Electromagnetic Wave Propagation, Radiation, and Scattering

Akira Ishimaru

Department of Electrical Engineering
University of Washington
Seattle, Washington

PRENTICE HALL
Upper Saddle River, NJ 07458

Library of Congress Cataloging-in-Publication Data

Ishimaru, Akira
Electomagnetic wave propagation, radiation, and scattering / Akira Ishimaru.
p. cm.
Includes bibliographical references and index.
ISBN 0-13-249053-6
1. Electromagnetic waves--Transmission. 2. Electromagnetic waves--Scattering. I. Title.
QC665.T7I74 1991
539.2--dc20 90-7645
CIP

Editorial/production supervision
and interior design: ***Kathleen Schiaparelli***
Manufacturing buyer: ***Linda Behrens/Patrice Fraccio***

To
Joyce, Karen, Jim, Jane,
Katie, Chuck, John,
and
Yuko

Printed in the United States of America

10 9 8 7 6 5 4 3 2

ISBN 0-13-249053-6

Prentice-Hall International (UK) Limited, London
Prentice-Hall of Australia Pty. Limited, Sydney
Prentice-Hall Canada Inc., Toronto
Prentice-Hall Hispanoamericana, S.A., Mexico
Prentice-Hall of India Private Limited, New Delhi
Prentice-Hall of Japan, Inc., Tokyo
Pearson Education Asia Pte. Ltd., Singapore
Editoria Prentice-Hall do Brasil, Ltda., Rio De Janeiro

Contents

APPENDICES

Preface

Numerous new advances have been made in electromagnetic theory in recent years. This is due, in part, to new applications of the theory to many practical problems. For example, in microwave and millimeter wave applications, there is an increasing need to investigate the electromagnetic problems of new guiding structures, phased arrays, microwave imaging, polarimetric radars, microwave hazards, frequency-sensitive surfaces, composite materials, and microwave remote sensing. In the field of optics, applications involving fiber optics, integrated optics, atmospheric optics, light diffusion in tissues, and optical oceanography are among many problems whose solutions require the use of electromagnetic theory as an essential element. The mathematical techniques used in electromagnetic theory are also equally applicable to ocean acoustics, acoustic scattering in the atmosphere, and ultrasound imaging of tissues. In this book we attempt to present a cohesive account of these advances with sufficient background material.

This book has three overall objectives. First, we present the fundamental concepts and formulations in electromagnetic theory. Second, advanced mathematical techniques and formulations such as the saddle-point technique, integral equations, GTD, and the *T*-matrix method are presented. Third, new topics, such as inverse scattering, diffraction tomography, vector radiative transfer, and polarimetry, are discussed.

The book includes several topics that are often not adequately covered in many electromagnetics texts. Examples are the Maxwell–Garnett and Polder–van Santern mixing formulas for the effective dielectric constant of mixtures and waves in chiral media. Also included are London's equations and the two-fluid model of the complex conductivity of superconductors at high frequencies. Other examples are

Radon transform, diffraction tomography, holographic and physical optics inverse problems, Abel's inversion formula, and polarimetry.

The book also includes more conventional but important topics such as the spectral method, strip lines, patches, apertures, Stokes vectors, modified saddle-point techniques, Watson transforms, residue series, *T*-matrix, GTD, and UTD. Also covered are Zenneck waves, plasmons, Goos–Hanchen effects, construction of Green's functions, Green's dyadic, radiometry, and noise temperatures. An important current topic that is not included due to space limitations is that of wave propagation and scattering in random media. However, this has been covered adequately in the author's previous book (Academic Press, 1978). Other topics that have been omitted are nonlinear electromagnetics, solitons, and chaos.

This book is based on a set of lecture notes prepared for a three-quarter graduate course on electromagnetic waves and applications given in the Department of Electrical Engineering at the University of Washington. The course has been offered yearly for students primarily from within the department of electrical engineering, but also including those from bioengineering, atmospheric sciences, geophysics, and oceanography. The course is also offered via cable television to students in remote sites located at nearby aerospace and bioengineering companies. The interest of the students vary considerably, from microwave antennas, radomes, and radar cross sections, to ultrasound imaging, optical scattering, ocean acoustics applications, microwave hazards, and bioelectromagnetics. This book is therefore intended to concentrate on fundamentals as well as recent analytical techniques, so that students may be prepared to handle a variety of old and new applications.

It is assumed that readers are familiar with material normally covered in undergraduate courses on electromagnetic theory, differential equations, Fourier and Laplace transforms, vector analysis, and linear algebra. However, brief reviews of these topics are given whenever needed. The time convention used in this book is $\exp(j\omega t)$. For some topics such as Stokes parameters, which are used extensively by the optics and acoustics community, the convention of $\exp(-i\omega t)$ is used and its use is clearly noted.

This book is intended as a text book for a first-year graduate course, and the material can be covered in two semesters or three quarters with appropriate selections depending on the students' interests. The book is also a useful reference for engineers working on electromagnetic, optical, or acoustics problems with aerospace, bioengineering, remote sensing, and ocean engineering applications.

I am grateful to Noel Henry for her expert typing and her assistance in organizing the manuscript. I also thank my graduate students and colleagues who have made many valuable comments. Some of the work in this book was supported by the National Science Foundation, the U.S. Army Research Office, the Office of Naval Research, and the U.S. Army Engineer Waterways Experiment Station. I also wish to thank Elizabeth Kaster, Editor, Prentice Hall, for her encouragement and editorial assistance.

Akira Ishimaru

1

Introduction

Many advances in electromagnetic theory were made in recent years in response to new applications of the theory to microwaves, millimeter waves, optics, and acoustics; as a result, there is a need to present a cohesive account of these advances with sufficient background material. In this book we present the fundamentals and the basic formulations of electromagnetic theory, as well as advanced analytical theory and mathematical techniques and current new topics and applications.

In Chapter 2 we review the fundamentals, starting with Maxwell's equations and covering such fundamental concepts and relationships as energy relations, potentials, Hertz vectors, and uniqueness and reciprocity theorems. The chapter concludes with linear acoustic-wave formulation. Plane-wave incidence on dielectric layers and wave propagation along layered media are often encountered in practice. Examples are microwaves in dielectric coatings, integrated optics, waves in the atmosphere, and acoustic waves in the ocean. Chapter 3 deals with these problems, starting with reviews of plane waves incident on layered media, Fresnel formulas, Brewster's angle, and total reflection. The concepts of complex waves, trapped surface waves, and leaky waves are presented with examples of surface-wave propagation along dielectric slabs, and this is followed by discussion on the relation between Zenneck waves and plasmons. The chapter concludes with WKB solutions and the Bremmer series for inhomogeneous media and turning points, and WKB solutions for the propagation constant of guided waves in inhomogeneous media such as graded-index fibers.

Chapter 4 deals with microwave waveguides, dielectric waveguides, and cavities. Formulations for TM, TE, and TEM waves, eigenfunctions, eigenvalues, and the k–β diagram are given, followed by pulse propagation in dispersive media. Dielectric waveguides, step-index fibers, and graded-index fibers are discussed next with due attention to dispersion. It concludes with radial and azimuthal waveguides, rectangular and cylindrical cavities, and spherical waveguides and cavities. This chapter introduces Green's identities, Green's theorem, special functions, Bessel and Legendre functions, eigenfunctions and eigenvalues, and orthogonality.

One of the most important and useful tools in electromagnetic theory is Green's functions. They are used extensively in the formulation of integral equations and radiation from various sources. Methods of constructing Green's functions are the topics in Chapter 5. First, the excitation of waves by electric and magnetic dipoles is reviewed. Three methods of expressing Green's functions are discussed. The first is the representation of Green's functions in a series of eigenfunctions. The second is to express them using the solutions of homogeneous equations. Here we discuss the important properties of Wronskians. The third is the Fourier transform representation of Green's functions. In actual problems, these three methods are often combined to obtain the most convenient representations. Examples are shown for Green's functions in rectangular waveguides and cylindrical and spherical structures.

Chapter 6 deals with the radiation field from apertures. We start with Green's theorem applied to the field produced by the sources and the fields on a surface. Here we discuss the extinction theorem and Huygens' formula. Next, we consider the Kirchhoff approximation and Fresnel and Frannhofer diffraction formulas. Spectral representations of the field are used to obtain Gaussian beam waves and the radiation from finite apertures. The interesting phenomenon of the Goos–Hanchen shift of a beam wave at an interface and higher-order beam waves are also discussed. The chapter concludes with the electromagnetic vector Green's theorem, Stratton–Chu formula, Franz formula, equivalence theorem, and electromagnetic Kirchhoff approximations.

The periodic structures discussed in Chapter 7 are used in many applications, such as optical gratings, phased arrays, and frequency-selective surfaces. We start with the Floquet-mode representation of waves in periodic structures. Guided waves along periodic structures and plane-wave incidence on periodic structures are discussed using integral equations and Green's function. An interesting question regarding the Rayleigh hypothesis for scattering from sinusoidal surfaces is discussed. Also included are the coupled-mode theory and co-directional and contra-directional couplers.

Chapter 8 deals with material characteristics. We start with the dispersive characteristics of dielectric material, the Sellmeier equation, plasma, and conductors. It also includes the Maxwell–Garnett and Polder–van Santern mixing formulas for the effective dielectric constant of mixtures. Wave propagation characteristics in magnetoplasma, which represents the ionosphere and ionized gas, and in ferrite, used in microwave networks, are discussed as well as Faraday rotation, group velocity, warm plasma, and reciprocity relations. This is followed by wave propagation in chiral material. The chapter concludes with London's equations and the two-fluid model of superconductors at high frequencies.

Chapter 9 presents selected topics on antennas, apertures, and arrays. Included in this chapter are radiation from current distributions, dipoles, slots, and loops. Also discussed are arrays with nonuniform spacings, microstrip antennas, mutual couplings, and the integral equation for current distributions on wire antennas. Chapter 10 starts with a general description of the scattering and absorption characteristics of waves by dielectric and conducting objects. Definitions of cross sections and scattering amplitudes are given, and Rayleigh scattering and Rayleigh–Debye approximations are discussed. Also included are the Stokes vector, the Mueller matrix, and the Poincaré sphere for a description of the complete and partial polarization states. Techniques discussed for obtaining the cross sections of conducting objects include the physical optics approximation and the moment method. Formal solutions for cylindrical structures, spheres, and wedges are presented in Chapter 11, including a discussion of branch points, the saddle-point technique, the Watson transform, the residue series, and Mie theory. Also discussed is diffraction by wedges, which will be used in Chapter 13.

Electromagnetic scattering by complex objects is the topic of Chapter 12. We present scalar and vector formulations of integral equations. Babinet's principle for scalar and electromagnetic fields, EFIE (electric field integral equation), and MFIE (magnetic field integral equation) are discussed. The T-matrix method, also called the extended boundary condition method, is discussed and applied to the problem of sinusoidal surfaces. In addition to the surface integral equation, also included are the volume integral equation for two- and three-dimensional dielectric bodies and Green's dyadic. Discussions of small apertures and slits are also included.

Geometric theory of diffraction (GTD) is one of the powerful techniques for dealing with high-frequency diffraction problems. GTD and UTD (uniform geometric theory of diffraction) are discussed in Chapter 13. Applications of GTD to diffraction by slits, knife edges, and wedges are presented, including slope diffraction, curved wedges, and vertex and surface diffractions.

Chapter 14 deals with excitation and scattering by sources, patches, and apertures embedded in planar structures. Excitation of a dielectric slab is discussed, followed by the WKB solution for the excitation of waves in inhomogeneous layers. An example of the latter is acoustic-wave excitation by a point source in the ocean. Next, we give general spectral formulations for waves in patches, strip lines, and apertures embedded in dielectric layers. Convenient equivalent network representations are presented that are applicable to strip lines and periodic patches and apertures.

The Sommerfeld dipole problem is that of finding the field when a dipole is located above the conducting earth. This classical problem, which dates back to 1907, when Zenneck investigated what is now known as the Zenneck wave, is discussed in Chapter 15, including a detailed study of the Sommerfeld pole, the modified saddle-point technique, lateral waves, layered media, and mode representations.

The inverse scattering problem in Chapter 16 is one of the important topics in recent years. It deals with the problem of obtaining the properties of an object by using the observed scattering data. First, we present the Radon transform, used in computed tomography or X-ray tomography. The inverse Radon transform is obtained by using the projection slice theorem and the back projection of the filtered projection. Also included is an alternative inverse Radon transform in terms of the Hilbert transform. For ultrasound and electromagnetic imaging problems, it is necessary to include the diffraction effect. This leads to diffraction tomography, which makes use of back propagation rather than back projection. Also discussed are physical optics inverse scattering and the holographic inverse problem. Abel's integral equations are frequently used in inverse problems. Here we illustrate this technique by using it to find the electron density profile in the ionosphere. Polarimetric radars are becoming increasingly more important because of the advances in polarimetric measurement techniques. We present the fundamentals of polarimetry and optimization as well as polarization signatures.

Chapter 17 presents fundamentals of radiometry and noise. The definitions of antenna temperature, radiative transfer theory, and the scattering cross section of surfaces are given followed by consideration of system noise temperature and minimum detectable temperature. Also included is a discussion on the determination of sky brightness distribution by interferometry used in radio astronomy. Here we discuss the Fourier transform and convolution relationships among the aperture distribution, the radiation pattern, and sky brightness distributions.

Numerical techniques are used in solving almost all electromagnetic problems. Basic ideas of the moment method, basis functions, weighting functions, and shape functions are reviewed in Chapter 18 together with discussions on the point-matching technique, Galerkin's method, Rayleigh–Ritz formula, finite-element method, and finite-difference method.

The appendices give many formulas and detailed derivations of equations that are too lengthy to be included in the text. They should be helpful in understanding the material in the text.

Useful reference books on electromagnetics at the intermediate undergraduate level include Ramo et al. (1965), Jordan and Balmain (1968), Wait (1986), Shen and Kong (1987), and Cheng (1983). Among books at the advanced level are Stratton (1941), Harrington (1961), Collin (1966), Felsen and Marcuvitz (1973), Schelkunoff (1965), Balanis (1989), Kong (1981, 1986), Jones (1964, 1979), and Van Bladel (1964).

2

Fundamental Field Equations

The fundamental field equations for electromagnetic fields are Maxwell's equations. In this chapter we review differential and integral forms of these equations and discuss boundary conditions, energy relations, the Poynting theorem, the uniqueness theorem, and the reciprocity theorem. Vector and scalar potentials and Hertz vectors are discussed, as they give alternative and often simpler formulations of the problem. Although electromagnetic waves are vector fields, they can sometimes be represented or approximated by scalar fields. We present formulations for scalar acoustic waves as examples of scalar fields.

It will be assumed that the reader is familiar with the electromagnetic field theory normally covered in undergraduate courses, and therefore, this book starts with a review of the fundamental field equations. A detailed historical development of electromagnetic theory is given in Elliott (1966) and Born and Wolf (1970).

2-1 MAXWELL'S EQUATIONS

The fundamental differential equations governing the behavior of electromagnetic fields were given by Maxwell in 1865:

$$\nabla \times \bar{H} = \frac{\partial \bar{D}}{\partial t} + \bar{J}, \tag{2-1}$$

$$\nabla \times \bar{E} = -\frac{\partial \bar{B}}{\partial t}, \tag{2-2}$$

$$\nabla \cdot \bar{D} = \rho, \tag{2-3}$$

$$\nabla \cdot \bar{B} = 0. \tag{2-4}$$

Here $\bar{E}$ is the electric field vector in volts/meter, $\bar{H}$ the magnetic field vector in amperes/meter, $\bar{D}$ the electric displacement vector in coulombs/meter^2, $\bar{B}$ the magnetic flux density vector in webers/meter^2, $\bar{J}$ the current density vector in amperes/meter^2, and ρ the volume charge density in coulombs/meter^3.

The physical meanings of Maxwell's equations are often easier to understand if expressed in alternative integral form. The first two equations, (2-1) and (2-2), can be converted into integral form by employing Stokes' theorem:

$$\int_a \nabla \times \bar{A} \cdot d\bar{a} = \oint_l \bar{A} \cdot d\bar{l}, \tag{2-5}$$

where $\bar{A}$ is a vector, $d\bar{a}$ is a differential surface element vector with magnitude da pointed in the normal direction $\hat{n}$ ($d\bar{a} = \hat{n}\, da$), and $d\bar{l}$ is a differential line element vector with magnitude dl pointed in the direction $\hat{l}$ (see Fig. 2-1). The directions of $\hat{n}$ and $\hat{l}$ are chosen so that $\hat{l}$ is in the direction of the rotation of the right-handed screw advancing in the direction $\hat{n}$.

Using (2-5), (2-1) and (2-2) are expressed as follows:

$$\oint_l \bar{H} \cdot d\bar{l} = \int_a \left(\frac{\partial \bar{D}}{\partial t} + \bar{J}\right) \cdot d\bar{a}, \tag{2-6}$$

$$\oint_l \bar{E} \cdot d\bar{l} = -\int_a \frac{\partial \bar{B}}{\partial t} \cdot d\bar{a}. \tag{2-7}$$

Equation (2-6) represents Ampère's law that the line integral of a magnetic field around a closed path is equal to the total current, including the displacement current, $\partial\bar{D}/\partial t$, through the loop. Equation (2-7) represents Faraday's law of induction that the line integral of an electric field around a closed path is equal to the negative of time rate of change of the total magnetic flux through the loop.

Equations (2-3) and (2-4) can be expressed in integral form by using the divergence theorem

$$\int_V \nabla \cdot \bar{A}\, dV = \int_S \bar{A} \cdot d\bar{a}, \tag{2-8}$$

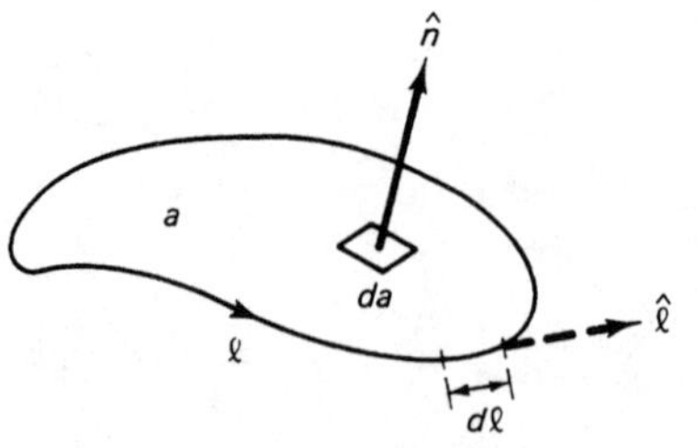

Figure 2-1 Stoke's theorem.

where S is the closed surface surrounding a volume V and $\hat{n}$ is pointed outward (Fig. 2-2). With (2-8), equations (2-3) and (2-4) can be written as

$$\int_S \bar{D} \cdot d\bar{a} = \int_V \rho \, dV, \tag{2-9}$$

$$\int_S \bar{B} \cdot d\bar{a} = 0. \tag{2-10}$$

Equation (2-9) is Gauss's law that the total electric flux flowing out of any closed surface is equal to the total charge enclosed by the surface. Equation (2-10) states that there is no magnetic charge and that there is no net magnetic flux flowing in or out of a closed surface. In addition to Maxwell's equations, the following force law holds concerning the force on a charge q moving with velocity $\bar{v}$ through an electric field $\bar{E}$ and a magnetic field $\bar{B}$:

$$\bar{F} = q(\bar{E} + \bar{v} \times \bar{B}). \tag{2-11}$$

The conservation of charge is embodied in the following continuity equation:

$$\nabla \cdot \bar{J} + \frac{\partial \rho}{\partial t} = 0. \tag{2-12}$$

The integral form of (2-12) is obtained by employing the divergence theorem:

$$\int_S \bar{J} \cdot d\bar{a} + \frac{\partial}{\partial t} \int_V \rho \, dV = 0. \tag{2-13}$$

This states that the outward flow of the current through a closed surface S must be accompanied by the decrease per unit time of the total charge inside the volume V.

The continuity equation (2-12) can be derived from Maxwell's equations by taking the divergence of (2-1) and using (2-3) and the identity

$$\nabla \cdot \nabla \times \bar{A} = 0. \tag{2-14}$$

Some vector formulas and theorems and the gradient, the divergence, the curl, and the Laplacian in Cartesian, cylindrical, and spherical coordinate systems are shown in the Appendix to Chapter 2, Section A.

Equations (2-1) to (2-4) exhibit a mathematical similarity. $\bar{E}$ and $\bar{H}$ appear on the left sides of (2-1) and (2-2) under the same operator, and $\bar{D}$ and $\bar{B}$ appear similarly on the right sides. This may appear to suggest that $\bar{E}$ and $\bar{H}$ belong to one class and $\bar{D}$ and $\bar{B}$ to another. However, (2-11) shows that the force depends on $\bar{E}$

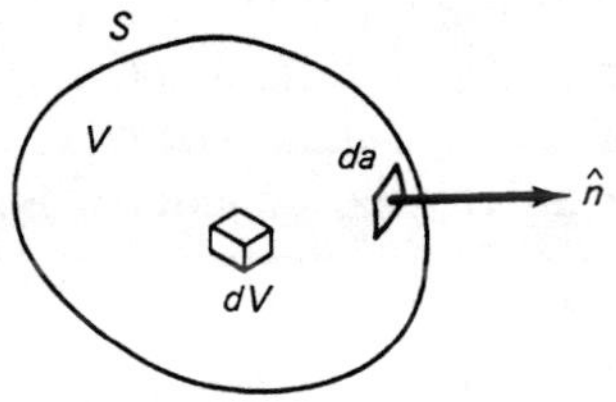

Figure 2-2 Divergence theorem.

and $\bar{B}$, not $\bar{E}$ and $\bar{H}$. In fact, physically, $\bar{E}$ and $\bar{B}$ are the fundamental fields and $\bar{D}$ and $\bar{H}$ are the derived fields related to $\bar{E}$ and $\bar{B}$ through constitutive relations (Section 2-3).

2-2 TIME-HARMONIC CASE

Since the general behavior of a wave as a function of time can always be expressed as a superposition of waves at different frequencies through a Fourier transform, it is sufficient to investigate the characteristics of a wave at a single frequency. The wave at a single frequency is often called a *time-harmonic* or *monochromatic wave* and is most conveniently described by the real part of the phasor field. For example, the vector field $\bar{E}(\bar{r}, t)$, a real function of position $\bar{r}$ and time t, is given by

$$\bar{E}(\bar{r}, t) = \mathrm{Re}[\bar{E}_{\mathrm{ph}}(\bar{r})e^{j\omega t}], \tag{2-15}$$

where $\bar{E}_{\mathrm{ph}}(\bar{r})$ is a phasor field and, in general, complex. Here we use the convention that $\bar{E}_{\mathrm{ph}}$ is the peak value rather than the rms value, and thus $\bar{E}_{\mathrm{ph}} = \sqrt{2}\bar{E}(\mathrm{rms})$. The x component of $\bar{E}(\bar{r}, t)$ is then given by

$$\begin{aligned} E_x(\bar{r}, t) &= \mathrm{Re}[E_{\mathrm{ph}x}(\bar{r})e^{j\omega t}] \\ &= A_x(\bar{r}) \cos[\omega t + \phi_x(\bar{r})], \end{aligned} \tag{2-16}$$

where A_x and ϕ_x are the amplitude and phase, respectively, of the x component of the phasor $E_{\mathrm{ph}x} = A_x \exp[j\phi_x]$. In the following, we omit the subscript ph for the phasor whenever no confusion is expected to arise.

Let us rewrite Maxwell's equations for the time-harmonic case:

$$\nabla \times \bar{H}(\bar{r}) = j\omega\bar{D}(\bar{r}) + \bar{J}(\bar{r}), \tag{2-17}$$

$$\nabla \times \bar{E}(\bar{r}) = -j\omega\bar{B}(\bar{r}), \tag{2-18}$$

$$\nabla \cdot \bar{D}(\bar{r}) = \rho(\bar{r}), \tag{2-19}$$

$$\nabla \cdot \bar{B}(\bar{r}) = 0, \tag{2-20}$$

with the continuity equation

$$\nabla \cdot \bar{J}(\bar{r}) + j\omega\rho(\bar{r}) = 0. \tag{2-21}$$

Note that $\exp(j\omega t)$ is dropped, as it is common to all terms and that all the field quantities in (2-17) to (2-21) are phasors. Equation (2-19) can be obtained by taking the divergence of (2-17) and using (2-21). Equation (2-20) is obtained by taking the divergence of (2-18). Thus, for the time-harmonic case, (2-17), (2-18), and (2-21) constitute the complete set of differential equations. For the static field, $\omega = 0$, however, we need (2-18) and (2-19) for electrostatics and (2-17) and (2-20) for magnetostatics.

Once the time-harmonic fields in terms of the phasor [e.g., $\bar{E}(\bar{r})$] are obtained, the general transient field $\bar{E}(\bar{r}, t)$ as a function of time can be obtained by the inverse Fourier transform

$$\bar{E}(\bar{r}, t) = \frac{1}{2\pi}\int \bar{E}(\bar{r})e^{j\omega t}\, d\omega. \tag{2-22}$$

Let us next consider the current $\bar{J}$ and charge ρ. It is often convenient to separate the current into the source current and the induced current. For example, in a radio broadcast the current on the transmitting antenna at a radio station is the *source current*, but the currents induced in receiver antennas and nearby metallic walls are considered the *induced current*. Similarly, we can separate the charges into the source charge and the induced charge. The current density $\bar{J}$ and the charge density ρ in Maxwell's equations refer to all currents and charges, source and induced. It is more convenient, however, to express separately the source currents $\bar{J}_s$ and the source charges ρ_s in the following manner:

$$\nabla \times \bar{H}(\bar{r}) = j\omega \bar{D}(\bar{r}) + \bar{J}(\bar{r}) + \bar{J}_s(\bar{r}), \tag{2-23}$$

$$\nabla \times \bar{E}(\bar{r}) = -j\omega \bar{B}(\bar{r}), \tag{2-24}$$

$$\nabla \cdot \bar{D}(\bar{r}) = \rho + \rho_s, \tag{2-25}$$

$$\nabla \cdot \bar{B}(\bar{r}) = 0. \tag{2-26}$$

In this form, $\bar{J}$ and ρ are the induced current density and the induced charge density, respectively, and as will be shown in the next section, $\bar{J}$ and ρ will be incorporated into the medium characteristics.

2-3 CONSTITUTIVE RELATIONS

Let us consider Maxwell's equations (2-23) to (2-26) for time-harmonic electromagnetic fields. The relationships among $\bar{D}$, $\bar{E}$, $\bar{B}$, $\bar{H}$, and $\bar{J}$ depend on the characteristics of the medium, and they are expressed by the *constitutive* parameters. In a linear passive medium, $\bar{D}$ and $\bar{B}$ are linearly related to $\bar{E}$ and $\bar{H}$, respectively. Furthermore, if the constitutive relationships between $\bar{D}$ and $\bar{E}$, and $\bar{B}$ and $\bar{H}$, do not depend on the direction of $\bar{E}$ and $\bar{H}$, the medium is called *isotropic*.

Let us first consider a lossless medium where $\bar{J} = 0$. For a linear, passive, isotropic medium, we have

$$\bar{D} = \epsilon \bar{E}, \tag{2-27}$$

$$\bar{B} = \mu \bar{H}, \tag{2-28}$$

where ϵ is the permittivity or dielectric constant (farads/meter), μ is the permeability (henries/meter), and both ϵ and μ are real and scalar. For free space, the dielectric constant and the permeability are

$$\begin{aligned} \epsilon_0 &= 8.854 \times 10^{-12} \simeq \frac{10^{-9}}{36\pi} \text{ farad/meter}, \\ \mu_0 &= 4\pi \times 10^{-7} \text{ henry/meter}. \end{aligned} \tag{2-29}$$

Note that the light velocity in free space is given by

$$c = \frac{1}{(\mu_0 \epsilon_0)^{1/2}} \simeq 3 \times 10^8 \text{ meters/s.} \tag{2-30}$$

It is often convenient to use the dimensionless quantities, the relative dielectric constant ϵ_r and the relative permeability μ_r.

$$\epsilon_r = \frac{\epsilon}{\epsilon_0} \quad \text{and} \quad \mu_r = \frac{\mu}{\mu_0}. \tag{2-31}$$

If ϵ and μ are constant from point to point, the medium is called *homogeneous*.

In an anisotropic medium, where the relationship between $\overline{D}$ and $\overline{E}$ (or $\overline{B}$ and $\overline{H}$) depends on the direction of $\overline{E}$ (or $\overline{H}$) and, therefore, $\overline{D}$ and $\overline{E}$ are in general not parallel, the constitutive parameters should be expressed by the tensor dielectric constant $\overline{\overline{\epsilon}}$ *(or permeability* $\overline{\overline{\mu}}$*)*:

$$\overline{D} = \overline{\overline{\epsilon}}\overline{E}, \tag{2-32}$$

$$\overline{B} = \overline{\overline{\mu}}\overline{H}. \tag{2-33}$$

For example, (2-32) in the Cartesian coordinate system can be expressed in the following matrix form:

$$\begin{bmatrix} D_x \\ D_y \\ D_z \end{bmatrix} = \begin{bmatrix} \epsilon_{11} & \epsilon_{12} & \epsilon_{13} \\ \epsilon_{21} & \epsilon_{22} & \epsilon_{23} \\ \epsilon_{31} & \epsilon_{32} & \epsilon_{33} \end{bmatrix} \begin{bmatrix} E_x \\ E_y \\ E_z \end{bmatrix}. \tag{2-34}$$

In (2-32) and (2-33), we called $\overline{\overline{\epsilon}}$ and $\overline{\overline{\mu}}$ the tensor without defining them. In Chapter 8 we present a detailed discussion of anisotropic media as well as chiral media in which $\overline{D}$ and $\overline{B}$ are coupled to both $\overline{E}$ and $\overline{H}$.

Let us consider a lossy medium. In a linear lossy medium, the current density $\overline{J}$ is proportional to the electric field $\overline{E}$, and this relationship is called *Ohm's law*:

$$\overline{J} = \sigma \overline{E}, \tag{2-35}$$

where σ is the conductivity (siemens/meter) of the medium. For low frequencies up to the microwave region, the conductivity is often essentially real and independent of frequency. We can then rewrite one of Maxwell's equations (2-23) in the following manner:

$$\begin{aligned} \nabla \times \overline{H} &= j\omega\epsilon\overline{E} + \sigma\overline{E} + \overline{J}_s \\ &= j\omega\epsilon_c\overline{E} + \overline{J}_s, \end{aligned} \tag{2-36}$$

where $\epsilon_c = \epsilon - j(\sigma/\omega)$ is called the complex dielectric constant. Note that in (2-36), the conductivity term $\sigma\overline{E}$ is absorbed into the dielectric constant as the imaginary part. The relative complex dielectric constant is then given by

$$\epsilon_r = \frac{\epsilon_c}{\epsilon_0} = \frac{\epsilon}{\epsilon_0} - j\frac{\sigma}{\omega\epsilon_0} = \epsilon' - j\epsilon''. \tag{2-37}$$

The ratio ϵ''/ϵ' is called the loss tangent,

$$\tan\delta = \frac{\epsilon''}{\epsilon'}, \tag{2-38}$$

and the complex index of refraction n is given by

$$n = (\epsilon_r)^{1/2} = n' - jn''. \tag{2-39}$$

For most material, the permeability μ is equal to that of free space ($\mu = \mu_0$). However, in magnetic material, μ is different from μ_0 and may be lossy. This will be discussed in more detail later when we discuss ferrite materials. We note here that in general the relative permeability $\mu_r = \mu/\mu_0$ can be complex ($\mu_r = \mu' - j\mu''$). The complex refractive index n is then given by $n = (\epsilon_r \mu_r)^{1/2}$. Some examples of the relative dielectric constant ϵ' and the conductivity σ are shown in Table 2-1. Note that using the relative dielectric constant ϵ_r in (2-37), the displacement vector $\bar{D}$ is now given by

$$\bar{D} = \epsilon_c \bar{E}.$$

In the frequency ranges above the microwave region, ϵ and σ in (2-37) are no longer independent of frequency and it is often more convenient simply to use the relative complex dielectric constant ϵ_c and not separate it into ϵ and σ. Therefore, whenever no confusion is expected to arise, we use ϵ to indicate the complex dielectric constant. Maxwell's equations can then be written as

$$\nabla \times \bar{E} = -j\omega\mu\bar{H}, \tag{2-40}$$

$$\nabla \times \bar{H} = j\omega\epsilon\bar{E} + \bar{J}_s, \tag{2-41}$$

$$\nabla \cdot \bar{D} = \rho_s, \tag{2-42}$$

$$\nabla \cdot \bar{B} = 0, \tag{2-43}$$

where $\bar{J}_s$ and $\bar{\rho}_s$ are the source current and charge density, respectively, and ϵ and μ are in general complex and dependent on frequency. The medium is called *dispersive* if ϵ or μ is dependent on frequency.

TABLE 2-1 Relative Dielectric Constants and Conductivities for Low Frequencies

	ϵ'	σ (S/m)
Wet earth	10	10^{-3}
Dry earth	5	10^{-5}
Fresh water	81	10^{-3}
Seawater	81	4
Copper	1	5.8×10^7
Silver	1	6.17×10^7
Brass	1	1.57×10^7

In (2-40) to (2-43), the constitutive equations should be written with the complex ϵ and μ:

$$\overline{D} = \epsilon\overline{E}, \qquad \overline{B} = \mu\overline{H}. \tag{2-44}$$

Equations (2-40) to (2-44) constitute a complete description of the fundamental field equations for a lossy medium. Note that the medium is now expressed in terms of the complex dielectric constant and the complex permeability. Alternatively, we can express the medium in terms of the real dielectric constant and the conductivity.

Instead of the dielectric constant and the permeability, we can use two vectors, the electric polarization $\overline{P}$ and the magnetic polarization $\overline{M}$ defined by

$$\begin{aligned} \overline{D} &= \epsilon_0\overline{E} + \overline{P}, \\ \overline{B} &= \mu_0(\overline{H} + \overline{M}). \end{aligned} \tag{2-45}$$

The vectors $\overline{P}$ and $\overline{M}$ represent the electric and magnetic dipole distributions in the medium and vanish in free space. In a linear medium, they are related to $\overline{E}$ and $\overline{H}$ by the electric and magnetic susceptibilities, χ_e and χ_m:

$$\begin{aligned} \overline{P} &= \chi_e\epsilon_0\overline{E}, \\ \overline{M} &= \chi_m\overline{H}. \end{aligned} \tag{2-46}$$

The dielectric constant ϵ in (2-44) is defined for a time-harmonic electromagnetic phasor field at a certain frequency ω. If the dielectric constant is independent of frequency, the relationship (2-44) in time is simply given by

$$\overline{D}(\overline{r}, t) = \epsilon\overline{E}(\overline{r}, t). \tag{2-47}$$

However, if the dielectric constant is a function of frequency, the temporal relationship is a Fourier transform of the product of $\epsilon(\omega)$ and $\overline{E}(\omega)$ and, therefore, is given by the following convolution integral:

$$\overline{D}(\overline{r}, t) = \int_{-\infty}^{t} h(t - t')\overline{E}(\overline{r}, t')\,dt', \tag{2-48}$$

where

$$h(t) = \frac{1}{2\pi}\int \epsilon(\omega)e^{j\omega t}\,d\omega, \tag{2-49}$$

$$\overline{E}(\overline{r}, t) = \frac{1}{2\pi}\int \overline{E}(\overline{r}, \omega)e^{j\omega t}\,d\omega. \tag{2-50}$$

The medium whose dielectric constant is a function of frequency $\epsilon(\omega)$ is called dispersive. Although strictly speaking, all media are dispersive, a medium can often be treated as nondispersive within a frequency range used for a particular problem.

If the medium is linear but time-varying, the relationship between $\overline{D}$ and $\overline{E}$ cannot be expressed as the convolution integral (2-48). The general relationship should then be given by

$$\overline{D}(\overline{r}, t) = \int_{-\infty}^{t} h(t, t - t')\overline{E}(\overline{r}, t')\,dt'. \tag{2-51}$$

Substituting (2-50) into (2-51), we get

$$\bar{D}(\bar{r}, t) = \frac{1}{2\pi} \int \epsilon(t, \omega) \bar{E}(\bar{r}, \omega) e^{j\omega t} \, d\omega, \tag{2-52}$$

where $\epsilon(t, \omega)$ is the time-varying dielectric constant given by

$$\epsilon(t, \omega) = \int_0^\infty h(t, t'') e^{-j\omega t''} \, dt''. \tag{2-53}$$

We will, however, not discuss time-varying media in this book.

2-4 BOUNDARY CONDITIONS

At an interface between two media, the field quantities must satisfy certain conditions. Consider an interface between two media with complex dielectric constants ϵ_1 and ϵ_2. In the absence of the source current $\bar{J}_s$, we write (2-41) in integral form,

$$\oint_l \bar{H} \cdot d\bar{l} = \int_a j\omega\epsilon \, \bar{E} \cdot d\bar{a}, \tag{2-54}$$

and apply it to a line integral shown in Fig. 2-3. As $dl \to 0$, the right side of (2-54) vanishes and the left side becomes

$$(H_{t1} - H_{t2}) \, \Delta L = 0, \tag{2-55}$$

where H_{t1} and H_{t2} are the tangential components of the magnetic field in media 1 and 2, respectively. Similarly, from (2-40) we get

$$(E_{t1} - E_{t2}) \, \Delta L = 0, \tag{2-56}$$

where E_{t1} and E_{t2} are the tangential components of the electric field in media 1 and 2. We therefore state that the tangential components of the electric field and magnetic fields must be continuous, respectively, across the boundary. Mathematically, we write

$$\hat{n} \times \bar{E}_1 = \hat{n} \times \bar{E}_2, \tag{2-57a}$$

$$\hat{n} \times \bar{H}_1 = \hat{n} \times \bar{H}_2, \tag{2-57b}$$

where $\hat{n}$ is the unit vector normal to the interface and (E_1, H_1) and (E_2, H_2) are the

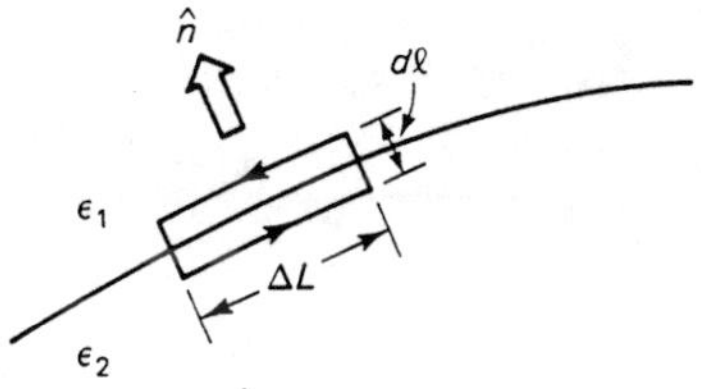

Figure 2-3 Boundary conditions for tangential components.

fields in the medium with ϵ_1 and ϵ_2, respectively. If a surface current J_{sf} (amperes/m) exists on the boundary, we have

$$\hat{n} \times \overline{H}_1 - \hat{n} \times \overline{H}_2 = \overline{J}_{sf}. \tag{2-58}$$

We can also obtain the conditions on the normal components of the electric and magnetic fields on the boundary. In the absence of the source ($\overline{J}_s = 0$ and $\rho_s = 0$), we apply the divergence theorem to (2-42) and (2-43) for a pillbox volume shown in Fig. 2-4, and in the limit $dl \to 0$ we get

$$\begin{aligned} (D_{n1} - D_{n2})\,\Delta a &= 0, \\ (B_{n1} - B_{n2})\,\Delta a &= 0. \end{aligned} \tag{2-59}$$

This states that the normal components of $\overline{D}$ and $\overline{B}$ must be continuous across the boundary. Mathematically, we write

$$\begin{aligned} \overline{D}_1 \cdot \hat{n} &= \overline{D}_2 \cdot \hat{n}, \\ \overline{B}_1 \cdot \hat{n} &= \overline{B}_2 \cdot \hat{n}. \end{aligned} \tag{2-60}$$

If a surface charge exists on the boundary, we have

$$(D_{n1} - D_{n2})\,\Delta a = \rho_{sf}\,\Delta a, \tag{2-61}$$

or

$$\overline{D}_1 \cdot \hat{n} - \overline{D}_2 \cdot \hat{n} = \rho_{sf}. \tag{2-62}$$

Here ρ_{sf} is the surface charge density (coulombs/m^2).

Let us examine how we use the boundary conditions (2-57), (2-58), (2-60), and (2-62) to solve electromagnetic problems. According to the uniqueness theorem, which is discussed in Section 2-10, only one unique solution exists among all the possible solutions of Maxwell's equations that satisfies both Maxwell's equations and the boundary conditions. Thus it is important to determine what constitutes the necessary and sufficient boundary conditions in order to yield the unique solution. (See Morse and Feshbach, 1953, Chapter 6, for more detail.)

Here, we first note that (2-60) is not independent of (2-57a) and (2-57b) for a time-harmonic case since (2-42) and (2-43) can be derived by taking the divergence of (2-40) and (2-41), respectively. Therefore, the necessary and sufficient boundary conditions are the continuity of the tangential electric and magnetic fields as given in (2-57a) and (2-57b).

If the second medium is a perfect conductor, the fields inside the conductor vanish, and the necessary and sufficient boundary condition is that the tangential component of the electric field is zero on the boundary:

$$\hat{n} \times \overline{E}_1 = 0. \tag{2-63}$$

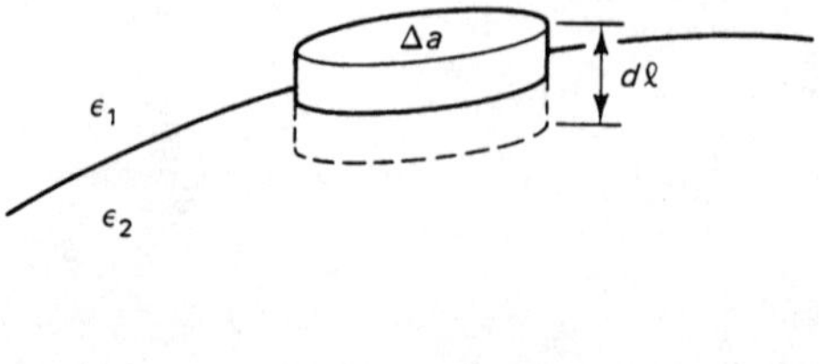

Figure 2-4 Boundary conditions for normal components.

The surface current density is then given by $\bar{J}_{sf} = \hat{n} \times \bar{H}_1$.

If the second medium is a good conductor, so that $\epsilon' << \epsilon''$, the wave can only penetrate a distance of the skin depth given by $(2/\omega\mu\sigma)^{1/2}$ (see Chapter 3). If the radius of curvature of the surface is much greater than the skin depth, the following condition, called the *Leontovich impedance boundary condition*, holds approximately (Brekhovskikh, 1960):

$$\bar{E}_t = Z_s(\hat{n} \times \bar{H}), \tag{2-64}$$

where $\bar{E}_t$ is the electric field tangential to the surface and $Z_s = (\mu/\epsilon)^{1/2} \approx (j\mu\omega/\sigma)^{1/2}$. This means that the ratio of the tangential electric field to the tangential magnetic field is constant at the surface. The Leontovich boundary condition eliminates the need to consider the field in the second medium and thus leads to considerable mathematical simplification.

In addition to the foregoing conditions at the boundary, if the region under consideration extends to infinity, the wave must be outgoing at infinity. This requirement at infinity is called the *Sommerfeld radiation condition* (Sommerfeld, 1949). For a scalar field ψ, the radiation condition is given by

$$\lim_{r \to \infty} r\left(\frac{\partial \psi}{\partial r} + jk\psi\right) = 0, \tag{2-65}$$

where $k = \omega/c$. For electromagnetic fields, they are given by (Collin and Zucker, 1969)

$$\begin{aligned} &\lim_{r \to \infty} r\left[\left(\frac{\mu}{\epsilon}\right)^{1/2} \hat{r} \times \bar{H} + \bar{E}\right] = 0, \\ &\lim_{r \to \infty} r\left[\hat{r} \times \bar{E} - \left(\frac{\mu}{\epsilon}\right)^{1/2} \bar{H}\right] = 0. \end{aligned} \tag{2-66}$$

This means that the field is outgoing and the Poynting vector is pointed outward and decreases as r^{-2}. The radial components E_r and H_r must decrease faster than r^{-1}.

If the region includes a sharp edge, the field can become infinite, but the energy stored around the edge must be finite. Thus the field must satisfy the edge condition, which is discussed in the Appendix to Chapter 7, Section C. In general, therefore, the complete mathematical description of the electromagnetic problem includes Maxwell's equations, boundary conditions, the radiation condition, and the edge condition.

2-5 ENERGY RELATIONS AND POYNTING THEOREM

Let us consider the general time-varying electromagnetic fields satisfying Maxwell's equations (2-1) to (2-4). To obtain the energy relations, we use the vector identity

$$\nabla \cdot (\bar{A} \times \bar{B}) = \bar{B} \cdot \nabla \times \bar{A} - \bar{A} \cdot \nabla \times \bar{B} \tag{2-67}$$

and let $\bar{A} = \bar{E}$ and $\bar{B} = \bar{H}$. We then substitute $\nabla \times \bar{E}$ from (2-2) and $\nabla \times \bar{H}$ from (2-1) into the right side of (2-67) and obtain

$$\nabla \cdot (\bar{E} \times \bar{H}) + \bar{H} \cdot \frac{\partial \bar{B}}{\partial t} + \bar{E} \cdot \frac{\partial \bar{D}}{\partial t} + \bar{E} \cdot \bar{J} = 0, \tag{2-68}$$

where all field quantities are real functions of position $\bar{r}$ and time t. The vector $\bar{S} = \bar{E} \times \bar{H}$ is called the *Poynting vector* and represents the flow of the power flux per unit area. It is the power flux density and its unit is watts/m^2.

Equation (2-68) is a mathematical representation of the Poynting theorem. Let us examine its physical meaning. In a nondispersive, lossless medium, $\bar{D} = \epsilon \bar{E}$ and $\bar{B} = \mu \bar{H}$ and ϵ and μ are real constants. We can then identify the electromagnetic energy density W:

$$W = W_e + W_m = \tfrac{1}{2}\epsilon \bar{E} \cdot \bar{E} + \tfrac{1}{2}\mu \bar{H} \cdot \bar{H}, \tag{2-69}$$

where W is the sum of the electric W_e and magnetic W_m energy densities. The Poynting theorem (2-68) can then be stated as

$$\nabla \cdot \bar{S} + \frac{\partial}{\partial t} W + \bar{E} \cdot \bar{J} = 0. \tag{2-70}$$

The physical meaning of this can be seen more clearly in the following integral form obtained by applying the divergence theorem to (2-70):

$$-\int_S \bar{S} \cdot d\bar{a} = \frac{\partial}{\partial t} \int_V W\,dV + \int_V \bar{E} \cdot \bar{J}\,dV. \tag{2-71}$$

Here the left side of (2-71) is the total power flow into the volume V. The first term on the right represents the time rate of increase of the total electromagnetic energy inside the volume and the second term represents the total power dissipation in the volume. Thus (2-71) states that the total energy flow into a volume per unit time is equal to the sum of the increase in the total electromagnetic energy and the energy dissipation per unit time in the volume.

In many practical problems, we deal with time-harmonic electromagnetic fields, and it is necessary to consider the Poynting theorem for the phasor field quantities. In a time-harmonic case, we use the complex Poynting vector

$$\bar{S} = \tfrac{1}{2} \bar{E} \times \bar{H}^*, \tag{2-72}$$

where $\bar{E}$ and $\bar{H}$ are phasors. The magnitudes $|\bar{E}|$ and $|\bar{H}|$ are peak values, and therefore the rms values are $(1/\sqrt{2})|\bar{E}|$ and $(1/\sqrt{2})|\bar{H}|$. $\bar{S}$ gives the direction and the rms value of the complex power flux density. The real part of $\bar{S}$ represents the real power flux density, and the imaginary part represents the reactive power flux density. Using the identity (2-67), we write

$$\nabla \cdot \bar{S} = \tfrac{1}{2} \bar{H}^* \cdot \nabla \times \bar{E} - \tfrac{1}{2} \bar{E} \cdot \nabla \times \bar{H}^*. \tag{2-73}$$

Substituting Maxwell's equations (2-40) and (2-41) into this, we get the Poynting theorem in the following complex form:

$$\nabla \cdot \bar{S} + 2j\omega(W_m - W_e) + L + \tfrac{1}{2} \bar{E} \cdot \bar{J}_s^* = 0, \tag{2-74}$$

where

$$W_e = \frac{\epsilon_0 \epsilon'}{4} |\bar{E}|^2,$$

$$W_m = \frac{\mu_0 \mu'}{4} |\bar{H}|^2,$$

$$L = \frac{\omega \epsilon_0 \epsilon''}{2} |\bar{E}|^2 + \frac{\omega \mu_0 \mu''}{2} |\bar{H}|^2,$$

$$\varepsilon = \epsilon_0(\epsilon' - j\epsilon''), \qquad \epsilon'' = \frac{\sigma}{\omega \epsilon_0}, \qquad \mu = \mu_0(\mu' - j\mu'').$$

Here ϵ and μ are complex dielectric constant and permeability, respectively (μ is generalized to be complex). W_e and W_m are the time-averaged electric and magnetic stored energy densities and are equal to the average of $\frac{1}{2}\epsilon_0\epsilon'|\bar{E}(t)|^2$ and $\frac{1}{2}\mu_0\mu'|\bar{H}(t)|^2$ over the period $T = 2\pi/\omega$.

$$\begin{aligned}\frac{1}{T}\int_0^T \frac{\epsilon_0 \epsilon'}{2} |\bar{E}(t)|^2\,dt &= \frac{1}{T}\frac{\epsilon_0 \epsilon'}{2}\int_0^T |\text{Re}[\bar{E}(\text{phasor})e^{j\omega t}]|^2\,dt \\ &= \frac{\epsilon_0 \epsilon'}{4} |\bar{E}(\text{phasor})|^2.\end{aligned} \tag{2-75}$$

The third term L in (2-74) is real and positive and represents the power dissipation per unit volume in a lossy medium. The last term in (2-74) is the power absorbed by the source current $\bar{J}_s$. The power emitted from the source $\bar{J}_s$ is therefore given by $-\frac{1}{2}\bar{E}\cdot\bar{J}_s^*$. This can be seen by taking the volume integral of (2-74) over the source volume only. Then W_e, W_m, and L are zero and the total power emitted is equal to the volume integral of $-\frac{1}{2}\bar{E}\cdot\bar{J}_s^*$.

The Poynting vector $\bar{S}$ is in general complex, and its real part represents the real power flow and its imaginary part represents the reactive power. Taking the real and imaginary part of (2-74), we write

$$-\nabla\cdot\bar{S}_r - \tfrac{1}{2}\,\text{Re}(\bar{E}\cdot\bar{J}_s^*) = L, \tag{2-76a}$$

$$-\nabla\cdot\bar{S}_i - \tfrac{1}{2}\,\text{Im}(\bar{E}\cdot\bar{J}_s^*) = 2\omega(W_m - W_e), \tag{2-76b}$$

where $\bar{S}_r$ and $\bar{S}_i$ are the real and imaginary parts of $\bar{S}$, respectively.

Equation (2-76a) states that the real power flowing into a unit volume ($-\nabla\cdot\bar{S}_r$) plus the power supplied by the source $-\frac{1}{2}\,\text{Re}(\bar{E}\cdot\bar{J}_s^*)$ per unit volume is equal to the power loss per unit volume L. Similarly, (2-76b) represents the reactive power per unit volume due to the power flow, the source, and the stored energy densities.

The specific absorption rate (SAR) is used to represent the power loss per unit mass of biological media when the incident power flux density is 1 mW/cm^2 (Table 2-2). If the density of the medium is given by ρ (kg/m^3), the SAR is given by

$$\text{SAR} = \frac{L}{\rho} \qquad \text{watts/kg.} \tag{2-77}$$

TABLE 2-2 Dielectric Constants and Conductivities of (a) Muscle, Skin, and Tissues with High Water Content and (b) Fat, Bone, and Tissues with Low Water Content

Frequency (MHz)	Dielectric constant (a)	Dielectric constant (b)	Conductivity (S/m) (a)	Conductivity (S/m) (b)
27.12	113	20	0.612	10.9–43.2
40.68	97.3	14.6	0.693	12.6–52.8
433	53	5.6	1.43	37.9–118
915	31	5.6	1.60	55.6–147
2450	47	5.5	2.21	96.4–213
5000	43.3	5.05	4.73	186–338

Source: Johnson and Guy (1972).

For biological media, $\mu'' = 0$, and therefore,

$$\text{SAR} = \frac{\omega\epsilon_0\epsilon''|E|^2}{2\rho} = \frac{\sigma|E|^2}{2\rho}. \tag{2-78}$$

The density ρ is usually taken to be approximately equal to that of water ($\rho = 10^3$ kg/m^3).

The definition of the time-averaged electric and magnetic stored energy densities as given in (2-74) and (2-75) is valid if ϵ and μ are independent of frequency. For a dispersive medium, the time-averaged electric and magnetic stored energy densities are given by

$$W_e = \frac{1}{4}\text{Re}\left[\frac{\partial}{\partial\omega}(\omega\epsilon)|\bar{E}|^2\right],$$
$$W_m = \frac{1}{4}\text{Re}\left[\frac{\partial}{\partial\omega}(\omega\mu)|\bar{H}|^2\right]. \tag{2-79}$$

Note that if ϵ and μ are constant, (2-79) reduces to those given in (2-74). The derivation of (2-79) requires consideration of $\omega\epsilon$ in the neighborhood of the operating frequency and is given in Landau and Lifshitz (1960) and Yeh and Liu (1972).

2-6 VECTOR AND SCALAR POTENTIALS

Maxwell's equations are vector differential equations and each equation represents three scalar equations for each of three orthogonal components. It would be more convenient, therefore, if the vector problem were reduced to a scalar problem with a fewer number of equations. This has been done in electrostatics and magnetostatics by using electrostatic and vector potentials to describe electric and magnetic fields, respectively. The concept of these potentials can be extended to electromagnetic fields in the following manner.

We assume that the medium is isotropic, homogeneous, and nondispersive, and therefore, μ and ϵ are scalar and constant. First, we note from (2-4) that the divergence of $\bar{B}$ is zero, and recalling that the divergence of the curl of any vector is zero, $\bar{B}$ can be expressed by the curl of an arbitrary vector $\bar{A}$, called the *vector potential.*

$$\nabla \cdot \bar{B} = 0, \tag{2-80}$$

$$\bar{B} = \nabla \times \bar{A}. \tag{2-81}$$

Then the second Maxwell's equation (2-2) becomes

$$\nabla \times \left(\bar{E} + \frac{\partial \bar{A}}{\partial t} \right) = 0. \tag{2-82}$$

Since the curl of the gradient of any scalar function is zero, the bracketed factor is represented by the gradient of an arbitrary scalar function ϕ, which is called the *scalar potential.*

$$\bar{E} + \frac{\partial \bar{A}}{\partial t} = -\nabla \phi. \tag{2-83}$$

Substituting (2-81) and (2-83) into the first of Maxwell's equation (2-1), we get

$$-\nabla \times \nabla \times \bar{A} - \mu\epsilon \frac{\partial^2 \bar{A}}{\partial t^2} - \mu\epsilon \nabla \frac{\partial \phi}{\partial t} = -\mu \bar{J}. \tag{2-84}$$

Now, substituting (2-83) into (2-3), we get

$$\nabla^2 \phi + \frac{\partial}{\partial t} \nabla \cdot \bar{A} = -\frac{\rho}{\epsilon}. \tag{2-85}$$

Alternatively, we can get (2-85) by taking the divergence of (2-84) and using the continuity equation (2-12). Equations (2-84) and (2-85) are the two equations the vector and scalar potentials must satisfy.

The vector potential $\bar{A}$ above is defined only through $\nabla \times \bar{A}$ in (2-81). In general, a vector field $\bar{A}$ consists of a curl-free component $\bar{A}_1$ and a divergence-free component $\bar{A}_2$.

$$\bar{A} = \bar{A}_1 + \bar{A}_2, \qquad \nabla \times \bar{A}_1 = 0, \qquad \text{and} \qquad \nabla \cdot \bar{A}_2 = 0. \tag{2-86}$$

Since $\nabla \times \bar{A} = \nabla \times \bar{A}_2$ and $\nabla \cdot \bar{A} = \nabla \cdot \bar{A}_1$, we can still choose any $\nabla \cdot \bar{A}$ without affecting $\bar{E}$ and $\bar{H}$. If we choose $\nabla \cdot \bar{A}$ so that it satisfies the *Lorentz condition,*

$$\nabla \cdot \bar{A} + \mu\epsilon \frac{\partial \phi}{\partial t} = 0, \tag{2-87}$$

(2-84) and (2-85) become

$$\nabla^2 \bar{A} - \mu\epsilon \frac{\partial^2 \bar{A}}{\partial t^2} = -\mu \bar{J}, \tag{2-88}$$

$$\nabla^2 \phi - \mu\epsilon \frac{\partial^2 \phi}{\partial t^2} = -\frac{\rho}{\epsilon}, \tag{2-89}$$

where $\nabla^2 \bar{A} = -\nabla \times \nabla \times \bar{A} + \nabla(\nabla \cdot \bar{A})$, and $\bar{J}$ and ρ are related through the continuity equation:

$$\nabla \cdot \bar{J} + \frac{\partial \rho}{\partial t} = 0. \tag{2-90}$$

Once (2-88) and (2-89) are solved for $\bar{A}$ and ϕ, the fields are given by (2-81) and (2-83).

$$\bar{E} = -\frac{\partial \bar{A}}{\partial t} - \nabla \phi, \tag{2-91}$$

$$\bar{B} = \nabla \times \bar{A}. \tag{2-92}$$

Another useful choice of $\nabla \cdot \bar{A}$ is the *Coulomb gauge*, for which $\nabla \cdot \bar{A} = 0$. This is particularly useful for a source-free region ($\bar{J} = 0, \rho = 0$). In this case, (2-84) and (2-85) become

$$\nabla^2 \bar{A} - \mu\epsilon \frac{\partial^2 \bar{A}}{\partial t^2} = 0 \qquad \text{and} \qquad \nabla^2 \phi = 0. \tag{2-93}$$

The fields are given by

$$\bar{E} = -\frac{\partial \bar{A}}{\partial t} \qquad \text{and} \qquad \bar{B} = \nabla \times \bar{A}. \tag{2-94}$$

For a dispersive, isotropic, and homogeneous medium, the time-harmonic Maxwell's equations (2-40) to (2-43) must be used. We then obtain

$$\begin{aligned} \nabla^2 \bar{A} + \omega^2 \mu\epsilon \bar{A} &= -\mu \bar{J}_s, \\ \nabla^2 \phi + \omega^2 \mu\epsilon\phi &= -\frac{\rho_s}{\epsilon_c}, \end{aligned} \tag{2-95}$$

and the Lorentz condition

$$\nabla \cdot \bar{A} + j\omega\mu\epsilon\phi = 0. \tag{2-96}$$

Equations (2-95) and (2-96) constitute the basic formulations of electromagnetic problems in terms of the vector and scalar potentials, and the fields are obtained by (2-91) and (2-92) from A and ϕ with $\partial/\partial t$ replaced by $j\omega$. If the medium is inhomogeneous, however, the equations above do not hold, and it is more convenient to go back and start with the original Maxwell's equations.

2-7 ELECTRIC HERTZ VECTOR

It is possible to combine the vector and scalar potentials and the Lorentz condition and form a single vector called the Hertz vector, from which all the field components can be derived. This useful formulation has been applied to many engineering problems.

Let us define the electric Hertz vector $\bar{\pi}$ such that

$$\bar{A} = \mu\epsilon\frac{\partial\bar{\pi}}{\partial t} \quad \text{and} \quad \phi = -\nabla\cdot\bar{\pi}. \tag{2-97}$$

Then the Lorentz condition is satisfied automatically. Furthermore, we combine $\bar{J}$ and ρ consistent with the continuity equation by using $\bar{P}$:

$$\bar{J} = \frac{\partial\bar{P}}{\partial t} \quad \text{and} \quad \rho = -\nabla\cdot\bar{P}. \tag{2-98}$$

Then we get a single vector equation

$$\nabla^2\bar{\pi} - \mu\epsilon\frac{\partial^2\bar{\pi}}{\partial t^2} = -\frac{\bar{P}}{\epsilon}, \tag{2-99}$$

from which all the electromagnetic fields can be derived.

$$\bar{E} = \nabla(\nabla\cdot\bar{\pi}) - \mu\epsilon\frac{\partial^2}{\partial t^2}\bar{\pi} = \nabla\times\nabla\times\bar{\pi} - \frac{\bar{P}}{\epsilon} \quad \text{and} \quad \bar{H} = \epsilon\nabla\times\frac{\partial\bar{\pi}}{\partial t}. \tag{2-100}$$

Equations (2-99) and (2-100) constitute the basic formulation in terms of the electric Hertz vector $\bar{\pi}$. The vector $\bar{P}$ is called the electric polarization vector and is equal to the dipole moment per unit volume of the exciting source.

For a time-harmonic case (2-40) to (2-43) applicable to a dispersive medium, we have

$$\nabla^2\bar{\pi} + \omega^2\mu\epsilon\bar{\pi} = -\frac{\bar{J}_s}{j\omega\epsilon}, \tag{2-101}$$

$$\bar{E} = \nabla(\nabla\cdot\bar{\pi}) + \omega^2\mu\epsilon\bar{\pi} = \nabla\times\nabla\times\bar{\pi} - \frac{\bar{J}_s}{j\omega\epsilon}, \tag{2-102}$$

$$\bar{H} = j\omega\epsilon\nabla\times\bar{\pi}, \tag{2-103}$$

where $\bar{J}_s$ is the source current density and ϵ is the complex dielectric constant. Equation (2-101) gives scalar wave equation for each Cartesian component π_x, π_y, and π_z. Care should be exercised in the use of (2-101) for the component of $\bar{\pi}$ in other coordinate systems. See Section 3-1.

2-8 DUALITY PRINCIPLE AND SYMMETRY OF MAXWELL'S EQUATIONS

At present, no magnetic charge has been found to exist in nature and Maxwell's equations contain only electric charges and currents. In practice, however, it is often convenient to use the concept of fictitious magnetic currents and charges. For example, we show later that a small current loop is equivalent to a magnetic current. If we include the fictitious magnetic current density $\bar{J}_m$ and charge density ρ_m, Maxwell's equations take the following symmetric form:

$$\nabla \times \bar{H} = \epsilon \frac{\partial \bar{E}}{\partial t} + \bar{J}, \tag{2-104}$$

$$\nabla \times \bar{E} = -\mu \frac{\partial \bar{H}}{\partial t} - \bar{J}_m, \tag{2-105}$$

$$\nabla \cdot \bar{B} = \rho_m. \tag{2-106}$$

$$\nabla \cdot \bar{D} = \rho. \tag{2-107}$$

Because of this symmetry, we can interchange $\bar{E}$ and $\bar{H}$, $\bar{J}$ and J_m, ρ and ρ_m, and ϵ and μ from the unprimed fields to the new primed fields in the following manner:

$$\begin{array}{llll} \bar{E} \to \bar{H}' & \bar{J} \to \bar{J}'_m & \rho \to \rho'_m & \mu \to \epsilon' \\ \bar{H} \to -\bar{E}' & \bar{J}_m \to -\bar{J}' & \rho_m \to -\rho' & \epsilon \to \mu' \end{array} \tag{2-108}$$

Then the primed fields satisfy the same Maxwell's equations. Using this duality principle, when a solution is known for the unprimed fields, the solution for the primed fields can easily be obtained.

The duality relations above are not the only ones to transform the unprimed fields to the primed fields. We can also use the following transformation without affecting Maxwell's equations:

$$\begin{array}{lll} \bar{E} \to \sqrt{\dfrac{\mu}{\epsilon}}\,\bar{H}' & \bar{J} \to \sqrt{\dfrac{\epsilon}{\mu}}\,\bar{J}'_m & \rho \to \sqrt{\dfrac{\epsilon}{\mu}}\,\rho'_m \\ \bar{H} \to -\sqrt{\dfrac{\epsilon}{\mu}}\,\bar{E}' & \bar{J}_m \to -\sqrt{\dfrac{\mu}{\epsilon}}\,\bar{J}' & \rho_m \to -\sqrt{\dfrac{\epsilon}{m}}\,\rho'. \end{array} \tag{2-109}$$

The transformation does not require the interchange of ϵ and μ, and therefore it is useful in dealing with the field relationship for the same medium.

2-9 MAGNETIC HERTZ VECTOR

The symmetric Maxwell's equations (2-104) to (2-107) contain the electric current $\bar{J}$ and the magnetic current $\bar{J}_m$. Therefore, the total field consists of the field due to $\bar{J}$ and the field due to $\bar{J}_m$. The field due to $\bar{J}$ was already obtained in Section 2-7 in terms of the electric Hertz vector $\bar{\pi}$. Similarly, the field due to $\bar{J}_m$ can be obtained by using the transformation (2-108) and by replacing the electric Hertz vector $\bar{\pi}$ with the magnetic Hertz vector $\bar{\pi}_m$. Thus we have the following vector equation corresponding to (2-99):

$$\nabla^2 \bar{\pi}_m - \mu\epsilon \frac{\partial^2}{\partial t^2} \bar{\pi}_m = -\bar{M} \quad \text{and} \quad \bar{M} = \frac{\bar{J}_{ms}}{j\omega\mu}, \tag{2-110}$$

where $\bar{M}$ is the magnetic polarization vector. The fields corresponding to (2-100) are given by

$$\bar{E} = -\mu \nabla \times \frac{\partial \bar{\pi}_m}{\partial t} \quad \text{and} \quad \bar{H} = \nabla(\nabla \cdot \bar{\pi}_m) - \mu\epsilon \frac{\partial^2}{\partial t^2} \bar{\pi}_m = \nabla \times \nabla \times \bar{\pi}_m - \bar{M}. \tag{2-111}$$

The magnetic polarization vector $\overline{M}$ is the magnetic dipole moment per unit volume.

For a time-harmonic case, we have

$$\nabla^2 \overline{\pi}_m + \omega^2 \mu\epsilon\overline{\pi}_m = -\frac{\overline{J}_{ms}}{j\omega\mu}, \tag{2-112}$$

$$\overline{E} = -j\omega\mu\nabla \times \overline{\pi}_m, \tag{2-113}$$

$$\overline{H} = \nabla(\nabla \cdot \overline{\pi}_m) + \omega^2 \mu\epsilon\overline{\pi}_m = \nabla \times \nabla \times \overline{\pi}_m - \frac{\overline{J}_{ms}}{j\omega\mu}, \tag{2-114}$$

where $\overline{J}_{ms}$ is the magnetic source current density.

2-10 UNIQUENESS THEOREM

For a passive network with N terminals, if N voltages $v_1, v_2, \ldots, v_N$ are applied at these N terminals, all the voltages and currents inside the network are uniquely determined. Similarly, if N currents $I_1, I_2, \ldots, I_N$ are applied at the terminals, this also uniquely determines all the voltages and currents. Or we can specify the voltages at some of the N terminals and specify the currents at the rest of the terminals. This also gives a unique distribution of voltages and currents. It is also obvious that we cannot specify *both* voltages $v_1 \cdots v_N$ and currents $I_1 \cdots I_N$ at the N terminals. These conditions appear to be obvious for network problems. However, in electromagnetic problems, these conditions are not obvious since we need to consider a volume V surrounded by a surface S and ask: What field quantities should be specified on the surface S in order to uniquely determine all fields inside?

The quantities could be tangential or normal, electric or magnetic fields, or displacement or flux vectors. Among all these quantities, which can give the unique field inside the surface S? We will show that one of the following three conditions is necessary and sufficient to determine uniquely all the fields inside.

1. The tangential electric field $\hat{n} \times \overline{E}$ is specified on S.
2. The tangential magnetic field $\hat{n} \times \overline{H}$ is specified on S.
3. The tangential electric field $\hat{n} \times \overline{E}$ is specified on a part of S and the tangential magnetic field $\hat{n} \times \overline{H}$ is specified on the rest of S. (2-115)

Note that these conditions correspond to the three conditions mentioned above for network problems.

We shall prove (2-115) for a time-harmonic case. Let us consider two different fields, $(\overline{E}_1, \overline{H}_1)$ and $(\overline{E}_2, \overline{H}_2)$, both of which satisfy Maxwell's equations. We will show that if both satisfy one of the conditions in (2-115) on S, these two fields are identical within V, and thus the field in V is unique. To show this, consider the fields $\overline{E}_d = E_1 - \overline{E}_2$ and $\overline{H}_d = \overline{H}_1 - \overline{H}_2$. Since both $(\overline{E}_1, \overline{H}_1)$ and $(\overline{E}_2, \overline{H}_2)$ satisfy Maxwell's equations, $(\overline{E}_d, \overline{H}_d)$ also satisfies Maxwell's equations and consequently, satisfies the Poynting theorem (2-74). Noting that in a passive medium, the source current $\overline{J}_s$ is zero, we write (2-74) in the following integral form for the volume V:

$$\int_S \frac{1}{2}\bar{E}_d \times \bar{H}_d^* \cdot d\bar{a} = -2j\omega \int_V \left(\frac{\epsilon_0 \epsilon'}{4}|\bar{E}_d|^2 - \frac{\mu_0 \mu'}{4}|\bar{H}_d|^2\right) dV$$
$$- \int_V \left(\frac{\omega\epsilon_0 \epsilon''}{2}|\bar{E}_d|^2 + \frac{\omega\mu_0 \mu''}{2}|\bar{H}_d|^2\right) dV. \qquad (2\text{-}116)$$

If both $(\bar{E}_1, \bar{H}_1)$ and $(\bar{E}_2, \bar{H}_2)$ satisfy one of the conditions (2-115), the left side of (2-116) is zero, since $\bar{E}_d \times \bar{H}_d^* \cdot \hat{n} = \hat{n} \times \bar{E}_d \cdot \bar{H}_d^* = 0$, where $d\bar{a} = \hat{n}\, da$. Therefore, the right side of equation (2-116) must be zero. The first integral is purely imaginary. The second integral on the right side of (2-116) is always positive and real whenever $\epsilon'' \neq 0$ and $\mu'' \neq 0$ unless $\bar{E}_d = 0$ and $\bar{H}_d = 0$. For any physical medium, ϵ'' and μ'' are always nonzero and positive, and therefore $\bar{E}_d$ and $\bar{H}_d$ must be zero, proving that $\bar{E}_1 = \bar{E}_2$ and $\bar{H}_1 = \bar{H}_2$ and the field inside the surface S is unique as long as one of (2-115) is satisfied.

Note that if a lossless cavity is inside S, the field inside the cavity is independent of the field on S and cannot be determined uniquely by the field on S. However, a completely lossless cavity does not exist and thus the uniqueness theorem can be applied to any physical medium. For a general time-varying case, a similar proof can be given to show that (2-115) is also a necessary and sufficient condition (Stratton, 1941, p. 486).

2-11 RECIPROCITY THEOREM

It is well understood that the reciprocity theorem holds for any linear passive network. For example, if a voltage V_a applied at the input terminal produces a short circuit current I_a at the output terminal, a voltage V_b applied at the output terminal produces the short circuit current I_b at the input satisfying the following reciprocity relationship:

$$\frac{I_a}{V_a} = \frac{I_b}{V_b}. \qquad (2\text{-}117)$$

In electromagnetics, the equivalent relationship is called the "Lorentz reciprocity theorem."

Let us consider time-harmonic electromagnetic fields. We consider the field $(\bar{E}_a, \bar{H}_a)$ produced by $\bar{J}_a$ and $\bar{J}_{ma}$. We also consider the field $(\bar{E}_b, \bar{H}_b)$ produced by $\bar{J}_b$ and $\bar{J}_{mb}$, in the same medium (Fig. 2-5). We first note the following:

$$\nabla \cdot (\bar{E}_a \times \bar{H}_b) = \bar{H}_b \cdot \nabla \times \bar{E}_a - \bar{E}_a \cdot \nabla \times \bar{H}_b. \qquad (2\text{-}118)$$

We then use Maxwell's equations for $(\bar{E}_a, \bar{H}_a)$:

$$\nabla \times \bar{E}_a = -j\omega\mu\bar{H}_a - \bar{J}_{ma},$$
$$\nabla \times \bar{H}_a = j\omega\epsilon\bar{E}_a + \bar{J}_a. \qquad (2\text{-}119)$$

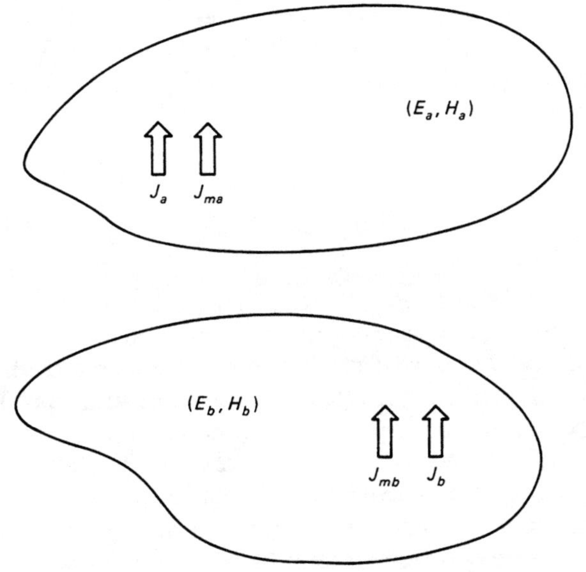

Figure 2-5 Lorentz reciprocity theorem.

We also use the similar equations for $(\bar{E}_b, \bar{H}_b)$. We then get

$$\nabla \cdot (\bar{E}_a \times \bar{H}_b) = -j\omega\mu\bar{H}_a \cdot \bar{H}_b - \bar{H}_b \cdot \bar{J}_{ma} - j\omega\epsilon\bar{E}_a \cdot \bar{E}_b - \bar{E}_a \cdot \bar{J}_b. \tag{2-120}$$

Interchanging a and b, we get a similar equation for $\nabla \cdot (\bar{E}_b \times \bar{H}_a)$. Subtracting one from another, we get the *Lorentz reciprocity theorem.*

$$\nabla \cdot (\bar{E}_a \times \bar{H}_b - \bar{E}_b \times \bar{H}_a) = -(\bar{E}_a \cdot \bar{J}_b - \bar{H}_a \cdot \bar{J}_{mb}) + \bar{E}_b \cdot \bar{J}_a - \bar{H}_b \cdot \bar{J}_{ma}. \tag{2-121}$$

In order to examine the meaning of the Lorentz reciprocity theorem, let us consider two special cases. If the volume V contains no exciting sources, $\bar{J} = \bar{J}_m = 0$, we have the following Lorentz reciprocity theorem applicable to any closed surface enclosing a source-free region:

$$\int_S (\bar{E}_a \times \bar{H}_b - \bar{E}_b \times \bar{H}_a) \cdot d\bar{a} = 0. \tag{2-122}$$

Next, we consider a case where the volume V extends to infinity. In this case, at a large distance r from the source, the field component behaves as a spherical wave and is proportional to $(1/r)\exp(-jkr)$ and since $d\bar{a}$ is proportional to r^2, the left side of (2-121) is proportional to $\exp(-j2kr)$. Since $k = k_0 n$ has a negative imaginary part for any physical medium, the left side of (2-121) tends to zero as $r \to \infty$. Therefore, the volume integral of the right side of (2-121) taken over the entire space extending to infinity is zero.

We rewrite this reciprocity relationship using the following form:

$$\langle a, b \rangle = \langle b, a \rangle, \tag{2-123}$$

where $\langle a, b \rangle = \int_\infty (\bar{E}_a \cdot \bar{J}_b - \bar{H}_a \cdot \bar{J}_{mb})\, dV$. This shows that the electric field $\bar{E}_a$ at a point

$\bar{r}_1$ due to the current $\bar{J}_a$ at $\bar{r}_2$ is the same as the electric field $\bar{E}_b$ at $\bar{r}_2$ due to the same current $\bar{J}_b = \bar{J}_a$ at $\bar{r}_1$. It should be easy to see the correspondence between (2-117) and (2-123). The quantity $\langle a, b \rangle$ given in (2-123) is called the *reaction* by Rumsey (1954) and is useful in solving boundary value problems.

2-12 ACOUSTIC WAVES

Although electromagnetic waves are vector fields, there are several reasons why we wish to examine scalar fields. First, electromagnetic problems can often be approximated by scalar problems because they reveal many important features of the problem without excessive mathematical complexities. Also, in cases such as two-dimensional problems, vector electromagnetic problems can be reduced to scalar problems. In addition, the study of scalar acoustic problems are of practical interest in many applications, including ultrasound in medicine and ocean acoustics.

For acoustic waves, the pressure variation $p(\bar{r}, t)$ replaces the electric field $\bar{E}(\bar{r}, t)$ in previous sections. In this section we present a brief description of basic formulations for acoustic waves propagating in material media. We may classify material media into fluids and solids. *Fluids* include both gases and liquids and are in general *viscous*. However, often the effects of viscosity are negligible, and we consider *nonviscous* fluids, which are also called *perfect* fluids. Fluids are in general *compressible*, but when the density of the medium can be assumed constant, they are called *incompressible* fluids. Propagation of an acoustic wave takes place in elastic solids, viscous compressible fluids, and perfect compressible fluids. In this section we outline the formulation for acoustic-wave propagation in perfect compressible fluids such as gases and liquids. Examples include fog particles in air and bubbles in water (Morse and Ingard, 1968).

There are two fundamental equations: the equation of motion and the equation of conservation of mass. The equation of motion for a small elementary volume of the medium is

$$\rho \frac{d\bar{V}}{dt} + \operatorname{grad} p = 0, \tag{2-124}$$

where ρ is density (kg/m^3), $\bar{V}$ is the particle velocity (m/s), and p is pressure (N/m^2). The particle velocity $\bar{V}$ is the velocity of the elementary volume of the fluid and should be distinguished from the velocity c of an acoustic wave in the medium. The derivative d/dt is the time rate of change in a coordinate system attached to a particular portion of a fluid and moving with the fluid. This is called the Lagrangian description. In contrast, $\partial/\partial t$ is the time derivative at a fixed point in space as the fluid flows past that point and is called the Eulerian description. They are related through

$$\frac{d}{dt} = \frac{\partial}{\partial t} + (\bar{V} \cdot \operatorname{grad}). \tag{2-125}$$

The conservation of mass is given by

$$\nabla \cdot (\rho \overline{V}) + \frac{\partial \rho}{\partial t} = 0. \tag{2-126}$$

Let us decompose p, ρ, and $\overline{V}$ into their average values p_0, ρ_0, and $\overline{V}_0$ and the small vibrating acoustic-wave components p_1, ρ_1, and $\overline{V}_1$.

First, we assume that the fluid is stationary, $\overline{V}_0 = 0$, and the magnitudes of acoustic pressure p_1 and density ρ_1 are small compared with the average values p_0 and ρ_0. Then keeping the linear terms, (2-124) and (2-126) become

$$\begin{aligned} \rho_0 \frac{\partial \overline{V}_1}{\partial t} + \operatorname{grad} p_1 = 0, \\ \rho_0 \nabla \cdot \overline{V}_1 + \frac{\partial \rho_1}{\partial t} = 0. \end{aligned} \tag{2-127}$$

In general, pressure p is a function of density ρ and therefore for small p_1 and ρ_1, we can expand p in Taylor's series and keep the first two terms.

$$p = p_0 + p_1 = p_0 + \left(\frac{\partial p}{\partial \rho}\right)_{\rho_0} \rho_1. \tag{2-128}$$

Since p_1 is linearly related to ρ_1, we write

$$p_1 = c^2 \rho_1, \qquad \text{where} \quad c^2 = \left(\frac{\partial p}{\partial \rho}\right)_{\rho_0}. \tag{2-129}$$

The value of the constant c depends on the material under consideration, and it is the velocity of an acoustic wave in that medium. Note the difference between the particle velocity $\overline{V}_1$ and the acoustic velocity c. Using (2-129), we express ρ_1 in (2-127) by p_1.

$$\nabla \cdot \overline{V}_1 + \frac{1}{c^2 \rho_0} \frac{\partial p_1}{\partial t} = 0. \tag{2-130}$$

Taking the divergence of $\overline{V}_1$ in (2-127) and substituting (2-130), we get

$$\nabla \cdot \left(\frac{1}{\rho_0} \nabla p_1\right) - \kappa \frac{\partial^2 p_1}{\partial t^2} = 0, \tag{2-131}$$

where $\kappa = 1/c^2 \rho_0$ is called the *compressibility*.

Equation (2-131) is the basic acoustic wave equation for a stationary medium. Note that $(1/\rho_0)$ is inside the divergence operation. For a uniform medium, this can be taken outside the divergence,

$$\nabla^2 p_1 - \frac{1}{c^2} \frac{\partial^2 p_1}{\partial t^2} = 0. \tag{2-132}$$

For a time-harmonic case $\exp(j\omega t)$, we have

$$\begin{aligned} (\nabla^2 + k^2) p_1 = 0, \\ \overline{V}_1 = -\frac{1}{j\omega\rho_0} \nabla p_1. \end{aligned} \tag{2-133}$$

The boundary conditions at an interface between two media are:

1. Continuity of pressure p_1
2. Continuity of the normal component of the particle velocity $\bar{V}_1$ [or $(1/\rho_0)(\partial p_1/\partial n)$, where $\partial/\partial n$ is the normal derivative]

The power flux density vector $\bar{S}$ is given by

$$\bar{S} = \tfrac{1}{2} p_1 \bar{V}_1^*. \tag{2-134}$$

Alternatively, we can use the velocity potential ψ. For a uniform medium, the curl of $\bar{V}_1$ is zero from (2-127), and thus it is possible to express $\bar{V}_1$ in terms of a scalar function ψ.

$$\bar{V}_1 = -\operatorname{grad} \psi. \tag{2-135}$$

The function ψ is called the *velocity potential*, and it satisfies the wave equation

$$\nabla^2 \psi - \frac{1}{c^2} \frac{\partial^2 \psi}{\partial t^2} = 0, \tag{2-136}$$

and p_1 and ρ_1 are expressed by

$$p_1 = \rho_0 \frac{\partial \psi}{\partial t}, \qquad \rho_1 = \frac{\rho_0}{c^2} \frac{\partial \psi}{\partial t}. \tag{2-137}$$

For a time-harmonic case $[\exp(j\omega t)]$, we have

$$\begin{aligned} &(\nabla^2 + k^2)\psi = 0, \qquad k = \frac{\omega}{c}, \\ &\bar{V}_1 = -\operatorname{grad} \psi, \qquad p_1 = j\omega\rho_0 \psi. \end{aligned} \tag{2-138}$$

The boundary conditions at an interface are:

1. Continuity of $\rho_0 \psi$
2. Continuity of $\partial\psi/\partial n$

For a plane acoustic wave propagating in the x direction, we have from (2-133)

$$\begin{aligned} p_1 &= A_0 \exp(-jkx) \\ \bar{V}_1 &= \hat{x} \frac{p_1}{\rho_0 c}. \end{aligned} \tag{2-139}$$

The ratio $\rho_0 c$ of p_1 to V_1 is called the characteristic impedance. Its MKS unit is rayl (after Rayleigh) and 1 rayl = 1 kg/m$^2 \cdot$ s. The MKS unit for acoustic pressure is Pa (pascal) and 1 Pa = 1 N/m^2 = 10 μbar.

In seawater at 13°C and with a salinity of 35 (in parts by weight per thousand), the standard sound velocity is taken to be 1500 m/s, the standard characteristic impedance is $\rho_0 c = 1.54 \times 10^6$ rayl, and the density $\rho_0 = 1026$ kg/m^3. Air at a temperature of 20°C and at standard atmospheric pressure (1 atm = 1.013×10^5 N/m^2 = 1013.25 mbar) has the density $\rho_0 = 1.21$ kg/m^3, the sound velocity $c = 343$ m/s, and the characteristic impedance $\rho_0 c = 415$ rayl. Oil has a density of 900 kg/m^3, sound velocity of 1300 m/s, and characteristic impedance of 1.117×10^6 rayl.

PROBLEMS

2-1. Consider a circular parallel-plate capacitor with dielectric material shown in Fig. P2-1. A dc current I_0 is switched on at $t = 0$. Find the fields $\bar{E}$, $\bar{H}$, and the Poynting vector $\bar{S} = \bar{E} \times \bar{H}$ as functions of position $\bar{r}$ and time inside the capacitor. Assume that the fringe field is negligible and that the fields are confined within the capacitor. Also find the electromagnetic energy density W defined in (2-69). Show that

$$\nabla \cdot \bar{S} + \frac{\partial}{\partial t} W = 0.$$

2-2. A microwave propagating in tissue is expressed by

$$E_x = E_0 e^{-jkz},$$

$$H_y = \frac{nE_0}{Z_0} e^{-jkz},$$

$$k = \frac{\omega}{c} n, \qquad Z_0 = \left(\frac{\mu_0}{\epsilon_0}\right)^{1/2}.$$

The power flux density at $z = 0$ is 1 mW/cm^2. Calculate E_0, W_e, W_m, L, and SAR as functions of z. Show that they satisfy (2-76a) and (2-76b). Use Table 2-2(a) and 915 MHz.

2-3. The magnetic Hertz vector $\bar{\pi}_m$ for a small loop antenna is given by

$$\bar{\pi}_m = \frac{Ae^{-jkr}}{4\pi r} \hat{z}, \qquad A = \text{constant}.$$

Find $\bar{E}$ and $\bar{H}$ and express all components (E_θ, E_ϕ, E_r, H_θ, H_ϕ, and H_r) in the spherical coordinate system.

2-4. The electric Hertz vector $\bar{\pi}$ for a small wire antenna in free space is given by

$$\bar{\pi} = \frac{Ae^{-jk_0 r}}{4\pi r} \hat{z}, \qquad A = \text{constant}.$$

Find $\bar{E}$ and $\bar{H}$ and express all components (E_θ, E_ϕ, E_r, H_θ, H_ϕ, and H_r) in the spherical coordinate systems.

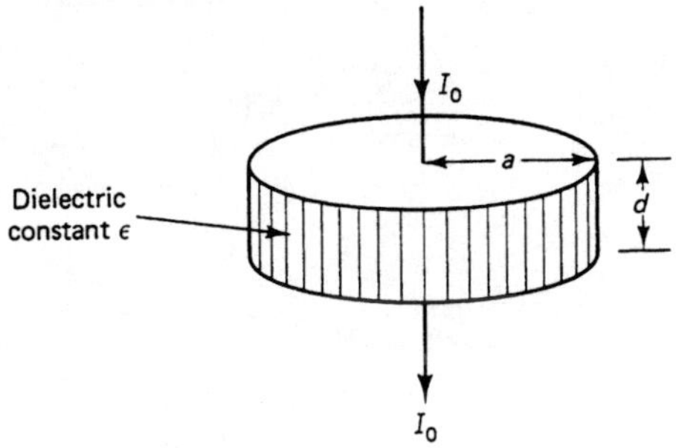

Figure P2-1 Circular capacitor.

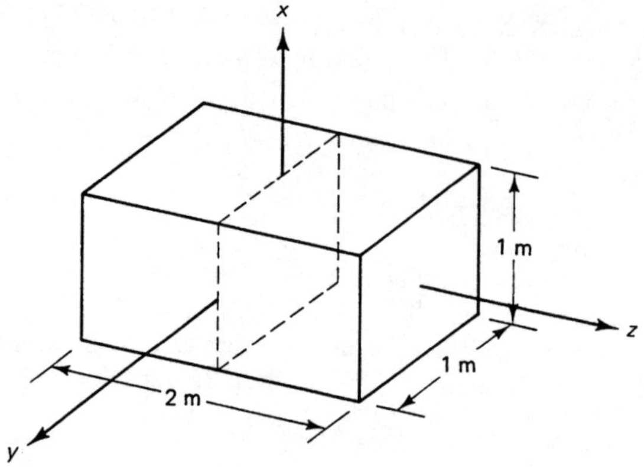

Figure P2-6 Current sheet is located at $z = 0$. The volume is $1\ \text{m} \times 1\ \text{m} \times 2\ \text{m}$.

2-5. Show that the following $\bar{\pi}$ satisfies the wave equation in free space:

$$\bar{\pi} = A \sin k_1 x \sin k_2 y \exp[-j\sqrt{k_0^2 - k_1^2 - k_2^2}z]\hat{z}.$$

Find all components of $\bar{E}$ and $\bar{H}$ derived from this $\bar{\pi}$.

2-6. A sheet of uniform current I_0 (A/m) flowing in the x direction is located in free space at $z = 0$. This is represented by the source current density $\bar{J}_s = \hat{x} I_0 \delta(z)$. The Hertz vector $\bar{\Pi} = \Pi_x \hat{x}$ is given by

$$\Pi_x = -\frac{I_0 \exp(-jk_0|z|)}{2k_0 \omega \epsilon_0}, \qquad k_0 = \omega(\mu_0 \epsilon_0)^{1/2}.$$

Find $\bar{E}$ and $\bar{H}$ at z.

Consider the volume shown in Fig. P2-6 and integrate (2-76) over this volume. Show that the real power flowing out of the volume is equal to the power supplied by the source. Also calculate the reactive power shown in (2-76b).

3

Waves in Inhomogeneous and Layered Media

In this chapter we start with a review of plane-wave propagation in homogeneous media and reflection and refraction of plane waves by layered media. We then consider the propagation characteristics of guided waves along layered media. This gives us an opportunity to discuss various types of guided waves and to introduce the idea of complex waves. The chapter concludes with the problem of wave propagation in a medium whose refractive index varies as a function of z only. Here we discuss the WKB approximation and the turning point.

3-1 WAVE EQUATION FOR A TIME-HARMONIC CASE

Maxwell's equations are the coupled first-order vector differential equations for the two field quantities $\overline{E}$ and $\overline{H}$. We can combine these two equations, eliminate one of the field quantities and obtain the uncoupled second-order vector differential equation for one field quantity in the following manner.

Let us write Maxwell's equations for a time-harmonic wave

$$\nabla \times \overline{H} = j\omega\epsilon\overline{E}, \tag{3-1}$$

$$\nabla \times \overline{E} = -j\omega\mu\overline{H}, \tag{3-2}$$

where $\epsilon = \epsilon_0 \epsilon_r$ is the complex dielectric constant (permittivity) and $\mu = \mu_0 \mu_r$ is the permeability. $\mu_r = 1$ for most materials except for magnetic material. We consider a homogeneous medium where ϵ and μ are constant.

To combine (3-1) and (3-2), we take the curl of (3-2) and substitute (3-1). We get

$$\nabla \times (\nabla \times \bar{E}) - \omega^2 \mu\epsilon \bar{E} = 0. \tag{3-3}$$

This is a vector wave equation and can be decomposed into three components in three orthogonal directions.

For the special case of a Cartesian coordinate system, we use the following identity:

$$-\nabla \times \nabla \times \bar{E} + \nabla(\nabla \cdot \bar{E}) = \nabla^2 \bar{E}. \tag{3-4}$$

Together with $\nabla \cdot \bar{D} = 0$, we get the following scalar wave equation for each component E_x, E_y, and E_z.

$$(\nabla^2 + k^2) E_x = 0, \tag{3-5}$$

where ∇^2 is the Laplacian, $k = k_0 n$ is the wave number in the medium, $k_0 = \omega(\mu_0 \epsilon_0)^{1/2} = \omega/c = (2\pi)/\lambda_0$ is the wave number in free space, c is the speed of light in free space, λ_0 is the wavelength in free space, and $n = (\mu_r \epsilon_r)^{1/2}$ is the refractive index of the medium.

The identity (3-4) is true for the Cartesian system, but care should be exercised in the use of (3-4) for other coordinate systems. For example, in the cylindrical coordinate system (ρ, ϕ, z), the z component of the left side of (3-4) is equal to $\nabla^2 E_z$, the right side. However, the ρ (or ϕ) component of the left side is *not* equal to $\nabla^2 E_\rho$ (or $\nabla^2 E_\phi$).

Once we solve (3-5) and obtain $\bar{E}$, we can easily obtain $\bar{H}$ by (3-2):

$$\bar{H} = -\frac{1}{j\omega\mu} \nabla \times \bar{E}. \tag{3-6}$$

3-2 TIME-HARMONIC PLANE-WAVE PROPAGATION IN HOMOGENEOUS MEDIA

Let us consider a time-harmonic plane wave propagating in the z direction. First, we note that all the fields are functions of z only and since $\nabla \cdot \bar{E} = \partial E_z/\partial z = 0$, E_z is independent of z. Therefore, E_z varying in z is zero, and we conclude that $E_z = 0$ for a plane wave propagating in the z direction. Similarly, we have $H_z = 0$ for a plane wave from $\nabla \cdot \bar{H} = 0$. Since all the components of $\bar{E}$ and $\bar{H}$ are transverse to the direction of the wave propagation, the plane electromagnetic field is called the *transverse electromagnetic* (TEM) wave.

Let us assume that the field is linearly polarized in the x direction ($E_x \neq 0, E_y = 0$). Solving (3-5) and using (3-6), we get

$$E_x = E_0 e^{-jkz}, \tag{3-7}$$

$$H_y = \frac{E_0}{\eta} e^{-jkz}, \tag{3-8}$$

where $\eta = (\mu/\epsilon)^{1/2} = \eta_0(\mu_r/\epsilon_r)^{1/2}$ is the characteristic impedance of the medium, and $\eta_0 = (\mu_0/\epsilon_0)^{1/2} \approx 120\pi$ ohms is the characteristic impedance of free space.

In a lossy medium, using $n = n' - jn''$, we get

$$\begin{aligned} E_x &= E_0 \exp(-jkz) = E_0 \exp(-j\beta z - \alpha z), \\ H_y &= \frac{E_x}{\eta}, \end{aligned} \tag{3-9}$$

where $\beta = k_0 n' = 2\pi/\lambda = \omega/V$ is called the *phase constant*, λ the wavelength in the medium, V the phase velocity, and α the *attenuation constant*. Also note that at the distance $\delta = \alpha^{-1}$, the wave attenuates to the value of $\exp(-1)$ of the magnitude at $z = 0$. This distance δ is called the *skin depth* of the medium, because the wave cannot penetrate in the medium much farther than this depth. For example, if the medium is highly conducting,

$$\epsilon_r = \epsilon' - j\epsilon'' \approx -j\frac{\sigma}{\omega\epsilon_0}. \tag{3-10}$$

Therefore, we get

$$\delta \approx \left(\frac{2}{\omega\mu_0\sigma}\right)^{1/2}, \tag{3-11}$$

In general, however, the skin depth δ is given by $\alpha^{-1} = (k_0 n'')^{-1}$.

A plane wave propagating in an arbitrary direction $\hat{i}$ is given by

$$\begin{aligned} \bar{E} &= \bar{E}_0 e^{-j\bar{K}\cdot\bar{r}}, \\ \bar{H} &= \frac{\hat{i} \times \bar{E}}{\eta}, \\ \bar{K} &= k\hat{i} \quad \text{and} \quad \bar{r} = x\hat{x} + y\hat{y} + z\hat{z}. \end{aligned} \tag{3-12}$$

We also note that for a plane wave,

$$\nabla e^{-j\bar{K}\cdot\bar{r}} = -j\bar{K}e^{-j\bar{K}\cdot\bar{r}}. \tag{3-13}$$

Therefore, $\nabla \cdot \bar{D} = 0$ means that $\bar{K} \cdot \bar{D} = 0$. Similarly $\bar{K} \cdot \bar{B} = 0$, showing that $\bar{D}$ and $\bar{B}$ are perpendicular to the direction of propagation $\hat{i}$.

3-3 POLARIZATION

In Section 3-2 we considered a linearly polarized plane wave polarized in the x direction ($E_x \neq 0$, $E_y = 0$, and $E_z = 0$) and propagating in the z direction. In general, however, the electric field vector $\bar{E}$ can be polarized in any direction ($E_x \neq 0$, $E_y \neq 0$, and $E_z = 0$). If E_x and E_y are in phase, the electric field vector $\overline{E}$ at a given location is always pointed in a fixed direction, and the wave is said to be *linearly polarized*. For a linearly polarized wave, the direction of the electric vector is called the *direction of polarization*, and the plane containing the electric vector and the direction of wave propagation is called the *plane of polarization* (Fig. 3-1a).

In general, when E_x and E_y are not in phase, we write

$$E_x = E_1 e^{j\delta_1},$$
$$E_y = E_2 e^{j\delta_2}, \tag{3-14}$$

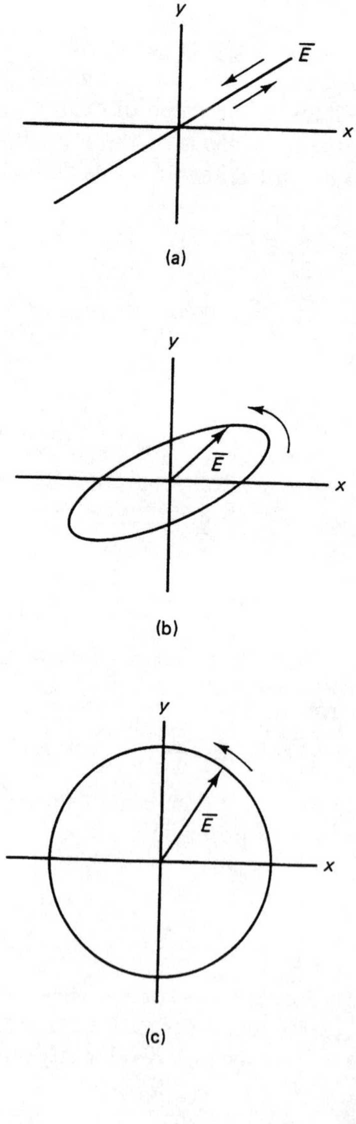

Figure 3-1 Polarizations: (a) linear; (b) elliptic; (c) circular (right-handed).

where $\delta = \delta_1 - \delta_2$ is the phase difference. As a function of time, E_x and E_y are given by

$$E_x = E_1 \cos(\omega t + \delta_1),$$
$$E_y = E_2 \cos(\omega t + \delta_2). \tag{3-15}$$

If we eliminate ωt from (3-15), we get an equation to describe the locus of the tip of the electric vector $\overline{E}$. To do so, we write

$$\frac{E_y}{E_2} = \cos A, \qquad A = \omega t + \delta_2, \tag{3-16}$$

$$\frac{E_x}{E_1} = \cos(A + \delta) = \cos A \ \cos\delta - \sin A \ \sin\delta. \tag{3-17}$$

Eliminating A from (3-16) and (3-17), we get

$$\left(\frac{E_x}{E_1}\right)^2 + \left(\frac{E_y}{E_2}\right)^2 - 2\left(\frac{E_x}{E_1}\right)\frac{E_y}{E_2}\cos\delta = \sin^2\delta. \tag{3-18}$$

This is an equation of an ellipse, and thus the wave is said to be *elliptically polarized* (Fig. 3-1b).

If the phase difference is exactly 90° or −90° and the magnitudes E_1 and E_2 are the same ($E_1 = E_2 = E_0$), then (3-18) becomes an equation of a circle, and the wave is said to be *circularly polarized* (Fig. 3-1c). If $\delta = +90°$, the rotation of the electric field vector and the direction of the wave propagation form a right-handed screw, and this is called the *right-handed circular polarization* (RHC), and E_x and E_y are related by

$$E_x = +jE_y. \tag{3-19}$$

If $\delta = -90°$, this is called the *left-handed circular polarization* (LHC) and E_x and E_y satisfy

$$E_x = -jE_y. \tag{3-20}$$

This definition of RHC and LHC is normally used in electrical engineering and IEEE publications. However, in physics, RHC refers to the wave whose electric vector is rotating in the clockwise direction as observed by the receiver, while the engineering definition of RHC refers to the wave whose electric vector is rotating in the clockwise direction as viewed by the transmitter. Therefore, RHC in engineering definition is LHC in physics.

The *polarized waves* described above are deterministic, which means that the field quantities are definite functions of time and position. In contrast, if the field quantities are completely random and E_x and E_y are uncorrelated, the wave is called *unpolarized*. The light from the sun can be considered unpolarized. In many situations the waves may be partially polarized. We include a detailed discussion of this topic, including Stokes' vectors and the Poincaré sphere, in Chapter 10.

3-4 PLANE-WAVE INCIDENCE ON A PLANE BOUNDARY: PERPENDICULAR POLARIZATION (*s* POLARIZATION)

Let us consider the reflection and refraction of a plane wave incident on a single boundary separating two media with dielectric constants ϵ_1 and ϵ_2 and permeabilities μ_1 and μ_2, respectively (Fig. 3-2). Both media can be lossy, and therefore, ϵ_1, ϵ_2, μ_1, and μ_2 can be complex. We define the *plane of incidence* as the plane including the direction of the wave propagation and the normal to the boundary, and we choose the x–z plane to be the plane of incidence. The *angle of incidence* θ_i is defined as the angle between the direction of propagation and the normal to the boundary (Fig. 3-2). The refractive indexes n_1 and n_2 are given by $n_1 = [(\epsilon_1 \mu_1)/(\epsilon_0 \mu_0)]^{1/2}$ and $n_2 = [(\epsilon_2 \mu_2)/(\epsilon_0 \mu_0)]^{1/2}$, and the characteristic impedance of the medium is $\eta_1 = (\mu_1/\epsilon_1)^{1/2}$ and $\eta_2 = (\mu_2/\epsilon_2)^{1/2}$, respectively (see Section 3-2).

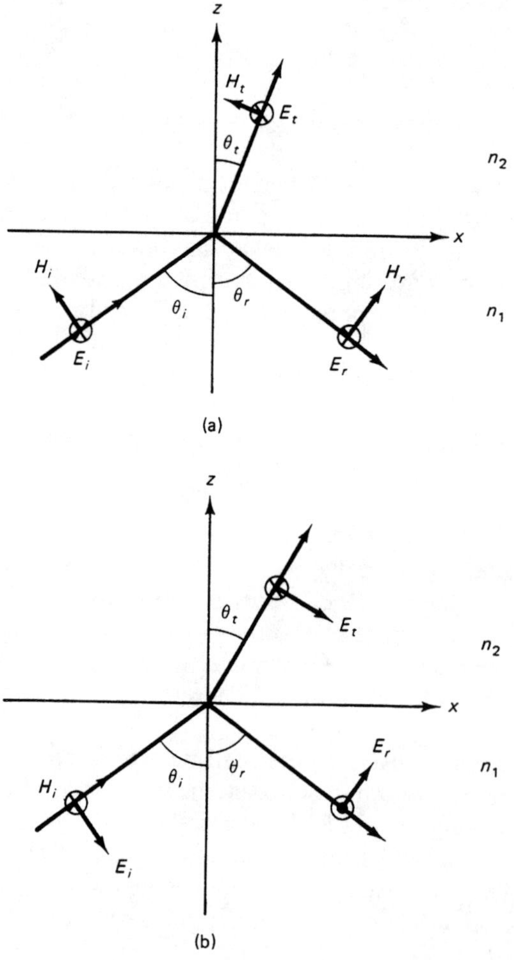

Figure 3-2 (a) Perpendicular and (b) parallel polarizations.

In Fig. 3-2 we show two cases: (a) perpendicular polarization (electric field perpendicular to the plane of incidence) and (b) parallel polarization (electric field parallel to the plane of incidence). Case (a) is also called the *s polarization* (*senkrecht*, "perpendicular" in German) or the *TE wave* (electric field is transverse to the direction of wave propagation). Case (b) is then called the *p polarization* (parallel) or the *TM wave* (magnetic field is transverse to the direction of wave propagation). These two waves are independent. In fact, for a two-dimensional problem where the dielectric constant and permeability are functions of x and z only ($\partial/\partial y = 0$), the electromagnetic field can be separated into two independent waves: TE and TM.

Let us consider the perpendicular polarization. We use (3-12) and note that $\hat{i} = \hat{x} \sin\theta_i + \hat{z} \cos\theta_i$. We then write the incident field E_{yi} with magnitude E_0.

$$E_{yi} = E_0 \exp(-jq_i z - j\beta_i x), \tag{3-21}$$

where $q_i = k_1 \cos\theta_i$, $\beta_i = k_1 \sin\theta_i$, $k_1 = \omega(\mu_1 \epsilon_1)^{1/2} = k_0 n_1$ is the wave number in medium I, and $k_0 = \omega(\mu_0 \epsilon_0)^{1/2}$ is a free-space wave number. The reflected field E_{yr} also satisfies the wave equation, and therefore we write

$$E_{yr} = R_s E_0 \exp(+jq_r z - j\beta_r x), \tag{3-22}$$

where R_s is the reflection coefficient for perpendicular polarization (s-polarization), $q_r = k_1 \cos\theta_r$ and $\beta_r = k_1 \sin\theta_r$, and θ_r is the angle of reflection. Similarly, for the transmitted field we have

$$E_{yt} = T_s E_0 \exp(-jq_t z - j\beta_t x), \tag{3-23}$$

where T_s is the transmission coefficient, $q_t = k_2 \cos\theta_t$, $\beta_t = k_2 \sin\theta_t$, $k_2 = k_0 n_2$, and θ_t is the angle of transmission.

Now we apply the boundary conditions at $z = 0$: the continuity of the tangential electric field E_y and the magnetic field H_x. First we apply the continuity of the tangential electric field:

$$E_{yi} + E_{yr} = E_{yt} \qquad \text{at } z = 0. \tag{3-24}$$

This yields the following:

$$\exp(-j\beta_i x) + R_s \exp(-j\beta_r x) = T_s \exp(-j\beta_t x). \tag{3-25}$$

For this to hold at all x, all exponents must be the same, thus we get the *phase-matching condition*:

$$\beta_i = \beta_r = \beta_t. \tag{3-26}$$

From this, we obtain the *law of reflection*,

$$\theta_i = \theta_r, \tag{3-27}$$

and *Snell's law*,

$$n_1 \sin\theta_i = n_2 \sin\theta_t. \tag{3-28}$$

With (3-26), (3-25) becomes

$$1 + R_s = T_s. \tag{3-29}$$

To apply the boundary condition for the tangential magnetic field H_x, we use

$$H_x = \frac{1}{j\omega\mu}\frac{\partial}{\partial z}E_y, \tag{3-30}$$

and write

$$\begin{aligned} H_{xi} &= -\frac{E_0}{Z_1}\exp(-jq_i z - j\beta_i x),\\ H_{xr} &= \frac{R_s E_0}{Z_1}\exp(+jq_i z - j\beta_i x),\\ H_{xt} &= -\frac{T_s E_0}{Z_2}\exp(-jq_t z - j\beta_i x), \end{aligned} \tag{3-31}$$

where $Z_1 = \omega\mu_1/q_i$ and $Z_2 = \omega\mu_2/q_t$ represent the ratio of the tangential components E_y and H_x and are called the *wave impedance*. We also use Snell's law and write

$$q_t = (k_2^2 - \beta_i^2)^{1/2} = k_2 \cos\theta_t$$

and

$$\cos\theta_t = \left[1 - \left(\frac{n_1}{n_2}\right)^2 \sin^2\theta_i\right]^{1/2}. \tag{3-32}$$

Since the transmitted wave is propagating in the $+z$ direction, the sign for the square root must be chosen such that $\operatorname{Re} q_t > 0$ and $\operatorname{Im} q_t < 0$.

We now apply the continuity of the tangential magnetic field

$$H_{xi} + H_{xr} = H_{xt} \qquad \text{at } z = 0. \tag{3-33}$$

We then get

$$\frac{1 - R_s}{Z_1} = \frac{T_s}{Z_2}. \tag{3-34}$$

From (3-29) and (3-34), we get the following *Fresnel formula*:

$$R_s = \frac{Z_2 - Z_1}{Z_2 + Z_1}, \qquad T_s = \frac{2Z_2}{Z_2 + Z_1}, \tag{3-35}$$

where Z_1 and Z_2 are as given in (3-31). Note that since $\exp(-jq_t z)$ in (3-23) should represent the wave propagating in the $+z$ direction, if $\cos\theta_t$ becomes complex, the choice of the plus or minus sign in the square root must be made such that the imaginary part of $\cos\theta_t$ is negative. If $\mu_1 = \mu_2 = \mu_0$, (3-35) is reduced to a more familiar form:

$$R_s = \frac{n_1\cos\theta_i - n_2\cos\theta_t}{n_1\cos\theta_i + n_2\cos\theta_t} \quad \text{and} \quad T_s = \frac{2n_1\cos\theta_i}{n_1\cos\theta_i + n_2\cos\theta_t}. \tag{3-36}$$

3-5 ELECTRIC FIELD PARALLEL TO A PLANE OF INCIDENCE: PARALLEL POLARIZATION (*p* POLARIZATION)

We now consider the parallel polarization. We let E_0 be the magnitude of the incident electric field. The x component of the incident electric field is given by

$$E_{xi} = E_0 \cos\theta_i \exp(-jq_i z - j\beta_i x). \tag{3-37}$$

The reflected and the transmitted waves are

$$E_{xr} = R_p E_0 \cos\theta_i \exp(+jq_i z - j\beta_i x), \tag{3-38}$$

$$E_{xt} = T_p E_0 \cos\theta_t \exp(-jq_t z - j\beta_i x), \tag{3-39}$$

where R_p and T_p are the reflection and transmission coefficients and the law of reflection $\theta_i = \theta_r$ has already been used. As explained in Section 3-4, Snell's law holds here:

$$n_1 \sin\theta_i = n_2 \sin\theta_t. \tag{3-40}$$

The magnetic field H_y is related to E_x by

$$E_x = -\frac{1}{j\omega\epsilon}\frac{\partial}{\partial z}H_y. \tag{3-41}$$

Now we apply the boundary conditions:

$$E_{xi} + E_{xr} = E_{xt} \qquad \text{and} \qquad H_{yi} + H_{yr} = H_{yt}, \tag{3-42}$$

and obtain the following *Fresnel formula*:

$$R_p = \frac{Z_2 - Z_1}{Z_2 + Z_1}, \qquad T_p = \frac{2Z_2}{Z_2 + Z_1}\frac{\cos\theta_i}{\cos\theta_t}. \tag{3-43}$$

Here the wave impedance Z_1 and Z_2 are given by

$$Z_1 = \frac{q_i}{\omega\epsilon_1} \qquad \text{and} \qquad Z_2 = \frac{q_t}{\omega\epsilon_2}. \tag{3-44}$$

If $\mu_1 = \mu_2 = \mu_0$, (3-43) is reduced to a more familiar form:

$$R_p = \frac{(1/n_2)\cos\theta_t - (1/n_1)\cos\theta_i}{(1/n_2)\cos\theta_t + (1/n_1)\cos\theta_i} \qquad \text{and} \qquad T_p = \frac{(2/n_2)\cos\theta_i}{(1/n_2)\cos\theta_t + (1/n_1)\cos\theta_i}, \tag{3-45}$$

where

$$\cos\theta_t = \left[1 - \left(\frac{n_1}{n_2}\right)^2 \sin^2\theta_i\right]^{1/2}.$$

The factor $\cos\theta_i/\cos\theta_t$ of T_p in (3-43) comes from the definition of T_p as the ratio of the total electric field. The total transmitted and incident electric fields are $E_{xt}/\cos\theta_t$ and $E_{xi}/\cos\theta_i$, respectively.

If we take the tangential components E_{xi}, E_{xr}, and E_{xt}, (3-43) should be identical to (3-35). Equations (3-43) and (3-35) can be recognized as the same as the

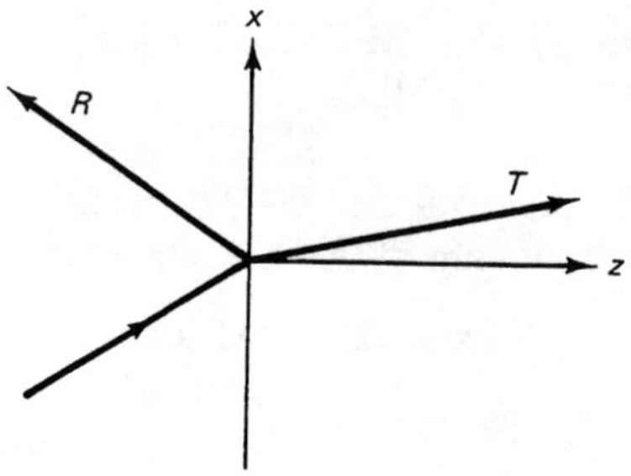

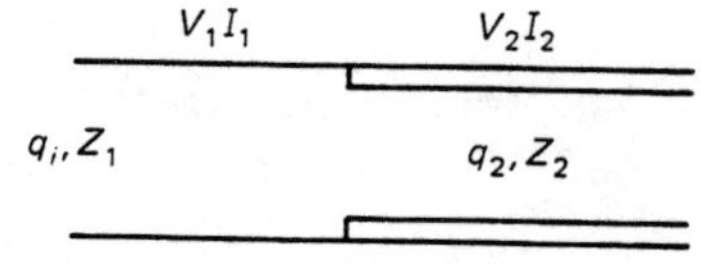

Figure 3-3 Equivalent transmission line.

reflection and transmission coefficients for a junction of two transmission lines. In fact, if we take the tangential electric field as the voltage and the tangential magnetic field as the current, we have a completely equivalent transmission-line problem (Fig. 3-3).

3-6 FRESNEL FORMULA, BREWSTER'S ANGLE, AND TOTAL REFLECTION

Let us examine the Fresnel formula (3-35) and (3-43) when n_1 and n_2 are real and positive. If medium 2 is denser than medium 1 ($n_1 < n_2$), the variations of these coefficients as functions of the incident angle θ_i are as shown in Fig. 3-4(a).

The reflection coefficient R_p for parallel polarization becomes zero when

$$\cos\theta_i = \frac{n_1}{n_2}\cos\theta_t = \frac{n_1}{n_2}\left[1 - \left(\frac{n_1}{n_2}\right)^2 \sin^2\theta_i\right]^{1/2}.$$

Solving this for θ_i, we get

$$\theta_i = \theta_b = \sin^{-1}\left(\frac{n_2^2}{n_1^2 + n_2^2}\right)^{1/2} = \tan^{-1}\frac{n_2}{n_1}. \tag{3-46}$$

This incident angle θ_b is called *Brewster's angle* (Fig. 3-5), and at this angle all the incident power passes into the second medium. Note that this occurs only when the incident field is polarized in the plane of incidence. If the wave consisting of both parallel and perpendicular polarizations is incident on the boundary at Brewster's angle, the component with parallel polarization is all transmitted into the second medium and the reflected wave consists only of the perpendicular polarization. Note also that the Brewster angle occurs for both $n_1 > n_2$ and $n_1 < n_2$. It is also easy to prove that at Brewster's angle, $\theta_b + \theta_t = \pi/2$. This means that the electric field in

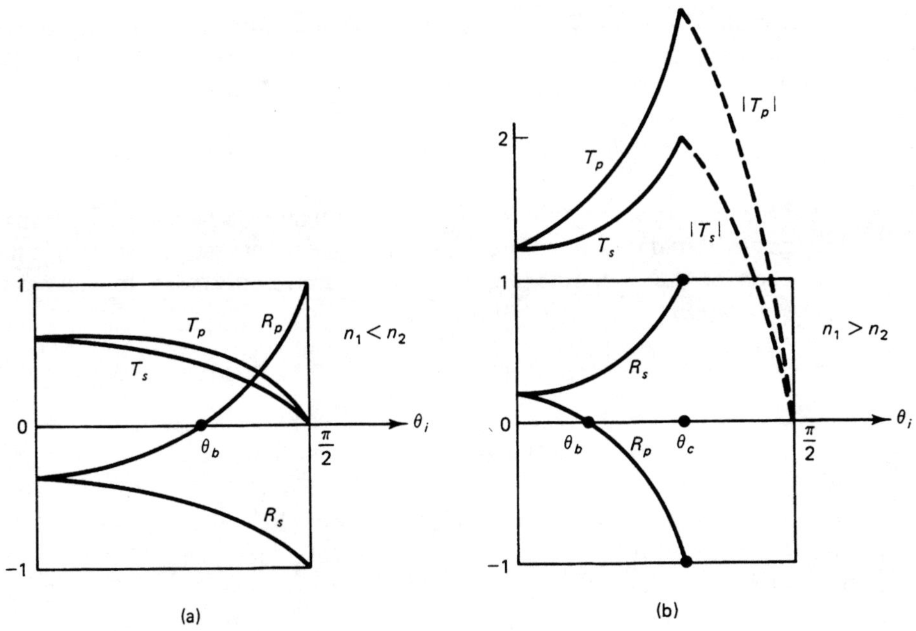

Figure 3-4 Reflection and transmission coefficients.

the second medium is polarized such that all the polarization vectors $\bar{P}$ are pointed in the direction of the reflected wave (Fig. 3-4). It will be shown that the far-field radiation from each electric dipole in the direction of the dipole axis is zero, and since the reflected wave is a sum of the radiation from all electric dipoles corresponding to the polarization vectors $\bar{P}$, the reflection should be zero at Brewster's angle (Sommerfeld, 1954).

Let us next examine the case when medium 2 is rarer than medium 1, $n_1 > n_2$. The reflection coefficients are shown in Fig. 3-4(b). Note that for the parallel

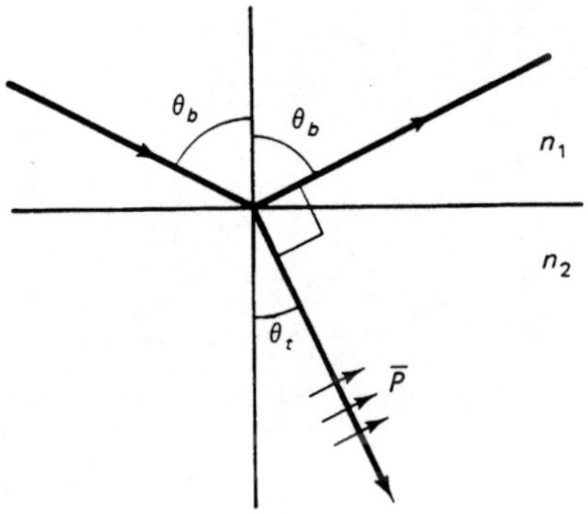

Figure 3-5 Brewster's angle.

polarization, there is a Brewster angle. Also, note that since $n_1 > n_2$, $\theta_t > \theta_i$, and there is a critical incident angle $\theta_i = \theta_c$ at which θ_t becomes $\pi/2$, and the incident wave $\theta_i > \theta_c$ is totally reflected. The critical angle θ_c is given by

$$\theta_c = \sin^{-1}\frac{n_2}{n_1}. \tag{3-47}$$

When the wave is totally reflected ($\theta_i > \theta_c$), there is no real power transmitted through the boundary. However, this does not mean that there is no field in medium 2. To examine this phenomenon, consider the transmitted wave for the perpendicular polarization.

$$E_{yt} = T_s E_0 \exp(-jk_2 z \cos\theta_t - jk_1 x \sin\theta_i), \tag{3-48}$$

where

$$\cos\theta_t = (1 - \sin^2\theta_t)^{1/2} = \left[1 - \left(\frac{n_1}{n_2}\right)^2 \sin^2\theta_i\right]^{1/2}.$$

When $\theta_i > \theta_c = \sin^{-1}(n_2/n_1)$, $\cos\theta_t$ becomes purely imaginary. In fact, as θ_i varies from 0 to θ_c, θ_t varies from 0 to $\pi/2$, and when θ_i becomes greater than θ_c, θ_t becomes complex $(\pi/2 + j\delta)$, $\delta > 0$ (Fig. 3-6).

$$\cos\theta_t = \cos\left(\frac{\pi}{2} + j\delta\right) = -j\sinh\delta = -j\left[\left(\frac{n_1}{n_2}\right)^2 \sin^2\theta_i - 1\right]^{1/2}.$$

Note that $\cos\theta_t$ must have a negative imaginary part, as noted in Section 3-4. The transmitted field is then given by

$$E_{yt} = T_0 E_0 \exp(-k_2 z \sinh\delta - jk_1 x \sin\theta_i). \tag{3-49}$$

This shows that the field decays exponentially in the positive z direction. This field, (3-49), carries no real power in the $+z$ direction. However, the reactive power given by the imaginary part of the Poynting vector in the $+z$ direction is not zero and represents the stored energy. This wave is called the *evanescent wave* and occurs in many situations involving total reflections. This is in contrast with the exponentially attenuating wave in a lossy medium, which does carry real power.

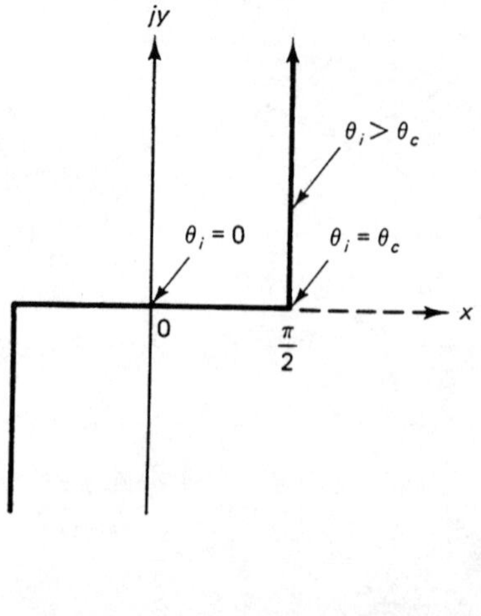

Figure 3-6 Complex θ_t plane ($\theta_t = x + jy$).

Let us next consider the conservation of power across a boundary. We showed at the end of Section 3-5 that the reflection and transmission of a wave at a boundary is equivalent to a transmission-line junction if the tangential electric and magnetic fields are defined as the voltage and current, respectively. Therefore, the Poynting vector normal to the surface is equal to $\frac{1}{2}VI^*$. Now we let the incident, reflected, and transmitted Poynting vectors be $\bar{S}_i$, $\bar{S}_r$, and $\bar{S}_t$, respectively. Then we have

$$\begin{aligned} P_i &= \bar{S}_i \cdot \hat{z} = \tfrac{1}{2}\bar{E}_i \times \bar{H}_i^* \cdot \hat{z} = \tfrac{1}{2}V_i I_i^*, \\ P_r &= \bar{S}_r \cdot (-\hat{z}) = \tfrac{1}{2}\bar{E}_r \times \bar{H}_r^* \cdot (-\hat{z}) = \tfrac{1}{2}V_r I_r^*, \\ P_t &= \bar{S}_t \cdot \hat{z} = \tfrac{1}{2}\bar{E}_t \times \bar{H}_t^* \cdot \hat{z} = \tfrac{1}{2}V_t I_t^* \end{aligned}$$

Using $1 + R = T$ and $(1 - R)/Z_1 = T/Z_2$, we can show the conservation of real power:

$$\text{Re}\,P_i = \text{Re}\,P_r + \text{Re}\,P_t, \tag{3-50}$$

if Z_1 or R is real.

3-7 WAVES IN LAYERED MEDIA

At the end of Section 3-5 we noted that if we take the tangential electric field to be the voltage and the tangential magnetic field to be the current, we get an equivalent transmission line. We can easily generalize this result to wave propagation in a layered medium.

Let us first consider the perpendicular polarization. If a wave is incident on a layer as shown in Fig. 3-7, we can obtain the equivalent transmission line as shown. Here we have, from Section 3-4,

$$\begin{aligned} &E_y = V(z)e^{-j\beta x}, \quad -H_x = I(z)e^{-j\beta x} \\ &q_m = k_m \cos\theta_m = k_m\left[1 - \left(\frac{n_1}{n_m}\right)^2 \sin^2\theta_i\right]^{1/2}, \\ &Z_m = \frac{\omega\mu_m}{q_m}, \qquad k_m = k_0 n_m, \\ &q_i = k_1 \cos\theta_i, \qquad q_t = k_t \cos\theta_t. \end{aligned} \tag{3-51}$$

The transmission-line problem can be solved by noting that the voltage and current at $z = 0$ and $z = l$ are related by (Fig. 3-7(b))

$$\begin{aligned} &\begin{bmatrix} V(0) \\ I(0) \end{bmatrix} = \begin{bmatrix} A_m & B_m \\ C_m & D_m \end{bmatrix} \begin{bmatrix} V(l) \\ I(l) \end{bmatrix}. \\ &A_m = D_m = \cos q_m l, \qquad B_m = jZ_m \sin q_m l, \\ &C_m = \frac{j \sin q_m l}{Z_m}, \qquad A_m D_m - B_m C_m = 1. \end{aligned} \tag{3-52}$$

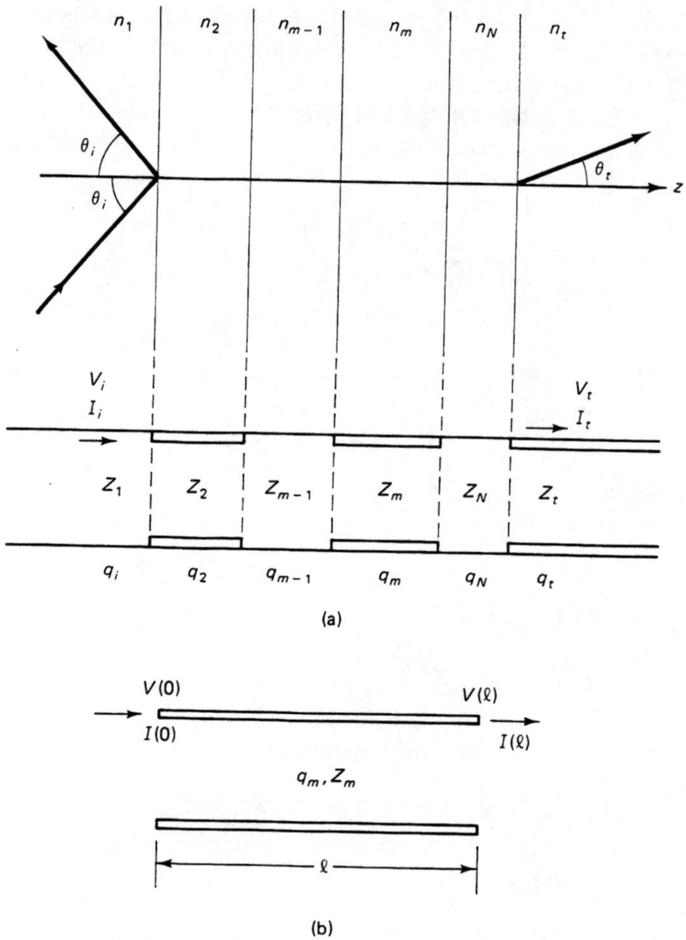

Figure 3-7 (a) Layered medium; (b) transmission line.

The total layer can be expressed by the total $ABCD$ matrix.

$$\begin{bmatrix} V_i \\ I_i \end{bmatrix} = \begin{bmatrix} A & B \\ C & D \end{bmatrix} \begin{bmatrix} V_t \\ I_t \end{bmatrix}, \tag{3-53}$$

where

$$\begin{bmatrix} A & B \\ C & D \end{bmatrix} = \begin{bmatrix} A_2 & B_2 \\ C_2 & D_2 \end{bmatrix} \begin{bmatrix} A_3 & B_3 \\ C_3 & D_3 \end{bmatrix} \cdots \begin{bmatrix} A_N & B_N \\ C_N & D_N \end{bmatrix}.$$

At the input we have

$$V_i = E_0(1 + R_s) \qquad \text{and} \qquad I_i = \frac{E_0}{Z_1}(1 - R_s). \tag{3-54}$$

At the other end we have

$$V_t = T_s E_0 \qquad \text{and} \qquad I_t = \frac{T_s E_0}{Z_t}. \tag{3-55}$$

From these we get the solution

$$R_s = \frac{A + B/Z_t - Z_1(C + D/Z_t)}{A + B/Z_t + Z_1(C + D/Z_t)},$$
$$T_s = \frac{2}{A + B/Z_t + Z_1(C + D/Z_t)}. \tag{3-56}$$

For the parallel polarization, we have the identical transmission line, but V, I, and Z are different.

$$V = E_x, \qquad I = H_y,$$
$$Z_m = \frac{q_m}{\omega\epsilon_m}, \qquad k_m = k_0 n_m, \tag{3-57}$$
$$q_m = k_m \cos\theta_m.$$

The total electric field E is, however, not E_x, but $E_x/\cos\theta_m$.

3-8 ACOUSTIC REFLECTION AND TRANSMISSION FROM A BOUNDARY

The fundamental equations for acoustic waves in fluid are given in Section 2-12. In terms of the acoustic pressure p_1 and the particle velocity $\bar{V}_1$, we have

$$(\nabla^2 + k^2)p_1 = 0,$$
$$\bar{V}_1 = -\frac{1}{j\omega\rho_0}\nabla p_1. \tag{3-58}$$

The boundary conditions are the continuity of p_1 and $(1/\rho_0)(\partial/\partial n)p_1$.

For the wave incident on a boundary at $z = 0$ from the medium 1 with the acoustic velocity c_1 and the density ρ_1 on the medium 2 with c_2 and ρ_2, we write the incident p_i, the reflected p_r, and the transmitted waves p_t.

$$p_i = A_0 \exp(-jq_i z - j\beta_i x),$$
$$p_r = RA_0 \exp(+jq_i z - j\beta_i x), \tag{3-59}$$
$$p_t = TA_0 \exp(-jq_t z - j\beta_i x),$$

where A_0 is a constant, R and T are the reflection and transmission coefficients, respectively, and

$$k_1 = \frac{\omega}{c_1}, \qquad k_2 = \frac{\omega}{c_2},$$

$$\beta_i = k_1 \sin\theta_i = k_2 \sin\theta_t \qquad \text{(Snell's law)},$$

$$q_i = k_1 \cos\theta_i,$$

$$q_t = k_2 \cos\theta_t = (k_2^2 - \beta_i^2)^{1/2},$$

$$\cos\theta_t = \left[1 - \left(\frac{c_2}{c_1}\right)^2 \sin^2\theta_i\right]^{1/2}.$$

Now we satisfy the boundary conditions at $z = 0$:

$$\begin{aligned} 1 + R &= T, \\ \frac{1}{Z_1}(1 - R) &= \frac{1}{Z_2}T, \end{aligned} \tag{3-60}$$

where $Z_1 = \rho_1 c_1/\cos\theta_i$ and $Z_2 = \rho_2 c_2/\cos\theta_t$, and $\rho_1 c_1$ and $\rho_2 c_2$ are the characteristic impedances of medium 1 and medium 2, respectively. From (3-60), we get

$$R = \frac{Z_2 - Z_1}{Z_2 + Z_1}, \qquad T = \frac{2Z_2}{Z_2 + Z_1}. \tag{3-61}$$

This is identical to electromagnetic problems. There are some differences, however. For the electromagnetic case, μ_1 and μ_2 for nonmagnetic materials are equal to μ_0, and therefore the medium is characterized by the refractive index $n = \sqrt{\epsilon}$ only. For the acoustic case, the medium is characterized by two parameters, the sound velocity and the density (or compressibility; see Section 2-12). This is equivalent to having different μ and ϵ for the electromagnetic case. Therefore, even though there is no Brewster's angle for electromagnetic perpendicular polarization, Brewster's angle θ_b is possible for the acoustic case and is given by setting $R = 0$.

$$\sin\theta_b = \left[\frac{(\rho_2/\rho_1)^2 - (c_1/c_2)^2}{(\rho_2/\rho_1)^2 - 1}\right]^{1/2}. \tag{3-62}$$

It is clear that Brewster's angle exists only when the right side of (3-62) is real and its magnitude is less than unity. Thus for Brewster's angle to exist, c_1, c_2, ρ_1, and ρ_2 must satisfy the condition that either $\rho_2/\rho_1 < c_1/c_2 < 1$ or $\rho_2/\rho_1 > c_1/c_2 > 1$.

3-9 COMPLEX WAVES

Up to this point we have discussed the reflection and transmission of a plane wave incident on plane boundaries. We also noted that the reflection coefficient becomes zero when the angle of incidence is equal to Brewster's angle. Can the reflection coefficient become infinite under certain conditions? The infinite reflection coefficient is equivalent to having a finite field with a vanishing incident field and thus represents a wave that is guided along the dielectric surface. This is often called the *surface wave* or the *trapped surface wave*. Important applications of this wave type include guided waves on optical fibers and thin films, microwaves on Goubau lines, and artificial dielectric, and thin dielectric coatings on metal surfaces.

In studying the surface-wave mode and other guided waves, it is important first to recognize various wave types and their mathematical representations. The general wave types are usually called *complex waves*, and the clear understanding of their characteristics is important in the study of surface waves, leaky waves, Zenneck waves, and many other wave types.

To clarify the relationships among various wave types, it is instructive to examine all possible waves propagating along a plane boundary and their physical significance. Let us examine a wave $u(x, z)$ propagating in the z direction in free space. We consider the two-dimensional problem $\partial/\partial y = 0$ and assume that the dielectric or other guiding structure is located below $x = 0$ and that free space extends to infinity in the $+x$ direction. This field $u(x, z)$ satisfies a scalar wave equation.

$$\left(\frac{\partial^2}{\partial x^2} + \frac{\partial^2}{\partial z^2} + k^2\right)u(x, z) = 0. \tag{3-63}$$

We write a solution in the following form:

$$u(x, z) = e^{-jpx - j\beta z}. \tag{3-64}$$

Substituting this into (3-63), we get the following condition:

$$p^2 + \beta^2 = k^2. \tag{3-65}$$

In general, p and β can be complex and thus we let

$$p = p_r - j\alpha_t \qquad \text{and} \qquad \beta = \beta_r - j\alpha. \tag{3-66}$$

Substituting (3-66) into (3-65) and equating the real and imaginary parts of both sides, we get

$$p_r^2 - \alpha_t^2 + \beta_r^2 - \alpha^2 = k^2, \tag{3-67}$$

$$p_r\alpha_t + \beta_r\alpha = 0. \tag{3-68}$$

Using (3-68), we can first show that the constant-amplitude and constant-phase planes are perpendicular to each other. To prove this, we write

$$u(x, z) = e^{-j(p_r x + \beta_r z) - (\alpha_t x + \alpha z)} = e^{-j\bar{k}_r \cdot \bar{r} - \bar{\alpha} \cdot \bar{r}} \tag{3-69}$$

The constant-phase plane is given by

$$\bar{k}_r \cdot \bar{r} = \text{constant}, \qquad \bar{k}_r = p_r\hat{x} + \beta_r\hat{z}, \qquad \hat{r} = x\hat{x} + z\hat{z}, \tag{3-70}$$

and the constant-amplitude plane by

$$\bar{\alpha} \cdot \bar{r} = \text{constant}, \qquad \bar{\alpha} = \alpha_t\hat{x} + \alpha\hat{z}. \tag{3-71}$$

Thus the constant-phase plane is perpendicular to $\bar{k}_r$ and the constant-amplitude plane is perpendicular to $\bar{\alpha}$. But according to (3-68), $\bar{k}_r$ is perpendicular to $\bar{\alpha}$.

$$\bar{k}_r \cdot \bar{\alpha} = p_r\alpha_t + \beta_r\alpha = 0, \tag{3-72}$$

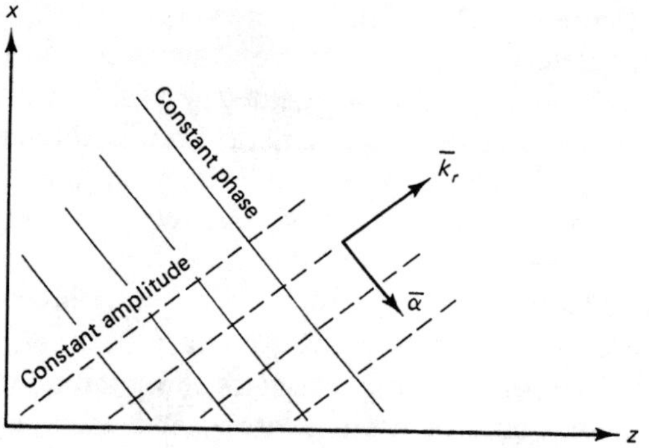

Figure 3-8 Complex waves.

and therefore the constant-amplitude plane is perpendicular to the constant-phase plane (Fig. 3-8).

Next, we consider the behavior of the wave as $x \to +\infty$. We take the wave whose phase progression along the surface is in the $+z$ direction ($\beta_r > 0$) and then consider the magnitude of the wave in the transverse (x) direction, which is given by

$$|e^{-jpx}| = e^{-\alpha_t x}. \tag{3-73}$$

From this we note that if $\alpha_t > 0$, the wave attenuates exponentially in the $+x$ direction and is called the *proper wave*. If $\alpha_t < 0$, the wave has the amplitude increasing exponentially in the $+x$ direction and is called the *improper wave*. It is obvious that depending on the signs of β_r, α, p_r, and α_t, a variety of wave types can result.

Let us start with the following combinations of p and β shown in Table 3-1. In terms of $\bar{k}_r = p_r\hat{x} + \beta_r\hat{z}$ and $\bar{\alpha} = \alpha_t\hat{x} + \alpha\hat{z}$, they are pictured in Fig. 3-9. In terms of the complex $p = p_r + jp_i$ and $\beta = \beta_r + j\beta_i$ planes, they are shown in Fig. 3-10. Note that for a given β, there can be two values of $p = \pm(k^2 - \beta^2)^{1/2}$. For example, two waves at D and H have the same β but different p. D is the Zenneck wave and H is the leaky wave. Those waves in the upper half-plane of the complex p plane are the improper waves, whereas those in the lower half are the proper waves (see Section 3-12 for Zenneck waves). A more complete discussion must be postponed to chapter 15 where these waves are interpreted as a portion of the complete spectrum representation of a wave. In this chapter we concentrate on trapped surface waves and leaky waves.

TABLE 3-1 Proper and Improper Waves

	Case	β_r	α	p_r	α_t	
	A	+	0	+	0	Fast wave (waveguide modes)
Proper wave	B	+	−	+	+	Backward leaky wave
	C	+	0	0	+	Trapped surface wave
	D	+	+	−	+	Zenneck wave
	E	+	0	−	0	Plane-wave incidence
Improper wave	F	+	−	−	−	
	G	+	0	0	−	Untrapped surface wave
	H	+	+	+	−	Forward leaky wave

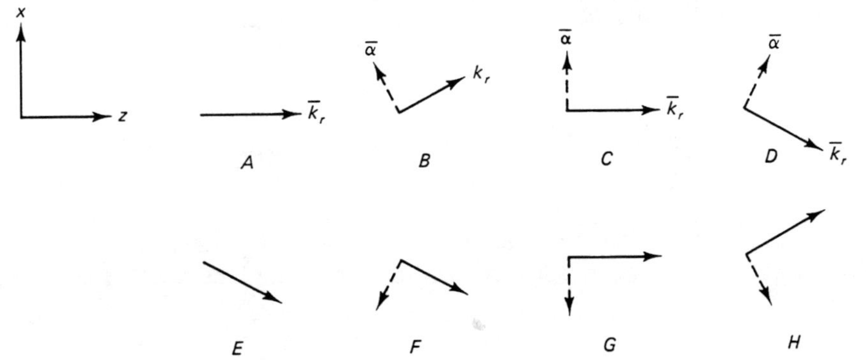

Figure 3-9 Proper and improper waves.

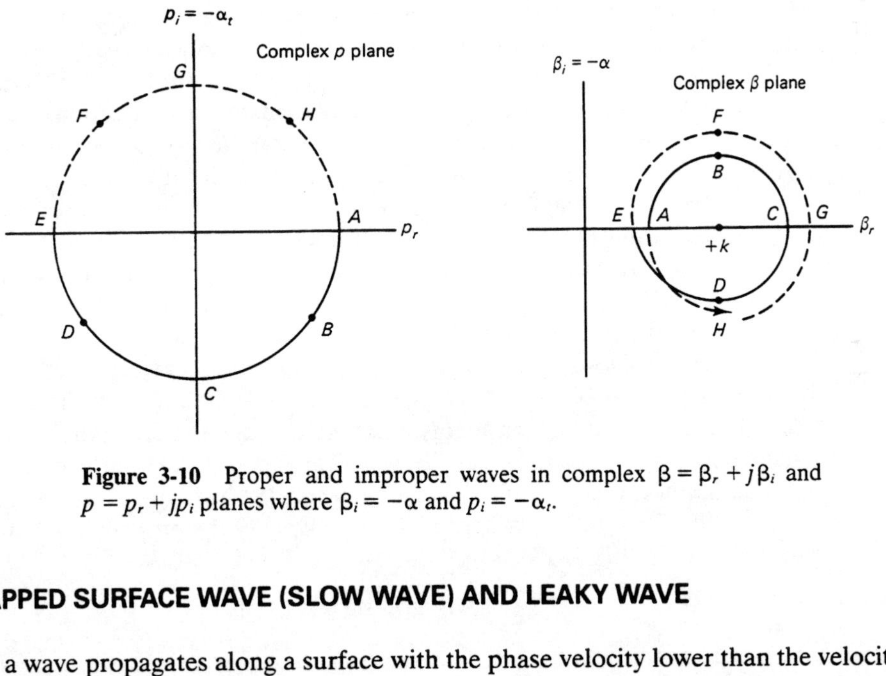

Figure 3-10 Proper and improper waves in complex $\beta = \beta_r + j\beta_i$ and $p = p_r + jp_i$ planes where $\beta_i = -\alpha$ and $p_i = -\alpha_t$.

3-10 TRAPPED SURFACE WAVE (SLOW WAVE) AND LEAKY WAVE

If a wave propagates along a surface with the phase velocity lower than the velocity of light, the wave can be trapped near the surface and can propagate without attenuation. This characteristic is useful for guiding waves over a long distance. Examples are optical communication through optical fiber, microwave transmission through Goubau wire, and surface waves on a dielectric coating. Other examples are waves along Yagi antennas and slow waves along helical structures in traveling-wave tubes.

Let us assume that a wave propagates along a surface without attenuation. If the phase velocity of this wave is slow, $\beta = \beta_r = \omega/v_p > k$. Thus, from (3-65), we get

$$p^2 = k^2 - \beta_r^2 < 0,$$

and therefore, p must be purely imaginary. Writing $p = -j\alpha_t$, the wave is expressed by

$$u(x,z) = e^{-\alpha_t x - j\beta_r z}, \tag{3-74}$$

where $\alpha_t = \sqrt{\beta_r^2 - k^2}$. Obviously, in this case, $\bar{k}_r$ is directed in the z direction and $\bar{\alpha}$ in the x direction (Fig. 3-11).

We note that the slower the wave, the greater the value of β, and therefore, the greater the value of α_t. Because of the attenuation due to α_t in the $+x$ direction, the wave is mostly concentrated near the surface and the total power is finite:

$$\int_0^\infty |u(x,y)|^2\,dx = \text{finite}. \tag{3-75}$$

Thus the wave is said to be *trapped* near the surface. The *trapped surface wave* propagates the finite amount of power along the surface without attenuation, and it decays exponentially in the transverse $+x$ direction. In this chapter we examine several slow-wave structures that support the trapped surface wave.

Let us consider a fast wave propagating along the surface, $\beta_r = \omega/v_p < k$. In this case, in general, the wave must attenuate in the $+z$ direction, $\alpha > 0$, and according to (3-68), $p_r\alpha_t$ must be negative. If we consider the outgoing wave in the $+x$ direction, p_r is positive, and therefore, α_t must be negative. Now $\alpha_t < 0$ represents the wave whose amplitude increases exponentially in the $+x$ direction, and thus this is the improper wave. Note that the amplitude decays in the z direction ($A_1 > A_2 > A_3 > A_4$), but in the x direction at a given z, the amplitude increases exponentially (Figure 3-12). This wave is called a *leaky wave*, as the energy is constantly leaked out from the surface.

Since the leaky wave is an improper wave, it cannot exist by itself, but it can exist within a portion of a space. A typical example is the radiation from a narrow slit cut along a waveguide. The energy leaks out through this slit, and thus the wave attenuates in the z direction. But the wave farther from the surface has greater amplitude because it is originated at the point on the waveguide where the amplitude is greater. The angle θ_c is approximately given by the propagation constant β_z for the waveguide (Fig. 3-13).

$$\beta_z = k \sin\theta_c.$$

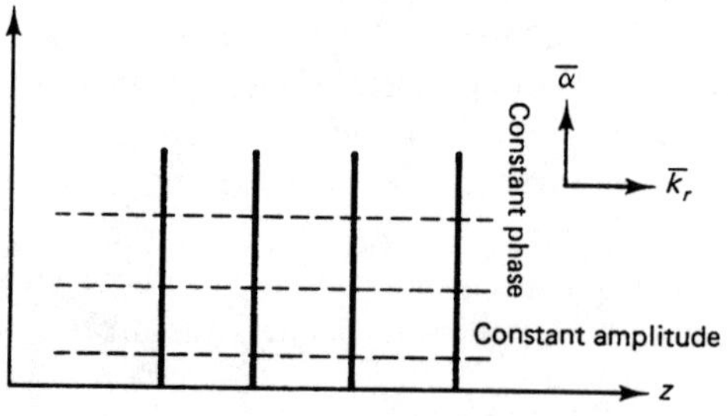

Figure 3-11 Trapped surface waves.

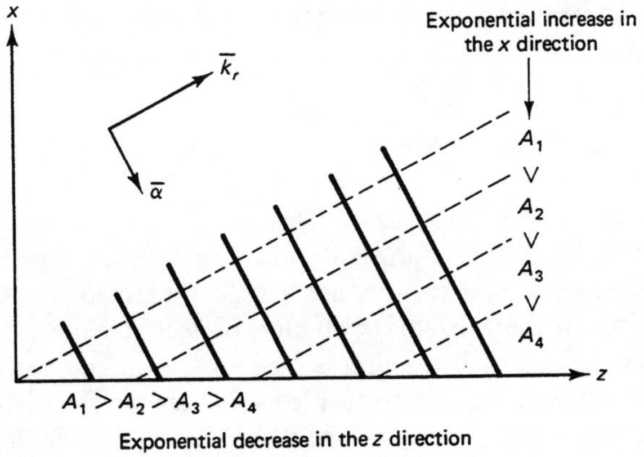

Figure 3-12 Leaky waves.

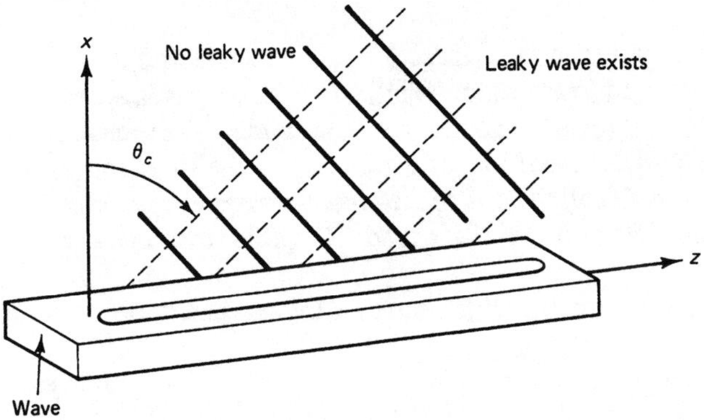

Figure 3-13 Leaky waveguides.

Note that the leaky wave exists only in a portion of the space, and thus in the transverse (x) direction the amplitude increases up to a point and then decreases. More rigorous analysis is necessary to analyze this problem completely and will be done later in terms of a Fourier transform technique (see Section 14-1).

Another important example is the coupling of an optical beam into a thin film as shown in Fig. 3-14. A slow wave propagates along a thin film. A prism is placed close to the thin film, and the wave in the prism becomes a leaky wave extracting the power from the surface wave. The leaky wave is then taken out from the prism at

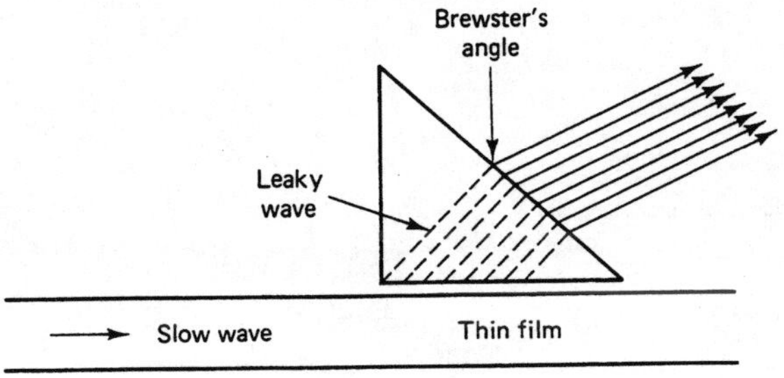

Figure 3-14 Coupling of a beam and surface wave.

Brewster's angle. This process can be reversed and a beam can enter into a prism in the opposite direction and the power can be converted into a surface wave.

3-11 SURFACE WAVES ALONG A DIELECTRIC SLAB

A dielectric slab can support a trapped surface wave. This occurs in a variety of situations, such as optical beam propagation along a thin film and optical fibers. Also, radar cross sections of spacecraft and rockets are severely affected by the existence of surface waves on the surface coated with thin dielectric materials. Furthermore, the study of this dielectric slab requires a basic mathematical technique that is common to other wave propagation problems for planar stratified media, such as the wave propagation along the earth–ionosphere waveguide. In this section we consider the characteristics of trapped surface-wave modes on a dielectric slab.

Let us consider a wave propagating in the z direction along a dielectric slab of thickness d placed on a perfectly conducting plane (Fig. 3-15). Since this is a two-dimensional problem with $\partial/\partial y = 0$, there are two independent modes. One consists of H_y, E_x, and E_z and since the magnetic field H_y is transverse to the direction of propagation (the z direction), we may call them TM (transverse magnetic) modes. The other consists of E_y, H_x, and H_z and may be called TE (transverse electric) modes.

The TM modes can be derived from H_y, which satisfies the wave equation

$$\left(\frac{\partial^2}{\partial x^2}+\frac{\partial^2}{\partial z^2}+k^2\right)H_y=0, \tag{3-76}$$

and E_x and E_z are given in terms of H_y:

$$E_x=j\frac{1}{\omega\epsilon}\frac{\partial}{\partial z}H_y, \qquad E_z=-j\frac{1}{\omega\epsilon}\frac{\partial}{\partial x}H_y. \tag{3-77}$$

Similarly, the TE mode is given by E_y:

$$\left(\frac{\partial^2}{\partial x^2}+\frac{\partial^2}{\partial z^2}+k^2\right)E_y=0, \tag{3-78}$$

$$H_x=-j\frac{1}{\omega\mu}\frac{\partial}{\partial z}E_y, \qquad H_z=j\frac{1}{\omega\mu}\frac{\partial}{\partial x}E_y, \tag{3-79}$$

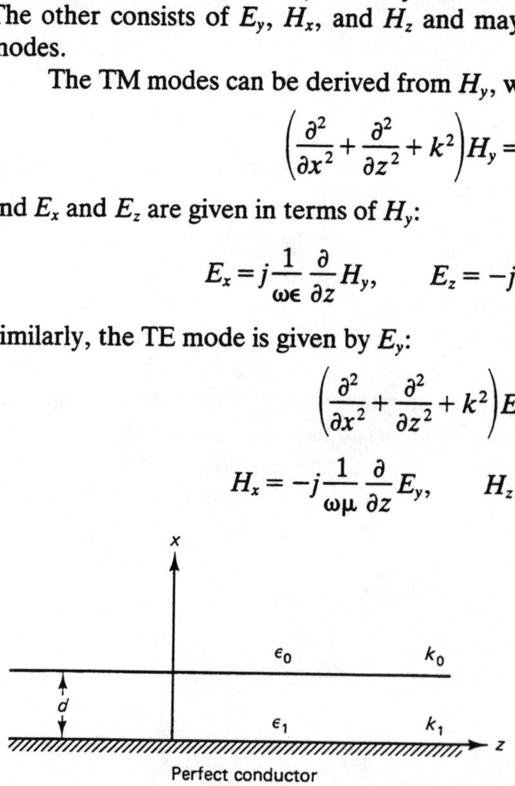

Figure 3-15 Dielectric slab on a conducting surface.

where $k = k_0$ and $\epsilon = \epsilon_0$ in free space and $k = k_1$ and $\epsilon = \epsilon_1$ in the dielectric slab.

Let us first consider the TM mode. This mode can easily be excited by a horn or a waveguide. We solve (3-76) in free space and in the dielectric slab.

$$H_{y0} = Ae^{-jp_0x - j\beta z} \qquad \text{for } x > d \quad \text{(free space)}, \tag{3-80}$$

$$H_{y1} = Be^{-jpx - j\beta z} + Ce^{+jpx - j\beta z} \qquad \text{for } d > x > 0 \quad \text{(in dielectric)} \tag{3-81}$$

where

$$p_0^2 + \beta^2 = k_0^2 \qquad \text{and} \qquad p^2 + \beta^2 = k_1^2.$$

We note that the propagation constant β along the surface must be the same for both free space and dielectric to ensure the same phase progression along the surface. This is necessary to satisfy the boundary condition. We also note that (3-80) has only one term because this region extends to infinity and only the outgoing wave exists. Equation (3-81), on the other hand, consists of two terms representing the waves propagating in the $+x$ and the $-x$ direction.

Now the boundary conditions at $x = 0$ and $x = d$ are applied. At $x = 0$, we have

$$E_{z1} = -j\frac{1}{\omega\epsilon_1}\frac{\partial}{\partial x}H_{y1} = 0, \tag{3-82}$$

and at $x = d$, E_z and H_y must be continuous.

$$E_{z0} = E_{z1} \qquad \text{and} \qquad H_{y0} = H_{y1}, \tag{3-83}$$

where

$$E_{z0} = -j\frac{1}{\omega\epsilon_0}\frac{\partial}{\partial x}H_{y0} \qquad \text{and} \qquad E_{z1} = -j\frac{1}{\omega\epsilon_1}\frac{\partial}{\partial x}H_{y1}.$$

Equations (3-82) and (3-83) may be written as

$$-B + C = 0, \qquad Be^{-jpd} + Ce^{jpd} = Ae^{-jp_0d},$$

$$\frac{p}{\epsilon_1}(Be^{-jpd} - Ce^{jpd}) = \frac{p_0}{\epsilon_0}Ae^{-jp_0d}.$$

From these, we eliminate A, B, and C and obtain the transcendental equation to determine the propagation constant β.

$$\frac{p}{\epsilon_1}\tan pd = j\frac{p_0}{\epsilon_0}. \tag{3-84}$$

Once this equation is solved for p, the propagation constant β is given by $\beta = (k_1^2 - p^2)^{1/2}$. Therefore, p is a number that is characteristic of the geometry of the problem and is called the *eigenvalue*. Equation (3-84) is the *eigenvalue equation*.

The ratio of the amplitudes can also be determined.

$$\frac{B}{A} = \frac{C}{A} = \frac{e^{-jp_0d}}{2\cos pd}. \tag{3-85}$$

Equations (3-84) and (3-85) should give the propagation constant and the field configurations for the TM mode.

Slow-Wave Solution for the TM Mode

Let us consider the most important case of a trapped surface-wave propagation. As noted in Section 3-10, we look for a solution of the form

$$H_{y0} = Ae^{-\alpha_t x - j\beta z}, \tag{3-86}$$

with

$$\beta > k_0 \qquad \text{and} \qquad p_0 = -j\alpha_t.$$

In this case, $\beta^2 - \alpha_t^2 = k_0^2$ and $\beta^2 + p^2 = k_1^2$ and thus, eliminating β from these two, we get

$$\alpha_t^2 = (k_1^2 - k_0^2) - p^2. \tag{3-87}$$

Thus the eigenvalue equation (3-84) becomes, for $\alpha_t > 0$,

$$\frac{\epsilon_0}{\epsilon_1} X \tan X = \sqrt{V^2 - X^2}, \tag{3-88}$$

where $X = pd$ and $V^2 = (k_1^2 - k_0^2)d^2$. If the dielectric constant ϵ of the material is greater than ϵ_0, V^2 is real and positive. If we plot the left side and the right side of (3-88) as functions of X, we get a curve Y_l for the left side and a circle Y_r for the right side. The solution is obtained at the intersection of Y_r and Y_l (Fig. 3-16). Note that when $0 < V < \pi$, there is only one solution. Similarly, when $(N-1)\pi < V < N\pi$, there are N different modes. Therefore, there exist N trapped surface-wave modes if the thickness of the slab d is given by

$$\frac{N-1}{2(\epsilon_r - 1)^{1/2}} < \frac{d}{\lambda_0} < \frac{N}{2(\epsilon_r - 1)^{1/2}}, \qquad \epsilon_r = \frac{\epsilon_1}{\epsilon_0}, \tag{3-89}$$

and in particular, there exists only one mode if

$$0 < \frac{d}{\lambda_0} < \frac{1}{2(\epsilon_r - 1)^{1/2}}. \tag{3-90}$$

The parameter $V = k_0 d(n_1^2 - 1)^{1/2}$ determines the number of modes and is called the *normalized frequency*. The total number of propagating modes N is given by the *mode volume formula*:

$$(N-1)\pi < V < N\pi. \tag{3-91}$$

Once we find $X = pd$ from (3-88), the propagation constant β is obtained by $\beta^2 = k_1^2 - p^2$. As is seen for $V = V_2$ (in Fig. 3-16), there are also some solutions for $\alpha_t < 0$. This is called *untrapped surface wave*. When this structure is excited by a source, this wave has some influence over the total field. In practice, however, this untrapped surface wave contributes very little to the total field and is insignificant.

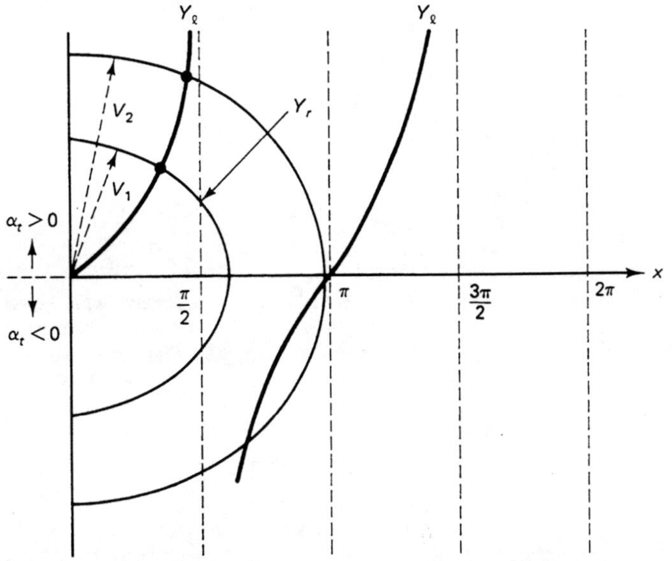

Figure 3-16 The plot of the left side Y_l and the right side Y_r of (3-88). The values of X at which Y_l and Y_r intersect are the solution. $V^2 = k_0^2(\epsilon_r - 1)\, d^2$.

The propagation constant β can be found from the value of X in (3-88) using $\beta^2 = k_1^2 - p^2$. A general shape of k_0–β diagram may be seen by examining a few important points (Fig. 3-17). For example, as $k_0 \to 0$, the curve approaches $k_0 = \beta$, and as $k_0 \to \infty$, the curve approaches $k_1 = \beta$. Also, note that at each cutoff frequency (k_c), the curve is tangent to $k_0 = \beta$. It is clear then that β is always between k_0 and k_1, and thus the phase velocity is always faster than that of a plane wave in dielectric $1/\sqrt{\mu_0 \epsilon_1}$, but slower than that in free space $1/\sqrt{\mu_0 \epsilon_0}$. At the cutoff frequency k_c, $V = X = N\pi$ in (3-88) and α_t becomes zero. This means that at the cutoff, the magnitude of the wave outside the slab becomes uniform, extending to infinity instead of decaying exponentially.

The field in dielectric is given using (3-85)

$$H_{y1} = Ae^{-\alpha_t d} \frac{\cos px}{\cos pd} e^{-j\beta z}, \tag{3-92}$$

and other field components are easily obtained from (3-77).

$$E_{x0} = \left(\frac{\beta}{k}\eta_0\right) Ae^{-\alpha_t x - j\beta z} \qquad \text{in free space}$$

$$E_{x1} = \left(\frac{\beta}{k_1}\eta_1\right) Ae^{-\alpha_t d} \frac{\cos px}{\cos pd} e^{-j\beta z} \qquad \text{in dielectric slab,}$$

where $\eta_0 = (\mu_0/\epsilon_0)^{1/2}$ and $\eta_1 = (\mu_0/\epsilon_1)^{1/2}$ are the characteristic impedances of free space and the dielectric medium. Note that the transverse electric and magnetic fields are perpendicular to each other and have the same configuration. The ratio of

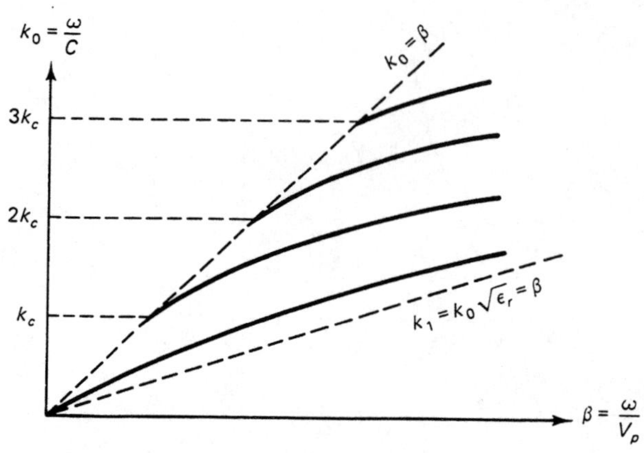

Figure 3-17 k_0–β diagram for the wave along a dielectric slab. The cutoff wavenumber $k_c = (\pi/d)(1/\sqrt{\epsilon_r - 1})$.

these two is the wave impedance $Z_e = (\beta/k)\eta$. Also, the field has sinusoidal distribution ($\cos px$) inside dielectric, but it has exponential behavior ($e^{-\alpha_t x}$) in free space (Fig. 3-18).

Next, let us examine the impedance looking down on the surface of the dielectric slab. Noting that the Poynting vector $E_z \hat{z} \times H_y^* \hat{y}$ is pointed into the dielectric, the impedance is given by

$$\frac{E_z}{H_y} = j\frac{\alpha_t}{k_0}\eta_0. \tag{3-93}$$

This is purely inductive. It is clear that the surface supporting the trapped surface wave must be purely reactive. The TM mode requires a purely inductive surface, whereas the TE mode requires a purely capacitive surface.

Slow-Wave Solution for the TE Mode

In a similar manner, we obtain the solution for the TE mode. We get

$$\begin{aligned} E_{y0} &= Ae^{-\alpha_t x - j\beta z}, \\ H_{x0} &= -\left(\frac{\beta}{k_0\eta_0}\right)Ae^{-\alpha_t x - j\beta z}, \\ H_{z0} &= -j\left(\frac{\alpha_t}{k_0\eta_0}\right)Ae^{-\alpha_t x - j\beta z}, \end{aligned} \tag{3-94}$$

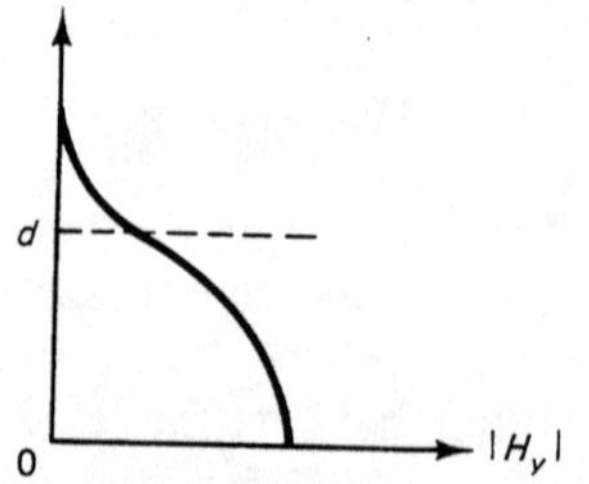

Figure 3-18 Field distribution for a trapped surface wave.

in free space $x > d$, and

$$E_{y1} = Ae^{-\alpha_t d}\frac{\sin px}{\sin pd}e^{-j\beta z},$$

$$H_{x1} = -\left(\frac{\beta}{k_1\eta_1}\right)Ae^{-\alpha_t d}\frac{\sin px}{\sin pd}e^{-j\beta z}, \tag{3-95}$$

$$H_{z1} = j\left(\frac{p}{k_1\eta_1}\right)Ae^{-\alpha_t d}\frac{\cos px}{\sin pd}e^{-j\beta z},$$

in the dielectric $0<x<d$, where $\eta_0 = (\mu_0/\epsilon_0)^{1/2}$ and $\eta_1 = (\mu_0/\epsilon_1)^{1/2}$. The eigenvalue equation is then, using $X = pd$,

$$-X\cot X = \sqrt{V^2 - X^2} \qquad \text{where} \quad V^2 = k_0^2(\epsilon_r - 1)d^2. \tag{3-96}$$

The left and right sides of (3-96) are pictured in Fig. 3-19. Note that there is no solution for $V < \pi/2$ and there are N solutions for $(N - \frac{1}{2})\pi < V < (N + \frac{1}{2})\pi$. This is the mode volume equation for TE modes. Thus the minimum thickness d required for a trapped wave is

$$\frac{d}{\lambda_0} > \frac{1}{4(\epsilon_r - 1)^{1/2}}. \tag{3-97}$$

The field distribution for the lowest mode is shown in Figure 3-20.

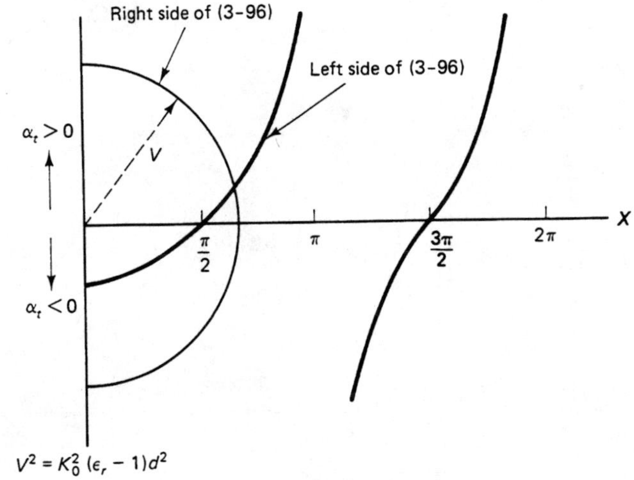

Figure 3-19 TE surface wave modes.

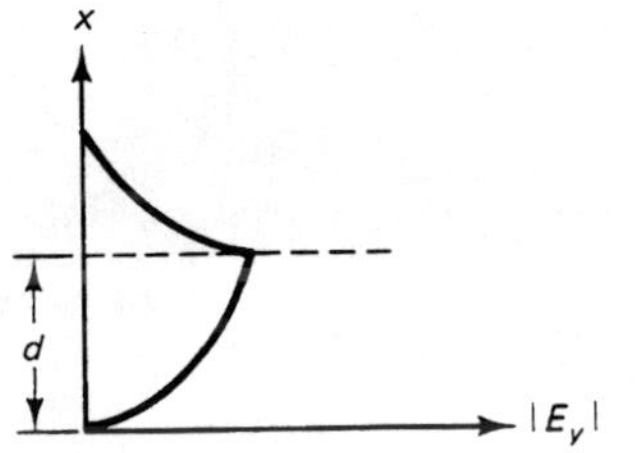

Figure 3-20 Lowest TE mode.

3-12 ZENNECK WAVES AND PLASMONS

In Section 3-9 we discussed various complex waves. In this section we show some examples. Consider a TM wave propagating along a plane boundary (Fig. 3-21). In air, $x > 0$, we have

$$H_{y0} = Ae^{-jp_0 x - j\beta z},$$
$$p_0^2 + \beta^2 = k_0^2. \tag{3-98}$$

In the lower half-space $x < 0$, we have

$$H_{y1} = Ae^{+jpx - j\beta z},$$
$$p^2 + \beta^2 = k_1^2 = k_0^2 \epsilon_r. \tag{3-99}$$

Here the continuity of H_y at $x = 0$ is already satisfied. The continuity of E_z yields the following:

$$\epsilon_r p_0 = -p. \tag{3-100}$$

From this, we get the following relating β to k_0 and k_1.

$$\frac{1}{\beta^2} = \frac{1}{k_1^2} + \frac{1}{k_0^2}. \tag{3-101}$$

Let us consider the case when the medium ($x < 0$) is highly conducting. Then we have

$$\epsilon_r = \epsilon' - j\epsilon'' \approx -j\epsilon'', \tag{3-102}$$
$$\epsilon'' = \frac{\sigma}{\omega\epsilon_0} >> 1.$$

We then get

$$\beta = \frac{k_0}{[1 + 1/\epsilon_r]^{1/2}} \approx k_0\left(1 - \frac{1}{2\epsilon_r}\right),$$
$$p_0 = -\frac{k_0}{(1 + \epsilon_r)^{1/2}}, \tag{3-103}$$
$$p = \frac{\epsilon_r k_0}{(1 + \epsilon_r)^{1/2}}.$$

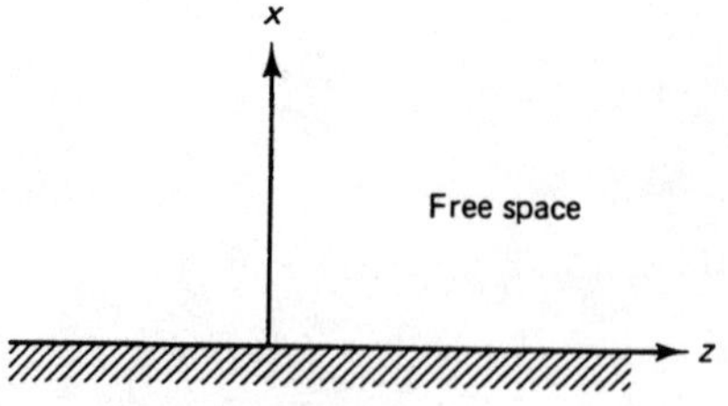

Figure 3-21 Zenneck waves and plasmons.

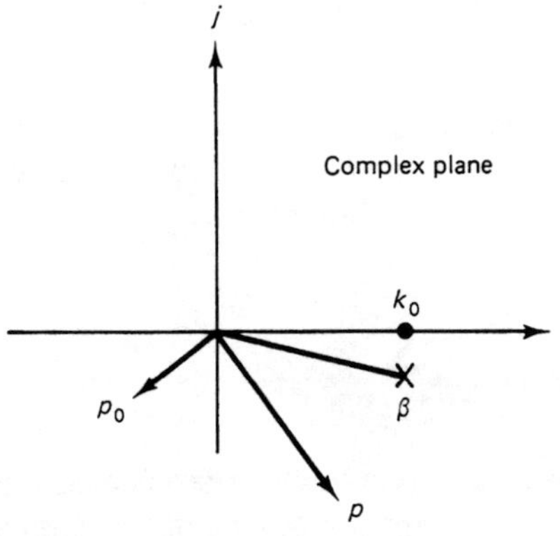

Figure 3-22 Zenneck waves.

Here $(1+\epsilon_r)^{1/2}$ is chosen such that its imaginary part is negative. These are shown in Fig. 3-22. β and p_0 may be compared with those shown in Fig. 3-10. The propagation constant β was obtained by Zenneck to explain the propagation characteristics of radio waves over the earth. The excitation of such a wave by a dipole source was investigated by Sommerfeld, and we discuss this historical Sommerfeld dipole problem in Chapter 15.

Next we consider the case when ϵ_r is real and negative. Ionized gas, called *plasma*, has this dielectric constant when the operating frequency is below plasma frequency and the loss is negligible. We let $\epsilon_r = -|\epsilon|$ and note that if $|\epsilon| > 1$, the propagation constant is real, $p_0 = -j\alpha_0$, and $p = -j\alpha$. This is a trapped surface wave propagating along the boundary.

Metal at optical frequencies can have a dielectric constant with a negative real part. For example, at $\lambda = 0.6\ \mu\text{m}$, silver has $\epsilon_r = -17.2 - j0.498$. In this case the wave is not trapped, but the wave propagating along the surface exhibits characteristics similar to surface waves (Figure 3-23). This is called the *plasmon*, and it has significant effects on the optical scattering characteristics of metal surfaces.

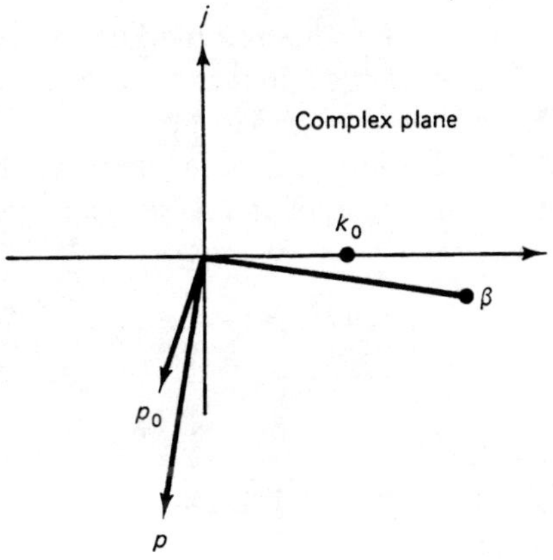

Figure 3-23 Plasmons.

Zenneck waves and plasmons cannot exist by themselves, but when the surface is illuminated by a localized source and the field is expressed in a Fourier transform, these waves appear as poles in the complex Fourier transform plane and the locations of the poles have significant effects on the total field. Above, we considered only TM waves. It can be shown, however, that these poles do not exist for TE waves.

3-13 WAVES IN INHOMOGENEOUS MEDIA

Up to this point we have discussed waves in homogeneous media separated by plane boundaries. Even though they represent a great many problems, there are many practical problems where the medium should be represented by a continuous function of position. For example, the ionosphere can be approximated by a medium with a sharply bounded lower edge at VLF, but for higher frequencies, the variation of the electron density over a wavelength is gradual and must be represented by a continuous function of height. The medium whose dielectric constant ϵ or permeability μ is a function of position is called the *inhomogeneous medium*.

In general, the characteristics of wave propagation and scattering in an inhomogeneous medium cannot be described in a simple form. However, relatively simple descriptions of the field are possible for two extreme cases: low-frequency and high-frequency fields. The high-frequency approximation is useful whenever the variation of the refractive index of the medium is negligibly small over the distance of a wavelength. This applies to ultrasound propagation in biological media, wave propagation in the ionosphere, microwaves and optical waves in large dielectric bodies, and sound propagation in ocean water. In contrast to this high-frequency approximation, the low-frequency approximation is applicable whenever the sizes of a dielectric body are much smaller than a wavelength. Examples are microwave radiation in a biological medium and scattering from small particles in the ocean, atmosphere, and biological media. In the following sections, we discuss the high-frequency approximation.

Let us consider a plane wave incident upon a medium whose dielectric constant is a function of height z. We choose the x axis so that the plane of incidence is in the x–z plane (Fig. 3-24). This is a two-dimensional problem ($\partial/\partial y = 0$), and thus there are two independent waves, TE (electric field is transverse to the z axis) and TM (magnetic field is transverse to the z axis). For TE (perpendicular polarization), we combine Maxwell's equations

$$\begin{aligned} \nabla \times \overline{E} &= -j\omega\mu\overline{H}, \\ \nabla \times \overline{H} &= j\omega\epsilon(z)\overline{E}, \end{aligned} \tag{3-104}$$

and get

$$\nabla \times \nabla \times \overline{E} = \nabla(\nabla \cdot \overline{E}) - \nabla^2\overline{E} = k_0^2 n^2(z)\overline{E}.$$

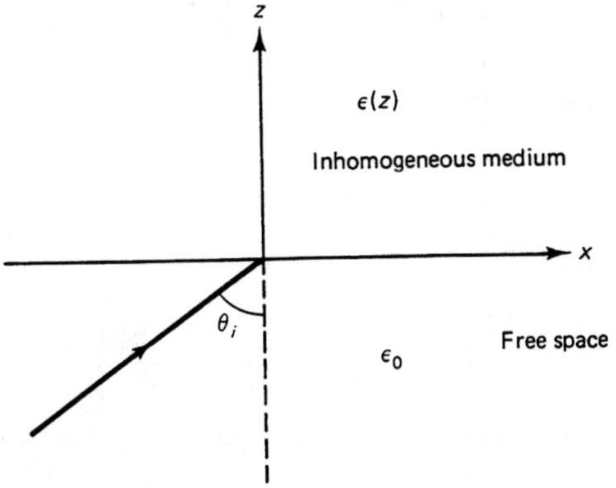

Figure 3-24 Wave incident in an inhomogeneous medium.

Noting that $\nabla \cdot \bar{E} = (\partial/\partial y)E_y = 0$, we get

$$\left[\frac{\partial^2}{\partial x^2} + \frac{\partial^2}{\partial z^2} + k_0^2 n^2(z)\right]E_y = 0. \tag{3-105}$$

For TM (parallel polarization), we eliminate $\bar{E}$ from (3-104) and get

$$\nabla \times \bar{E} = \nabla \times \left(\frac{1}{j\omega\epsilon}\nabla \times \bar{H}\right) = -j\omega\mu\bar{H}. \tag{3-106}$$

Rewriting this using $\bar{H} = H_y\hat{y}$ and $\partial/\partial y = 0$, we get

$$\frac{\partial^2}{\partial x^2}H_y + \epsilon(z)\frac{\partial}{\partial z}\left[\frac{1}{\epsilon(z)}\frac{\partial}{\partial z}H_y\right] + k_0^2 n^2(z)H_y = 0. \tag{3-107}$$

We may simplify this by using

$$U = \frac{H_y}{n(z)}, \tag{3-108}$$

and get

$$\left[\frac{\partial^2}{\partial x^2} + \frac{\partial^2}{\partial z^2} + k_0^2 N^2(z)\right]U = 0, \tag{3-109}$$

where

$$k_0^2 N^2(z) = k_0^2 n^2 + \frac{n''}{n} - \frac{2n'^2}{n^2}.$$

Note that (3-105) and (3-109) have the same mathematical form. Thus the study of the equation for the TE wave may be easily extended to that of the TM wave with changes from n to N.

Let us now consider the TE plane wave obliquely incident upon the medium (Fig. 3-24). We write

$$E_y(x, z) = u(z)e^{-j\beta_i x}, \tag{3-110}$$

where $\beta_i = k_0 \sin \theta_i$. Then (3-105) becomes

$$\left[\frac{d^2}{dz^2} + q^2(z)\right] u(z) = 0, \tag{3-111}$$

where $q^2(z) = k_0^2[n^2(z) - \sin^2 \theta_i]$. Similarly, for the TM wave, we let

$$U(x, z) = u(z)e^{-j\beta_i x}. \tag{3-112}$$

We get the same equation, (3-111), with

$$q^2(z) = k_0^2[N^2(z) - \sin^2 \theta_i]. \tag{3-113}$$

In the next section we discuss the WKB solution of (3-111).

3-14 WKB METHOD

In general, the exact solution of (3-111) may be obtained only for a few special functions $q(z)$. Here we consider an approximate solution of (3-111) called the *WKB solution*, named after Wentzel, Kramers, and Brillouin. Sometimes, this is also called the *WKBJ method*, adding Jeffreys.

The WKB approximation can be thought of as the first term of an asymptotic expansion. There are two ways of obtaining the asymptotic series. One is to expand the phase of $u(z)$ in a series

$$u(z) = \exp\left(\sum_n \psi_n\right), \tag{3-114}$$

and the other is to expand the field $u(z)$ in a series

$$u(z) = \sum_n u_n(z). \tag{3-115}$$

The first approach, (3-114), is convenient when the total phase is important, whereas the approach (3-115) is convenient for identifying various waves propagating in the medium. In either case the first term gives the identical WKB solution.

Let us consider (3-111). Noting that q is proportional to the wave number k_0, we may write $q(z) = k_0 n_e(z)$ and $n_e(z)$ as the equivalent refractive index.

Let us use (3-114), which is called the *phase integral approach*,

$$u(z) = e^{\psi(z)} \quad \text{and} \quad \psi(z) = \int^z \phi(z)\, dz. \tag{3-116}$$

Then substituting this into (3-111) we have

$$\frac{d\phi}{dz} + \phi^2 + k_0^2 n_e^2 = 0. \tag{3-117}$$

Note that (3-117) is a first-order nonlinear differential equation of the Riccati type.

In general, it is always possible to transform a second-order linear differential equation,

$$\left[P_0(z)\frac{d^2}{dz^2} + P_1(z)\frac{d}{dz} + P_2(z)\right]u(z) = 0, \tag{3-118}$$

into a first-order nonlinear differential equation of the Riccati type by using the transformation

$$u(z) = e^{\psi(z)} = e^{\int Q(z)\phi(z)dz}. \tag{3-119}$$

The resulting Riccati equation is

$$\frac{d\phi}{dz} + P(z)\phi + Q(z)\phi^2 = R(z), \tag{3-120}$$

where

$$P(z) = \frac{P_1}{P_0} + \frac{1}{Q}\frac{dQ}{dz} \qquad \text{and} \qquad R(z) = -\frac{P_2}{P_0 Q}.$$

For our problem (3-111), $P_0 = 1$, $P_1 = 0$, $P_2 = q^2$, and $Q = 1$.

Let us examine (3-117). We seek an expansion of ϕ in inverse powers of k. We write

$$\phi = \phi_0 k_0 + \phi_1 + \frac{\phi_2}{k_0} + \frac{\phi_3}{k_0^2} + \cdots. \tag{3-121}$$

The first term of (3-121) is proportional to k_0, as is clear from (3-117). Substituting (3-121) into (3-117) and expanding (3-117) in inverse powers of k_0, we get

$$\begin{aligned}\frac{d\phi}{dz} + \phi^2 + k_0^2 n_e^2 = (\phi_0^2 + n_e^2)k_0^2 &+ (2\phi_0\phi_1 + \phi_0')k_0 + (\phi_1^2 + 2\phi_0\phi_2 + \phi_1') \\ &+ (2\phi_1\phi_2 + 2\phi_0\phi_3 + \phi_2')\frac{1}{k_0} + \cdots.\end{aligned} \tag{3-122}$$

Equating the coefficients of each power of k_0 to zero, we get an infinite number of equations. The first three are

$$\begin{aligned}\phi_0^2 + n_e^2 &= 0, \\ 2\phi_0\phi_1 + \phi_0' &= 0, \\ \phi_1^2 + 2\phi_0\phi_2 + \phi_1' &= 0.\end{aligned} \tag{3-123}$$

The first equation of (3-123) gives

$$\phi_0 = \pm jn_e. \tag{3-124}$$

From the second equation, we have

$$\phi_1 = -\frac{\phi_0'}{2\phi_0}. \tag{3-125}$$

Therefore, the solution $u(z)$ is

$$\begin{aligned} u(z) &= e^{\int \phi dz} \\ &= e^{\int(\phi_0 k_0 + \phi_1 + \cdots)dz} \\ &= \exp(\pm j \int k_0 n_e \, dz + \int \phi_1 \, dz + \cdots). \end{aligned} \tag{3-126}$$

From (3-125), we get

$$\int \phi_1 \, dz = -\int \frac{\phi_0'}{2\phi_0} dz = -\frac{1}{2} \ln \phi_0,$$

and thus if we continue to the terms ϕ_0 and ϕ_1, we get

$$u(z) = \frac{1}{\sqrt{\phi_0}} \exp(\pm j \int k_0 n_e \, dz).$$

The general solution to (3-111) is, therefore,

$$u(z) = \frac{1}{q^{1/2}} (A e^{-j\int q dz} + B e^{+j\int q dz}), \tag{3-127}$$

where A and B are arbitrary constants and $q = k_0 n_e$. The derivative du/dz, consistent with the approximation above, is

$$\frac{du(z)}{dz} = -j q^{1/2} (A e^{-j\int q dz} - B e^{+j\int q dz}). \tag{3-128}$$

Here since we kept only the terms with $q^{-1/2}$ in (3-127), we keep only the terms with $q^{1/2}$ in (3-128). The term with $q^{-3/2}$ should not be kept as it is comparable to the term arising from ϕ_2. Equations (3-127) and (3-128) constitute the WKB solution to the original equation (3-111).

The WKB solution (3-127) consists of the wave u_1 traveling in the $+z$ direction and the wave u_2 traveling in the $-z$ direction.

$$u_1 = \frac{1}{q^{1/2}} e^{-j\int q dz} \quad \text{and} \quad u_2 = \frac{1}{q^{1/2}} e^{+j\int q dz}. \tag{3-129}$$

The Wronskian of u_1 and u_2 is then given by (see Section 5-5)

$$\Delta = u_1 u_2' - u_2 u_1' = 2j. \tag{3-130}$$

Let us examine the condition for the validity of the WKB solution. If we substitute u_1 in (3-129) into the original equation (3-111), we get

$$\left(\frac{d^2}{dz^2} + q^2\right) u_1 = f \neq 0.$$

Therefore, the range of validity of the WKB solution is

$$|f| << |q^2 u_1|. \tag{3-131}$$

In terms of $q = k_0 n_e$, (3-131) is

$$\frac{1}{k_0^2}\left|\frac{3}{4n_e^4}\left(\frac{dn_e}{dz}\right)^2 - \frac{1}{2n_e^3}\frac{d^2 n_e}{dz^2}\right| << 1. \tag{3-132}$$

It is clear that the medium must be slowly varying, but because of k_0^2, it is a better approximation for higher frequencies. However, whenever $q = 0$ (or $n_e = 0$), the WKB solution is not valid.

Let us next consider the physical meaning of the WKB solution. The total phase $\int q\,dz$ is the integral of the wave number q along the path. The amplitude $q^{-1/2}$ is such that the total power flux in a lossless medium is kept constant. For a scalar field $u(\bar{r})$, the real power flux density $\bar{F}(\bar{r})$ is given by [see equation (2-134)]

$$\bar{F}(\bar{r}) = \mathrm{Im}[u \nabla u^*]. \tag{3-133}$$

For our problem, the wave traveling in the $+z$ direction is $u(z) = Aq^{-1/2} \exp(-j\int q\,dz)$, and therefore, for a lossless medium (q is real), the power flux is kept constant.

$$\bar{F}(z) = |A|^2 \hat{z} = \text{constant}. \tag{3-134}$$

The WKB solution is identical to the geometric optics approximation, in which the phase is given by an eikonal equation and the amplitude is obtained so as to conserve power. The WKB solution is similar to the Rytov solution (see the Appendix to Chapter 16, Section B).

3-15 BREMMER SERIES

The WKB solution is the first term of the high-frequency series representation of a wave. It is applicable to the wave in a slowly varying inhomogeneous medium at high frequencies. The next term is usually difficult to obtain, but it is important to examine the rest of the series in order to determine the correction to the WKB method and to establish the convergence of the series. The Bremmer series provides such a representation (Wait, 1962).

Let us consider a series representation of a field $u(z)$:

$$u(z) = \sum_n u_n(z). \tag{3-135}$$

This is convenient because each term can be identified with the actual wave reflected by the inhomogeneities. This is done by Bremmer in the following manner.

We seek a solution to the differential equation

$$\left(\frac{d^2}{dz^2} + q^2\right)u(z) = 0, \tag{3-136}$$

in a form of the WKB solution, except that we allow A and B in (3-127) to be a function of z and try to find $A(z)$ and $B(z)$ to satisfy the differential equation. Thus we write

$$u(z) = \frac{1}{q^{1/2}}[A(z)e^{-j\int q dz} + B(z)e^{+j\int q dz}]. \tag{3-137}$$

Furthermore, we impose the condition between $A(z)$ and $B(z)$, such that the derivative du/dz has the same form as (3-128):

$$\frac{du(z)}{dz} = -jq^{1/2}[A(z)e^{-j\int q dz} - B(z)e^{+j\int q dz}]. \tag{3-138}$$

This requires that

$$\left(A' - \frac{1}{2}\frac{q'}{q}A\right)e^{-j\int q dz} + \left(B' - \frac{1}{2}\frac{q'}{q}B\right)e^{+j\int q dz} = 0. \tag{3-139}$$

Now, substituting (3-138) into (3-136), we get

$$\left(A' + \frac{1}{2}\frac{q'}{q}A\right)e^{-j\int q dz} + \left(B' + \frac{1}{2}\frac{q'}{q}B\right)e^{+j\int q dz} = 0. \tag{3-140}$$

We rewrite (3-139) and (3-140) in the following form:

$$\left(A' - \frac{1}{2}\frac{q'}{q}Be^{+j\int 2q dz}\right) + \left(B' - \frac{1}{2}\frac{q'}{q}Ae^{-j\int 2q dz}\right)e^{+j\int 2q dz} = 0. \tag{3-141}$$

and

$$\left(A' - \frac{1}{2}\frac{q'}{q}Be^{+j\int 2q dz}\right) - \left(B' - \frac{1}{2}\frac{q'}{q}Ae^{-j\int 2q dz}\right)e^{+j\int 2q dz} = 0. \tag{3-142}$$

Adding and subtracting these two, we get

$$A' - \left(\frac{1}{2}\frac{q'}{q}e^{+j\int 2q dz}\right)B = 0, \tag{3-143}$$

$$B' - \left(\frac{1}{2}\frac{q'}{q}e^{-j\int 2q dz}\right)A = 0. \tag{3-144}$$

These two equations constitute the coupling between two waves $A(z)$ and $B(z)$. Let us first recognize that these couplings depend on how the medium varies. If the medium is a slowly varying function, then q'/q, and therefore the coupling, is small. If the coupling is small, we may write

$$\begin{aligned} \frac{1}{2}\frac{q'}{q}e^{j\int 2q dz} &= \epsilon\lambda_1, \\ \frac{1}{2}\frac{q'}{q}e^{-j\int 2q dz} &= \epsilon\lambda_2, \end{aligned} \tag{3-145}$$

where ϵ is a small parameter and (3-143) and (3-144) become

$$\frac{dA}{dz} = \epsilon\lambda_1 B,$$
$$\frac{dB}{dz} = \epsilon\lambda_2 A, \tag{3-146}$$

Now we seek a perturbation solution in the following form:

$$A = A_0 + \epsilon A_1 + \epsilon^2 A_2 + \cdots,$$
$$B = B_0 + \epsilon B_1 + \epsilon^2 B_2 + \cdots. \tag{3-147}$$

Substituting (3-147) into (3-146) and equating equal powers of ϵ, we get

$$\frac{dA_{n+1}}{dz} = \lambda_1 B_n \quad \text{and} \quad \frac{dA_0}{dz} = 0,$$
$$\frac{dB_{n+1}}{dz} = \lambda_2 A_n \quad \text{and} \quad \frac{dB_0}{dz} = 0. \tag{3-148}$$

From these we obtain

$$A_{n+1} = \int_{-\infty}^{z} \lambda_1 B_n \, dz \quad \text{and} \quad A_0 = \text{constant},$$
$$B_{n+1} = \int_{\infty}^{z} \lambda_2 A_n \, dz \quad \text{and} \quad B_0 = \text{constant}. \tag{3-149}$$

The limit of integration in (3-149) is chosen to represent the physical picture as follows. We first write the complete solution

$$u(z) = \frac{1}{q^{1/2}}[(A_0 + \epsilon A_1 + \epsilon^2 A_2 + \cdots)e^{-j\int q dz} + (B_0 + \epsilon B_1 + \epsilon^2 B_2 + \cdots)e^{+j\int q dz}]$$
$$= u_{0+} + u_{1+} + u_{2+} + u_{3+} + \cdots + u_{0-} + u_{1-} + u_{2-} + u_{3-} + \cdots, \tag{3-150}$$

where

$$u_{0+} = \frac{A_0}{q^{1/2}} e^{-j\int q dz},$$
$$u_{0-} = \frac{B_0}{q^{1/2}} e^{+j\int q dz},$$

are the WKB solution. u_{0+} then generates the wave u_{1-} given by

$$u_{1-}(z) = \frac{e^{+j\int q dz}}{q^{1/2}} \int_{\infty}^{z} \epsilon\lambda_2 A_0 \, dz'$$
$$= \frac{1}{q(z)^{1/2}} \int_{\infty}^{z} \frac{1}{2} \frac{q'(z')}{q(z')} A_0 \exp\left(j\int_{z'}^{z} q\, dz'' - j\int_{z_0}^{z'} q\, dz''\right) dz'. \tag{3-151}$$

This gives the first correction term to the WKB solution u_{0+}. This wave $u_{1-}(z)$ then generates the wave u_{2+}:

$$u_{2+}(z) = \frac{e^{-j\int q\,dz}}{q^{1/2}} \int_{-\infty}^{z} \epsilon\lambda_1\,dz' \int_{\infty}^{z'} \epsilon\lambda_2 A_0\,dz''. \tag{3-152}$$

Similarly, u_{1+} is generated by u_{0-}, and u_{2-} is generated by u_{1+} (Fig. 3-25).

The Bremmer series (3-150) gives a complete series representation of the solution to (3-136) if the series is convergent. This requires that, considering u_{1-} of (3-151),

$$\left| \int_{\infty}^{z} \left(\frac{1}{2}\frac{q'}{q} e^{-j\int^{z} 2q\,dz} \right) dz \right| << 1. \tag{3-153}$$

Now the exponential $e^{-j\int 2q\,dz}$ oscillates with an approximate period of π/q. Thus if

$$\left| \frac{1}{2}\frac{q'}{q}\frac{\pi}{2q} \right| << 1, \tag{3-154}$$

then the integral (3-153) in a half-period ($\pi/2q$) cancels the integral in the next half-period, and thus (3-153) is approximately satisfied. Therefore, (3-154) is the necessary condition for the WKB solution to be valid.

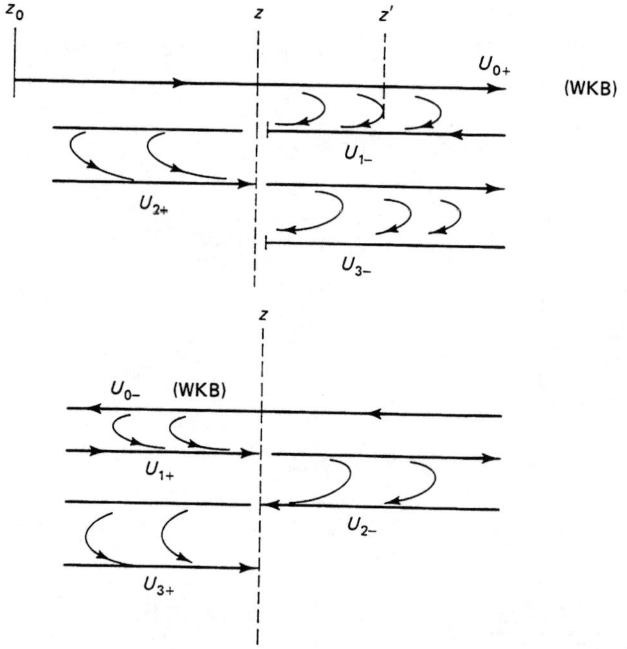

Figure 3-25 Bremmer series.

3-16 WKB SOLUTION FOR THE TURNING POINT

It is clear that the WKB solution breaks down in the neighborhood of $z = z_0$, where

$$q(z_0) = 0. \tag{3-155}$$

Consider the profile of $q(z)$ shown in Fig. 3-26. The wave propagating in the $+z$ direction cannot propagate beyond the point z_0, and thus the wave must be reflected back. This point z_0 is called the *turning point*.

Let us consider the three regions approximately divided at z_1 and z_2, as shown in Fig. 3-26. In region I the WKB solution is applicable, and thus the incident wave can be written as

$$u_i(z) = \frac{A_0}{q^{1/2}} e^{-j\int_0^z q\,dz}, \tag{3-156}$$

where $A_0/q(0)^{1/2}$ is the amplitude at $z = 0$.

The reflected wave in region I can also be written in the WKB form. It is shown in the Appendix to Chapter 3, Section A, that the WKB reflected wave is given by

$$u_r(z) = \frac{A_0}{q^{1/2}} e^{-j\int_0^{z_0} q(z)dz + j\int_{z_0}^{z} q(z)dz + j(\pi/2)}. \tag{3-157}$$

Note that the phase of the reflected wave is the integral of $q(z)$ from 0 to z_0, and then from z_0 back to z with the additional phase jump of $\pi/2$. Thus the reflected wave u_r behaves as if the WKB solution is applicable to z_0 and back, except for the phase jump. The WKB solution is, of course, not applicable near z_0, as shown in the Appendix, but the total effect of the turning point to the WKB solution far from the turning point is simply the phase jump of $\pi/2$.

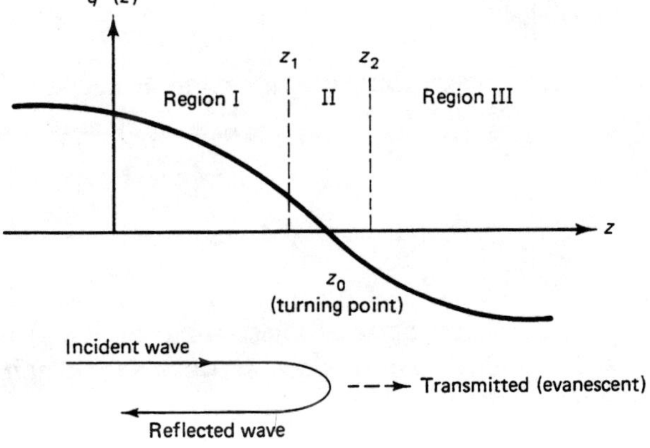

Figure 3-26 Turning point.

The transmitted wave is evanescent since q^2 is negative. The WKB transmitted wave u_t far from the turning point is given by

$$\begin{aligned} u_t(z) &= \frac{A_0}{q^{1/2}} e^{-j\int_0^{z_0} q\,dz - j\int_{z_0}^{z} q\,dz} \\ &= \frac{A_0 e^{j(\pi/4)}}{\alpha(z)^{1/2}} e^{-j\int_0^{z_0} q\,dz - \int_{z_0}^{z} \alpha\,dz}, \end{aligned} \tag{3-158}$$

where $q(z) = -j\alpha(z)$. The derivation of (3-158) is also given in the Appendix.

3-17 TRAPPED SURFACE-WAVE MODES IN AN INHOMOGENEOUS SLAB

The WKB method can conveniently deal with the propagation constant of the guided-wave modes in an inhomogeneous slab. This approach is useful for thin-film, graded-index optical fibers (GRIN) and guided acoustic waves in the ocean.

Consider a dielectric slab whose refractive index n (or relative dielectric constant ϵ_r) is a function of z.

$$\epsilon_r(z) = n^2(z). \tag{3-159}$$

The wave $U(x, z)$ is propagating in the x direction and we wish to find the propagation constant β in the x direction. The field $U(x, z)$ satisfies

$$\left[\frac{\partial^2}{\partial x^2} + \frac{\partial^2}{\partial z^2} + k_0^2 n^2(z)\right] U(x, z) = 0. \tag{3-160}$$

Now we let

$$U(x, z) = u(z) e^{-j\beta x} \tag{3-161}$$

and obtain

$$\left[\frac{d^2}{dz^2} + q^2(z)\right] u(z) = 0, \tag{3-162}$$

where $q^2(z) = k_0^2 n^2(z) - \beta^2$. The refractive index profile $n(z)$ is maximum near $z = 0$ and slopes down on both sides of $z = 0$ (Fig. 3-27).

The WKB solution starting at $z = 0$ is given by

$$u(z) = \frac{A_0}{q^{1/2}} \exp\left[-j\int_0^{z_2} q\,dz + j\int_{z_2}^{z_1} q\,dz - j\int_{z_1}^{z} q\,dz + j\pi\right], \tag{3-163}$$

where the turning points z_1 and z_2 are the roots of $q^2(z) = 0$. The total wave is reflected at two turning points z_1 and z_2, and the total phase jump of $2(\pi/2)$ is included. If this wave represents the guided-wave mode, this wave should join smoothly with the wave at $z = 0$. Thus we should have

$$\exp\left(-j\int_{z_1}^{z_2} q\,dz + j\int_{z_2}^{z_1} q\,dz + j\pi\right) = 1. \tag{3-164}$$

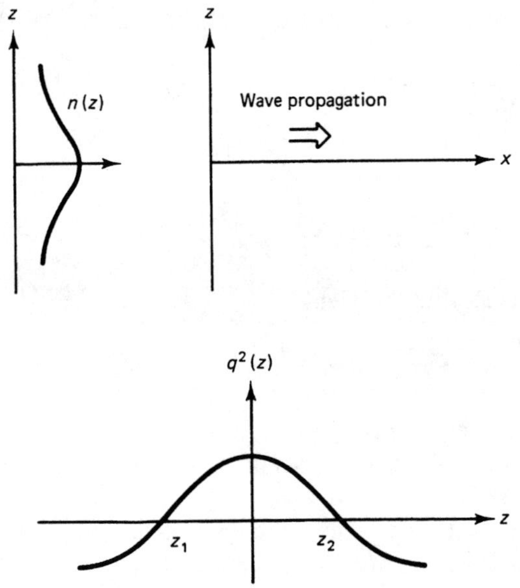

Figure 3-27 Trapped surface wave.

From this, we get

$$\int_{z_1}^{z_2} q(z)\,dz = \left(m + \frac{1}{2}\right)\pi, \qquad m = \text{integer}, \tag{3-165}$$

where $q(z) = [k^2 n^2(z) - \beta^2]^{1/2}$ and $q(z_1) = q(z_2) = 0$. Equation (3-165) gives the propagation constant β for each mode with $m = 0, 1, 2, \ldots$. Since the WKB solution is applicable to the region far from the turning points, (3-165) gives a better approximation for large m. If the refractive index is bounded by n_1 and n_2, the propagation constant β must also be bounded (Fig. 3-28):

$$k_0 n_2 < \beta < k_0 n_1. \tag{3-166}$$

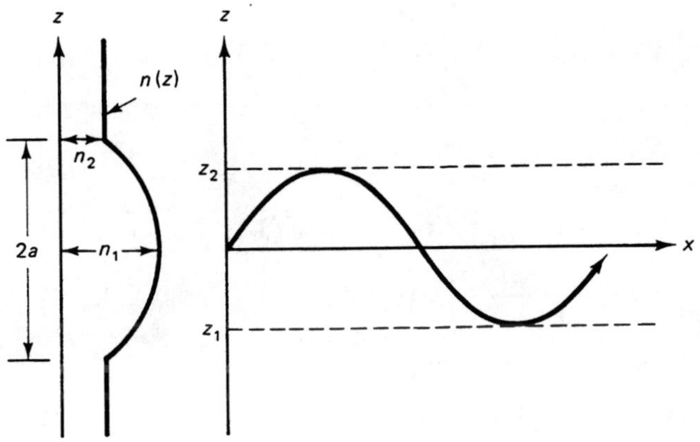

Figure 3-28 Square-law profile.

Let us consider an example of the square-law profile:

$$n^2(z) = \begin{cases} n_1^2\left[1 - 2\left(\frac{z}{a}\right)^2 \Delta\right] & \text{for } |z| < a \\ n_2^2 = n_1^2(1 - 2\Delta) & \text{for } |z| > a \end{cases}. \tag{3-167}$$

where

$$\Delta = \frac{n_1^2 - n_2^2}{2n_1^2} \approx \frac{n_1 - n_2}{n_1}.$$

Integration in (3-165) can be performed to yield

$$\beta^2 = k_0^2 n_1^2\left[1 - \frac{2\sqrt{2\Delta}}{k_0 n_1 a}\left(m + \frac{1}{2}\right)\right], \tag{3-168}$$

where m = integers and the turning points z_1 and z_2 are assumed to be within the region $|z| < a$ (Fig. 3-28). The propagation constant β is therefore bounded by $k_0 n_2 < \beta < k_0 n_1$. The WKB solution (3-165) is valid for large m. However, for this square-law profile, (3-168) can be shown to give the exact propagation constant β for the finite number of $m = 0, 1, 2, \ldots, M$.

3-18 MEDIUM WITH PRESCRIBED PROFILE

It is often possible to approximate the profile of the medium $n(z)$ by using a particular functional form such that the resulting equation has a well-known exact solution. Here we discuss one such example.

There are a great number of differential equations, which may be converted into Bessel's differential equation (see Jahnke et al., 1960, p. 156). For example,

$$W'' + \left(\beta^2 - \frac{4\nu^2 - 1}{4z^2}\right)W = 0, \tag{3-169}$$

has a solution

$$W = \sqrt{z}Z_\nu(\beta z),$$

where Z_ν is a Bessel function. We may convert (3-111) into the form above if the medium profile is given by

$$n^2 = a^2 - \frac{b^2}{(z + z_0)^2}, \tag{3-170}$$

where a, b, and z_0 are constants to be chosen arbitrarily. Then (3-111) becomes

$$\left\{\frac{d^2}{dz^2} + k^2\left[a^2 - \sin^2\theta_i - \frac{b^2}{(z + z_0)^2}\right]\right\}u(z) = 0. \tag{3-171}$$

Comparing (3-171) with (3-169), we get the solution

$$u(z) = (z + z_0)^{1/2} Z_\nu[\beta(z + z_0)], \tag{3-172}$$

where $\nu = \sqrt{k^2 b^2 + \frac{1}{4}}$ and $\beta = k\sqrt{a^2 - \sin^2 \theta_i}$.

The choice of the Bessel function Z must be made to satisfy the boundary conditions. For example, if the medium extends to $z \to \infty$, then to satisfy the radiation condition as $z \to \infty$, we must have (see p. 90)

$$Z_\nu = H_\nu^{(2)}[\beta(z + z_0)]. \tag{3-173}$$

Let us now calculate the reflected wave from such an inhomogeneous medium. Let the incident wave be

$$E_y^i = E_0 e^{-j(k \sin \theta_i)x - j(k \cos \theta_i)z}, \qquad z < 0, \tag{3-174}$$

and the reflected wave be

$$E_y^r = RE_0 e^{-j(k \sin \theta_i)x + j(k \cos \theta_i)z}, \qquad z < 0. \tag{3-175}$$

The transmitted wave is given by (3-172),

$$E_y^t = TE_0(z + z_0)^{1/2} H_\nu^{(2)}[\beta(z + z_0)]e^{-j(k \sin \theta_i)x}, \qquad z > 0, \tag{3-176}$$

where R and T are constants to be determined.

By applying the boundary conditions

$$\begin{aligned} E_y^i + E_y^r &= E_y^t, \\ H_x^i + H_x^r &= H_x^t \qquad \text{at } z = 0, \end{aligned} \tag{3-177}$$

we obtain

$$R = \frac{k \cos \theta_i - j\left[\dfrac{1}{2z_0} + \beta \dfrac{H_\nu^{(2)\prime}(\beta z_0)}{H_\nu^{(2)}(\beta z_0)}\right]}{k \cos \theta_i + j\left[\dfrac{1}{2z_0} + \beta \dfrac{H_\nu^{(2)\prime}(\beta z_0)}{H_\nu^{(2)}(\beta z_0)}\right]}. \tag{3-178}$$

The amplitude of the transmitted wave T is then given by

$$T = \frac{1 + R}{z_0^{1/2} H_\nu^{(2)}(\beta z_0)}. \tag{3-179}$$

PROBLEMS

3-1. Microwaves at 915 MHz are normally incident from air to muscle ($\epsilon' = 51, \sigma = 1.6$). Calculate the SAR (specific absorption rate) in watts/kg as a function of depth. The incident power flux density is 1 mW/cm^2. Assume that the density of muscle is approximately equal to that of water. SAR is the power absorbed per unit mass of the medium and is commonly used in bioelectromagnetics.

3-2. Consider an electromagnetic wave at 10 GHz normally incident from air on a semi-infinite ferromagnetic material. The relative dielectric constant ϵ_r is $2 - j10$.

(a) What should be the permeability of the medium to reduce the reflection?

(b) If the wave is obliquely incident on the medium given in part (a), find and plot the reflection coefficient as a function of the angle of incidence for the TE and TM cases.

(c) If a thickness of 1 cm of the material is placed on a conducting surface, what is the reflection coefficient (in dB) at normal incidence?

3-3. A plane radio wave is incident on the ionosphere from the air. The frequency is 6 MHz, the plasma frequency of the ionosphere is 5 MHz, and the collision frequency is negligibly small.

(a) Calculate and plot the reflection and transmission coefficients as functions of the incident angle for TE and TM waves.

(b) What is the critical angle for total reflection and the Brewster angle?

(c) Discuss the conservation of power at the boundary. The relative dielectric constant of lossless plasma (ionosphere) is given by

$$\epsilon_r = 1 - \frac{\omega_p^2}{\omega^2},$$

where $f_p = \omega_p/2\pi$ is the plasma frequency.

3-4. A TE wave with λ_0 (free space) = 3 cm is incident on the dielectric layers as shown in Fig. P3-4. Calculate the ABCD parameters for each layer and for the total layers. Calculate the reflection and transmission coefficients. Find the angle of transmission.

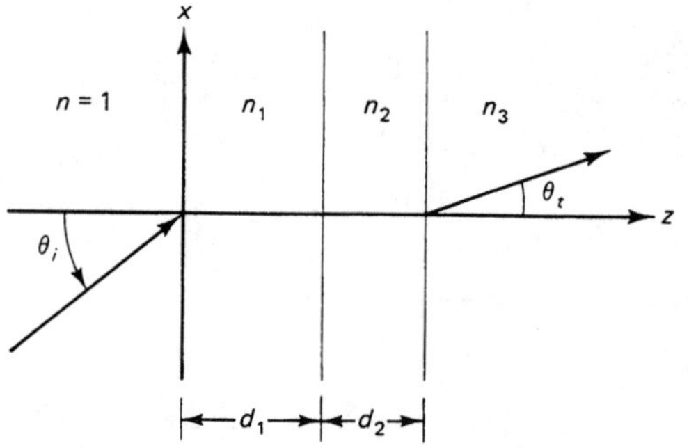

Figure P3-4 TE wave is incident on a dielectric layer. $n_1 = 1.5$, $n_2 = 2.0$, $n_3 = 2.5$, $d_1 = 1$ cm, $d_2 = 1.5$ cm, and $\theta_i = 30°$.

3-5. A wave with unit power flux density is normally incident on a dielectric film as shown in Fig. P3-5. Calculate the reflected and the transmitted power flux densities as functions of $n_2 d/\lambda_0$ (λ_0 in free space) for the following two cases: ($0 \le n_2 d/\lambda_0 \le 1$):

(a) $n_1 = 1, n_2 = 2, n_3 = 4$

(b) $n_1 = 1, n_2 = 3, n_3 = 4$

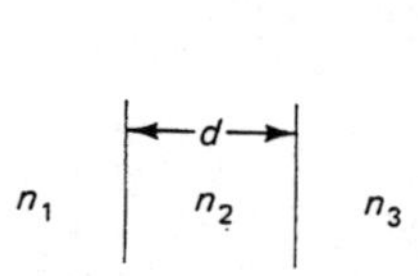

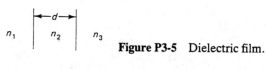

Figure P3-5 Dielectric film.

3-6. An ultrasound wave at 1 MHz is incident from fat ($C_0 = 1.44 \times 10^5$ cm/s, $\rho_0 = 0.97$ g/cm^3) to muscle ($C_0 = 1.57 \times 10^5$, $\rho_0 = 1.07$). Assuming that the incident wave is plane and the boundary is plane, calculate and plot the reflection and transmission coefficients as functions of the incident angle. C_0 = sound velocity and ρ_0 = density.

3-7. A TE wave with unit power flux density is incident on an air gap between two prisms as shown in Fig. P3-7. Calculate the power flux density transmitted as a function of the air gap d/λ_0.

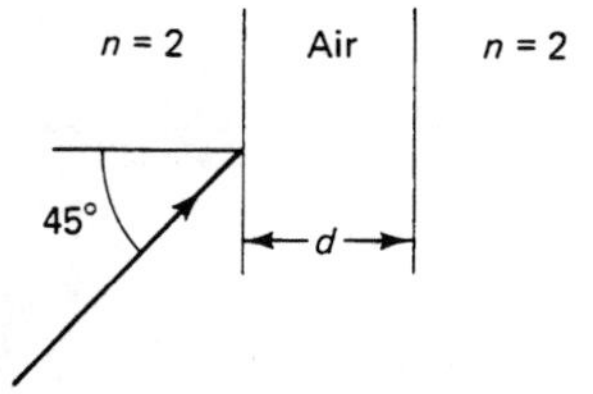

Figure P3-7 Air gap between two prisms.

3-8. Consider TM modes along a dielectric slab lying on a perfect conductor. Let ϵ of the slab be $2.56\epsilon_0$, and the medium above the waveguide is air. The thickness of the slab is 0.5 cm. What are the cutoff frequencies of the first six modes? At a frequency of 30 GHz, what are the propagation constants of the propagating modes? What is the ratio of the power outside the slab to the power inside the slab for each propagating mode at 30 GHz?

3-9. A trapped surface wave may propagate along the interface between the air and plasma (see Fig. 3-21). If the plasma frequency is 1 MHz, find the frequency range in which a trapped surface wave propagates along the interface. Do this for TM and TE modes. Find the propagation constant β, p_0, and p when the operating frequency is 500 kHz as shown in Fig. 3-23.

3-10. Calculate β, p_0, and p for a Zenneck wave at 100 kHz for wet earth, dry earth, fresh water, and seawater (see Table 2-1).

3-11. Calculate β, p_0, and p for plasmons at an air–silver interface with $\lambda = 0.6\ \mu\text{m}$.

3-12. Calculate the reflection coefficient when a TM wave at 0.6 μm is incident on silver from air as a function of the incident angle.

3-13. Consider a TE trapped surface wave ($E_y \neq 0, E_x = E_z = 0$) propagating in the x direction along an inhomogeneous dielectric slab whose refractive index is given by

$$n^2(z) = \begin{cases} n_1^2\left[1 - 2\Delta\left(\dfrac{z}{a}\right)^2\right] & \text{for } |z| < a, \\ n_2^2 = n_1^2(1 - 2\Delta) & \text{for } |z| > a, \end{cases}$$

where $n_1 = 1.48$, $n_2 = 1.46$, $a = 10\ \mu\text{m}$, and $\lambda_0 = 0.82\ \mu\text{m}$. How many surface wave modes can exist? Calculate the propagation constant for each mode, using the WKB method.

4 Waveguides and Cavities

In Chapter 3 we discussed wave propagation along layered media. In this chapter we concentrate on hollow waveguides, dielectric waveguides, and cavities commonly used in microwave, millimeter wave, and optical fibers. In discussing these problems, we present basic formulations and solutions in terms of special functions. Waveguide problems are treated extensively in Montgomery et al. (1948) and Marcuvitz (1951); dielectric waveguides are discussed in Marcuse (1982). For special functions, see Gradshteyn and Ryzhik (1965), Jahnke et al. (1960), Abramowitz and Stegun (1964), and Magnus and Oberhettinger (1949).

4-1 UNIFORM ELECTROMAGNETIC WAVEGUIDES

We start with a general expression for the electromagnetic field in the source-free region in terms of the electric Hertz vector $\bar{\pi}$ and the magnetic Hertz vector $\bar{\pi}_m$ shown in Sections 2-7 and 2-9.

$$\begin{aligned} \bar{E} &= \nabla \times \nabla \times \bar{\pi} - j\omega\mu\nabla \times \bar{\pi}_m, \\ \bar{H} &= j\omega\epsilon\nabla \times \bar{\pi} + \nabla \times \nabla \times \bar{\pi}_m. \end{aligned} \tag{4-1}$$

It has been shown (Stratton, 1941, p. 392) that a completely general solution to

Maxwell's equations can be given by (4-1) when $\bar{\pi}$ and $\bar{\pi}_m$ are pointed in a constant direction $\hat{a}$.

$$\bar{\pi} = \pi(x, y, z)\hat{a}, \qquad \bar{\pi}_m = \pi_m(x, y, z)\hat{a}. \tag{4-2}$$

As noted in Sections 2-7 and 2-9, $\bar{\pi}$ and $\bar{\pi}_m$ satisfy the scalar wave equation.

For a waveguide whose cross section is uniform along the z direction (Fig. 4-1), it is convenient to choose $\hat{a} = \hat{z}$. It will be shown that the fields generated by $\bar{\pi} = \pi_z \hat{z}$ do not have H_z, the z component of the magnetic field, and this is called the *transverse magnetic* (TM) mode. It is also called the E mode since E_z, the z component of the electric field, is proportional to π_z, which generates this mode. Similarly, the field generated by $\bar{\pi}_m = \pi_{mz}\hat{z}$ does not have E_z and is called the *transverse electric* (TE) mode. It is also called the H mode. We now show that the TM and TE modes correspond to the solutions to Dirichlet's and Neumann's eigenvalue problems, respectively.

4-2 TM MODES OR E MODES

The TM modes are generated by $\bar{\pi} = \pi_z \hat{z}$, which satisfies the scalar wave equation:

$$(\nabla^2 + k^2)\pi_z = 0, \tag{4-3}$$

where k is the wave number. If the waveguide is hollow, it is given in terms of the velocity of light c and wavelength λ_0 in a vacuum,

$$k = k_0 = \frac{\omega}{c} = \frac{2\pi}{\lambda_0}. \tag{4-4}$$

If the waveguide is filled with a material with an index of refraction n, k is given by

$$k = k_0 n = \frac{\omega}{c} n = \frac{2\pi}{\lambda_0} n. \tag{4-5}$$

All field components are then given in terms of π_z,

$$\bar{E} = \nabla(\nabla \cdot \bar{\pi}) + k^2\bar{\pi} = \nabla\frac{\partial \pi_z}{\partial z} + k^2\bar{\pi}, \tag{4-6}$$

$$\bar{H} = j\omega\epsilon\nabla \times \bar{\pi} = j\omega\epsilon\left(\hat{z}\frac{\partial}{\partial z} + \nabla_t\right) \times \bar{\pi} = j\omega\epsilon(\nabla_t \pi_z \times \hat{z}), \tag{4-7}$$

where ∇_t is a "del" operator in the transverse plane,

$$\nabla_t = \hat{x}\frac{\partial}{\partial x} + \hat{y}\frac{\partial}{\partial y}. \tag{4-8}$$

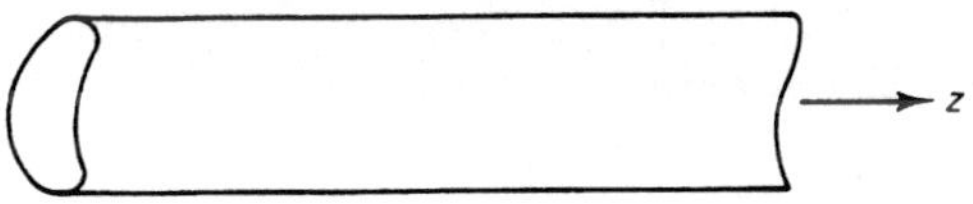

Figure 4-1 Uniform waveguide.

Also note that $\overline{H}$ is transverse to the z direction.

Let us consider the propagation constant β of a wave propagating in the z direction. The Hertz vector is given by

$$\pi_z(x, y, z) = \phi(x, y)e^{-j\beta z}. \tag{4-9}$$

Substituting this into (4-3), we get

$$\begin{aligned} (\nabla_t^2 + k_c^2)\phi(x, y) &= 0, \\ \beta^2 &= k^2 - k_c^2. \end{aligned} \tag{4-10}$$

For a given waveguide, k_c is a constant and depends only on the geometry of the guide. This constant k_c is called the *cutoff wave number*. Once k_c is known, the propagation constant β for a given frequency is determined from (4-10).

The field components are then given by (4-6) and (4-7) using (4-9),

$$\begin{aligned} E_z &= k_c^2\, \phi(x, y)e^{-j\beta z}, \\ \overline{E}_t &= -j\beta\nabla_t\phi(x, y)e^{-j\beta z}, \\ H_z &= 0, \\ \overline{H}_t &= \frac{\hat{z} \times \overline{E}_t}{Z_e} = -j\omega\epsilon\hat{z} \times \nabla_t\phi(x, y)e^{-j\beta z}, \end{aligned} \tag{4-11}$$

where $Z_e = \beta/\omega\epsilon$.

The significance of the formulation above is that the waveguide problem is now reduced to solving a scalar wave equation (4-10) for a scalar function $\phi(x, y)$ over the waveguide cross section. Also, we note from (4-11) that in the cross section the electric and magnetic field distributions have the same form except that the magnetic field is rotated by 90° from the electric field and the ratio of the electric field to the magnetic field is Z_e. The quantity Z_e has the dimension of impedance and is called the *wave impedance*.

4-3 TE MODES OR *H* MODES

The dual of the TM modes are the TE (transverse electric) modes or H modes. This can be derived from the magnetic Hertz vector $\pi_m = \pi_{mz}\hat{z}$ and π_{mz} satisfies the wave equation

$$(\nabla^2 + k^2)\pi_{mz} = 0. \tag{4-12}$$

For a wave propagating in the z direction with propagation constant β, we write

$$\pi_{mz} = \psi(x, y)e^{-j\beta z}. \tag{4-13}$$

We then get the scalar wave equation for $\psi(x, y)$

$$(\nabla_t^2 + k_c^2)\psi(x, y) = 0. \tag{4-14}$$

The propagation constant β at a given frequency is obtained once the cutoff wave number k_c is known,

$$\beta^2 = k^2 - k_c^2. \tag{4-15}$$

The field components are given in the following form, similar to (4-11):

$$\begin{aligned} H_z &= k_c^2\, \psi(x,y)e^{-j\beta z}, \\ \overline{H}_t &= -j\beta\nabla_t\psi(x,y)e^{-j\beta z}, \\ E_z &= 0, \\ \overline{E}_t &= Z_h(\overline{H}_t \times \hat{z}) = j\omega\mu\hat{z} \times \nabla_t\psi(x,y)e^{-j\beta z}, \end{aligned} \tag{4-16}$$

where Z_h is the wave impedance given by

$$Z_h = \frac{\omega\mu}{\beta}.$$

Note that if the waveguide is empty, $k = k_0$, but if the waveguide is filled with a material with an index of refraction n, $k = k_0 n$.

The boundary condition for a perfectly conducting wall is that the tangential electric field must be zero. For TM modes, this means that

$$E_z = 0 \quad \text{and} \quad \overline{E}_t \times \hat{n} = 0, \tag{4-17}$$

where $\hat{n}$ is the unit vector normal to the boundary (Fig. 4-2). From (4-11), the first condition requires that either $k_c = 0$ or $\phi(x,y) = 0$ on the wall. If $k_c = 0$, E_z as well as H_z are identically zero at all points in the guide, and therefore this is a TEM mode (transverse electromagnetic mode). It will be shown later that TEM modes cannot exist inside a waveguide with a single wall. Therefore, for TM modes, $k_c \neq 0$ and

$$\phi(x,y) = 0 \qquad \text{on the wall.} \tag{4-18}$$

If (4-18) is satisfied, we can show that $\overline{E}_t \times \hat{n}$ is also zero on the wall. To show this, write $E_t = (\overline{E}_t \cdot \hat{n})\hat{n} + (\overline{E}_t \cdot \hat{t})\hat{t}$, where $\hat{t}$ is the unit vector tangential to the wall and $\hat{n}$ is the unit vector normal to the wall (Fig. 4-2). Then $\overline{E}_t \times \hat{n} = -(\overline{E}_t \cdot \hat{t})\hat{z}$, but $(\overline{E}_t \cdot \hat{t})$ is equal to $-j\beta(\partial)/(\partial t)\phi(x,y)e^{-j\beta z}$ from (4-11), where $\partial/\partial t$ is the derivative in the direction of $\hat{t}$. If $\phi(x,y) = 0$ on the wall, then $\partial/(\partial t)\phi$ is also zero, and thus $\overline{E}_t \times \hat{n} = 0$ on the wall.

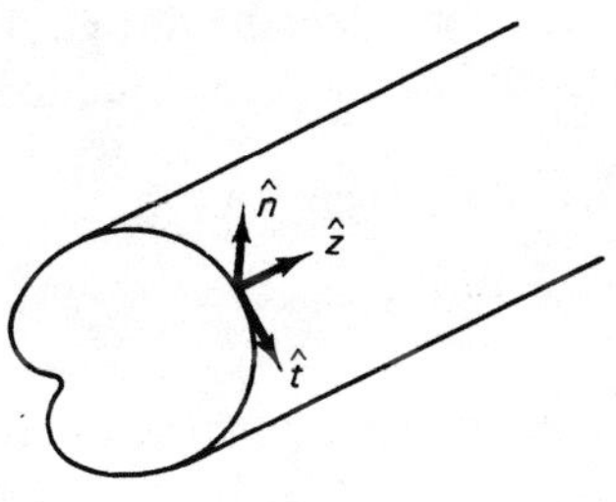

Figure 4-2 Three orthogonal unit vectors, $\hat{n}$, $\hat{t}$, and $\hat{z}$. $\hat{n}$ is normal to the wall and $\hat{t}$ is tangential to the wall and transverse to $\hat{z}$.

For TE modes, the boundary condition is $\bar{E}_t \times \hat{n} = 0$. From (4-16), this condition can be shown to be equivalent to

$$\frac{\partial \psi}{\partial n} = 0 \qquad \text{on the wall.} \tag{4-19}$$

These two boundary conditions, (4-18) and (4-19), are called *Dirichlet's* and *Neumann's conditions*, respectively, and boundary value problems with these boundary conditions are often encountered in many branches of science and engineering.

4-4 EIGENFUNCTIONS AND EIGENVALUES

As shown in (4-10), for TM modes, the problem is reduced to that of finding the function $\phi(x, y)$ and the constant k_c satisfying the following equation:

$$(\nabla_t^2 + k_c^2)\phi(x, y) = 0,$$

over the cross section of the guide and the boundary condition

$$\phi(x, y) = 0 \qquad \text{on the boundary.} \tag{4-20}$$

Similarly, for TE modes, we have, from (4-14),

$$\begin{aligned} &(\nabla_t^2 + k_c^2)\psi(x, y) = 0, \\ &\frac{\partial \psi}{\partial n} = 0 \qquad \text{on the boundary.} \end{aligned} \tag{4-21}$$

The function $\phi(x, y)$ or $\psi(x, y)$ satisfying (4-20) or (4-21) is called the *eigenfunction* and k_c is called the *eigenvalue*. Therefore, the problem of solving for electromagnetic wave propagation in a uniform guide with a perfectly conducting wall is reduced to the problem of finding eigenfunctions and eigenvalues for the two-dimensional Dirichlet's or Neumann's problems (4-20) or (4-21), and once k_c is obtained, the propagation constant β is given by

$$\beta = (k^2 - k_c^2)^{1/2}. \tag{4-22}$$

In general, to satisfy (4-20) or (4-21), the eigenvalue k_c cannot be an arbitrary number, and it can take only specific values. There is an infinite number of discrete eigenvalues. As an example, consider a rectangular guide with sides a and b (Fig. 4-3). The eigenfunction $\phi(x, y)$ satisfies the wave equation:

$$\left(\frac{\partial^2}{\partial x^2} + \frac{\partial^2}{\partial y^2} + k_c^2\right)\phi(x, y) = 0. \tag{4-23}$$

The boundary conditions are

$$\phi(x, y) = 0 \qquad \text{at } x = 0 \text{ and } a \quad \text{and} \quad \text{at } y = 0 \text{ and } b. \tag{4-24}$$

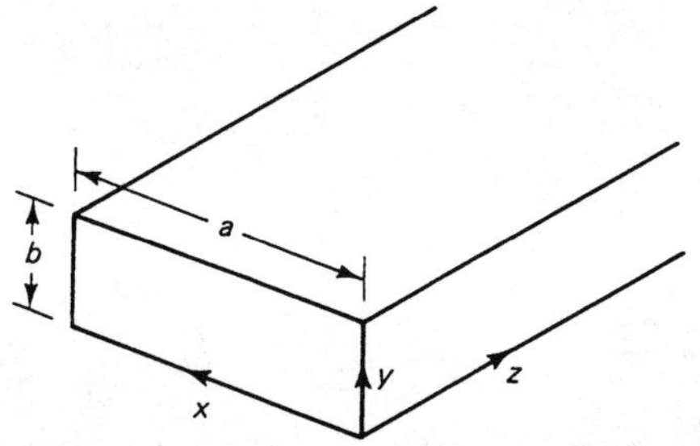

Figure 4-3 Rectangular waveguide.

Equations (4-23) and (4-24) can be solved by assuming that $\phi(x, y)$ is a product of a function $X(x)$ of x only and a function $Y(y)$ of y only. This technique is called the *method of separation of variables* and is possible only when the boundary coincides with the coordinate system. Let us write

$$\phi(x, y) = X(x)Y(y). \tag{4-25}$$

Substituting this into (4-23) and multiplying the resulting equation by $[X(x)Y(y)]^{-1}$, we get

$$\frac{X''}{X} + \frac{Y''}{Y} + k_c^2 = 0, \tag{4-26}$$

where the double prime means the second derivative. Since the first term depends only on x and the second term depends only on y, (4-26) can be satisfied for arbitrary x and y only when the first term and the second term, respectively, are constant and the constants are chosen such that the sum of the constants and k_c^2 equals zero. Let these two constants be $-k_x^2$ and $-k_y^2$. Then we have

$$\left(\frac{d^2}{dx^2} + k_x^2\right)X(x) = 0,$$
$$\left(\frac{d^2}{dy^2} + k_y^2\right)Y(y) = 0, \tag{4-27}$$

where $k_x^2 + k_y^2 = k_c^2$. Since each of (4-27) is a second-order differential equation, $X(x)$ is given by a linear combination of two independent solutions,

$$X(x) = A \sin k_x x + B \cos k_x x, \tag{4-28}$$

where A and B are constant. Now we apply the boundary condition that $X(0) = X(a) = 0$. This gives $B = 0$ and $\sin k_x a = 0$. From this we get

$$k_x = \frac{m\pi}{a}, \qquad m = 1, 2, 3, \ldots, \infty. \tag{4-29}$$

Let us choose A such that $X(x)$ satisfies the following *normalization* condition:

$$\int_0^a X(x)^2\, dx = 1, \tag{4-30}$$

yielding $A = (2/a)^{1/2}$.

Similarly, the normalized $Y(y)$ is given by

$$Y(y) = \left(\frac{2}{b}\right)^{1/2} \sin k_y y, \tag{4-31}$$

where

$$k_y = \frac{n\pi}{b}, \qquad n = 1, 2, 3, \ldots, \infty.$$

Since the eigenvalue k_c depends on m and n, we write $k_c = k_{mn}$ and from (4-27), we get

$$k_{mn}^2 = \left(\frac{m\pi}{a}\right)^2 + \left(\frac{n\pi}{b}\right)^2, \tag{4-32}$$

where $m = 1, 2, 3, \ldots, \infty$ and $n = 1, 2, 3, \ldots, \infty$. For given m and n, the corresponding normalized eigenfunction $\phi(x, y) = \phi_{mn}(x, y)$ is given by

$$\phi_{mn}(x, y) = \frac{2}{\sqrt{ab}} \sin\frac{m\pi x}{a} \sin\frac{n\pi y}{b}. \tag{4-33}$$

4-5 GENERAL PROPERTIES OF EIGENFUNCTIONS FOR CLOSED REGIONS

As we discussed in the preceding section, waveguide problems are a two-dimensional eigenfunction problem, since we only need to consider the waveguide cross-sectional area. Cavity resonators can be regarded as a three-dimensional eigenfunction problem, because we need to consider waves in a volume. In either case we are considering the region completely surrounded by the boundary and the region does not extend to infinity. This is called the *closed region*. If a part of the boundary extends to infinity, this is called the *open region*.

Let us consider the general properties of eigenfunctions in a closed region as defined by (4-20) and (4-21). It was indicated in Section 4-4 that there exists a set of a doubly infinite number of eigenfunctions

$$\phi_{mn}, \qquad m = 0, 1, 2, \ldots, \infty$$
$$n = 0, 1, 2, \ldots, \infty \tag{4-34}$$

and the corresponding eigenvalues k_{mn}. The eigenfunction ϕ_{mn} satisfies the wave equation

$$(\nabla_t^2 + k_{mn}^2)\phi_{mn} = 0 \tag{4-35}$$

and boundary conditions. The boundary conditions can be either one of the following two types:

$$\phi_{mn} = 0 \qquad \text{Dirichlet's condition} \tag{4-36}$$

$$\frac{\partial \phi_{mn}}{\partial n} = 0 \qquad \text{Neumann's condition.} \tag{4-37}$$

Let us consider the general properties of the eigenfunction.

1. *ϕ_{mn} are orthogonal to each other*. The eigenfunctions ϕ_{mn} satisfy the orthogonality condition:

$$\int_a \phi_{mn}\phi_{m'n'}\,da = N_{mn}\,\delta_{mm'}\,\delta_{nn'}, \tag{4-38}$$

where the left side is the surface integral over the cross-sectional area a, $\delta_{mm'}$ is a Kronecker delta defined by

$$\delta_{mm'} = \begin{cases} 1 & \text{for } m = m' \\ 0 & \text{for } m \neq m'' \end{cases} \tag{4-39}$$

and N_{mn} is the normalizing factor

$$N_{mn} = \int_a \phi_{mn}^2\,da. \tag{4-40}$$

To prove (4-38), let us start with the following:

$$\begin{aligned} (\nabla_t^2 + k_{mn}^2)\phi_{mn} &= 0 \\ (\nabla_t^2 + k_{m'n'}^2)\phi_{m'n'} &= 0. \end{aligned} \tag{4-41}$$

We multiply the first equation by $\phi_{m'n'}$ and the second by ϕ_{mn} and subtract one from the other. We get

$$\phi_{m'n'}\nabla_t^2\phi_{mn} - \phi_{mn}\nabla_t^2\phi_{m'n'} = (k_{m'n'}^2 - k_{mn}^2)\phi_{mn}\phi_{m'n'}. \tag{4-42}$$

Integrating both sides of (4-42) and noting Green's second identity (see the Appendix to Chapter 4, Section A),

$$\int_a (u\nabla_t^2 v - v\nabla_t^2 u)\,da = \int_l \left(u\frac{\partial v}{\partial n} - v\frac{\partial u}{\partial n}\right) dl, \tag{4-43}$$

where $\partial/\partial n$ is the normal derivative with n outward from a, we obtain

$$\int_l \left(\phi_{m'n'}\frac{\partial\phi_{mn}}{\partial n} - \phi_{mn}\frac{\partial\phi_{m'n'}}{\partial n}\right) dl = (k_{m'n'}^2 - k_{mn}^2)\int_a \phi_{m'n'}\phi_{mn}\,da. \tag{4-44}$$

Since ϕ_{mn} and $\phi_{m'n'}$ satisfy either Dirichlet's or Neumann's boundary condition, the integrand on the left side vanishes, and we obtain

$$(k_{m'n'}^2 - k_{mn}^2)\int_a \phi_{m'n'}\phi_{mn}\,da = 0. \tag{4-45}$$

Thus if $m \neq m'$ and $n \neq n'$, $k_{m'n'} \neq k_{mn}$, and therefore the integral must be zero, proving the orthogonality (4-38).

2. *ϕ_n forms a complete set, and thus any continuous function $f(\bar{r})$ can be expanded in a series of eigenfunctions*. Let us write

$$f(\bar{r}) = \sum_m^\infty \sum_n^\infty A_{mn}\phi_{mn}(\bar{r}), \tag{4-46}$$

where $\bar{r} = x\hat{x} + y\hat{y}$. Then A_{mn} is given by multiplying both sides of (4-46) by $\phi_{mn}(\bar{r})$ and integrating over the cross-sectional area.

$$A_{mn} = \frac{\int_a f(\bar{r})\phi_{mn}(\bar{r})\,da}{\int_a \phi_{mn}(\bar{r})^2\,da}. \tag{4-47}$$

In particular, a delta function $\delta(\bar{r} - \bar{r}')$ can be expanded in a series of eigenfunctions.

$$\delta(\bar{r} - \bar{r}') = \sum_m^{\infty}\sum_n^{\infty} \frac{\phi_{mn}(\bar{r})\phi_{mn}(\bar{r}')}{\int_a \phi_{mn}(\bar{r})^2\,da}. \tag{4-48}$$

As an example consider the rectangular waveguide with Dirichlet boundary condition as shown in (4-23) and (4-24). The eigenfunction $\phi_{mn}(x, y)$ is given in (4-33), and therefore,

$$\delta(\bar{r} - \bar{r}') = \sum_{m=1}^{\infty}\sum_{n=1}^{\infty} \frac{4}{ab} \sin\frac{m\pi x}{a} \sin\frac{m\pi x'}{a} \sin\frac{n\pi y}{b} \sin\frac{n\pi y'}{b}, \tag{4-49}$$

where $\bar{r} = x\hat{x} + y\hat{y}$ and $\bar{r}' = x'\hat{x} + y'\hat{y}$.

3. *k_{mn}^2 is real for Dirichlet and Neumann boundary conditions*. To prove this, we consider $\phi_{mn}(\bar{r})$ and its complex conjugate $\phi_{mn}^*(\bar{r})$:

$$\begin{aligned} (\nabla_t^2 + k_{mn}^2)\phi_{mn}(\bar{r}) &= 0, \\ (\nabla_t^2 + k_{mn}^{*2})\phi_{mn}^*(\bar{r}) &= 0. \end{aligned} \tag{4-50}$$

Using Green's second identity (4-43) and letting $u = \phi_{mn}$ and $v = \phi_{mn}^*$, we get

$$(k_{mn}^2 - k_{mn}^{*2}) \int_a \phi_{mn}\phi_{mn}^*\,da = \int_l \left(\phi_{mn}\frac{\partial \phi_{mn}^*}{\partial n} - \phi_{mn}^*\frac{\partial \phi_{mn}}{\partial n}\right) dl. \tag{4-51}$$

The right side is zero because of the boundary conditions. Therefore, we get $k_{mn}^2 = k_{mn}^{*2}$, proving that k_{mn}^2 is real.

Furthermore, we can show that k_{mn}^2 is not only real but positive. Recall Green's first identity,

$$\int_a (\nabla_t v \cdot \nabla_t u + u\nabla_t^2 v)\,da = \int_l u\frac{\partial v}{\partial n}\,dl. \tag{4-52}$$

Letting $u = v = \phi_{mn}$ and noting

$$\nabla^2 \phi_{mn} = -k_{mn}^2\,\phi_{mn}, \tag{4-53}$$

we get

$$k_{mn}^2 = \frac{\int_a \nabla_t\phi_{mn} \cdot \nabla_t\phi_{mn}\,da}{\int_a \phi_{mn}^2\,da}. \tag{4-54}$$

All the integrands in (4-54) are real and positive, and therefore k_{mn}^2 is also real and positive. We can also rewrite (4-54) as follows:

$$k_{mn}^2 = \frac{\int_a \phi_{mn}(-\nabla_t^2 \phi_{mn})\, da}{\int_a \phi_{mn}^2\, da}. \tag{4-55}$$

Equations (4-54) and (4-55) will be used later for numerical solutions of waveguides with complex cross sections.

4-6 k–β DIAGRAM AND PHASE AND GROUP VELOCITIES

Let us consider a hollow waveguide. Once the eigenvalue k_c is determined, the propagation constant β is given by

$$\beta = (k^2 - k_c^2)^{1/2}. \tag{4-56}$$

The eigenvalue k_c depends only on the geometry of the guide and is called the cutoff wave number. It is related to the cutoff wavelength λ_c and the cutoff frequency f_c by the following:

$$k_c = \frac{\omega_c}{c} = \frac{2\pi f_c}{c} = \frac{2\pi}{\lambda_c}. \tag{4-57}$$

When the frequency of operation is higher than the cutoff frequency, β is real and the wave behaves as $\exp(-j\beta z)$, exhibiting the propagation of the wave. On the other hand, if the operating frequency is below cutoff, the wave is evanescent, and the wave attenuates as $\exp(-\alpha z)$, where $\alpha = (k_c^2 - k^2)^{1/2}$ is real.

The frequency characteristics of the propagation constant β and the attenuation constant α are conventionally expressed in a k–β diagram. For the case of waveguides, the k–β diagram is given by a hyperbola. Also, the attenuation constant α can be shown in the same diagram as a circle (Fig. 4-4).

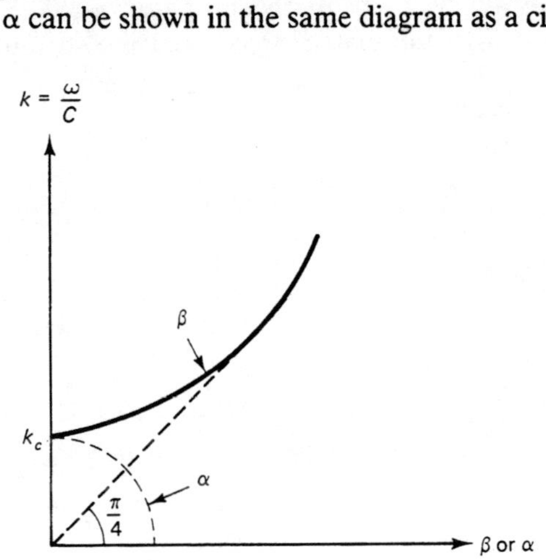

Figure 4-4 k–β diagram for a waveguide.

The phase velocity V_p is given by

$$V_p = \frac{\omega}{\beta} \quad \text{or} \quad \frac{V_p}{c} = \frac{k}{\beta}. \tag{4-58}$$

If $k < \beta$, $V_p < c$, it is called a *slow wave*, and if $k > \beta$, it is called a *fast wave*. As seen in the k–β diagram (Fig. 4-5), the waveguide modes are fast waves.

The relative phase velocity (V_p/c) is given in the k–β diagram as a slope of a straight line connecting the origin and the operating point.

$$\frac{V_p}{c} = \tan\theta_1. \tag{4-59}$$

The group velocity V_g is given by

$$V_g = \frac{d\omega}{d\beta}, \tag{4-60}$$

or its normalized form,

$$\frac{V_g}{c} = \frac{dk}{d\beta}.$$

Thus the relative group velocity is given by the slope of the k–β curve at the operating point:

$$\frac{V_g}{c} = \tan\theta_2. \tag{4-61}$$

It is clear that for the waveguide problem, the wave is fast and

$$V_p V_g = c^2. \tag{4-62}$$

Note that this relation (4-62) is valid only for the form of the propagation constant given in (4-56) and does not hold for other waveguides, such as dielectric guides.

The group velocity V_g is the velocity of the propagation of the signal as represented by the envelope of the modulated signal, and it also represents the

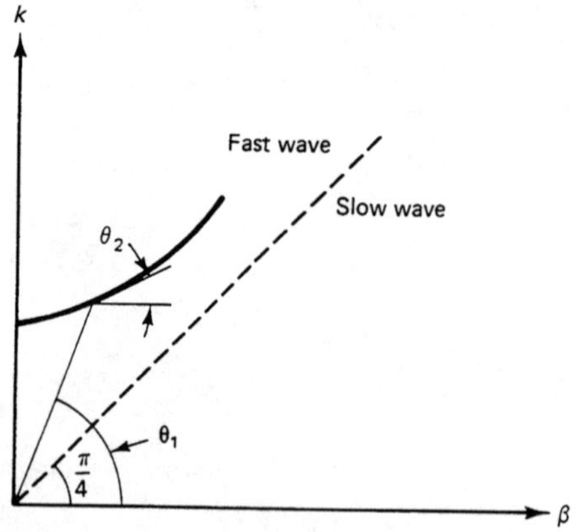

Figure 4-5 Phase and group velocities.

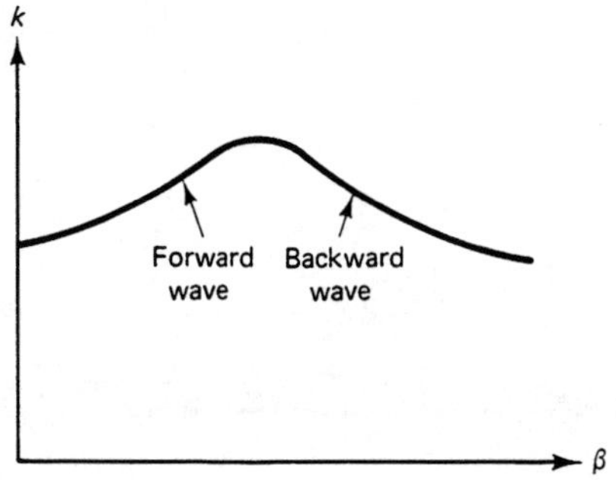

Figure 4-6 Backward and forward waves.

velocity of the energy transport. In some cases the phase velocity and the group velocity can be directed in opposite directions (Fig. 4-6). This is called the *backward wave*. In contrast, when the phase and group velocities are directed in the same direction, it is called the *forward wave*. The backward waves are important in such applications as backward wave tubes and log-periodic antennas.

If the waveguide is filled with material with refractive index n, the wave number k in (4-56) becomes $k = k_0 n = (\omega/c)n$. If the refractive index n is constant, the velocity of light c in (4-56) to (4-62) should be changed to c/n, the velocity of light in the medium. For dispersive media, $n = n(\omega)$, (4-56) to (4-61) are valid if c is replaced by c/n, but (4-62) does not hold.

4-7 RECTANGULAR WAVEGUIDES

Let us consider TM modes in a rectangular guide with sides a and b (Fig. 4-3). As shown in Section 4-4, for TM modes, we have the normalized eigenfunction

$$\phi_{mn}(x,y) = \frac{2}{\sqrt{ab}} \sin\frac{m\pi x}{a} \sin\frac{n\pi y}{b}, \tag{4-63}$$

and the eigenvalue

$$k_{mn}^2 = \left(\frac{m\pi}{a}\right)^2 + \left(\frac{n\pi}{b}\right)^2,$$

where $m = 1, 2, 3, \ldots$ and $n = 1, 2, 3, \ldots$. This mode is called the TM_{mn} mode or the E_{mn} mode. The propagation constant β_{mn} is given by

$$\beta_{mn} = (k^2 - k_{mn}^2)^{1/2}. \tag{4-64}$$

The actual Hertz vector corresponding to a given wave with a certain power is given by

$$\pi_z(x,y,z) = A_0 \phi_{mn}(x,y) e^{-j\beta_{mn} z}, \tag{4-65}$$

where A_0 is constant. The field components are given by (4-6) and (4-7).

The total power propagating through the guide when the frequency is above cutoff is given by

$$\begin{aligned} P &= \int_0^a dx \int_0^b dy \frac{1}{2} \overline{E} \times \overline{H}^* \cdot \hat{z} \\ &= \int_0^a dx \int_0^b dy \frac{|\overline{E}|^2}{2Z_e} \\ &= \frac{|A_0|^2 \beta_{mn}^2}{2Z_e} \left[\left(\frac{m}{a}\right)^2 + \left(\frac{n}{b}\right)^2 \right] \left(\frac{\pi^2 ab}{4}\right) \end{aligned} \tag{4-66}$$

In a similar manner, TE modes can be given by the normalized eigenfunction

$$\psi_{mn}(x,y) = \frac{2}{\sqrt{ab}} \cos\frac{m\pi x}{a} \cos\frac{n\pi y}{b}, \tag{4-67}$$

where $m = 0, 1, 2, \ldots, \infty$ and $n = 0, 1, 2, \ldots, \infty$, but $m = n = 0$ must be excluded as it leads to zero field. The normalizing constant is $\sqrt{2}/\sqrt{ab}$ if $m = 0$ or $n = 0$. The eigenvalue k_{mn}^2 is the same as the TM case, and this is called the TE_{mn} (or H_{mn}) mode.

As an important special case, consider the TE_{10} mode. This mode usually has the lowest cutoff frequency and is the fundamental mode used for most microwave waveguides. We can easily obtain the following field components:

$$\begin{aligned} E_y &= E_0 \sin\frac{\pi x}{a} e^{-j\beta z}, \\ H_x &= -\frac{E_0}{Z_h} \sin\frac{\pi x}{a} e^{-j\beta z}, \\ H_z &= \frac{jk_c E_0}{\beta Z_h} \cos\frac{\pi x}{a} e^{-j\beta z}, \\ \beta &= \left[k^2 - \left(\frac{\pi}{a}\right)^2\right]^{1/2}. \end{aligned} \tag{4-68}$$

The total power propagating through the guide is

$$P = \frac{|E_0|^2 ab}{4Z_h}. \tag{4-69}$$

The cutoff wavelength is $\lambda_c = 2a$, and the cutoff frequency is $f_c = c/2a$ if the guide is empty. The cutoff frequency for a guide filled with material with relative dielectric constant ϵ_r is $f_c = c/2a(\epsilon_r)^{1/2}$.

In general, in a waveguide, there may be many propagating and evanescent modes. Therefore, a completely general electromagnetic field in a waveguide should consist of all possible TM and TE modes, and it can be expressed as follows:

$$\begin{aligned} \overline{E} &= \nabla(\nabla \cdot \overline{\pi}) + k^2 \overline{\pi} - j\omega\mu \nabla \times \overline{\pi}_m, \\ \overline{H} &= j\omega\epsilon \nabla \times \overline{\pi} + \nabla(\nabla \cdot \overline{\pi}_m) + k^2 \overline{\pi}_m, \end{aligned} \tag{4-70}$$

where

$$\bar{\pi} = \hat{z} \sum_{m,n} A_{mn} \phi_{mn}(x, y) e^{-j\beta_{mn} z},$$
$$\bar{\pi}_m = \hat{z} \sum_{m,n} B_{mn} \psi_{mn}(x, y) e^{-j\beta_{mn} z}. \tag{4-71}$$

A_{mn} and B_{mn} are constants representing the amount of each mode.

In the Cartesian coordinate system, (4-70) may be written

$$E_z = \left(\frac{\partial^2}{\partial z^2} + k^2\right)\pi_z,$$
$$E_x = \frac{\partial^2}{\partial x\, \partial z}\pi_z - j\omega\mu \frac{\partial}{\partial y}\pi_{mz},$$
$$E_y = \frac{\partial^2}{\partial y\, \partial z}\pi_z + j\omega\mu \frac{\partial}{\partial x}\pi_{mz},$$
$$H_z = \left(\frac{\partial^2}{\partial z^2} + k^2\right)\pi_{mz}, \tag{4-72}$$
$$H_x = j\omega\epsilon \frac{\partial}{\partial y}\pi_z + \frac{\partial^2}{\partial x\, \partial z}\pi_{mz},$$
$$H_y = -j\omega\epsilon \frac{\partial}{\partial x}\pi_z + \frac{\partial^2}{\partial y\, \partial z}\pi_{mz}.$$

4-8 CYLINDRICAL WAVEGUIDES

Let us represent a TM mode in the cylindrical coordinate system. We have

$$\pi_z(\rho, \phi, z) = \phi(\rho, \phi) e^{-j\beta z}. \tag{4-73}$$

where the eigenfunction $\phi(\rho, \phi)$ satisfies the wave equation

$$\left[\frac{1}{\rho}\frac{\partial}{\partial \rho}\left(\rho \frac{\partial}{\partial \rho}\right) + \frac{1}{\rho^2}\frac{\partial^2}{\partial \phi^2} + k_c^2\right]\phi(\rho, \phi) = 0. \tag{4-74}$$

By the method of separation of variables, we write

$$\phi(\rho, \phi) = X_1(\rho) X_2(\phi), \tag{4-75}$$

and obtain

$$\left[\frac{1}{\rho}\frac{d}{d\rho}\left(\rho \frac{d}{d\rho}\right) + k_c^2 - \frac{\nu^2}{\rho^2}\right] X_1(\rho) = 0, \tag{4-76}$$

and

$$\left(\frac{d^2}{d\phi^2} + \nu^2\right) X_2(\phi) = 0. \tag{4-77}$$

Equations (4-76) and (4-77) are the second-order ordinary differential equations and a general solution is given by a linear combination of two independent solutions. First, we get from (4-77),

$$X_2(\phi) = c_1 \sin \nu\phi + c_2 \cos \nu\phi, \tag{4-78}$$

where c_1 and c_2 are constants.

Equation (4-76) is Bessel's differential equation, and a general solution is given by

$$X_1(\rho) = c_3 J_\nu(k_c \rho) + c_4 N_\nu(k_c \rho), \tag{4-79}$$

where J_ν and N_ν are called the Bessel function and the Neumann function, respectively. It is sometimes convenient to use the Hankel function of the first kind $H_\nu^{(1)}$ and the second kind $H_\nu^{(2)}$:

$$X_1(\rho) = c_5 H_\nu^{(1)}(k_c \rho) + c_6 H_\nu^{(2)}(k_c \rho). \tag{4-80}$$

The Hankel functions are related to the Bessel and Neumann functions by (see the Appendix to Chapter 4, Section B)

$$H_\nu^{(1)} = J_\nu + jN_\nu \quad \text{and} \quad H_\nu^{(2)} = J_\nu - jN_\nu. \tag{4-81}$$

The general solution for the Hertz vector is given by (4-73) with the propagation constant $\beta = (k^2 - k_c^2)^{1/2}$. Similarly for TE modes, the magnetic Hertz vector $\pi_m = \psi e^{-j\beta z}$ satisfies the same equation, (4-74).

Let us consider a TM mode propagating in a circular waveguide of radius a (Fig. 4-7). First, we note that the field must be a periodic function of ϕ with period 2π, and therefore $X_2(\phi)$ is $\sin n\phi$ or $\cos n\phi$, where n is an integer. $X_1(\rho)$ is given by (4-79) with $J_n(k_c\rho)$ and $N_n(k_c\rho)$, but $N_n(k_c\rho)$ diverges as $\rho \to 0$. Therefore, only $J_n(k_c\rho)$ is acceptable. Thus we write

$$\pi_z(\rho, \phi, z) = A_0 J_n(k_c \rho) \cos n\phi \, e^{-j\beta z}. \tag{4-82}$$

Here we used $\cos n\phi$ since $\sin n\phi$ gives an identical field distribution except for the rotation of the axis around the z axis. The boundary condition at $\rho = a$ requires that

$$J_n(k_c a) = 0. \tag{4-83}$$

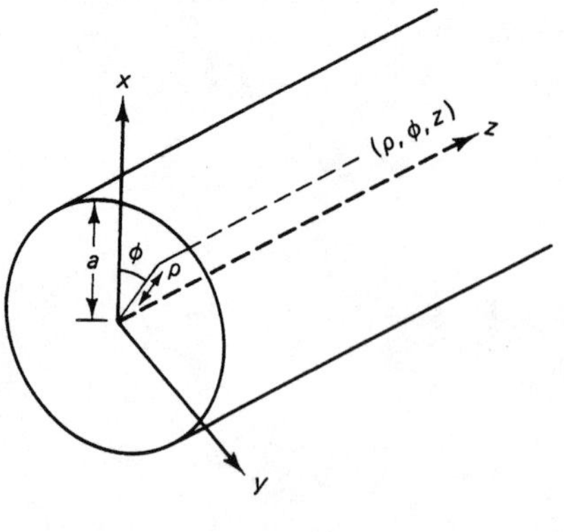

Figure 4-7 Cylindrical waveguide.

This determines an infinite number of eigenvalues k_c. Using the roots χ_{nl} of the equation

$$J_n(\chi_{nl}) = 0, \tag{4-84}$$

we write k_c as

$$k_{nl} = \frac{\chi_{nl}}{a}, \qquad l = 1, 2, 3, \ldots, \tag{4-85}$$

and the propagation constant β is given by

$$\beta = (k^2 - k_{nl}^2)^{1/2}. \tag{4-86}$$

For each k_{nl}, the field is given by the eigenfunction in (4-82) and this is called the TM_{nl} mode.

Similarly, for the TE_{nl} mode, we have

$$\pi_{mz}(\rho, \phi, z) = B_0 J_n(k_{nl}\rho) \cos n\phi\, e^{-j\beta z}, \tag{4-87}$$

where k_{nl} is given by

$$J_n'(k_{nl} a) = 0. \tag{4-88}$$

Table 4-1 gives some of the roots of (4-84) and (4-88).

It should be noted that all eigenfunctions are orthogonal as indicated in (4-38). For the cylindrical waveguide, we have the following orthogonality condition:

$$\int_0^a J_n(k_{nl}\rho) J_n(k_{nl'}\rho)\rho\, d\rho = \begin{cases} 0 & \text{for } l \neq l', \\ N_{nl}^2 & \text{for } l = l'. \end{cases} \tag{4-89}$$

The constant N_{nl}^2 for the TM case is

$$N_{nl}^2 = \frac{a^2}{2}[J_n'(k_{nl} a)]^2. \tag{4-90}$$

TABLE 4-1

l	n = 0	1	2	3
	Roots of $J_n(\chi_{nl}) = 0$			
1	2.405	3.832	5.136	6.380
2	5.520	7.016	8.417	9.761
3	8.654	10.173	11.620	13.015
	Roots of $J_n'(\chi_{nl}) = 0$			
1	3.832	1.841	3.054	4.201
2	7.016	5.331	6.706	8.015
3	10.173	8.563	9.969	11.346

For the TE case (Abramowitz and Stegun, 1964, p. 485),

$$N_{nl}^2 = \frac{1}{2k_{nl}^2}(k_{nl}^2 a^2 - n^2)[J_n(k_{nl} a)]^2. \tag{4-91}$$

If the cross section is a sector (Fig. 4-8), the field is not a periodic function of ϕ, and thus $\sin n\phi$ and $\cos n\phi$ cannot be used. For the TM modes, we have

$$\phi(\rho, \phi) = J_\nu(k_c \rho)(c_1 \sin \nu\phi + c_2 \cos \nu\phi). \tag{4-92}$$

To satisfy the Dirichlet boundary condition, we let

$$\phi(\rho, \phi) = 0 \qquad \text{at } \phi = 0 \quad \text{and} \quad \phi = \alpha, \tag{4-93}$$

and get $c_2 = 0$ and $\nu = n\pi/\alpha$. Thus we obtain

$$\begin{aligned} \pi_z &= \phi(\rho, \phi) e^{-j\beta z}, \\ \phi(\rho, \phi) &= A_0 J_{n\pi/\alpha}(k_{nl} \rho) \sin \frac{n\pi}{\alpha}\phi, \qquad n = 1, 2, 3, \ldots, \end{aligned} \tag{4-94}$$

where the eigenvalue k_{nl} is given by

$$J_{n\pi/\alpha}(k_{nl} a) = 0. \tag{4-95}$$

Similarly, TE modes are given by

$$\begin{aligned} \pi_{mz} &= \psi(\rho, \phi) e^{-j\beta z}, \\ \psi(\rho, \phi) &= B_0 J_{n\pi/\alpha}(k_{nl} \rho) \cos \frac{n\pi}{\alpha}\phi, \qquad n = 0, 1, 2, \ldots, \end{aligned} \tag{4-96a}$$

and the eigenvalue k_{nl} is given by

$$J'_{n\pi/\alpha}(k_{nl} a) = 0. \tag{4-96b}$$

General electromagnetic fields in the cylindrical system are given by

$$E_z = \left(\frac{\partial^2}{\partial z^2} + k^2\right)\pi_z,$$

$$E_\rho = \frac{\partial^2}{\partial\rho\,\partial z}\pi_z - j\omega\mu\frac{1}{\rho}\frac{\partial}{\partial\phi}\pi_{mz},$$

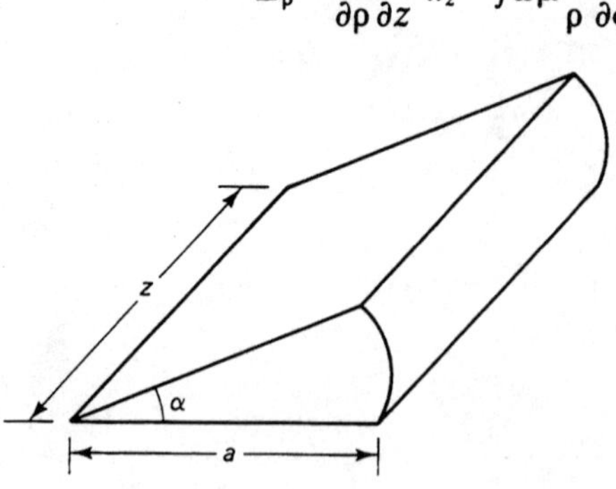

Figure 4-8 Sector waveguide.

$$E_\phi = \frac{1}{\rho}\frac{\partial^2}{\partial\phi\,\partial z}\pi_z + j\omega\mu\frac{\partial}{\partial\rho}\pi_{mz},$$

$$H_z = \left(\frac{\partial^2}{\partial z^2} + k^2\right)\pi_{mz}, \tag{4-97}$$

$$H_\rho = j\omega\epsilon\frac{1}{\rho}\frac{\partial}{\partial\phi}\pi_z + \frac{\partial^2}{\partial\rho\,\partial z}\pi_{mz},$$

$$H_\phi = -j\omega\epsilon\frac{\partial}{\partial\rho}\pi_z + \frac{1}{\rho}\frac{\partial^2}{\partial\phi\,\partial z}\pi_{mz},$$

where

$$\pi_z = A_0\phi(\rho,\phi),$$

$$\pi_{mz} = B_0\psi(\rho,\phi),$$

and ϕ and ψ are given by $X_1(\rho)X_2(\phi)$, shown in (4-75).

4-9 TEM MODES

A hollow waveguide whose cross section is enclosed by a single wall can propagate only TE and TM modes: It cannot support a TEM mode. However, a waveguide consisting of two separate walls, such as coaxial lines and two wire lines, can support a TEM mode as well as TE and TM modes (Fig. 4-9). Since a TEM mode means that $E_z = 0$ and $H_z = 0$, we get from (4-11) and (4-16)

$$k_c = 0. \tag{4-98}$$

From this we get

$$\beta = (k^2 - k_c^2)^{1/2} = k. \tag{4-99}$$

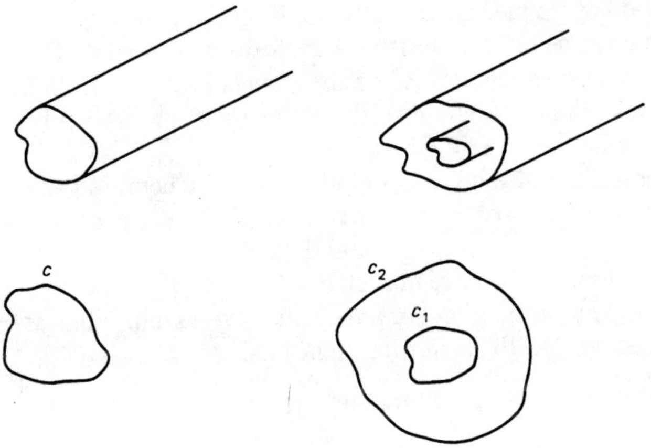

Figure 4-9 TEM guides.

Therefore, the propagation constant of the TEM mode is equal to that of the free-space propagation constant.

Also since $k_c = 0$, ϕ and ψ satisfy the Laplace equation,

$$\nabla^2 \phi = 0,$$
$$\nabla^2 \psi = 0. \tag{4-100}$$

If the waveguide is hollow and enclosed by a single wall, the solution to the Laplace equation with either Dirichlet's or Neumann's boundary condition is identically zero or constant, and therefore the field is identically zero from (4-11) and (4-16).

If the waveguide consists of two surfaces, the potential ϕ can be a constant $\phi = \phi_1$ on one surface c_1 and a different constant $\phi = \phi_2$ on the other surface c_2 and the solution exists. The TEM mode is given by the solution of the Laplace equation (4-100), and $\phi(x, y) = \phi_1$ on c_1 and ϕ_2 on c_2. From (4-11), letting $V = jk\phi$,

$$\overline{E}_t(x, y, z) = -\nabla V(x, y)e^{-jkz},$$
$$\overline{H}_t = \frac{1}{\eta}\hat{z} \times \overline{E}_t, \tag{4-101}$$
$$\nabla^2 V = 0.$$

Note that $V(x, y)$ satisfies the Laplace equation and is identical to a two-dimensional electrostatic potential function, and the electric field distribution $\overline{E}_t$ is identical to the electrostatic field.

4-10 DISPERSION OF A PULSE IN A WAVEGUIDE

The propagation constant β for a hollow waveguide is given by $\beta = (k^2 - k_c^2)^{1/2}$, where $k = \omega/c$. If β is equal to $k = \omega/c$, the phase velocity and the group velocity are identical and when a pulse propagates, all the frequency components propagate with the same velocity and no distortion of the pulse shape takes place. However, when β is a more general function of frequency, the different frequency components of a pulse propagate with different velocities, and the waveform is distorted. The variation of the phase velocity with frequency is called the *dispersion*. To calculate the output waveform, we first take the Fourier transform of the input, multiply it by the transfer function, and then take the inverse Fourier transform.

In many practical problems, the pulse input is a modulated wave centered at a carrier frequency. In this section we first present a general formulation of the propagation of a modulated pulse in a dispersive waveguide and then give an approximate and useful solution for the broadening of a pulse in such a waveguide.

Let us consider a TE_{10} mode in a rectangular waveguide. We assume that at $z = 0$, E_y is given as a modulated wave with carrier frequency ω_0.

$$E_y(0, t) = A(t) \cos[\omega_0 t + \phi(t)] = \text{Re}[u_0(t)e^{j\omega_0 t}], \tag{4-102}$$

where $A(t)$ and $\phi(t)$ are the slowly varying amplitude and phase, $u_0(t)$ is the *complex envelope* $u(z,t)$ evaluated at $z=0$, and $u_0 = A\ \exp(j\phi)$. The complex envelope $u(z,t)$ is related to the *analytical signal* $u_a(z,t)$ by $u_a = u\ \exp(j\omega_0 t)$ (see Born and Wolf, 1970). In the above, the variation $\sin(\pi x/a)$ of E_y in the waveguide is independent of z and t and is included in $A(t)$.

Let us first express the analytical signal $u_a(0,t)$ at $z=0$ in a Fourier integral.

$$u_a(0,t) = u_0(t)e^{j\omega_0 t} = \frac{1}{2\pi}\int U_a(\omega)e^{j\omega t}\,d\omega. \tag{4-103}$$

The Fourier component U_a is therefore given by

$$\begin{aligned} U_a(\omega) &= \int u_0(t)e^{j\omega_0 t - j\omega t}\,dt \\ &= U(\omega-\omega_0), \end{aligned} \tag{4-104}$$

where $U(\omega')$ is the Fourier transform of the complex envelope.

$$U(\omega') = \int u_0(t)e^{-j\omega' t}\,dt. \tag{4-105}$$

Equation (4-104) shows that the Fourier component of the modulated pulse is given by the Fourier component of the complex envelope centered at the carrier frequency.

Each Fourier component at a point $z \neq 0$ satisfies the wave equation and its propagation characteristics are given by the dependence $\exp(-j\beta z)$. Therefore, adding all the frequency components, the field at $z \neq 0$ is given by

$$E_y(z,t) = \mathrm{Re}\left[\frac{1}{2\pi}\int U(\omega-\omega_0)e^{j\omega t - j\beta z}\,d\omega\right]. \tag{4-106}$$

This is a general expression for the wave $E_y(z,t)$ propagating in a dispersive guide. Note that (4-106) is equivalent to an inverse Laplace transform ($j\omega = s$), and therefore the evaluation of the integral for a general transient problem requires a detailed study of an integral in a complex plane by allowing ω to be a complex variable.

Next we consider a simpler, special case when the pulse is narrow band and $\beta(\omega)$ is a slowly varying function of ω. Since $\omega - \omega_0 << \omega_0$, we expand the exponent in Taylor's series and keep its first three terms.

$$\beta(\omega) = \beta(\omega_0) + (\omega-\omega_0)\left.\frac{\partial\beta}{\partial\omega}\right|_{\omega_0} + \frac{1}{2}(\omega-\omega_0)^2\left.\frac{\partial^2\beta}{\partial\omega^2}\right|_{\omega_0}. \tag{4-107}$$

Substituting this into (4-106), we get

$$E_y(z,t) = \mathrm{Re}[u(z,t)e^{j\omega_0 t}], \tag{4-108}$$

$$u(z,t) = \frac{1}{2\pi}\int U(\omega')\exp\left[-j\beta(\omega_0)z + j\omega'(t-t_0) - j\frac{\omega'^2}{2}\frac{\partial^2\beta}{\partial\omega^2}z\right]d\omega',$$

where

$$t_0 = z\frac{\partial\beta}{\partial\omega} \quad \text{and} \quad \frac{\partial^2\beta}{\partial\omega^2} \text{ is evaluated at } \omega_0. \tag{4-109}$$

In general, the propagation constant $\beta(\omega)$ for a lossy guide is complex, and therefore t_0 is also complex and the physical meaning of the complex t_0 is difficult to determine. For a lossless guide, however, the meaning of t_0 is clear. If we write $t_0 = (z/V_g)$, $V_g = (\partial\beta/\partial\omega)^{-1}$ is real for a lossless guide and is called the *group velocity*, representing the velocity of the signal. t_0 is called the *group delay*, and $\hat{t}_0 = \partial\beta/\partial\omega = V_g^{-1}$ is called the *specific group delay* (group delay per unit distance).

The second term in the exponent of (4-109) involving $\partial^2\beta/\partial\omega^2$ represents the broadening of the pulse. If this term is negligibly small, the field is given by

$$E_y(z,t) = \text{Re}[u_0(t-t_0)e^{j\omega_0 t - j\beta(\omega_0)z}]. \tag{4-110}$$

This shows that in this case, if the medium is lossless, the wave propagates without distortion, with the group velocity.

Let us consider an example. The input pulse is assumed to have a Gaussian envelope

$$E_y(0,t) = A_0 \exp\left(-\frac{t^2}{T_0^2}\right) \cos\omega_0 t. \tag{4-111}$$

The complex envelope at $z = 0$ is then

$$u_0(t) = A_0 \exp\left(-\frac{t^2}{T_0^2}\right). \tag{4-112}$$

The Fourier transform of $u_0(t)$ is given by

$$U(\omega') = A_0(\pi)^{1/2}\, T_0 \exp\left(-\frac{T_0^2\omega'^2}{4}\right). \tag{4-113}$$

Substituting this into (4-108), we get

$$u(z,t) = \frac{A_0 e^{-j\beta(\omega_0)z}}{(1+jS/T_0)^{1/2}} \exp\left[-\frac{(t-t_0)^2(1-jS/T_0)}{T_0^2+S^2}\right], \tag{4-114}$$

where $S = (2/T_0)(\partial^2\beta/\partial\omega^2)|_{\omega_0} z$. Note that the pulse width is broadened from the initial value of T_0 to

$$T = (T_0^2 + S^2)^{1/2}. \tag{4-115}$$

For a lossless guide, we can obtain the physical meaning of S. Noting that the group delay is given by $t_0 = (\partial\beta/\partial\omega)z$, S may be written as $\Delta\omega\ \partial t_0/\partial\omega$, where $\Delta\omega = (2/T_0)$ is the bandwidth of the Gaussian pulse at $z = 0$. $\partial t_0/\partial\omega$ is the group delay per unit frequency, and therefore S is the total group delay caused by all the frequency components of the pulse. In some cases the pulse broadening parameter S is expressed in terms of the free-space wavelength $\lambda_0 = 2\pi/k_0$. Noting that

$$\omega\lambda_0 = 2\pi c \quad \text{and} \quad \frac{\Delta\omega}{\omega} = -\frac{\Delta\lambda_0}{\lambda_0}, \tag{4-116}$$

we get

$$|S| = \left|\Delta\omega \frac{\partial t_0}{\partial \omega}\right| = \left|\Delta\lambda_0 \frac{\partial t_0}{\partial \lambda_0}\right|. \tag{4-117}$$

The pulse broadening $|\hat{S}|$ per unit distance is given by

$$|\hat{S}| = \Delta\hat{\tau} = \left|\Delta\lambda_0 \frac{\partial \hat{\tau}}{\partial \lambda_0}\right|. \tag{4-118}$$

For example, in optical fibers, the pulse broadening is normally given by $|\partial\hat{\tau}/\partial\lambda_0|$ in units of (ns/km · nm), and when multiplied by $\Delta\lambda$ (in nm of source bandwidth), it gives the spread $|\hat{S}|$ in nanoseconds per kilometer of fiber. $D = \partial\hat{\tau}/\partial\lambda_0$ is called the *dispersive parameter* or simply the *dispersion*.

For the TE_{10} mode with the propagation constant $\beta = [k^2 - (\pi/a)^2]^{1/2}$, we get the specific group delay $\hat{\tau}$ (group delay per unit length of the guide)

$$\hat{\tau} = \frac{\partial \beta}{\partial \omega} = \frac{\omega}{\beta c^2}. \tag{4-119}$$

The pulse broadening $|\hat{S}|$ per unit length of the guide when the bandwidth of the pulse at $z = 0$ is $\Delta\omega$ is

$$\begin{aligned} |\hat{S}| = \Delta\hat{\tau} &= \left|\Delta\omega \frac{\partial \hat{\tau}}{\partial \omega}\right| = \left|\Delta\lambda_0 \frac{\partial \hat{\tau}}{\partial \lambda_0}\right| \\ &= \left|\Delta\omega \frac{(\pi/a)^2}{c^2\beta^3}\right|. \end{aligned} \tag{4-120}$$

The propagation constant β in (4-119) and (4-120) is evaluated at the carrier frequency ω_0.

4-11 STEP-INDEX OPTICAL FIBERS

Optical fibers commonly used in optical communications are divided into step-index fibers and graded-index fibers. A *step-index fiber* consists of the circular center *core* with refractive index n_1 and a surrounding *cladding* with n_2. A *graded-index fiber*, also called GRIN, has a core with a slowly varying refractive index (Fig. 4-10). In this section we examine the propagation characteristics of the wave along the step-index fibers. It is clear from Section 3-11 that in order to support the trapped surface wave along the core, the refractive index of the core must be higher than that of the cladding. In our analysis we assume that the cladding extends to infinity since the wave in the cladding decays exponentially in the radial direction, and the outer boundary of the cladding has little effect on the propagation characteristics (see Marcuse, 1982).

Let us consider a wave propagating along a dielectric cylinder of radius a (Fig. 4-11). It was shown in Section 4-1 that a general electromagnetic field consists of TM and TE modes. For the TM (transverse magnetic) modes, the magnetic fields are

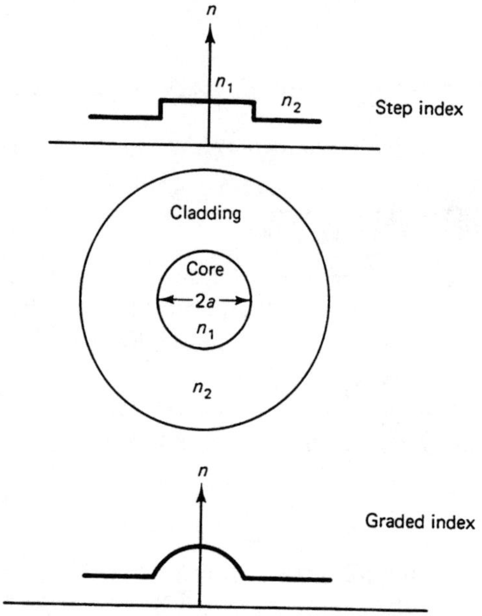

Figure 4-10 Step-index and graded-index fibers.

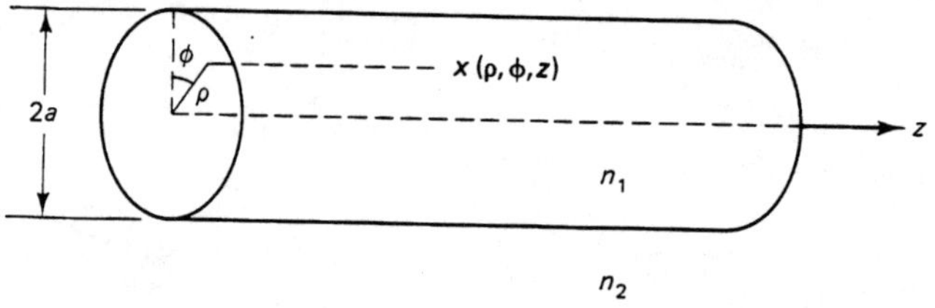

Figure 4-11 Round optical fiber.

transverse to a fixed direction denoted by a unit vector $\hat{a}$. For the TE (transverse electric) modes, the electric fields are all transverse to $\hat{a}$. All the TE modes can be derived from a magnetic Hertz vector $\overline{\pi}_m = \pi_m \hat{a}$, all the TM modes can be derived from an electric Hertz vector $\overline{\pi} = \pi \hat{a}$, and π_m and π satisfy a scalar wave equation. For cylindrical geometry, (ρ, ϕ, z), we take $\hat{a} = \hat{z}$ and write $\overline{\pi} = \pi_z \hat{z}$ and $\overline{\pi}_m = \pi_{mz} \hat{z}$. The solution to the scalar wave equation is given by a product of Bessel functions $Z_n(p\rho) \exp(\pm jn\phi)$ and $\exp(\pm j\beta z)$, where p and β are constant and satisfy the condition $p^2 + \beta^2 = k^2$ (see Section 4-8). In general, for a dielectric waveguide, TE and TM modes are coupled and both are needed to satisfy the boundary conditions, except for the case where there is no azimuthal variation ($\partial/\partial\phi = 0$). Let us write the Hertz potentials for the TM and TE modes:

$$\left.\begin{aligned} \pi_z &= AJ_n(p_1\rho)e^{-jn\phi - j\beta z} \\ \pi_{mz} &= BJ_n(p_1\rho)e^{-jn\phi - j\beta z} \end{aligned}\right\} \quad \text{for } \rho < a$$
$$\left.\begin{aligned} \pi_z &= CK_n(\alpha_2\rho)e^{-jn\phi - j\beta z} \\ \pi_{mz} &= DK_n(\alpha_2\rho)e^{-jn\phi - j\beta z} \end{aligned}\right\} \quad \text{for } \rho > a, \tag{4-121}$$

where

$$p_1^2 + \beta^2 = k_0^2 n_1^2,$$
$$\alpha_2^2 = \beta^2 - k_0^2 n_2^2 = k_0^2(n_1^2 - n_2^2) - p_1^2.$$

Note that for $\rho > a$, we used $K_n(\alpha_2 \rho)$. We may also use $H_n^{(2)}(p_2 \rho)$, where $p_2^2 + \beta^2 = k_0^2 n_2^2$. However, since we are interested in the trapped surface-wave solution, β must be real and greater than $k_0 n_2$. Therefore, $p_2^2 = k_0^2 n_2^2 - \beta^2$ is negative, and thus p_2 is purely imaginary. It is, therefore, more convenient to use the modified Bessel function $K_n(z)$ with $z = \alpha_2 \rho = j p_2 \rho$.

$$K_n(z) = -\frac{\pi j}{2} e^{-jn(\pi/2)} H_n^{(2)}(-jz). \tag{4-122}$$

The modified Bessel function exhibits exponentially decaying behavior as $\rho \to \infty$.

$$K_n(z) \sim \left(\frac{\pi}{2z}\right)^{1/2} e^{-z} \qquad \text{as } z \to \infty. \tag{4-123}$$

Now we consider the boundary conditions that E_z, E_ϕ, H_z, and H_ϕ be continuous at $\rho = a$.

$$\begin{aligned} E_z &= \left(\frac{\partial^2}{\partial z^2} + k^2\right)\pi_z, \qquad & E_\phi &= \frac{1}{\rho}\frac{\partial^2}{\partial \phi\, \partial z}\pi_z + j\omega\mu \frac{\partial}{\partial \rho}\pi_{mz}, \\ H_z &= \left(\frac{\partial^2}{\partial z^2} + k^2\right)\pi_{mz}, \qquad & H_\phi &= -j\omega\epsilon \frac{\partial}{\partial \rho}\pi_z + \frac{1}{\rho}\frac{\partial^2}{\partial \phi\, \partial z}\pi_{mz}, \end{aligned} \tag{4-124}$$

where $k = k_0 n_1$ and $\epsilon = \epsilon_1$ in the core and $k = k_0 n_2$ and $\epsilon = \epsilon_2$ outside. Applying the continuity of E_z and H_z at $\rho = a$, we obtain

$$\begin{aligned} p_1^2 A J_n(p_1 a) &= -\alpha_2^2 C K_n(\alpha_2 a), \\ p_1^2 B J_n(p_1 a) &= -\alpha_2^2 D K_n(\alpha_2 a). \end{aligned} \tag{4-125}$$

This determines the ratio $A/C = B/D$, and using this, we rewrite (4-121) in the following more convenient form:

$$\left.\begin{aligned} \pi_z &= \frac{A_0}{p_1^2}\frac{J_n(p_1\rho)}{J_n(p_1 a)} e^{-jn\phi - j\beta z} \\ \pi_{mz} &= \frac{B_0}{p_1^2}\frac{J_n(p_1\rho)}{J_n(p_1 a)} e^{-jn\phi - j\beta z} \end{aligned}\right\} \text{ for } \rho < a$$
$$\left.\begin{aligned} \pi_z &= \frac{(-A_0)}{\alpha_2^2}\frac{K_n(\alpha_2\rho)}{K_n(\alpha_2 a)} e^{-jn\phi - j\beta z} \\ \pi_{mz} &= \frac{(-B_0)}{\alpha_2^2}\frac{K_n(\alpha_2\rho)}{K_n(\alpha_2 a)} e^{-jn\phi - j\beta z} \end{aligned}\right\} \text{ for } \rho > a, \tag{4-126}$$

where A_0 and B_0 are constant.

Now we apply the boundary condition that E_ϕ and H_ϕ must be continuous at $\rho = a$ and obtain

$$\left[\frac{n_1^2}{u_1}\frac{J_n'(u_1)}{J_n(u_1)} + \frac{n_2^2}{u_2}\frac{K_n'(u_2)}{K_n(u_2)}\right]\left[\frac{1}{u_1}\frac{J_n'(u_1)}{J_n(u_1)} + \frac{1}{u_2}\frac{K_n'(u_2)}{K_n(u_2)}\right] = \left(\frac{n\beta}{k_0}\right)^2\left(\frac{1}{u_1^2} + \frac{1}{u_2^2}\right)^2, \quad (4\text{-}127)$$

where $u_1 = p_1 a$ and $u_2 = \alpha_2 a$, $u_2^2 = V^2 - u_1^2$, and $V^2 = k_0^2 a^2(n_1^2 - n_2^2)$.

We note that only when $n = 0$, (4-127) separates into two equations corresponding to TM and TE modes. But, in general, TM and TE modes are mixed, and they are called *hybrid* modes and are designated as EH_{nm} or HE_{nm} modes, depending on whether TM or TE modes are dominant. The HE_{11} mode is the only mode that does not have a low-frequency cutoff and is often called the HE_{11} dipole mode. This is equivalent to the lowest TM mode on a dielectric slab (Fig. 4-12).

The frequency dependence of the propagation constant β is conveniently expressed in terms of the normalized propagation constant b as functions of the normalized frequency V.

$$b = \frac{\beta^2 - k_0^2 n_2^2}{k_0^2(n_1^2 - n_2^2)} = \frac{\alpha_2^2 a^2}{V^2} = \frac{u_2^2}{V^2},$$
$$V^2 = k_0^2 a^2(n_1^2 - n_2^2). \quad (4\text{-}128)$$

Typical curves showing the frequency dependence are sketched in Fig. 4-13.

The cutoff frequency of the HE_{11} mode is zero. The modes with the next-lowest cutoff frequencies are the TE_{01} and TM_{01} modes ($n = 0$). Their cutoff frequencies are obtained by letting $n = 0$ in (4-127), thus separating the equation into TE and TM modes, and then by recognizing that the cutoff requires that $u_2 \to 0$. Noting that as $u_2 \to 0$,

$$K_0(u_2) \to \ln\frac{2}{\gamma u_2}$$
$$\gamma = \text{Euler's constant} = 1.781, \quad (4\text{-}129)$$

we get the cutoff condition for both TE_{01} and TM_{01}:

$$J_0(u_1) = 0. \quad (4\text{-}130)$$

The lowest root is $u_1 = 2.4048$. Therefore, the cutoff normalized frequency is

$$V_c = k_0 a(n_1^2 - n_2^2)^{1/2} = 2.4048$$
$$\sim \frac{2\pi a}{\lambda_c} n_1 (2\Delta)^{1/2}, \quad (4\text{-}131)$$

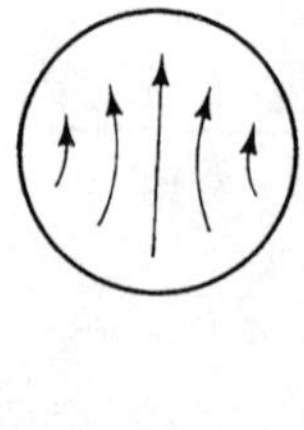

Figure 4-12 HE_{11} dipole mode.

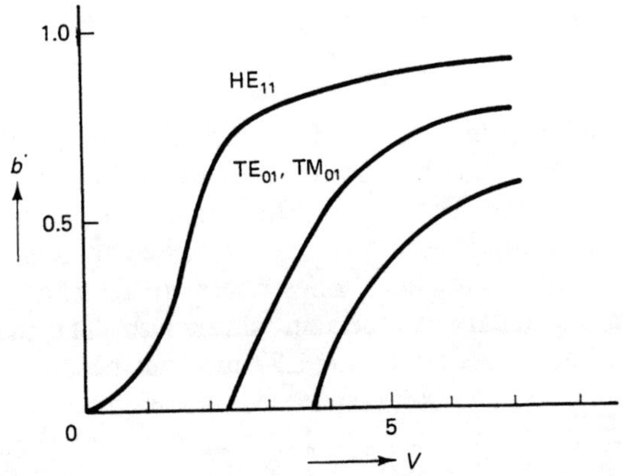

Figure 4-13 Normalized propagation constant b as a function of normalized frequency V.

where λ_c is the cutoff wavelength and $\Delta \sim (n_1 - n_2)/n_1$. We conclude that as long as $V < 2.4048$, only a single mode can propagate along the optical fiber. For example, for typical values of $n_1 = 1.48$, $n_2 = 1.46$, and $\lambda_0 = 0.82\ \mu\text{m}$, the maximum fiber diameter that supports the *single mode* is $2a = 2.59\ \mu\text{m}$.

In contrast to the single-mode fiber, a fiber with a large diameter (50 μm, for example) supports many modes and is called the *multimode fiber*. In this case the number of modes for a given normalized frequency V is approximately given by the mode volume formula

$$N \approx \tfrac{1}{2}V^2. \tag{4-132}$$

This may be compared with the number of modes for a dielectric slab, $N \approx 2V/\pi$ for the total TE and TM modes.

The specific group delay (group delay per unit length) $\hat{\tau}$ is given by

$$\hat{\tau} = \frac{d\beta}{d\omega}, \tag{4-133}$$

and the increase of the pulse width per unit length of the fiber is given by

$$\Delta\hat{\tau} = \frac{d\hat{\tau}}{d\lambda}\Delta\lambda. \tag{4-134}$$

The dispersion is then expressed by

$$D = \frac{d\hat{\tau}}{d\lambda}, \tag{4-135}$$

measured in ns/km · nm. When multiplied by the source spectrum width $\Delta\lambda_s$, (4-135) gives the pulse broadening in ns/km.

The dispersion over a distance L can also be expressed in terms of the bandwidth B:

$$B = (\Delta\hat{\tau}L)^{-1}. \tag{4-136}$$

We can then define the *bandwidth–distance product*:

$$BL = (\Delta\hat{\tau})^{-1} \qquad \text{Hz}\cdot\text{km}. \tag{4-137}$$

The dispersion of a single-mode fiber consists of *material dispersion* and *mode dispersion*. The mode dispersion depends on the mode structure and therefore depends also on the radius of the fiber. The material dispersion is caused by the variation of refractive index with frequency and can be described by Sellmeier's equation (Chapter 8). Dispersion curves for fused silica fibers are sketched in Fig. 4-14. Note that the dispersion disappears at a certain wavelength. The material dispersion ($a \to \infty$) for fused silica disappears at $\lambda = 1.27$ μm, but because of the mode dispersion, the vanishing dispersion point is shifted. It is sometimes desirable to shift this point to the minimum-loss wavelength. For example, fibers made of germanium-doped fused silica have a minimum loss of 0.2 dB/km near $\lambda = 1.55$ μm. Recently, long-wavelength systems with 1.3-μm wavelength with minimum dispersion have been developed. A typical bandwidth–distance product can be $BL \geq 100$ GHz · km at 1.3 μm for the source spectrum width $\Delta\lambda_s = 10$ nm and the loss may be 0.27 dB/km. At 1.5 to 1.65 μm, the dispersion is much greater than at 1.3 μm and BL is only about 5 GHz · km with $\Delta\lambda_s = 10$ nm.

The dispersion of a multimode fiber consists of the *material dispersion* and the *intermodal dispersion*. The intermodal dispersion is caused by the different group delays of many modes. An estimate of the pulse spread of step-index multimode fibers can be made by noting the difference in travel time between the longest ray path and the shortest ray path. The shortest ray path is a straight line inside the fiber, and the travel time over the distance z along the fiber is $(n_1/c)z$. The largest path is when the path makes the maximum angle with the z axis, which corresponds to the critical angle θ_c for total reflection on the fiber surface. Beyond this maximum angle, the rays escape from the fiber. Thus the travel time along this longest path is $(n_1/c)(z/\sin\theta_c) = (n_1/c)(n_1/n_2)z$. Therefore, the pulse spread $\Delta\hat{\tau}$ per unit distance along the fiber is given by

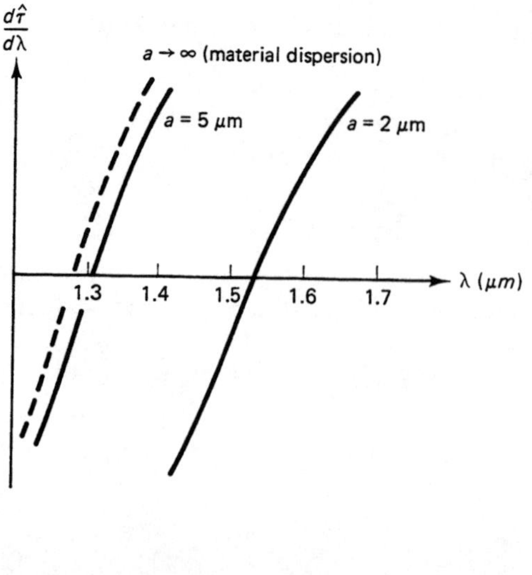

Figure 4-14 Dispersion curves (Marcuse, p. 504).

$$\Delta\hat{\tau} \sim \frac{(n_1/c)(n_1 - n_2)}{n_2} = (n_1/c)(n_1/n_2)\Delta,$$
$$\Delta = \frac{n_1 - n_2}{n_1}. \tag{4-138}$$

Note that the dispersion for a step-index fiber is proportional to Δ. It will be shown in the next section that the dispersion of the graded-index fiber is much less than (4-138) and is proportional to Δ^2. Multimode short-wavelength systems operating at a 0.85-μm wavelength are commercially available with a typical loss of 2.5 dB/km and a bandwidth–distance product of several hundred MHz · km.

One of the important parameters of a multimode fiber is its *numerical aperture* (NA). It is a measure of the ability to gather light into the fiber and is given by the sine of the maximum angle of the entrance of a ray that is trapped in the core (Fig. 4-15). Noting that the maximum angle θ occurs when the angle θ_c is the critical angle for total reflection, we get

$$\mathrm{NA} = \sin\theta = (n_1^2 - n_2^2)^{1/2}. \tag{4-139}$$

4-12 DISPERSION OF GRADED-INDEX FIBERS

In Section 3-17 we discussed the surface-wave modes in an inhomogeneous slab using the WKB method. The propagation constant β is obtained by (3-165). By differentiating (3-165) with respect to ω and noting that even though z_1 and z_2 depend on β and therefore on ω, $(\partial z_1/\partial\omega)q(z_1) = (\partial z_2/\partial\omega)q(z_2) = 0$, we get the specific group delay $\hat{\tau}$.

$$\hat{\tau} = \frac{d\beta}{d\omega} = \frac{k_0}{\beta c}\,\frac{\int_{z_1}^{z_2}\{[n^2(z) + k_0 n(z)(dn/dk_0)]/q(z)\}\,dz}{\int_{z_1}^{z_2} dz/q(z)} \tag{4-140}$$

The material dispersion $dn/d\omega = c\,dn/dk_0$ is included in the above. The maximum and minimum values of $\hat{\tau}$ are obtained by letting $\beta = k_0 n_2$ and $k_0 n_1$, respectively, and therefore the pulse spread $\Delta\hat{\tau}$ per unit length is given by

$$\Delta\hat{\tau} = \hat{\tau}(\beta = k_0 n_2) - \hat{\tau}(\beta = k_0 n_1). \tag{4-141}$$

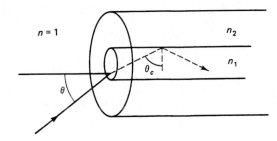

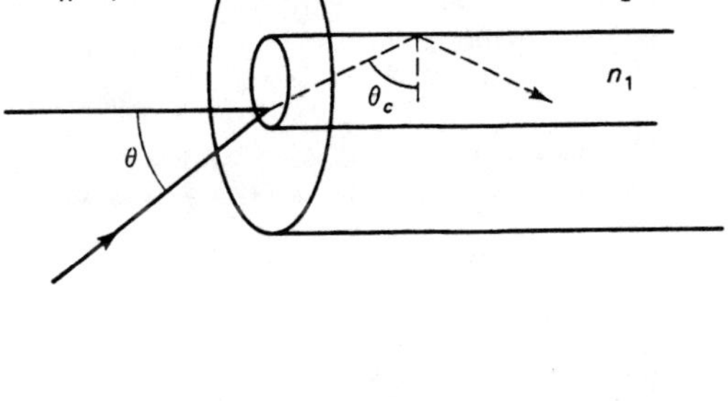

Figure 4-15 Numerical aperture NA = sin θ.

For example, for the square-law profile, we have (3-167):

$$n^2(z) = \begin{cases} n_1^2\left[1 - 2\Delta\left(\frac{z}{a}\right)^2\right], & z < a \\ n_2^2 = n_1^2(1 - 2\Delta), & z > a \end{cases}. \tag{4-142}$$

Because of symmetry, we have $z_2 = -z_1$ and $q(z_2) = 0$. Substituting (4-142) into (4-140), we get, assuming no material dispersion,

$$\hat{\tau} = \frac{n_1[1 - \Delta(z_2/a)^2]}{c[1 - 2\Delta(z_2/a)^2]^{1/2}}. \tag{4-143}$$

The minimum $\hat{\tau}$ is obtained when $z_2 = 0$ and the maximum $\hat{\tau}$ is obtained when $z_2 = a$. From these we get

$$\hat{\tau}(\beta = k_0 n_2) = \frac{n_1^2 + n_2^2}{2n_2 c}$$

$$\hat{\tau}(\beta = k_0 n_1) = \frac{n_1}{c} \tag{4-144}$$

$$\Delta\hat{\tau} = \frac{n_1 \Delta^2}{2c}.$$

In Section 4-11 it was shown that the pulse spread $\Delta\hat{\tau}$ for the step-index fiber is proportional to Δ [equation (4-138)]. Equation (4-144) shows that the pulse spread for the graded-index fiber is proportional to Δ^2 and is much smaller than that for the step-index fiber.

4-13 RADIAL AND AZIMUTHAL WAVEGUIDES

The formulations given in Section 4-8 on cylindrical structures are applicable not only to a uniform waveguide in which the wave propagates in the z direction, but also to nonuniform waveguides in which the wave propagates in the radial (ρ) or the azimuthal directions (ϕ). Before we discuss the radial and azimuthal waveguides, let us summarize the boundary conditions in terms of the electric and magnetic Hertz vectors.

$$\bar{\pi} = \pi_z \hat{z} \quad \text{and} \quad \bar{\pi}_m = \pi_{mz} \hat{z}. \tag{4-145}$$

The boundary conditions for a perfectly conducting wall (electric wall) parallel to the z axis are

$$\pi_z = 0 \quad \text{and} \quad \frac{\partial \pi_{mz}}{\partial n} = 0. \tag{4-146}$$

These correspond to the TM and TE modes in metallic waveguides. The boundary conditions for a perfectly conducting wall perpendicular to the z axis are

$$\frac{\partial \pi_z}{\partial n} = 0 \quad \text{and} \quad \pi_{mz} = 0. \tag{4-147}$$

Similarly, for a magnetic wall where the tangential magnetic field vanishes, we have

$$\frac{\partial \pi_z}{\partial n} = 0 \qquad \text{and} \qquad \pi_{mz} = 0, \tag{4-148}$$

on a wall parallel to the z axis, and

$$\pi_z = 0 \qquad \text{and} \qquad \frac{\partial \pi_{mz}}{\partial n} = 0, \tag{4-149}$$

on a wall perpendicular to the z axis (Fig. 4-16).

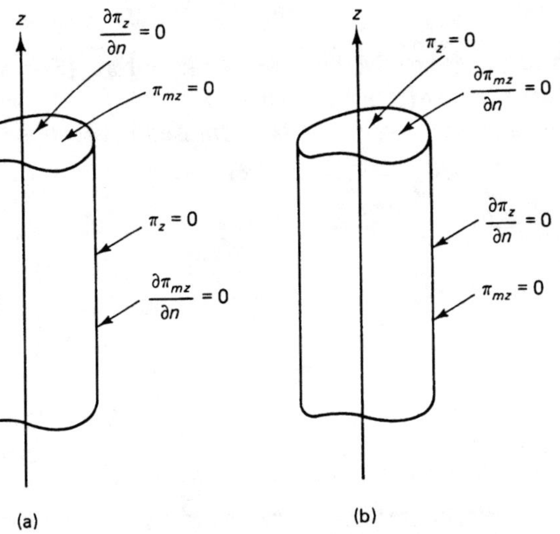

Figure 4-16 Boundary conditions for Hertz vectors: (a) electric wall; (b) magnetic wall.

Radial Waveguides

As an example, consider a sectoral horn (Fig. 4-17). The TM modes (the magnetic field perpendicular to the z axis) are given by

$$\pi_z = H^{(2)}_{m\pi/\alpha}(k_\rho \rho) \sin\frac{m\pi\phi}{\alpha} \cos\frac{n\pi z}{l}, \qquad m = 1, 2, 3, \ldots, \quad n = 0, 1, 2, \ldots. \tag{4-150}$$

This satisfies the boundary condition that $\pi_z = 0$ at $\phi = 0$ and α and $(\partial \pi_z)/(\partial z) = 0$ at $z = 0$ and $z = l$. The radial dependence of $H^{(2)}_{m\pi}/a$ is chosen to represent the outgoing wave. The propagation constant in the radial direction is given by

$$k_\rho^2 = k^2 - \left(\frac{n\pi}{l}\right)^2. \tag{4-151}$$

Similarly, the TE modes are given by

$$\pi_{mz} = H^{(2)}_{m\pi/\alpha}(k_\rho \rho) \cos\frac{m\pi\phi}{\alpha} \sin\frac{n\pi}{l}z, \qquad m = 0, 1, 2, \ldots, \quad n = 1, 2, 3, \ldots. \tag{4-152}$$

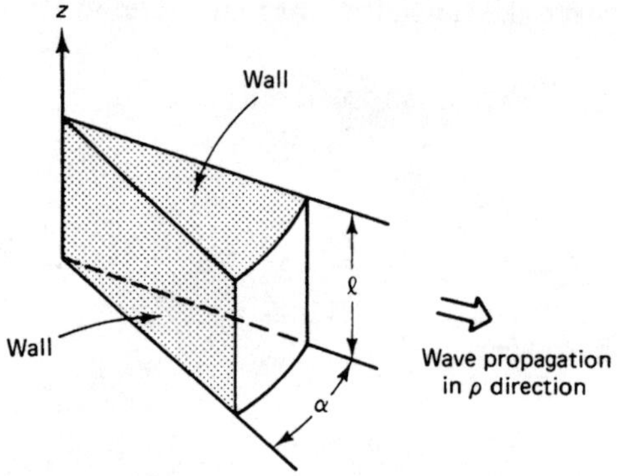

Figure 4-17 Sectoral horn.

Note that for TM modes, the lowest n is zero. In this case $k_\rho = k$, but for TE modes, the lowest n is one, and thus there is a low-frequency cutoff.

A special TM mode with no azimuthal variation is often used in practice (Fig. 4-18). This is given by

$$\pi_z = H_0^{(2)}(K_\rho \rho) \cos\frac{n\pi z}{l}, \qquad n = 0, 1, 2, \ldots, \tag{4-153}$$

where $k_\rho^2 = k^2 - (n\pi/l)^2$ and the electric and magnetic fields are obtained by (4-97). In particular, when $n = 0$,

$$\begin{aligned} E_z &= k^2 H_0^{(2)}(k\rho), \\ H_\phi &= -j\sqrt{\frac{\epsilon}{\mu}}k^2 H_0^{(2)\prime}(k\rho) \\ &= j\sqrt{\frac{\epsilon}{\mu}}k^2 H_1^{(2)}(k\rho). \end{aligned} \tag{4-154}$$

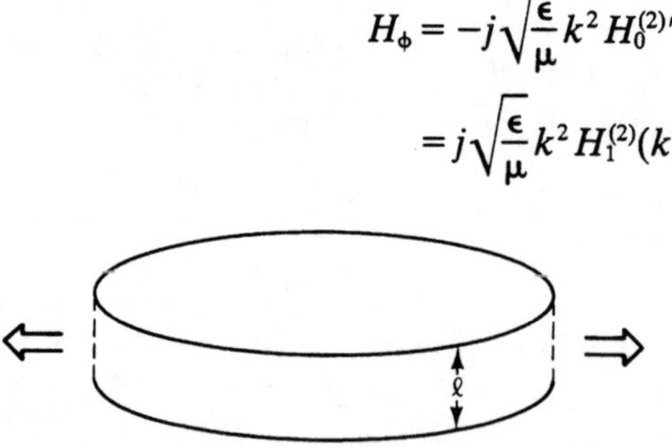

Figure 4-18 Radial waveguides.

Azimuthal Waveguides

An example of the azimuthal wave propagation is a waveguide bend (Fig. 4-19). Let us consider a rectangular waveguide with a TE_{10} mode. If the waveguide is bent in the H plane, we can express π_z in the following manner:

$$\pi_z = Z_\nu(k\rho)e^{-j\nu\phi}, \tag{4-155}$$

where

$$Z_\nu(k\rho) = J_\nu(k\rho)N_\nu(ka) - J_\nu(ka)N_\nu(k\rho),$$

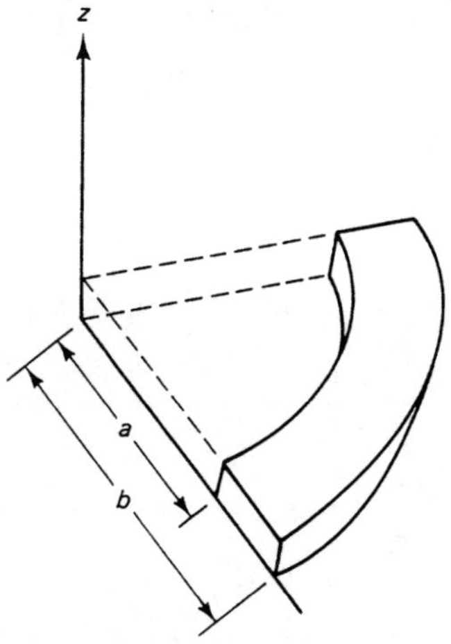

Figure 4-19 Azimuthal waveguides.

and thus $Z_\nu(ka) = 0$, satisfying the boundary conditions at $\rho = a$,

$$\pi_z|_{\rho=a} = 0.$$

The azimuthal propagation constant ν is given by satisfying the boundary condition at $\rho = b$.

$$Z_\nu(kb) = 0,$$

or

$$\frac{J_\nu(ka)}{N_\nu(ka)} = \frac{J_\nu(kb)}{N_\nu(kb)}. \tag{4-156}$$

This equation determines the propagation constant ν in the azimuthal direction.

4-14 CAVITY RESONATORS

The rectangular cavity is formed by closing the ends of a rectangular waveguide with perfectly conducting walls. For a rectangular cavity with dimensions $a \times b \times h$, the eigenfunction for TM modes is

$$\pi_z = \sin\frac{m\pi x}{a}\sin\frac{n\pi y}{b}\cos\frac{l\pi z}{h}, \tag{4-157}$$

where

$$m = 1, 2, \ldots,$$
$$n = 1, 2, \ldots,$$
$$l = 0, 1, 2, 3, \ldots.$$

The eigenvalue is the wave number for the resonant frequency f_r, and for the TM_{mnl} mode, it is given by

$$k_c = \frac{2\pi f_r}{c} = \left[\left(\frac{m\pi}{a}\right)^2 + \left(\frac{n\pi}{b}\right)^2 + \left(\frac{l\pi}{h}\right)^2\right]^{1/2}. \tag{4-158}$$

Similarly, for TE modes, we have

$$\pi_{mz} = \cos\frac{m\pi x}{a} \cos\frac{n\pi y}{b} \sin\frac{l\pi z}{h}, \tag{4-159}$$

where

$$m = 0, 1, 2, \ldots,$$

$$n = 0, 1, 2, \ldots,$$

$$l = 1, 2, 3, \ldots,$$

and $m = n = 0$ is excluded. The resonant frequency for the TE_{mnl} mode is given by the same formula, (4-158). If the cavity is filled with dielectric material, then $2\pi f_r/c$ in (4-158) must be replaced by $2\pi f_r \epsilon_r^{1/2}/c$, where ϵ_r is the relative dielectric constant of the material.

The cylindrical cavity is formed by closing the ends of a cylindrical waveguide by perfectly conducting walls. For a cylindrical cavity with radius a and height h, the eigenfunction for the TM_{nlp} mode is

$$\pi_z = J_n(k_{\rho nl}\rho) \cos n\phi \cos k_{zp} z, \tag{4-160}$$

where

$$J_n(k_{\rho nl} a) = 0, \qquad k_{zp} = \frac{p\pi}{h},$$

$$n = 0, 1, 2, \ldots,$$

$$p = 0, 1, 2, \ldots.$$

The resonant frequency f_r for the TM_{nlp} mode is

$$\frac{2\pi f_r \epsilon_r^{1/2}}{c} = (k_{\rho nl}^2 + k_{zp}^2)^{1/2}, \tag{4-161}$$

where ϵ_r is the relative dielectric constant of the medium inside the cavity.

Similarly, for the TE_{nlp} mode, we have

$$\pi_{mz} = J_n(k_{\rho nl}\rho) \cos n\phi \sin k_{zp} z, \tag{4-162}$$

where

$$J_n'(k_{\rho nl} a) = 0, \qquad k_{zp} = \frac{p\pi}{h},$$

$$n = 0, 1, 2, \ldots,$$

$$p = 1, 2, 3, \ldots.$$

The resonant frequency is given by (4-161).

As an example of the cylindrical cavity, consider a circular microstrip antenna (Fig. 4-20). This antenna is fed by a coaxial line located off the center of the patch. The current on the patch is then similar to that of a horizontal dipole, and the radiation is in the broadside direction. Microstrip antennas are therefore useful as low-profile antennas on the surface of aircraft. The microstrip antenna is high Q and narrow band and is operated near the resonant frequency of the cylindrical cavity of radius a and height h. The top and bottom surfaces are conducting (electric wall), while the side is open. The radiating power is transmitted through the side and some reactive power is stored near the fringe of the patch. However, as a first approximation, we assume that on the side, the tangential magnetic field is negligibly small (magnetic wall). Noting Fig. 4-16, the boundary conditions for TM modes are

$$\begin{aligned} \frac{\partial \pi_z}{\partial z} &= 0 \qquad \text{on the top and bottom} \\ \frac{\partial \pi_z}{\partial \rho} &= 0 \qquad \text{on the side.} \end{aligned} \tag{4-163}$$

For TE modes, the boundary conditions are

$$\begin{aligned} \pi_{mz} &= 0 \qquad \text{on the top and bottom} \\ \pi_{mz} &= 0 \qquad \text{on the side.} \end{aligned} \tag{4-164}$$

From Table 4-1 it is seen that the lowest eigenvalue $\chi_{11} = 1.841$ is obtained from $J_n'(\chi_{nl}) = 0$ when $n = 1$ and $l = 1$. Thus the lowest resonant frequency is obtained for the TM_{11} mode with the boundary condition (4-163). The field for this cavity is, therefore, given by

$$\begin{aligned} E_z &= A_0 J_1(k_{11}\rho) \cos\phi, \\ H_\phi &= -j\omega\epsilon k_{11} A_0 J_1'(k_{11}\rho) \cos\phi, \\ H_\rho &= -j\omega\frac{1}{\rho} A_0 J_1(k_{11}\rho) \sin\phi, \\ k_{11} &= \frac{1.841}{a}. \end{aligned} \tag{4-165}$$

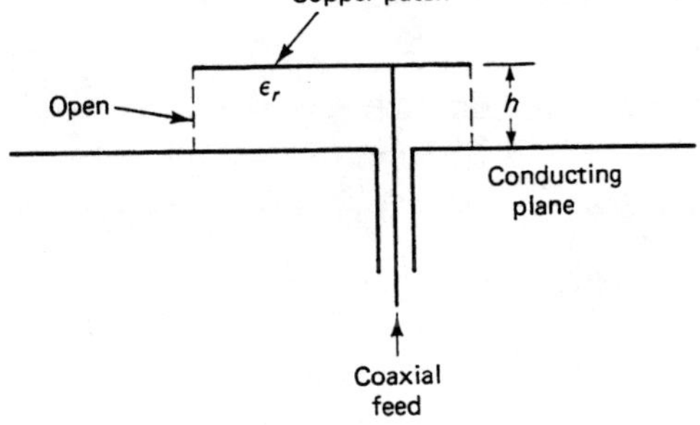

Figure 4-20 Microstrip antennas.

The resonant frequency f_r is then given by

$$\frac{2\pi f_r\sqrt{\epsilon_r}}{c}=\frac{1.841}{a}. \tag{4-166}$$

The microstrip antenna is operated at a frequency close to this frequency.

4-15 WAVES IN SPHERICAL STRUCTURES

Unlike the cylindrical coordinate system, the spherical coordinate system does not possess an axis with constant direction, and the formulation based on $\pi\hat{a}$ and $\pi_m\hat{a}$ in (4-2) cannot be used. Therefore, it is necessary to devise a different method by which all the field components can be derived from scalar functions satisfying a well-known differential equation. This can be done by a special choice of the Hertz vector, and the complete electromagnetic fields can be described by two scalar functions that satisfy a scalar wave equation.

Let us reexamine the derivation of the Hertz vector. We need a condition similar to but different from the Lorentz condition, which yields the desired result. We consider the time-harmonic case and start with the vector potential $\bar{A}$ and scalar potential Φ. Rewriting (2-84) for the time-harmonic case, we get

$$-\nabla\times\nabla\times\bar{A}+k^2\bar{A}=j\omega\mu\epsilon\nabla\Phi-\mu\bar{J}. \tag{4-167}$$

We need to find the relationship between $\bar{A}$ and Φ such that (4-167) reduces to a simpler scalar equation in a spherical coordinate system. Let us assume that $\bar{A}$ and $\bar{J}$ have only the radial component A_r and J_r.

$$\bar{A}=A_r\hat{r} \qquad \text{and} \qquad \bar{J}=J_r\hat{r}. \tag{4-168}$$

The radial component of (4-167) is then given by

$$\left[\frac{1}{r^2\sin\theta}\frac{\partial}{\partial\theta}\left(\sin\theta\frac{\partial}{\partial\theta}\right)+\frac{1}{r^2\sin^2\theta}\frac{\partial^2}{\partial\phi^2}+k^2\right]A_r=j\omega\mu\epsilon\frac{\partial}{\partial r}\Phi-\mu J_r. \tag{4-169}$$

The θ and ϕ components are

$$\begin{aligned}-\frac{1}{r}\frac{\partial^2}{\partial r\,\partial\theta}A_r&=\frac{j\omega\mu\epsilon}{r}\frac{\partial}{\partial\theta}\Phi,\\ -\frac{1}{r\sin\theta}\frac{\partial^2}{\partial r\,\partial\phi}A_r&=\frac{j\omega\mu\epsilon}{r\sin\theta}\frac{\partial}{\partial\phi}\Phi.\end{aligned} \tag{4-170}$$

The Lorentz condition $\nabla\cdot\bar{A}+j\omega\mu\epsilon\,\Phi=0$ in (2-87) cannot simplify (4-170). Instead, we use

$$\frac{\partial A_r}{\partial r}+j\omega\mu\epsilon\,\Phi=0. \tag{4-171}$$

Then (4-170) is automatically satisfied, and (4-169) becomes

$$\left[\frac{\partial^2}{\partial r^2}+\frac{1}{r^2 \sin\theta}\frac{\partial}{\partial\theta}\left(\sin\theta\frac{\partial}{\partial\theta}\right)+\frac{1}{r^2\sin^2\theta}\frac{\partial^2}{\partial\phi^2}+k^2\right]A_r=-\mu J_r. \tag{4-172}$$

This is a simple differential equation. However, it would be more convenient if this were transformed into a scalar wave equation. This can be accomplished by choosing

$$j\omega\mu\epsilon r\pi_r = A_r. \tag{4-173}$$

Then (4-172) becomes

$$(\nabla^2+k^2)\pi_r=-\frac{J_r}{j\omega\epsilon r}, \tag{4-174}$$

where ∇^2 is the Laplacian operator

$$\nabla^2=\frac{1}{r^2}\frac{\partial}{\partial r}\left(r^2\frac{\partial}{\partial r}\right)+\frac{1}{r^2\sin\theta}\frac{\partial}{\partial\theta}\left(\sin\theta\frac{\partial}{\partial\theta}\right)+\frac{1}{r^2\sin^2\theta}\frac{\partial^2}{\partial\phi^2}. \tag{4-175}$$

A definite advantage of using the scalar wave equation is that its solution has been extensively studied and is well documented. The fields derived from π_r are called E modes or TM modes because the only radial component is the electric field and all the magnetic fields are transverse to the radial vector.

By the duality principle, we can obtain H modes or TE modes from π_{mr}. Thus we get

$$(\nabla^2+k^2)\pi_r=-\frac{J_r}{j\omega\epsilon r},$$

$$(\nabla^2+k^2)\pi_{mr}=-\frac{J_{mr}}{j\omega\mu r}.$$

$$\begin{aligned}\bar{E}&=\nabla\times\nabla\times(r\pi_r\hat{r})-j\omega\mu\nabla\times(r\pi_{mr}\hat{r})-\frac{J_r\hat{r}}{j\omega\epsilon}\\&=\nabla\frac{\partial}{\partial r}(r\pi_r\hat{r})+k^2r\pi_r\hat{r}-j\omega\mu\nabla\times(r\pi_{mr}\hat{r}),\\\bar{H}&=j\omega\epsilon\nabla\times(r\pi_r\hat{r})+\nabla\times\nabla\times(r\pi_{mr}\hat{r})-\frac{J_{mr}\hat{r}}{j\omega\mu}\\&=j\omega\epsilon\nabla\times(r\pi_r\hat{r})+\nabla\frac{\partial}{\partial r}(r\pi_{mr}\hat{r})+k^2r\pi_{mr}\hat{r}.\end{aligned} \tag{4-176}$$

In terms of the field components, we have

$$E_r=\frac{\partial^2}{\partial r^2}(r\pi_r)+k^2r\pi_r,$$

$$E_\theta=\frac{1}{r}\frac{\partial^2}{\partial r\,\partial\theta}(r\pi_r)-j\omega\mu\frac{1}{\sin\theta}\frac{\partial}{\partial\phi}\pi_{mr},$$

$$E_\phi = \frac{1}{r\sin\theta}\frac{\partial^2}{\partial r\,\partial\phi}(r\pi_r) + j\omega\mu\frac{\partial}{\partial\theta}\pi_{mr},$$

$$H_r = \frac{\partial^2}{\partial r^2}(r\pi_{mr}) + k^2 r\pi_{mr}, \tag{4-177}$$

$$H_\theta = j\omega\epsilon\frac{1}{\sin\theta}\frac{\partial}{\partial\phi}\pi_r + \frac{1}{r}\frac{\partial^2}{\partial r\,\partial\theta}(r\pi_{mr}),$$

$$H_\phi = -j\omega\epsilon\frac{\partial}{\partial\theta}\pi_r + \frac{1}{r\sin\theta}\frac{\partial^2}{\partial r\,\partial\phi}(r\pi_{mr}).$$

In the above, we used the source term of the radial electric current J_r and the radial magnetic current J_{mr}. For those radial current sources, π_r and π_{mr} are simply related to J_r and J_{mr}, as shown in (4-174). For a more general source current with θ and ϕ components, the relationships are more complicated. It should, however, be noted that the completely general electromagnetic field in the spherical coordinate system can be expressed by two scalar functions π_r and π_{mr}.

Let us now consider a formal solution of the homogeneous wave equation in the spherical coordinate system.

$$(\nabla^2 + k^2)\pi_r = 0. \tag{4-178}$$

Assume the product solution

$$\pi_r = X_1(r)X_2(\theta)X_3(\phi). \tag{4-179}$$

Then we get

$$\left\{\frac{1}{r^2}\frac{d}{dr}\left(r^2\frac{d}{dr}\right) + \left[k^2 - \frac{\nu(\nu+1)}{r^2}\right]\right\}X_1(r) = 0, \tag{4-180}$$

$$\left\{\frac{1}{\sin\theta}\frac{d}{d\theta}\left(\sin\theta\frac{d}{d\theta}\right) + \left[\nu(\nu+1) - \frac{\mu^2}{\sin^2\theta}\right]\right\}X_2(\theta) = 0, \tag{4-181}$$

$$\left(\frac{d^2}{d\phi^2} + \mu^2\right)X_3(\phi) = 0, \tag{4-182}$$

where ν and μ are constant.

Each of the equations above is a second-order differential equation, and the solution is in general represented by a linear combination of two independent solutions. Let us first note that (4-182) can easily be solved and that $X_3(\phi)$ is given by

$$\begin{aligned} X_3(\phi) &= C_1 e^{j\mu\phi} + C_2 e^{-j\mu\phi} \\ &= C_3 \sin\mu\phi + C_4\cos\mu\phi, \end{aligned} \tag{4-183}$$

where all C are constant.

Next consider $X_1(r)$ in (4-180). This is a spherical Bessel's equation, and its solution, the spherical Bessel functions $z_\nu(x)$, is related to the ordinary Bessel functions by

$$z_\nu(x) = \sqrt{\frac{\pi}{2x}}\, Z_{\nu+1/2}(x). \tag{4-184}$$

Thus, corresponding to $J_\nu(x)$, $N_\nu(x)$, $H_\nu^{(1)}(x)$, and $H_\nu^{(2)}(x)$, we have

$$\begin{aligned} j_\nu(x) &= \sqrt{\frac{\pi}{2x}}\, J_{\nu+1/2}(x), \\ n_\nu(x) &= \sqrt{\frac{\pi}{2x}}\, N_{\nu+1/2}(x), \\ h_\nu^{(1)}(x) &= \sqrt{\frac{\pi}{2x}}\, H_{\nu+1/2}^{(1)}(x), \\ h_\nu^{(2)}(x) &= \sqrt{\frac{\pi}{2x}}\, H_{\nu+1/2}^{(2)}(x). \end{aligned} \tag{4-185}$$

The behavior of spherical Bessel functions can readily be understood by that of the corresponding Bessel functions.

Note that (see Jahnke et al. 1960, p. 142)

$$\begin{aligned} j_0(x) &= \frac{\sin x}{x} \quad \text{and} \quad n_0(x) = -\frac{\cos x}{x}, \\ h_0^{(1)}(x) &= -j\frac{e^{jx}}{x} \quad \text{and} \quad h_0^{(2)}(x) = j\frac{e^{-jx}}{x}. \end{aligned} \tag{4-186}$$

In general,

$$\begin{aligned} j_n(x) &= x^n\left(-\frac{1}{x}\frac{d}{dx}\right)^n \frac{\sin x}{x}, \\ n_n(x) &= -x^n\left(-\frac{1}{x}\frac{d}{dx}\right)^n \frac{\cos x}{x} \end{aligned}$$

Let us consider $X_2(\theta)$ in (4-181). This is the Legendre differential equation, and X_2 is given by a linear combination of its two independent solutions. We consider the following three cases.

1. $\mu = 0, \nu = n$ *integer*. In this case, $X_2(\theta)$ is given by

$$X_2(\theta) = C_1 P_n(\cos\theta) + C_2 Q_n(\cos\theta). \tag{4-187}$$

$P_n(x), x = \cos\theta$, is a polynomial of degree n.

$$\begin{aligned} P_0(x) &= 1, \\ P_1(x) &= x, \\ P_2(x) &= \tfrac{1}{2}(3x^2 - 1). \end{aligned} \tag{4-188}$$

$P_n(x)$ is an even function of x when n is even and an odd function when n is odd. $P_n(x)$ is also regular in the range, $-1 \le x \le 1$ $(\pi \ge \theta \ge 0)$. $Q_n(x)$ is called the Legendre function of the second kind and becomes infinite at $\theta = 0$ and π.

$$Q_0(x) = \frac{1}{2} \ln \frac{1+x}{1-x},$$

$$Q_1(x) = \frac{x}{2} \ln \frac{1+x}{1-x} - 1, \tag{4-189}$$

$$Q_2(x) = \frac{3x^2 - 1}{4} \ln \frac{1+x}{1-x} - \frac{3x}{2}.$$

2. $\mu = m, \nu = n, m, n$ *are integers*. In this case $X_2(\theta)$ is given by

$$X_2(\theta) = C_1 P_n^m(\cos\theta) + C_2 Q_n^m(\cos\theta). \tag{4-190}$$

P_n^m and Q_n^m are called the associated Legendre function of the first kind and the second kind, respectively. Both functions vanish if $m > n$, and therefore, m ranges only over $0, 1, 2, \ldots, n$. Q_n^m becomes infinite at $\theta = 0$ and $\theta = \pi$ while P_n^m is regular in $0 \le \theta \le \pi$. $P_n^m(x)$ is also orthogonal in this range.

$$\int_{-1}^{1} P_n^m(x) P_{n'}^m(x)\, dx = \begin{cases} 0, & n \ne n', \\ \dfrac{2}{2n+1} \dfrac{(n+m)!}{(n-m)!}, & n = n'. \end{cases} \tag{4-191}$$

The function $P_n^m(\cos\theta)e^{jm\phi}$ is regular and orthogonal over the range $0 \le \theta \le \pi$ and $0 \le \phi \le 2\pi$ and is called the *spherical harmonics*.

3. $\mu = m$ *integer*, $\nu =$ *noninteger*. This case arises in connection with a problem of a cone. Two independent solutions are:

$$P_\nu^m(\cos\theta) \quad \text{and} \quad P_\nu^m(-\cos\theta).$$

The first becomes infinite at $\theta = \pi$ and the second becomes infinite at $\theta = 0$. $P_\nu^m(\cos\theta)$ is not zero for $m > \nu$ when ν is a noninteger.

4-16 SPHERICAL WAVEGUIDES AND CAVITIES

Let us consider waves propagating in each of the three orthogonal directions in the spherical coordinate system.

Wave Propagation in the Radial Direction

As an example, consider a wave propagating in the radial direction in a conical waveguide (Fig. 4-21). TM waves can be expressed by the following Hertz potential:

$$\pi_r = \sum_m \sum_n A_{mn} P_{\nu_n}^m(\cos\theta) h_{\nu_n}^{(2)}(kr) \begin{pmatrix} \cos m\phi \\ \sin m\phi \end{pmatrix}. \tag{4-192}$$

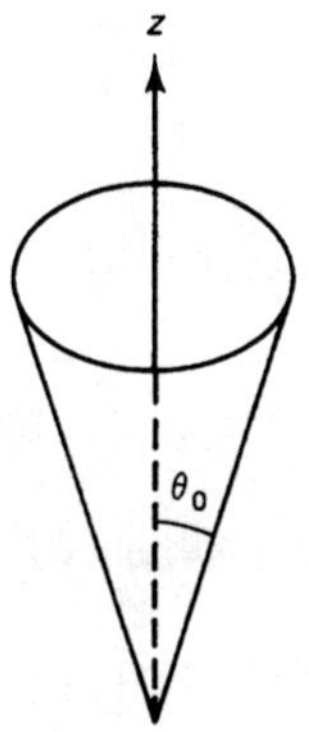

Figure 4-21 Conical waveguide.

where ν_n is determined by

$$P_{\nu_n}^m(\cos\theta_0) = 0.$$

Note that $h_\nu^{(2)}$ represents the outgoing wave in the radial direction.

Wave Propagation in the θ Direction

An example of this case is the VLF propagation between the spherical earth and the ionosphere. For example, we may consider a resonance in a spherical cavity formed by the earth and the ionosphere (Fig. 4-22). For TM modes we write

$$\pi_r = P_n^m(\cos\theta)Z_n(kr)\begin{pmatrix}\cos m\phi\\ \sin m\phi\end{pmatrix}, \tag{4-193}$$

where $Z_n(kr)$ is a linear combination of two spherical Bessel functions that satisfies the boundary condition at the earth surface and the ionosphere. If we assume that the earth surface ($r = a$) and the lower edge of the ionosphere ($r = b$) are *conducting*, the boundary conditions are

$$\frac{\partial}{\partial r}(r\pi_r) = 0 \qquad \text{at } r = a \text{ and } b. \tag{4-194}$$

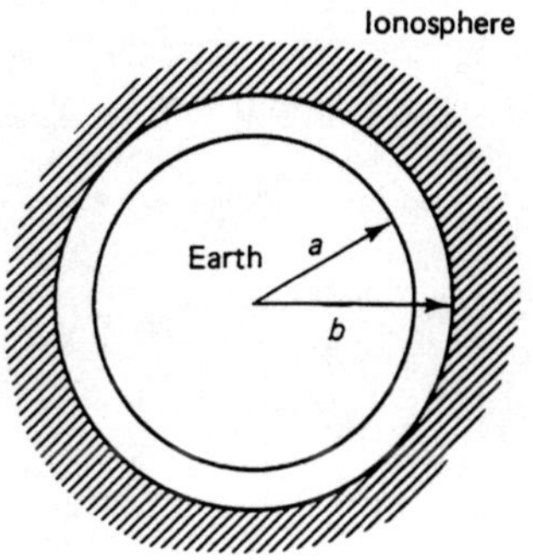

Figure 4-22 Earth–ionosphere cavity.

Thus we have

$$
\begin{aligned}
Z_n(kr) &= C_1 j_n(kr) + C_2 h_n^{(2)}(kr), \\
C_1 &= \frac{\partial}{\partial r}[r h_n^{(2)}(kr)]\Big|_{r=a}, \\
C_2 &= -\frac{\partial}{\partial r}[r j_n(kr)]\Big|_{r=a}.
\end{aligned}
\tag{4-195}
$$

The resonant frequency f_r is then given by $k = 2\pi f_r/c$, which satisfies

$$
\frac{\partial}{\partial r}[rZ_n(kr)]\Big|_{r=b} = 0. \tag{4-196}
$$

This yields the following equation to determine the resonant frequency:

$$
\left.\frac{\frac{\partial}{\partial r}[r j_n(kr)]}{\frac{\partial}{\partial r}[r h_n^{(2)}(kr)]}\right|_{r=a} = \left.\frac{\frac{\partial}{\partial r}[r j_n(kr)}{\frac{\partial}{\partial r}[r h_n^{(2)}(kr)]}\right|_{r=b}. \tag{4-197}
$$

The resonance phenomena above were first studied by Schumann and are now called the Schumann resonance. Although (4-197) gives the exact equation for the resonant frequency when both the ionosphere and the earth are assumed conducting, it is instructive to obtain an approximate solution for the Schumann resonance. We first note that the boundary condition (4-194) requires the derivative of $r\pi_r$. Thus we let $rX_1 = u$ in (4-180) and obtain

$$
\left[\frac{d^2}{dr^2} + k^2 - \frac{n(n+1)}{r^2}\right]u(r) = 0. \tag{4-198}
$$

Now since the distance between the ionosphere and the earth is very much smaller than the earth's radius, we let $r \sim a + z \sim a$ and approximate (4-198) by

$$
\left[\frac{d^2}{dz^2} + k^2 - \frac{n(n+1)}{a^2}\right]u(z) = 0, \tag{4-199}
$$

with the boundary condition

$$
\frac{\partial u}{\partial z} = 0 \qquad \text{at } z = 0 \text{ and } z = h. \tag{4-200}
$$

The solution is

$$
u = \cos\frac{m\pi z}{h}, \qquad m = 0, 1, 2, \ldots
$$

$$
k^2 - \frac{n(n+1)}{a^2} = \left(\frac{m\pi}{h}\right)^2.
$$

The lowest resonant frequencies f_r are

$$f_r = \frac{c}{2\pi a}\sqrt{n(n+1)}, \qquad n = 1, 2, 3, \ldots. \tag{4-201}$$

These values are close to the observed peaks in the frequency spectrum of the noise power generated by lightning around the earth.

Wave Propagation in the Azimuthal Direction

An example may be a leaky waveguide around a sphere. A general form of π_r is then

$$\pi_r = z_\nu(kr)P_\nu^\mu(\cos\theta)e^{-j\mu\phi}, \tag{4-202}$$

where μ is the azimuthal propagation constant.

Let us now consider the boundary conditions for a perfectly conducting wall (Fig. 4-23). Consider a surface containing a radial vector. At ϕ = constant, E_r and E_θ must be zero. At θ = constant, E_r and E_ϕ must be zero. They are satisfied if

$$\pi_r = 0 \qquad \text{and} \qquad \frac{\partial}{\partial n}\pi_{mr} = 0. \tag{4-203}$$

$\partial/\partial n$ is the normal derivative, and in the two cases above, this can also be expressed by $\partial/\partial\phi$ and $\partial/\partial\theta$, respectively.

The boundary condition at the surface perpendicular to the radial vector is that E_θ and E_ϕ be zero, which leads to the condition

$$\frac{\partial}{\partial n}(r\pi_r) = \frac{\partial}{\partial r}(r\pi_r) = 0 \qquad \text{and} \qquad \pi_{mr} = 0. \tag{4-204}$$

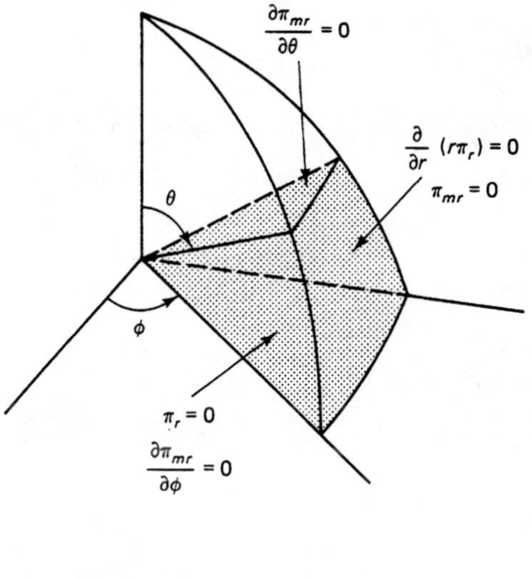

Figure 4-23 Boundary conditions for Hertz vectors in a spherical system.

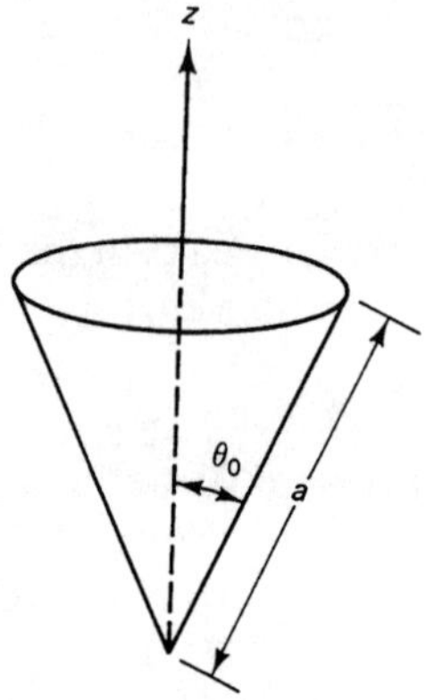

Figure 4-24 Conical cavity.

As an example, let us consider a conical cavity shown in Fig. 4-24. The TM mode is given by

$$\pi_r = j_\nu(kr)P_\nu^m(\cos\theta)\begin{pmatrix}\cos m\phi \\ \sin m\phi\end{pmatrix}, \tag{4-205}$$

where $j_\nu(kr)$ is used because $n_\nu(kr)$ becomes infinite at the origin.

For a given m, ν must be determined by the boundary condition at $\theta = \theta_0$, which is given by

$$P_{\nu_n}^m(\cos\theta_0) = 0. \tag{4-206}$$

Once ν_n is determined, the boundary condition at $r = a$ is applied and yields the resonant frequency f_r (or wavelength λ_r), given by $k_r = \omega_r/c = 2\pi/\lambda_r$, $\omega_r = 2\pi f_r$, where $X_r = k_r a$ is the root of

$$\frac{d}{dX}[Xj_{\nu_n}(X)] = 0. \tag{4-207}$$

There is an infinite number of roots for (4-207) for a given ν_n, and thus we should write

$$X_r = X_{mnl}. \tag{4-208}$$

The TM_{mnl} mode is therefore given by

$$\pi_r = j_{\nu_n}(k_{mnl}r)P_{\nu_n}^m(\cos\theta)\begin{pmatrix}\cos m\phi \\ \sin m\phi\end{pmatrix}, \tag{4-209}$$

where $k_{mnl}a = X_{mnl}$.

PROBLEMS

4-1. (a) Find the normalized eigenfunctions $\phi_n(x)$ and eigenvalues k_n^2 that satisfy the following:

$$\left(\frac{d^2}{dx^2} + k_n^2\right)\phi_n(x) = 0,$$

$$\phi_n(x) = 0 \text{ at } x = 0,$$

$$\frac{d\phi_n}{dx} + h\phi_n = 0, \qquad \text{at } x = a.$$

(b) Find the lowest two eigenvalues for $a = 1$ and $h = 2$.

4-2. Consider a TE_{10} mode in a rectangular waveguide. At $x = 0$, $E_y = 0$, and at $x = a$, $E_y/H_z = j100$. Find the propagation constant β at 10 GHz. $a = 2.5$ cm and $b = 1.25$ cm.

4-3. A TM_{11} mode is propagating in a rectangular waveguide with $a = 0.2$ m and $b = 0.1$ m.
(a) What is the cutoff frequency f_c?
(b) If the operating frequency f is $2f_c$, calculate the phase velocity and the group velocity.
(c) At $f = 2f_c$, if the maximum value of $|E_z| = 5$ V/m, calculate the total power propagated through the guide.

4-4. A TE_{11} wave is propagating in a cylindrical guide with radius 1 cm. The operating frequency is 10 GHz and the total transmitted power is 100 mW.
(a) Find the phase and group velocities.
(b) What is the cutoff frequency for this guide?
(c) If the guide is filled with dielectric material with $\epsilon_r = 2$, do the phase and group velocities and the cutoff frequencies change? If so, find their values.
(d) Assume that the guide is hollow. Find expressions for the electric and magnetic fields.
(e) Find an expression for the surface current.

4-5. Consider a coaxial line with $a = 1$ cm and $b = 0.5$ cm. Calculate the characteristic impedance of this line. If the total power of 1 mW is propagated through the line, calculate the maximum electric field intensity.

4-6. Find the lowest two cutoff frequencies for the sector waveguide shown in Fig. 4-8. $a =$ 1 cm and $\alpha = 90°$.

4-7. A pulse wave in the TE_{10} mode is propagating in a rectangular guide with $a = 2.5$ cm and $b = 1$ cm. The carrier frequency is $f_0 = 8$ GHz and the waveform of the electric field at $z = 0$ is given by

$$E_y(t, z = 0) = E_0 \sin\frac{\pi x}{a} \exp\left(-\frac{t^2}{T_0^2}\right) \cos\omega_0 t,$$

where $E_0 = 0.5$ V/m, $T_0 = 1$ ns, and $\omega_0 = 2\pi f_0$. Calculate the phase velocity, group velocity, group delay per meter, and pulse spread per meter of the guide.

4-8. Consider a multimode step-index fiber with $n_1 = 1.48$ and $n_2 = 1.46$. The diameter is 50 μm, and the wavelength is $\lambda = 0.82$ μm. Find the number of modes, pulse spread per unit length in ns/km, and numerical aperture (NA).

4-9. A graded-index fiber has the square-law refractive index profile given by (4-142) with $n_1 = 1.48$, $n_2 = 1.46$, and $2a = 50$ μm. Find the pulse spread in ns/km.

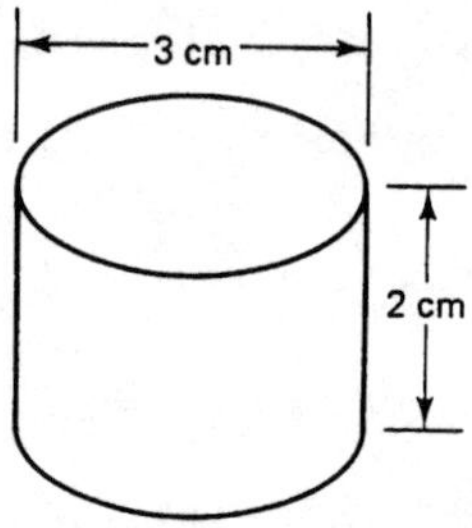

Figure P4-11 Cylindrical cavity.

4-10. Consider TM modes propagating in the radial direction inside the sectorial horn in Fig. 4-17 with $\alpha = 45°$ and $l = 3$ cm. Find the two lowest cutoff frequencies. Find the expressions for $\bar{E}$ and $\bar{H}$ for the propagating modes $m = 1$ and $n = 1$ at 10 GHz.

4-11. Find the three lowest resonant frequencies of the cylindrical cavity shown in Fig. P4-11.

4-12. Find the resonant frequency of the microstrip antenna shown in Fig. 4-20 with $h = 0.3$ cm, $a = 4$ cm, and $\epsilon_r = 2.5$.

4-13. Use Rodrigues's formula,

$$P_n(x) = \frac{1}{2^n n!} \frac{d^n}{dx^n}(x^2 - 1)^n,$$

to show the following:

$$\int_{-1}^{1} x^m P_n(x)\, dx = 0 \qquad \text{if } m < n.$$

4-14. Expand the following function in a series of Legendre functions $P_n(x)$:

$$F(x) = x^3 + x^2 + x + 1.$$

4-15. Find the lowest five Schumann resonance frequencies.

4-16. Consider a spherical cavity of radius a. For TM modes, show that the lowest nonzero mode is TM_{101}, where the subscripts are variations on the r, ϕ, and θ directions, and the Hertz vector is $\Pi_r = j_1(kr)P_1(\cos\theta)$. Show that the resonant frequency is given by

$$\tan x = \frac{x}{1 - x^2}, \qquad x = ka.$$

Its solution is $x = 2.744$. Find expressions for $\bar{E}$ and $\bar{H}$.

For TE modes, the lowest mode is TE_{101} and the resonant frequency is given by

$$\tan x = x, \qquad x = ka.$$

Its solution is $x = 4.493$. Find expressions for $\bar{E}$ and $\bar{H}$.

4-17. Consider the cylindrical cavity shown in Fig. P4-17. Find the lowest two resonant frequencies for (a) the TM modes and (b) the TE modes.

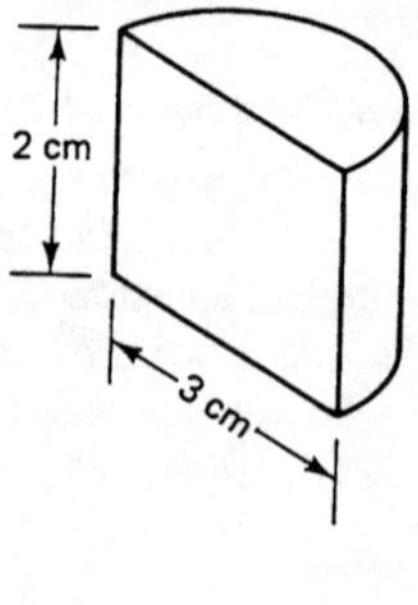

Figure P4-17 Cylindrical sector cavity.

5

Green's Functions

In Chapter 3 we discussed the reflection and transmission of plane waves incident on layers of dielectric medium and the propagation characteristics of guided waves along the layered medium. In Chapter 4 we discussed the wave propagation in waveguides and the wave modes in cavities. In this chapter we discuss the problems of the excitation of waves and Green's function, which is the field excited by a point source.

5-1 ELECTRIC AND MAGNETIC DIPOLES IN HOMOGENEOUS MEDIA

There are two basic sources of excitation: electric and magnetic. The simplest forms of these two sources are the *electric dipole* and the *magnetic dipole*. An arbitrary source can always be represented by a distribution of electric and magnetic dipole sources.

A short wire antenna located at $\bar{r}'$ which is fed at the center with the current I_0 oscillating at a frequency $f = \omega/2\pi$, and whose length L is much smaller than a wavelength, can represent the electric dipole whose current density $\bar{J}$ is given by

$$\bar{J}(\bar{r}) = \hat{i} I_0 L_0 \delta(\bar{r} - \bar{r}'), \tag{5-1}$$

where $\hat{i}$ is a unit vector in the direction of the current, and L_0 is the effective length of the wire antenna given by

$$I_0 L_0 = \int_{\Delta V} \bar{J}\, dV = \int_{-L/2}^{L/2} I(z)\, dz. \tag{5-2}$$

For a short wire antenna, $I(z)$ is known to have a triangular shape, and therefore $L_0 = L/2$ (see Fig. 5-1).

Let us next consider the magnetic dipole. If the current I_0 flows in a small loop with the area A, the magnetic dipole moment $\bar{m}$ is defined by

$$\bar{m} = I_0 A \hat{i}, \tag{5-3}$$

where $\hat{i}$ is the unit vector normal to the area A (Fig. 5-2). Noting that the magnetic current density $\bar{J}_m$ is related to $\bar{M}$ by (see Section 2-9)

$$\bar{J}_m = j\omega\mu_0 \bar{M}, \tag{5-4}$$

the magnetic current density $\bar{J}_m$ for a small current loop shown in Fig. 5-2 is represented by

$$\bar{J}_m = j\omega\mu_0 A I_0 \delta(\bar{r} - \bar{r}')\hat{i}. \tag{5-5}$$

The electric current density $\bar{J}$ given in (5-1) and the magnetic current density $\bar{J}_m$ given in (5-5) are the two basic sources of electromagnetic fields. In general, many other practical sources can be represented by continuous distributions of the electric and magnetic current sources. For example, a long wire antenna can be regarded as a continuous distribution of the electric dipole source. Another common source is

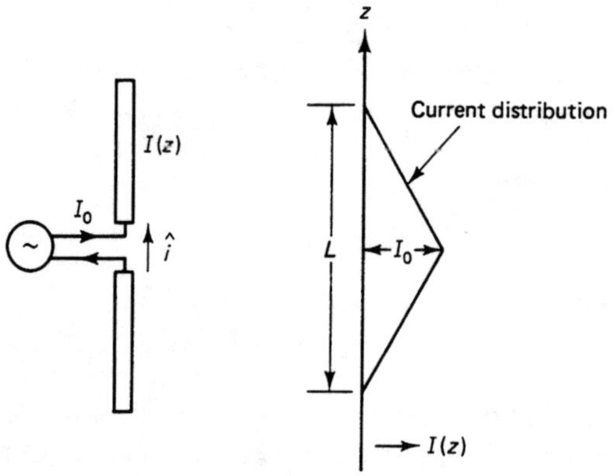

Figure 5-1 Electric dipole.

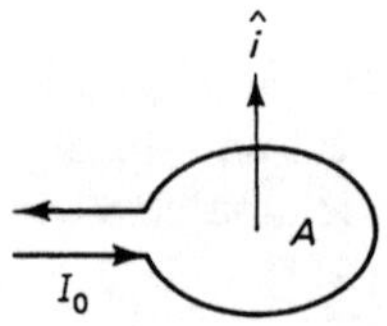

Figure 5-2 Magnetic dipole.

aperture fields for parabolic dish antennas and slot antennas. They can be represented by the electric or magnetic current distribution over the surface.

5-2 ELECTROMAGNETIC FIELDS EXCITED BY AN ELECTRIC DIPOLE IN A HOMOGENEOUS MEDIUM

The time-harmonic electromagnetic wave excited by the current source $\bar{J}_s$ is given in terms of the Hertz vector $\bar{\pi}$ (Section 2-7):

$$\bar{E} = \nabla(\nabla \cdot \bar{\pi}) + k^2 \bar{\pi},$$

$$\bar{H} = j\omega\epsilon \nabla \times \bar{\pi},$$

and

$$(\nabla^2 + k^2)\bar{\pi} = -\frac{\bar{J}_s}{j\omega\epsilon}, \tag{5-6}$$

where $k = \omega\sqrt{\mu\epsilon}$ and μ and ϵ are, in general, complex.

Let us consider the electromagnetic field excited by an electric dipole pointed in the z direction and located at the origin of a Cartesian coordinate system. The electric dipole is represented by

$$\bar{J}_s = \hat{z} I_0 L_0 \delta(\bar{r}). \tag{5-7}$$

The z component of $\bar{\pi}$ then satisfies

$$(\nabla^2 + k^2)\pi_z = -\frac{I_0 L_0}{j\omega\epsilon} \delta(r). \tag{5-8}$$

We write (5-8) using a function $G(r)$ which satisfies the equation

$$(\nabla^2 + k^2) G(\bar{r}, \bar{r}') = -\delta(\bar{r} - \bar{r}'). \tag{5-9}$$

Then π_z is given by

$$\pi_z = \frac{I_0 L_0}{j\omega\epsilon} G(\bar{r}, 0). \tag{5-10}$$

Note that once we solve (5-9), the solution $\bar{\pi}$ is given by (5-10) and the fields are given by (5-6).

The function $G(\bar{r}, \bar{r}')$ is called Green's function and represents the response of a physical system in space due to a point exciting source $\delta(\bar{r} - \bar{r}')$. The differential equation (5-9) alone is insufficient to determine Green's function uniquely, and it is necessary to apply additional conditions. These conditions are (1) the radiation condition at infinity whenever the region extends to infinity and (2) the boundary conditions. Therefore, Green's function $G(\bar{r}, \bar{r}')$ must satisfy the inhomogeneous differential equation of the type (5-9) and the radiation and boundary conditions.

Let us solve (5-9) in free space satisfying the radiation condition. We choose $\bar{r}'$

to be at the origin, and thus $G(\bar{r})$ is spherically symmetric and a function of r only. We write (5-9) as

$$\left[\frac{1}{r^2}\frac{d}{dr}\left(r^2\frac{d}{dr}\right)+k^2\right]G(r)=-\delta(r). \tag{5-11}$$

The solution is given by

$$G(r)=\frac{\exp(-jkr)}{4\pi r}. \tag{5-12}$$

To show this, first we consider $G(\bar{r})$ when $\bar{r}\neq 0$. Here $\delta(r)=0$. We then let $G=u/r$ and obtain

$$\left(\frac{d^2}{dr^2}+k^2\right)u(r)=0,$$

whose general solution is

$$u(r)=c_1 e^{-jkr}+c_2 e^{+jkr}, \qquad c_1, c_2 \text{ constants.}$$

Therefore, for $r\neq 0$, G is given by

$$G(r)=\frac{1}{r}(c_1 e^{-jkr}+c_2 e^{+jkr}). \tag{5-13}$$

Now $G(r)$ is the wave originating at $r=0$, and at infinity, the wave must be outgoing. The first term of (5-13) represents the outgoing wave, but the second term represents the incoming wave. Even though both terms vanish at $r\to\infty$, the second term must be discarded because there should be no incoming wave from infinity. This is called the *radiation condition* or *Sommerfeld radiation condition*, and its mathematical expression is given by (Section 2-4)

$$\lim_{r\to\infty} r\left(\frac{\partial G}{\partial r}+jkG\right)=0. \tag{5-14}$$

Only the first term of (5-13) satisfies (5-14) as expected, and therefore c_2 must be zero.

Next we determine the constant c_1 by considering the behavior of G near the origin. To do this, let us write (5-9) as follows:

$$\nabla\cdot(\nabla G)+k^2 G=-\delta(r), \tag{5-15}$$

and integrate both sides over a small spherical volume of radius r_0 and let the radius r_0 approach zero:

$$\lim_{r_0\to 0}\int_V[\nabla\cdot(\nabla G)+k^2 G]\,dV=-\lim_{r_0\to 0}\int_V \delta(r)\,dV. \tag{5-16}$$

The right side is -1. The first term in the left side becomes, using the divergence theorem, $-4\pi c_1$. The second term in the left becomes zero. Therefore, (5-16) becomes

$$-4\pi c_1 = -1,$$

from which we obtain $c_1 = 1/4\pi$.

Thus Green's function $G(\bar{r}, \bar{r}')$ satisfying

$$(\nabla^2 + k^2)G(\bar{r}, \bar{r}') = -\delta(\bar{r} - \bar{r}'), \tag{5-17}$$

and the radiation condition at infinity is given by

$$G(\bar{r}, \bar{r}') = \frac{e^{-jk|\bar{r}-\bar{r}'|}}{4\pi|\bar{r}-\bar{r}'|}. \tag{5-18}$$

Using this Green's function, the Hertz vector due to an electric dipole located at the origin pointed in the $\hat{z}$ direction is given by

$$\begin{aligned}\bar{\pi} &= \hat{z}\pi_z \\ &= \hat{z}\frac{I_0 L_0}{j\omega\epsilon}G(\bar{r}) \\ &= \hat{z}\frac{I_0 L_0}{j\omega\epsilon}\frac{e^{-jkr}}{4\pi r}.\end{aligned} \tag{5-19}$$

The electric and magnetic fields are given by (5-6). We may use the spherical coordinate system, writing

$$\bar{\pi} = \pi_r\hat{r} + \pi_\theta\hat{\theta} + \pi_\phi\hat{\phi}, \tag{5-20}$$

where

$$\pi_r = \pi_z\hat{z}\cdot\hat{r} = \pi_z\cos\theta,$$

$$\pi_\theta = \pi_z\hat{z}\cdot\hat{\theta} = \pi_z(-\sin\theta), \qquad \pi_\phi = 0.$$

and obtain

$$\begin{aligned}E_r &= \frac{\partial}{\partial r}\left[\frac{1}{r^2}\frac{\partial}{\partial r}(r^2\pi_r) + \frac{1}{r\sin\theta}\frac{\partial}{\partial\theta}(\sin\theta\pi_\theta)\right] + k^2\pi_r \\ &= \frac{I_0 L_0}{j\omega\epsilon}\frac{e^{-jkr}}{4\pi}\left(\frac{j2k}{r^2} + \frac{2}{r^3}\right)\cos\theta, \\ E_\theta &= \frac{I_0 L_0}{j\omega\epsilon}\frac{e^{-jkr}}{4\pi}\left(-\frac{k^2}{r} + \frac{jk}{r^2} + \frac{1}{r^3}\right)\sin\theta, \\ E_\phi &= 0, \\ H_r &= H_\theta = 0, \\ H_\phi &= I_0 L_0\frac{e^{-jkr}}{4\pi}\left(\frac{jk}{r} + \frac{1}{r^2}\right)\sin\theta.\end{aligned} \tag{5-21}$$

We note that the radial component E_r is proportional to r^{-2} and r^{-3}, while E_θ and H_ϕ contain the term with r^{-1}. Thus at a large distance from the dipole, the radial component E_r vanishes much faster than E_θ and H_ϕ. The range $|k_0 r| >> 1$ is called the *far zone* and the field in this range is called the *far field*. Similarly, the field in the *near zone* $|k_0 r| << 1$ is called the *near field*.

In the far zone, the field components are given by

$$E_\theta = j(I_0 L_0)\omega\mu \frac{e^{-jkr}}{4\pi r} \sin\theta,$$

$$H_\phi = j(I_0 L_0)k \frac{e^{-jkr}}{4\pi r} \sin\theta, \tag{5-22}$$

and therefore, the radiation pattern is proportional to $\sin\theta$ (see Fig. 5-3). The electric field E_θ and the magnetic field H_ϕ are perpendicular to each other, and both are perpendicular to the direction of wave propagation $\hat{r}$. The ratio of E_θ to H_ϕ is the characteristic impedance

$$\frac{E_\theta}{H_\phi} = \eta = \left(\frac{\mu}{\epsilon}\right)^{1/2}. \tag{5-23}$$

Note that these are also the characteristics of a plane wave, and thus the spherical wave radiating from a dipole behaves like a plane wave at a large distance except that the amplitude decreases as $1/r$.

Let us examine the total real power radiated from the dipole. If the medium is lossless, μ and ϵ are real and all the power radiated from the dipole must be equal to the power flowing out from any surface enclosing the dipole. In particular, we may take the surface to be that of a sphere with a large radius so that (5-22) holds. The total real power P_t is given by integrating the Poynting vector over this sphere.

$$P_t = \int_0^{2\pi} d\phi \int_0^{\pi} \sin\theta \, d\theta \, r^2 \bar{S} \cdot \hat{r}, \tag{5-24}$$

where

$$\bar{S} = \tfrac{1}{2}\bar{E} \times \bar{H}^*.$$

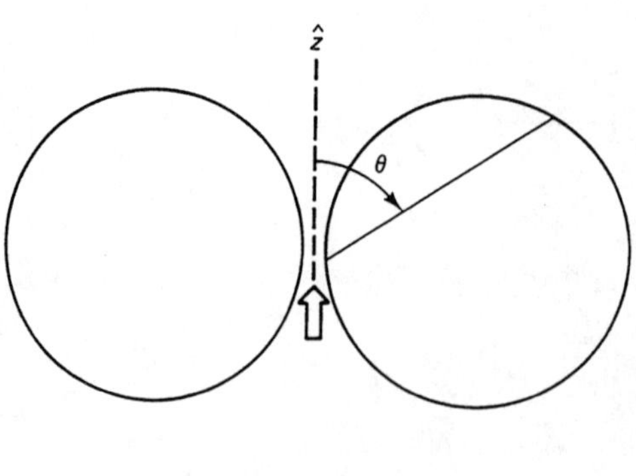

Figure 5-3 Dipole radiation pattern.

Substituting (5-22) into (5-24), we get

$$
\begin{aligned}
P_t &= \int_0^{2\pi} d\phi \int_0^{\pi} \sin\theta\, d\theta\, (I_0 L_0)^2 \frac{k^2\eta}{2(4\pi)^2} \sin^2\theta \\
&= (I_0 L_0)^2 \frac{k^2\eta}{12\pi} = \left(I_0 \frac{L_0}{\lambda}\right)^2 \frac{\pi}{3}\eta.
\end{aligned}
\tag{5-25}
$$

The effectiveness of an antenna may be expressed by the amount of real power the current I_0 can radiate. This is expressed by the radiation resistance R_{rad} defined by

$$P_t = \tfrac{1}{2} I_0^2 R_{\text{rad}}. \tag{5-26}$$

The radiation resistance of a short dipole of length L (which is equal to $2L_0$) is then given by

$$
\begin{aligned}
R_{\text{rad}} &= \eta \frac{2\pi}{3}\left(\frac{L_0}{\lambda}\right)^2 \\
&= \eta \frac{\pi}{6}\left(\frac{L}{\lambda}\right)^2 = 20\pi^2 \left(\frac{L}{\lambda}\right)^2.
\end{aligned}
\tag{5-27}
$$

The expression (5-27) is applicable only to a short dipole ($L \ll \lambda$), but it is a good approximation for dipoles of lengths up to a quarter-wavelength.

The radiation resistance (5-27) is the equivalent resistance of the input impedance of the dipole antenna representing the radiation and is independent of the wire radius. In addition to the radiation resistance, there is the resistance R_0 representing the ohmic loss of the wire and the reactance X (Fig. 5-4). The reactance X represents the stored energy in the near field of the antenna and depends on the geometry and the radius of the wire and requires a detailed study of the field near the dipole.

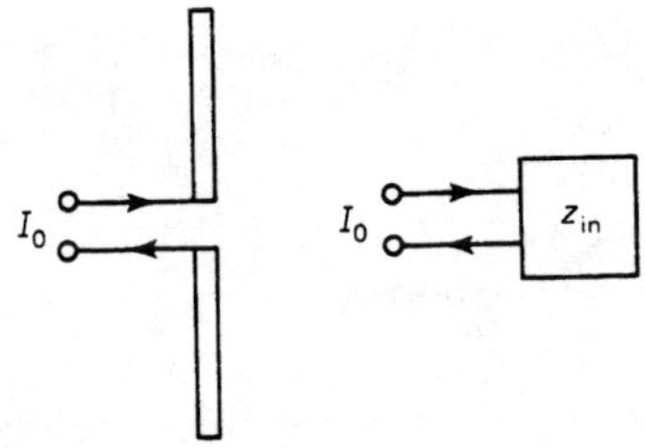

Figure 5-4 Input impedance of a dipole antenna.

5-3 ELECTROMAGNETIC FIELDS EXCITED BY A MAGNETIC DIPOLE IN A HOMOGENEOUS MEDIUM

As we noted in Section 5-1, a small loop antenna can be represented by a magnetic current dipole

$$\bar{J}_m = \hat{z} j\omega\mu_0 I_0 A\, \delta(\bar{r} - \bar{r}'). \tag{5-28}$$

We can derive electromagnetic fields due to the magnetic dipole by using the duality principle. If $\bar{E}$, $\bar{H}$, $\bar{J}$, $\bar{\pi}$, μ, and ϵ for the case of an electric dipole are replaced by $-\bar{H}$, $\bar{E}$, $-\bar{J}_m$, $-\bar{\pi}_m$, ϵ, and μ, then Maxwell's equations and the formulations in terms of the Hertz vectors are unchanged, and we have

$$\bar{E} = -j\omega\mu\nabla \times \bar{\pi}_m,$$
$$\bar{H} = \nabla(\nabla \cdot \bar{\pi}_m) + k^2\bar{\pi}_m,$$

and

$$(\nabla^2 + k^2)\bar{\pi}_m = -\frac{\bar{J}_m}{j\omega\mu}. \tag{5-29}$$

Therefore, we obtain

$$\begin{aligned}
H_r &= I_0 A \frac{e^{-jkr}}{4\pi}\left(\frac{j2k}{r^2} + \frac{2}{r^3}\right) \cos\theta, \\
H_\theta &= I_0 A \frac{e^{-jkr}}{4\pi}\left(-\frac{k^2}{r} + \frac{jk}{r^2} + \frac{1}{r^3}\right) \sin\theta, \\
H_\phi &= 0, \\
E_r &= E_\theta = 0, \\
E_\phi &= (-j\omega\mu I_0 A)\frac{e^{-jkr}}{4\pi}\left(\frac{jk}{r} + \frac{1}{r^2}\right) \sin\theta,
\end{aligned} \tag{5-30}$$

and the radiation pattern is given by

$$\begin{aligned}
H_\theta &= -(I_0 A)k^2 \frac{e^{-jkr}}{4\pi r} \sin\theta, \\
E_\phi &= (I_0 A)k^2 \eta \frac{e^{-jkr}}{4\pi r} \sin\theta.
\end{aligned} \tag{5-31}$$

The radiation resistance R_{rad} is given by

$$R_{\text{rad}} = \eta\left(\frac{2\pi}{\lambda}\right)^4 \frac{A^2}{6\pi} = 20(2\pi)^4\frac{A^2}{\lambda^4}. \tag{5-32}$$

When an electric dipole is located near a conducting planar surface, the total field can conveniently be expressed as a sum of the field due to the dipole and the field due to an image of the dipole. For a vertical electric dipole located above a conducting plane as shown in Fig. 5-5, the field due to the image must be such that the total electric field tangential to the conducting surface vanishes. From (5-21), since $r_1 = r_2$ and $\theta_1 = \pi - \theta_2$, we can easily show that the total electric field due to the vertical electric dipole and its image dipole oriented in the same direction has no tangential component on the surface of the conductor. Similarly, the image of a horizontal electric dipole is directed opposite to the dipole (Fig. 5-5). We can

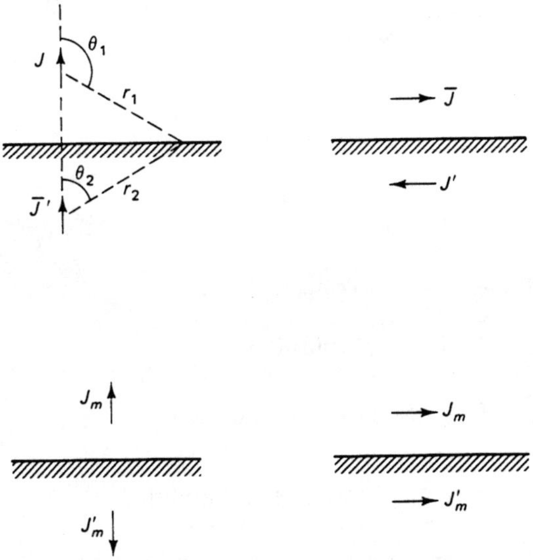

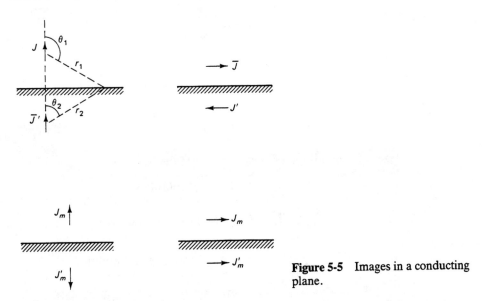

Figure 5-5 Images in a conducting plane.

also follow the similar reasoning to show that the image of a vertical-magnetic dipole is directed opposite while the image of a horizontal magnetic dipole is directed parallel to the dipole.

5-4 SCALAR GREEN'S FUNCTION FOR CLOSED REGIONS AND EXPANSION OF GREEN'S FUNCTION IN A SERIES OF EIGENFUNCTIONS

In Sections 5-1 to 5-3, we discussed the excitation by an electric or magnetic dipole in a homogeneous medium and expressed electromagnetic fields in terms of Green's function in a homogeneous medium. In general, however, Green's function must satisfy appropriate boundary condition.

If the region under consideration extends to infinity, it is called the *open* region. Green's function in the open region can be represented by Fourier transform, and this will be discussed later. In this section we consider the finite *closed* region V surrounded by a surface S (Fig. 5-6). In this case it is possible to construct Green's function in terms of a series of eigenfunctions. In addition to the foregoing two representations, Fourier transform and eigenfunctions, we can express a one-dimensional Green's function in terms of solutions of homogeneous differential equations. We will discuss these three representations of Green's functions.

Let us consider a volume V surrounded by a surface S (Fig. 5-6). Green's function is a solution to an inhomogeneous wave equation with the delta function as an exciting source and satisfies an appropriate boundary condition on S.

$$(\nabla^2 + k^2)G(\bar{r}, \bar{r}') = -\delta(\bar{r} - \bar{r}') \qquad \text{in } V, \tag{5-33}$$

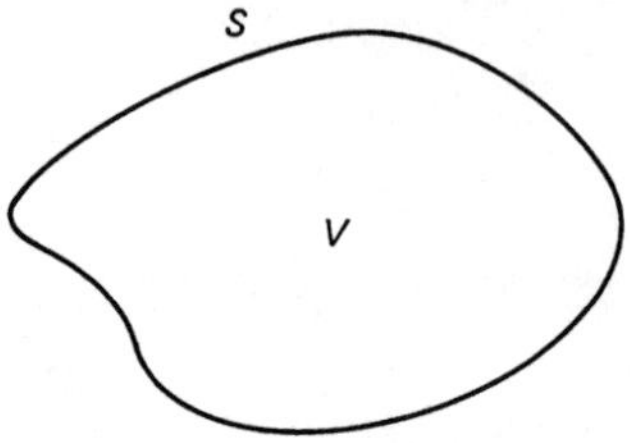

Figure 5-6 Closed region.

and the boundary condition on S can be one of the following:

$$G = 0 \qquad \text{Dirichlet's condition,}$$
$$\frac{\partial G}{\partial n} = 0 \qquad \text{Neumann's condition,} \tag{5-34}$$
$$G + h\frac{\partial G}{\partial n} = 0 \qquad \text{general homogeneous condition.}$$

Since any function can be expressed by a series of eigenfunctions, we seek to represent G in the following form:

$$G = \sum_n A_n \phi_n(\bar{r}), \tag{5-35}$$

where A_n is an unknown coefficient and ϕ_n is an eigenfunction. Here the summation over n means triple summation for the three-dimensional case, double summation for the two-dimensional case, and single summation for the one-dimensional case.

We also expand the delta function in a series of eigenfunctions

$$\delta(\bar{r} - \bar{r}') = \sum_n B_n \phi_n(\bar{r}), \tag{5-36}$$

where B_n is given by multiplying both sides of (5-36) by $\phi_m(\bar{r})$ and integrating over the volume. Noting the orthogonality of eigenfunctions,

$$\int_V \phi_n(\bar{r})\phi_m(\bar{r})\, dV = 0 \qquad \text{for } n \neq m, \tag{5-37}$$

we get

$$\begin{aligned} B_n &= \frac{\int_V \delta(\bar{r} - \bar{r}')\phi_n(\bar{r})\, dV}{\int_V [\phi_n(\bar{r})]^2\, dV} \\ &= \frac{\phi_n(\bar{r}')}{\int_V [\phi_n(\bar{r})]^2\, dV}. \end{aligned} \tag{5-38}$$

Substituting (5-35) and (5-36) in the wave equation (5-33), we get

$$\sum_n (\nabla^2 + k^2) A_n \phi_n(\bar{r}) = -\sum_n B_n \phi_n(\bar{r}). \tag{5-39}$$

Since

$$(\nabla^2 + k_n^2)\phi_n(\bar{r}) = 0,$$

the left side of (5-39) becomes

$$\sum_n (k^2 - k_n^2) A_n \phi_n(\bar{r}).$$

Now both sides of (5-39) are expansions in terms of the same orthogonal functions ϕ_n, and therefore the coefficient of each term must be equal. Thus

$$(k^2 - k_n^2) A_n = -B_n,$$

from which the unknown coefficients A_n are obtained.

Green's function is then given by

$$G(\bar{r}, \bar{r}') = \sum_n \frac{\phi_n(\bar{r})\phi_n(\bar{r}')}{(k_n^2 - k^2)} \frac{1}{N_n^2}, \tag{5-40}$$

where

$$N_n^2 = \int_V [\phi_n(\bar{r})]^2 \, dV$$

is a normalization factor. Equation (5-40) holds as long as $k_n \neq k$. If for a particular $n = n_0$, $k_{n_0} = k$, Green's function diverges unless the source is located at a zero of the eigenfunction $\phi_{n_0}(\bar{r}') = 0$. In this case the term for this particular n_0 should be excluded.

Let us apply the general expression (5-40) to the one-dimensional problem. Let us assume that Green's function satisfies Dirichlet's condition at $z = 0$ and $z = a$, and $0 < z' < a$.

$$\begin{aligned} \left(\frac{d^2}{dz^2} + k^2\right) G(z, z') &= -\delta(z - z'), \\ G(0, z') &= 0, \\ G(a, z') &= 0. \end{aligned} \tag{5-41}$$

The normalized eigenfunctions are

$$\phi_n(z) = \sqrt{\frac{2}{a}} \sin \frac{n\pi z}{a}, \tag{5-42}$$

and the eigenvalues are

$$k_n = \frac{n\pi}{a}, \qquad n = 1, 2, \ldots, \infty.$$

Therefore, Green's function is given by

$$G(z,z') = \sum_{n=1}^{\infty} \frac{2}{a} \frac{\sin(n\pi z/a)\sin(n\pi z'/a)}{(n\pi/a)^2 - k^2}. \tag{5-43}$$

Let us next consider a two-dimensional Green's function that satisfies Dirichlet's condition at $\rho = a$ in the cylindrical coordinte system, and $0 \le \rho' < a$.

$$\left[\frac{1}{\rho}\frac{\partial}{\partial\rho}\left(\rho\frac{\partial}{\partial\rho}\right) + \frac{1}{\rho^2}\frac{\partial^2}{\partial\phi^2} + k^2\right] G = -\frac{\delta(\rho-\rho')\delta(\phi-\phi')}{\rho}, \tag{5-44}$$

satisfying the condition

$$G = 0 \quad \text{at} \quad \rho = a.$$

The denominator ρ on the right-hand side is due to the definition of the delta functions (see the Appendix to Chapter 5, Section A). We choose the right side in such a manner that the integral with respect to the area is unity:

$$\int \frac{\delta(\rho-\rho')\delta(\phi-\phi')}{\rho}\, dS = 1, \tag{5-45}$$

where

$$dS = \rho\, d\rho\, d\phi.$$

Let us first note that in the ϕ direction, the solution must be periodic, and therefore we write G in a series of normalized eigenfunctions

$$\phi_n(\phi) = \frac{1}{\sqrt{2\pi}} e^{-jn\phi}. \tag{5-46}$$

Then we write

$$G = \sum_{n=-\infty}^{\infty} \phi_n(\phi)\phi_n^*(\phi')G_n(\rho,\rho'). \tag{5-47}$$

Noting that

$$\delta(\phi-\phi') = \sum_{n=-\infty}^{\infty} \phi_n(\phi)\phi_n^*(\phi'), \tag{5-48}$$

we obtain

$$\left[\frac{1}{\rho}\frac{\partial}{\partial\rho}\left(\rho\frac{\partial}{\partial\rho}\right) - \frac{n^2}{\rho^2} + k^2\right] G_n = -\frac{\delta(\rho-\rho')}{\rho}. \tag{5-49}$$

Now we expand G_n in a series of eigenfunctions.

$$G_n = \sum_{m=1}^{\infty} A_m J_n(k_{nm}\rho), \tag{5-50}$$

where $J_n(k_{nm}a) = 0$ and k_{nm} is the eigenvalue. Also the delta function is expressed as

$$\frac{\delta(\rho-\rho')}{\rho} = \sum_{m=1}^{\infty} \frac{J_n(k_{nm}\rho)J_n(k_{nm}\rho')}{N_{nm}^2}, \tag{5-51}$$

where the normalization factor N_{nm}^2 is given by

$$N_{nm}^2 = \int_0^a J_n(k_{nm}\rho)^2 \rho\, d\rho = \frac{a^2}{2}[J_n'(k_{nm}a)]^2. \tag{5-52}$$

Substituting (5-50) into (5-49) and using (5-51), we get

$$G_n = \sum_{m=1}^{\infty} \frac{J_n(k_{nm}\rho)J_n(k_{nm}\rho')}{(k_{nm}^2 - k^2)N_{nm}^2}. \tag{5-53}$$

Finally, we substitute (5-53) into (5-47) and obtain the two-dimensional Green's function

$$G(\rho,\phi;\rho',\phi') = \sum_{n=-\infty}^{\infty}\sum_{m=1}^{\infty} \frac{J_n(k_{nm}\rho)J_n(k_{nm}\rho')e^{-jn(\phi-\phi')}}{2\pi(k_{nm}^2 - k^2)N_{nm}^2}. \tag{5-54}$$

5-5 GREEN'S FUNCTION IN TERMS OF SOLUTIONS OF THE HOMOGENEOUS EQUATION

The representation of Green's function in terms of a series of eigenfunctions is in fact an expression in the form of resonant cavity modes, but this is not the only representation. Green's function can also be expressed in terms of the solutions of a homogeneous equation, as discussed in this section.

Consider a general second-order differential equation

$$\left[\frac{1}{f(z)}\frac{d}{dz}\left(f(z)\frac{d}{dz}\right) + q(z)\right]G(z,z') = -\frac{\delta(z-z')}{f(z)}. \tag{5-55}$$

The function $f(z)$ on the right side is introduced so that Green's function becomes symmetric with respect to z and z', as will be seen shortly. This is consistent with Green's function for wave equations in cylindrical and spherical coordinate systems.

Let us write (5-55) as

$$\left[\frac{d^2}{dz^2} + p(z)\frac{d}{dz} + q(z)\right]G(z,z') = -\frac{\delta(z-z')}{f(z)}, \tag{5-56}$$

where

$$p(z) = \frac{1}{f(z)}\frac{df(z)}{dz}, \tag{5-57}$$

or

$$f(z) = \exp[\int p(z)\, dz]. \tag{5-58}$$

For the region $z > z'$, $G(z,z')$ is a solution of the homogeneous differential equation.

$$\left[\frac{1}{f(z)}\frac{d}{dz}\left(f(z)\frac{d}{dz}\right) + q(z)\right]G(z,z') = 0. \tag{5-59}$$

We write this solution as

$$G(z,z') = A_0 y_1(z). \tag{5-60}$$

For the region $z < z'$, we write

$$G(z,z') = B_0 y_2(z). \tag{5-61}$$

Let us now use the following properties of Green's function.

1. The first derivative of Green's function has a jump at $z = z'$. From the differential equation, note that the second derivative of G should have the behavior of the delta function. Thus the first derivative has a discontinuity. The function itself, however, is continuous being an integral of the first derivative. Thus we have the second property.
2. Green's function is a continuous function of z when z' is fixed, including $z = z'$.
3. We also note that, in general, Green's function is symmetric with respect to $\bar{r}$ and $\bar{r}'$ as a direct consequence of the reciprocity theorem.

$$G(\bar{r},\bar{r}') = G(\bar{r}',\bar{r}). \tag{5-62}$$

We make use of the foregoing properties to obtain the constants A_0 and B_0. First we note that G is continuous at $z = z'$. Then we get

$$A_0 y_1(z') - B_0 y_2(z') = 0. \tag{5-63}$$

Next we note that the first derivative of G is discontinuous. To make use of this, let us integrate the original differential equation with a weighting function $f(z)$ from $z' - \epsilon$ to $z' + \epsilon$ and let ϵ approach zero. Thus we get

$$\lim_{\epsilon \to 0} \int_{z'-\epsilon}^{z'+\epsilon} \left[\frac{d}{dz}\left(f \frac{d}{dz}\right) + fq \right] G\, dz = -\int_{z'-\epsilon}^{z'+\epsilon} \frac{\delta(z-z')}{f(z)} f(z)\, dz. \tag{5-64}$$

The right side becomes -1, and therefore,

$$\lim_{\epsilon \to 0} \left[f(z) \frac{d}{dz} G \right]_{z'-\epsilon}^{z'+\epsilon} = -1.$$

Here we must use $A_0 y_1'$ for $z' + \epsilon$ and $B_0 y_2'$ for $z' - \epsilon$. We write this using y_1 and y_2:

$$A_0 y_1'(z') - B_0 y_2'(z') = -\frac{1}{f(z')}. \tag{5-65}$$

where $y_1' = dy_1/dz$ and $y_2' = dy_2/dz$.

Solving (5-63) and (5-65) for A_0 and B_0, we get

$$A_0 = \frac{y_2(z')}{f(z')\Delta(z')}, \tag{5-66}$$

$$B_0 = \frac{y_1(z')}{f(z')\Delta(z')}, \tag{5-67}$$

$$\Delta(z') = \begin{vmatrix} y_1(z') & y_2(z') \\ y_1'(z') & y_2'(z') \end{vmatrix}. \tag{5-68}$$

Thus Green's function is given by

$$G(z,z') = \begin{cases} \dfrac{y_1(z)y_2(z')}{f(z')\Delta(z')}, & z > z', \quad (5\text{-}69) \\ \dfrac{y_1(z')y_2(z)}{f(z')\Delta(z')}, & z < z', \quad (5\text{-}70) \end{cases}$$

Note that both the numerator and the denominator contain the product of y_1 and y_2 or its derivative such as y_1y_2, y_1y_2', and so on. Thus any constant in front of y_1 and y_2 cancels out, and therefore, only the form of the function for y_1 and y_2 is necessary. Its magnitude does not affect the final form. $\Delta(z')$ is called the *Wronskian of* y_1 *and* y_2, and since it has some useful characteristics, some of the details are given below. The denominator $f(z')\Delta(z')$ of (5-69) and (5-70) appears to be a function of z', and if so, $G(z,z')$ is not symmetric. It will be shown shortly, however, that $f(z')\Delta(z')$ is constant and is independent of z'. The final form of Green's function $G(z,z')$ is therefore given by

$$G(z,z') = \begin{cases} \dfrac{y_1(z)y_2(z')}{D}, & z > z', \quad (5\text{-}71) \\ \dfrac{y_1(z')y_2(z)}{D}, & z < z', \quad (5\text{-}72) \end{cases}$$

where $D = f(z')\Delta(z') = \text{constant}$.

Equations (5-71) and (5-72) are often combined in the following convenient form:

$$G(z,z') = \frac{y_1(z_>)y_2(z_<)}{D}, \tag{5-73}$$

where $z_>$ and $z_<$ denote the greater or lesser of z and z'.

Let us consider the Wronskian $\Delta(z)$, which we write in the following form:

$$\Delta(y_1, y_2) = y_1y_2' - y_1'y_2, \tag{5-74}$$

where $y_1 = y_1(z)$ and $y_2 = y_2(z)$, and $y_1' = dy_1/dz$ and $y_2' = dy_2/dz$. First, we note that if y_1 and y_2 are two independent solutions of the homogeneous differential equation, then $\Delta(y_1, y_2) \neq 0$, but if y_1 and y_2 are dependent, then $\Delta(y_1, y_2) = 0$, as can be verified by (5-74).

Let us next find the form of the Wronskian. To do this, we first show that Δ satisfies a simple first-order differential equation. Taking the derivative of Δ, we get

$$\begin{aligned} \frac{d\Delta}{dz} &= \frac{d}{dz}(y_1y_2' - y_1'y_2) \\ &= y_1y_2'' - y_2y_1''. \end{aligned} \tag{5-75}$$

But y_1 and y_2 satisfy the differential equation

$$y_1'' + py_1' + qy_1 = 0,$$
$$y_2'' + py_2' + qy_2 = 0. \tag{5-76}$$

Thus

$$\frac{d\Delta}{dz} = y_1(-py_2' - qy_2) - y_2(-py_1' - qy_1),$$

which reduces to

$$\frac{d\Delta}{dz} = -p\Delta. \tag{5-77}$$

This is the differential equation that Δ satisfies. The solution can be easily obtained:

$$\Delta = (\text{constant})e^{-\int p\,dz},$$

or

$$f(z)\Delta = \text{constant}. \tag{5-78}$$

This is a very important and useful relationship. This shows that whatever y_1 and y_2 are, $f(z)\Delta(y_1, y_2)$ is always constant and independent of z. This property was used to obtain (5-73). Since this relationship (5-78) holds for any z, the constant $D = f(z)\Delta(y_1, y_2)$ can be determined by choosing any convenient z.

As an example, let us consider the one-dimensional problem we discussed in Section 5-4.

$$\left(\frac{d^2}{dz^2} + k^2\right)G(z, z') = -\delta(z - z'),$$
$$G(0, z') = G(a, z') = 0. \tag{5-79}$$

We let

$$y_1 = \sin k(a - z),$$
$$y_2 = \sin kz,$$

where the constants in front of these functions are immaterial.

Now we know that the Wronskian must be constant. To find this constant, we see that

$$\Delta(y_1, y_2) = k\ \sin k(a - z)\ \cos kz + k\ \cos k(a - z)\ \sin kz.$$

If we expand sine and cosine terms, all terms containing z cancel out, and the result is a constant. But it is not necessary to do this. Since we know that Δ is constant, we can simply choose any appropriate z. Choosing $z = 0$, we immediately obtain

$$\Delta(y_1, y_2) = k\ \sin ka,$$

and we can write down the results immediately.

$$G = \begin{cases} \dfrac{\sin kz' \sin k(a-z)}{k \sin ka}, & z > z', \\ \dfrac{\sin kz \sin k(a-z')}{k \sin ka}, & z < z'. \end{cases} \tag{5-80}$$

Equation (5-80) is identical to (5-43), but these are two different representations of the same Green's function.

So far, we have two representations of Green's function. The representation in terms of an infinite series of eigenfunctions is valid for the closed region and applicable to one-, two-, and three-dimensional cases. On the other hand, the representation in terms of two solutions of homogeneous differential equations is valid for both closed and open regions, but it is applicable only to the one-dimensional case.

In addition to the two representations above, there is a method using integral transforms that will be discussed in the next section. For an actual problem, we combine these three representations in an appropriate manner.

5-6 FOURIER TRANSFORM METHOD

In addition to the above two methods, Green's function may be expressed in terms of a Fourier transform. In this section, we examine this procedure.

As an example, let us consider the following problem:

$$\left(\frac{\partial^2}{\partial x^2} + \frac{\partial^2}{\partial z^2} + k^2\right) G = -\delta(x - x')\delta(z - z'), \tag{5-81}$$

and $G = 0$ at $x = 0$ and a. Here we start with the Fourier transform of both sides of the differential equation. We let

$$g(x, h) = \int_{-\infty}^{\infty} G(x, z) e^{jhz}\, dz, \tag{5-82}$$

and the inverse Fourier transform yields

$$G(x, z) = \frac{1}{2\pi} \int_{-\infty}^{\infty} g(x, h) e^{-jhz}\, dh. \tag{5-83}$$

Using

$$\int_{-\infty}^{\infty} \left(\frac{\partial^2}{\partial z^2} G\right) e^{jhz}\, dz = -h^2 g(x, h),$$

$$\int_{-\infty}^{\infty} \delta(z - z') e^{jhz}\, dz = e^{jhz'}, \tag{5-84}$$

$$\delta(z - z') = \frac{1}{2\pi} \int_{-\infty}^{\infty} e^{-jh(z - z')}\, dh,$$

we write (5-81) in the following form:

$$\left(\frac{d^2}{dx^2}+k^2-h^2\right)g(x,h)=-\delta(x-x')e^{jhz'}. \tag{5-85}$$

This is solved by noting that $G=0$ at $x=0$ and $x=a$, we get

$$g(x,h)=\sum_{n=1}^{\infty}\frac{\phi_n(x)\phi_n(x')}{k_n^2-k^2+h^2}e^{jhz'}, \tag{5-86}$$

where

$$\phi_n(x)=\sqrt{\frac{2}{a}}\sin\frac{n\pi x}{a}, \qquad k_n=\frac{n\pi}{a}.$$

Thus the complete solution is

$$G(x,z;x',z')=\frac{1}{2\pi}\int_{-\infty}^{\infty}\sum_{n=1}^{\infty}\frac{\phi_n(x)\phi_n(x')}{k_n^2-k^2+h^2}e^{-jh(z-z')}\,dh. \tag{5-87}$$

Alternatively, we can solve (5-85) using the solutions of the homogeneous equation. We get

$$g(x,h)=\frac{y_1(x_>)y_2(x_<)}{D}e^{jhz'},$$

where $y_1(x)=\sin p(a-x)$, $y_2(x)=\sin px$, $p=(k^2-h^2)^{1/2}$, and $D=p\sin pa$.
We, therefore, get

$$G(x,z;x',z')=\frac{1}{2\pi}\int_{-\infty}^{\infty}\frac{\sin px'\,\sin p(a-x)}{p\sin pa}e^{-jh(z-z')}\,dh, \qquad x>x', \tag{5-88}$$

where $p=(k^2-h^2)^{1/2}$ and for $x<x'$, x and x' are interchanged. Equation (5-88) is an alternative expression for Green's function (5-87).

We can still obtain another form of Green's function by first expanding G in a series of eigenfunctions $\phi_n(x)$.

$$G=\sum_{n=1}^{\infty}\phi_n(x)\phi_n(x')G_n(z,z'). \tag{5-89}$$

We then get

$$\left(\frac{d^2}{dz^2}+k^2-k_n^2\right)G_n(z,z')=-\delta(z-z').$$

The solution for G_n is given by

$$G_n(z,z')=\frac{y_1(z_>)y_2(z_<)}{D}, \tag{5-90}$$

where

$$y_1(z)=e^{-jq_nz},$$

$$y_2(z)=e^{+jq_nz},$$

$$D=2jq_n, \qquad q_n=(k^2-k_n^2)^{1/2}.$$

Substituting (5-90) into (5-89), we get

$$G(x,z;x',z') = \sum_{n=1}^{\infty} \frac{\phi_n(x)\phi_n(x')}{2jq_n} e^{-jq_n|z-z'|}. \tag{5-91}$$

Equations (5-87), (5-88), and (5-91) are the same Green's function as expressed in three different representations.

5-7 EXCITATION OF A RECTANGULAR WAVEGUIDE

As an example, let us consider a rectangular waveguide excited by a current element of small length dl carrying I_0 as shown in Fig. 5-7. We choose Π_y that satisfies the wave equation

$$(\nabla^2 + k^2)\Pi_y = -\frac{J_y}{j\omega\epsilon}, \tag{5-92}$$

where

$$J_y = I\,dl\,\delta(x-x')\delta(y-y')\delta(z-z').$$

The boundary conditions at $x = 0$ and $x = a$ are

$$\Pi_y = 0, \tag{5-93}$$

and the boundary conditions at $y = 0$ and $y = b$ are

$$\frac{\partial \Pi_y}{\partial y} = 0. \tag{5-94}$$

Let us rewrite the problem using Green's function G, which satisfies

$$\left(\frac{\partial^2}{\partial x^2} + \frac{\partial^2}{\partial y^2} + \frac{\partial^2}{\partial z^2} + k^2\right) G(x,y,z;x',y',z') = -\delta(x-x')\delta(y-y')\delta(z-z') \tag{5-95}$$

$$G = 0 \quad \text{at } x = 0 \text{ and } a,$$

$$\frac{\partial G}{\partial y} = 0 \quad \text{at } y = 0 \text{ and } b.$$

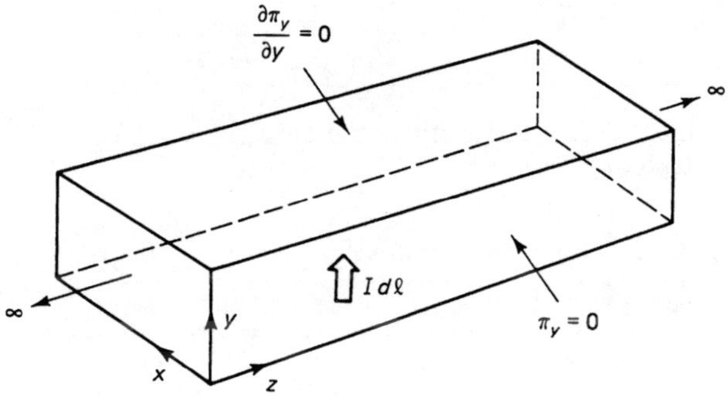

Figure 5-7 Excitation of a rectangular waveguide.

The Hertz vector Π_y is then given by

$$\Pi_y = \frac{I\,dl}{j\omega\epsilon} G. \tag{5-96}$$

To find Green's function from (5-95), we first note that the region is closed in the x and y directions. Thus G can be represented by a series of eigenfunctions.

$$\left(\frac{\partial^2}{\partial x^2} + \frac{\partial^2}{\partial y^2} + k_{mn}^2\right)\phi_{mn}(x,y) = 0, \tag{5-97}$$

$$\phi_{mn} = 0 \qquad \text{at } x = 0 \quad \text{and} \quad a,$$

$$\frac{\partial \phi_{mn}}{\partial y} = 0 \qquad \text{at } y = 0 \quad \text{and} \quad b.$$

We normalize ϕ_{mn} by

$$\int_0^a dx \int_0^b dy\,[\phi_{mn}(x,y)]^2 = 1. \tag{5-98}$$

The eigenfunctions $\phi_{mn}(x,y)$ are, therefore, given by

$$\begin{aligned} \phi_{mn}(x,y) &= \phi_m(x)\phi_n(y), \\ \phi_m(x) &= \left(\frac{2}{a}\right)^{1/2} \sin\frac{m\pi x}{a}, \qquad m = 1, 2, \ldots, \\ \phi_n(y) &= \begin{cases} \left(\dfrac{1}{b}\right)^{1/2}, & n = 0, \\ \left(\dfrac{2}{b}\right)^{1/2} \cos\dfrac{n\pi y}{b}, & n = 1, 2, \ldots. \end{cases} \end{aligned} \tag{5-99}$$

Let us write Green's function in a series of eigenfunctions:

$$G = \sum_{m=1}^{\infty} \sum_{n=0}^{\infty} \phi_{mn}(x,y)\phi_{mn}(x',y')G_{mn}(z,z'), \tag{5-100}$$

where $G_{mn}(z,z')$ is still an unknown function. We also write the right side of (5-95) using the following:

$$\delta(x - x')\delta(y - y') = \sum_{m=1}^{\infty} \sum_{n=1}^{\infty} \phi_{mn}(x,y)\phi_{mn}(x',y'). \tag{5-101}$$

Substituting (5-100) and (5-101) into (5-95) and noting that this is an orthogonal expansion in a series of $\phi_{mn}(x,y)$, we get

$$\left(\frac{d^2}{dz^2} + k^2 - k_{mn}^2\right)G_{mn}(z,z') = -\delta(z - z'). \tag{5-102}$$

Now, we note that in the z direction, the region is open, so we use the representation of the two solutions of the homogeneous equation.

The solution for $z > z'$ is

$$y_1(z) = e^{-j\sqrt{k^2 - k_{mn}^2}(z - z')},$$

and for $z < z'$,

$$y_2(z) = e^{j\sqrt{k^2 - k_{mn}^2}(z - z')}, \tag{5-103}$$

and the Wronskian is simply constant:

$$\Delta(y_1, y_2) = j \cdot 2\sqrt{k^2 - k_{mn}^2}. \tag{5-104}$$

Thus we finally get

$$G = \sum_{m=1}^{\infty} \sum_{n=0}^{\infty} \frac{\phi_{mn}(x, y)\phi_{mn}(x', y')e^{-j\sqrt{k^2 - k_{mn}^2}|z - z'|}}{j \cdot 2\sqrt{k^2 - k_{mn}^2}}. \tag{5-105}$$

5-8 EXCITATION OF A CONDUCTING CYLINDER

As another example, let us consider the problem of exciting a conducting cylinder with an electric current element of small length dl carrying the current I_0 pointed in the z direction (Fig. 5-8). We have a scalar wave equation

$$(\nabla^2 + k^2)\Pi_z = -\frac{I_0\, dl}{j\omega\epsilon} \frac{\delta(\phi - \phi')\delta(\rho - \rho')\delta(z - z')}{\rho}. \tag{5-106}$$

Considering that $\Pi_z = 0$ at $\rho = a$, the problem is reduced to solving

$$(\nabla^2 + k^2)G = -\frac{\delta(\phi - \phi')\delta(\rho - \rho')\delta(z - z')}{\rho}. \tag{5-107}$$

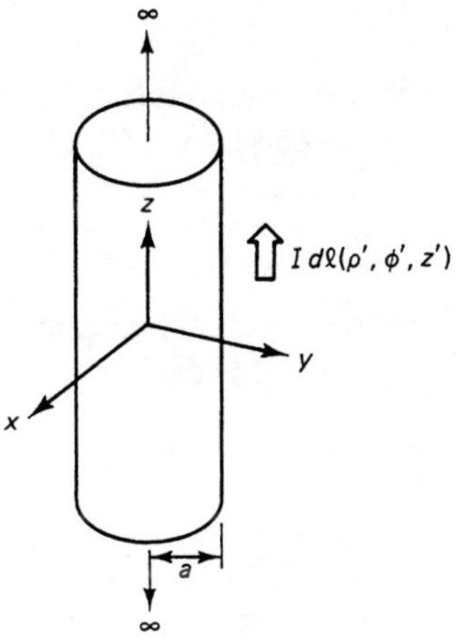

Figure 5-8 Excitation of a conducting cylinder.

$G = 0$ at $\rho = a$. First, we take the Fourier transform in z.

$$G = \frac{1}{2\pi}\int g(\rho, \phi, h)e^{-jh(z-z')}\,dh. \tag{5-108}$$

We then get

$$\left[\frac{1}{\rho}\frac{\partial}{\partial\rho}\left(\rho\frac{\partial}{\partial\rho}\right) + \frac{1}{\rho^2}\frac{\partial^2}{\partial\phi^2} + k^2 - h^2\right]g = -\frac{\delta(\rho-\rho')\delta(\phi-\phi')}{\rho}. \tag{5-109}$$

Expanding g in a series of eigenfunctions $\phi_n(\phi)$ around the cylinder, we get

$$g = \sum_{n=-\infty}^{\infty}\frac{e^{-jn(\phi-\phi')}}{2\pi}g_n(\rho, \rho'),$$

$$\left[\frac{1}{\rho}\frac{d}{d\rho}\left(\rho\frac{d}{d\rho}\right) - \frac{n^2}{\rho^2} + k^2 - h^2\right]g_n(\rho, \rho') = -\frac{\delta(\rho-\rho')}{\rho}. \tag{5-110}$$

Now we express g_n in terms of solutions to the homogeneous equation, which is Bessel's differential equation in this case. We choose appropriate Bessel functions for the two different regions. In the range $\rho > \rho'$, the solution should be an outgoing wave to satisfy the radiation condition. Thus

$$y_1(\rho) = H_n^{(2)}(q\rho) \quad \text{for } \rho > \rho', \qquad q = (k^2 - h^2)^{1/2}. \tag{5-111}$$

In the range $a < \rho < \rho'$, the solution should be a linear combination of two independent Bessel functions because here the wave travels back and forth forming a standing wave. Two independent functions can be $J_n(q\rho)$ and $N_n(q\rho)$, for example. But here we choose $J_n(q\rho)$ and $H_n^{(2)}(q\rho)$. The reason for this choice is explained later. Thus

$$y_2(\rho) = AJ_n(q\rho) + BH_n^{(2)}(q\rho) \qquad \text{for } a < \rho < \rho'. \tag{5-112}$$

Using the boundary condition that $G = 0$ at $\rho = a$, the ratio of the constants A to B is determined, and we obtain

$$y_2(\rho) = J_n(q\rho)H_n^{(2)}(qa) - J_n(qa)H_n^{(2)}(q\rho). \tag{5-113}$$

To find the Wronskian of y_1 and y_2, we note that the second term of $y_2(\rho)$ has the same form as $y_1(\rho)$, and they are dependent. The Wronskian between them is, therefore, zero. Thus we get

$$\Delta(y_1 y_2) = H_n^{(2)}(qa)q\,\Delta(H_n^{(2)}(x), J_n(x)),$$

where q is due to the fact that the derivative is with respect to ρ for $\Delta(y_1, y_2)$, but it is with respect to x for $\Delta(H_n(x), J_n(x))$. Now consider

$$\Delta(H_n^{(2)}(x), J_n(x)).$$

$J_n(x)$ is written as

$$J_n(x) = \tfrac{1}{2}(H_n^{(1)}(x) + H_n^{(2)}(x)).$$

Noting that $\Delta(H_n^{(2)}(x), H_n^{(2)}(x)) = 0$, we get

$$\Delta(H_n^{(2)}(x), J_n(x)) = \tfrac{1}{2}\Delta(H_n^{(2)}(x), H_n^{(1)}(x)).$$

To evaluate the last Wronskian, we first note that $f\Delta$ should be constant. $f(x)$ in this cylindrical case is simply x, and therefore Wronskian Δ should be equal to (constant)/x. To determine this constant, we can choose any convenient x and evaluate Δ. We let $x \to \infty$ and noting that

$$\begin{aligned}
H_n^{(2)}(x) &\to \left(\frac{2}{\pi x}\right)^{1/2} \exp\left[-j\left(x - \frac{n\pi}{2} - \frac{\pi}{4}\right)\right],\\
H_n^{(1)}(x) &\to \left(\frac{2}{\pi x}\right)^{1/2} \exp\left[+j\left(x - \frac{n\pi}{2} - \frac{\pi}{4}\right)\right],\\
H_n^{(2)\prime}(x) &\to \left(\frac{2}{\pi x}\right)^{1/2} \exp\left[-j\left(x - \frac{n\pi}{2} + \frac{\pi}{4}\right)\right],\\
H_n^{(1)\prime}(x) &\to \left(\frac{2}{\pi x}\right)^{1/2} \exp\left[+j\left(x - \frac{n\pi}{2} + \frac{\pi}{4}\right)\right],
\end{aligned} \tag{5-114}$$

we get

$$\Delta(y_1, y_2) = j\frac{2}{\pi\rho'} H_n^{(2)}(qa). \tag{5-115}$$

Therefore,

$$g_n(\rho, \rho') = \frac{y_1(\rho_>)y_2(\rho_<)}{j(2/\pi)H_n^{(2)}(qa)}. \tag{5-116}$$

The complete solution is then

$$\begin{aligned}
G(\rho, \phi, z; \rho', \phi', z') &= \frac{1}{2\pi}\int g(\rho, \phi, h)e^{-jh(z-z')}\,dh,\\
g(\rho, \phi, h) &= -j\frac{1}{4}\sum_{n=-\infty}^{\infty} e^{-jn(\phi-\phi')}\\
&\quad \times \frac{[J_n(q\rho_<)H_n^{(2)}(qa) - J_n(qa)H_n^{(2)}(q\rho_<)]H_n^{(2)}(q\rho_>)}{H_n^{(2)}(qa)},
\end{aligned} \tag{5-117}$$

where $\rho_>$ and $\rho_<$ are the larger or smaller of ρ and ρ'.

Note that (5-117) consists of two terms.

$$g = g_i + g_s, \tag{5-118}$$

where

$$g_i = -j\frac{1}{4}\sum_{n=-\infty}^{\infty} e^{-jn(\phi-\phi')}J_n(q\rho_<)H_n^{(2)}(q\rho_>).$$

This term q_i does not contain a, the radius of the cylinder, and this represents the incident or primary wave. The second term contains the effects of the radius of the cylinder and is called the scattered wave or secondary wave.

We also note from (5-118) that if $\rho_<$ becomes zero, all the terms with $n \neq 0$ disappear because of J_n and thus

$$g_i = -j\frac{1}{4}H_0^{(2)}(q\rho), \tag{5-119}$$

represents the primary wave at a distance ρ from the source.

The final solution (5-117) is given as an inverse Fourier transform. Here the contour of integration in the complex h plane must be chosen such that the radiation condition is satisfied. Also, if the observation point is far from the dipole, the radiation field can be obtained simply by the saddle-point technique. These problems are discussed fully in Chapter 11.

5-9 EXCITATION OF A CONDUCTING SPHERE

Let us next consider the problem of exciting a conducting sphere with a radial dipole (Fig. 5-9). As shown in Section 4-15, the electric Hertz vector $\bar{\Pi} = \Pi\hat{r}$ is pointed in the radial direction and satisfies the equation

$$(\nabla^2 + k^2)\Pi = -\frac{J_r}{j\omega\epsilon r}. \tag{5-120}$$

The boundary condition at $r = a$ is (Section 4-16)

$$\frac{\partial}{\partial r}(r\Pi) = 0. \tag{5-121}$$

For a short electric dipole of length dl carrying the current I_0, we have

$$J_r = I_0\, dl\, \delta(\bar{r} - \bar{r}').$$

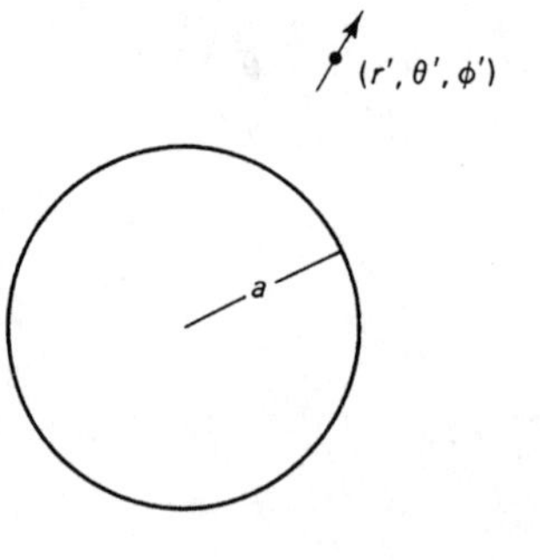

Figure 5-9 Excitation of a conducting sphere by a radial dipole.

Therefore, the problem is reduced to that of finding Green's function G satisfying

$$(\nabla^2 + k^2)G = -\delta(\bar{r} - \bar{r}'),$$
$$\frac{\partial}{\partial r}(rG) = 0 \qquad \text{at } r = a. \tag{5-122}$$

Let us first recognize that Green's function G consists of the primary wave G_p and the scattered wave G_s. The primary wave G_p is Green's function in free space in the absence of the sphere and G_s represents the effect of the sphere. Let us first find G_p. This is simply given by $G_p = \exp(-jk|\bar{r} - \bar{r}'|)/4\pi|\bar{r} - \bar{r}'|$. However, to satisfy the boundary conditions at $r = a$, it is necessary to expand G_p in spherical harmonics. We expand G_p in Fourier series in the ϕ direction and in a series of Legendre functions in the θ direction. Thus we write

$$G_p = \sum_{n=0}^{\infty} \sum_{m=-n}^{n} G_{nm}(r, r')Y_{nm}(\theta, \phi)Y^*_{nm}(\theta', \phi'), \tag{5-123}$$

where

$$Y_{nm}(\theta, \phi) = \left[\frac{2n+1}{2}\frac{(n-m)!}{(n+m)!}\right]^{1/2} P_n^m(\cos\theta)\frac{e^{jm\phi}}{(2\pi)^{1/2}}$$

The functions $Y_{nm}(\theta, \phi)$ are the spherical harmonics, combining the θ and the ϕ dependence, and are orthogonal and normalized (see Section 4-15).

$$\int_0^{\pi} \sin\theta d\theta \int_0^{2\pi} d\phi Y_{nm} Y^*_{n'm'} = \begin{cases} 0 & \text{if } n \neq n' \quad \text{or} \quad m \neq m', \\ 1 & \text{if } n = n' \quad \text{and} \quad m = m'. \end{cases} \tag{5-124}$$

Note also that $P_n^m(\cos\theta) = 0$ for $m > n$. The delta function on the right side of (5-122) is given by

$$\delta(\bar{r} - \bar{r}') = \frac{\delta(r - r')\delta(\theta - \theta')\delta(\phi - \phi')}{r^2 \sin\theta} \tag{5-125}$$

We now expand the delta function in series of spherical harmonics:

$$\frac{\delta(\theta - \theta')\delta(\phi - \phi')}{\sin\theta} = \sum_{n=0}^{\infty} \sum_{m=-n}^{n} Y_{nm}(\theta, \phi)Y^*_{nm}(\theta', \phi'). \tag{5-126}$$

Substituting (5-123) and (5-126) into (5-122), we get

$$\left[\frac{1}{r^2}\frac{d}{dr}\left(r^2\frac{d}{dr}\right) + k^2 - \frac{n(n+1)}{r^2}\right]G_n = -\frac{\delta(r - r')}{r^2}, \tag{5-127}$$

where we used G_n instead of G_{nm} since the differential equation contains only n. Using the technique shown in Section 5-5, the solution is given by

$$G_n = \frac{y_1(r_>)y_2(r_<)}{D},$$

where $D = r^2\Delta = r^2(y_1 y_2' - y_1' y_2)$, and $y_1 = h_n^{(2)}(kr)$ and $y_2 = j_n(kr)$.

Now the Wronskian Δ can be evaluated noting that as $kr \to$ large,

$$h_n^{(2)}(kr) \to \frac{1}{kr} \exp\left[-j\left(kr - \frac{n+1}{2}\pi\right)\right],$$

$$j_n(kr) \to \frac{1}{kr} \cos\left(kr - \frac{n+1}{2}\pi\right).$$

We then get $\Delta = j/(kr^2)$ and $D = j/k$. The final expression for Green's function G_p for the primary wave is then given by

$$G_p = \sum_{n=0}^{\infty} \sum_{m=-n}^{n} G_n(r, r') Y_{nm}(\theta, \phi) Y_{nm}^*(\theta', \phi'),$$

$$G_n = \begin{cases} -jkj_n(kr')h_n^{(2)}(kr) & \text{if } r' < r \\ -jkj_n(kr)h_n^{(2)}(kr') & \text{if } r' > r. \end{cases} \tag{5-128}$$

We can rewrite (5-128) using

$$P_n^{-m}(x) = (-1)^m \frac{(n-m)!}{(n+m)!} P_n^m(x) \tag{5-129}$$

and the addition theorem

$$P_n(\cos\gamma) = \sum_{m=0}^{n} \epsilon_m \frac{(n-m)!}{(n+m)!} P_n^m(\cos\theta) P_n^m(\cos\theta') \cos m(\phi - \phi'), \tag{5-130}$$

$$\cos\gamma = \hat{r} \cdot \hat{r}' = \cos\theta \cos\theta' + \sin\theta \sin\theta' \cos m(\phi - \phi'),$$

$$\epsilon_0 = 1, \qquad \epsilon_m = 2, \qquad \text{for } m \geq 1.$$

Then we have

$$G_p = \sum_{n=0}^{\infty} \frac{2n+1}{4\pi} G_n(r, r') P_n(\cos\gamma). \tag{5-131}$$

Now let us go back to our problem (5-122) and find G_s. We expand G_s in the same spherical harmonics. We write

$$G_s = \sum_{n=0}^{\infty} A_n h_n^{(2)}(kr) P_n(\cos\gamma), \tag{5-132}$$

where A_n are unknown coefficients to be determined by the boundary condition

$$\frac{\partial}{\partial r}[r(G_p + G_s)] = 0 \qquad \text{at } r = a. \tag{5-133}$$

Noting that $G_n = -jkj_n(kr)h_n^{(2)}(kr')$ for $r < r'$, we get

$$A_n = \frac{(2n+1)jkh_n^{(2)}(kr')}{4\pi} \left\{ \frac{\frac{\partial}{\partial r}[rj_n(kr)]}{\frac{\partial}{\partial r}[rh_n^{(2)}(kr)]} \right\}_{r=a}. \tag{5-134}$$

The final expression for Green's function is then $G = G_p + G_s$, where G_p and G_s are given by (5-131) and (5-132), respectively.

PROBLEMS

5-1. Find the radiation fields $\bar{E}$ and $\bar{H}$ from a vertical electric dipole at a height h above a conducting plane.

5-2. Find the radiation resistance of the monopole shown in Fig. P5-2.

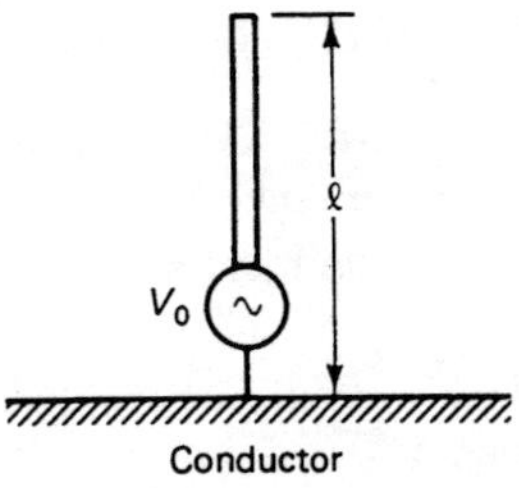

Figure P5-2 Monopole.

5-3. Find the radiation field from a horizontal electric dipole located at a height h above a conducting plane.

5-4. Show that one-dimensional Green's function in free space is given by

$$G(x, x') = \frac{\exp(-jk|x - x'|)}{2jk}.$$

5-5. Show that a two-dimensional Green's function in free space is given by

$$G(x, x') = -\frac{j}{4} H_0^{(2)}(k|\bar{r} - \bar{r}'|).$$

5-6. A sheet of uniform current I_0 (A/m) flowing in the x direction is located in free space at $z = 0$ (see Problem 2-6). Find the Hertz vector $\bar{\Pi}$, $\bar{E}$, and $\bar{H}$.

5-7. The current sheet shown in Problem 5-6 is located at $z = z'$, and perfectly conducting sheets are placed at $z = 0$ and $z = d$ as shown in Fig. P5-7. Find $\bar{E}$ and $\bar{H}$. Find the real and reactive power supplied by the current [see equation (2-76)].

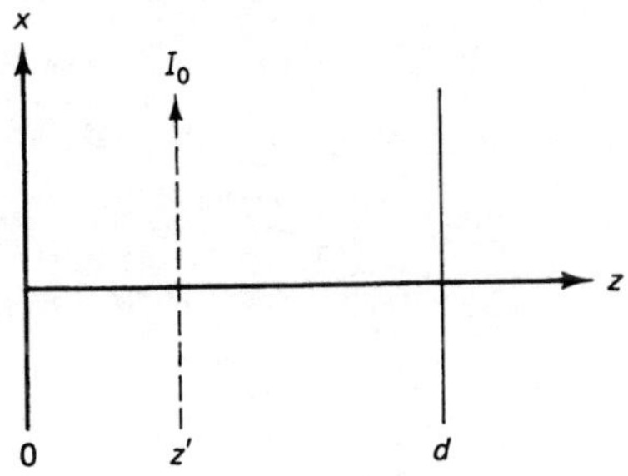

Figure P5-7 Current sheet I_0 at $z = z'$.

5-8. If an impedance sheet is placed at $z = d$ for Problem 5-7, find $\overline{E}$, $\overline{H}$, and the power supplied by the current. Assume that at $z = d$, $E_x/H_y = 100\ \Omega$.

5-9. Show that the normalization constant for spherical Bessel functions are given by (see Abramowitz and Stegun, 1964, p. 485)

$$\int_0^a j_\nu(\alpha r) j_\nu(\alpha r) r^2\, dr = \begin{cases} \dfrac{a^3}{2}[j_{\nu+1}(\alpha a)]^2 & \text{if } j_\nu(\alpha a) = 0, \\ \dfrac{a^3}{2}\left[\dfrac{1}{4(\alpha a)^2} + 1 - \dfrac{\nu + \frac{1}{2}}{(\alpha a)^2}\right][j_\nu(\alpha a)]^2 & \text{if } \dfrac{\partial}{\partial r}(r j_\nu) = 0 \text{ at } r = a. \end{cases}$$

5-10. Find the Green's function $G(r, r')$ inside a spherical cavity that satisfies the boundary condition

$$\frac{\partial}{\partial r}[rG] = 0 \qquad \text{at } r = a.$$

5-11. Find the Green's function for a rectangular cavity with width a, length b, and height c satisfying Dirichlet's boundary condition.

5-12. Find the Green's function for a cylindrical cavity of radius a and height h with Neumann's boundary condition on the wall.

5-13. Find the Green's function inside an infinitely long cylindrical waveguide of radius a with Dirichlet's boundary condition.

5-14. Show that three-dimensional Green's function $G(\bar{r} - \bar{r}') = \exp(-jk|\bar{r} - \bar{r}'|)/4\pi|\bar{r} - \bar{r}'|$ in free space can be expressed as

$$G = \frac{1}{(2\pi)^2}\iint \frac{\exp[-jq_1(x - x') - jq_2(y - y') - jq|z - z'|]}{2jq}\, dq_1\, dq_2,$$

where

$$q = \begin{cases} (k^2 - q_1^2 - q_2^2)^{1/2} & \text{if } |k| > (q_1^2 + q_2^2)^{1/2}, \\ -j(q_1^2 + q_2^2 - k^2)^{1/2} & \text{if } |k| < (q_1^2 + q_2^2)^{1/2}. \end{cases}$$

5-15. Consider the Green's function for the wave equation

$$\left(\nabla^2 - \frac{1}{c^2}\frac{\partial^2}{\partial t^2}\right) g(\bar{r}, t) = -\delta(\bar{r})\delta(t).$$

(a) Define the Fourier transform of $g(\bar{r}, t)$ to be $G(\bar{r}, \omega)$, so that

$$\bar{g}(r, t) = \frac{1}{2\pi}\int d\omega\, e^{j\omega t} G(\bar{r}, \omega).$$

Show that

$$G(r, \omega) = \frac{e^{-jkr}}{4\pi r}.$$

(b) By performing the inverse Fourier transform, show that

$$g(\bar{r}, t) = \frac{\delta(t - r/c)}{4\pi r}.$$

6

Radiation from Apertures and Beam Waves

In Chapter 5 we discussed Green's functions and their representations for different problems. In this chapter we make use of Green's functions and apply them to obtain the waves radiated from apertures. We also discuss the spectral domain method applied to beam wave propagation, the Goos–Hanchen shift, and higher order beam waves (Tamir and Blok, 1986). Electromagnetic aperture problems are then discussed, including the Stratton–Chu formula, the equivalence theorem, and Kirchhoff approximation.

6-1 HUYGENS' PRINCIPLE AND EXTINCTION THEOREM

According to Huygens' principle, the field at a point is the superposition of the spherical wavelets originating from a surface located between the observation point and the source (Fig. 6-1). In this section we discuss the mathematical formulation of Huygens' principle for scalar waves.

Let us consider the field $\psi(\bar{r})$ generated by the source $f(\bar{r})$ located in the volume V_f surrounded by the surface S_f (Fig. 6-2):

$$(\nabla^2 + k^2)\psi(\bar{r}) = -f(\bar{r}). \tag{6-1}$$

Let us now apply Green's theorem to the volume V_1 surrounded by S, S_∞, and S_1. The surface S is arbitrary, and S_∞ is the surface at infinity. The surface S_1 is the

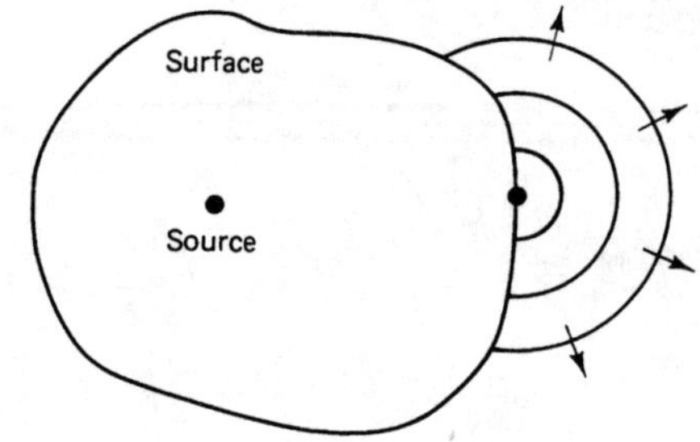

Figure 6-1 Huygens' principle.

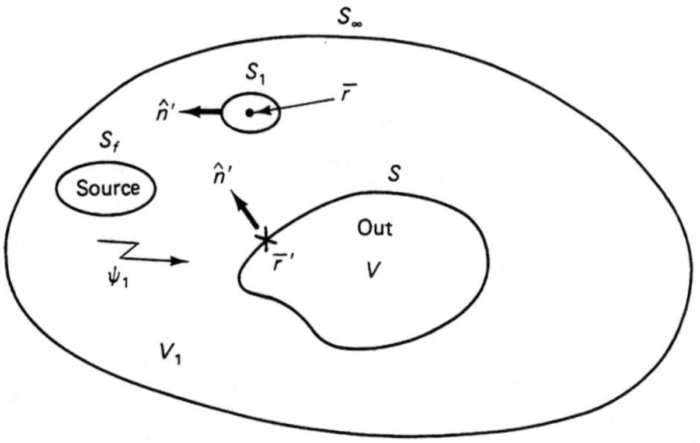

Figure 6-2 Volume V_1 surrounded by S_f, S_1, S, and S_∞. The observation point $\bar{r}$ is outside S.

surface of a small sphere centered at $\bar{r}$ with radius ϵ, and $\bar{r}$ will be identified later as the observation point. This procedure of using S_1 may appear complicated, but it is necessary when we consider the case when $\bar{r}$ is on the surface.

Let us apply Green's theorem to V_1.

$$\int_{V_1} (u\nabla^2 v - v\nabla^2 u)\, dV = \int_{S_t} \left(u\frac{\partial v}{\partial n} - v\frac{\partial u}{\partial n}\right) dS, \tag{6-2}$$

and $S_t = S + S_\infty + S_1$, and $\partial/\partial n$ is the derivative in the outward normal direction. We first let $\bar{r}'$ denote the integration point and let $\partial/\partial n'$ be the normal derivative "into" the volume V_1 rather than "outward." We then let

$$\begin{aligned} u(\bar{r}') &= \psi(\bar{r}'), \\ v(\bar{r}') &= G(\bar{r}', \bar{r}) = G(\bar{r}, \bar{r}') = \text{Green's function.} \end{aligned} \tag{6-3}$$

Now the source point $\bar{r}$ for Green's function is inside S_1, and therefore it is outside the volume V_1. Thus ψ and G satisfy (6-1) and the homogeneous wave equation in V_1, respectively.

$$(\nabla'^2 + k^2)G(\bar{r}', \bar{r}) = 0, \tag{6-4}$$

where ∇'^2 is the Laplacian with respect to $\bar{r}'$. Now we have

$$\psi\nabla'^2 G - G\nabla'^2\psi = \psi(\nabla'^2 + k^2)G - G(\nabla'^2 + k^2)\psi = Gf \quad \text{in } V_1.$$

Therefore, (6-2) becomes, noting that $\partial/\partial n = -\partial/\partial n'$,

$$\psi_i(\bar{r}) = -\int_{S_t} \left[\psi(\bar{r}') \frac{\partial G(\bar{r}, \bar{r}')}{\partial n'} - G(\bar{r}, \bar{r}') \frac{\partial \psi(\bar{r}')}{\partial n'} \right] dS', \tag{6-5}$$

where $\psi_i(\bar{r}) = \int_{V_f} G(\bar{r}, \bar{r}') f(\bar{r}')\, dV'$.

Now we consider the integral for each surface, S_∞, S_1, and S. The field at infinity on S_∞ is an outgoing spherical wave with propagation constant k, and since k in any physical medium has some small negative imaginary part, the integral on S_∞ vanishes. This is called the *radiation condition* and is expressed by

$$\lim_{r \to \infty} r \left(\frac{\partial}{\partial r} + jk \right) \psi(\bar{r}) = 0. \tag{6-6}$$

Next consider the integral on S_1. We note that $G(\bar{r}, \bar{r}')$ has a singularity at $\bar{r} = \bar{r}'$, and we can write

$$G(\bar{r}, \bar{r}') = \frac{e^{-jk\epsilon}}{4\pi\epsilon} + G_1(\bar{r}'),$$

where $\epsilon = |\bar{r} - \bar{r}'|$ and G_1 has no singularity at $\bar{r} = \bar{r}'$ and is regular. We then get

$$\begin{aligned} \lim_{\epsilon \to 0} \int_{S_1} \psi(\bar{r}') \frac{\partial G(\bar{r}, \bar{r}')}{\partial n'}\, dS' &= \lim_{\epsilon \to 0} \left[\psi(\bar{r}) \frac{\partial}{\partial \epsilon} \left(\frac{e^{-jk\epsilon}}{4\pi\epsilon} \right) 4\pi\epsilon^2 \right] \\ &= -\psi(\bar{r}). \end{aligned} \tag{6-7}$$

Also, we have

$$\lim_{\epsilon \to 0} \int_{S_1} G(\bar{r}, \bar{r}') \frac{\partial \psi(\bar{r}')}{\partial n'}\, dS' = 0. \tag{6-8}$$

Therefore, the integration on S_1 is equal to $-\psi(\bar{r})$. Equation (6-5) then becomes

$$\psi_i(\bar{r}) + \int_S \left[\psi(\bar{r}') \frac{\partial G(\bar{r}, \bar{r}')}{\partial n'} - G(\bar{r}, \bar{r}') \frac{\partial \psi(\bar{r}')}{\partial n'} \right] dS' = \psi(\bar{r}), \qquad \text{if } \bar{r} \text{ is outside } S, \tag{6-9}$$

where the field ψ and $\partial\psi/\partial n'$ are evaluated as $\bar{r}'$ approaches the surface from outside. The field $\psi_i(\bar{r})$ is given by (6-5). It is the field in the absence of the surface S and is equal to the *incident field*. At this point, we have not imposed any boundary conditions on Green's function. The most common choice for G is simply the free-space Green's function $G(\bar{r}, \bar{r}') = G_0(\bar{r}, \bar{r}') = \exp(-jk|\bar{r} - \bar{r}'|)/(4\pi|\bar{r} - \bar{r}'|)$. Other choices are discussed later. Equation (6-9) gives the fundamental expression for the field $\psi(\bar{r})$ outside the object consisting of the incident field $\psi_i(\bar{r})$ and the scattered field $\psi_s(\bar{r})$, which is the contribution from the field on the surface S.

Next, we consider the case when the observation point $\bar{r}$ approaches the surface S from outside (Fig. 6-3). We then consider the semispherical surface S_2 as shown. Then considering that the normal direction $\hat{n}'$ for S_2 is opposite to the normal direction $\hat{n}'$ for S_1 on the semisphere, we get

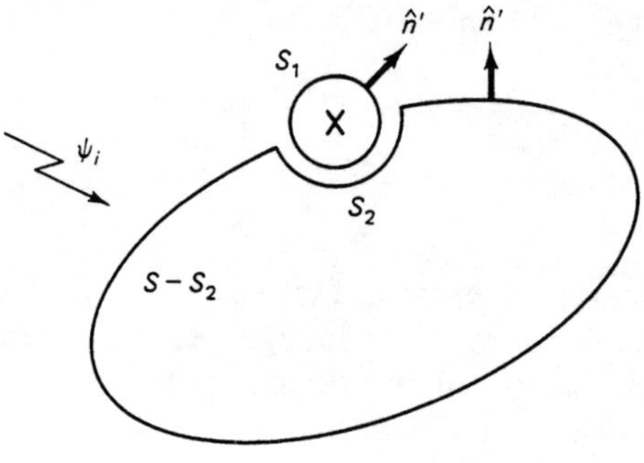

Figure 6-3 Observation point $\bar{r}$ on the surface S.

$$\int_{S_1} dS' = -\psi(\bar{r}),$$
$$\int_{S_2} dS' = \tfrac{1}{2}\psi(\bar{r}). \tag{6-10}$$

We therefore get

$$\psi_i(\bar{r}) + \fint_S \left[\psi(\bar{r}')\frac{\partial G(\bar{r},\bar{r}')}{\partial n'} - G(\bar{r},\bar{r}')\frac{\partial \psi(\bar{r}')}{\partial n'}\right] dS' = \frac{1}{2}\psi(\bar{r}) \qquad \text{if } \bar{r} \text{ is on the surface } S, \tag{6-11}$$

where the surface integral is called Cauchy's principal value and it indicates the integration over the surface S excluding the small circular area of radius ϵ.

Finally, we consider the case when the observation point $\bar{r}$ is inside the surface S (Fig. 6-4). The integration does not include the surface S_1, and therefore we get

$$\psi_i(\bar{r}) + \int_S \left[\psi(\bar{r}')\frac{\partial G(\bar{r},\bar{r}')}{\partial n'} - G(\bar{r},\bar{r}')\frac{\partial \psi(\bar{r}')}{\partial n'}\right] dS' = 0 \qquad \text{if } \bar{r} \text{ is inside } S. \tag{6-12}$$

The three equations (6-9), (6-11), and (6-12) are the fundamental equations that are important and useful for the formulation of many problems. They are the basis for constructing the surface integral equations, as we discuss in later sections. In this section, however, we consider the physical meaning of (6-12). It states that inside the surface S, the incident field $\psi_i(\bar{r})$ and the contribution from the surface field combine to produce the null field. The incident field $\psi_i(\bar{r})$ inside the surface is therefore "extinguished" by the surface contribution. For this reason, (6-12) is

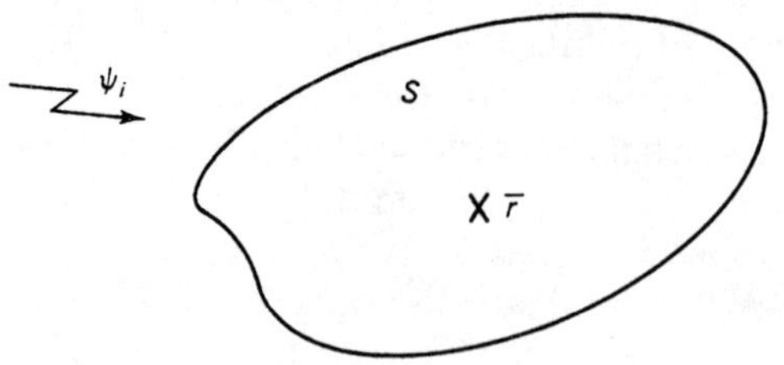

Figure 6-4 Observation point $\bar{r}$ inside the surface S.

called the *extinction theorem*, the *Ewald–Oseen extinction theorem*, or the *null field theorem*. This theorem can be used as a boundary condition to obtain an extended integral equation, as will be shown later, and this method is called the *extended boundary condition method* or the *T-matrix method*.

Let us next apply the extinction theorem to a special case when there is no object inside S. The field everywhere is the incident wave $\psi_i(\bar{r})$ itself. We then write the extinction theorem as follows:

$$\psi_i(\bar{r}) = \int_S \left[\psi_i(\bar{r}')\frac{\partial G_0(\bar{r},\bar{r}')}{\partial n} - G_0(\bar{r},\bar{r}')\frac{\partial \psi_i(\bar{r}')}{\partial n}\right] dS', \tag{6-13}$$

where $G_0(\bar{r},\bar{r}') = \exp(-k|\bar{r}-\bar{r}'|)/(4\pi|\bar{r}-\bar{r}'|)$ is a free-space Green's function and $\partial/\partial n$ is now taken as the normal derivative pointed inside S. Equation (6-13) shows that the field at $\bar{r}$ can be calculated by knowing the field ψ and $\partial\psi/\partial n$ on a surface S, which act as the secondary source for spherical waves. This is the mathematical statement of Huygens' principle.

We note that the three fundamental equations (6-9), (6-11), and (6-12) are valid for two-dimensional problems ($\partial/\partial z = 0$) where Green's function is the two-dimensional Green's function $G_0(\bar{r},\bar{r}') = -(j/4)H_0^{(2)}(k|\bar{r}-\bar{r}'|)$, $\bar{r} = x\hat{x} + y\hat{y}$, $\bar{r}' = x'\hat{x} + y'\hat{y}$, and the surface integral is replaced by the line integral.

6-2 FIELDS DUE TO THE SURFACE FIELD DISTRIBUTION

In Section 6-1 we obtained the scalar field $\psi(\bar{r})$ in (6-9) when $\bar{r}$ is outside the surface S. The field $\psi(\bar{r})$ consists of the incident field $\psi_i(\bar{r})$ and the field $\psi_s(\bar{r})$ scattered from the surface S.

$$\begin{aligned}\psi(\bar{r}) &= \psi_i(\bar{r}) + \psi_s(\bar{r}),\\ \psi_s(\bar{r}) &= \int_S \left[\psi(\bar{r}')\frac{\partial G(\bar{r},\bar{r}')}{\partial n'} - G(\bar{r},\bar{r}')\frac{\partial \psi(\bar{r}')}{\partial n'}\right] dS'.\end{aligned} \tag{6-14}$$

To derive this we used Green's function $G(\bar{r},\bar{r}')$ which has a singularity at $\bar{r} = \bar{r}'$ and satisfies the equation

$$(\nabla^2 + k^2)G(\bar{r},\bar{r}') = -\delta(\bar{r}-\bar{r}'). \tag{6-15}$$

However, we have not imposed any boundary conditions on G.

The simplest Green's function is the free-space Green's function:

$$G(\bar{r},\bar{r}') = G_0(\bar{r},\bar{r}') = \frac{\exp(-jk|\bar{r},\bar{r}'|)}{4\pi|\bar{r}-\bar{r}'|}. \tag{6-16}$$

With the free-space Green's function, the scattered field is calculated if both ψ and $\partial\psi/\partial n'$ are known on the surface.

$$\psi_s(\bar{r}) = \int_S \left[\psi(\bar{r}')\frac{\partial G_0(\bar{r},\bar{r}')}{\partial n'} - G_0(\bar{r},\bar{r}')\frac{\partial \psi(\bar{r}')}{\partial n'}\right] dS'. \tag{6-17}$$

This is called the *Helmholtz–Kirchhoff formula*. It should, however, be recognized that it is not necessary to know both ψ and $\partial\psi/\partial n$ on the surface, because according to the uniqueness theorem (Section 2-10), if the field ψ (or $\partial\psi/\partial n$) is known on the surface, the field should be uniquely determined everywhere outside the surface.

To obtain the expression for ψ_s in terms of the surface field only, we use Green's function $G_1(\bar{r}, \bar{r}')$, which satisfies the boundary condition

$$G_1(\bar{r}, \bar{r}') = 0 \qquad \text{when } \bar{r}' \text{ is on } S. \tag{6-18}$$

We then get

$$\psi_s(\bar{r}) = \int_S \psi(\bar{r}') \frac{\partial}{\partial n'} G_1(\bar{r}, \bar{r}')\, dS'. \tag{6-19}$$

We can also obtain the field $\psi_s(\bar{r})$ in terms of $\partial\psi/\partial n'$ on the surface using $G_2(\bar{r}, \bar{r}')$, which satisfies Neumann's boundary condition on S.

$$\frac{\partial}{\partial n'} G_2(\bar{r}, \bar{r}') = 0 \qquad \text{when } \bar{r}' \text{ is on } S. \tag{6-20}$$

Then we get

$$\psi_s(\bar{r}) = -\int_S G_2(\bar{r}, \bar{r}') \frac{\partial \psi(\bar{r}')}{\partial n'}\, dS'. \tag{6-21}$$

The three equations (6-17), (6-19), and (6-21) are exact and should yield the identical exact field $\psi_s(\bar{r})$. In practice, however, the exact field may not be known on S, and therefore one of them may be preferred, depending on the problems.

Consider a field $\psi_s(\bar{r})$ behind a plane screen with an aperture where the field is given (Fig. 6-5). The field $\psi_s(\bar{r})$ can be expressed using the preceding three formulas.

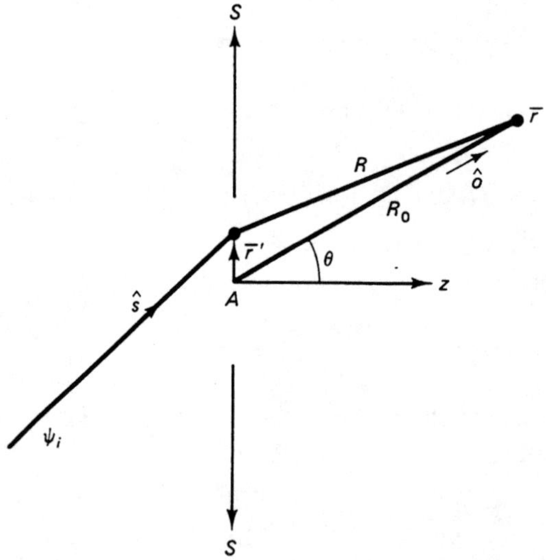

Figure 6-5 Aperture A on screen S.

Let us first find Green's function that satisfies Dirichlet's condition on the screen. Green's function in this case is easily found to be the difference between free-space Green's function due to $\bar{r}'$ and its image position $\bar{r}''$ (Fig. 6-6).

$$G_1(\bar{r}, \bar{r}') = \frac{\exp(-jkr_1)}{4\pi r_1} - \frac{\exp(-jkr_2)}{4\pi r_2}, \tag{6-22}$$

where $r_1 = |\bar{r} - \bar{r}'|$ and $r_2 = |\bar{r} - \bar{r}''|$. To calculate $\partial/(\partial n')G$, note that

$$\frac{\partial}{\partial n'}\left[\frac{\exp(-jkr_1)}{4\pi r_1}\right] = \frac{\partial}{\partial r_1}\left[\frac{\exp(-jkr_1)}{4\pi r_1}\right]\frac{\partial r_1}{\partial n'}$$

$$= \frac{\exp(-jkr_1)}{4\pi r_1}\left(-jk - \frac{1}{r_1}\right)\frac{z' - z}{r_1}, \tag{6-23}$$

$$\frac{\partial}{\partial n'}\left[\frac{\exp(-jkr_2)}{4\pi r_2}\right] = \frac{\exp(-jkr_2)}{4\pi r_2}\left(-jk - \frac{1}{r_2}\right)\frac{z' + z}{r_2},$$

where

$$r_1^2 = (x - x')^2 + (y - y')^2 + (z - z')^2,$$

$$r_2^2 = (x - x')^2 + (y - y')^2 + (z + z')^2.$$

Therefore, we get

$$\frac{\partial}{\partial n'}G(\bar{r}, \bar{r}') = \frac{\exp(-jkR)}{2\pi R}\left(jk + \frac{1}{R}\right)\frac{z}{R}, \tag{6-24}$$

where $R^2 = (x - x')^2 + (y - y')^2 + z^2$.

The field $\psi(\bar{r})$ is then given by

$$\psi_s(\bar{r}) = \int_s \frac{\exp(-jkR)}{2\pi R}\left(jk + \frac{1}{R}\right)\frac{z}{R}\,\psi(\bar{r}')\,dS'. \tag{6-25}$$

If $\bar{r}$ is close to the z axis ($z/R \approx 1$) and $kR >> 1$, we get

$$\psi_s(\bar{r}) = \frac{jk}{2\pi}\int_S \frac{\exp(-jkR)}{R}\,\psi(\bar{r}')\,dS'. \tag{6-26}$$

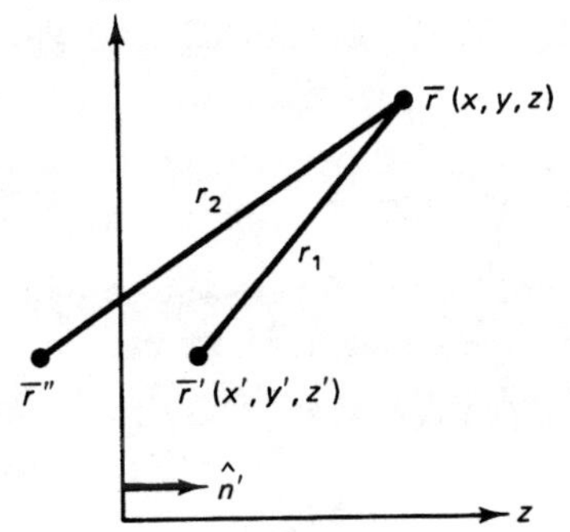

Figure 6-6 Green's function $G_1(\bar{r}, \bar{r}')$ which satisfies Dirichlet's condition.

This expression is often used to calculate the field $\psi_s(\bar{r})$ due to the aperture field $\psi(\bar{r}')$ on a screen.

Next consider Green's function G_2, which satisfies Neumann's condition on a plane screen. We have

$$G_2(\bar{r}, \bar{r}') = \frac{e^{-jkr_1}}{4\pi r_1} + \frac{e^{-jkr_2}}{4\pi r_2}. \tag{6-27}$$

Therefore, we have

$$\psi_s(\bar{r}) = -\frac{1}{2\pi}\int_S \frac{e^{-jkR}}{R}\frac{\partial\psi(\bar{r}')}{\partial n'}\,dS'. \tag{6-28}$$

Equations (6-25) and (6-26) are applicable when the aperture field is known on a plane surface. Equation (6-28) can be used when the normal derivative of the field is known on a plane surface. When both the field and its normal derivative are known, we can use (6-17).

6-3 KIRCHHOFF APPROXIMATION

In (6-17) we showed that if the field and its normal derivative are known on a surface, the field at any point can be calculated. However, the exact evaluation of the field on a surface, such as the aperture field on a screen (Fig. 6-5), is difficult and requires solution of the complete boundary value problem. If the aperture size is large in terms of wavelength, the aperture field may be approximately equal to those of the incident field. This is called the *Kirchhoff approximation* and can be stated mathematically as follows.

On the aperture A (Fig. 6-5), we assume that

$$\psi = \psi_i \quad \text{and} \quad \frac{\partial\psi}{\partial n} = \frac{\partial\psi_i}{\partial n}. \tag{6-29}$$

On the screen S, we assume that

$$\psi = 0 \quad \text{and} \quad \frac{\partial\psi}{\partial n} = 0. \tag{6-30}$$

Making use of (6-23), we get the Kirchhoff approximation for the aperture problem shown in Fig. 6-5:

$$\psi(\bar{r}) = \int_S \left[\psi_i(\bar{r}')\left(jk + \frac{1}{R}\right)\frac{z}{R} - \frac{\partial\psi_i}{\partial n'}\right]\frac{\exp(-jkR)}{4\pi R}\,dS'. \tag{6-31}$$

For example, if the incident field is a plane wave propagating in the $\hat{s}$ direction and the aperture is in the plane $z = 0$ (Fig. 6-5), we have

$$\begin{aligned} \psi_i(\bar{r}') &= A_0 e^{-jk\hat{s}\cdot\bar{r}'}, \\ \frac{\partial\psi_i(\bar{r}')}{\partial n'} &= -jk\hat{s}\cdot\hat{z}A_0 e^{-jk\hat{s}\cdot\bar{r}'} \end{aligned} \tag{6-32}$$

If we consider the field at a large distance from the aperture, we can neglect $1/R$ in $(jk + 1/R)$, let z/R be $\cos\theta$, and $1/R = 1/R_0$. The phase kR cannot be equated to kR_0, however, as we need to consider the difference in R and R_0 in terms of wavelength. We then should use $R = R_0 - \bar{r}' \cdot \hat{o}$. Finally, we get the field far from the aperture:

$$\psi(\bar{r}) = \frac{\exp(-jkR_0)}{4\pi R_0} \int_S jk\psi_i(\bar{r}')(\cos\theta + \hat{s}\cdot\hat{z})e^{jk\bar{r}'\cdot\bar{o}}\, dS'. \tag{6-33}$$

The Kirchhoff approximation can be applied to the problem of scattering from an object (Fig. 6-7). At $\bar{r}'$, we assume that the field is approximately equal to the field if the surface is plane and tangential to the surface at $\bar{r}'$. If the incident field is a plane wave propagating in the $\hat{s}$ direction, we have

$$\psi_i = A_0 \exp(-jk\hat{s}\cdot\bar{r}). \tag{6-34}$$

If the reflection coefficient at $\bar{r}'$ for a plane surface is R, we have in the illuminated part of the surface

$$\psi = (1+R)\psi_i,$$
$$\frac{\partial\psi}{\partial n'} = (-jk\hat{s}\cdot\hat{n}')(1-R)\psi_i. \tag{6-35}$$

In the shadow region, we assume that

$$\psi = 0 \quad \text{and} \quad \frac{\partial\psi}{\partial n'} = 0. \tag{6-36}$$

Equations (6-34), (6-35), and (6-36) are then substituted in (6-17) to obtain the *Kirchhoff approximation*. Since we use the approximation that the field at any point on the surface is equal to the field for a plane tangent to the surface, we sometimes call this the *tangent approximation*. Electromagnetic equivalent of this approximation is called the *physical optics approximation*.

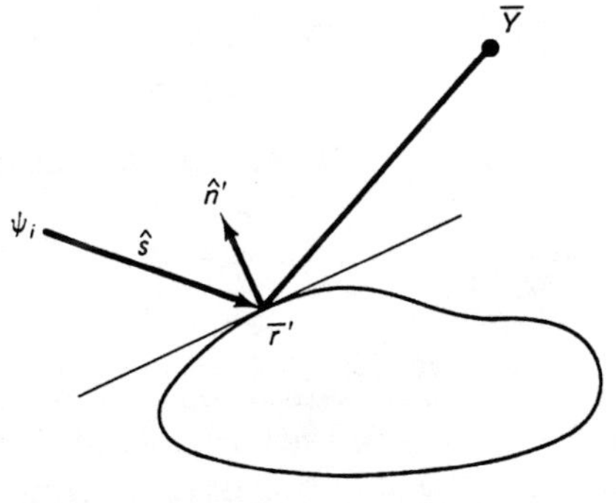

Figure 6-7 Kirchhoff approximation.

6-4 FRESNEL AND FRAUNHOFER DIFFRACTION

Let us next consider the field u at $\bar{r}$ where the field $u_0(\bar{r}')$ is given on the aperture at $z = 0$ (Fig. 6-8). This is given in (6-25). Consider an observation point (x, y, z) and let r_0 be the distance from the origin $x' = y' = 0$ of the aperture. Then we assume

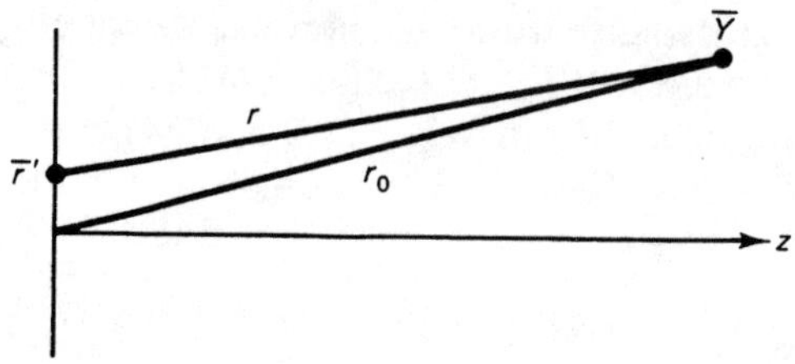

Figure 6-8 Fresnel and Fraunhofer diffraction.

that $kr >> 1$, $z/r \approx 1$, and $1/r \approx 1/r_0$. The phase kr cannot be approximated by kr_0 since the difference can be of the order of wavelengths. We use the following approximation:

$$\begin{aligned} r &= [z^2 + (x - x')^2 + (y - y')^2]^{1/2} \\ &\approx z + \frac{(x - x')^2 + (y - y')^2}{2z}, \end{aligned} \tag{6-37}$$

Then we get the following "Fresnel diffraction" formula:

$$u(x, y, z) = \frac{jk}{2\pi z} e^{-jkz} \iint u_0(x', y') \exp\left[-jk \frac{(x - x')^2 + (y - y')^2}{2z}\right] dx'\, dy' \tag{6-38}$$

If we rewrite (6-37) as

$$r \approx r_0 - \frac{xx' + yy'}{z} + \frac{x'^2 + y'^2}{2z}, \tag{6-39}$$

where

$$r_0 = z + \frac{x^2 + y^2}{2z},$$

and note that $kx'^2/z < 1$ at large distance z, we can approximate r by

$$r \approx r_0 - \frac{xx'}{z} - \frac{yy'}{z}. \tag{6-40}$$

Then we get the *Fraunhofer diffraction formula*, valid at large distance r_0 from the aperture, where (6-40) holds.

$$u(x, y, z) = \frac{jk}{2\pi r_0} e^{-jkr_0} \iint u_0(x', y') e^{+jk(xx'/z + yy'/z)}\, dx'\, dy'. \tag{6-41}$$

Note the difference between the Fresnel diffraction formula (6-38) and the Fraunhofer diffraction formula (6-41). The Fresnel formula contains the quadratic terms in x' and y' and is applicable to the near field of the aperture $z < a^2/\lambda$, while the Fraunhofer formula contains only the linear terms in x' and y' and is applicable to the far field $z >> a^2/\lambda$, where a is the aperture size. We also note that (6-41) is in a form of Fourier transform, and therefore we can state that the Fraunhofer field is the Fourier transform of the aperture field. In antenna theory, we state that the radiation pattern of the aperture antenna is the Fourier transform of the aperture field.

Let us consider the Fraunhofer diffraction from a circular aperture of radius a with constant field:

$$u_0(x',y') = \begin{cases} A_0 & \text{if } \rho' \le a \\ 0 & \text{if } \rho' > a \end{cases}. \tag{6-42}$$

Then from (6-41), letting

$$x' = \rho' \cos\phi', \qquad y' = \rho' \sin\phi'$$

$$\frac{x}{z} = \sin\theta \cos\phi, \qquad \frac{y}{z} = \sin\theta \sin\phi$$

we get

$$u(x,y,z) = \frac{jk}{2\pi r_0} e^{-jkr_0} A_0 \int_0^a \int_0^{2\pi} \exp[jk\rho' \sin\theta \cos(\phi - \phi')]\, d\phi' \rho' \, d\rho'. \tag{6-43}$$

Now we make use of the following integral representation of Bessel's function and the integral formula:

$$J_0(x) = \frac{1}{2\pi} \int_0^{2\pi} e^{jx\cos\phi}\, d\phi,$$

$$\int J_0(x)\, x\, dx = xJ_1(x). \tag{6-44}$$

We then get

$$u(x,y,z) = u(r_0, \theta)$$

$$= \frac{jk}{2\pi r_0} e^{-jkr_0} A_0 \pi a^2 \frac{2J_1(ka \sin\theta)}{ka \sin\theta}. \tag{6-45}$$

Normalizing u with respect to u at $\theta = 0$, we get

$$\frac{u(r_0, \theta)}{u(r_0, 0)} = \frac{2J_1(ka \sin\theta)}{ka \sin\theta}. \tag{6-46}$$

The angle θ_a of the first zero of $u(r_0, \theta)$ is given by

$$ka \sin\theta_a = 3.832 \qquad \text{or} \qquad \sin\theta_a = 0.610\lambda/a. \tag{6-47}$$

The diffraction pattern (6-46) is called the *Airy pattern* and the main lobe in $0 \le \theta \le \theta_a$ is called the *Airy disk*.

As an example of the Fresnel diffraction, consider the uniform field in the square aperture.

$$U_0 = \begin{cases} A_0 & \text{for } |x'| < a \text{ and } |y'| < b \\ 0 & \text{outside} \end{cases}. \tag{6-48}$$

The field on the axis ($x = y = 0$) is then given by

$$U(z) = \frac{jk}{2\pi z} e^{-jkz} \int_{-a}^{a} dx' \int_{-b}^{b} dy' A_0 \exp\left[-\frac{jk(x'^2 + y'^2)}{2z}\right]. \tag{6-49}$$

We make use of the Fresnel integral, defined by

$$F(Z) = \int_0^Z \exp\left(-j\frac{\pi}{2}t^2\right) dt = C(Z) - jS(Z). \tag{6-50}$$

The field in (6-49) is then given by

$$U(Z) = j2A_0 e^{-jkz} F(Z_1)F(Z_2), \tag{6-51}$$

where $Z_1 = (k/\pi z)^{1/2} a$ and $Z_2 = (k/\pi z)^{1/2} b$.

Let us consider one more example of Fresnel diffraction. Consider a circular aperture on which the field has the field distribution $A_0(x', y')$ with the quadratic phase front with the radius of curvature R_0 at $z = 0$.

$$U_0(x', y') = A_0(x', y') \exp\left[+j\frac{k(x'^2 + y'^2)}{2R_0}\right]. \tag{6-52}$$

The phase distribution above focuses the wave at $z = R_0$ since the wave propagated from (x', y') to $(x = 0, y = 0, z = R_0)$ has the phase given by

$$\begin{aligned} \exp(-jkR) &= \exp[-jk(R_0^2 + x'^2 + y'^2)^{1/2}] \\ &\simeq \exp\left[-jkR_0 - \frac{jk(x'^2 + y'^2)}{2R_0}\right] \end{aligned}$$

and the quadratic phase distribution in (6-52) exactly compensates for the quadratic term in $\exp(-jkR)$. We substitute (6-52) into (6-38). If we consider the field at the focal plane $z = R_0$, we get

$$U(x, y, z) = \frac{jk}{2\pi z} e^{-jkr_0} \iint A_0 e^{+jk(xx' + yy')/R_0} dx' \, dy'. \tag{6-53}$$

This is identical to (6-41) for Fraunhofer diffraction. If A_0 is constant, we get (6-46), where $\sin\theta = (x^2 + y^2)^{1/2}/R_0$.

This last example shows that if the aperture field has the quadratic phase distribution such that the wave is focused at $z = R_0$, the field distribution at the focal plane $z = R_0$ is identical to the Fraunhofer diffraction. We can also state that the field at the focal plane is the Fourier transform of the aperture field. This fact is often used in optical signal processing when Fourier transform is needed.

6-5 FOURIER TRANSFORM (SPECTRAL) REPRESENTATION

Let us consider a scalar wave $u(x, y, z)$ generated by the field $u_0(x', y')$ at the plane $z = 0$. There are two ways to obtain $u(x, y, z)$ from $u_0(x', y')$. One is to find the Green's function that is the field at (x, y, z) due to a delta function at (x', y'), and then to multiply it with $u_0(x',y')$ and integrate over the aperture at $z=0$. This is similar to finding the output voltage of a network due to an input by using the impulse response, and was used in Section 6.4. The other method is the Fourier transform technique. The Fourier transform of the aperture field $u_0(x',y')$ is first taken, multiplied by the transfer function, and then the inverse Fourier transform is taken to produce $u(x,y,z)$. This is similar to solving the network problem by using the frequency spectrum of the input, multiplying it by the transfer function, and taking the inverse transform to obtain the output. We first discuss the Fourier transform technique in this section.

Let us start with the scalar wave equation,

$$(\nabla^2 + k^2)u(\bar{r}) = 0. \tag{6-54}$$

We first take a Fourier transform in x and y:

$$U(q_1, q_2, z) = \iint u(x,y,z)e^{jq_1x + jq_2y}\,dx\,dy. \tag{6-55}$$

We then get

$$\left(\frac{d^2}{dz^2} + k^2 - q_1^2 - q_2^2\right)U(q_1,q_2,z) = 0. \tag{6-56}$$

The general solution of this consists of two waves traveling in the $+z$ and $-z$ directions, respectively.

$$U(q_1,q_2,z) = A_+(q_1,q_2)e^{-jqz} + A_-(q_1,q_2)e^{+jqz}, \tag{6-57}$$

where

$$q = \begin{cases} (k^2 - q_1^2 - q_2^2)^{1/2} & \text{if } k^2 > q_1^2 + q_2^2 \\ -j(q_1^2 + q_2^2 - k^2)^{1/2} & \text{if } k^2 < q_1^2 + q_2^2. \end{cases}$$

Note that for a positive-going wave with $\exp(-jqz)$, $q = q_r + jq_i$ must be in the fourth quadrant in the complex plane ($q_r \geq 0$ and $q_i \leq 0$), because $\exp(-jqz) = \exp(-jq_rz + q_iz)$ must decay exponentially as $z \to +\infty$. The same condition also ensures that the negative-going wave $\exp(+jqz)$ decays exponentially as $z \to -\infty$.

Let us consider the wave propagating in the $+z$ direction in the region $z \geq 0$, with $A_- = 0$. Taking the inverse Fourier transform, we get

$$u(x,y,z) = \frac{1}{(2\pi)^2}\iint A_+(q_1,q_2)e^{-jq_1x - jq_2y - jqz}\,dq_1\,dq_2. \tag{6-58}$$

The function $A_+(q_1,q_2)$ can be given by letting $u(x',y',z) = u_0(x',y')$ at $z=0$ in (6-58). We then get

$$A_+(q_1,q_2) = \iint u_0(x',y')e^{jq_1x' + jq_2y'}\,dx'\,dy'. \tag{6-59}$$

Equations (6-58) and (6-59) therefore give the field at an arbitrary point (x,y,z) due to the given field u_0 at the plane $z=0$.

6-6 BEAM WAVES

As an example of the spectral method described in Section 6-5, we discuss an important class of waves called *beam waves*. Typical examples are a laser beam and a beam of millimeter wave propagating in the atmosphere. The characteristics of a beam wave are also exhibited in the near field in front of a parabolic reflector antenna, lens waveguides, and open resonator (laser resonator) (see Fig. 6-9).

Equation (6-58) is exact, but further simplification is possible for a beam wave. If the wave is assumed to be propagating mostly in the z direction, as in the case of a beam wave, the propagation constants q_1 and q_2 in the x and y directions are much smaller than the propagation constant q in the z direction. Therefore, the major contribution to the Fourier integral in (6-58) comes from the region where $|q|$ is close to k and $|q_1|$ and $|q_2|$ are much smaller than k. Thus we make the following *para-axial approximation* for q:

$$\begin{aligned} q &= (k^2 - q_1^2 - q_2^2)^{1/2} \\ &\approx k - \frac{q_1^2 + q_2^2}{2k}. \end{aligned} \tag{6-60}$$

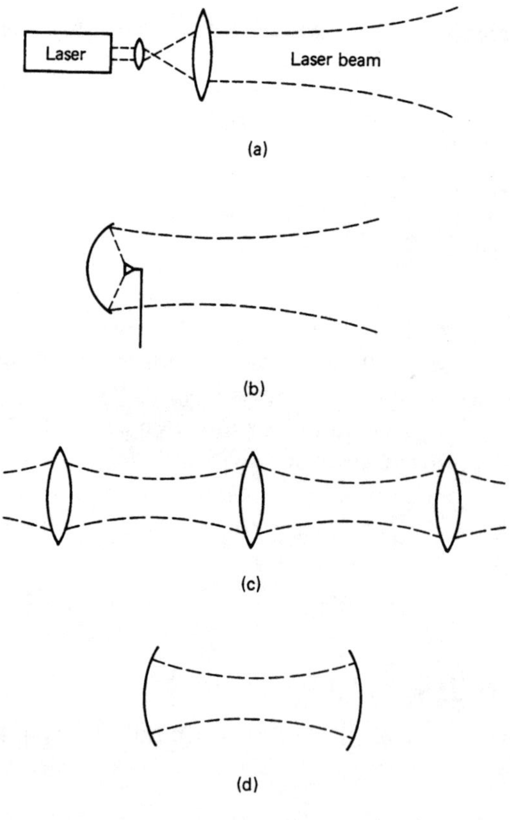

Figure 6-9 Beam waves: (a) laser beam; (b) parabolic reflector antenna; (c) lens waveguides; (d) laser resonator.

We now make use of the para-axial approximation (6-60) to obtain the expression for a beam wave. Let us assume that the field at $z = 0$ has the Gaussian amplitude distribution and the quadratic phase front with the radius of curvature R_0 (Fig. 6-10).

$$u_0(x, y) = A_0 \exp\left(-\frac{\rho^2}{W_0^2} + j\frac{k\rho^2}{2R_0}\right), \tag{6-61}$$

where $\rho^2 = x^2 + y^2$ and W_0 is the beam width at $z = 0$. The quadratic phase shift compensates for the difference between the actual distance $(R_0^2 + \rho^2)^{1/2}$ to the focal point and the focal distance R_0.

$$(R_0^2 + \rho^2)^{1/2} - R_0 \approx \frac{\rho^2}{2R_0}. \tag{6-62}$$

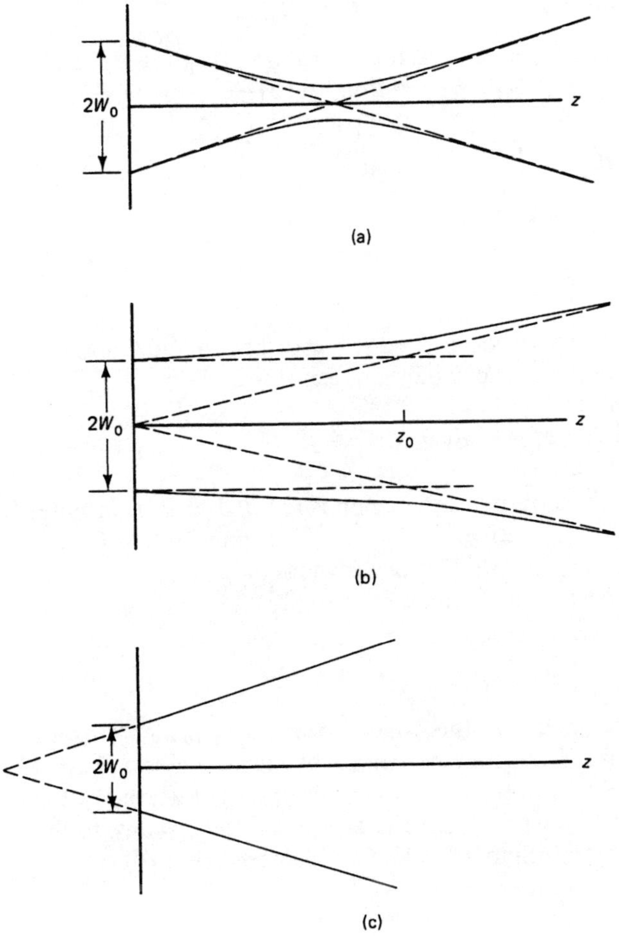

Figure 6-10 (a) Focused $R_0 > 0$; (b) collimated $R_0 \to \infty$; (c) diverging beams $R_0 < 0$.

We substitute (6-61) into (6-59) and use the following useful formula:

$$\int_{-\infty}^{\infty} \exp(-at^2 + bt)\, dt = \left(\frac{\pi}{a}\right)^{1/2} \exp\left(\frac{b^2}{4a}\right), \qquad \operatorname{Re} a > 0. \tag{6-63}$$

We then get

$$A_+(q_1, q_2) = \frac{2\pi A_0}{k\alpha} \exp\left(-\frac{q_1^2 + q_2^2}{2k\alpha}\right), \tag{6-64}$$

where

$$\alpha = \frac{1}{z_0} - j\frac{1}{R_0}, \qquad z_0 = \frac{kW_0^2}{2}.$$

Substituting (6-64) to (6-58) and using the para-axial approximation (6-60), we get

$$u(x, y, z) = \frac{A_0}{1 - j\alpha z} \exp\left(-jkz - \frac{k\alpha}{2} \frac{\rho^2}{1 - j\alpha z}\right). \tag{6-65}$$

This is the expression for a Gaussian beam when the field at $z = 0$ is given by (6-61).

Let us consider the intensity $I = |u|^2$. We get from (6-65)

$$I(x, y, z) = I_0 \frac{W_0^2}{W^2} \exp\left(-\frac{2\rho^2}{W^2}\right), \tag{6-66}$$

where $I_0 = |A_0|^2$ is the intensity at $z = 0$, and

$$W^2 = W_0^2\left[\left(1 - \frac{z}{R_0}\right)^2 + \left(\frac{z}{z_0}\right)^2\right]. \tag{6-67}$$

Equation (6-66) indicates that the beam width W varies according to (6-67) and that the total intensity I_t is independent of the distance z as expected.

$$I_t = \iint I(x, y, z)\, dx\, dy = I_0 W_0^2\left(\frac{\pi}{2}\right). \tag{6-68}$$

If the radius of curvature R_0 is positive, the beam is focused, if R_0 is infinite, the beam is collimated, and if R_0 is negative, the beam is diverging (Fig. 6-10). Let us first examine the focused beam $R_0 > 0$. The beam width W at the focal point $z = R_0$ is called the *beam spot size* and is obtained from (6-67).

$$\frac{W_s}{W_0} = \frac{\lambda R_0}{\pi W_0^2}. \tag{6-69}$$

The beam spot size W_s becomes smaller as the original beam size W_0 increases for given λ and R_0. This relationship (6-69) may be viewed by examining the Fresnel numbers N_f. At a given distance z from the center of the aperture of width W_0, the difference between the distance to the edge of the aperture and the distance to the center may be measured in the multiple of a half-wavelength (Fig. 6-11). The number of the multiple is the Fresnel number N_f.

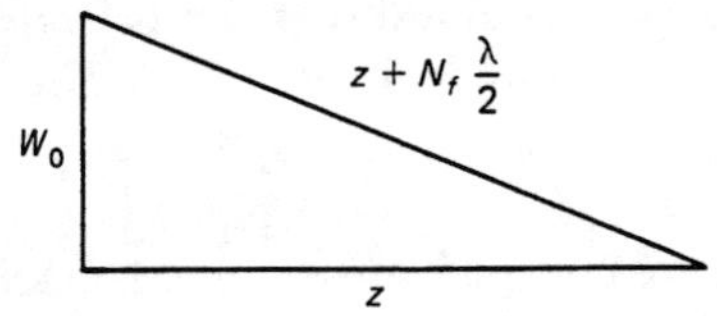

Figure 6-11 Fresnel number $N_f = W_0^2/\lambda z$.

To obtain N_f, we let

$$z + N_f \frac{\lambda}{2} = (z^2 + W_0^2)^{1/2},$$

and we get

$$N_f \approx \frac{W_0^2}{\lambda z} \qquad \text{for } W_0 \ll z. \tag{6-70}$$

At $z = R_0$, therefore, the spot size W_s is given by

$$\frac{W_s}{W_0} = \frac{1}{\pi N_f}. \tag{6-71}$$

The ratio of spot size to the aperture size is inversely proportional to the Fresnel number N_f.

Let us consider a collimated beam $R_0 \to \infty$. We get

$$u(x,y,z) = \frac{A_0}{1 - j\alpha_1 z} \exp\left(-jkz - \frac{\rho^2}{W_0^2} \frac{1}{1 - j\alpha_1 z}\right),$$

where

$$\alpha_1 = \frac{\lambda}{\pi W_0^2}, \qquad \alpha_1 z = \frac{1}{\pi N_f}. \tag{6-72}$$

The half-power beam width θ_b as $z \to \infty$ is obtained by letting

$$\exp\left(-\frac{2\rho^2}{W^2}\right) = \frac{1}{2}$$

$$\rho = \frac{z\theta_b}{2}, \qquad W^2 \approx W_0^2(\alpha_1 z)^2.$$

We then get

$$\theta_b = \frac{\lambda\,(2 \ln 2)^{1/2}}{\pi W_0}. \tag{6-73}$$

6-7 GOOS–HANCHEN EFFECT

When a beam wave is incident on a plane boundary, the transmission and reflection normally take place. If a beam wave is incident from a dense medium on a less dense medium, the total reflection takes place if the angle of incidence is greater than the critical angle. When this happens it is noted that the reflected beam is shifted in space and this is called the *Goos–Hanchen shift*. This phenomenon gives some interesting physical insight into the total reflection mechanism.

Let us consider a beam wave incident on a plane boundary as shown in Fig. 6-12. To simplify the analysis, we consider a two-dimensional ($\partial/\partial y = 0$) problem. The incident beam wave $u_i(x', z')$ is collimated with amplitude distribution $A(x')$:

$$u_i(x', z') = A(x') \exp(-jk_1 z'). \tag{6-74}$$

We then use the coordinate transformation from (x', z') to (x, z)

$$\begin{aligned} x' &= x \cos\theta_0 + (z - h)\sin\theta_0, \\ z' &= x \sin\theta_0 - (z - h)\cos\theta_0. \end{aligned} \tag{6-75}$$

The incident field at $z = h$ is, therefore, given by

$$u_i(x, h) = A(x \cos\theta_0) \exp(-j\beta_0 x), \tag{6-76}$$

where $\beta_0 = k_1 \sin\theta_0$.

We take a Fourier transform of (6-76):

$$\begin{aligned} U(\beta - \beta_0) &= \int u_i(x, h) e^{j\beta x}\, dx \\ &= \int A(x \cos\theta_0) e^{j(\beta - \beta_0)x}\, dx. \end{aligned} \tag{6-77}$$

The incident field $u_i(x, z)$ satisfies the wave equation

$$\left(\frac{\partial^2}{\partial x^2} + \frac{\partial^2}{\partial z^2} + k_1^2\right) u_i(x, z) = 0. \tag{6-78}$$

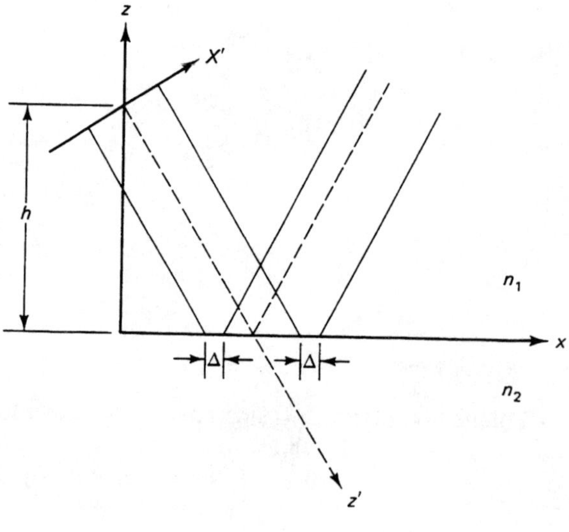

Figure 6-12 Goos–Hanchen shift.

Taking the Fourier transform in the x direction and letting $u_i(x, z)$ be equal to (6-76) at $z = h$, we obtain

$$u_i(x, z) = \frac{1}{2\pi} \int U(\beta - \beta_0) \exp[-j\beta x + jq(z - h)]\, d\beta, \tag{6-79}$$

where $q = (k_1^2 - \beta^2)^{1/2}$. Note that the incident wave u_i consists of the spectrum $U(\beta - \beta_0)$ centered at β_0. For example, if u_i is a Gaussian beam, we have

$$A(x') = A_0 \exp\left[-\left(\frac{x'}{W_0}\right)^2\right],$$

$$U(\beta - \beta_0) = \frac{A_0 W_0 \sqrt{\pi}}{\cos\theta_0} \exp\left[-\frac{W_0^2(\beta - \beta_0)^2}{4\cos^2\theta_0}\right]. \tag{6-80}$$

The wave number β in the x direction extends from $-\infty$ to $+\infty$. The range $-k_1 < \beta < k_1$ corresponds to the plane wave with real angle θ_i of incidence $\beta = k_1 \sin\theta_i (-\pi/2 < \theta_i < \pi/2)$. However, the range $\beta > k_1$ corresponds to the complex angle of incidence $\theta_i = \pi/2 + j\theta'', 0 \le \theta'' \le \infty$, and the range $\beta < -k_1$ corresponds to $\theta_i = -\pi/2 - j\theta'', 0 \le \theta'' \le \infty$. These ranges $|\beta| > k_1$ represent the evanescent wave. In general, the finite beam or the wave from a localized source excites both the propagating $|\beta| < k_1$ and the evanescent waves $|\beta| > k_1$.

The reflected wave $u_r(x, z)$ can be written as

$$u_r(x, z) = \frac{1}{2\pi} \int R(\beta) U(\beta - \beta_0) \exp[-j\beta x - jq(z + h)]\, d\beta, \tag{6-81}$$

where $R(\beta)$ is the reflection coefficient to be determined by applying the boundary condition.

When we apply the boundary condition at $z = 0$ using (6-79), (6-81), and the transmitted wave,

$$u_t(x, z) = \frac{1}{2\pi} \int T(\beta) U(\beta - \beta_0) e^{-j\beta x + jq_t z - jqh}\, d\beta, \tag{6-82}$$

where $q_t = (k_2^2 - \beta^2)^{1/2}$, we obtain the relationship identical to the plane wave case except that $k_1 \sin\theta_i$ for a plane wave is replaced by β. Thus each Fourier component $U(\beta - \beta_0)$ behaves as if it is a plane wave with $\beta = k_1 \sin\theta_i$. The difference is that the Fourier components include all β, both propagating and evanescent waves, while the plane wave is limited to the real angle of incidence.

If u represents a TE wave in Section 3-4, we get (for $\mu_1 = \mu_2 = \mu_0$)

$$R(\beta) = \frac{q - q_t}{q + q_t}, \tag{6-83}$$

where $q = (k_1^2 - \beta^2)^{1/2}$ and $q_t = (k_2^2 - \beta^2)^{1/2}$. If u represents a TM wave in Section 3-5, we get

$$R(\beta) = \frac{(q_t/n_2^2) - (q/n_1^2)}{(q_t/n_2^2) + (q/n_1^2)}. \tag{6-84}$$

If u represents an acoustic wave in Section 3-8,

$$R(\beta) = \frac{(\rho_2/q_t) - (\rho_1/q)}{(\rho_2/q_t) + (\rho_1/q)}. \tag{6-85}$$

Let us examine the reflected field (6-81). First, we note that for a beam wave, $U(\beta - \beta_0)$ has a peak at $\beta = \beta_0$ and decays away from $\beta = \beta_0$, as seen in (6-80). If the beam width W_0 is many wavelengths wide, U is very much concentrated near $\beta = \beta_0$. If the reflection coefficient $R(\beta)$ is a slowly varying function of β, then as a first approximation, we can let $R(\beta) \sim R(\beta_0)$. We then get from (6-81) the reflected wave that is equal to the beam wave $u_{r0}(x, z)$ originated at the image point $z = -h$ except for the reflection coefficient.

$$\begin{aligned} u_r(x, z) &\sim R(\beta_0) u_{r0}(x, z), \\ u_{r0}(x, z) &= \frac{1}{2\pi} \int U(\beta - \beta_0) \exp[-j\beta x - jq(z + h)]\, d\beta. \end{aligned} \tag{6-86}$$

The approximation $R(\beta) \approx R(\beta_0)$ is no longer valid, however, when the angles of incidence θ_0 and n_1 and n_2 are such that the total reflection takes place, $\sin\theta_0 > (n_2/n_1)$. Then the beam is not only totally reflected, but the reflected beam is shifted in space because of the additional phase in the reflection coefficient. As an example, consider the TE wave (6-83). Since in the neighborhood of $\beta \approx \beta_0$, $q_t = (k_2^2 - \beta^2)^{1/2} \sim k(n_2^2 - n_1^2 \sin^2\theta_0)^{1/2} \sim -jk(n_1^2 \sin^2\theta_0 - n_2^2)^{1/2}$, we let $q_t = -j\alpha_t$. Then we write

$$R(\beta) = \frac{q - q_t}{q + q_t} = \frac{q + j\alpha_t}{q - j\alpha_t} = \exp[j\phi(\beta)], \tag{6-87}$$

where $\phi(\beta) = 2\tan^{-1}(\alpha_t/q)$. Since the major contribution to the integral (6-81) comes from the neighborhood of $\beta = \beta_0$, we expand $\phi(\beta)$ in Taylor's series about β_0 and keep the first two terms:

$$\phi(\beta) = \phi(\beta_0) + (\beta - \beta_0)\phi'(\beta_0). \tag{6-88}$$

Substituting (6-87) and (6-88) into (6-81), we get

$$u_r(x, z) = R(\beta_0) e^{-j\beta_0 \phi'(\beta_0)} u_{r0}(x - \phi'(\beta_0), z) \tag{6-89}$$

This shows that the reflected beam $u_r(x, z)$ is proportional to the beam u_{r0} from the image point with the lateral shift of $\Delta = \phi'(\beta_0)$ (Fig. 6-12). This shift Δ is called the Goos–Hanchen effect and it becomes greater the closer the incident angle θ_0 be-

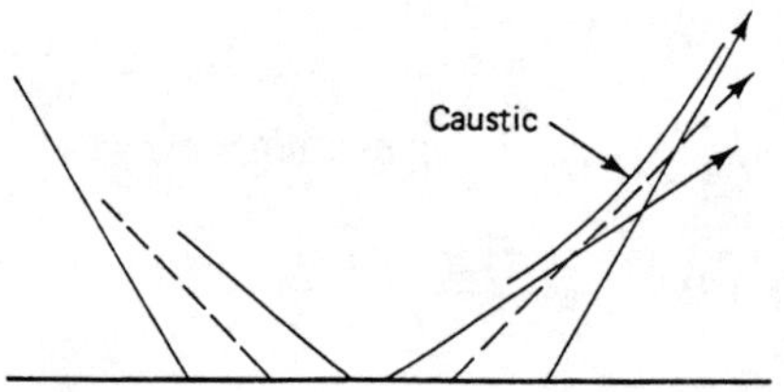

Figure 6-13 Caustic when the beam wave is incident at an angle close to the critical angle.

comes to the critical angle $\sin^{-1}(n_2/n_1)$. As θ_0 approaches the critical angle, the first-order approximation (6-88) becomes insufficient, and in the neighborhood of the critical angle, the reflected rays form a caustic (Fig. 6-13).

6-8 HIGHER-ORDER BEAM-WAVE MODES

In Sections 6-6 and 6-7 we discussed a beam whose amplitude distribution is Gaussian. This is the most important practical beam wave. However, this constitutes the fundamental beam-wave mode, and there are an infinite number of higher-order beam-wave modes in addition to the fundamental mode. To investigate these higher-order modes, it is convenient to start with the parabolic approximation to the wave equation. Consider the field $u(x, y, z)$ that satisfies the wave equation

$$(\nabla^2 + k^2)u(x, y, z) = 0. \tag{6-90}$$

For a beam wave, the field is mostly propagating in the z direction, and thus we write

$$u(x, y, z) = U(x, y, z)e^{-jkz}. \tag{6-91}$$

The function U should be a slowly varying function of z. Substituting (6-91) into (6-90), and noting that

$$\nabla u = (\nabla U)e^{-jkz} - jk\hat{z}Ue^{-jkz},$$

$$\nabla^2 u = \left(\nabla^2 U - j2k\frac{\partial}{\partial z}U - k^2 U\right)e^{-jkz},$$

we get

$$\left(\nabla^2 - j2k\frac{\partial}{\partial z}\right)U = 0. \tag{6-92}$$

This is exact. Now if U is slowly varying in z such that the variation of U over a wavelength is negligibly small, we have

$$\left|\frac{\partial U}{\partial z}\right| \sim \left|\frac{\Delta U}{\Delta z}\right| < \left|\frac{\Delta U}{\lambda}\right| << \left|\frac{U}{\lambda}\right| \sim |kU|. \tag{6-93}$$

Thus we can approximate (6-92) by

$$\left(\nabla_t^2 - j2k\frac{\partial}{\partial z}\right)U = 0. \tag{6-94}$$

where ∇_t^2 is the Laplacian in x and y (transverse to $\hat{z}$). This is the parabolic approximation to the wave equation (6-92) and greatly simplifies the mathematical analysis for a beam wave. This parabolic approximation is equivalent to the para-axial approximation (6-60).

Since (6-94) is identical to the para-axial approximation, it is easy to show that the Gaussian beam wave shown in (6-65)

$$U_g(x,y,z)=\frac{A_0}{1-j\alpha z}\exp\left[-\frac{\rho^2}{W_0^2(1-j\alpha z)}\right],\qquad \alpha=\frac{\lambda}{\pi W_0^2},\tag{6-95}$$

satisfies (6-94).

To obtain the higher-order modes, we let

$$U(x,y,z)=U_g(x,y,z)f(t)g(\tau)e^{j\phi(z)},\tag{6-96}$$

where

$$t=\sqrt{2}\frac{x}{W},\qquad \tau=\sqrt{2}\frac{y}{W}.$$

$$W=W_0[1+(\alpha z)^2]^{1/2}.$$

Substituting (6-96) into (6-94), and noting that U_g also satisfies (6-94), we get

$$\frac{1}{f}\left(\frac{d^2f}{dt^2}-2t\frac{df}{dt}\right)+\frac{1}{g}\left(\frac{d^2g}{d\tau^2}-2\tau\frac{dg}{d\tau}\right)-kW^2\frac{d\phi}{dz}=0.\tag{6-97}$$

Noting that x and y enter into the first and second terms only, respectively, we require that each term be constant to satisfy (6-97) at all x and y. Thus we let

$$\begin{aligned}\frac{1}{f}\left(\frac{d^2f}{dt^2}-2t\frac{df}{dt}\right)&=2m,\\ \frac{1}{g}\left(\frac{d^2g}{d\tau^2}-2\tau\frac{df}{d\tau}\right)&=2n,\\ kW^2\frac{d\phi}{dz}&=2(m+n),\end{aligned}\tag{6-98}$$

where m and n are constant.

The first two equations of (6-98) are Hermite differential equations:

$$\left(\frac{d^2}{dt^2}-2t\frac{d}{dt}+2m\right)H_m(t)=0,\tag{6-99}$$

whose solutions $H_m(t)$ are Hermite polynomials. Equation (6-98) can be integrated to obtain

$$\begin{aligned}\phi&=\int_0^z\frac{2(m+n)}{kW^2}dz\\ &=\frac{2(m+n)}{kW_0^2}\int_0^z\frac{dz}{1+(\alpha z)^2}\\ &=(m+n)\tan^{-1}\alpha z.\end{aligned}\tag{6-100}$$

We therefore obtain the general representation of the higher-order beam-wave modes:

$$U_{mn}(x,y,z) = U_g(x,y,z)H_m(t)H_n(\tau)\exp[j(m+n)\tan^{-1}\alpha z], \qquad (6\text{-}101)$$

where t and τ are defined in (6-96) and α is defined in (6-95).

Hermite polynomials $H_m(t)$ are orthogonal:

$$\int_{-\infty}^{\infty} e^{-t^2} H_m(t)H_{m'}(t)\,dt = \begin{cases} 0 & \text{if } m \neq m' \\ 2^m\sqrt{\pi}\,m & \text{if } m = m'. \end{cases} \qquad (6\text{-}102)$$

Therefore, noting that

$$|U_g|^2 = \frac{A_0^2}{W^2}\exp[-(t^2+\tau^2)], \qquad (6\text{-}103)$$

all modes $U_{mn}(x,y,z)$ are orthogonal in a constant z plane.

$$\iint U_{mn}U^*_{m'n'}\,dx\,dy = 0 \quad \text{if } m \neq n' \quad \text{or} \quad n \neq n'. \qquad (6\text{-}104)$$

The first few Hermite polynomials $H_m(t)$ are

$$H_0(t) = 1, \qquad H_1(t) = 2t,$$

$$H_2(t) = 4t^2 - 2, \qquad H_3(t) = 8t^3 - 12t.$$

The functions $H_m(t)\exp(-t^2/2)$ are sketched in Fig. 6-14. The beam wave radiating out of a He–Ne laser consists of the fundamental Gaussian beam and the higher-order modes described above.

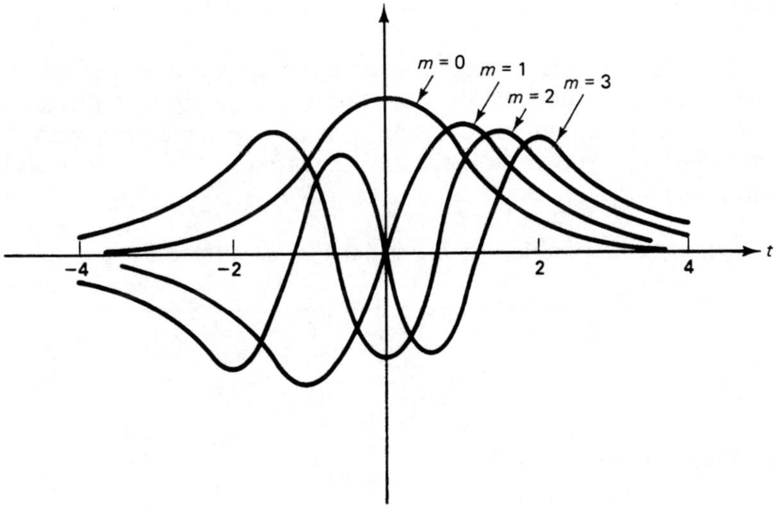

Figure 6-14 Function $H_m(t)\exp(-t^2/2)$ normalized by $(2^m\sqrt{\pi}\,m)^{1/2}$.

6-9 VECTOR GREEN'S THEOREM, STRATTON–CHU FORMULA, AND FRANZ FORMULA

We have discussed scalar Green's theorem in Section 6-1. In this section we consider the equivalent theorem for vector fields. Let us consider the vector fields $\bar{P}(\bar{r})$ and $\bar{Q}(\bar{r})$ in the volume V surrounded by the surface S. We assume that $\bar{P}$ and $\bar{Q}$ and their first and second derivatives are continuous in V and on S. Using the divergence theorem, we have

$$\int_V \nabla\cdot(\bar{P}\times\nabla\times\bar{Q})\,dV = \int_S \bar{P}\times\nabla\times\bar{Q}\cdot d\bar{S}. \tag{6-105}$$

We now use the identity

$$\nabla\cdot(\bar{A}\times\bar{B}) = \bar{B}\cdot\nabla\times\bar{A} - \bar{A}\cdot\nabla\times\bar{B}. \tag{6-106}$$

Letting $\bar{A}=\bar{P}$ and $\bar{B}=\nabla\times\bar{Q}$, we get the vector Green's first identity.

$$\int_V (\nabla\times\bar{Q}\cdot\nabla\times\bar{P} - \bar{P}\cdot\nabla\times\nabla\times\bar{Q})\,dV = \int_S \bar{P}\times\nabla\times\bar{Q}\cdot d\bar{S}. \tag{6-107}$$

To get the second identity or Green's theorem, we interchange $\bar{P}$ and $\bar{Q}$ in (6-107).

$$\int_V (\nabla\times\bar{P}\cdot\nabla\times\bar{Q} - \bar{Q}\cdot\nabla\times\nabla\times\bar{P})\,dV = \int_S \bar{Q}\times\nabla\times\bar{P}\cdot d\bar{S}. \tag{6-108}$$

Subtracting (6-108) from (6-107), we get the vector Green's theorem or the vector Green's second identity.

$$\int_V (\bar{Q}\cdot\nabla\times\nabla\times\bar{P} - \bar{P}\cdot\nabla\times\nabla\times\bar{Q})\,dV = \int_S (\bar{P}\times\nabla\times\bar{Q} - \bar{Q}\times\nabla\times\bar{P})\cdot d\bar{S}. \tag{6-109}$$

The vector Green's theorem developed above is now applied to the electromagnetic field problem. We consider the scattering problem discussed in Section 6-1. The electromagnetic field $\bar{E}_i$ and $\bar{H}_i$ is incident on a body with the volume V surrounded by the surface S (Fig. 6-15). The Stratton–Chu formula can be stated for the following three cases.

1. When the observation point $\bar{r}$ is outside the surface S,

$$\begin{aligned} \bar{E}_i(\bar{r}) + \int_S \bar{E}_s\,dS' &= \bar{E}(\bar{r}),\\ \bar{H}_i(\bar{r}) + \int_S \bar{H}_s\,dS' &= \bar{H}(\bar{r}). \end{aligned} \tag{6-110}$$

2. When $\bar{r}$ is on the surface S,

$$\begin{aligned} \bar{E}_i(\bar{r}) + \fint_S \bar{E}_s\,dS' &= \tfrac{1}{2}\bar{E}(\bar{r}),\\ \bar{H}_i(\bar{r}) + \fint_S \bar{H}_s\,dS' &= \tfrac{1}{2}\bar{H}(\bar{r}). \end{aligned} \tag{6-111}$$

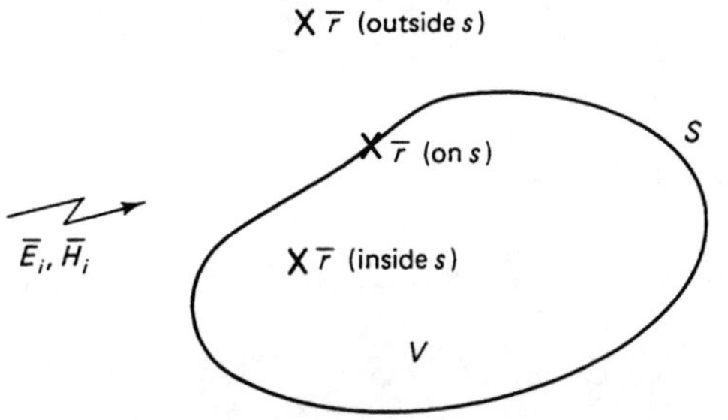

Figure 6-15 Stratton–Chu formula.

3. When $\bar{r}$ is inside S,

$$\begin{aligned} \bar{E}_i(\bar{r}) + \int_S \bar{E}_s\, dS' &= 0, \\ H_i(\bar{r}) + \int_S \bar{E}_s\, dS' &= 0, \end{aligned} \tag{6-112}$$

where

$$\begin{aligned} \bar{E}_s &= -[j\omega\mu G\hat{n}' \times \bar{H} - (\hat{n}' \times \bar{E}) \times \nabla' G - (\hat{n}' \cdot \bar{E})\nabla' G], \\ \bar{H}_s &= j\omega\mu G\hat{n}' \times \bar{E} + (\hat{n}' \times \bar{H}) \times \nabla' G + (\hat{n}' \cdot \bar{H})\nabla' G], \\ \bar{E} &= E(\bar{r}'), \qquad \bar{H} = \bar{H}(\bar{r}'), \end{aligned} \tag{6-113}$$

$G(\bar{r} - \bar{r}') = \exp(-jk|\bar{r} - \bar{r}'|)/(4\pi|\bar{r} - \bar{r}'|)$ is the scalar free-space Green's function, and ∇' is the gradient with respect to $\bar{r}'$. The surface integral in (6-111) is the Cauchy principal value of the integral, and the fields $\bar{E}(\bar{r}')$ and $\bar{H}(\bar{r}')$ are those as $\bar{r}'$ approaches S from outside.

Equation (6-111) will be used later to construct an EFIE (electric field integral equation) and an MFIE (magnetic field integral equation). Equation (6-112) is the vector extinction theorem, also called the vector Ewald–Oseen extinction theorem or the vector null field theorem. Inside the surface S, the incident fields $\bar{E}_i$ and $\bar{H}_i$ are extinguished by the contribution from the surface fields. Proof of formula (6-113) is given in the Appendix to Chapter 6, Section A.

In addition to the Stratton–Chu formula, there are other equivalent representations of the electromagnetic field. Consider the field $\bar{E}(\bar{r})$ and $\bar{H}(\bar{r})$ outside the surface S and the volume V (Fig. 6-16). The Stratton–Chu formula is given by

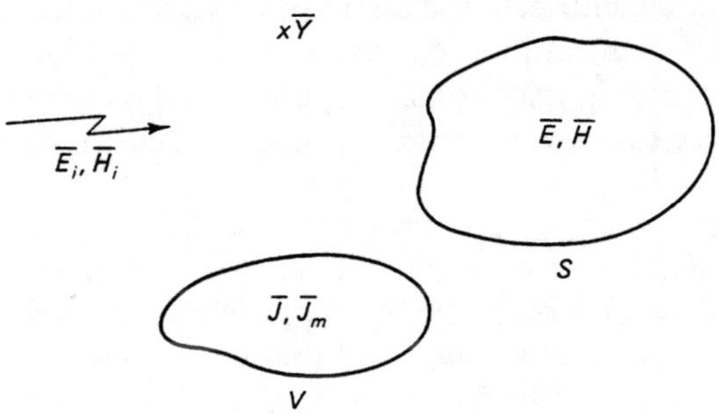

Figure 6-16 Franz formula.

$$\bar{E}(\bar{r}) = \bar{E}_i(\bar{r}) + \int_S \bar{E}_s\, dS' + \int_V \bar{E}_v\, dV',$$
$$\bar{H}(\bar{r}) = \bar{H}_i(\bar{r}) + \int_S \bar{H}_s\, dS' + \int_V \bar{H}_v\, dV', \tag{6-114}$$

where $\bar{E}_s$ and $\bar{H}_s$ are given in (6-113) and $\bar{E}_v$ and $\bar{H}_v$ are given by

$$\bar{E}_v(\bar{r}) = -(j\omega\mu G\bar{J} + \bar{J}_m \times \nabla' G - \frac{\rho}{\epsilon}\nabla' G),$$
$$\bar{H}_v(\bar{r}) = -(j\omega\epsilon G\bar{J}_m - \bar{J} \times \nabla' G - \frac{\rho_m}{\mu}\nabla' G). \tag{6-115}$$

The following equivalent representation is called the *Franz formula* (Tai, 1972):

$$\bar{E}(\bar{r}) = \bar{E}_i(\bar{r}) + \nabla \times \nabla \times \bar{\pi} - j\omega\mu\nabla \times \bar{\pi}_m,$$
$$\bar{H}(\bar{r}) = \bar{H}_i(\bar{r}) + j\omega\epsilon\nabla \times \bar{\pi} + \nabla \times \nabla \times \bar{\pi}_m, \tag{6-116}$$

where

$$\pi(\bar{r}) = \frac{1}{j\omega\epsilon}\left(\int_V \bar{J}G\, dV' + \int_S \hat{n}' \times \bar{H}G\, dS'\right),$$
$$\pi_m(\bar{r}) = \frac{1}{j\omega\mu}\left(\int_V \bar{J}_m G\, dV' - \int_S \hat{n}' \times \bar{E}G\, dS'\right).$$

The equivalence between the Stratton–Chu formula and the Franz formula is shown by Tai.

6-10 EQUIVALENCE THEOREM

In Section 6-9 we discussed the Stratton–Chu formula and its equivalent Franz formula. According to the Franz formula, the field outside the surface S is given by the incident field and the surface integral of the tangential electric field $\hat{n}' \times \bar{E}$ and the tangential magnetic field $\hat{n}' \times \bar{H}$. Inside the surface S, the field is zero as shown in the extinction theorem. Since the surface S is arbitrary, we can state that the actual field outside the surface S is identical to the field generated by the equivalent surface magnetic current $\bar{J}_{ms} = \bar{E} \times \hat{n}'$ and the equivalent surface electric current $\bar{J}_s = \hat{n}' \times \bar{H}$. These fictitious currents $\bar{J}_{ms}$ and $\bar{J}_s$ on S produce a field identical to the original field outside S, but they extinguish the field inside S, producing the null field (Harrington, 1968, Chap. 3) (Fig. 6-17).

In the above we used both $\bar{J}_s$ and $\bar{J}_{ms}$ on the surface. However, according to the uniqueness theorem, if we specify either $\bar{J}_s$ or $\bar{J}_{ms}$ on S, the field outside S is uniquely determined, and therefore we may use $\bar{J}_s$ or $\bar{J}_{ms}$. For example, consider a conducting body with the aperture A on its surface S. The tangential electric field over the aperture is $\bar{E} \times \hat{n}'$. As far as the field outside S is concerned, the field is identical to the field produced by the conducting body with the aperture closed but with the

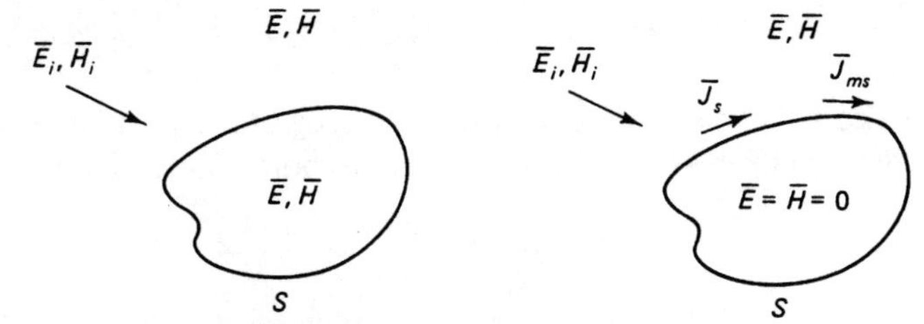

Figure 6-17 Equivalence theorem.

magnetic surface current $\bar{J}_{ms} = \bar{E} \times \hat{n}'$ placed in front of the original location of the aperture (Fig. 6-18). Note that $\bar{E} \times \hat{n}'$ is zero just behind $\bar{J}_{ms}$ on S, but $\bar{E} \times \hat{n}'$ is identical to the original field just in front of $\bar{J}_{ms}$.

6-11 KIRCHHOFF APPROXIMATION FOR ELECTROMAGNETIC WAVES

In Section 6-3 we discussed the Kirchhoff approximation for the scalar field. The electromagnetic Kirchhoff approximation can be obtained using the same technique. Let us consider a large aperture on a screen. According to the Kirchhoff approximation, we assume that the electric and magnetic fields tangential to the aperture are equal to those of the incident field:

$$\begin{aligned} \hat{n}' \times \bar{H} &= \hat{n}' \times \bar{H}_i, \\ \bar{E} \times \hat{n}' &= \bar{E}_i \times \bar{n}'. \end{aligned} \tag{6-117}$$

Then using the Franz formula (6-116), we get

$$\begin{aligned} \bar{E}(\bar{r}) &= \nabla \times \nabla \times \bar{\Pi} - j\omega\mu \nabla \times \bar{\Pi}_m, \\ \bar{\Pi} &= \frac{1}{j\omega\epsilon} \int_S \hat{n}' \times \bar{H}_i G \, dS', \\ \bar{\Pi}_m &= \frac{1}{j\omega\mu} \int_S \bar{E}_i \times \hat{n}' G \, dS', \\ G &= \frac{\exp(-jk|\bar{r} - \bar{r}')}{4\pi|\bar{r} - \bar{r}'|}. \end{aligned} \tag{6-118}$$

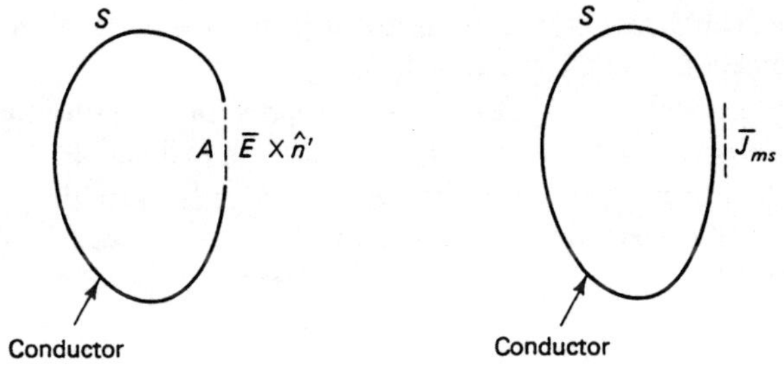

Figure 6-18 Aperture on a conducting surface.

Here all the quantities are known and therefore $\bar{E}(\bar{r})$ at any point $\bar{r}$ can be calculated.

If we consider the field $\bar{E}(\bar{r})$ in the far zone of the aperture where $|\bar{r}| >> D^2/\lambda$ and D is the size of the aperture, we can approximate G by the following (Fig. 6-5):

$$G = \frac{\exp(-jkR_0 + jk\hat{o} \cdot \bar{r}')}{4\pi R_0}. \tag{6-119}$$

Also note that in the far zone, the field is propagating in the direction $\hat{o}$ and is thus proportional to $\exp(-jk\hat{o} \cdot \bar{r})$. Since $\nabla \exp(-jk\hat{o} \cdot \bar{r}) = -jk\hat{o} \exp(-jk\hat{o} \cdot \bar{r})$, the operator ∇ is equal to $-jk\hat{o}$. Thus we have $\nabla \times \nabla \times = -k^2 \hat{o} \times \hat{o} \times$ and $\nabla \times = -jk\hat{o} \times$.

Also, we consider the components E_θ and E_ϕ. Then we finally get the following Kirchhoff approximation for the radiation field when the incident field $\bar{E}_i$ and $\bar{H}_i$ is known over the aperture S:

$$\begin{aligned}
E_\theta &= -\frac{j\omega\mu}{4\pi R_0} e^{-jkR_0} \int \hat{\theta} \cdot (\hat{n}' \times \bar{H}_i) e^{jk\hat{o}\cdot\bar{r}'}\, dS' \\
&\quad - \frac{jk}{4\pi R_0} e^{-jkR_0} \int \hat{\phi} \cdot (\bar{E}_i \times \hat{n}') e^{jk\hat{o}\cdot\bar{r}'}\, dS', \\
E_\phi &= -\frac{j\omega\mu}{4\pi R_0} e^{-jkR_0} \int \hat{\phi} \cdot (\hat{n}' \times \bar{H}_i) e^{jk\hat{o}\cdot\bar{r}'}\, dS' \\
&\quad + \frac{jk}{4\pi R_0} e^{-jkR_0} \int \hat{\theta} \cdot (\bar{E}_i \times \hat{n}') e^{jk\hat{o}\cdot\bar{r}'}\, dS'.
\end{aligned} \tag{6-120}$$

For example, if the incident wave is a plane wave normally incident on a rectangular aperture ($2a \times 2b$) in the x–y plane, and is polarized in the x direction, we have $\bar{E}_i = E_0 \hat{x}$ and $\bar{H}_i = (E_0/\eta)\hat{y}$ with $\eta = \sqrt{\mu_0/\epsilon_0}$. Then the radiaton field $\bar{E}$ is given by

$$\bar{E} = \frac{jke^{-jkR_0}}{4\pi R_0}(1 + \cos\theta)[\hat{\theta} \cos\phi - \hat{\phi} \sin\phi] F(\theta, \phi), \tag{6-121}$$

$$F(\theta, \phi) = 4ab \frac{\sin K_1 a}{K_1 a} \frac{\sin K_2 b}{K_2 b},$$

$$K_1 = k \sin\theta \cos\phi, \qquad K_2 = k \sin\theta \sin\phi.$$

PROBLEMS

6-1. A plane scalar wave is normally incident on a large square aperture of a (m) $\times$ b (m). Using the Kirchhoff approximation, find the radiation field.

6-2. If a plane wave is normally incident on a square plate of $a \times b$ with the reflection coefficient of $R = -1$, find the scattered field using the Kirchhoff approximation.

6-3. Find the radiation pattern, half-power beamwidth, and first sidelobe level (in dB) in the x–z plane and y–z plane for the aperture field distribution $U_0(x, y) = A_0 \cos^2(\pi x/a)$ in a square aperture of $a = 3$ m $\times$ $b = 1$ m at a frequency of 10 GHz.

6-4. Is it possible to send a light beam ($\lambda = 0.5\ \mu m$) from the earth to illuminate an area of 500 m diameter between half-power points on the surface of the moon? If so, what should be the size of the aperture of the transmitter? The distance between the moon and the earth is approximately 384,400 km.

6-5. It is planned to collect the solar energy by a large solar panel on a synchronous satellite, convert it to microwaves, and send it down to the earth. The total power to be transmitted should be 5 GW. The power density on the ground should be less than 10 mW/cm^2. A microwave at 2.45 GHz is to be used. Assume that the transmitter is a circular aperture of 1 km diameter and that the aperture distribution is uniform. Assume that the altitude of the satellite in geosynchronous equatorial orbit is 35,800 km. Find the intensity distribution on the ground.

6-6. Plot the beam size W of a focused optical beam with $W_0 = 1$ cm at $\lambda = 0.6\ \mu m$ as a function of distance. The focal distance is 1 m. What is the half-power beamwidth (in degrees) at a large distance?

6-7. Calculate the Goos–Hanchen shift Δx for an optical beam with $\lambda = 0.6\ \mu m$ incident from the medium $n = 2.5$ to air. The wave is polarized in the plane of incidence. Plot Δx as a function of the incident angle.

6-8. Find an expression for $\bar{E}$ in (6-121) when both the x and y components of the aperture field are known.

7

Periodic Structures and Coupled-Mode Theory

There are many important structures whose characteristics are periodic in space. Examples are three-dimensional lattice structures for crystals, artificial dielectric consisting of periodically placed conducting pieces, Yagi antennas that have periodically spaced elements, corrugated surfaces, and waveguides with periodic loadings. In addition, systems of lenses that are placed with equal spacings and open resonators can be considered periodic structures. Guided waves along these structures exhibit a unique frequency dependence often characterized by stop bands and passbands. The scattered waves from these periodic structures are characterized by grating modes resulting from periodic interference of waves in different directions. The starting point in solving the problems of periodic structures is Floquet's theorem, which is described in Section 7-1.

In this chapter we discuss guided waves propagating along a periodic structure, waves propagating through periodic layers, and plane waves incident on periodic structures. In discussing these problems, we present integral equation formulations that are used extensively in many electromagnetic problems. We also discuss a classical problem of scattering by sinusoidal surfaces and include a short description of coupled-mode theory.

7-1 FLOQUET'S THEOREM

Let us consider a wave propagating in periodic structures, which may be characterized by periodic boundary conditions or a periodically varied dielectric constant (Fig. 7-1). We note that the fields at a point z in an infinite periodic structure differ from the fields one period L away by a complex constant. This is obviously true because in an infinite periodic structure, there should be no difference between the fields at z and at $z + L$ except for the constant attenuation and phase shift. Let a function $u(z)$ represent a wave. Then a wave $u(z)$ at z and a wave $u(z + L)$ at $z + L$ are related in the same manner as a wave $u(z + L)$ at $z + L$ and a wave $u(z + 2L)$ at $z + 2L$.

Mathematically, we write

$$\frac{u(z+L)}{u(z)} = \frac{u(z+2L)}{u(z+L)} = \frac{u(z+mL)}{u[z+(m-1)L]} = C = \text{constant}. \tag{7-1}$$

From this we obtain

$$u(z+mL) = C^m u(z). \tag{7-2}$$

The constant C is in general complex, which we write

$$C = e^{-j\beta L}, \qquad \beta = \text{complex}, \tag{7-3}$$

and β represents the propagation constant.

Now let us consider a function

$$R(z) = e^{j\beta z} u(z). \tag{7-4}$$

Then $R(z + L) = e^{j\beta(z+L)} u(z + L) = R(z)$. Therefore, $R(z)$ is a periodic function of z with the period L, and thus can be represented in a Fourier series.

$$R(z) = \sum_{n=-\infty}^{\infty} A_n e^{-j(2n\pi/L)z}. \tag{7-5}$$

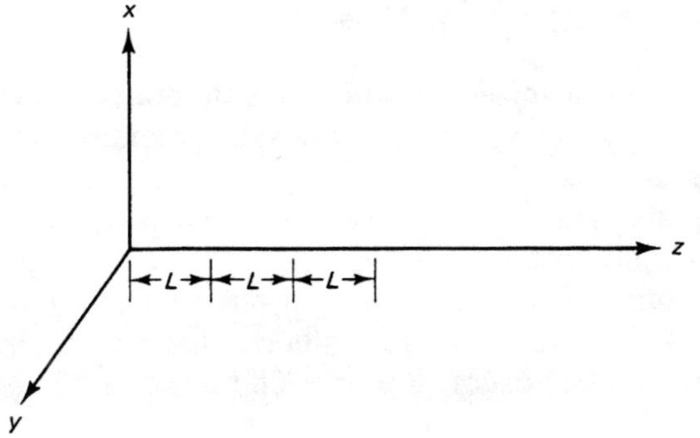

Figure 7-1 Periodic structures.

Using (7-4), we finally obtain a general expression for a wave in a periodic structure with the period L.

$$u(z) = \sum_{n=-\infty}^{\infty} A_n e^{-j(\beta + 2n\pi/L)z}$$
$$= \sum_{n=-\infty}^{\infty} A_n e^{-j\beta_n z}, \qquad \beta_n = \beta + \frac{2n\pi}{L}. \tag{7-6}$$

Noting that, in general, the wave consists of both positive-going and negative-going waves, we write

$$u(z) = \sum_{n=-\infty}^{\infty} A_n e^{-j\beta_n z} + \sum_{n=-\infty}^{\infty} B_n e^{+j\beta_n z}. \tag{7-7}$$

This is a representation of a wave in periodic structures in a form of an infinite series, resembling harmonic representation ($e^{-j\omega_n t}$) in time. The nth term in (7-6) is called the nth *space harmonic* or *Hartree harmonic*. Equation (7-7) is the mathematical representation of Floquet's theorem, which states that the wave in periodic structures consists of an infinite number of *space harmonics*. In this chapter we consider two cases. One is guided waves along a periodic structure and the other is the scattering of plane waves from periodic structures.

7-2 GUIDED WAVES ALONG PERIODIC STRUCTURES

Consider a wave propagating along a periodic structure with period L. The positive-going wave is given by (7-6), and when the wave is propagating, β is real and the phase velocity is different for each harmonic.

$$v_{pn} = \frac{\omega}{\beta_n} = \frac{\omega}{\beta + 2n\pi/L}, \tag{7-8}$$

but the group velocity is the same for all harmonics

$$v_{gn} = \frac{1}{d\beta_n/d\omega} = \frac{1}{d\beta/d\omega} = v_{g0}. \tag{7-9}$$

The propagation constant β is real in some frequency ranges, and this is called the *pass band*. The frequency range where β is purely imaginary and the wave is evanescent is called the *stop band*.

We now examine the k–β diagram for periodic structures. We first note that if β is increased by $2\pi/L$, this is equivalent to changing β_n to β_{n+1}, and the general expression for the fields is unaltered. Therefore, k is a periodic function of β with period $2\pi/L$. It is also clear that since the wave propagation in the $+z$ and the $-z$ direction should have the same characteristics, k is an even function of β. As an example, the k–β diagram for a tape helix is shown in Fig. 7-2. The labeling of each mode is usually made to identify the lowest β with positive slope ($v_g > 0$) by the

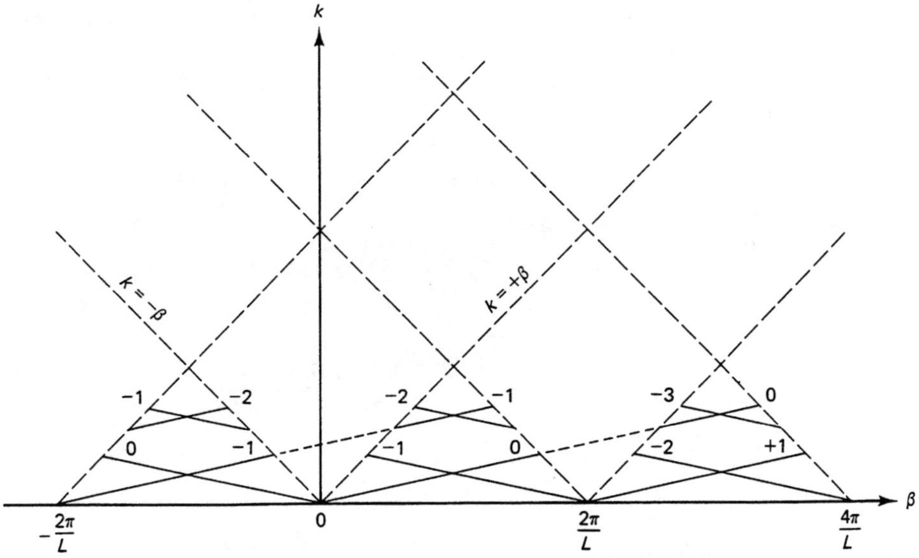

Figure 7-2 k–β diagram for periodic structures.

zeroth mode and the $\beta + 2n\pi/L$ by the nth mode. Similarly, for the mode with negative slopes ($v_g < 0$), $-(\beta + 2n\pi/L)$ is labeled by the nth mode.

As an example, let us consider TM modes propagating along a corrugated surface (Fig. 7-3). We look for a trapped surface-wave solution. In this section we employ the *integral equation formulation* for the boundary value problem. In the integral equation formulation, the differential equation and the boundary conditions are combined to obtain an integral equation. The integral equation contains an unknown function under an integral operator, just as the differential equation contains an unknown function under a differential operator. Let us examine this technique for the problem of a corrugated surface.

In this problem we first express the magnetic field H_1 in the region $x > 0$ due to the unknown tangential electric field at $x = 0$. Next we express the magnetic field H_2 in the region $x < 0$ due to the "same" tangential electric field at $x = 0$. Since we used

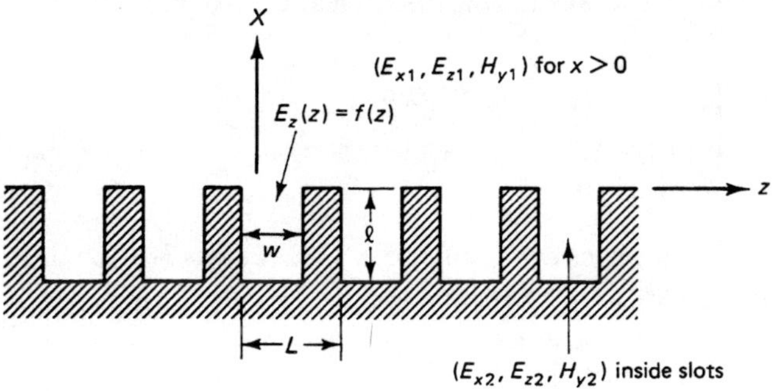

Figure 7-3 Corrugated surface.

the same electric field at $x=0$, although it is still unknown, we automatically satisfied the boundary condition that the tangential electric field is continuous at $x=0$. We then satisfy the boundary condition that the tangential magnetic field is continuous by equating H_1 to H_2, yielding an integral equation for the unknown tangential electric field at $x=0$.

To follow the foregoing procedure, let us first note that all TM modes are given by H_y satisfying a scalar wave equation,

$$\left(\frac{\partial^2}{\partial x^2}+\frac{\partial^2}{\partial z^2}+k^2\right)H_y=0, \tag{7-10}$$

with

$$E_x=j\frac{1}{\omega\epsilon}\frac{\partial}{\partial z}H_y \quad \text{and} \quad E_z=-j\frac{1}{\omega\epsilon}\frac{\partial}{\partial x}H_y.$$

Next we express H_1 for $x>0$. According to Floquet's theorem, we write H_{y1} for $x>0$ in a series of space harmonics.

$$H_{y1}(x,z)=\sum_{n=-\infty}^{\infty} f_n(x)e^{-j\beta_n z}, \qquad \beta_n=\beta+\frac{2n\pi}{L}. \tag{7-11}$$

Since this must satisfy a wave equation (7-10), $f_n(x)$ should have a form $f_n(x)\approx e^{\pm jq_n x}$, where $q_n^2+\beta_n^2=k^2$. Furthermore, this structure is open in the $+x$ direction and H_y must satisfy the radiation condition as $x\to+\infty$, and therefore, $f_n(x)$ must have the form $e^{-jq_n x}$.

Since we are only interested in a trapped slow-wave solution, q_n must be purely imaginary: $q_n=-j\alpha_n$. Therefore, we write

$$H_{y1}(x,z)=\sum_{-\infty}^{\infty} A_n e^{-\alpha_n x-j\beta_n z}, \qquad x>0, \tag{7-12}$$

where $\alpha_n^2=\beta_n^2-k^2$, and A_n is the amplitude for each space harmonic.

To express H_{y1} in terms of the tangential electric field E_z at $x=0$, we write E_{z1} using (7-10) and (7-12),

$$E_{z1}(x,z)=j\frac{1}{\omega\epsilon}\sum_{-\infty}^{\infty}\alpha_n A_n e^{-\alpha_n x-j\beta_n z}. \tag{7-13}$$

Now at $x=0$, $E_z(0,z)=0$ on the top surface of the corrugation and $E_z(0,z)$ is equal to the field in the slot $f(z)$.

$$E_z(0,z)=\begin{cases}0 & \text{for } \frac{W}{2}<|z|<\frac{L}{2} \\ f(z) & \text{for } |z|<\frac{W}{2}\end{cases}. \tag{7-14}$$

The coefficients A_n can be expressed in terms of $f(z)$ by using (7-13) and (7-14) and recognizing that the space harmonics are orthogonal.

$$\int_{-L/2}^{L/2}(e^{-j\beta_n z})(e^{-j\beta_m z})^*\,dz=\int_{-L/2}^{L/2}e^{-j(\beta_n-\beta_m)z}\,dz=\int_{-L/2}^{L/2}e^{-j(n-m)(2\pi/L)z}\,dz=L\delta_{mn}, \tag{7-15}$$

where Kronecker's delta $\delta_{mn} = 1$ when $m = n$ and $\delta_{mn} = 0$ when $m \neq n$. Recognizing that (7-13) is an expansion of $f(z)$ in an orthogonal series, we can obtain the coefficients of the series by multiplying both sides of (7-14) by $e^{j\beta_n z}$ and integrating over a period L.

$$\frac{j\alpha_n A_n}{\omega\epsilon} = \frac{1}{L}\int_{-W/2}^{W/2} f(z)e^{j\beta_n z}\,dz. \tag{7-16}$$

Substituting this into (7-12), we write H_{y1} in the form

$$H_{y1}(x,z) = -j\omega\epsilon\int_{-W/2}^{W/2} f(z')G_1(z,x;z',0)\,dz' \qquad \text{for } x > 0, \tag{7-17}$$

where

$$G_1(z,x;z',0) = \sum_{-\infty}^{\infty} \frac{e^{-\alpha_n x - j\beta_n(z-z')}}{L\alpha_n}.$$

This form (7-17) is written to conform to Green's function formulation, and G_1 is Green's function for periodic structures (see the Appendix to Chapter 7, Section A).

Next, consider the field inside the slot due to the field $f(z)$ at $x = 0$. H_y in the slot should consist of TEM modes and higher-order TM modes. Noting that $E_x = 0$ at $z = \pm W/2$ and $E_z = 0$ at $x = -l$, we write

$$H_{y2} = \sum_{n=0}^{\infty} B_n \cos\frac{n\pi(z + W/2)}{W}\cos[k_n(x+l)], \tag{7-18}$$

where $k_n^2 + (n\pi/W)^2 = k^2$ and B_0 represents the TEM mode and $B_n, n \neq 0$, represents all the higher-order modes. We obtain E_z from (7-18) using $E_z = -j(1/\omega\epsilon)(\partial/\partial x)H_y$. Equating this to $f(z)$ at $x = 0$, we get B_n, and substituting B_n into (7-18), we get

$$H_{y2}(x,z) = -j\omega\epsilon\int_{-W/2}^{W/2} f(z')G_2(z,x;z',0)\,dz', \tag{7-19}$$

where

$$G_2(z,x;z',0) = \frac{\cos k(l+x)}{Wk\sin kl} + \sum_{n=1}^{\infty}\frac{2\psi_n(z)\psi_n(z')\cos k_n(l+x)}{Wk_n \sin k_n l},$$

$$\psi_n(z) = \cos\left[\frac{n\pi}{W}\left(z + \frac{W}{2}\right)\right].$$

Equations (7-17) and (7-19) give $H_y(x,z)$ in the region $x > 0$ and $x < 0$, respectively, in terms of the same tangential electric field $f(z)$ in the slot. Now let us consider the boundary conditions at $x = 0$, which requires the continuity of tangential electric and magnetic fields at $x = 0$. We note that the tangential electric field over the slot is continuous at $x = 0$ because we used the same field $f(z)$. The continuity of the tangential magnetic field requires that (7-17) and (7-19) be equal over the slot at $x = 0$. Thus we get

$$\int_{-W/2}^{W/2} f(z')G(z,0;z',0)\,dz' = 0 \qquad \text{over the slot } |z| < \frac{W}{2}, \tag{7-20}$$

where $G(z,0;z',0) = G_1(z,0;z',0) - G_2(z,0;z',0)$.

If we can solve (7-20) for the unknown function $f(z')$ and the propagation constant β, all the other fields can be obtained by (7-17) and (7-19). But the analytical solution of (7-20) is, in general, not available. It is possible to solve (7-20) by a numerical technique such as the moment method. In this section, however, we are not concerned with a detailed description of $f(z)$ but are interested in obtaining the propagation constant β.

In this case we can obtain a convenient solution by equating the total complex power at both sides of the slot opening (Fig. 7-4). We note that $P_1 = P_2$, where

$$P_1 = \int_{-W/2}^{W/2} E_z^*(z)H_{y1}(z)\,dz,$$

$$P_2 = \int_{-W/2}^{W/2} E_z^*(z)H_{y2}(z)\,dz.$$

With (7-20), this is expressed as

$$\int_{-W/2}^{W/2} dz \int_{-W/2}^{W/2} dz' f^*(z)G(z,0;z',0)f(z') = 0. \tag{7-21}$$

This form, (7-21), can be shown to be a variational form for the propagation constant β (see the Appendix to Chapter 7, Section B).

Let us use the simplest trial function, $f(z) = \text{constant}$. This gives

$$\sum_{-\infty}^{\infty} \frac{1}{\alpha_n}\left[\frac{\sin \beta_n(W/2)}{\beta_n(W/2)}\right]^2 = \frac{L}{kW}\cot kl. \tag{7-22}$$

The k–β diagram for this structure is shown in Fig. 7-5.

The choice of $f(z)$ above is made for mathematical convenience. The best choice would be the one that satisfies the *edge condition*. The field behavior near the edge is not arbitrary and it must satisfy a certain condition called the edge condition

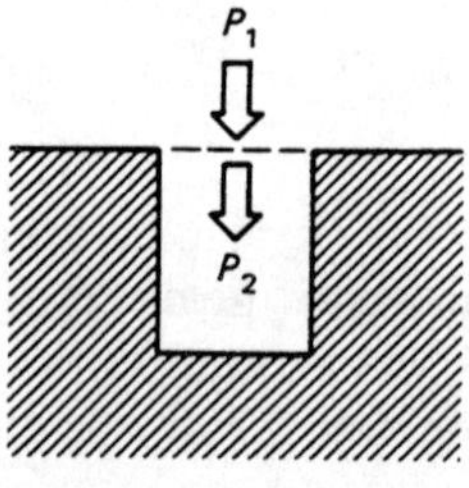

Figure 7-4 Conservation of the total power.

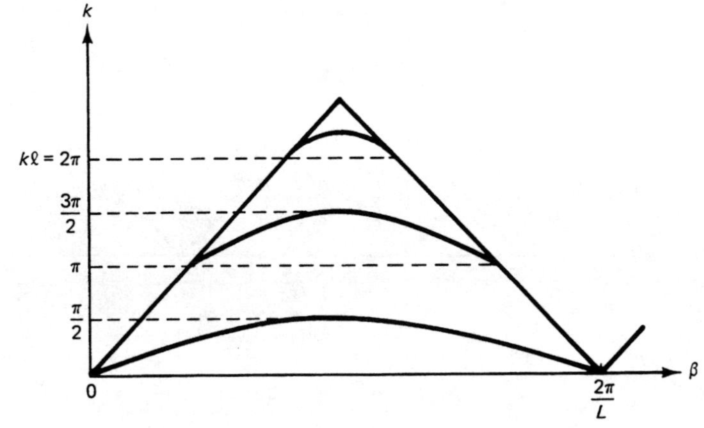

Figure 7-5 k–β diagram for a corrugated surface.

(see the Appendix to Chapter 7, Section C). For example, the electric field normal to the edge with the angle ϕ_0 shown in Fig. 7-6 should behave as

$$E_z \approx \left(\frac{W}{2} - z\right)^{(\pi/\phi_0)-1}. \tag{7-23}$$

Since $\phi_0 = 3\pi/2$, an appropriate choice for the field $f(z)$ should be

$$f(z) = (\text{const.})\left[\left(\frac{W}{2}\right)^2 - z^2\right]^{-1/3}. \tag{7-24}$$

If a periodic structure is open in the transverse direction, the trapped surface wave can propagate along this structure without attenuation. Since the trapped surface wave is a slow wave, the wave exists only in the region $|\beta| > k$ in the k–β diagram. Recognizing that the diagram is a periodic function of β with period $2\pi/L$, we note that the trapped surface wave can exist only in a series of triangles in Fig. 7-7. Outside these triangles, the wave is fast and the energy leaks out of the structure, resulting in attenuation. This region is called *forbidden* because the structure cannot support waves without attenuation. For example, a helix used in traveling-wave tubes is designed to support the slow wave and obviously should not be operated in the forbidden region. On the other hand, if the structure is designed as an antenna, the leakage of the energy represents the radiation, and therefore the

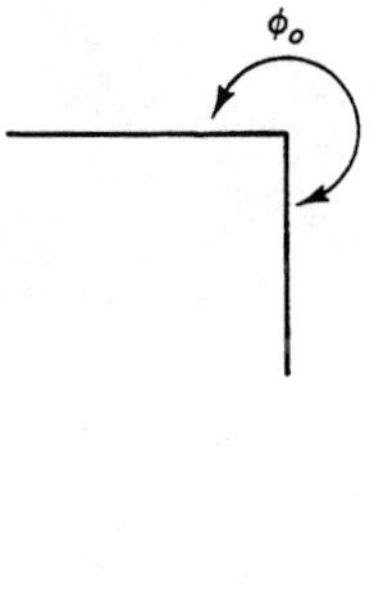

Figure 7-6 Edge condition.

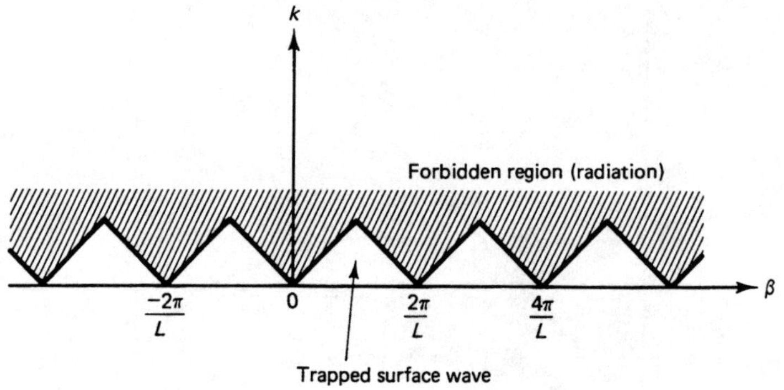

Figure 7-7 Forbidden region.

characteristics in this forbidden region are of great interest. Log-periodic antennas can be analyzed in terms of the behavior of the wave in the forbidden region.

If the structure is closed in the transverse direction, however, a propagating wave need not be slow, and thus there is no forbidden region. For example, if the corrugated surface discussed previously is closed (Fig. 7-8), the k–β diagram may appear as shown. Obviously, an exact diagram depends on the relative sizes of L, W, l, and the wavelength.

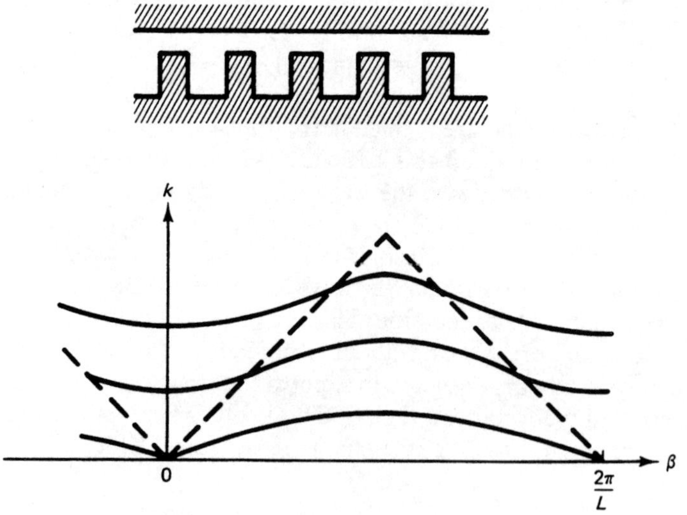

Figure 7-8 Closed corrugated guide.

7-3 PERIODIC LAYERS

Periodic stratified media are important in many optical and microwave applications. Let us consider the simplest case of a periodic medium consisting of alternating layers of two different indices of refraction (Fig. 7-9). We wish to find the propagation constant of a wave as it propagates through this structure. We first note that in an infinite periodic structure, there should be no difference between the field $U(x)$ at one point and the field $U(x+d)$ at another point separated by the period d except for a constant attenuation and phase shift. Mathematically, we state that

$$U(x+d) = U(x)\exp(-jqd), \tag{7-25}$$

where q is a complex propagation constant. This concept, expressed in (7-25), is Floquet's theorem, which was discussed previously.

Let us consider one period of the structure (Fig. 7-10). If we choose the voltages and currents V_1, I_1, V_2, and I_2 as shown, the Floquet theorem yields

$$\begin{aligned} V_2 &= V_1\exp(-jqd), \\ I_2 &= I_1\exp(-jqd), \end{aligned} \tag{7-26}$$

where q is a complex propagation constant and $d = d_1 + d_2$.

We also note that since this is a linear passive network, the input V_1 and I_1 and the output V_2 and I_2 are related through the $ABCD$ parameters, as discussed in Sections 3-7 and 4-6.

$$\begin{bmatrix} V_1 \\ I_1 \end{bmatrix} = \begin{bmatrix} A & B \\ C & D \end{bmatrix}\begin{bmatrix} V_2 \\ I_2 \end{bmatrix}, \tag{7-27}$$

where the $ABCD$ matrix is the product of the $ABCD$ matrices for each layer.

$$\begin{bmatrix} A & B \\ C & D \end{bmatrix} = \begin{bmatrix} A_1 & B_1 \\ C_1 & D_1 \end{bmatrix}\begin{bmatrix} A_2 & B_2 \\ C_2 & D_2 \end{bmatrix}. \tag{7-28}$$

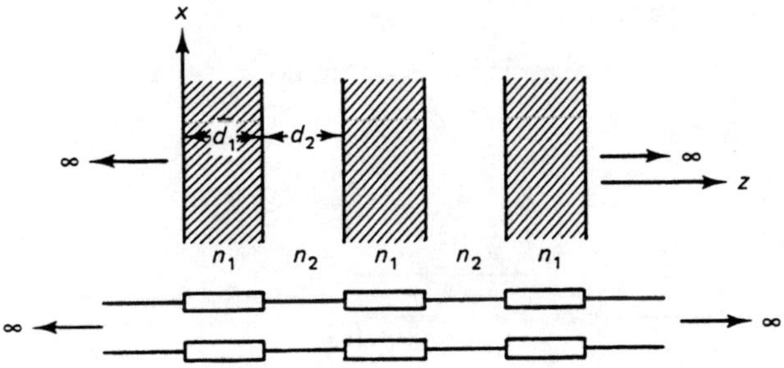

Figure 7-9 Periodic layers.

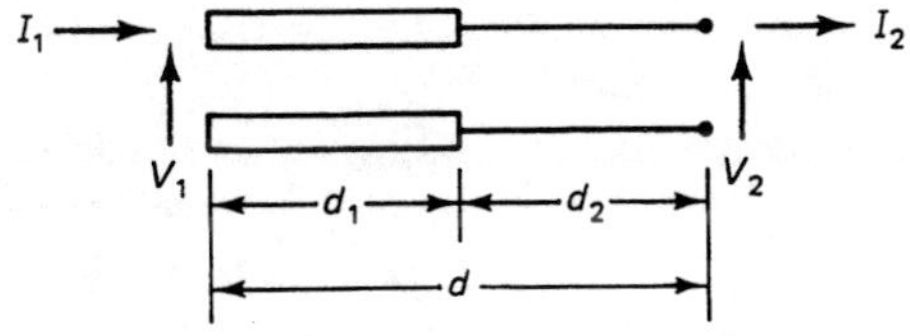

Figure 7-10 One period of layers.

The matrix elements are given by

$$\begin{aligned} A_i &= D_i = \cos q_i d_i, \\ B_i &= jZ_i \sin q_i d_i, \\ C_i &= \frac{j \sin q_i d_i}{Z_i}, \end{aligned} \tag{7-29}$$

where $i = 1, 2$ represents the first and second layers, and

$$q_i^2 + \beta^2 = k_0^2 n_i^2,$$

$$Z_i = \begin{cases} \dfrac{\omega\mu}{q_i} & \text{for perpendicular polarization } E_y \neq 0 \\ \dfrac{q_i}{\omega\epsilon} & \text{for parallel polarization } H_y \neq 0 \end{cases}$$

For a reciprocal network, A, B, C, and D satisfy the condition

$$AD - BC = 1. \tag{7-30}$$

The constant β is the phase constant along the x direction (Fig. 7-9) and is constant for all layers according to Snell's law.

$$\beta = k_0 n_1 \sin\theta_1 = k_0 n_2 \sin\theta_2, \tag{7-31}$$

where θ_1 and θ_2 are the angles between the direction of the wave in each layer and the z direction.

To find the propagation constant q in the z direction, we combine (7-26) and (7-27) and obtain the following eigenvalue equation:

$$\begin{bmatrix} A & B \\ C & D \end{bmatrix} \begin{bmatrix} V_2 \\ I_2 \end{bmatrix} = \lambda \begin{bmatrix} V_2 \\ I_2 \end{bmatrix}, \tag{7-32}$$

where $\lambda = \exp(jqd)$ is the eigenvalue. The eigenvalue is obtained by solving

$$\begin{vmatrix} A - \lambda & B \\ C & D - \lambda \end{vmatrix} = 0. \tag{7-33}$$

We get

$$\begin{aligned} qd &= -j \ln \left[\frac{A + D}{2} \pm i\sqrt{1 - \left(\frac{A + D}{2}\right)^2} \right] \\ &= \cos^{-1} \frac{A + D}{2}. \end{aligned} \tag{7-34}$$

The two eigenvalues for λ correspond to $\exp(\pm jqd)$ for positive- and negative-going waves.

Equation (7-34) gives the fundamental expression for the propagation constant q. We note that if $|(A + D)/2| < 1$, q is real, corresponding to the propagating wave, and the frequency range for this condition is called the pass band. If $|(A + D)/2| > 1$,

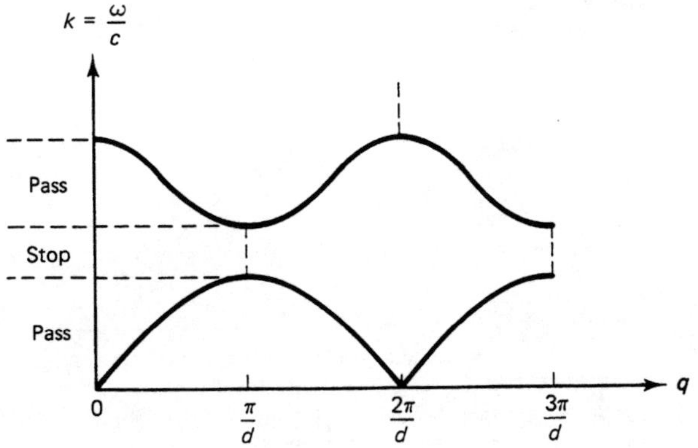

Figure 7-11 Pass-bands and stop bands.

then $q = m\pi + j$ (real) corresponding to the evanescent wave, and the frequency range under this condition is called the stop band. The frequency dependence of the propagation constant q has the general shape shown in Fig. 7-11. Note that $qd = \pi$ corresponds to the band edge $|(A + D)/2| = 1$. Periodic dielectric media have been used in distributed feedback lasers and distributed Bragg reflection lasers.

7-4 PLANE WAVE INCIDENCE ON A PERIODIC STRUCTURE

The reflection and transmission of a plane wave incident on periodic structures are of great importance in many areas of engineering and physics. Examples are microwave mesh reflectors, optical gratings, and crystal structures. In this section we outline the basic approaches to this problem.

Let a periodic structure be located at $z = 0$ and a plane wave incident on this surface from the direction defined by (θ_p, ϕ_p) (see Fig. 7-12). The incident wave, whether it is an electric field, magnetic field, or Hertz potential, can be written as

$$U_i = A_i e^{-jk_x x - jk_y y + jk_z z}, \tag{7-35}$$

where $k_x = k \sin\theta_p \cos\phi_p$, $k_y = k \sin\theta_p \sin\phi_p$, and $k_z = k \cos\theta_p$.

Now the reflected wave U_r should be written in a series of space harmonics in the x and y directions. Thus at $z = 0$, we write

$$U_r = \sum_{m=-\infty}^{\infty} \sum_{n=-\infty}^{\infty} B_{mn} e^{-jk_{xm}x - jk_{yn}y}, \tag{7-36}$$

where

$$k_{xm} = k_x + \frac{2m\pi}{L_x} \quad \text{and} \quad k_{yn} = k_y + \frac{2n\pi}{L_y}.$$

Considering that U satisfies the wave equation, we write the reflected wave

$$U_r(x, y, z) = \sum_{m=-\infty}^{\infty} \sum_{n=-\infty}^{\infty} B_{mn} \exp[-jk_{xm}x - jk_{yn}y - j(k^2 - k_{xm}^2 - k_{yn}^2)^{1/2} z]. \tag{7-37}$$

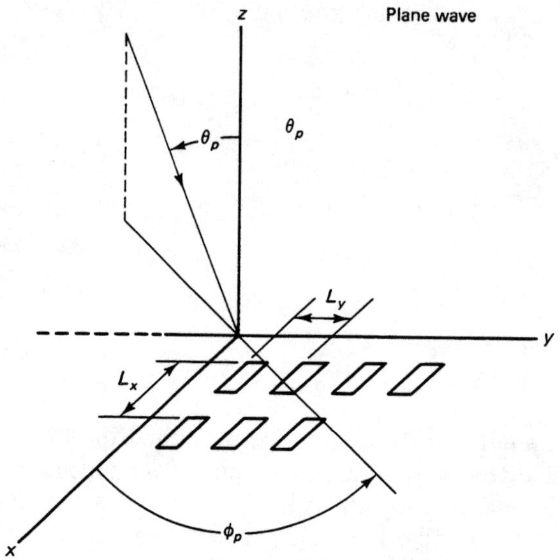

Figure 7-12 Plane wave incident on a periodic structure.

The determination of B_{mn} can be made by applying the boundary conditions. Let us note that the propagation constant for each mode in the z direction $\beta_{mn} = (k^2 - k_{xm}^2 - k_{yn}^2)^{1/2}$ can be real or purely imaginary depending on the incident direction (θ_p, ϕ_p) and m and n. If β_{mn} is real, the wave propagates away from the surface carrying real power and is called the *grating mode*. If β_{mn} is purely imaginary, the wave does not carry real power away from the surface and is *evanescent*.

As an example, let us consider a wave incident on a periodic conducting grating as shown in Fig. 7-13. We assume that the plane of incidence is in the x–z plane and all the gratings are parallel to the y axis, and therefore this is a two-dimensional problem. For a TE wave, $E_x = E_z = 0$ and E_y satisfies the wave equation and Dirichlet's boundary condition ($E_y = 0$) on the conducting tapes. For a TM wave, $H_x = H_z = 0$ and H_y satisfies the wave equation and Neumann's boundary condition on the conducting tapes.

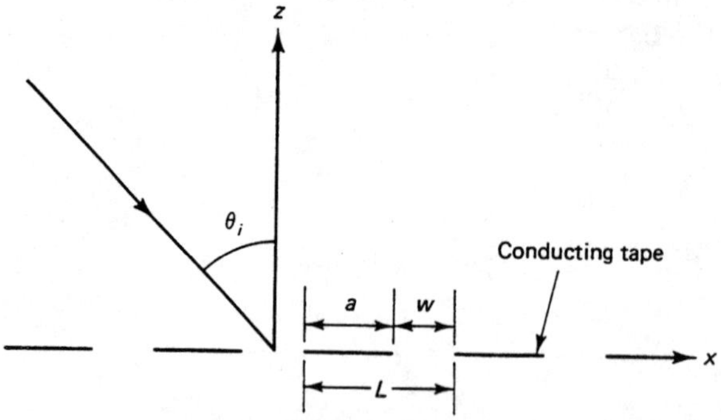

Figure 7-13 Plane wave incident on grating.

Let us consider the TM case. The magnetic field H_y in the region $z > 0$ consists of the incident wave H_{yi}, the reflected wave H_{yr} when the aperture is completely closed, and the scattered wave H_{ys} generated by the field in the aperture.

$$\begin{aligned} H_{yi} &= A_0 e^{+jqz - j\beta x}, \\ H_{yr} &= A_0 e^{-jqz - j\beta x}, \\ H_{ys1} &= \sum_{n=-\infty}^{\infty} B_n e^{-jq_n z - j\beta_n x}, \end{aligned} \tag{7-38}$$

where

$$\beta = k \sin \theta_i, \qquad q = k \cos \theta_i,$$

$$\beta_n = \beta + \frac{2n\pi}{L}, \qquad q_n^2 + \beta_n^2 = k^2.$$

Note that $E_x = (j/\omega\epsilon)(\partial/\partial z)H_y$, and therefore $E_{xi} + E_{xr} = 0$ at $z = 0$.

The field component E_x at $z = 0$ is equal to the unknown function $f(x)$ in the aperture $|x| \le (w/2)$.

$$E_x(x, z = 0) = \begin{cases} 0 & \text{for } \frac{w}{2} < |x| < \frac{L}{2} \\ f(x) & \text{for } |x| < \frac{w}{2}. \end{cases} \tag{7-39}$$

Following the procedure in Section 7-2, we get

$$H_{ys1}(x, z) = j\omega\epsilon \int_{-w/2}^{w/2} f(x')G_1(x, z; x', 0)\, dx' \qquad \text{for } z > 0, \tag{7-40}$$

where

$$G_1(x, z; x', 0) = \sum_{n=-\infty}^{\infty} \frac{e^{-jq_n z - j\beta_n(x - x')}}{jq_n L}.$$

Similarly, for $z < 0$, we get

$$H_{ys2}(x, z) = -j\omega\epsilon \int_{-w/2}^{w/2} f(x')G_2(x, z; x', 0)\, dx' \qquad \text{for } z < 0, \tag{7-41}$$

where

$$G_2(x, z; x', 0) = \sum_{n=-\infty}^{\infty} \frac{e^{+jq_n z - j\beta_n(x - x')}}{jq_n L}.$$

Now let us consider the boundary conditions at $z = 0$. The continuity of the tangential electric field E_x is already satisfied as we used the same function $f(x')$. The continuity of the tangential magnetic field requires that

$$H_{yi} + H_{yr} + H_{ys1} = H_{ys2} \qquad \text{at } z = 0. \tag{7-42}$$

Substituting (7-38), (7-40), and (7-41) in (7-42), we get

$$A_0 = -j\omega\epsilon \int_{-w/2}^{w/2} f(x')G(x,x')\,dx', \tag{7-43}$$

where

$$G(x,x') = \sum_{n=-\infty}^{\infty} \frac{e^{-j\beta_n(x-x')}}{jq_n L}.$$

An approximate solution of (7-43) may be obtained by assuming that

$$f(x) = \frac{Ce^{-j\beta x}}{[(w/2)^2 - x^2]^{1/2}}. \tag{7-44}$$

Here we included the edge condition for E_z and C is constant. We then multiply both sides of (7-43) by $f^*(x)$ and integrate over the aperture.

$$A_0 \int_{-w/2}^{w/2} f^*(x)\,dx = -j\omega\epsilon \int_{-w/2}^{w/2} dx \int_{-w/2}^{w/2} dx' f^*(x) f(x')G(x,x').$$

We then obtain

$$C = A_0\left[-j\omega\epsilon \sum_{n=-\infty}^{\infty} \frac{\pi}{jq_n L} J_0^2\left(\frac{n\pi w}{L}\right)\right]^{-1}. \tag{7-45}$$

Once we obtain the solution $f(x)$, we use (7-40) and (7-41) to calculate the scattered field. For example, in the region $z > 0$, we get H_{ys1} from (7-40) using the approximate solution (7-45).

Note that far from the surface, all the evanescent modes are negligibly small and only the propagating modes exist. For the propagating modes, q_n is real, and therefore $k > |\beta_n|$. This is the forbidden (radiation) region shown in Fig. 7-7.

Let us consider the conservation of power. The incident power is in the $-z$ direction and is obtained from (7-38):

$$\bar{P}_i = \text{Re}\left(\frac{1}{2} E_{xi} H_{yi}^*\right)\hat{z} = -P_i\hat{z},$$
$$P_i = \frac{1}{2\omega\epsilon}|A_0|^2 q. \tag{7-46}$$

The scattered wave consists of H_{yr} and H_{ys1} in (7-38). The scattered power per unit period L is given by

$$\bar{P}_s = P_s\hat{z},$$
$$P_s = \frac{1}{L}\int_0^L \text{Re}\left(\frac{1}{2} E_{xs} H_{ys}^*\right) dx, \tag{7-47}$$

where

$$H_{ys} = H_{yr} + H_{ys1} \quad \text{and} \quad E_{xs} = \frac{1}{\omega\epsilon}\frac{\partial}{\partial z} H_{ys}.$$

Substituting (7-38) into (7-47) and noting that all space harmonics are orthogonal and that $E_{xs} H_{ys}^*$ for all evanescent modes are purely imaginary, we get

$$P_s = \frac{1}{2\omega\epsilon}\left(q|A_0 + B_0|^2 + \sum_{\substack{n=N_1 \\ n\neq 0}}^{N_2} q_n |B_n|^2\right), \tag{7-48}$$

where $N_1 \le n \le N_2$ includes all the propagating modes (q_n is real).

The transmitted power is similarly given by

$$\bar{P}_t = \frac{1}{L}\int_0^L \text{Re}\left(\frac{1}{2}E_{xt}H_{yt}^*\right)\hat{z}\,dx = -P_t\hat{z},$$
$$P_t = \frac{1}{2\omega\epsilon}\left(\sum_{N_1}^{N_2} q_n |C_n|^2\right), \tag{7-49}$$

where

$$H_{ys2} = \sum_{n=-\infty}^{\infty} C_n e^{+jq_n z - j\beta_n x}.$$

The conservation of power is then given by

$$P_i = P_s + P_t. \tag{7-50}$$

Let us next consider a TE wave. The electric field E_y in the region $z > 0$ consists of the incident wave E_{yi}, the wave E_{yr} reflected from the conducting plane at $z = 0$ and the wave E_{ys1} produced by the aperture field at $z = 0$.

$$\begin{aligned} E_y &= E_{yi} + E_{yr} + E_{ys1}, \\ E_{yi} &= A_0 e^{+jqz - j\beta x}, \\ E_{yr} &= -A_0 e^{-jqz - j\beta x}, \\ E_{ys1} &= \sum_{n=-\infty}^{\infty} B_n e^{-jq_n z - j\beta_n x}, \end{aligned} \tag{7-51}$$

where

$$q = k\cos\theta_i, \qquad \beta = k\sin\theta_i,$$
$$\beta_n = \beta + \frac{2n\pi}{L}, \qquad q_n^2 + \beta_n^2 = k^2.$$

Now we let the aperture field be $f(x)$.

$$E_y(x, z=0) = \begin{cases} 0 & \text{for } \frac{w}{2} < |x| < \frac{L}{2} \\ f(x) & \text{for } |x| < \frac{w}{2}. \end{cases} \tag{7-52}$$

We can then express B_n in terms of $f(x)$.

$$B_n = \frac{1}{L}\int_{-w/2}^{w/2} f(x')e^{+j\beta_n x'}\,dx'. \tag{7-53}$$

The magnetic field component H_{x1} is given by

$$H_{x1}(x,z) = \frac{1}{j\omega\mu}\frac{\partial}{\partial z}E_y = H_{x0}(x,z) + \frac{1}{j\omega\mu}\int_{-w/2}^{w/2} Kf(x')\,dx', \tag{7-54}$$

where

$$H_{x0}(x,z) = H_{xi}(x,z) + H_{xr}(x,z)$$

$$= \frac{1}{j\omega\mu}2jqA_0 \cos qz e^{-j\beta x},$$

$$K = K(x,z;x') = \sum_{n=-\infty}^{\infty} \frac{-jq_n}{L} e^{-jq_n z - j\beta_n(x-x')}.$$

Similarly, in the region $z < 0$, we have

$$E_y = E_{s2} = \sum_{n=-\infty}^{\infty} C_n e^{+jq_n z - j\beta_n x},$$

where using (7-52) and (7-53), $B_n = C_n$. The magnetic field is then given by

$$H_{x2}(x,z) = \frac{-1}{j\omega\mu}\int_{-w/2}^{w/2} Kf(x')\,dx'. \tag{7-55}$$

Equating H_{x1} to H_{x2} in the aperture at $z = 0$, we get

$$\int_{-w/2}^{w/2} K(x;x')f(x')\,dx' = -2qA_0 e^{-j\beta x}, \tag{7-56}$$

where

$$K(x;x') = \sum_{n=-\infty}^{\infty} \frac{-jq_n}{L} e^{-j\beta_n(x-x')}.$$

As an approximate solution to (7-56), we let $f(x') = C[(w/2)^2 - x'^2]^{1/2} e^{-j\beta x}$ using the edge condition (see the Appendix to Chapter 7, Section C) and integrate over the aperture. We then get

$$\int_{-w/2}^{w/2} dx \int_{-w/2}^{w/2} dx' K(x,x')f(x)f(x') = -jqA_0 \int_{-w/2}^{w/2} f(x)e^{-j\beta x}\,dx. \tag{7-57}$$

From this we get C (see Gradshteyn and Ryzhik, 1965, p. 482).

$$C = qA_0\left(\frac{\pi}{2}\right)\left\{\sum_{n=-\infty}^{\infty} \frac{q_n}{L}\left[\frac{\pi J_1(n\pi w/L)}{n\pi w/L}\right]^2\right\}^{-1}. \tag{7-58}$$

The conservation of power is satisfied by

$$P_i = P_s + P_t, \tag{7-59}$$

where P_i, P_s, and P_t are given by the same equations as (7-46), (7-48), and (7-49) except that $1/2\omega\epsilon$ is replaced by $1/2\omega\mu$.

7-5 SCATTERING FROM PERIODIC SURFACES BASED ON THE RAYLEIGH HYPOTHESIS

Let us consider a wave scattered by a sinusoidally varying surface illuminated by a plane wave (Fig. 7-14). We will consider a two-dimensional problem where the surface does not vary in the y direction and the plane of incidence is in the x–z plane. The surface is given by $z = \zeta$,

$$\zeta = -h \cos\frac{2\pi x}{L}. \tag{7-60}$$

This problem was first solved by Rayleigh using what is now called the Rayleigh hypothesis.

The Dirichlet Problem

$$(\nabla^2 + k^2)\psi = 0, \qquad \psi = 0 \text{ on surface.} \tag{7-61}$$

The incident wave ψ_i can be written as

$$\psi_i = A_0 e^{+jqz - j\beta x}, \tag{7-62}$$

where $q = k \cos\theta_i$ and $\beta = k \sin\theta_i$. The scattered wave ψ_s can be expressed in terms of the space harmonics.

$$\psi_s = \sum_{n=-\infty}^{\infty} B_n e^{-jq_n z - j\beta_n x}, \tag{7-63}$$

where $\beta_n = \beta + 2n\pi/L$ and $q_n^2 = k^2 - \beta_n^2$. Note that in (7-63), the scattered field ψ_s is expressed in terms of the outgoing wave with $\exp(-jq_n z)$ only and the incoming wave with $\exp(+jq_n z)$ is not included. Whereas the wave is outgoing in the region above the highest point of the surface ($z > h$), the wave in the region $-h < z < h$ should consist of both the outgoing and incoming waves. The approach that the

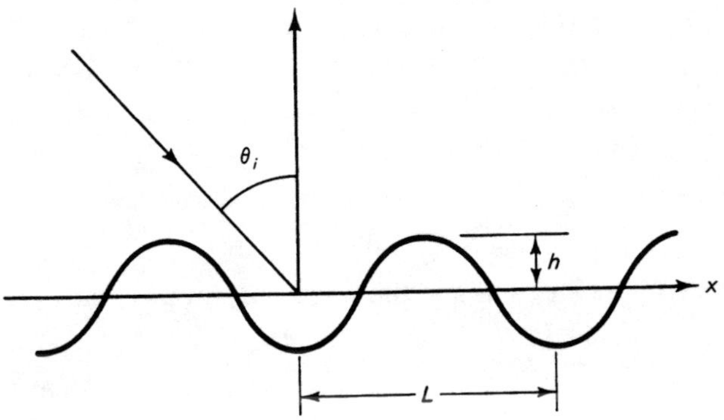

Figure 7-14 Sinusoidally varying surface illuminated by a plane wave.

scattered wave can be represented by the outgoing wave only, even in the region $-h < z < h$, is called the *Rayleigh hypothesis*. Although the Rayleigh hypothesis is not valid for a general periodic surface, it has been shown that it is valid for the sinusoidal surface if the maximum slope of the surface $(\partial\zeta/\partial x)_{\max} = (2\pi h)/L$ is less than 0.448.

Assuming the Rayleigh hypothesis, we apply the boundary condition that $\psi_i + \psi_s = 0$ on the surface $(z = \zeta)$. Noting that $\exp(-j\beta x)$ is common to all terms, we get

$$A_0 \exp(+jq\zeta) + \sum_{n=-\infty}^{\infty} B_n \exp\left(-jq_n\zeta - j\frac{2n\pi}{L}x\right) = 0. \qquad (7\text{-}64)$$

The relationships among B_n and A_0 are obtained by expanding (7-64) in Fourier series and equating all Fourier coefficients to zero. To do this we multiply (7-64) by $\exp(j2m\pi x/L)$ and integrate the result over L. Noting that

$$\frac{1}{2\pi}\int_0^{2\pi} e^{jz\cos\beta + jn\beta - jn(\pi/2)}\, d\beta = J_n(z), \qquad (7\text{-}65)$$

we get, for $m = -\infty \cdots +\infty$,

$$A_0 e^{-j(\pi/2)|m|} J_{|m|}(qh) + \sum_{n=-\infty}^{\infty} B_n e^{-j(\pi/2)|m-n|} J_{|m-n|}(q_n h) = 0. \qquad (7\text{-}66)$$

This may be arranged in the following matrix form:

$$[K_{mn}][B_n] = [A_m]A_0, \qquad (7\text{-}67)$$

where

$$K_{mn} = e^{-j(\pi/2)|m-n|} J_{|m-n|}(q_n h),$$
$$A_m = -e^{-j(\pi/2)|m|} J_{|m|}(qh).$$

It can be truncated and $[B_n]$ may be obtained by $[K]^{-1}[A]A_0$. For example, if the height is small compared with a wavelength and the period L is much greater than a wavelength, $qh \ll 1$ and $|q_n h| \ll 1$ for propagating modes, and therefore, $J_{m-n}(q_n h)$ is on the order of $|q_n h|^{m-n}$ and the matrix can be truncated without too much error. The conservation of power should be a good check for the convergence of the solution.

Neumann Problem

For the Neumann problem, we write the incident and the scattered wave as

$$\begin{aligned} \psi_i &= A_0 e^{-jqz - j\beta x}, \\ \psi_s &= \sum_{n=-\infty}^{\infty} B_n e^{-jq_n z - j\beta_n x}, \end{aligned} \qquad (7\text{-}68)$$

where β, q, β_n, and q_n are as defined in Section 7.4. Now we impose the boundary condition

$$\frac{\partial}{\partial n}(\psi_i + \psi_s) = \hat{n} \cdot \nabla(\psi_i + \psi_s) = 0 \qquad \text{at } z = \zeta, \tag{7-69}$$

where $\hat{n}$ is the unit vector normal to the surface and is given by

$$\hat{n} = \frac{-(\partial\zeta/\partial x)\hat{x} - (\partial\zeta/\partial y)\hat{y} + \hat{z}}{[1 + (\partial\zeta/\partial x)^2 + (\partial\zeta/\partial y)^2]^{1/2}}, \tag{7-70}$$

Substituting (7-68) into (7-69), we get

$$\begin{aligned} -\frac{\partial\zeta}{\partial x}\left[-j\beta A_0 e^{jq\zeta} + \sum_{n=-\infty}^{\infty}(-j\beta_n)B_n e^{-jq_n\zeta - j(2n\pi x/L)}\right] \\ + \left[jqA_0 e^{jq\zeta} + \sum_{n=-\infty}^{\infty}(-jq_n)B_n e^{-jq_n\zeta - j(2n\pi x/L)}\right] = 0. \end{aligned} \tag{7-71}$$

Noting that $\zeta = -h\cos(2\pi x/L)$ and $\partial\zeta/\partial x = (2\pi h/L)\sin(2\pi x/L)$, we can multiply (7-71) by $\exp[j(2m\pi x/L)]$ and integrate it with respect to x over L. We then get the following matrix equation:

$$[H_{mn}][B_n] = [D_m]A_0, \tag{7-72}$$

where

$$\begin{aligned} H_{mn} &= \frac{\pi h \beta_n}{L}[e^{-j(\pi/2)|m-n+1|}J_{|m-n+1|}(q_n h) - e^{-j(\pi/2)|m-n-1|}J_{|m-n-1|}(q_n h)] \\ &\quad - jq_n e^{-j(\pi/2)|m-n|}J_{|m-n|}(q_n h), \\ D_m &= -\frac{\pi h \beta}{L}[e^{-j(\pi/2)|m+1|}J_{|m+1|}(qh) - e^{-j(\pi/2)|m-1|}J_{|m-1|}(qh)] \\ &\quad - jq\, e^{-j(\pi/2)|m|}J_{|m|}(qh). \end{aligned}$$

Two-Media Problem

Above, we discussed Dirichlet's and Neumann's problems using the Rayleigh hypothesis. Similar techniques can be used to solve the two-media problem. We write the incident ψ_i and the scattered wave ψ_s in medium 1, and the transmitted wave ψ_t in medium 2 as

$$\begin{aligned} \psi_i &= A_0 e^{+jqz - j\beta x}, \\ \psi_s &= \sum_{n=-\infty}^{\infty} B_n e^{-jq_n z - j\beta_n x}, \\ \psi_t &= \sum_{n=-\infty}^{\infty} C_n e^{+jq_{tn} z - j\beta_n x}, \end{aligned} \tag{7-73}$$

where

$$\beta = k_i \sin\theta_i, \qquad q = k_i \cos\theta_i,$$

$$\beta_n = \beta + \frac{2n\pi}{L},$$

$$q_n = (k_1^2 - \beta_n^2)^{1/2},$$

$$q_{tn} = (k_2^2 - \beta_n^2)^{1/2},$$

$$k_1 = \frac{\omega}{c} n_1, \qquad k_2 = \frac{\omega}{c} n_2.$$

The boundary conditions are

$$\rho_1 \psi_1 = \rho_2 \psi_2 \quad \text{and} \quad \frac{\partial \psi_1}{\partial n} = \frac{\partial \psi_2}{\partial n} \quad \text{on the surface.}$$

Following the procedure for Dirichlet's and Neumann's problems, we have

$$\begin{aligned} [K_{mn}][B_m] - [K_{tmn}][C_n] &= [A_m]A_0, \\ [H_{mn}][B_m] - [H_{tmn}][C_n] &= [D_m]A_0, \end{aligned} \tag{7-74}$$

where $[K_{mn}]$, $[H_{mn}]$, $[A_m]$, and $[D_m]$ are already given. $[K_{tmn}]$ and $[H_{tmn}]$ have the same form as $[K_{mn}]$ and $[H_{mn}]$, respectively, except that all q_n are replaced by $-q_{tn}$.

The power conservation is checked as follows: For Dirichlet's and Neumann's problems, we should have

$$q|A_0|^2 = \sum_{n=N_1}^{N_2} q_n |B_n|^2, \tag{7-75}$$

where $N_1 \leq N \leq N_2$ includes all the propagating modes. For two-media problems, we should have

$$\rho_1 q |A_0|^2 = \rho_1 \sum_{n=N_1}^{N_2} q_n |B_n|^2 + \rho_2 \sum_{n=N_3}^{N_4} q_{tn} |C_n|^2. \tag{7-76}$$

In this section we discussed the scattering from a sinusoidal surface using the Rayleigh hypothesis. This is valid if $(2\pi h/L) < 0.448$. If the surface slope is higher than this or if the surface has a more general shape than a sinusoidal surface, the Rayleigh hypothesis is, in general, not valid, and a more rigorous method should be used. One such technique is the T-matrix method, which will be discussed later.

We note here that each mode of the scattered wave, $B_n \exp(-jq_n z - j\beta_n x)$, propagates in the direction with the angle θ_n from the z axis given by

$$k \sin\theta_n = \beta_n = \beta + \frac{2n\pi}{L} = k \sin\theta_i + \frac{2n\pi}{L}. \tag{7-77}$$

If this angle θ_n becomes close to $\pm\pi/2$, the scattered mode propagates along the surface and a rapid redistribution of the power in all modes takes place within a small variation of the wavelength or the angle. This effect, known as the *Wood anomalies*, has been studied extensively. The condition for the Wood anomalies is, therefore,

$$\lambda = \frac{L}{n}(\pm 1 - \sin\theta_i). \tag{7-78}$$

This wavelength is sometimes called the *Rayleigh wavelength*.

7-6 COUPLED-MODE THEORY

Consider two wave guiding structures, such as two waveguides, two optical fibers, or two strip lines. The wave in each guide propagates with a definite propagation constant. Suppose that these two guides are close to each other, so that some coupling of the power takes place. For example, for waveguides, these may be a series of holes or slits between two guides as in the case of directional couplers. For two strip lines, the two lines are closely located and the power of the two lines may be coupled. If the coupling is weak, the coupling perturbs the original guide modes slightly and transfers the power from one guide to the other. The coupled-mode theory described in this section provides a mathematical formulation of this coupling process. It is clear from the above that coupled-mode theory is applicable only to a weakly coupled system. For a system with strong coupling, the improved theory should be used (Hardy and Streifer, 1986; Tsang and Chuang, 1988).

Consider two modes a_1 and a_2 representing the waves propagating in the two guides, respectively (Fig. 7-15). If two guides are isolated, each mode is assumed to propagate with the propagation constant β_{10} and β_{20}. Therefore, a_1 and a_2 in the isolated guides satisfy the following:

$$\frac{da_1}{dz} = -j\beta_{10}a_1, \qquad \frac{da_2}{dz} = -j\beta_{20}a_2. \tag{7-79}$$

If these two guides are coupled, the coupling terms between a_1 and a_2 should be included in (7-79). We thus write the following coupled equations:

$$\begin{aligned} \frac{da_1}{dz} &= -j\beta_{10}a_1 - jc_{12}a_2, \\ \frac{da_2}{dz} &= -jc_{21}a_1 - j\beta_{20}a_2. \end{aligned} \tag{7-80}$$

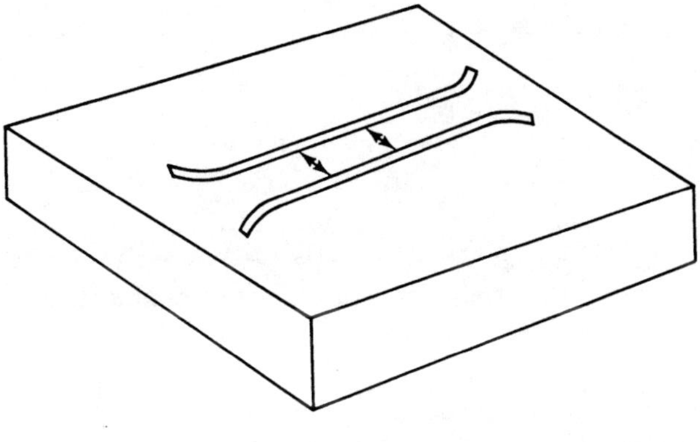

Figure 7-15 Coupling between two strip lines.

The constants c_{12} and c_{21} are the mutual coupling coefficients per unit length. Note that β_{10} and β_{20} are real in a lossless system, and that if $\beta_{10} > 0$ and $\beta_{20} > 0$, the phase velocities for the modes a_1 and a_2 are in the positive z direction, and if $\beta_{10} < 0$, and $\beta_{20} < 0$, the phase velocities are in the negative z direction.

Let us next consider the power P_1 carried by the mode a_1. We get

$$\frac{dP_1}{dz} = \frac{d}{dz}(a_1 a_1^*) = a_1 \frac{da_1^*}{dz} + \frac{da_1}{dz} a_1^*. \tag{7-81}$$

Substituting (7-80), we get

$$\frac{dP_1}{dz} = -ja_1(\beta_{10} - \beta_{10}^*)a_1^* + 2\,\mathrm{Re}(ja_1 c_{12}^* a_2^*). \tag{7-82}$$

Similarly for $P_2 = |a_2|^2$, we get

$$\frac{dP_2}{dz} = -ja_2(\beta_{20} - \beta_{20}^*)a_2^* + 2\,\mathrm{Re}(-ja_1 c_{21} a_2^*), \tag{7-83}$$

where Re denotes "real part of."

If the system is lossless, β_{10} and β_{20} are real, and therefore $\beta_{10} - \beta_{10}^* = 0$ and $\beta_{20} - \beta_{20}^* = 0$. If both powers P_1 and P_2 are propagated in the same direction, and thus the group velocities for a_1 and a_2 are in the same direction, the conservation of power requires that

$$\frac{d}{dz}(P_1 + P_2) = 0. \tag{7-84}$$

Using (7-82) and (7-83), this means that for a lossless system,

$$c_{12} = c_{21}^*. \tag{7-85}$$

This is called a *codirectional coupler*. If P_1 and P_2 are propagated in the opposite direction, and thus the group velocities are in the opposite directions, we require that

$$\frac{d}{dz}(P_1 - P_2) = 0, \tag{7-86}$$

and therefore, for a lossless system,

$$c_{12} = -c_{21}^*. \tag{7-87}$$

This is called a *contradirectional coupler*.

Codirectional Coupler

Consider a lossless system consisting of two guides weakly coupled to each other. We assume that the phase velocities and the group velocities of waves a_1 and a_2 are both in the positive z direction. Thus we have

$$\beta_{10} > 0, \qquad \beta_{20} > 0, \qquad c_{12} = c_{21}^*. \tag{7-88}$$

To solve the coupled equation (7-80), we let

$$a_1(z) = A_1 \exp(-j\beta z),$$
$$a_2(z) = A_2 \exp(-j\beta z). \tag{7-89}$$

Substituting this into (7-80), we get the following eigenvalue problem:

$$\begin{bmatrix} \beta_{10} & c_{12} \\ c_{21} & \beta_{20} \end{bmatrix} \begin{bmatrix} A_1 \\ A_2 \end{bmatrix} = \beta \begin{bmatrix} A_1 \\ A_2 \end{bmatrix}, \tag{7-90}$$

where the propagation constant β is the eigenvalue and $[A_1, A_2]$ is the eigenvector.

The propagation constant β can be obtained by equating the determinant of the matrix in (7-90) to zero,

$$\begin{vmatrix} \beta_{10} - \beta & c_{12} \\ c_{21} & \beta_{20} - \beta \end{vmatrix} = 0. \tag{7-91}$$

From this we get two values for β. We arrange these two propagation constants β_1 and β_2 in the following form:

$$\beta_1 = \beta_a + \beta_b,$$
$$\beta_2 = \beta_a - \beta_b, \tag{7-92}$$

where

$$\beta_a = \frac{1}{2}(\beta_{10} + \beta_{20}),$$
$$\beta_b = (\beta_d^2 + c_{12}c_{21})^{1/2},$$
$$\beta_d = \frac{1}{2}(\beta_{10} - \beta_{20}).$$

Since the two powers are propagating in the same direction, this is a codirectional coupler, and therefore $c_{12}\, c_{21} = |c_{12}|^2$ and β_a and β_b are both real. When β_a and β_b are real, the two modes are said to be *passively* coupled.

The eigenvectors $[A_1, A_2]$ for β_1 and β_2 are obtained from (7-90):

$$\frac{A_2}{A_1} = \frac{\beta - \beta_{10}}{c_{12}} = \frac{c_{21}}{\beta - \beta_{20}}. \tag{7-93}$$

The general solutions for a_1 and a_2 are then given by

$$a_1(z) = C_1 e^{-j\beta_1 z} + C_2 e^{-j\beta_2 z},$$
$$a_2(z) = C_1 \frac{\beta_1 - \beta_{10}}{c_{12}} e^{-j\beta_1 z} + C_2 \frac{\beta_2 - \beta_{10}}{c_{12}} e^{-j\beta_2 z}, \tag{7-94}$$

where C_1 and C_2 are constant.

The constants C_1 and C_2 are determined by the boundary conditions. Suppose that the wave is incident in guide 1 at $z = 0$ and no wave is incident in guide 2 at $z = 0$. Thus we have

$$a_1(0) = a_0 \quad \text{and} \quad a_2(0) = 0. \tag{7-95}$$

Using this, we get

$$a_1(z) = a_0\left(\cos\beta_b z - j\frac{\beta_d}{\beta_b}\sin\beta_b z\right)\exp(-j\beta_a z),$$
$$a_2(z) = -ja_0\frac{c_{21}}{\beta_b}\sin\beta_b z\ \exp(-j\beta_a z). \tag{7-96}$$

The power in each guide is given by

$$P_1(z) = |a_1(z)|^2 \quad \text{and} \quad P_2(z) = |a_2(z)|^2. \tag{7-97}$$

Substituting (7-96) into (7-97), it is easily verified that $P_1(z) + P_2(z) = \text{constant}$ as expected. The power is then periodically transferred between two guides (Fig. 7-16). Note that the maximum transferred power $P_2(z)$ is

$$P_{2\max} = \frac{|C_{12}|^2}{|\beta_d|^2 + |C_{12}|^2}|a_0|^2. \tag{7-98}$$

Thus if $\beta_{10} = \beta_{20}$, the power transfer is 100%.

Contradirectional Coupler

Consider a lossless contradirectional coupler. In this case the phase velocities for both modes are in the same direction, but the group velocities are in the opposite direction. Thus we have

$$\beta_{10} > 0, \qquad \beta_{20} > 0, \qquad c_{12} = -c_{21}^*. \tag{7-99}$$

Following the procedure described in Section 7-6, we get

$$\beta_1 = \beta_a + \beta_b,$$
$$\beta_2 = \beta_a - \beta_b, \tag{7-100}$$

where

$$\beta_a = \tfrac{1}{2}(\beta_{10} + \beta_{20}),$$
$$\beta_b = [\beta_d^2 - |c_{12}|^2]^{1/2},$$
$$\beta_d = \tfrac{1}{2}(\beta_{10} - \beta_{20}).$$

Note that here we used $c_{12}c_{21} = -|c_{12}|^2$.

The expression for β_b indicates that if $|\beta_d| < |c_{12}|$, then β_b is purely imaginary

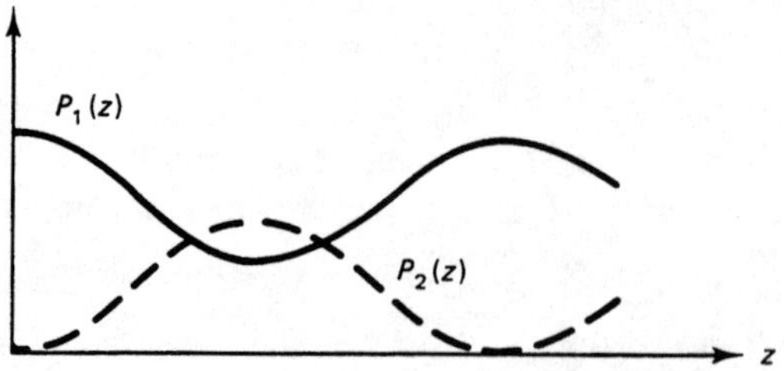

Figure 7-16 Periodic transfer of power for a codirectional coupler.

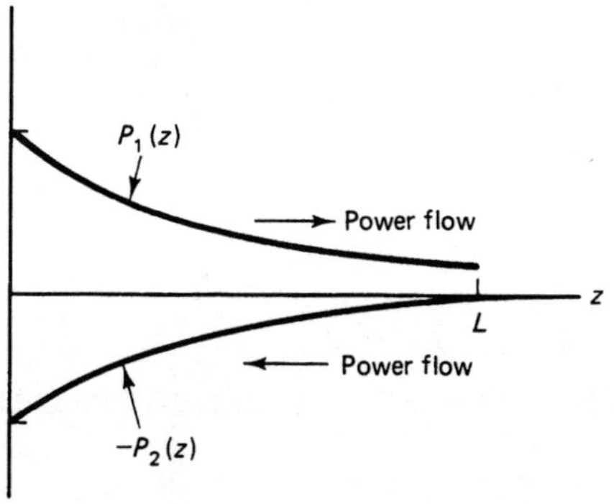

Figure 7-17 Contradirectional coupler.

and the wave will be exponentially growing or decaying. When β_1 and β_2 are complex, the two modes are said to be "actively" coupled.

If the power is injected to guide 1 at $z = 0$, this power is coupled into guide 2 and propagates in the negative z direction. If no power is injected in guide 2 at $z = L$, we have the boundary condition:

$$\begin{aligned} a_1(0) &= a_0, \\ a_2(L) &= 0. \end{aligned} \tag{7-101}$$

The solution can be obtained following the procedure given for the codirectional case. The powers $P_1(z)$ and $P_2(z)$ are sketched in Fig. 7-17. Note that $P_1(z) - P_2(z)$ is constant.

In this and previous sections, we discussed lossless coupled systems where there is no net power gain. If the guides contain an active medium, the wave may be amplified, as in the case of coupling between a beam of electrons and a circuit in a traveling-wave tube. The determination of coupling coefficients for dielectric waveguides are discussed in Tamir (1975), Hardy and Streifer (1986), and Tsang and Chuang (1988).

PROBLEMS

7-1. Find an equation to determine the propagation constant of a wave propagating in a waveguide with the wall structure shown in Fig. P7-1. Assume no variation in y and that the wave is polarized in the x–z plane. Consider the limit as $l \to 0$ or $w \to 0$.

7-2. If the period L in Problem 7-1 is much smaller than a wavelength, the surface $x = 0$ may be approximated by the average surface impedance Z_s given by

$$\frac{E_z}{H_y} = Z_s = j\frac{W}{L}\sqrt{\frac{\mu_0}{\epsilon_0}}\tan kl.$$

Find the propagation constant for this case and compare the results with those of Problem 7-1.

7-3. A wave is propagating in the z direction through the periodic structure shown in Fig. P7-3. Plot the k–q diagram in the frequency range 0 to 10 GHz.

7-4. Consider a TM plane wave normally incident on the periodic grating shown in Fig. 7-13. $a = W = 5$ cm and frequency is 10 GHz. Identify all propagating modes and find

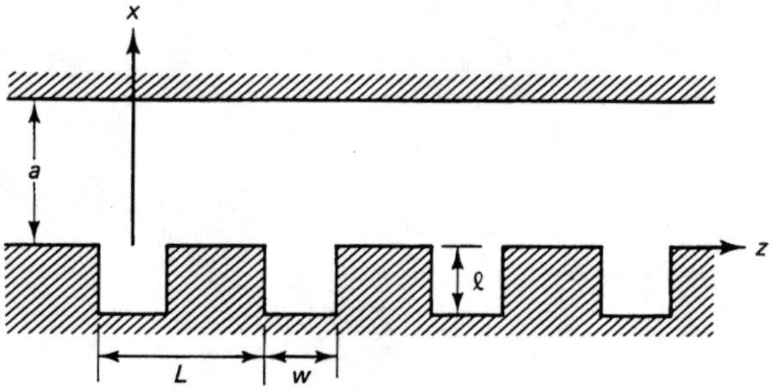

Figure P7-1 Periodic waveguide.

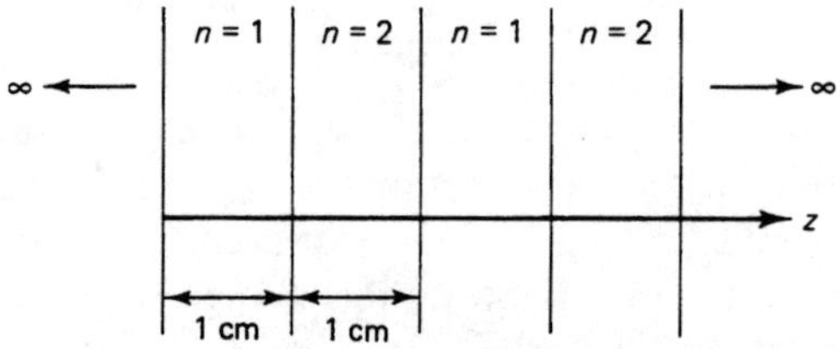

Figure P7-3 Periodic layers.

the direction of propagation for each mode. Find the expressions for the magnitude of each mode and check the conservation of power.

7-5. Consider the periodic Dirichlet surface given in (7-60). If the wave is normally incident on the surface, and $kh = 0.1$ and $L = 1.5\lambda$, find the amplitudes of the propagating modes and their direction of propagation, and check the power conservation.

7-6. In the k–β diagram Fig. 7-2, let $\beta = k \sin\theta_i$ and locate the points on the diagram that satisfy the Rayleigh wavelength condition.

7-7. Consider the TE_{10} modes in two rectangular waveguides with $a = 1$ in. and $b = \frac{1}{2}$ in. at 10 GHz. If these two modes are weakly coupled and the maximum power transfer from one to the other waveguide is 20 dB over a distance of 20 cm, find the coupling coefficient C_{12}. Assume that C_{12} is real and positive.

7-8. Consider a contradirectional coupler with $\beta_{10} = 1$, $\beta_{20} = 1.1$, and $C_{12} = -C_{21} = 0.1$. Find the eigenvalues and eigenvectors. Calculate and plot $P_1(z)$ and $P_2(z)$ for the boundary conditions $a_1(0) = 1$ and $a_2(5) = 0$.

7-9. Two pendulums of length l_1 and l_2 and mass m_1 and m_2 are coupled by a weightless spring with spring constant k as shown in Fig. P7-9. The amplitudes of the oscillations are assumed to be small and l_1 is not very different from l_2.

(a) Derive the coupled-mode equations for x_1 and x_2.

(b) Find solutions if $x_1 = dx_1/dt = dx_2/dt = 0$ and $x_2 = x_0$ at $t = 0$ when $l_1 = l_2 = 1m$, $m_1 = m_2 = 1$ g, and $k = 10^{-3}$ N/m.

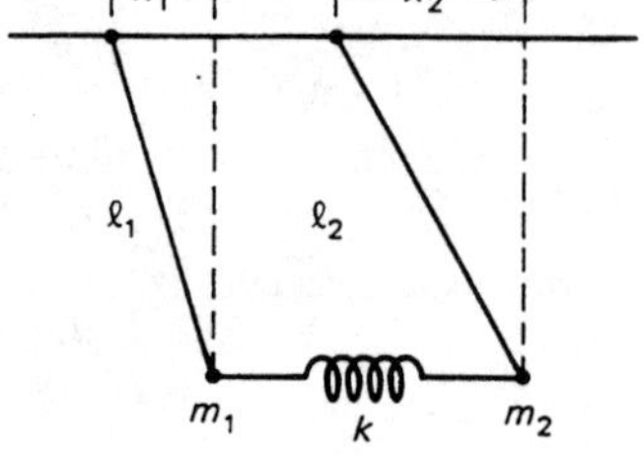

Figure P7-9 Coupled pendulums.

8

Dispersion and Anisotropic Media

In Section 2-3 we discussed the constitutive relations $\bar{D} = \epsilon\bar{E}$ and $\bar{B} = \mu\bar{H}$. They are valid for a "linear" medium, where $\bar{D}$ and $\bar{B}$ are proportional to $\bar{E}$ and $\bar{H}$, respectively. If $\bar{D}$ or $\bar{B}$ is a more general function of $\bar{E}$ [$\bar{D} = \bar{D}(\bar{E})$], this is the "nonlinear" medium. For a time-harmonic case, ϵ and μ are in general functions of frequency $\epsilon(\omega)$ and $\mu(\omega)$, and this is the dispersive medium. For a nondispersive medium, ϵ and μ are independent of frequency. If ϵ and μ are functions of position, this is called the inhomogeneous medium; for a homogeneous medium, ϵ and μ are constant. In an isotropic medium, ϵ and μ are scalar and therefore $\bar{D}$ and $\bar{B}$ are proportional to $\bar{E}$ and $\bar{H}$, respectively. In an anisotropic medium, as shown in Section 8-7, $\bar{D}$ and $\bar{E}$, and $\bar{B}$ and $\bar{H}$ are in general not parallel. In a bianisotropic medium, $\bar{D}$ depends on both $\bar{E}$ and $\bar{B}$, and $\bar{H}$ depends on both $\bar{E}$ and $\bar{B}$. Chiral medium is an example of a biisotropic medium, and these are discussed in Section 8-22. Two-fluid model of superconductors at high frequencies is discussed in Sections 8-23 and 8-24.

8-1 DIELECTRIC MATERIAL AND POLARIZABILITY

In Section 2-3 we discussed the constitutive relations for a medium in terms of the dielectric constant ϵ, the electric susceptibility χ_e, or the electric polarization $\bar{P}$. They are related by [equations (2-45) and (2-46)]

$$\bar{P} = (\epsilon - \epsilon_0)\bar{E} = \chi_e \epsilon_0 \bar{E}. \tag{8-1}$$

Alternatively, the polarization vector $\bar{P}$ can be viewed as the dipole moments per unit volume of the medium. In this interpretation we can write

$$\bar{P} = N\bar{p} = N\alpha\bar{E}', \tag{8-2}$$

where N is the number of dipoles per unit volume contributing to $\bar{P}$, and $\bar{p}$ is the moment of each elementary dipole. The dipole moment $\bar{p}$ is, in turn, produced by the local electric field $\bar{E}'$ and α is called the *polarizability*. Note that the local field $\bar{E}'$ is not equal to the applied field $\bar{E}$.

There are four major mechanisms of producing the dipole moment in a material. The *electronic polarization*, expressed by the polarizability α_e, is caused by a slight displacement of electrons surrounding positively charged atomic nuclei under the influence of the field $\bar{E}'$, forming a dipole. The *atomic polarization* α_a is caused by displacement of differently charged atoms with respect to each other. The *dipole polarization* α_d, also called the *orientation polarization*, is caused by the change of orientation of equivalent dipoles in a medium. Polarizations α_e, α_a, and α_d are due to the locally bound charges in the atoms or molecules. The fourth polarization, α_s, is called the *space charge* or *interfacial polarization*. We discuss the dispersion properties of these polarizations in the following sections.

In (8-1) and (8-2) we noted that the external applied field $\bar{E}$ is, in general, different from the local field $\bar{E}'$ that causes the polarization. They are almost identical for low-pressure gases but are different for solids, liquids, and high-pressure gases. The relationship between $\bar{E}'$ and $\bar{E}$ can be obtained by considering a fictitious sphere surrounding a molecule in the medium (Fig. 8-1). The local field $\bar{E}'$ acting on the molecule at the center of the spherical cavity of radius r_0 is the sum of the applied field $\bar{E}$ and the field $\bar{E}_p$ due to the polarization vector $\bar{P}$ surrounding the cavity.

$$\bar{E}' = \bar{E} + \bar{E}_p. \tag{8-3}$$

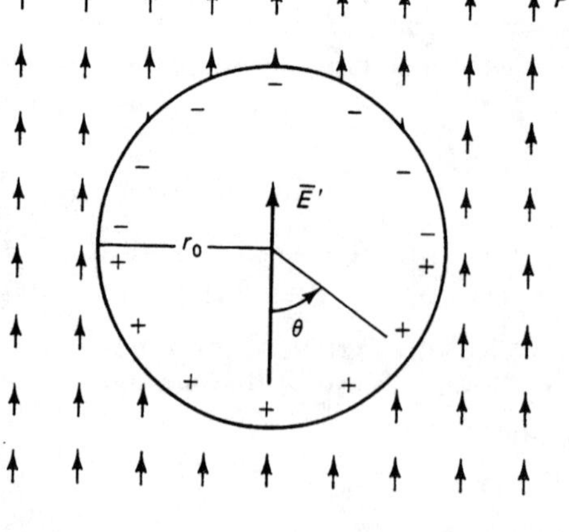

Figure 8-1 Local field $\bar{E}'$ and polarization $\bar{P}$.

The polarization $\bar{P}$ creates equivalent charges on the wall of the sphere, and the charge over the elementary area $d\bar{a}$ is given by $\bar{P}\cdot d\bar{a} = P\cos\theta\, da$, $da = 2\pi r\,\sin\theta\, r\, d\theta$.

The field $\bar{E}_p$ is obtained by summing the contributions from the charges $\bar{P}\cdot d\bar{a}$. It is pointed in the z direction and its magnitude is equal to

$$E_p = \int_0 \frac{P\cos^2\theta}{4\pi\epsilon_0 r_0^2} 2\pi r_0\,\sin\theta\, r_0\, d\theta = \frac{P}{3\epsilon_0}. \tag{8-4}$$

Using this, we get the local field $\bar{E}'$, called the *Mossotti field*, in terms of the applied field $\bar{E}$.

$$\bar{E}' = \bar{E} + \frac{\bar{P}}{3\epsilon_0} = \frac{\epsilon_r + 2}{3}\bar{E}, \qquad \epsilon_r = \frac{\epsilon}{\epsilon_0}. \tag{8-5}$$

Using (8-1), (8-2), and (8-5), we can express the electric susceptibility χ_e in terms of the polarizability α:

$$\chi_e = \frac{N\alpha/\epsilon_0}{1 - N\alpha/3\epsilon_0} \qquad \text{or} \qquad \frac{\epsilon}{\epsilon_0} = \frac{1 + 2N\alpha/3\epsilon_0}{1 - N\alpha/3\epsilon_0}. \tag{8-6}$$

Similarly, we can relate the polarizability α to the relative dielectric constant ϵ_r:

$$\alpha = \frac{3\epsilon_0}{N}\frac{\epsilon_r - 1}{\epsilon_r + 2}. \tag{8-7}$$

This is called the *Clausius–Mossotti formula* or *Lorentz–Lorenz formula*.

8-2 DISPERSION OF DIELECTRIC MATERIAL

The dielectric constant of any material is in general dependent on frequency, and it can be considered constant only within a narrow frequency band. If a broadband pulse is propagated through such a medium, however, the frequency dependence of the medium cannot be ignored. The variation of the dielectric constant with frequency is called *dispersion*. In this section we discuss some simple examples of dispersive media.

Let us consider the dispersion characteristics of dielectric material. We assume a simplified model of molecules with electrons bound elastically to the heavy nuclei. The equation of motion for an electron is

$$m\frac{d^2\bar{r}}{dt^2} = -m\omega_0^2\bar{r} - m\nu\frac{d\bar{r}}{dt} + \bar{F}, \tag{8-8}$$

where m is the mass of the electron, $\bar{r}$ the displacement of the electron, $-m\omega_0^2\bar{r}$ the elastic restoring force, $-m\nu\, d\bar{r}/dt$ is the damping force, ν is the collision frequency, and $\bar{F}$ the Lorentz force acting on the electron. The restoring force is assumed to be proportional to the displacement of the electron, and the constant ω_0 is equal to the frequency of the free oscillations of the electron under the influence of the restoring force alone. The Lorentz force is given by

$$\bar{F} = e(\bar{E}' + \bar{v} \times \bar{B}'), \tag{8-9}$$

where e is the charge of an electron, $\bar{E}'$ and $\bar{B}'$ the local Mossotti field (8-5), and $\bar{v}$ the velocity of the electron. Since $\bar{B}' = \mu_0 \bar{H}'$ and $|\bar{H}'|$ is on the order of $(\epsilon_0/\mu_0)^{1/2}|\bar{E}'|$, $|\bar{B}'|$ is on the order of $(1/c)|\bar{E}'|$, and therefore, assuming that $|\bar{v}| << c$, the second term of (8-9) is negligible compared with the first term.

Consider a time-harmonic field with $\exp(j\omega t)$. Assume that there are N bound electrons per unit volume. The polarization vector $\bar{P}$ is then given by

$$\bar{P} = Ne\bar{r}. \tag{8-10}$$

Equation (8-8) for the time-harmonic field is

$$-m\omega^2 \bar{r} = -m\omega_0^2 \bar{r} - j\omega m \nu \bar{r} + e\left(\bar{E} + \frac{Ne\bar{r}}{3\epsilon_0}\right). \tag{8-11}$$

Noting that $\bar{D} = \epsilon_0 \epsilon_r \bar{E} = \epsilon_0 \bar{E} + \bar{P}$, we get the relative dielectric constant ϵ_r as a function of frequency.

$$\epsilon_r = 1 + \frac{Ne^2}{m\epsilon_0(\omega_1^2 - \omega^2 + j\omega\nu)}, \tag{8-12}$$

where $\omega_1^2 = \omega_0^2 - Ne^2/3\epsilon_0 m$. Figure 8-2 shows the general shape of ϵ_r as a function of frequency.

In more general cases, there is more than one resonance, and we need to generalize (8-12) to

$$\epsilon_r = 1 + \sum_s \frac{N_s e^2}{m_s \epsilon_0(\omega_s^2 - \omega^2 + j\omega\nu_s)}. \tag{8-13}$$

For lossless material, we write (8-13) as

$$\begin{aligned} \epsilon_r &= 1 + \sum_s \frac{N_s e^2}{m_s \epsilon_0(\omega_s^2 - \omega^2)} \\ &= 1 + \sum_s \frac{\lambda^2 B_s}{\lambda^2 - \lambda_s^2} \end{aligned} \tag{8-14}$$

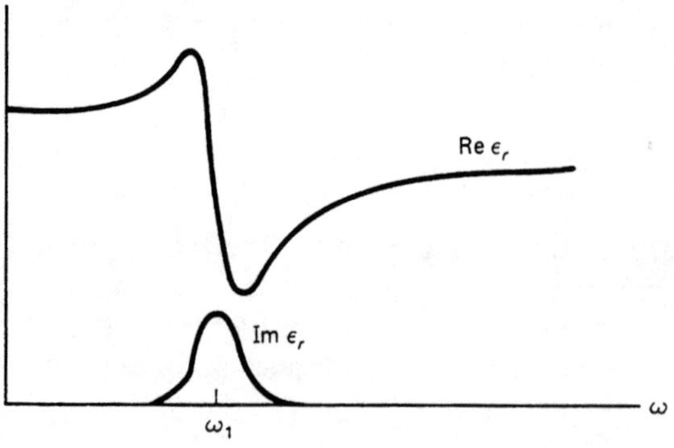

Figure 8-2 Dispersion.

where $\omega/c = 2\pi/\lambda$ and B_s are constants to be determined experimentally. Equation (8-14), called the *Sellmeier equation*, is often used in the study of dispersion in optical fibers. For example, the refractive index of fused silica (SiO_2) used for fibers in the wavelength $\lambda = 0.5$ to 2.0 μm can be given by (8-14) with $\lambda_1 = 0.1$ μm, $B_1 = 1.0955$ and $\lambda_2 = 9$ μm, $B_2 = 0.9$ (Marcuse, 1982, p. 485).

8-3 DISPERSION OF CONDUCTOR AND ISOTROPIC PLASMA

In dielectric material, the resonant frequency ω_1 in (8-12) is nonzero, and at low frequency $\omega \to 0$, ϵ_r in (8-12) approaches the static dielectric constant. However, in a conductor, there are free electrons that are not bound to molecules, and therefore the restoring force $(-m\omega_0^2\bar{r})$ in (8-8) is absent. Also, the interaction between the molecules can be neglected, and the local field $\overline{E}'$ is equal to the applied field $\overline{E}$. Equation (8-12) is therefore

$$\epsilon_r = 1 + \frac{\omega_p^2}{-\omega^2 + j\omega\nu,} \tag{8-15}$$

where $\omega_p = (Ne^2/m\epsilon_0)^{1/2}$ is called the plasma frequency. N is the number of free electrons per unit volume and is called the electron density. The damping is caused by the collisions between the electron and other molecules, and ν is called the collision frequency. If we compare (8-15) with the expression for a conducting medium (note that $\epsilon' = 1$, Table 2-1),

$$\epsilon_r = 1 - j\frac{\sigma}{\omega\epsilon_0}, \tag{8-16}$$

we get the equivalent conductivity σ:

$$\frac{\sigma}{\epsilon_0} = \frac{\omega_p^2}{\nu + j\omega}. \tag{8-17}$$

At low frequencies $\omega << \nu$, the conductivity σ is, therefore, almost constant. In general, however, the conductivity σ is a function of frequency.

The dielectric constant of metal in optical wavelengths can be approximately given by (8-15). For example, at $\lambda = 0.6$ μm, silver has the plasma frequency $f_p = 2 \times 10^{15}$ (ultraviolet), the collision frequency $f_\nu = 5.7 \times 10^{13}$ (infrared), and $\epsilon_r = -17.2 - j0.498$. If the frequency is increased beyond the plasma frequency, the dielectric constant becomes almost real and positive and the wave can propagate through metal, which is called the *ultraviolet transparency* of metals (Jackson, 1975).

Electromagnetic wave propagation through ionized gas has received considerable attention for many years. In particular, the reflection of radio waves and the transmission from and through the ionosphere have been studied extensively. The ionosphere was postulated as the Kenelly–Heaviside layer in 1902, and the formula for its index of refraction, now known as the Appleton–Hartree formula, was obtained around 1930. Such an ionized gas in which electron and ion densities are substantially the same is electrically neutral and is called the *plasma*. The problem of reentry of high-speed vehicles such as missiles and rockets has generated considerable interest in plasma problems. When high-speed vehicles enter the atmosphere, high temperature and pressure in front of the vehicle ionize the air molecules and produce the so-called *plasma sheath*. The problems of antenna characteristics, wave propagation through the plasma, and the radar cross section are of considerable importance. Also, the antenna and wave propagation characteristics of artificial satellites in the ionosphere are important in communication between the vehicle and the earth station.

If a dc magnetic field is present, the plasma becomes anisotropic and this is normally called the *magnetoplasma*. In the absence of dc magnetic fields, the plasma is isotropic and the equivalent dielectric constant is given by (8-15). Thus the refraction index n depends on the operating frequency ω, the plasma frequency ω_p, and the collision frequency ν. The *electron plasma frequency* plays a most important role in magnetic-ionic theory. Substituting the values of m, e, and ϵ_0, we get

$$f_p = \begin{cases} 8.98 N_e^{1/2} & (N_e \text{ in m}^{-3}) \\ 8.98 \times 10^3 N_e^{1/2} & (N_e \text{ in cm}^{-3}) \end{cases}. \tag{8-18}$$

The propagation constant β for a plane wave propagating in a lossless isotropic plasma is given by

$$\beta = k_0 n = (k_0^2 - k_p^2)^{1/2}, \qquad k_p = \frac{\omega_p}{c} \tag{8-19}$$

Mathematically, this is identical to the propagation constant for a hollow waveguide:

$$\beta = (k_0^2 - k_c^2)^{1/2}, \qquad k_c = \text{cutoff wave number} \tag{8-20}$$

In a waveguide, if the frequency is above the cutoff frequency, the wave propagates, and if the frequency is below the cutoff frequency, the wave becomes evanescent. In exactly the same manner, the plasma frequency plays the role of the cutoff frequency.

A well-known example of the cutoff phenomenon is wave propagation through the ionosphere. When the operating frequency is higher than the plasma frequency, radio waves can penetrate through the ionosphere, but at lower frequencies, radio waves are bounced off the ionosphere, thus contributing long-distance radio-wave propagation. Typical characteristics of the lower ionosphere are shown in Fig. 8-3. Some typical values of the electron density are shown in Table 8-1.

8-4 DEBYE RELAXATION EQUATION AND DIELECTRIC CONSTANT OF WATER

The dielectric constant of water at microwave frequencies is governed primarily by the relaxation phenomenon. Water molecules have permanent dipole moments, and when microwaves are applied, the polar molecules tend to rotate as if they are

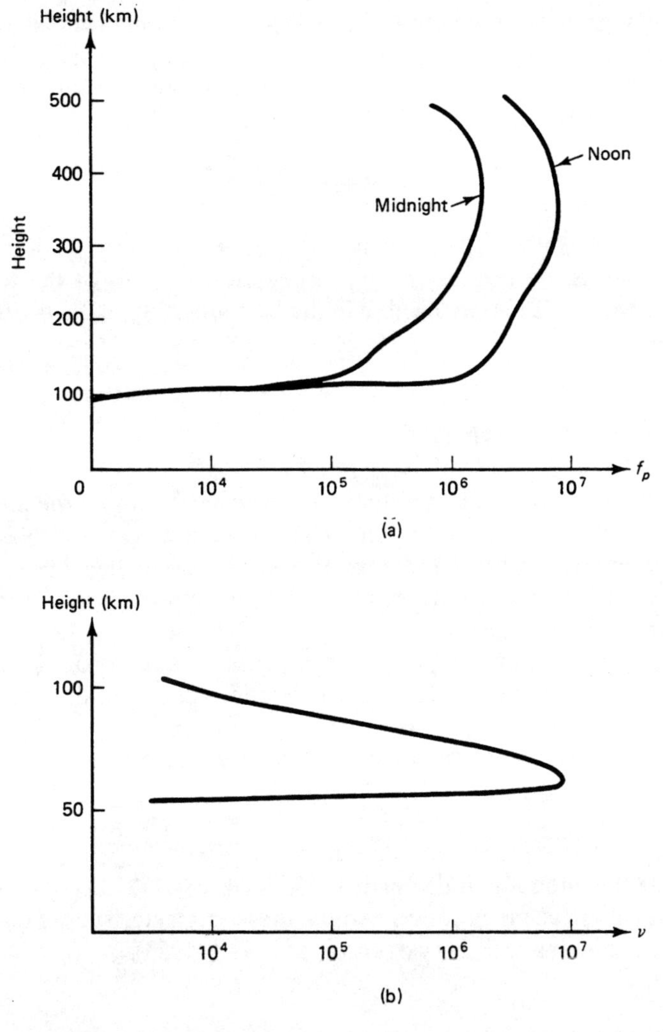

Figure 8-3 Typical plasma frequency (a) and collision frequency (b) of the ionosphere.

TABLE 8-1 Typical Values of Electron Densities

	N_e (cm^{-3})	Temperature (°K)
Ionosphere	$10^3 \sim 3 \times 10^6$	$300 \sim 3000$
Interplanetary space	$1 \sim 10^4$	
Solar Corona	$10^4 \sim 3 \times 10^8$	10^6
Interstellar space	$10^{-3} \sim 10$	$100 \sim 10^4$
Thermonuclear reaction	10^{15}	$10^6 \sim 10^7$
Gas discharge device	10^{12}	
In metals	3×10^{22}	

in a damping frictional medium. In (8-12) this frictional force is represented by ν. However, the acceleration term $-m\omega^2 \bar{r}$ in (8-11) may be negligibly small compared with other terms. Thus the dielectric constant for the medium of polar molecules may be expressed as

$$\epsilon_r = \epsilon_\infty + \frac{\epsilon_s - \epsilon_\infty}{1 + j\omega\tau}, \tag{8-21}$$

where ϵ_s is the static dielectric constant as $\omega \to 0$, ϵ_∞ is the high-frequency limit as $\omega \to \infty$, and τ is the relaxation time. They are functions of the temperature. The Debye formula (8-21) is applicable in the frequency range 0.3 to 300 GHz (Oguchi, 1983; Ray, 1972).

8-5 INTERFACIAL POLARIZATION

In Section 8-1 we discussed three polarization mechanisms: the electronic, atomic, and dipole orientation polarizations. They are caused by the displacement of bound or free electrons or by the change in orientation of the dipole moment of the molecule. In addition to these three, there is another process, called *interfacial polarization* or *space-charge polarization*. This is due to the large-scale field distortions caused by the piling up of space charges in the volume or of the surface charges at the interfaces between different small portions of materials with different characteristics.

The complex dielectric constant ϵ_r is written in the form

$$\epsilon_r = -j\frac{\sigma_0}{\omega\epsilon_0} + \sum_{m=1}^{M}\left(a_m + \frac{b_m - a_m}{1 + j\omega\tau_m}\right). \tag{8-22}$$

Note that this model is indistinguishable from (8-21) except for the conductivity term. Geophysical media often exhibit these characteristics. For a complete discussion on complex resistivity of earth, scc Wait (1989).

8-6 MIXING FORMULA

In Section 8-1 we discussed the Clausius–Mossotti formula, relating the dielectric constant to the polarizability. The dielectric material was viewed as consisting of many equivalent dipoles in free space created by the local Mossotti field. The Clausius–Mossotti formula can be used to obtain the effective dielectric constant of a mixture of two or more materials with different dielectric constants. The formula that gives the effective dielectric constant is called the *mixing formula*.

Let us first consider a simple example of a dielectric material with relative dielectric constant ϵ_1 in which many spheres of radius a and relative dielectric constant ϵ_2 are embedded (Fig. 8-4). If the dimension a is comparable to or greater than a wavelength, substantial scattering can take place. Also, if the fractional

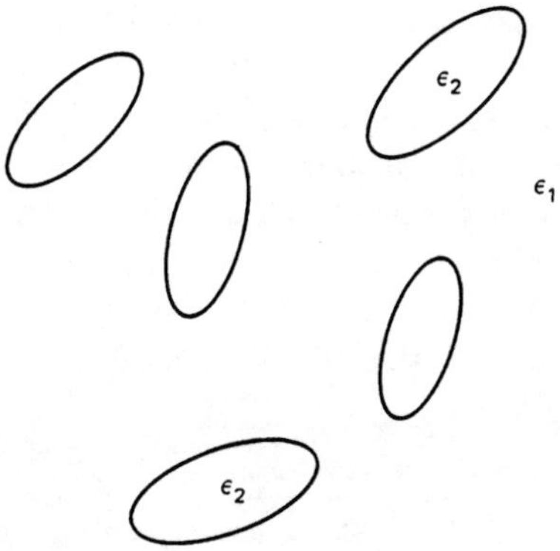

Figure 8-4 Effective dielectric constant of a mixture.

volume f, which is the fraction of the volume occupied by the spheres, is a few percent or higher, a correlation between the spheres needs to be considered. Here we limit ourselves to the case where dimensions of the spheres are much smaller than a wavelength, and the spheres are sparsely distributed. The situation described here is therefore similar to that discussed in Section 8-1, where the relative dielectric constant ϵ_r of the material consisting of many dipoles is given by

$$\epsilon_r = 1 + \chi_e = \frac{1 + 2N\alpha/3\epsilon_0}{1 - N\alpha/3\epsilon_0}, \tag{8-23}$$

where N is the number of dipoles per unit volume and α is the polarizability of the dipole.

In the case shown in Fig. 8-4, the background dielectric constant is $\epsilon_1 \epsilon_0$, and thus we have the effective dielectric constant ϵ_e:

$$\frac{\epsilon_e}{\epsilon_1} = \frac{1 + 2N\alpha/3\epsilon_1}{1 - N\alpha/3\epsilon_1}, \tag{8-24}$$

where N is the number of spheres per unit volume. The polarizability α of the sphere is given by (see Section 10-5)

$$\alpha = \frac{3(\epsilon_2 - \epsilon_1)}{\epsilon_2 + 2\epsilon_1} \epsilon_1 V, \tag{8-25}$$

where V is the volume of the sphere. The fractional volume f is then given by

$$f = NV. \tag{8-26}$$

Substituting (8-25) and (8-26) into (8-24), we get the effective dielectric constant ϵ_e:

$$\epsilon_e = \epsilon_1 \frac{1 + 2fy}{1 - fy}$$
$$y = \frac{\epsilon_2 - \epsilon_1}{\epsilon_2 + 2\epsilon_1}. \tag{8-27}$$

This is called the *Maxwell–Garnett mixing formula*. Note that even though we expect that the formula is valid only for a small fractional volume $f << 1$, the effective dielectric constant ϵ_e reduces to ϵ_1 when $f = 0$ and ϵ_2 when $f = 1$. Thus we may expect that the formula may be a reasonable approximation even when f is not small. However, if the inhomogeneity is not spherical, the polarizability is different and although ϵ_e reduces to ϵ_1 when $f = 0$, it does not reduce to ϵ_2 when $f = 1$. Therefore, the Maxwell–Garnett formula is in general applicable only when f is small. We can also rearrange (8-27) in the following form, known as the *Rayleigh mixing formula*.

$$\frac{\epsilon_e - \epsilon_1}{\epsilon_e + 2\epsilon_1} = f\frac{\epsilon_2 - \epsilon_1}{\epsilon_2 + 2\epsilon_1}. \tag{8-28}$$

If the inhomogeneity has a nonspherical shape, the appropriate polarizability for that shape should be used in place of (8-25).

The Maxwell–Garnett formula (8-27) is based on the idea that the inhomogeneity with ϵ_2 is embedded in the background ϵ_1. However, more generally, when two inhomogeneities are mixed, there should be no distinction between the background and the inhomogeneities. Thus both the inhomogeneity with ϵ_1 and f_1 and the inhomogeneity with ϵ_2 and $f_2(f_1 + f_2 = 1)$ are embedded in the artificial background with the effective dielectric constant ϵ_e.

Here we assume that the inhomogeneities are isotropic and have no preferred shape or direction. These inhomogeneities, which are the differences between ϵ_1 and ϵ_e and between ϵ_2 and ϵ_e, create equivalent dipole moments per unit volume $(N_1\alpha_1 + N_2\alpha_2)\overline{E}_e$, where $\overline{E}_e$ is the average field for the background medium with the effective dielectric constant ϵ_e. The effective dielectric constant ϵ_e is chosen such that the average of these dipole moments is zero. Thus we have

$$N_1\alpha_1 + N_2\alpha_2 = 0, \tag{8-29}$$

Since the inhomogeneities are isotropic, on the average, the polarizability should be equal to that of a sphere.

$$\begin{aligned} \alpha_1 &= \frac{3(\epsilon_1 - \epsilon_e)}{\epsilon_1 + 2\epsilon_e} V_1, \\ \alpha_2 &= \frac{3(\epsilon_2 - \epsilon_e)}{\epsilon_2 + 2\epsilon_e} V_2, \end{aligned} \tag{8-30}$$

$N_1 V_1 = f_1$, $N_2 V_2 = f_2$, and $f_1 + f_2 = 1$. Rearranging these, we get

$$f_1\frac{\epsilon_1 - \epsilon_e}{\epsilon_1 + 2\epsilon_e} + f_2\frac{\epsilon_2 - \epsilon_e}{\epsilon_2 + 2\epsilon_e} = 0. \tag{8-31}$$

This is known as the *Polder–van Santern mixing formula* and can be rearranged to give the following form:

$$f_1\frac{\epsilon_1 - \epsilon_0}{\epsilon_1 + 2\epsilon_e} + f_2\frac{\epsilon_2 - \epsilon_0}{\epsilon_2 + 2\epsilon_e} = \frac{\epsilon_e - \epsilon_0}{3\epsilon_e}. \tag{8-32}$$

Note that the Polder–van Santern form is completely symmetric and ϵ_1, f_1 can be interchanged with ϵ_2, f_2, giving the same formula; the Maxwell–Garnett formula is not symmetrical. The Polder–van Santern formula can be extended to many species with ϵ_n and f_n.

$$\sum_{n=1}^{M} \frac{\epsilon_n - \epsilon_0}{\epsilon_n + 2\epsilon_e} f_n = \frac{\epsilon_e - \epsilon_0}{3\epsilon_e}, \qquad \sum_{n=1}^{M} f_n = 1. \tag{8-33}$$

Note that the mixing formulas above are for low-frequency cases where the scattering is negligible. More exact formulas, including scattering and correlations between particles, must be obtained by considering the propagation constant K of the coherent wave. Then the effective dielectric constant ϵ_e is related to K by $K^2 = k^2 \epsilon_e$, where k is the free space wave number. Extensive studies have been reported on this topic (Tsang et al., 1985).

8-7 DIELECTRIC CONSTANT AND PERMEABILITY FOR ANISOTROPIC MEDIA

The interactions of electromagnetic fields with materials are characterized by the constitutive parameters: complex dielectric constant ϵ and permeability μ. In an isotropic medium, the property of the material does not depend on the direction of electric or magnetic field polarizations. Thus ϵ and μ are scalar quantities.

In anisotropic media, however, the material characteristics depend on the direction of the electric or magnetic field vectors and thus in general, the displacement vector $\bar{D}$ and magnetic flux density vector $\bar{B}$ are not in the same direction as the electric field $\bar{E}$ and magnetic field vectors $\bar{H}$, respectively. The dielectric constant ϵ must then be represented by a tensor ϵ_{ij}:

$$D_i = \sum_{j=1}^{3} \epsilon_{ij} E_j, \qquad i = 1, 2, 3, \tag{8-34}$$

where $i, j = 1$, 2, and 3 denote the x, y, and z components, respectively. We may write (8-34) in the following form:

$$\bar{D} = \bar{\bar{\epsilon}} \bar{E}. \tag{8-35}$$

Using matrix notation in the rectangular system, (8-34) is expressed by

$$\begin{bmatrix} D_x \\ D_y \\ D_z \end{bmatrix} = \begin{bmatrix} \epsilon_{11} & \epsilon_{12} & \epsilon_{13} \\ \epsilon_{21} & \epsilon_{22} & \epsilon_{23} \\ \epsilon_{31} & \epsilon_{32} & \epsilon_{33} \end{bmatrix} \begin{bmatrix} E_x \\ E_y \\ E_z \end{bmatrix}. \tag{8-36}$$

Similarly, we have the tensor permeability $\bar{\bar{\mu}}$ *relating* $\bar{B}$ to $\bar{H}$.

$$\bar{B} = \bar{\bar{\mu}} \bar{H}. \tag{8-37}$$

As will be shown shortly, in general, the reciprocity theorem does not hold for anisotropic media, and for a plane wave, $\bar{E}$ and $\bar{H}$ are not necessarily transverse to the direction of wave propagation.

8-8 MAGNETOIONIC THEORY FOR ANISOTROPIC PLASMA

A dc magnetic field is often present in plasma. Examples are the earth magnetic field in the ionosphere and a dc magnetic field applied to a laboratory plasma. The presence of the dc magnetic field makes the plasma anisotropic. In this section, we examine the characteristics of such anisotropic plasma (Yeh and Liu, 1972).

The equation of motion for an electron in electromagnetic fields $(\bar{E}, \bar{H})$ in the presence of a dc magnetic field $\bar{H}_{dc}$ is given by

$$m\frac{d\bar{v}}{dt} = e\bar{E} + e[\bar{v} \times (\bar{B} + \bar{B}_{dc})] - m\nu\bar{v}, \tag{8-38}$$

where $\bar{B} = \mu_0 \bar{H}$ and $\bar{B}_{dc} = \mu_0 \bar{H}_{dc}$. As shown in Section 8-2, the term with $\bar{B}$ is negligible compared with $e\bar{E}$.

For a time-harmonic electromagnetic field with time dependence $\exp(j\omega t)$, neglecting the term with $\bar{B}$, we get

$$j\omega m\bar{v} = e\bar{E} + \mu_0 e(\bar{v} \times \bar{H}_{dc}) - m\nu\bar{v}. \tag{8-39}$$

We rewrite this equation using the plasma frequency ω_p:

$$\omega_p^2 = \frac{N_e e^2}{m\epsilon_0} \tag{8-40}$$

and the cyclotron frequency ω_c:

$$\omega_c = \frac{|e|\mu_0 H_{dc}}{m}. \tag{8-41}$$

Note that ω_c is the frequency of a circular motion of an electron in a plane perpendicular to the dc magnetic field. This is obtained by equating the centrifugal force $(mv^2)/r$ to the force $ev\mu_0 H_{dc}$ due to the magnetic field and noting that $\omega_c = 2\pi/T$ and $T = 2\pi r/v$.

The dc magnetic field $\bar{H}_{dc}$ is pointed in the direction (θ, ϕ) and its rectangular components are (Fig. 8-5)

$$\begin{aligned} \bar{H}_{dc} &= H_{dc}[\sin\theta_d \cos\phi_d \hat{x} + \sin\theta_d \sin\phi_d \hat{y} + \cos\theta_d \hat{z}] \\ &= H_{dcx}\hat{x} + H_{dcy}\hat{y} + H_{dcz}\hat{z}. \end{aligned} \tag{8-42}$$

We also note that the polarization vector $\bar{P}$ is given by

$$\bar{P} = N_0 e\bar{r}. \tag{8-43}$$

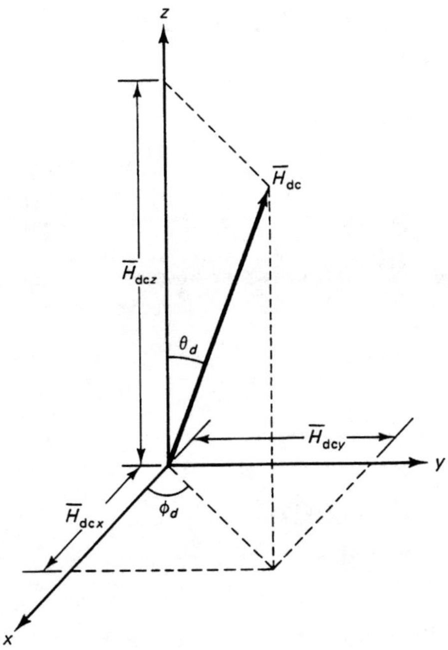

Figure 8-5 The dc magnetic field pointed in the direction (θ_d, ϕ_d) in plasma.

We now rewrite (8-39) in the following form:

$$-\overline{P}U = \epsilon_0 X\overline{E} + j\overline{P} \times \overline{Y}, \tag{8-44}$$

where $Z = j(\nu/\omega)$, $X = \omega_p^2/\omega^2$, $U = 1 - j(\nu/\omega)$, and $\overline{Y} = e\mu_0 \overline{H}_{dc}/m\omega$. Note that since e is negative for electrons, $\overline{Y}$ is pointed in the opposite direction to $\overline{H}_{dc}$.

Now let us write (8-44) in the following matrix form (see the Appendix to Chapter 8, Section A):

$$-U[P] = \epsilon_0 X[E] + j[y][P],$$

where

$$[P] = \begin{bmatrix} P_x \\ P_y \\ P_z \end{bmatrix}, \quad [E] = \begin{bmatrix} E_x \\ E_y \\ E_z \end{bmatrix}, \quad [y] = \begin{bmatrix} 0 & Y_z & -Y_y \\ -Y_z & 0 & Y_x \\ Y_y & -Y_x & 0 \end{bmatrix}.$$

This can be rearranged to yield

$$\epsilon_0[E] = [\sigma][P], \tag{8-45}$$

where

$$[\sigma] = -\frac{1}{X}\begin{bmatrix} U & jY_z & -jY_y \\ -jY_z & U & jY_x \\ jY_y & -jY_x & U \end{bmatrix}.$$

Inverting the matrix, we get the tensor electric susceptibility $[\chi_e]$:

$$
\begin{aligned}
[P] &= \epsilon_0[\chi_e][E], \\
[\chi_e] &= [\sigma]^{-1} \\
&= -\frac{X}{U(U^2 - Y^2)}\begin{bmatrix} U^2 - Y_x^2 & -jY_zU - Y_xY_y & jY_yU - Y_xY_z \\ jY_zU - Y_xY_y & U^2 - Y_y^2 & -jY_xU - Y_yY_z \\ -jY_yU - Y_xY_z & jY_xU - Y_yY_z & U^2 - Y_z^2 \end{bmatrix},
\end{aligned} \tag{8-46}
$$

where $Y^2 = Y_x^2 + Y_y^2 + Y_z^2 = \omega_c^2/\omega^2$. The relative tensor dielectric constant $[\epsilon_r]$ is then given by

$$[\epsilon_r] = [1] + [\chi_e], \tag{8-47}$$

where [1] is a 3×3 unit matrix.

Note that if the dc magnetic field is reversed, all Y_x, Y_y, and Y_z change the sign, and as seen in (8-46), this is equivalent to transposing the matrix $[\chi_e]$ and $[\epsilon_r]$:

$$
\begin{aligned}
[\chi_e] &\to [\tilde{\chi}_e], \\
[\epsilon_r] &\to [\tilde{\epsilon}_r].
\end{aligned} \tag{8-48}
$$

Note also that the anisotropy is produced by the cyclotron frquency ω_c. The cyclotron frequency $f_c = \omega_c/2\pi$ of the earth magnetic field is approximately $f_c = 1.42$ MHz.

8-9 PLANE-WAVE PROPAGATION IN ANISOTROPIC MEDIA

Let us consider the characteristics of a plane wave propagating in an anisotropic medium. We let $\bar{k} = k\hat{i}$, where k is the propagation constant and $\hat{i}$ is the unit vector in the direction of wave propagation. In general, the propagation constant k depends on the direction $\hat{i}$.

We seek a plane-wave solution that has the following general form:

$$e^{j(\omega t - \bar{k}\cdot\bar{r})}. \tag{8-49}$$

First, we note that in general, $\bar{E}$ and $\bar{H}$ are not necessarily perpendicular to $\bar{k}$, but $\bar{D}$ and $\bar{B}$ are always perpendicular to $\bar{k}$. To prove this, we note that for a plane wave

$$\frac{\partial}{\partial x}(e^{-j\bar{k}\cdot\bar{r}}) = -jk_x(e^{-j\bar{k}\cdot\bar{r}}),$$

$$\bar{k} = k_x\hat{x} + k_y\hat{y} + k_z\hat{z},$$

and therefore

$$
\begin{aligned}
\nabla &= \hat{x}\frac{\partial}{\partial x} + \hat{y}\frac{\partial}{\partial y} + \hat{z}\frac{\partial}{\partial z} \\
&= -jk_x\hat{x} - jk_y\hat{y} - jk_z\hat{z} = -j\bar{k}.
\end{aligned} \tag{8-50}
$$

Thus the divergence equations

$$\nabla \cdot \bar{B} = 0 \qquad \text{and} \qquad \nabla \cdot \bar{D} = 0$$

become

$$-j\bar{k} \cdot \bar{B} = 0 \qquad \text{and} \qquad -j\bar{k} \cdot \bar{D} = 0, \tag{8-51}$$

which proves that $\bar{B}$ and $\bar{D}$ are perpendicular to $\bar{k}$. This, however, does not show that $\bar{E}$ and $\bar{H}$ should be perpendicular to $\bar{k}$ since $\bar{E}$ and $\bar{D}$ (or $\bar{H}$ and $\bar{B}$) are not parallel in anisotropic media.

8-10 PLANE-WAVE PROPAGATION IN MAGNETOPLASMA

Let us write Maxwell's equations for a plane wave using (8-50):

$$\begin{aligned} -j\bar{k} \times \bar{E} &= -j\omega\bar{B}, \\ -j\bar{k} \times \bar{H} &= j\omega\bar{D}. \end{aligned} \tag{8-52}$$

In magnetoplasma, we have

$$\bar{B} = \mu_0 \bar{H} \qquad \text{and} \qquad \bar{D} = \epsilon_0 \bar{\bar{\epsilon}}_r \bar{E}, \tag{8-53}$$

and thus substituting (8-53) into (8-52), we obtain the equation for $\bar{E}$:

$$\bar{k} \times \bar{k} \times \bar{E} + \omega^2 \mu_0 \epsilon_0 \bar{\bar{\epsilon}}_r \bar{E} = 0 \tag{8-54}$$

and $\bar{H}$ is given by

$$\bar{H} = \frac{\bar{k} \times \bar{E}}{\omega\mu_0}. \tag{8-55}$$

It is now possible to obtain the propagation constant k from (8-54). Let us first write (8-54) in matrix form. Noting that

$$\begin{aligned} \bar{k} \times \bar{k} \times \bar{E} &= \bar{k}(\bar{k} \cdot \bar{E}) - (\bar{k} \cdot \bar{k})\bar{E}, \\ \bar{k} &= k_x\hat{x} + k_y\hat{y} + k_z\hat{z}, \end{aligned} \tag{8-56}$$

we write (8-54) in the following matrix form:

$$\{K\tilde{K} - k^2[1] + k_0^2[\epsilon_r]\}[E] = 0, \tag{8-57}$$

where

$$K = \begin{bmatrix} k_x \\ k_y \\ k_z \end{bmatrix} \qquad \text{and} \qquad K\tilde{K} = \begin{bmatrix} k_x k_x & k_x k_y & k_x k_z \\ k_y k_x & k_y k_y & k_y k_z \\ k_z k_x & k_z k_y & k_z k_z \end{bmatrix}$$

$k = |\bar{k}|$, $k_0^2 = \omega^2 \mu_0 \epsilon_0$, [1] is a 3×3 unit matrix, $[\epsilon_r]$ is a 3×3 matrix given by (8-47), and $[E]$ is a column matrix given in (8-45).

Equation (8-57) is the fundamental matrix equation for an anisotropic medium with the tensor dielectric constant $[\epsilon_r]$. Since this is a homogeneous linear equation for $[E]$, the nonzero solution for $[E]$ is obtained when the following determinant is zero.

$$|K\tilde{K} - k^2[1] + k_0^2[\epsilon_r]| = 0. \tag{8-58}$$

The solution of this equation gives the propagation constant k.

It will be shown in the next section that for the wave propagating along the dc magnetic field, there are two circularly polarized waves with different propagation constants, and for the wave propagating in the direction perpendicular to the dc magnetic field, there are two linearly polarized waves with different propagation constants. In general, there are two elliptically polarized waves for the wave propagating in an arbitrary direction.

8-11 PROPAGATION ALONG THE DC MAGNETIC FIELD

Let us take the z axis along the direction of propagation $\bar{k}$ and the dc magnetic field $\bar{H}_{dc}$:

$$\bar{k} = k\hat{z}, \qquad \bar{H}_{dc} = H_{dc}\hat{z}. \tag{8-59}$$

Noting that $Y_z = Y$, $Y_x = Y_y = 0$ in (8-46), we get

$$[\chi_e] = -\frac{X}{U(U^2 - Y^2)}\begin{bmatrix} U^2 & -jYU & 0 \\ jYU & U^2 & 0 \\ 0 & 0 & U^2 - Y^2 \end{bmatrix}. \tag{8-60}$$

Thus we get the tensor relative dielectric constant $[\epsilon_r]$:

$$[\epsilon_r] = \begin{bmatrix} \epsilon & ja & 0 \\ -ja & \epsilon & 0 \\ 0 & 0 & \epsilon_z \end{bmatrix}, \tag{8-61}$$

where

$$\epsilon = 1 - \frac{XU}{U^2 - Y^2} = 1 - \frac{(\omega_p/\omega)^2[1 - j(\nu/\omega)]}{[1 - j(\nu/\omega)]^2 - (\omega_c/\omega)^2},$$

$$a = \frac{XY}{U^2 - Y^2} = -\frac{(\omega_p/\omega)^2(\omega_c/\omega)}{[1 - j(\nu/\omega)]^2 - (\omega_c/\omega)^2},$$

$$\epsilon_z = 1 - \frac{X}{U} = 1 - \frac{(\omega_p/\omega)^2}{1 - j(\nu/\omega)}.$$

Note that Y is negative for electrons and therefore $Y = -(\omega_c/\omega)$.

Equation (8-57) then becomes

$$\begin{bmatrix} -k^2 + k_0^2\epsilon & jk_0^2 a & 0 \\ -jk_0^2 a & -k^2 + k_0^2\epsilon & 0 \\ 0 & 0 & k_0^2\epsilon_z \end{bmatrix}\begin{bmatrix} E_x \\ E_y \\ E_z \end{bmatrix} = 0. \tag{8-62}$$

From this, we get the following two values of the propagation constant k:

$$k_+ = k_0 n_+ \quad \text{and} \quad k_- = k_0 n_-, \tag{8-63}$$

where

$$n_+ = \sqrt{\epsilon_+} = \sqrt{\epsilon - a},$$
$$n_- = \sqrt{\epsilon_-} = \sqrt{\epsilon + a}.$$

The equivalent dielectric constant ϵ_+ and ϵ_- are plotted in Fig. 8-6. It is obvious that the behavior of ϵ_+ is similar to that for the isotropic case, and for this reason, the wave for ϵ_+ is called the *ordinary wave* and the other with ϵ_- is called the *extraordinary wave*.

The behavior of the electric fields can be studied by noting (8-62). The first equation is

$$(k_0^2 \epsilon - k^2)E_x + jk_0^2 aE_y = 0. \tag{8-64}$$

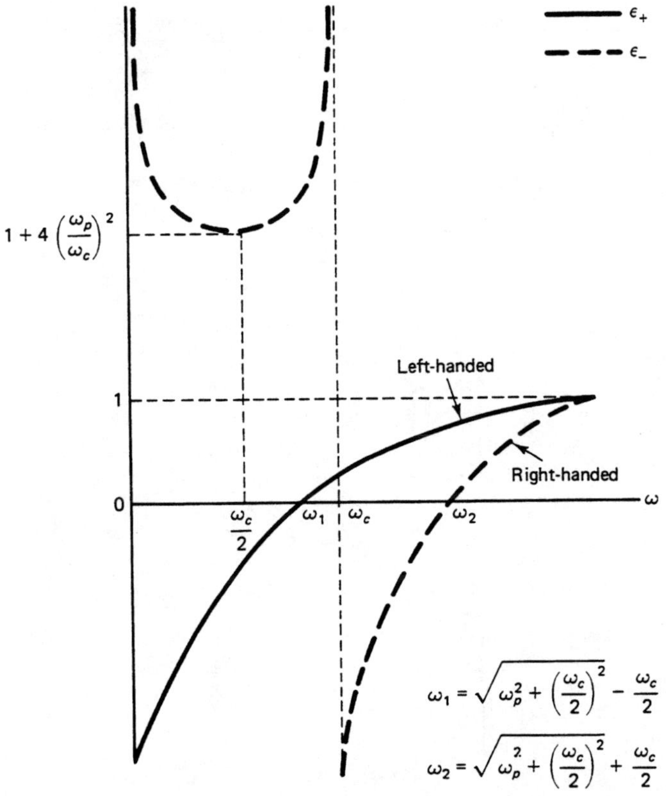

Figure 8-6 Equivalent dielectric constant ϵ_+ (left-handed) for the ordinary wave (a) and ϵ_- (right-handed) for the extraordinary wave (b). The medium is assumed lossless.

For the ordinary wave, $k = k_+$ and therefore we get

$$k_0^2 aE_x + jk_0^2 aE_y = 0,$$

which yields

$$E_x = -jE_y. \tag{8-65}$$

This is a left-handed circularly polarized wave (LHC) whose electric field vector rotates clockwise in the x–y plane (see Fig. 8-7). Also, from (8-56), we note that

$$E_z = 0. \tag{8-66}$$

The displacement vector D is given by

$$D_x = \epsilon_+ E_x, \qquad D_y = \epsilon_+ E_y, \qquad D_z = 0. \tag{8-67}$$

The magnetic field is perpendicular to $\bar{k}$ and $\bar{E}$ and is given by

$$\frac{E_x}{H_y} = -\frac{E_y}{H_x} = Z_+ = \frac{Z_0}{\sqrt{\epsilon - a}}, \qquad Z_0 = \left(\frac{\mu_0}{\epsilon_0}\right)^{1/2}. \tag{8-68}$$

When the propagation is in the $-z$ direction, we write

$$\bar{k} = -k\hat{z}, \tag{8-69}$$

which does not change the electric field E nor D, but the magnetic field is reversed as shown in Fig. 8-7.

Similarly, for the extraordinary wave, $k = k_-$, we get

$$E_x = +jE_y,$$
$$E_z = 0, \tag{8-70}$$

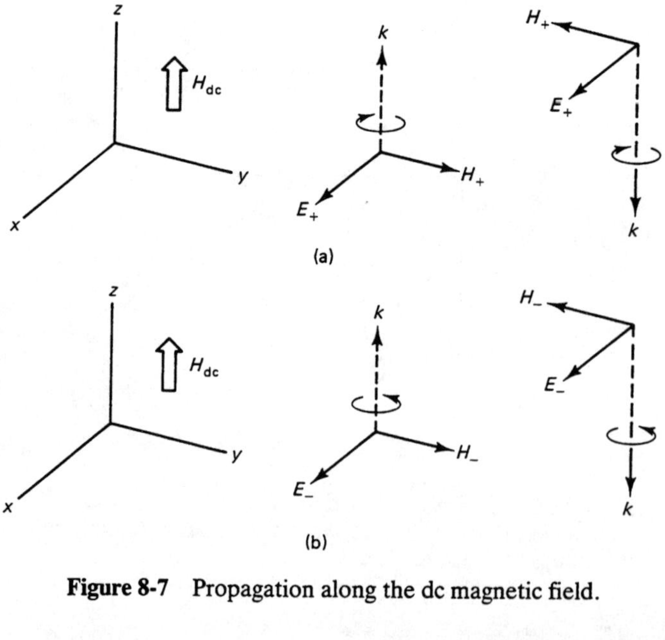

Figure 8-7 Propagation along the dc magnetic field.

which is a right-handed circularly polarized wave (RHC) whose vector rotates counterclockwise in the x–y plane. Also, we get

$$D_x = \epsilon_- E_x, \qquad D_y = \epsilon_- E_y, \qquad D_z = 0, \tag{8-71}$$

and the equivalent wave impedance is

$$Z_- = \frac{Z_0}{\sqrt{\epsilon + a}}. \tag{8-72}$$

We note from Fig. 8-7 that ϵ_+ is negative when $\omega < \omega_1$ and thus the ordinary wave does not propagate. However, in this frequency range, ϵ_- is positive and the extraordinary wave propagates. The VLF waves cannot penetrate the ionosphere in the absence of the dc magnetic field because the frequency is below the plasma frequency. However, in the presence of the earth magnetic field, the extraordinary VLF wave can propagate in the direction of the dc magnetic field. This is the main mechanism of the whistler mode.

The *whistler* is a form of radio noise in the audio-frequency range (1 to 20 kHz) characterized by a whistling tone. The VLF components of the short electromagnetic pulse due to a lightning stroke can penetrate the ionosphere by means of the extraordinary mode and propagate along the earth geomagnetic field as shown in Fig. 8-8. The signal may then be reflected and returned back propagating along the magnetic field. The whistling effect results from the different group velocities and the time delay for different frequency components (Fig. 8-8). The time T required for the signal at a certain frequency with the group velocity v_g is

$$T = \int_{\text{path}} \frac{ds}{v_g} = \int_{\text{path}} \frac{\partial k_-}{\partial \omega} ds. \tag{8-73}$$

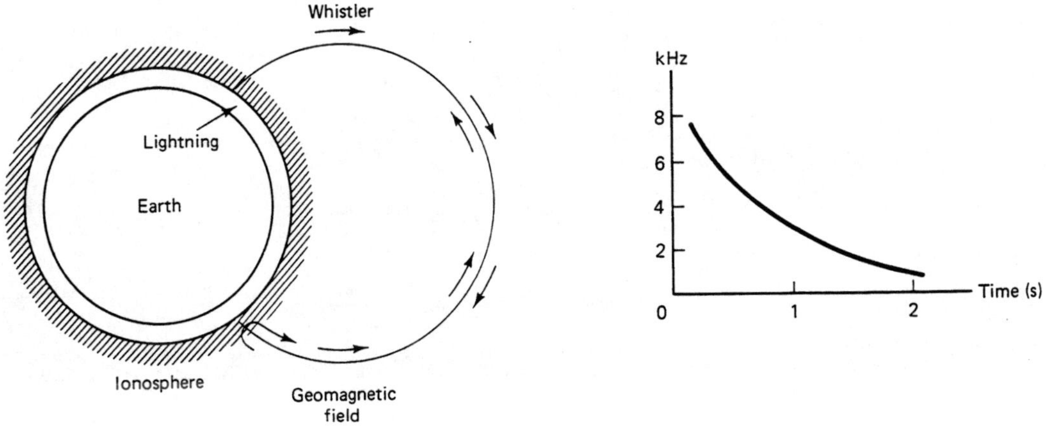

Figure 8-8 Whistler mode.

8-12 FARADAY ROTATION

As discussed in Section 8-11, when a wave is propagating through an anisotropic medium in the direction of the dc magnetic field, two circularly polarized waves can propagate with different propagation constants. As a result, if these two circularly polarized waves are properly combined so as to produce a linearly polarized wave at one point, then as the wave propagates, the plane of polarization rotates and the angle of rotation is proportional to the distance. This is called the *Faraday rotation*.

To show this, let us assume that at $z = 0$, two circularly polarized waves are combined to give

$$E_x = E_0, \qquad E_y = 0. \tag{8-74}$$

At any other point, in general

$$E_x = E_{x+} e^{-jk_+ z} + E_{x-} e^{-jk_- z}, \tag{8-75}$$

where E_{x+} and E_{x-} are the magnitude of two circularly polarized waves given in Section 8-11. Each E_{x+} and E_{x-} must be accompanied by its E_{y+} and E_{y-} as given by (8-65) and (8-70). Thus

$$E_y = jE_{x+} e^{-jk_+ z} - jE_{x-} e^{-jk_- z}. \tag{8-76}$$

At $z = 0$, $E_x = E_{x+} + E_{x-} = E_0$ and $E_y = j(E_{x+} - E_{x-}) = 0$, as given by (8-74), and thus

$$E_{x+} = E_{x-} = \frac{E_0}{2}.$$

Therefore, we have

$$\begin{aligned} E_x &= \frac{E_0}{2}(e^{-jk_+ z} + e^{-jk_- z}), \\ E_y &= j\frac{E_0}{2}(e^{-jk_+ z} - e^{-jk_- z}). \end{aligned} \tag{8-77}$$

By writing

$$k_{\pm} = \frac{k_+ + k_-}{2} \pm \frac{k_+ - k_-}{2},$$

we express E_x and E_y by

$$\begin{aligned} E_x &= E_0 e^{-j[(k_+ + k_-)/2]z} \cos\frac{k_+ - k_-}{2} z, \\ E_y &= E_0 e^{-j[(k_+ + k_-)/2]z} \sin\frac{k_+ - k_-}{2} z. \end{aligned} \tag{8-78}$$

This represents a linearly polarized wave that propagates with the propagation constant

$$k_f = \frac{k_+ + k_-}{2} \tag{8-79}$$

and whose plane of polarization rotates with the angle

$$\theta_f = \frac{k_+ - k_-}{2} z. \tag{8-80}$$

The angle θ_f is proportional to the distance z.

If the wave is propagating in the negative z direction, the formulas above are valid with the change $k_+ \to -k_+$, $k_- \to -k_-$, and $z \to -z$. Thus if the wave is propagated in the positive z direction and then at the end of the path, reflected back and propagated in the negative direction, the Faraday rotation is doubled. This is an important characteristic of *magnetic rotation*. In contrast with this, *natural rotation* is canceled as the wave is propagated forward and then reflected back. The natural rotation is the rotation of the plane of polarization in liquids such as sugar solutions, which have an asymmetrically bound carbon atom. It also occurs in crystals such as quartz and sodium chlorate, which have helical structure. They are characterized by two types of structures, which are related to each other like right-handed and left-handed screws. These two screws are otherwise identical, but they cannot be brought into coincidence by any rotation in three-dimensional space (see Sommerfeld, 1954, pp. 106 and 164). This natural rotation is also called *optical activity* and takes place in bianisotropic medium such as chiral medium. See Section 8-22.

8-13 PROPAGATION PERPENDICULAR TO THE DC MAGNETIC FIELD

Consider the propagation in the x direction and the dc magnetic field in the z direction:

$$\bar{k} = k\hat{x}, \qquad k = k_x = k_0 n, \qquad \bar{H}_{\text{dc}} = H_{\text{dc}}\hat{z}.$$

Then (8-57) becomes

$$\begin{bmatrix} k_0^2 \epsilon & jk_0^2 a & 0 \\ -jk_0^2 a & k_0^2 \epsilon - k^2 & 0 \\ 0 & 0 & k_0^2 \epsilon_z - k^2 \end{bmatrix} \begin{bmatrix} E_x \\ E_y \\ E_z \end{bmatrix} = 0. \tag{8-81}$$

From this, we get two solutions

$$k = k_0 n_0 = k_0 \sqrt{\epsilon_z} \quad \text{and} \quad k = k_0 n_e = k_0 \left(\frac{\epsilon^2 - a^2}{\epsilon} \right)^{1/2}. \tag{8-82}$$

For the wave with the propagation constant $k_0 n_0$, we get

$$\begin{aligned} &E_x = E_y = 0, \qquad E_z \neq 0, \\ &D_x = D_y = 0, \qquad D_z = \epsilon_z E_z, \\ &H_x = H_z = 0, \\ &-\frac{E_z}{H_y} = Z_1 = \frac{Z_0}{\sqrt{\epsilon_z}}, \\ &B_y = \mu_0 H_y. \end{aligned} \tag{8-83}$$

This wave is the ordinary wave because ϵ_z is the same as the case of isotropic plasma, and the dc magnetic field has no effect on this plane-wave propagation. This is expected because both electric and displacement vectors are in the z direction, and electrons move along the dc magnetic field and their motions are not affected by the presence of the dc magnetic field.

For the wave with the propagation constant

$$k = k_0 n_e = k_0 \sqrt{\epsilon_t}, \qquad \epsilon_t = \frac{\epsilon^2 - a^2}{\epsilon},$$

we get

$$\begin{aligned} \epsilon E_x &= -jaE_y, \\ E_z &= 0, \\ D_x &= D_z = 0, \\ D_y &= \epsilon_0 \epsilon_t E_y, \\ \frac{E_y}{H_z} &= Z_2 = \frac{Z_0}{\sqrt{\epsilon_t}}. \end{aligned} \tag{8-84}$$

H_z, D_y, and B_z behave as if the medium has the equivalent dielectric constant $\epsilon_0 \epsilon_t$. But in addition to these fields, E_x, the component in the direction of the propagation, appears. This is produced by the coupling between E_x and E_y in the anisotropic medium.

8-14 THE HEIGHT OF THE IONOSPHERE

Consider a radio wave pulse that is sent vertically toward the ionosphere. The electron density of the ionosphere depends on the height, and its typical profile is shown in Fig. 8-3. The electron density distribution with height may be explained by observing that the rate of production of electrons depends on the sun's radiation and the air density, but the sun's radiation increases with height while the air density decreases with height, and therefore there is a maximum electron density at a certain height.

Let us assume that the effects of the earth magnetic field and the collision frequencies are negligible. The radio wave with a certain frequency $f = \omega/2\pi$ propagates up to the height z_0, where the plasma frequency $f_p = \omega_p/2\pi$ becomes equal to f (Fig. 8-9). This height z_0 is called the *true height*. However, the time τ required for a radio pulse to travel from the ground to z_0 and back is, assuming that v_g is a slowly varying function of z,

$$\tau = 2\int_0^{z_0} \frac{dz}{v_g}, \tag{8-85}$$

where v_g is the group velocity,

$$\frac{1}{v_g} = \frac{\partial k(\omega, z)}{\partial \omega} = \frac{1}{c}\frac{\partial}{\partial \omega}[\omega n(\omega, z)],$$

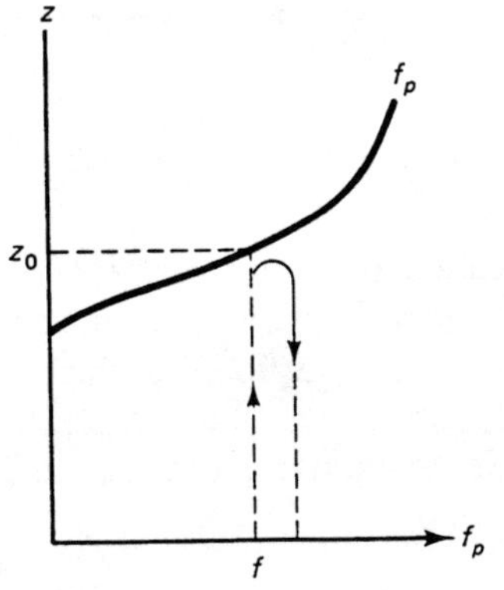

Figure 8-9 Height of the ionosphere.

where the refractive index n varies with height. The *equivalent height* h_e is the fictitious distance over which a pulse propagates in free space, during the time $\tau/2$.

$$h_e = \frac{c\tau}{2} = \int_0^{z_0} \frac{\partial}{\partial \omega}[\omega n(\omega, z)]\, dz. \tag{8-86}$$

The *phase height* h_p is the fictitious distance in free space corresponding to the total phase from $z = 0$ to $z = z_0$.

$$h_p = \int_0^{z_0} n(\omega, z)\, dz. \tag{8-87}$$

8-15 GROUP VELOCITY IN ANISOTROPIC MEDIUM

We defined the group velocity v_g of the wave propagating with propagation constant k by the following:

$$v_g = \frac{\partial \omega}{\partial k}. \tag{8-88}$$

In an anisotropic medium, the propagation constant $k = (\omega/c)n$ depends on the direction of the wave propagation and therefore, (8-88) should hold for each component of the group velocity $\bar{v}_g$.

$$\begin{aligned} \bar{v}_g &= v_{gx}\hat{x} + v_{gy}\hat{y} + v_{gz}\hat{z} \\ &= \left(\frac{\partial}{\partial k_x}\hat{x} + \frac{\partial}{\partial k_y}\hat{y} + \frac{\partial}{\partial k_z}\hat{z}\right)\omega = \nabla_k \omega, \\ \bar{k} &= k_x\hat{x} + k_y\hat{y} + k_z\hat{z}. \end{aligned} \tag{8-89}$$

This is the general expression for $\bar{v}_g$.

Equation (8-89) indicates that the group velocity $\bar{v}_g$ is perpendicular to surfaces of constant ω. This is the surface representing the dispersion relation for a fixed frequency (Fig. 8-10):

$$k(\theta, \phi) = \frac{\omega}{c} n(\theta, \phi). \tag{8-90}$$

The surface $k =$ constant is called the *wave vector surface*. Often it is useful to introduce the *refractive index surface*, where the refractive index n is constant. Obviously, these two surfaces are proportional to each other and carry the same information.

If the refractive index n is symmetric about the z axis and therefore $n = n(\theta)$ as in Fig. 8-10, it is easily seen that

$$\tan \alpha = -\frac{1}{n} \frac{\partial n}{\partial \theta}. \tag{8-91}$$

More generally, the group velocity $\bar{v}_g$ is given by

$$\begin{aligned}
\bar{v}_g &= v_{gk} \hat{k} + v_{g\theta} \hat{\theta} + v_{g\phi} \hat{\phi}, \\
v_{gk} &= \frac{c}{\partial(n\omega)/\partial\omega}, \\
v_{g\theta} &= -\frac{c}{\partial(n\omega)/\partial\omega} \frac{1}{n} \frac{\partial n}{\partial \theta}, \\
v_{g\phi} &= -\frac{c}{\partial(n\omega)/\partial\omega} \frac{1}{n \sin\theta} \frac{\partial n}{\partial \phi},
\end{aligned} \tag{8-92}$$

where $\hat{k}$, $\hat{\theta}$, and $\hat{\phi}$ are unit vectors in the directions of $\bar{k}$, θ, and ϕ, respectively (Yeh and Liu, 1972).

8-16 WARM PLASMA

In the magnetoionic theory discussed in previous sections, it is assumed that the motions of electrons are caused by the Lorentz force. However, since the temperature of the electron gas is finite, the motions of electrons are also affected by the

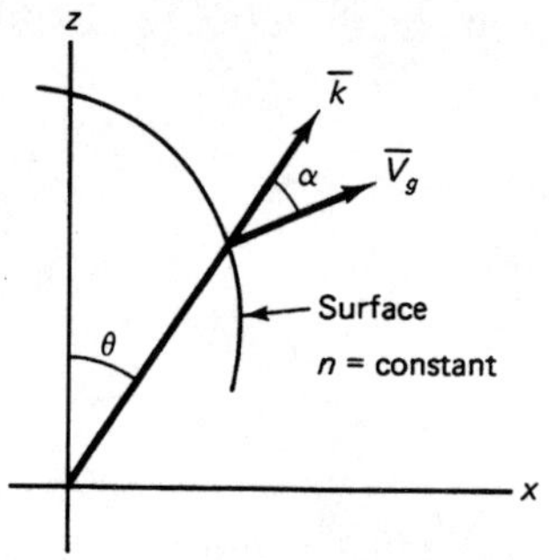

Figure 8-10 Group velocity in an anisotropic medium.

pressure variations of the electron gas. The magnetoionic theory ignores the effects of the finite temperature and the pressure variations, as they are normally small, and therefore it is equivalent to dealing with the electron gas at negligible temperature. Thus the magnetoionic theory may be considered as the theory dealing with *cold plasma*. However, recent studies on the impedance of antennas in the ionosphere indicate that there is a need to include the effects of the finite temperature and *warm plasma*. The pressure variation of warm plasma may be considered as an acoustic wave, and therefore inclusion of the finite temperature means the study of the interaction of the electromagnetic wave and the acoustic wave in the electron gas.

Let us start with the following basic equations for warm plasma.

MAXWELL'S EQUATIONS

$$\nabla \times \bar{E} = -\mu_0 \frac{\partial \bar{H}}{\partial t}$$

$$\nabla \times \bar{H} = \epsilon_0 \frac{\partial \bar{E}}{\partial t} + Ne\bar{v}; \tag{8-93}$$

HYDRODYNAMIC EQUATION

$$m\frac{\partial \bar{v}}{\partial t} = e(\bar{E} + \bar{v} \times \bar{B}) - \frac{1}{N}\nabla p; \tag{8-94}$$

CONTINUITY EQUATION

$$\nabla \cdot (N\bar{v}) + \frac{\partial}{\partial t}N = 0; \tag{8-95}$$

EQUATION OF STATE

$$P = k_b NT; \tag{8-96}$$

where

$\bar{v}$ = velocity of electron
e, m = charge and mass of electron
N = electron density, number of electrons per unit volume
P = pressure
k_b = Boltzmann's constant
T = temperature, Kelvin

Note that the current density $\bar{J}$ is expressed in (8-93) as $Ne\bar{v}$. The second term on the right side of (8-94) is the force due to the pressure gradient, and (8-95) represents the change of the number of electrons per unit volume.

We assume that there is no dc magnetic field $H_{dc} = 0$. We also write N and P as the sum of the average N_0 and P_0 and the ac (acoustic) components n and p:

$$N = N_0 + n,$$
$$P = P_0 + p. \tag{8-97}$$

Now we assume that the ac components are small compared with the average values, and therefore all nonlinear terms containing the product of two ac components are negligible.

Under the foregoing assumptions, (8-93) through (8-95) become

$$\nabla \times \bar{E} = -\mu_0 \frac{\partial \bar{H}}{\partial t}, \tag{8-98}$$

$$\nabla \times \bar{H} = \epsilon_0 \frac{\partial \bar{E}}{\partial t} + N_0 e \bar{v}, \tag{8-99}$$

$$\frac{\partial \bar{v}}{\partial t} = \frac{e}{m}\bar{E} - \frac{1}{mN_0}\nabla p, \tag{8-100}$$

$$N_0 \nabla \cdot \bar{v} + \frac{\partial n}{\partial t} = 0. \tag{8-101}$$

The equation of state (8-96) for the "isothermal" case (T = constant) becomes

$$p = nk_b T. \tag{8-102}$$

For acoustic waves, however, the adiabatic process, where no heat transfer is taking place, is more appropriate. In this case, P and N satisfy the following:

$$\frac{P}{N^\gamma} = \frac{P_0}{N_0^\gamma} = \text{constant}, \tag{8-103}$$

where γ is the ratio of specific heat at constant pressure and constant volume. From this we get

$$p = \gamma n k_b T. \tag{8-104}$$

Equations (8-98) to (8-101) and (8-104) constitute the fundamental equations for isotropic warm plasma. The value of γ for plasma is approximately equal to 3. (Note that $\gamma = \frac{5}{3}$ for perfect monoatomic gas, $\gamma = \frac{7}{5}$ for diatomic gas such as air, and $\gamma = \frac{4}{3}$ for polyatomic gas) (Yeh and Liu, 1972, p. 94).

8-17 WAVE EQUATIONS FOR WARM PLASMA

In order to obtain the wave equation for acoustic waves, we take the divergence of (8-100) and use (8-101) and (8-104) to eliminate $\bar{v}$ and n. We get

$$\nabla^2 p - \frac{1}{u^2}\frac{\partial^2}{\partial t^2} p - N_0 e \nabla \cdot \bar{E} = 0, \tag{8-105}$$

where $u = (\gamma k_b T/m)^{1/2}$ is the sound velocity of the electron gas if there were no charge $e = 0$ and is called *Laplace's sound velocity* for the adiabatic case. [In contrast, for the isothermal case, it is called *Newton's sound velocity* and is equal to $(k_b T/m)^{1/2}$.] The last term of (8-105) is proportional to $\nabla \cdot \bar{v}$ through (8-99), which is in turn proportional to p through (8-101) and (8-104). Taking the divergence of (8-99), we get

$$\begin{aligned}\nabla\cdot\nabla\times\overline{H}=0&=\epsilon_0\frac{\partial}{\partial t}\nabla\cdot\overline{E}+N_0e\nabla\cdot\overline{v}\\&=\epsilon_0\frac{\partial}{\partial t}\nabla\cdot\overline{E}-\frac{e}{\gamma k_bT}\frac{\partial p}{\partial t}.\end{aligned}\tag{8-106}$$

Thus we get

$$N_0e\nabla\cdot\overline{E}=\frac{\omega_p^2}{u^2}p,\qquad \omega_p^2=\frac{N_0e^2}{m\epsilon_0}.$$

Substituting this into (8-105), we get the following wave equation:

$$\nabla^2p-\frac{1}{u^2}\frac{\partial^2}{\partial t^2}p-\frac{\omega_p^2}{u^2}p=0.\tag{8-107}$$

For a time-harmonic case [exp($j\omega t$)], we get

$$\begin{aligned}(\nabla^2+k_p^2)p&=0,\\k_p^2&=\frac{\omega^2-\omega_p^2}{u^2}.\end{aligned}\tag{8-108}$$

The phase velocity is therefore

$$v_p=\frac{u}{[1-(\omega_p/\omega)^2]^{1/2}}.\tag{8-109}$$

Note that the acoustic velocity v_p is modified by the plasma frequency ω_p.

The wave equation for the electromagnetic field is obtained by taking the curl of (8-98) and substituting (8-99). We then get

$$-\nabla\times\nabla\times\overline{E}-\mu_0\epsilon_0\frac{\partial^2}{\partial t^2}\overline{E}-\mu_0\epsilon_0\omega_p^2\overline{E}+\mu_0\epsilon_0u^2\nabla(\nabla\cdot\overline{E})=0.\tag{8-110}$$

Let us consider a time-harmonic case and assume that a plane wave is propagating in a direction given by the propagation vector $\overline{k}$. Thus all electric and magnetic vectors have $\exp(-j\overline{k}\cdot\overline{r})$ dependence, and therefore the operator ∇ can be replaced by $-j\overline{k}$. Noting that

$$\overline{\nabla}\times\overline{\nabla}\times\overline{E}=-\overline{k}\times\overline{k}\times\overline{E}=-[\overline{k}(\overline{k}\cdot\overline{E})-\overline{E}k^2],$$

with $\overline{k}\cdot\overline{k}=k^2$, we get

$$\overline{k}(\overline{k}\cdot\overline{E})(1-\mu_0\epsilon_0u^2)-\left[k^2-k_0^2\left(1-\frac{\omega_p^2}{\omega^2}\right)\right]\overline{E}=0.\tag{8-111}$$

Now we examine the component of $\overline{E}$ along and perpendicular to $\overline{k}$:

$$\overline{E}=\overline{E}_{\parallel}+\overline{E}_{\perp}.\tag{8-112}$$

Taking the component of (8-111) along $\bar{k}$, we get

$$k^2(1-\mu_0\epsilon_0 u^2) - \left[k^2 - k_0^2\left(1-\frac{\omega_p^2}{\omega^2}\right)\right] = 0, \tag{8-113}$$

from which we obtain the propagation constant k for $\bar{E}_\parallel$,

$$k_\parallel^2 = \frac{\omega^2-\omega_p^2}{u^2} = k_p^2. \tag{8-14}$$

This component $\bar{E}_\parallel$ therefore propagates with the propagation constant identical to that of the pressure wave k_p; thus this is called the *acoustic wave*.

On the other hand, the perpendicular component $\bar{E}_\perp$ can be determined from (8-111) to have the propagation constant

$$k^2 = k_0^2\left(1-\frac{\omega_p^2}{\omega^2}\right), \tag{8-115}$$

which is identical to that of a cold plasma.

Since $\bar{H} = (1/\omega\mu_0)\bar{k}\times\bar{E}$, no magnetic field is associated with $\bar{E}_\parallel$, but $\bar{H}_\perp$ associated with $\bar{E}_\perp$ is given in the same manner as that of a cold plasma. These two waves $\bar{E}_\parallel$ and $\bar{E}_\perp$ can exist in an infinite space independently, with two different propagation constants. However, these two components are coupled together at a boundary or at an exciting source. For example, a plane wave incident on a warm plasma excites both components, and the amount of excitation for each component depends on boundary conditions. Also, a dipole source in a warm plasma excites an acoustic wave, and thus the impedance of the antenna is affected by the acoustic wave.

8-18 FERRITE AND THE DERIVATION OF ITS PERMEABILITY TENSOR

In 1845, Faraday discovered the rotation of the plane of polarization of light when propagating through various materials under the influence of the dc magnetic field, now called the *Faraday rotation*. Until about 1946, this effect could not be utilized at microwave frequencies due to the high loss in ferromagnetic materials. But the discovery of low-loss ferrite materials made it possible to employ this material for a variety of microwave applications. Polder developed the general tensor permeability for the ferrite in 1949, and Tellegen in 1948 and Hogan in 1952 developed a microwave network element called the *gyrator* utilizing ferrite. Commercial ferrite devices have been available since about 1953.

Let us first derive the tensor permeability for the ferrite based on a simplified model of a spinning electron. A spinning electron can be considered as a gyromagnetic top. Let the angular momentum be $\bar{J}$. Then the magnetic moment $\bar{m}$ is parallel to $\bar{J}$ and in the opposite direction. The magnitude of m is proportional to J. Thus

$$\bar{m} = \gamma\bar{J}. \tag{8-116}$$

The proportionality constant γ is negative and is found to be

$$\gamma = -\frac{e}{m_e} = -1.7592 \times 10^{11} \text{ C/kg}$$

$$= \text{gyromagnetic ratio.}$$

The equation of angular motion for a single electron is then

$$\frac{d\bar{J}}{dt} = \text{torque.} \tag{8-117}$$

The torque on the magnetic moment m is given by

$$\bar{m} \times \bar{B}.$$

Thus we get

$$\frac{d\bar{m}}{dt} = \gamma\bar{m} \times \bar{B}. \tag{8-118}$$

We first note that when this magnetic top is placed in the dc magnetic field, the top precesses around the direction of the magnetic field with the angular velocity

$$\omega_0 = -\gamma B = -\gamma\mu_0 H, \tag{8-119}$$

which is known as the *Larmor precessional frequency*. This precessional motion is described by

$$\frac{dJ}{dt} = J \times \omega_0. \tag{8-120}$$

Eventually, however, this precession dies down due to the damping, and all the magnetic tops become aligned with the dc magnetic field. Then the ferrite is said to be *saturated*. Let H_0 be the applied external dc magnetic field. Then the field in the ferrite is given by

$$H_i = H_0 - H_{\text{dem}}, \tag{8-121}$$

where H_{dem} is the demagnetizing field. In general, H_i is not in the same direction as H_0 and depends on the shape of the material. However, if the ferrite is an ellipsoid, H_i is parallel to H_0. It must be kept in mind that the internal field H_i is not the same as the applied field and may not even be in parallel direction.

Let us take a unit volume and consider the magnetic polarization

$$\bar{M} = N_e\bar{m}, \tag{8-122}$$

where N_e is the number of effective electrons per unit volume. In terms of M, we write (8-118)

$$\frac{d\bar{M}}{dt} = \gamma\bar{M} \times \bar{B}. \tag{8-123}$$

Next we express each of B, H, and M as a sum of the dc component and small ac component. Thus we write

$$B = \mu_0(H + M),$$
$$H = H_i + H_a, \tag{8-124}$$
$$M = M_i + M_a,$$

where H_i and M_i are the dc components and H_a and M_a are the small ac components.

Noting that $M \times M = 0$, we get

$$\frac{d\overline{M}}{dt} = \mu_0 \gamma(\overline{M} \times \overline{H}). \tag{8-125}$$

We then obtain

$$\frac{dM_i}{dt} + \frac{dM_a}{dt} = \mu_0 \gamma(M_i \times H_i + M_i \times H_a + M_a \times H_i + M_a \times H_a). \tag{8-126}$$

Note that M_i is constant and M_i and H_i are in parallel, and thus $dM_i/dt = 0$ and $M_i \times H_i = 0$.

We assume that the ac component is small compared with the dc component, and thus the last term is negligibly small compared with the other terms. We then obtain the following linearized equation:

$$\frac{dM_a}{dt} = \mu_0 \gamma(M_i \times H_a + M_a \times H_i). \tag{8-127}$$

We choose the z axis along the direction of the internal magnetic field H_i and M_i. We also consider the time-harmonic case, with $\exp(j\omega t)$ dependence. Then we can represent (8-127) in the following matrix form:

$$\begin{bmatrix} j\omega & \omega_0 \\ -\omega_0 & j\omega \end{bmatrix} \begin{bmatrix} M_x \\ M_y \end{bmatrix} = \begin{bmatrix} 0 & \omega_M \\ -\omega_M & 0 \end{bmatrix} \begin{bmatrix} H_x \\ H_y \end{bmatrix} \tag{8-128}$$

and $M_z = 0$, where

$$M_a = \begin{bmatrix} M_x \\ M_y \\ M_z \end{bmatrix}, \qquad H_a = \begin{bmatrix} H_x \\ H_y \\ H_z \end{bmatrix},$$

$$\omega_0 = -\gamma\mu_0 H_i$$

$$= \text{gyromagnetic response frequency},$$

$$\omega_M = -\gamma\mu_0 M_i$$

$$= \text{saturation magnetization frequency}.$$

From this, M_a can be expressed by

$$\begin{bmatrix} M_x \\ M_y \\ M_z \end{bmatrix} = \frac{\omega_M}{\omega_0^2 - \omega^2} \begin{bmatrix} \omega_0 & j\omega & 0 \\ -j\omega & \omega_0 & 0 \\ 0 & 0 & 0 \end{bmatrix} \begin{bmatrix} H_x \\ H_y \\ H_z \end{bmatrix}, \tag{8-129}$$

which we can write as

$$M_a = \bar{\bar{\chi}} H_a, \tag{8-130}$$

and therefore, for the ac components, we obtain

$$\begin{aligned} B_a &= \mu_0(H_a + M_a) \\ &= \bar{\bar{\mu}} H_a, \end{aligned} \tag{8-131}$$

where the permeability tensor $\bar{\bar{\mu}}$ can be written as

$$\begin{aligned} \bar{\bar{\mu}} &= \mu_0(1 + \bar{\bar{\chi}}) = \begin{bmatrix} \mu & -j\kappa & 0 \\ j\kappa & \mu & 0 \\ 0 & 0 & \mu_0 \end{bmatrix}, \\ \frac{\mu}{\mu_0} &= 1 - \frac{\omega_0 \omega_M}{\omega^2 - \omega_0^2}, \\ \frac{\kappa}{\mu_0} &= \frac{\omega \omega_M}{\omega^2 - \omega_0^2}. \end{aligned} \tag{8-132}$$

8-19 PLANE-WAVE PROPAGATION IN FERRITE

In ferrite media, the dielectric constant is scalar, and thus we write

$$\begin{aligned} \bar{k} \times \bar{k} \times \bar{H} &= \omega^2 \epsilon \bar{\bar{\mu}} \bar{H}, \\ \bar{E} &= -\frac{1}{\omega\epsilon} \bar{k} \times \bar{H}. \end{aligned} \tag{8-133}$$

This can also be obtained by noting the duality principle, in which E, H, $\bar{\bar{\epsilon}}$, and $\bar{\bar{\mu}}$ are replaced by H, $-E$, $\bar{\bar{\epsilon}}$, and $\bar{\bar{\mu}}$. Since this is the same form as the magnetoplasma case, only the results are shown here.

Propagation Along the DC Magnetic Field

The propagation constant $k = k_z$ is given by

$$\begin{aligned} k_+ &= \omega\sqrt{\epsilon(\mu + \kappa)}, \\ k_- &= \omega\sqrt{\epsilon(\mu - \kappa)}. \end{aligned} \tag{8-134}$$

When $k = k_+$, we get

$$H_x = -jH_y, \tag{8-135}$$

which gives a circularly polarized wave rotating clockwise in the x–y plane. The electric field E is given by

$$\begin{aligned} E_y &= -Z_+ H_x, \\ E_x &= Z_+ H_y, \\ Z_+ &= Z_0 \frac{k_+}{k_0} = Z_0 \sqrt{\frac{\mu + \kappa}{\mu_0}}. \end{aligned} \tag{8-136}$$

When $k = k_-$, we have

$$H_x = jH_y, \tag{8-137}$$

and the wave impedance is

$$Z_- = Z_0\sqrt{\frac{\mu - \kappa}{\mu_0}}. \tag{8-138}$$

Propagation Perpendicular to the DC Magnetic Field

We let

$$\bar{k} = k_x\hat{x} = k\hat{x} \tag{8-139}$$

and we get two cases. The case when

$$k = k_1 = \omega\sqrt{\mu_0 \epsilon} \tag{8-140}$$

is the same as that of isotropic medium. The only magnetic field component is H_z, and there is no coupling between H_x and H_y. When $k = k_2 = \omega\sqrt{\epsilon\mu_t}$, $\mu_t = (\mu^2 - \kappa^2)/\mu$, we have

$$\begin{aligned} \mu H_x - j\kappa H_y &= 0, \\ B_y &= \mu_t H_y, \\ E_z = -Z_2 H_y, \qquad Z_2 &= Z_0\sqrt{\frac{\mu_t}{\mu_0}}. \end{aligned} \tag{8-141}$$

8-20 MICROWAVE DEVICES USING FERRITES

Faraday Rotation and Circulators

The Faraday rotation can be used to construct nonreciprocal microwave networks. For example, the device pictured here contains a ferrite that produces 90° rotation of the polarization. If a wave is incident from the left, the phase is reversed on the other side. But if a wave is incident from the right, there is no phase shift (see Fig. 8-11). Thus this device is represented by a schematic diagram in Fig. 8-11c, showing the phase shift of 180° for a wave propagating to the right and no phase shift for a wave propagating to the left.

It is possible to make use of this device and combine it with two magic tees to construct a circulator (see Fig. 8-12). A wave entering terminal a divides equally, but because of the π phase shift, these two waves do not appear in d, but they are combined into b. Similarly, the wave entering b appears only in c. The wave entering c appears only in d, and the wave entering d appears only in a.

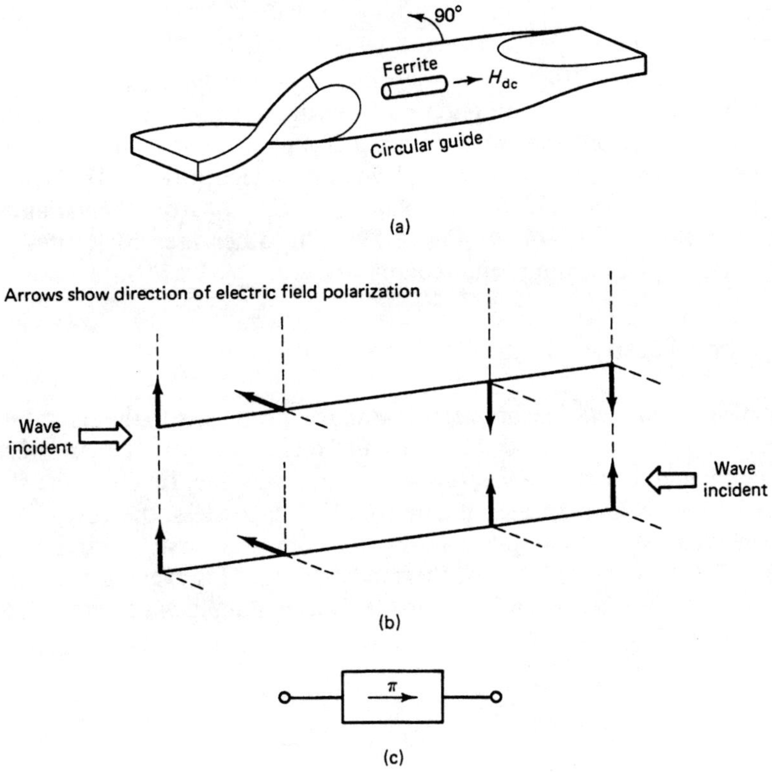

Figure 8-11 Nonreciprocal device.

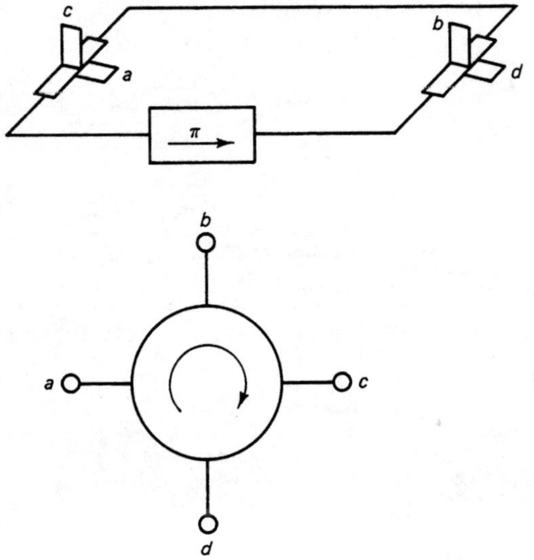

Figure 8-12 Circulator.

One-Way Line

It is also possible to construct a one-way line using the Faraday rotation. One scheme may be as follows. A wave entering from the left goes through this device with little attenuation, but the wave entering from the right is absorbed in the resistive sheet, which is placed in parallel with the electric field. This device is placed at the output of an oscillator to isolate the oscillator from the variations in the load and is called the *isolator* (Fig. 8-13). The other uses of ferrite-loaded waveguides include phase shifters and modulators.

Thermodynamic Paradox of One-Way Line

In the example of a one-way line above, the wave can propagate to the right, but the wave propagating to the left is absorbed by the resistive sheet. If there were no resistive sheet, it could be shown that this is not a one-way line. In fact, there should be no lossless one-way line, because if a one-way line is lossless, the energy on one side can transfer to the other side, causing a temperature rise without external work. This violates the second law of thermodynamics. However, in the study of microwave propagation through a ferrite-loaded waveguide, it was found that Max-

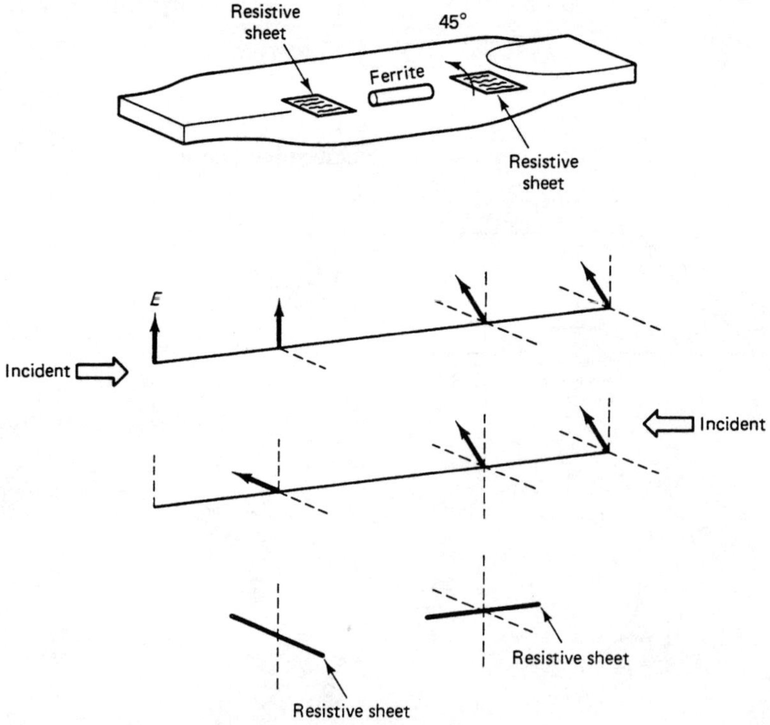

Figure 8-13 One-way line.

well's equations can be solved for lossless waveguide with lossless ferrite, and a one-way propagation constant can be obtained within some ranges of frequency and parameters. This prompted the question of whether lossless Maxwell's equations violate the law of thermodynamics.

The answer to this so-called "thermodynamic paradox" can be found in the concept that mathematical formulations of any physical problems must satisfy three conditions: (1) uniqueness, (2) existence, and (3) the solution must depend continuously on the variation of physical parameters. Problems that satisfy these three conditions are called *properly posed.* For the microwave problem above, it can be shown that if we solve Maxwell's equations for lossy ferrite and then take the limit as the resistivity goes to zero, the absorbed power is correctly accounted for and there is no conflict with the law of thermodynamics. Only if we start with lossless Maxwell's equations can the second law of thermodynamics be violated, but the problem in this case is improperly posed and therefore does not represent a physical problem.

8-21 LORENTZ RECIPROCITY THEOREM FOR ANISOTROPIC MEDIA

Electromagnetic fields in anisotropic media do not in general obey the reciprocity relationship. It is, however, possible to find the condition under which the reciprocal relationship holds for anisotropic media. This section deals with this question.

Let us consider the time-harmonic case ($e^{j\omega t}$). Consider the fields E_1 and H_1 in the medium $\bar{\bar{\epsilon}}_1$ and $\bar{\bar{\mu}}_1$, and E_2 and H_2 in the medium $\bar{\bar{\epsilon}}_2$ and $\bar{\bar{\mu}}_2$. We write two sets of Maxwell's equations

$$\nabla \times E_1 = -j\omega B_1 - J_{m1},$$

$$\nabla \times H_1 = j\omega D_1 + J_1$$

and (8-142)

$$\nabla \times E_2 = -j\omega B_2 - J_{m2},$$

$$\nabla \times H_2 = j\omega D_2 + J_2.$$

Now consider the identity

$$\nabla \cdot (E_1 \times H_2) = H_2 \cdot \nabla \times E_1 - E_1 \cdot \nabla \times H_2$$

and substituting from Maxwell's equations, we get

$$\nabla \cdot (E_1 \times H_2) = -j\omega[H_2 \cdot B_1 + E_1 \cdot D_2] - H_2 \cdot J_{m1} - E_1 \cdot J_2. \quad (8\text{-}143)$$

Similarly,

$$\nabla \cdot (E_2 \times H_1) = -j\omega[H_1 \cdot B_2 + E_2 \cdot D_1] - H_1 \cdot J_{m2} - E_2 \cdot J_1.$$

In isotropic media, we subtract the second equation from the first and setting $\epsilon_1 = \epsilon_2$ and $\mu_1 = \mu_2$, we obtain the Lorentz reciprocity theorem.

$$\nabla \cdot (E_1 \times H_2) - \nabla \cdot (E_2 \times H_1) = H_1 \cdot J_{m2} - H_2 \cdot J_{m1} + E_2 \cdot J_1 - E_1 \cdot J_2. \quad (8\text{-}144)$$

We made use of the fact that the first term of the right side of both equations (8-143) are the same and cancel out:

$$H_2 \cdot B_1 - H_1 \cdot B_2 = (\mu_1 - \mu_2) H_1 \cdot H_2 = 0.$$

However, in anisotropic media, these terms do not cancel each other. Expressing in matrix form, we get

$$\tilde{H}_2 \bar{\bar{\mu}}_1 H_1 - \tilde{H}_1 \bar{\bar{\mu}}_2 H_2 = \tilde{H}_1 (\tilde{\bar{\bar{\mu}}}_1 - \bar{\bar{\mu}}_2) H_2. \tag{8-145}$$

This is not zero when $\bar{\bar{\mu}}_1 = \bar{\bar{\mu}}_2$. For this term to be zero, $\bar{\bar{\mu}}_2$ must be a transpose of $\bar{\bar{\mu}}_1$.

$$\tilde{\bar{\bar{\mu}}}_1 = \bar{\bar{\mu}}_2 \qquad \text{and also} \qquad \tilde{\bar{\bar{\epsilon}}}_1 = \bar{\bar{\epsilon}}_2. \tag{8-146}$$

This occurs when the dc magnetic field is reversed, or $\bar{\bar{\epsilon}}$ *and* $\bar{\bar{\mu}}$ are symmetric tensors. Then we have the same reciprocity theorem.

Under the condition (8-146), let us consider a case when $J_{m1} = J_{m2} = 0$. We integrate over a volume V, and obtain

$$\int_S (E_1 \times H_2 - E_2 \times H_1) \cdot dS = \int_V (E_2 \cdot J_1 - E_1 \cdot J_2)\, dV. \tag{8-147}$$

Now we let the surface S expand to infinity. At a large distance R, E, and H behave as a plane wave and E and H are perpendicular to R and to each other. Thus

$$E_1 = A\frac{e^{-jkR}}{R}, \qquad H_1 = \sqrt{\frac{\epsilon_0}{\mu_0}}\, i_R \times E_1,$$

$$E_2 = B\frac{e^{-jkR}}{R}, \qquad H_2 = \sqrt{\frac{\epsilon_0}{\mu_0}}\, i_R \times E_2.$$

Then

$$E_1 \times H_2 \cdot dS = E_1 \times H_2 \cdot i_R\, dS = \sqrt{\frac{\epsilon_0}{\mu_0}}\, H_1 \cdot H_2\, dS$$

and

$$E_2 \times H_1 \cdot dS = \sqrt{\frac{\epsilon_0}{\mu_0}}\, H_1 \cdot H_2\, dS.$$

Therefore, the integral over S then becomes zero. We then obtain

$$\int_V E_2 \cdot J_1\, dV = \int_V E_1 \cdot J_2\, dV, \tag{8-148}$$

where V is taken as an entire space.

Let us consider the physical significance of (8-148). We let $J_1 = J_1(r_1)$ and $J_2 = J_2(r_2)$. Then the equation states that the field E_1 at r_2 due to the source J_1 at r_1 is equal to the field E_2 at r_1 due to the source J_2 at r_2, *provided that* $\bar{\bar{\epsilon}}$ *and* $\bar{\bar{\mu}}$ *are transposed*. Thus we can say that in isotropic media, the reciprocity theorem holds, but in anisotropic media, the reciprocity holds for the following cases: (1) when $\bar{\bar{\epsilon}}$ and $\bar{\bar{\mu}}$ are symmetric as in the case of crystal, or (2) when the dc magnetic field is reversed for the case of ferrite and magnetoplasma. When ferrite material is used in antennas, the transmitting pattern is not equal to the receiving pattern unless the dc magnetic field is reversed. This is an important consideration when making measurement of antenna radiation pattern.

8-22 BIANISOTROPIC MEDIA AND CHIRAL MEDIA

At the end of section 8-12, we mentioned the difference between Faraday rotation and Natural rotation. If a wave is propagated in plasma or ferrite along the dc magnetic field, and then reflected back and propagated in the reverse direction, the rotation of the plane of polarization is doubled, while the rotation is cancelled for natural rotation. This natural rotation of the plane of polarization is called the "optical activity," and is caused by the right-handed or left-handed property of the medium such as right-handed helix or left-handed helix. The optical activity has been known for many years and its mathematical formulations have been investigated by many workers (see Sommerfeld, 1954; Kong, 1972, 1974; Bassiri et al., 1988; Lakhtakia et al., 1988).

In a preceding section, we have discussed anisotropic media characterized by the constitutive relations

$$\bar{D} = \bar{\bar{\epsilon}}\bar{E} \qquad \text{and} \qquad \bar{B} = \bar{\bar{\mu}}\bar{H}. \tag{8-149}$$

We can generalize these relations to the bianisotropic media:

$$\begin{bmatrix} \bar{D} \\ \bar{H} \end{bmatrix} = \begin{bmatrix} \bar{P} & \bar{L} \\ \bar{M} & \bar{Q} \end{bmatrix} \begin{bmatrix} \bar{E} \\ \bar{B} \end{bmatrix}, \tag{8-150}$$

where in general, $\bar{P}$, $\bar{L}$, $\bar{M}$, and $\bar{Q}$ are 3×3 matrices. Note that in (8-150), $\bar{D}$ and $\bar{H}$ are given in terms of $\bar{E}$ and $\bar{B}$. As indicated at the end of Secton 2-1, the force depends on $\bar{E}$ and $\bar{B}$, and therefore $\bar{E}$ and $\bar{B}$ are the fundamental field quantities and $\bar{D}$ and $\bar{H}$ are the derived fields through the constitutive relations.

If the medium is lossless, then $\text{Re}(\nabla \cdot \bar{S})$ must be zero, where $\bar{S} = \frac{1}{2}\bar{E} \times \bar{H}^*$ is the complex Poynting vector (Section 2-5). Noting that

$$\nabla \cdot \bar{S} = -\frac{j\omega}{2}[\bar{H}^* \cdot \bar{B} - \bar{E} \cdot \bar{D}^*], \tag{8-151}$$

$\text{Re}(\nabla \cdot \bar{S}) = 0$ is equivalent to

$$\bar{H}^* \cdot \bar{B} - \bar{H} \cdot \bar{B}^* - \bar{E} \cdot \bar{D}^* + \bar{E}^* \cdot \bar{D} = 0. \tag{8-152}$$

Substituting (8-150) into this, we get

$$\begin{aligned} \bar{P} &= \bar{P}^+, \\ \bar{Q} &= \bar{Q}^+, \\ \bar{M} &= -\bar{L}^+. \end{aligned} \tag{8-153}$$

where $\overline{P}^{+}$ means the complex conjugate of the transpose of $\overline{P}$. The detailed exposition on this topic is given by Kong (1972).

If $\overline{P}$, $\overline{L}$, $\overline{M}$, and $\overline{Q}$ are scalar, this is called the *biisotropic medium*, which is also called *chiral medium*. Noting the symmetry relations (8-153), we write the constitutive relations for a lossless chiral medium:

$$\begin{aligned} \overline{D} &= \epsilon \overline{E} - j\gamma \overline{B}, \\ \overline{H} &= -j\gamma \overline{E} + \frac{1}{\mu}\overline{B}, \end{aligned} \tag{8-154}$$

where ϵ, μ, and γ are real scalar constants. Note that γ has the dimension of admittance, $(\epsilon/\mu)^{1/2}$. It is also possible to rewrite (8-154) in the form

$$\begin{aligned} \overline{D} &= (\epsilon + \gamma^2 \mu)\overline{E} - j\gamma\mu \overline{H}, \\ \overline{B} &= j\gamma\mu \overline{E} + \mu \overline{H}. \end{aligned} \tag{8-155}$$

Furthermore, using $\nabla \times \overline{H} = +j\omega \overline{D}$ and $\nabla \times \overline{E} = -j\omega \overline{B}$, we can rewrite (8-155) in the following form:

$$\begin{aligned} \overline{D} &= \epsilon_1[\overline{E} + \beta \nabla \times \overline{E}], \\ \overline{B} &= \mu_1[\overline{H} + \beta \nabla \times \overline{H}]. \end{aligned} \tag{8-156}$$

This form shows that for biisotropic medium, $\overline{D}$ depends not only on $\overline{E}$ at a point, but also on the behavior of $\overline{E}$ in the neighborhood of that point represented by $\nabla \times \overline{E}$. This nonlocal behavior of $\overline{D}$ is called the *spatial dispersion*.

Let us combine Maxwell's equations with the constitutive relations (8-155).

$$\begin{aligned} \nabla \times \overline{E} &= -j\omega \overline{B} = -j\omega(\mu \overline{H} + j\gamma\mu \overline{E}), \\ \nabla \times \overline{H} &= j\omega \overline{D} = j\omega[(\epsilon + \gamma^2 \mu)\overline{E} - j\gamma\mu \overline{H}]. \end{aligned} \tag{8-157}$$

We first substitute $\overline{H}$ from the first equation into the second equation and express $\nabla \times \overline{H}$ in terms of $\nabla \times \overline{E}$ and $\overline{E}$. Then we take the curl of the first equation and substitute $\nabla \times \overline{H}$. We then get the equation for $\overline{E}$.

$$-\nabla \times \nabla \times \overline{E} + 2\omega\gamma\mu \nabla \times \overline{E} + \omega^2 \mu\epsilon \overline{E} = 0 \tag{8-158}$$

and we get the identical equation for $\overline{H}$.

Let us find propagation constant K for a plane wave propagating in the z direction. Since $\overline{E}$ behaves as $\exp(-jKz)$, we get $\nabla = -jK\hat{z}$ and therefore (8-158) becomes

$$\begin{bmatrix} -K^2 + k^2 & jK2\omega\gamma\mu \\ -jK2\omega\gamma\mu & -K^2 + k^2 \end{bmatrix} \begin{bmatrix} E_x \\ E_y \end{bmatrix} = 0, \tag{8-159}$$

where $k^2 = \omega^2 \mu\epsilon$.

A nonzero solution to (8-159) is obtained by letting the determinant of coefficients be zero:

$$\begin{vmatrix} -K^2 + k^2 & jK2\omega\gamma\mu \\ -jK2\omega\gamma\mu & -K^2 + k^2 \end{vmatrix} = 0 \tag{8-160}$$

This can be solved to obtain two propagation constants,

$$\begin{aligned} K_1 &= \omega\mu\gamma + [(\omega\mu\gamma)^2 + k^2]^{1/2}, \\ K_2 &= -\omega\mu\gamma + [(\omega\mu\gamma)^2 + k^2]^{1/2}. \end{aligned} \tag{8-161}$$

Substituting K_1 into one of the equations (8-159), we get

$$(-K_1^2 + k^2)E_x + jK_1 2\omega\gamma\mu E_y = 0,$$

from which we get

$$E_x = jE_y. \tag{8-162}$$

This is a right-hand circularly polarized wave (RHC). The corresponding magnetic field is obtained from (8-157):

$$\overline{H} = \frac{j}{\omega\mu}\nabla \times \overline{E} - j\gamma\overline{E}. \tag{8-163}$$

Using $\nabla = -jK\hat{z}$ and (8-162), we get

$$\frac{E_x}{H_y} = -\frac{E_y}{H_x} = \frac{\omega\mu}{[(\omega\mu\gamma)^2 + k^2]^{1/2}}; \tag{8-164}$$

similarly, for K_2, we get a left-handed circularly polarized wave (LHC):

$$E_x = -jE_y. \tag{8-165}$$

The ratio of E to H is the same as given in (8-164). It is clear from (8-161) that if $\gamma > 0$, $K_1 > k > K_2$ and thus the phase velocity for RHC is slower than for LHC. If $\gamma < 0$, $K_1 < k < K_2$ and thus the LHC wave has a slower phase velocity than for the RHC wave. From (8-157) we note that taking the divergence, $\nabla \cdot \overline{B} = 0$ and $\nabla \cdot \overline{D} = 0$. Also, from (8-158), we get $\nabla \cdot \overline{E} = 0$ and $\nabla \cdot \overline{H} = 0$. Therefore, using $\nabla = -j\overline{K}$ for a plane wave, we conclude that the plane wave in chiral medium is a TEM wave.

All the analysis shown above can be derived from (8-156) with the following correspondence:

$$\mu = \frac{\mu_1}{1 - k_1^2\beta^2}, \qquad \epsilon = \epsilon_1, \qquad \gamma = \omega\epsilon_1\beta, \qquad k_1^2 = \omega^2\mu_1\epsilon_1. \tag{8-166}$$

The wave equation becomes

$$\nabla \times \nabla \times \overline{E} = 2\gamma_1^2\beta\nabla \times \overline{E} + \gamma_1^2\overline{E}, \tag{8-167}$$

where $\gamma_1^2 = k_1^2/(1 - k_1^2 \beta^2)$. The propagation constant then becomes

$$K_1 = \frac{k_1}{1 - k_1 \beta}, \qquad K_2 = \frac{k_1}{1 + k_1 \beta}. \tag{8-168}$$

8-23 SUPERCONDUCTORS, LONDON EQUATION, AND THE MEISSNER EFFECTS

Superconductors are used in transmission lines, waveguides, and resonant cavities because they have low losses, their dispersion is small, and they can be used for broad band transmission. In this and the next section we discuss the derivation of London equations; the Meissner effect; and the complex conductivity, penetration depth, and surface impedance of superconductors at high frequencies. We will not discuss the physics of superconductors or the historical development, such as the discovery of superconductivity by Kamerlingh Onnes in 1911, Meissner's work, Fritz and Heinz London's work, BCS theory (Bardeen, Cooper, and Schrieffer), Ginzburg–Landau theory, and work on high-temperature superconductors. They are clearly outside the scope of this book. For the topics discussed in this section, readers are referred to a very informative book by Mendelssohn (1966), Van Duzer and Turner (1981), Ghoshal and Smith (1988), and Lee and Itoh (1989).

In normal conductors, free electrons are assumed to move under the influence of the electric field and experience collisions as described in Section 8-3. In superconductors, pairs of electrons are involved and their behavior is quite different from that of single electrons. The electron pairs are immune from collisions.

Consider the electron-pair fluid. We write the equation of motion

$$m^* \frac{\partial \bar{v}_s}{\partial t} = e^* \bar{E}, \tag{8-169}$$

where $m^* = 2m$ and $e^* = -2e$ are the pair effective mass and pair effective charge, and m and e are electron mass and charge. The pair-current density J_s is given by

$$\bar{J}_s = n_s^* \, e^* \bar{v}_s, \tag{8-170}$$

where n_s^* is the number density of pair. Combining these two, we get the first London equation:

$$\Lambda \frac{\partial \bar{J}_s}{\partial t} = \bar{E}, \qquad \Lambda = \frac{m^*}{n_s^* \, e^{*2}}. \tag{8-171}$$

Next we use one of the Maxwell equations,

$$\nabla \times \bar{E} = -\frac{\partial \bar{B}_s}{\partial t}. \tag{8-172}$$

Substituting the first London equation, we get

$$\Lambda \nabla \times \bar{J}_s + \bar{B}_s = 0. \tag{8-173}$$

This is the second London equation. It is interesting to note that in ordinary electromagnetic theory the steady current produces the magnetic field through $\nabla \times \bar{H} = \bar{J}$, but steady magnetic field does not produce the current in a conductor. However, in the superconductor, steady magnetic field causes current as shown in (8-173).

Let us consider the dc case. We then have

$$\nabla \times \bar{H} = \bar{J}_s. \tag{8-174}$$

Taking the curl of this equation and substituting (8-173) and noting that $\mu \simeq \mu_0$ and $\nabla \cdot \bar{B} = 0$, we get

$$\nabla^2 \bar{B} = \frac{\bar{B}}{\lambda^2}, \tag{8-175}$$

where $\lambda^2 = \Lambda/\mu_0 = m^*/n_s^* e^{*2} \mu_0$. If $z = 0$ is the surface of superconductor and $\bar{B} = B(z)\hat{x}$, we get, from (8-175),

$$B(z) = B_0 \exp\left(-\frac{z}{\lambda}\right). \tag{8-176}$$

This shows that the static magnetic field penetrates the superconductor only a small distance given by λ. This λ is called the *penetration distance* and is on the order of 0.05 to 0.1 μm. The phenomenon that the superconductor tends to exclude the static magnetic flux is called the *Meissner effect.*

When a small magnet is dropped onto a superconductive plate, the magnetic flux cannot penetrate the surface and induces current on the surface. This, in effect, creates the image of magnet with the same polarity, which lies below the surface, and therefore the magnet is repelled by the image, cannot approach the superconductor, and is suspended in air.

The penetration distance λ depends on the pair density n_s^* as shown in (8-175). The number of paired electron $n_s = 2n_s^*$ varies as a function of temperature T.

$$\frac{n_s}{n} = 1 - \left(\frac{T}{T_c}\right)^4, \tag{8-177}$$

where n is the number of conducting electrons and T_c is critical temperature. Therefore, the penetration distance λ also varies as a function of temperature:

$$\lambda(T) = \lambda(0)\left[1 - \left(\frac{T}{T_c}\right)^4\right]^{-1/2}. \tag{8-178}$$

8-24 TWO-FLUID MODEL OF SUPERCONDUCTORS AT HIGH FREQUENCIES

Let us next consider the high-frequency behavior of the superconductor. In general, only a fraction of the conducting electrons is in superconductive state; the remainder is in the normal state. For electron pairs in superconductive state, we have, from (8-169),

$$m\frac{d\bar{v}_s}{dt} = -e\bar{E}, \tag{8-179}$$

where $m^* = 2m$ and $e^* = -2e$ are used. For a normal state, we have

$$m\frac{d\bar{v}_n}{dt} + m\frac{\bar{v}_n}{\tau} = -e\bar{E}, \tag{8-180}$$

where τ is the momentum relaxation time and is normally much shorter than 10^{-11} s. The total current density $\bar{J}$ is given by the sum of the superconductive current $\bar{J}_s$ and the normal current $\bar{J}_n$:

$$\begin{aligned} \bar{J} &= \bar{J}_s + \bar{J}_n, \\ \bar{J}_s &= -n_s e\bar{v}_s, \\ \bar{J}_n &= -n_n e\bar{v}_n, \\ n &= n_s + n_n. \end{aligned} \tag{8-181}$$

For a time-harmonic field with $\exp(j\omega t)$, we get from (8-179) and (8-180),

$$\begin{aligned} \bar{J}_s &= -j\frac{e^2 n_s}{m\omega}\bar{E}, \\ \bar{J}_n &= -j\frac{e^2 n_n}{m\omega}\frac{\bar{E}}{1 - j(1/\omega\tau)}. \end{aligned} \tag{8-182}$$

Thus we get complex conductivity σ:

$$\begin{aligned} \bar{J} &= \sigma\bar{E}, \\ \sigma &= \sigma_1 - j\sigma_2, \\ \sigma_1 &= \frac{e^2 n_n \tau}{m(1 + \omega^2\tau^2)}, \\ \sigma_2 &= \frac{e^2 n_s}{m\omega} + \frac{e^2 n_n(\omega\tau)^2}{m\omega(1 + \omega^2\tau^2)}. \end{aligned} \tag{8-183}$$

In most applications, $\omega^2\tau^2 << 1$ for frequency $<10^{11}$ Hz, and therefore

$$\begin{aligned} \sigma_1 &= \frac{e^2 n_n \tau}{m} \\ &= \sigma_n\left(\frac{n_n}{n}\right) = \sigma_n\left(\frac{T}{T_c}\right)^4 \\ \sigma_2 &= \frac{e^2 n_s}{m\omega} = \frac{1}{\omega\mu_0\lambda^2}, \end{aligned} \tag{8-184}$$

where σ_n is the conductivity in the normal state ($\sigma_n = e^2 n\tau/m$).

Let us next consider the penetration for electromagnetic wave into the superconductor. The wave is propagating into the superconductor ($z > 0$). Then we have the field $\bar{E} = E_x\hat{x}$, given by

$$E_x(z) = E_x(0)\,\exp(-jKz). \tag{8-185}$$

Noting that the displacement current is negligible compared with the conduction current, we have

$$\begin{aligned} K &= \omega\left[\mu_0\left(\epsilon - j\frac{\sigma}{\omega}\right)\right]^{1/2} \\ &\simeq \omega\left[\mu_0\left(-j\frac{\sigma}{\omega}\right)\right]^{1/2}. \end{aligned} \tag{8-186}$$

Using $\sigma = \sigma_1 - j\sigma_2$, we get

$$\begin{aligned} \exp(-jKz) &= \exp\left[-\frac{1}{\lambda}\left(1 + j\tau\omega\frac{n_n}{n_s}\right)^{1/2}\right]z \\ &= \exp(-\alpha - j\beta)z. \end{aligned} \tag{8-187}$$

The penetration depth is then given by α^{-1} and is approximately equal to λ. Note that as $\omega \rightarrow 0$, the penetration distance reduces to λ, but as the frequency is increased, the penetration depth decreases.

Next, let us consider the surface impedance Z_s. This is given by

$$Z_s = \left[\frac{\mu_0}{\epsilon - j(\sigma/\omega)}\right]^{1/2} \approx \left(\frac{j\mu_0\omega}{\sigma}\right)^{1/2}.$$

Noting that $\sigma = \sigma_1 - j\sigma_2$, we get

$$\begin{aligned} Z_s &= R_s + jX_s, \\ R_s &= \frac{1}{2}\sigma_n\left(\frac{n_n}{n}\right)(\omega\mu_0)^2\lambda^3, \\ X_s &= \omega\mu_0\lambda. \end{aligned} \tag{8-188}$$

Note that R_s increases with ω^2, in contrast with normal conductors, whose R_s increase only as $\omega^{1/2}$.

Microstrip lines using a high-temperature superconductor give low loss and virtually no dispersion (Lee and Itoh, 1989; Ghoshal and Smith, 1988). Typical values used are $T_c = 92.5$ K, $T = 77$ K (liquid nitrogen), $\lambda(0) = 0.14$ μm and $\sigma_n = 0.5$ S/μm. Thus $\sigma_1 = 0.24$ S/μm and $\sigma_2 = 336$ S/μm at 10 GHz. The strip line has the attenuation of 10^{-3} dB/cm, compared with an aluminum line with 1 dB/cm attenuation. The phase velocity for the superconducting line is almost constant over a wide frequency range (Lee and Itoh, 1989).

PROBLEMS

8-1. Calculate and plot the refraction index of fused silica as a function of wavelength (0.5 to 2.0 μm) using the Sellmeier equation.

8-2. Assuming that the plasma frequency and the collision frequency of silver given in Section 8-3 are constant, calculate and plot the dielectric constant of silver (real and imaginary parts) as a function of wavelength ($\lambda = 0.01$ to 1.0 μm).

8-3. For an isotropic plasma, assume that the electron density is 10^6 cm^{-3} and that the collision is negligible. A plane wave with $|E| = 1$ V/m at $f = 10$ MHz is propagating through this plasma. Find the phase velocity, group velocity, magnitudes of the magnetic field H, displacement vector D and magnetic flux B, and power flux density. Also find the maximum displacement of an electron.

8-4. Calculate and plot the effective dielectric constant of a mixture of two media with $\epsilon_1 = 2$ and $\epsilon_2 = 3$ as a function of the fractional volume $f_1(0 - 1)$, using the Maxwell–Garnett and Polder–van Santern formulas.

8-5. A radio wave is propagating through the ionosphere along the dc magnetic field. Assume that the electron density is $N_e = 10^3$ cm^{-3}, the geomagnetic field is 0.5 G, the collision frequency is zero, and the radio frequency is 10 kHz. Calculate the propagation constant, phase velocity, and group velocity. If the magnitude of the electric field is 0.1 V/m, describe the motion of an electron (1 tesla = 1 weber/m^2 = 10^4 Gauss).

8-6. A plane wave is normally incident on a plasma with the dc magnetic field normal to the surface. The incident wave is polarized in the x direction. Find the reflected waves E_x and E_y (Fig. P8-6).

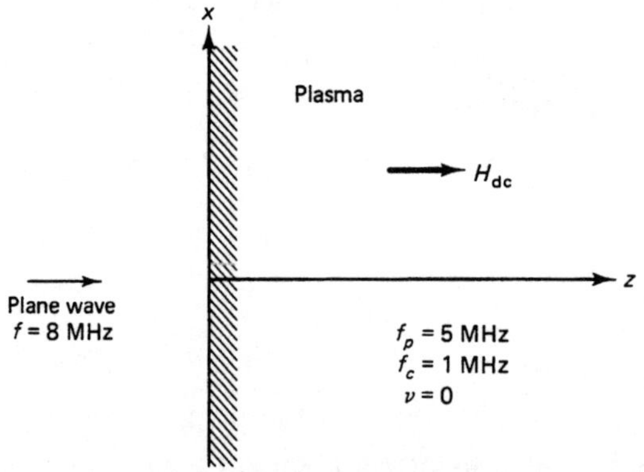

Figure P8-6 Plane wave incident on plasma.

8-7. Consider a wave propagating in a lossless plasma perpendicular to the dc magnetic field. Plot the equivalent dielectric constants ϵ_z and ϵ_t as a function of frequency and show the frequency ranges where the dielectric constant is positive.

8-8. Calculate the reflection coefficient of a radio wave of 1 MHz incident on the ionosphere from the air as shown in Fig. P8-8. The dc magnetic field is pointed in the z direction. The ionosphere characteristics are the same as those in Problem 8-5. Plot the reflection coefficient as a function of the incident angle $\theta(-\pi/2 < \theta < \pi/2)$. Do this for the TM and TE waves.

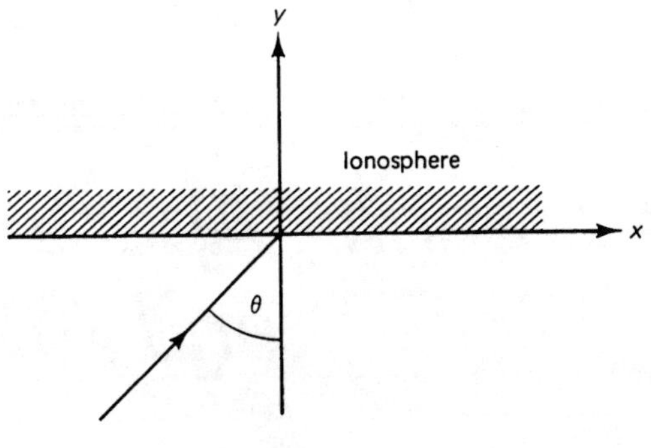

Figure P8-8 Wave incident on the ionosphere.

8-9. Assume that the ionosphere has the following plasma frequency profile:

$$f_p^2 = f_0^2\left[1 - \left(\frac{z - z_1}{h_0}\right)^2\right] \qquad \text{for } |z - z_1| < h_0,$$

where $f_0 = 1$ MHz, $z_1 = 200$ km, $h_0 = 100$ km, and $f_p = 0$ for $|z - z_1| > h_0$. For a radio wave of 500 kHz, find the true height, phase height, and equivalent height.

8-10. For a whistler mode, if $\omega << \omega_p$, $\omega << \omega_c$, show that the group velocity is given approximately by

$$v_g \sim 2c\frac{(ff_c)^{1/2}}{f_p}$$

and that the time required for wave propagation along the magnetic field is proportional to $f^{-1/2}$.

8-11. Assume that the cyclotron frequency is 1.42 MHz and that the plasma frequency is 0.5 MHz. Calculate the time required for the whistler mode to propagate over a path length of 50,000 km in the frequency range 1 to 10 kHz.

8-12. Find the phase velocity and group velocity of the acoustic wave and electromagnetic wave of magnetoplasma at $T = 3000$ K. The plasma frequency is 8 MHz and the operating frequency is 10 MHz. Assume that the collision frequency is zero.

8-13. Consider ferrite placed in a dc magnetic field (the internal dc magnetic field $H_i = 1000$ Oe and the saturation magnetization $M_i = 1700$ G). The relative dielectric constant is $\epsilon_r = 10$. Find the frequency range in which Faraday rotation takes place (1 A/m = $4\pi \times 10^{-3}$ Oe; 1 Wb/m^2 = 10^4 G).

8-14. For the ferrite in Problem 8-13, plot $\mu + k$, $\mu - k$, and μ_t as functions of frequency.

8-15. Show that ϵ_1, μ_1, and β in (8-156) are related to ϵ, μ, and γ through (8-166).

8-16. A linearly polarized plane wave is normally incident on a chiral medium from the air. Find the reflected and transmitted waves. The chiral medium has $\epsilon = 9\epsilon_0$ and $\mu = \mu_0$. Calculate the above for $\gamma(\mu_0/\epsilon_0)^{1/2} = 0.1$, 1.0, and 10.

8-17. A microwave at 10 GHz is normally incident on a superconductor shown at the end of Section 8-24 at a liquid nitrogen temperature of 77 K. First derive (8-188), taking into account that $\sigma_1 << \sigma_2$. Next, if the incident power flux density is 1 mW/cm^2, calculate the power absorbed by the surface. Compare this with the power absorbed by the copper surface with $\sigma = 450$ S/μm at 77 K and with $\sigma = 58$ S/μm at room temperature.

9

Antennas, Apertures, and Arrays

In this chapter we review the fundamental definitions and formulations related to antennas, aperture antennas, linear antennas, and arrays. For a more complete treatment of these topics, see Elliott (1981), Jull (1981), Mittra (1973), Wait (1986), Balanis (1982), Ma (1974), Hansen (1966), Stutzman and Thiele (1981), Stark (1974), Mailloux (1982), and Lo and Lee (1988).

9-1 ANTENNA FUNDAMENTALS

An antenna is usually designed to produce a desired radiation pattern in various directions. The most useful measures of antenna characteristics are *directivity* and *gain*. If the power radiated per unit solid angle in the direction (θ, ϕ) is $P(\theta, \phi)$ and the total power radiated is P_t, the directivity D is given by

$$D(\theta, \phi) = \frac{4\pi P(\theta, \phi)}{P_t}. \tag{9-1}$$

Note that $P_t/4\pi$ is the average power per unit solid angle if the antenna is an isotropic radiator, and therefore the directivity represents the ability of the antenna to produce more power in a certain direction and less power in other directions than the isotropic radiator. It is clear from the definition (9-1) that the integral of the directivity over all the solid angle is 4π:

$$\int_{4\pi} D(\theta, \phi)\, d\Omega = 4\pi. \tag{9-2}$$

The gain function $G(\theta, \phi)$ is defined by

$$G(\theta, \phi) = \frac{4\pi P(\theta, \phi)}{P_i}, \tag{9-3}$$

where P_i is the power input to the antenna.

If the antenna is lossless, $P_i = P_t$ and therefore $G = D$. The transmitting pattern is the normalized directivity defined by $D(\theta, \phi)/D_m$, where D_m is the maximum value of D.

A receiving antenna absorbs power from incident waves. If the power flux density of the incident wave coming from the direction (θ, ϕ) is $S(\theta, \phi)$ and the received power is P_r, the receiving cross section $A_r(\theta, \phi)$ is defined by

$$P_r = SA_r. \tag{9-4}$$

The receiving pattern is given by $A_r(\theta, \phi)/A_{rm}$, where A_{rm} is the maximum value of the receiving cross section.

Next we show that the transmitting and receiving patterns of an antenna are the same:

$$\frac{D(\theta, \phi)}{D_m} = \frac{A_r(\theta, \phi)}{A_{rm}}. \tag{9-5}$$

Also, we show that the receiving cross section $A_r(\theta, \phi)$ and the gain $G(\theta, \phi)$ of any matched antennas are related by

$$A_r(\theta, \phi) = \frac{\lambda^2}{4\pi} G(\theta, \phi). \tag{9-6}$$

To show these relationships, consider two matched lossless antennas 1 and 2 (Fig. 9-1). If the power P_1 is supplied to antenna 1, the received power P_{r2} at antenna 2 is given by

$$P_{r2} = \frac{P_1 G_1 A_{r2}}{4\pi R^2}. \tag{9-7}$$

Similarly, if the power P_2 is applied to antenna 2, the received power at antenna 1 is

$$P_{r1} = \frac{P_2 G_2 A_{r1}}{4\pi R^2}. \tag{9-8}$$

Now we can show the relationship between P_{r2} and P_{r1} by examining the equivalent network shown in Fig. 9-1. The impedances are matched with the conjugate impedances. Then P_1 and P_{r2} are given by

$$P_1 = \frac{|V_1|^2}{8R_1}, \qquad P_{r2} = \frac{R_2|i_2|^2}{2}, \tag{9-9}$$

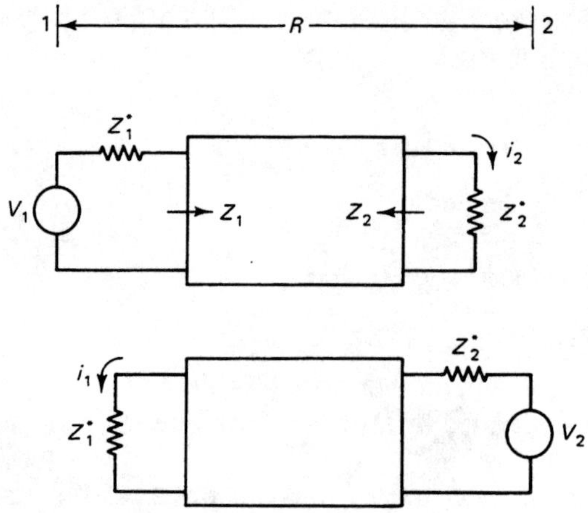

Figure 9-1 Reciprocity.

where $Z_1 = R_1 + jX_1$ and $Z_2 = R_2 + jX_2$ are the input impedances shown in Fig. 9-1. Similarly, we have

$$P_2 = \frac{|V_2|^2}{8R_2}, \qquad P_{r1} = \frac{R_1|i_1|^2}{2}. \tag{9-10}$$

Therefore,

$$\begin{aligned} \frac{P_{r2}}{P_1} &= \frac{4R_1 R_2|i_2|^2}{|V_1|^2}, \\ \frac{P_{r1}}{P_2} &= \frac{4R_1 R_2|i_1|^2}{|V_2|^2}. \end{aligned} \tag{9-11}$$

But by the reciprocity theorem, we have

$$\frac{i_2}{V_1} = \frac{i_1}{V_2}. \tag{9-12}$$

Therefore, we get

$$\frac{P_{r2}}{P_1} = \frac{P_{r1}}{P_2} = \frac{G_1 A_{r2}}{4\pi R^2} = \frac{G_2 A_{r1}}{4\pi R^2}. \tag{9-13}$$

From this we get

$$\frac{G_1}{A_{r1}} = \frac{G_2}{A_{r2}}. \tag{9-14}$$

Since the foregoing antennas are arbitrary, we conclude that

$$\frac{G}{A_r} = \text{universal constant} = C. \tag{9-15}$$

This also applies to the direction of maximum gain, and thus

$$\frac{G(\theta,\phi)}{A_r(\theta,\phi)}=\frac{G_m}{A_{rm}}, \tag{9-16}$$

which gives (9-5).

To obtain the universal constant for a lossless matched antenna in (9-15), we consider the average receiving cross section:

$$\frac{1}{4\pi}\int A_r(\theta,\phi)\,d\Omega=\frac{1}{4\pi C}\int G\,d\Omega=\frac{1}{C}. \tag{9-17}$$

Since this applies to any antenna, the constant C is obtained by calculating the average receiving cross section of a convenient simple antenna. If we take a short dipole antenna terminated with the matched conjugate impedance with the radiation resistance R_0, the received power P_r, when a plane wave polarized in the plane containing the dipole is incident at angle θ, is given by (Fig. 9-2)

$$P_r=\frac{V^2}{8R_0}=\frac{E_0^2 l^2 \sin^2\theta}{8R_0}=A_r(\theta)\frac{|E_0|^2}{2\eta}. \tag{9-18}$$

We therefore get

$$A_r(\theta)=\frac{\eta l^2}{4R_0}\sin^2\theta. \tag{9-19}$$

The average receiving cross section is

$$\frac{1}{C}=\frac{1}{4\pi}\int A_r(\theta)\,d\Omega=\frac{\eta l^2}{6R_0}. \tag{9-20}$$

However, the radiation resistance of a short dipole is [equation (5-27)]

$$R_0=\eta\left(\frac{l}{\lambda}\right)^2\frac{2\pi}{3}. \tag{9-21}$$

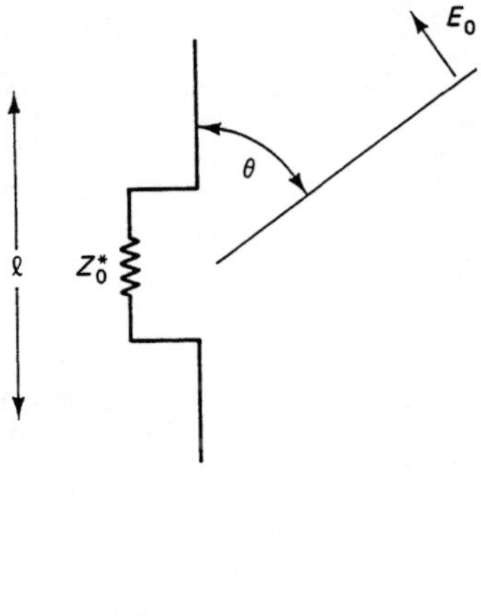

Figure 9-2 Short dipole terminated with conjugate impedance Z_0^*.

Therefore, we get

$$\frac{1}{C} = \frac{\lambda^2}{4\pi}, \tag{9-22}$$

which proves (9-6).

9-2 RADIATION FIELDS OF GIVEN ELECTRIC AND MAGNETIC CURRENT DISTRIBUTIONS

The exact current distributions on an antenna such as a wire antenna must be determined by solving the complete boundary value problem. This is discussed in Section 9-8. However, in many practical antennas, the approximate current distributions are known based on more exact calculations or on experimental data. In this section we present the radiation field due to given current distributions.

Let us consider the given distribution of the electric current $\bar{J}$ and the magnetic current $\bar{J}_m$ in space (Fig. 9-3). The electric field outside the source region is given by the Hertz vector formulations in Sections 5-2 and 5-3.

$$\begin{aligned}
\bar{E} &= \nabla \times \nabla \times \bar{\pi} - j\omega\mu_0 \nabla \times \bar{\pi}_m, \\
\bar{\pi} &= \int G_0(\bar{r}, \bar{r}') \frac{\bar{J}(\bar{r}')}{j\omega\epsilon_0} dV', \\
\bar{\pi}_m &= \int G_0(\bar{r}, \bar{r}') \frac{\bar{J}_m(\bar{r}')}{j\omega\mu_0} dV',
\end{aligned} \tag{9-23}$$

where

$$G_0(\bar{r}, \bar{r}') = \frac{\exp(-jk|\bar{r} - \bar{r}'|)}{4\pi|\bar{r} - \bar{r}'|}.$$

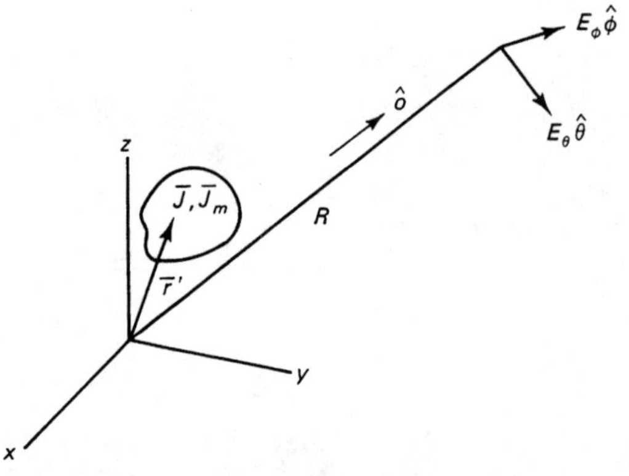

Figure 9-3 Radiation field E_θ and E_ϕ due to $\bar{J}(\bar{r}')$ and $\bar{J}_m(\bar{r}')$.

In the far field, Green's function is approximated by (6-119):

$$G_0(\bar{r}, \bar{r}') = \frac{e^{-jkR + jk\hat{o}\cdot\bar{r}'}}{4\pi R},$$

and the operator ∇ is equal to $\nabla = -jk\hat{o}$. Therefore, in the far zone, we get

$$\bar{E} = \frac{j\omega\mu_0}{4\pi R} e^{-jkR}\,\hat{o} \times \hat{o} \times \bar{I} + \frac{jk}{4\pi R} e^{-jkR}\,\hat{o} \times \bar{I}_m, \tag{9-24}$$

where

$$I = \int \bar{J}(\bar{r}') e^{jk\hat{o}\cdot\bar{r}'}\, dV',$$

$$I_m = \int \bar{J}_m(\bar{r}') e^{jk\hat{o}\cdot\bar{r}'}\, dV'.$$

Now the components E_θ and E_ϕ can be obtained by

$$\begin{aligned} E_\theta &= \hat{\theta} \cdot \bar{E}, \\ E_\phi &= \hat{\phi} \cdot \bar{E}. \end{aligned} \tag{9-25}$$

Noting the identities

$$\begin{aligned} \hat{o} \times \hat{o} \times \bar{J} &= \hat{o}(\hat{o} \cdot \bar{J}) - \bar{J}, \\ \hat{\theta} \cdot (\hat{o} \times \bar{J}_m) &= \bar{J}_m \cdot (\hat{\theta} \times \hat{o}) = -\hat{\phi} \cdot \bar{J}_m, \\ \hat{\phi} \cdot (\hat{o} \times \bar{J}_m) &= \bar{J}_m \cdot (\hat{\phi} \times \hat{o}) = -\hat{\theta} \cdot \bar{J}_m, \end{aligned}$$

we get

$$E_\theta = -\frac{j\omega\mu_0}{4\pi R} e^{-jkR} \int \hat{\theta} \cdot \bar{J}(\bar{r}') e^{jk\hat{o}\cdot\bar{r}'}\, dV' - \frac{jk}{4\pi R} e^{-jkR} \int \hat{\phi} \cdot \bar{J}_m(\bar{r}') e^{jk\hat{o}\cdot\bar{r}'}\, dV',$$

$$E_\phi = -\frac{j\omega\mu_0}{4\pi R} e^{-jkR} \int \hat{\phi} \cdot \bar{J}(\bar{r}') e^{jk\hat{o}\cdot\bar{r}'}\, dV' + \frac{jk}{4\pi R} e^{-jkR} \int \hat{\theta} \cdot \bar{J}_m(\bar{r}') e^{jk\hat{o}\cdot\bar{r}'}\, dV'. \tag{9-26}$$

The magnetic field $\bar{H}$ in the far zone is simply related to $\bar{E}$:

$$\bar{H} = \frac{1}{\eta}\hat{o} \times \bar{E}, \qquad \eta = \left(\frac{\mu_0}{\epsilon_0}\right)^{1/2}. \tag{9-27}$$

Therefore,

$$H_\theta = -\frac{E_\phi}{\eta}, \qquad H_\phi = \frac{E_\theta}{\eta}. \tag{9-28}$$

This is the fundamental formula for the far-field components E_θ and E_ϕ, due to the given current distributions $\bar{J}$ and $\bar{J}_m$. Note that the θ and ϕ components of $\bar{J}$ contribute to E_θ and E_ϕ and the ϕ and θ components of $\bar{J}_m$ contribute to E_θ and E_ϕ. The actual calculations can be done in different coordinate systems. For example, in a Cartesian system, we have

$$
\begin{aligned}
\hat{\theta} &= \hat{x}\cos\theta\cos\phi + \hat{y}\cos\theta\sin\phi - \hat{z}\sin\theta, \\
\hat{\phi} &= -\hat{x}\sin\phi + \hat{y}\cos\phi, \\
\hat{o} &= \hat{x}\sin\theta\cos\phi + \hat{y}\sin\theta\sin\phi + \hat{z}\cos\theta, \\
\bar{r}' &= \hat{x}x' + \hat{y}y' + \hat{z}z'.
\end{aligned}
\tag{9-29}
$$

Therefore,

$$
\begin{aligned}
\hat{\theta}\cdot\bar{J} &= \cos\theta\cos\phi J_x + \cos\theta\sin\phi J_y - \sin\theta J_z. \\
\hat{\phi}\cdot\bar{J} &= -\sin\phi J_x + \cos\phi J_y, \\
\hat{o}\cdot\bar{r}' &= x'\sin\theta\cos\phi + y'\sin\theta\sin\phi + z'\cos\theta.
\end{aligned}
\tag{9-30}
$$

In a cylindrical system, we have

$$
\begin{aligned}
\bar{J}(r', \phi', z') &= \hat{r}'J_{r'} + \hat{\phi}'J_{\phi'} + \hat{z}'J_{z'}, \\
\bar{r}' &= \hat{x}\rho'\cos\phi' + \hat{y}\rho'\sin\phi' + \hat{z}z'.
\end{aligned}
\tag{9-31}
$$

Therefore, we get

$$
\begin{aligned}
\hat{\theta}\cdot\bar{J} &= \cos\theta\sin(\phi - \phi')J_{\phi'} + \cos\theta\cos(\phi - \phi')J_{r'} - \sin\theta J_{z'}, \\
\hat{\phi}\cdot\bar{J} &= \cos(\phi - \phi')J_{\phi'} - \sin(\phi - \phi')J_{r'}, \\
\hat{o}\cdot\bar{r}' &= \rho'\sin\theta\cos(\phi - \phi') + z'\cos\theta.
\end{aligned}
\tag{9-32}
$$

9-3 RADIATION FIELDS OF DIPOLES, SLOTS, AND LOOPS

In this section we give some examples of the radiation fields of known current distributions based on the formulation (9-26). It is known (Elliott, 1981, Chap. 2) that the current distribution on a wire antenna is approximately sinusoidal. For the wire antenna shown in Fig. 9-4, the current distribution is given by

$$
I(z') = I_0 \frac{\sin[k(l - |z'|)]}{\sin kl}. \tag{9-33}
$$

Substituting this into (9-26), we get

$$
E_\theta = \frac{j\omega\mu_0 \sin\theta e^{-jkR}}{4\pi R} \int_{-l}^{l} I(z') e^{jkz'\cos\theta}\, dz'. \tag{9-34}
$$

The integration can be simplified by noting that $I(z')$ is an even function of z', and therefore the integral is equal to

$$
2\int_0^l I(z')\cos(kz'\cos\theta)\, dz'.
$$

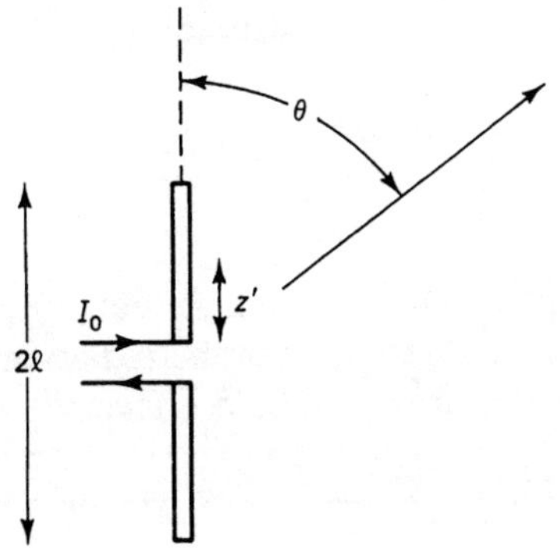

Figure 9-4 Radiation from a wire antenna.

After integration, we get

$$E_\theta = \frac{j\eta I_0}{2\pi R} e^{-jkR} \frac{\cos(kl \cos\theta) - \cos kl}{\sin kl \sin\theta},$$
$$E_\phi = 0. \tag{9-35}$$

For a half-wave dipole, $2l = \lambda/2$ and we get

$$E_\theta = \frac{j\eta I_0}{2\pi R} e^{-jkR} \frac{\cos[(\pi/2)\cos\theta]}{\sin\theta}. \tag{9-36}$$

The total radiated power is given by

$$P_t = \int_0^{2\pi} \int^{\pi} \frac{|E_\theta|^2}{2\eta} R^2 \sin\theta \, d\theta \, d\phi. \tag{9-37}$$

The radiation resistance R_r is defined by

$$P_t = \tfrac{1}{2} I_0^2 R_r. \tag{9-38}$$

For a half-wave dipole, the integral in (9-37) can be evaluated either numerically or analytically using sine and cosine integrals, and we get

$$R_r = 73.09 \text{ ohms}. \tag{9-39}$$

Note that the resistive component of the input impedance $Z_i = R_r + jX$ is given by the radiated power and is independent of the wire diameter, but the reactive component X depends on the stored energy in the near field. Therefore, X depends on the wire diameter and must be obtained by solving the boundary value problem. The maximum gain of a half-wave dipole antenna is in the direction $\theta = \pi/2$ and is given by

$$G = \frac{4\pi R^2 |E_\theta|^2 / 2\eta}{P_t} = 1.64. \tag{9-40}$$

For a short dipole $2l << \lambda$, the radiation resistance R_r and the maximum gain are

$$R_r = 20\left(\frac{\pi 2l}{\lambda}\right)^2,$$
$$G = 1.5. \tag{9-41}$$

A wire antenna is often placed in front of a conducting ground plane as shown in Fig. 9-5. The radiation field is then the sum of the radiation from the antenna and the radiation from its image. Therefore, the field is given by

$$E_\theta = \frac{j\eta I_0}{2\pi R} e^{-jkR} \frac{\cos(kl\ \cos\theta) - \cos kl}{\sin kl\ \sin\theta} 2j\ \sin(kh\ \sin\theta\ \sin\phi). \tag{9-42}$$

Let us next consider the radiation from a center-fed slot in the ground plane (Fig. 9-6). The electric field in the slot is given by

$$E_x = \frac{V_0}{w} \frac{\sin k(l - |z|)}{\sin kl}. \tag{9-43}$$

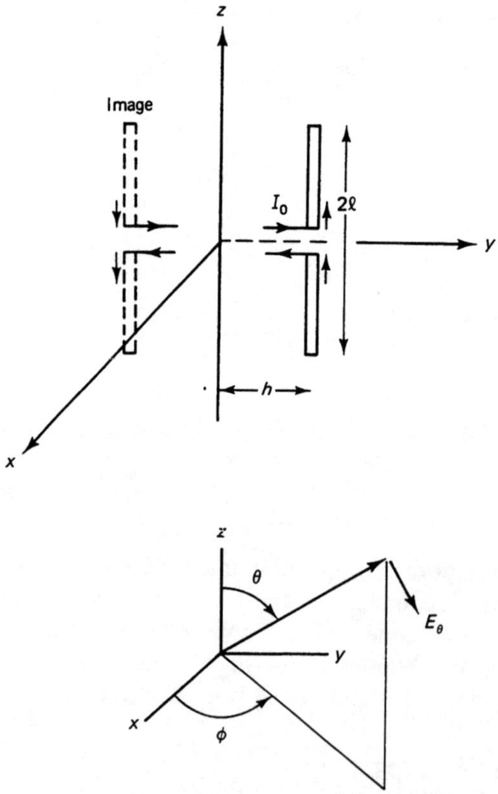

Figure 9-5 Radiation from a dipole in front of a ground plane (x–z).

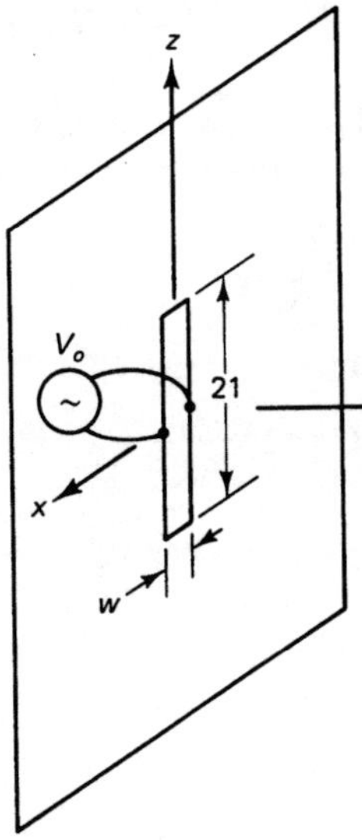

Figure 9-6 Center-fed-slot.

As shown in Section 6-10, this is identical to the magnetic current J_m in front of the ground plane.

$$\bar{J}_m = \bar{E} \times \hat{n} = E_x \hat{x} \times \hat{y} = E_x \hat{z}. \tag{9-44}$$

This is equivalent to the magnetic current $\bar{J}_m$ and its image in free space. Therefore, the field is equal to the field generated by $2\bar{J}_m$. Using (9-26), we get

$$\begin{aligned} E_\theta &= 0, \\ E_\phi &= -j\frac{e^{-jkR}}{\pi R} V_0 \frac{\cos(kl \cos\theta) - \cos kl}{\sin kl \sin\theta}. \end{aligned} \tag{9-45}$$

Next let us consider the radiation from a small loop of radius a. The current I_ϕ in the loop is assumed to be constant. Then using (9-31), (9-32), and

$$\begin{aligned} \hat{o} &= \hat{x} \sin\theta \cos\phi + \hat{y} \sin\theta \sin\phi + \hat{z} \cos\theta, \\ \hat{o} \cdot \bar{r}' &= a \sin\theta \cos(\phi - \phi'), \end{aligned} \tag{9-46}$$

we get

$$\begin{aligned} E_\theta &= 0, \\ E_\phi &= -\frac{j\omega\mu_0 e^{-jkR}}{4\pi R} (jk\pi a^2 I_\phi \sin\theta). \end{aligned} \tag{9-47}$$

The radiation resistance is then given by

$$R_r = \frac{\pi\eta}{6}(ka)^4 = 320\pi^6\left(\frac{a}{\lambda}\right)^4. \tag{9-48}$$

9-4 ANTENNA ARRAYS WITH EQUAL AND NONEQUAL SPACINGS

Let us consider the radiation field from a wire antenna excited by the current I_n. As can be seen from (9-35), we can write the radiation field $\bar{E}_n$ as

$$\bar{E}_n = \frac{I_n}{R_n}\bar{f}_n(\theta,\phi)e^{-jkR_n}, \tag{9-49}$$

where $\bar{f}_n(\theta,\phi)$ represents the radiation pattern (Fig. 9-7). For the far field, we have

$$\begin{aligned} \frac{1}{R_1} \approx \frac{1}{R_2} \approx \frac{1}{R_n} \approx \frac{1}{R_0}, \\ kR_n = kR_0 - \phi_n, \end{aligned} \tag{9-50}$$

where ϕ_n is the phase difference shown in Fig. 9-7. We also assume that all the antennas are identical.

$$\bar{f}_1 = \bar{f}_2 = \cdots = \bar{f}(\theta,\phi). \tag{9-51}$$

Then the total field is given by

$$\begin{aligned} \bar{E} &= \frac{\bar{f}(\theta,\phi)}{R_0}e^{-jkR_0}F(\theta,\phi), \\ F(\theta,\phi) &= \sum_{n=1}^{N} I_n e^{j\phi_n}. \end{aligned} \tag{9-52}$$

The function $F(\theta,\phi)$ is called the *array factor*. For the case shown in Fig. 9-7, where all the antennas are located on the y axis with equal spacing d, we have

$$F(\theta,\phi) = \sum_{n=1}^{N} I_n e^{jkd(n-1)\sin\theta\sin\phi}. \tag{9-53}$$

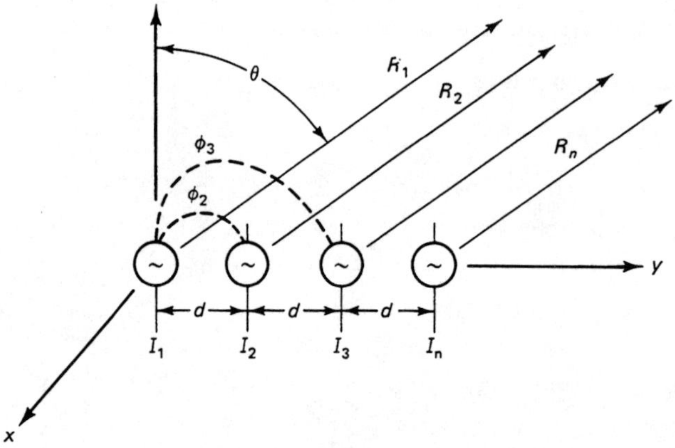

Figure 9-7 Radiation from an antenna array.

If all the antennas are located on a line as shown in Fig. 9-7, they are called *linear arrays*. If the antennas are located on a circle or on the surface of a sphere, they are called *circular arrays* or *spherical arrays*.

If all the currents in a linear array are identical, we get

$$I_n = I_0,$$

$$F(\theta, \phi) = I_0 \sum_{n=1}^{N} e^{j(n-1)\gamma}, \tag{9-54}$$

$$= NI_0 \exp\left[\frac{j(N-1)\gamma}{2}\right] \frac{\sin(N\gamma/2)}{\sin(\gamma/2)},$$

where $\gamma = kd \sin\theta \sin\phi$.

The magnitude of the array factor $|F(\theta, \phi)|$ normalized with NI_0 is pictured in Fig. 9-8. Note the main lobe, grating lobes, and the visible region $(-\pi/2 < \theta < \pi/2)$.

Extensive studies have been made on antenna array problems (Elliott, 1981; Ma, 1974). In this section we present some interesting ideas about linear arrays. Consider the array factor (9-53) in the y–z plane ($\phi = \pi/2$). If we let

$$Z = \exp(jkd \sin\theta), \tag{9-55}$$

we get

$$F = \sum_{n=1}^{N} I_n Z^{n-1}. \tag{9-56}$$

This is a polynomial of the degree $N - 1$, and therefore we write

$$F = A(Z - Z_1)(Z - Z_2) \cdots (Z - Z_{N-1}), \tag{9-57}$$

where Z_n is the zeros of the polynomial and A is constant. Note that the array factor $|F|$ is given by the magnitude of F evaluated on a unit circle in the complex Z plane, and that the desired array factor can be obtained by placing the zeros Z_n appro-

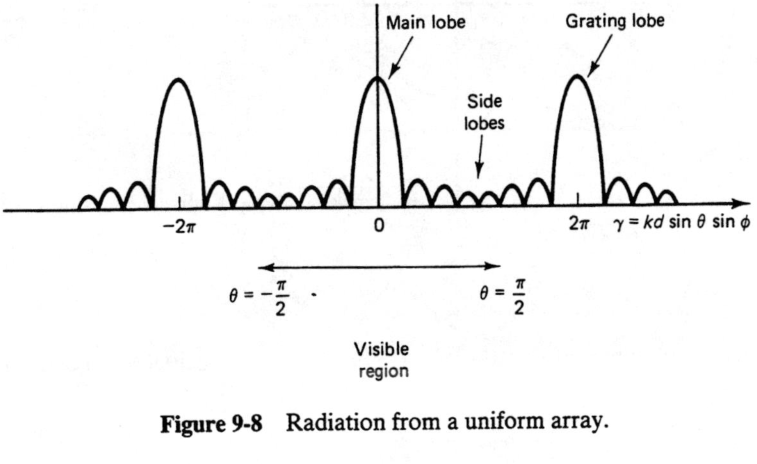

Figure 9-8 Radiation from a uniform array.

priately in the complex Z plane (Fig. 9-9). This method was first proposed by Shelkunoff in 1943.

Next let us consider a method to synthesize a desired radiation pattern. Suppose that we place $2N + 1$ antennas as shown in Fig. 9-10. The array factor is then given by

$$F(\theta) = I_0 + I_1 e^{j\gamma} + I_2 e^{j2\gamma} + \cdots + I_n e^{jN\gamma} + I_{-1} e^{-j\gamma} + I_{-2} e^{-j2\gamma} + \cdots + I_{-N} e^{-jN\gamma}. \tag{9-58}$$

This is a Fourier series representation of $F(\theta) = F(\gamma)$, and therefore for a desired radiation pattern $F(\gamma)$, we expand $F(\gamma)$ in Fourier series and the Fourier coefficient can be identified as the current in each antenna element. For a more detailed discussion on antenna pattern synthesis, see Collin and Zucker (1969) and Ma (1974).

In the above we considered only equally spaced arrays. They are mathematically the most convenient and are normally used in practical applications. There are some applications, however, where unequally spaced arrays are more desirable. For example, the resolution capability of an antenna depends on the total size: $\Delta\theta = \lambda/\text{size}$. However, if the spacing d is greater than a wavelength and is uniform, there will be grating lobes in the visible region (Fig. 9-8). We can, however, eliminate these grating lobes even when the spacing is much greater than a wavelength if we use unequally spaced arrays. Therefore, unequally spaced arrays can be used to obtain a better resolution with a smaller number of antenna elements.

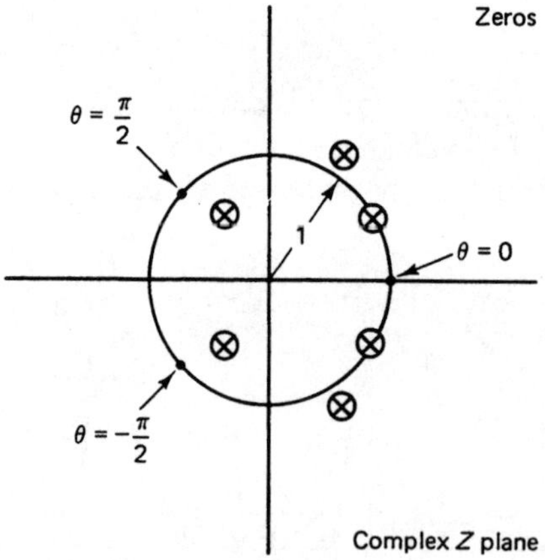

Figure 9-9 Schelkunoff's unit circle.

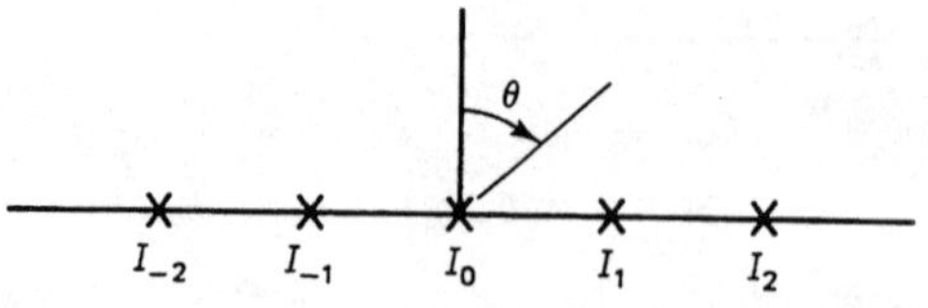

Figure 9-10 Fourier series synthesis.

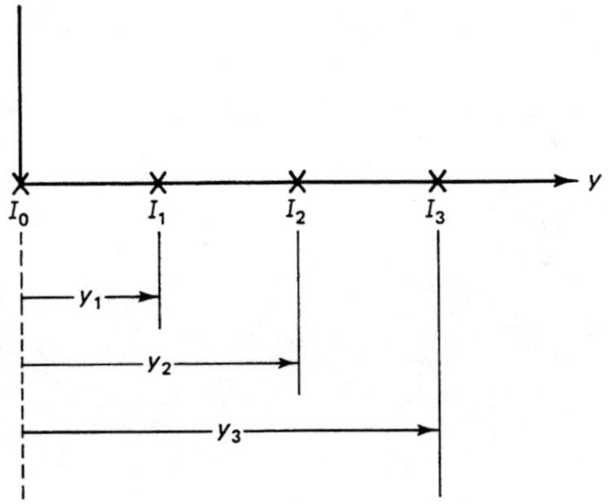

Figure 9-11 Unequally spaced array.

Let us consider the array factor for an unequally spaced array (Fig. 9-11):

$$F(\theta) = \sum_{n=0}^{N-1} I_n e^{j(k \sin\theta) y_n}. \tag{9-59}$$

We may write this as

$$F(\theta) = \sum_{n=0}^{N-1} I(n) e^{j(k \sin\theta) y(n)}, \tag{9-60}$$

where $I(n)$ and $y(n)$ are written as functions of n (Fig. 9-12). Now we make use of Poisson's sum formula:

$$\sum_n f(n) = \sum_m \int f(v) e^{-j2m\pi v}\, dv. \tag{9-61}$$

We then get

$$\begin{aligned} F(\theta) &= \sum_{m=-\infty}^{\infty} F_m(\theta), \\ F_m(\theta) &= \int_0^N I(v) e^{j(k \sin\theta) y(v) - j2m\pi v}\, dv \\ &= \int_0^L I(y) e^{j(k \sin\theta) y - j2m\pi v(y)} \left(\frac{dv}{dy}\right) dy. \end{aligned} \tag{9-62}$$

Here we converted the finite sum (9-60) into the infinite sum (9-62). However, this infinite sum reveals some interesting characteristics. For example, consider the $m = 0$ term:

$$F_0(\theta) = \int_0^L I(y) \left(\frac{dv}{dy}\right) e^{jky \sin\theta}\, dy. \tag{9-63}$$

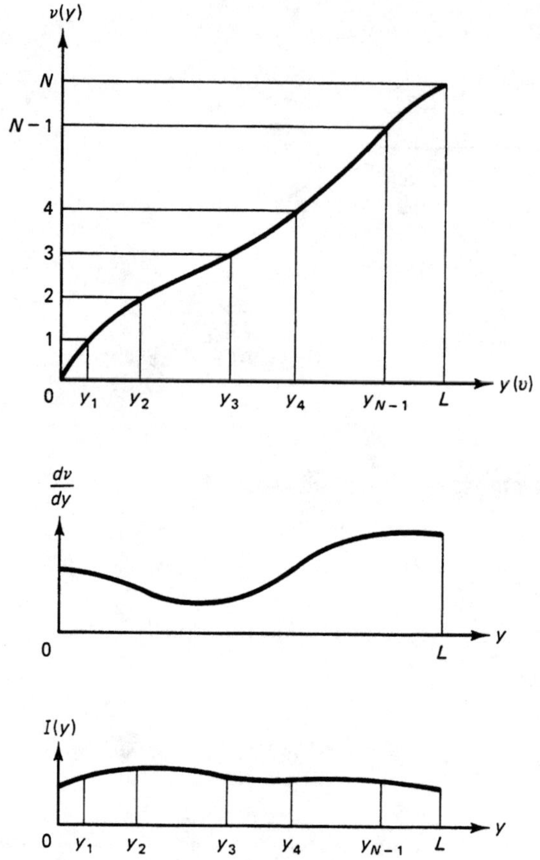

Figure 9-12 Element positions for an unequally spaced array.

This is identical to the radiation from an equivalent continuous source distribution $S_0(y)$:

$$S_0(y) = I(y)\frac{d\nu}{dy}. \tag{9-64}$$

This means that this equivalent source distribution is equal to the actual current distribution $I(y)$ modified by the *density function* $d\nu/dy$. The equivalent source distribution $S_m(y)$ for F_m is given by

$$S_m(y) = S_0(y)e^{-j2m\pi\nu(y)}. \tag{9-65}$$

This has the additional phase shift across the aperture. If the spacings are uniform, $\nu(y) = Ny/L$ and this creates the grating lobe at $k \sin\theta = 2m\pi N/L$. However, if the spacing is nonuniform, the grating lobe at this angle disappears and the power contained in this lobe is spread out in the angle.

We will not go into a detailed discussion on several important array problems, such as Dolph–Chebyshev arrays, circular arrays, spherical arrays, conformal arrays, three-dimensional arrays, randomly spaced arrays, and phased arrays, as they are discussed extensively in the literature (Hansen, 1966; Lo and Lee, 1988).

9-5 RADIATION FIELDS FROM A GIVEN APERTURE FIELD DISTRIBUTION

If the exact field $(\bar{E}_s, \bar{H}_s)$ on a surface S is known, we can obtain the field $(\bar{E}, \bar{H})$ at any point in terms of the tangential fields $\bar{E}_s \times \hat{n}$ and $\hat{n} \times \bar{H}_s$ by using the Franz formula (Section 6-9).

$$\begin{aligned}
\bar{E} &= \nabla \times \nabla \times \bar{\pi} - j\omega\mu_0 \nabla \times \bar{\pi}_m, \\
\bar{\pi} &= \int G_0(\bar{r}, \bar{r}') \frac{\hat{n} \times \bar{H}_s(\bar{r}')}{j\omega\epsilon_0}\, dS', \\
\bar{\pi}_m &= \int G_0(\bar{r}, \bar{r}') \frac{\bar{E}_s(\bar{r}') \times \hat{n}}{j\omega\mu_0}\, dS',
\end{aligned} \tag{9-66}$$

where

$$G_0(\bar{r}, \bar{r}') = \frac{\exp(-jk|\bar{r} - \bar{r}'|)}{4\pi|\bar{r} - \bar{r}'|},$$

and $\hat{n}$ is the unit vector normal to the surface S pointed toward the observation point. Following the procedure in Section 9-2, we get the far field from the known field distribution on a surface S:

$$\begin{aligned}
E_\theta &= -\frac{j\omega\mu_0}{4\pi R} e^{-jkR} \int \hat{\theta} \cdot (\hat{n} \times \bar{H}_s(\bar{r}')) e^{jk\hat{o}\cdot\bar{r}'}\, dS' \\
&\quad - \frac{jk}{4\pi R} e^{-jkR} \int \hat{\phi} \cdot (\bar{E}_s(\bar{r}') \times \hat{n}) e^{jk\hat{o}\cdot\bar{r}'}\, dS', \\
E_\phi &= -\frac{j\omega\mu_0}{4\pi R} e^{-jkR} \int \hat{\phi} \cdot (\hat{n} \times \bar{H}_s(\bar{r}')) e^{jk\hat{o}\cdot\bar{r}'}\, dS' \\
&\quad + \frac{jk}{4\pi R} e^{-jkR} \int \hat{\theta} \cdot (\bar{E}_s(\bar{r}') \times \hat{n}) e^{jk\hat{o}\cdot\bar{r}'}\, dS'.
\end{aligned} \tag{9-67}$$

This gives the exact field (E_θ, E_ϕ) if the exact surface fields $\hat{n} \times \bar{H}_s$ and $\bar{E}_s \times \hat{n}$ are known. In practice, the exact surface fields $\bar{E}_s$ and $\bar{H}_s$ are rarely known and we need to use approximate surface fields.

For example, if the aperture is large, as in the case of parabolic reflector antennas, the field $\bar{E}_s, \bar{H}_s$ on the conducting reflector surface S_1 can be well approximated by the field that would exist if the reflector surface were locally plane (Fig. 9-13). Thus we have

$$\begin{aligned}
&\bar{E}_s \times \hat{n} = 0, \\
&\hat{n} \times \bar{H}_s = 2\hat{n} \times \bar{H}_i \qquad \text{on } S_1,
\end{aligned} \tag{9-68}$$

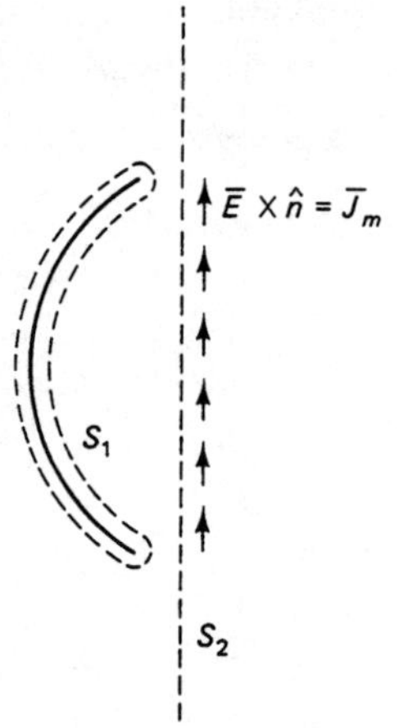

Figure 9-13 Reflector surface S_1 and the plane aperture surface S_2.

where $\bar{H}_i$ is the field from the feed horn in the absence of the reflector. The use of the approximation (9-68) is called the *Kirchhoff approximation* and is identical to the *physical optics approximation.*

Instead of using the field (9-68) on the surface S_1 of the reflector, we can also use the field on the plane aperture S_2 in front of the reflector. We can use (9-67) and approximate $\bar{E}_s$ and $\bar{H}_s$ by the field on S_2 from the feed reflected from the reflector, or we can use the *equivalence theorem* (Section 6-10) and note that the field on the right of S_2 is given by the field $2\bar{E} \times \hat{n}$ due to the magnetic current $\bar{J}_m = \bar{E} \times \hat{n}$ and its image $\bar{E} \times \hat{n}$ in free space. Then the far field is given by

$$E_\theta = -\frac{jk}{2\pi R} e^{-jkR} \int_{S_2} \hat{\phi} \cdot (\bar{E}_s \times \hat{n}) e^{jk\hat{o}\cdot\bar{r}'} \, dS',$$
$$E_\phi = \frac{jk}{2\pi R} e^{-jkR} \int_{S_2} \hat{\theta} \cdot (\bar{E}_s \times \hat{n}) e^{jk\hat{o}\cdot\bar{r}'} \, dS'. \qquad (9\text{-}69)$$

In principle, (9-67) and (9-69) should give the same correct far field if the appropriate surface fields $\bar{E}_s$ and $\bar{H}_s$ are exactly known and if the surface completely surrounds the object. In practice, however, this is not possible. The fields on S_1 or S_2 are only approximately known, and therefore there are slight differences between (9-67) and (9-69) in the far-sidelobe region. In the main lobe region, both are almost identical and give excellent results.

As an example, consider the radiation from a rectangular aperture (Fig. 9-14) with the field distribution:

$$\bar{E}_s = \hat{x} E_{sx}(x', y') \text{ in the aperture } \begin{array}{c} -a/2 \le x' \le a/2 \\ -b/2 \le y' \le b/2 \end{array}. \qquad (9\text{-}70)$$

Using (9-69) and noting (9-29) and that $\bar{E}_s \times \hat{n} = -\hat{y}E_0$, we get

$$E_\theta = \frac{jk}{2\pi R} e^{-jkR} \cos\phi \, F(k \sin\theta \cos\phi, k \sin\theta \sin\phi),$$
$$E_\phi = \frac{jk}{2\pi R} e^{-jkR} (-\cos\theta \sin\phi) F(k \sin\theta \cos\phi, k \sin\theta \sin\phi), \qquad (9\text{-}71)$$

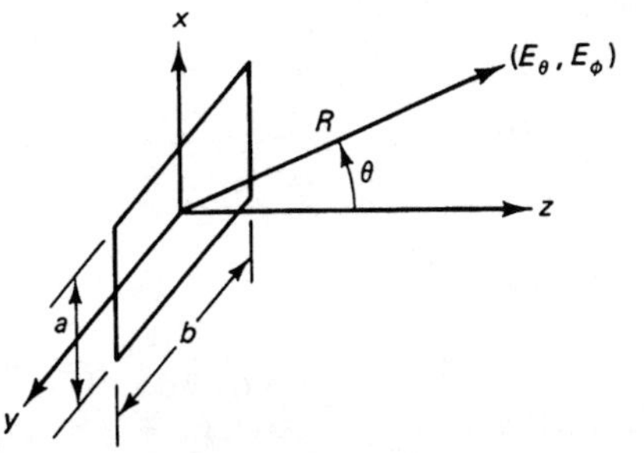

Figure 9-14 Radiation from a rectangular aperture.

where

$$F(k_x, k_y) = \iint E_{sx}(x', y')e^{jk_x x' + jk_y y'}\, dx'\, dy'.$$

Note that the radiation pattern is proportional to the Fourier transform $F(k_x, k_y)$ of the aperture field $E_{sx}(x', y')$ evaluated at $k_x = k \sin\theta \cos\phi$ and $k_y = k \sin\theta \sin\phi$. This Fourier transform relationship between the aperture field distribution and the radiation pattern is an important and useful result. If the aperture distribution is uniform, $E_{sx} = E_0 =$ constant, and we get

$$F(k_x, k_y) = ab\frac{\sin(k_x a/2)}{k_x a/2}\frac{\sin(k_y b/2)}{k_y b/2}. \tag{9-72}$$

If the aperture is circular and the aperture field $\bar{E}_s$ is given in a cylindrical system,

$$\bar{E}_s = \hat{x}E_{sx}(\rho', \phi') \qquad \text{in the aperture } \rho' \le a, \tag{9-73}$$

we have the same formula (9-71) except that F should be expressed in the cylindrical system:

$$F(\theta, \phi) = \iint E_{sx}(\rho', \phi')e^{jk\rho' \sin\theta \cos(\phi - \phi')}\rho'\, d\rho'\, d\phi'. \tag{9-74}$$

Furthermore, if E_{sx} is independent of ϕ', we get

$$F(\theta) = 2\pi \int_0^a E_{sx}(\rho')J_0(k\rho' \sin\theta)\rho'\, d\rho'. \tag{9-75}$$

For a uniform aperture field distribution $E_{sx} = E_0 =$ constant, using the formula

$$\int xJ_0(x)\, dx = xJ_1(x),$$

we get

$$F(\theta) = E_0 \pi a^2 \frac{J_1(ka \sin\theta)}{(ka \sin\theta)/2}. \tag{9-76}$$

If a slot is fed by a waveguide (Fig. 9-15) and the slot length is equal to a half-wavelength, the aperture field is known to be approximately sinusoidal and is given by

$$E_x = E_0 \cos kz. \tag{9-77}$$

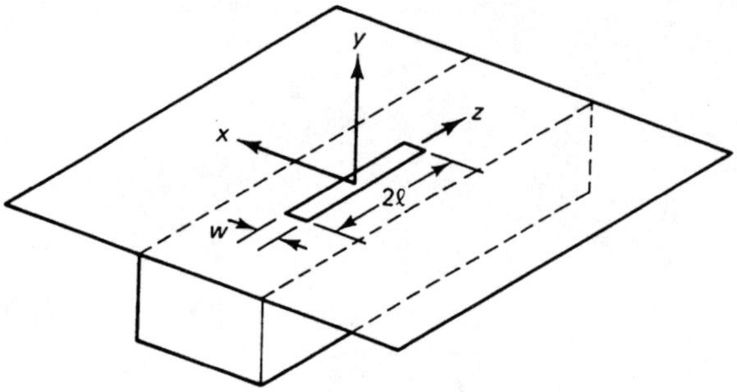

Figure 9-15 Half-wavelength slot antenna ($2l = \lambda/2$) fed by a waveguide.

The radiation field is then given by (9-45) with $kl = \pi/2$ and $V_0 = E_0 w$:

$$E_\theta = 0,$$
$$E_\phi = -j\frac{e^{-jkR}}{\pi R} E_0 w \frac{\cos[(\pi/2)\cos\theta]}{\sin\theta}. \tag{9-78}$$

9-6 RADIATION FROM MICROSTRIP ANTENNAS

Microstrop antennas are made of a thin patch of metal on a grounded dielectric slab as shown in Fig. 9-16. The metal patch is approximately a half-wavelength long and is fed by a coaxial line or a strip line. Thus it can be viewed as a half-wavelength transmission-line cavity with open terminations at both ends where the power leaks out. It is a narrow-band antenna, but it has a low profile and is lightweight. Its radiation pattern is close to that of a half-wave dipole radiating normal to the surface, and thus an array of microstrip antennas can easily be formed for high-gain scanning operations.

Let us first consider a rectangular patch antenna (Fig. 9-16). This can be considered as a rectangular cavity of dimensions $a \times b \times t$ with the boundary condition at the edge that the tangential magnetic field is zero (magnetic wall) (see Section 4-14). Therefore, we write the electric field inside the cavity as

$$E_z(x, y) = E_0 \cos\frac{m\pi x}{a} \cos\frac{n\pi y}{b}. \tag{9-79}$$

The resonant frequency $f_r = \omega_r/2\pi$ is then given by

$$\frac{\omega_r}{c}\sqrt{\epsilon_r} = \left[\left(\frac{m\pi}{a}\right)^2 + \left(\frac{n\pi}{b}\right)^2\right]^{1/2}. \tag{9-80}$$

The antenna is operated at a frequency close to this resonant frequency. For $a > b$, the lowest resonant frequency occurs when $m = 1$ and $n = 0$. (The static field $m = n = 0$ is excluded.)

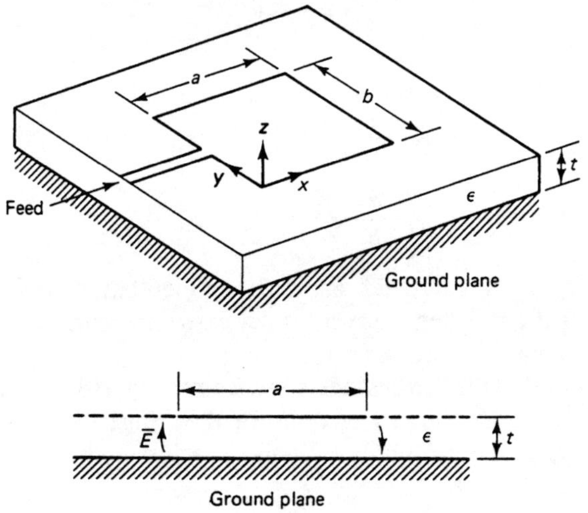

Figure 9-16 Rectangular microstrip antennas.

The radiation field can be calculated using several equivalent techniques (Lo et al., 1979; Bahl and Bhartia, 1980, p. 8; Elliott, 1981, p. 105). Here we use the equivalent magnetic current at the edge of the patch and its image due to the ground plane. We then get the following magnetic currents at the edge of the patch:

$$\begin{aligned}
\bar{E}_s \times \hat{n} &= \hat{y}\left(-E_0 \cos\frac{n\pi y}{b}\right), & x &= 0, \\
&= \hat{y}(-1)^m E_0 \cos\frac{n\pi y}{b}, & x &= a, \\
&= \hat{x}E_0 \cos\frac{m\pi x}{a}, & y &= 0, \\
&= \hat{x}\left[-(-1)^n E_0 \cos\frac{m\pi x}{a}\right], & y &= b.
\end{aligned} \tag{9-81}$$

We substitute these into (9-69) and note the following:

$$\begin{aligned}
\hat{\phi}\cdot(\bar{E}_s \times \hat{n}) &= -\sin\phi\, E_x + \cos\phi\, E_y, \\
\hat{\theta}\cdot(\bar{E}_s \times \hat{n}) &= \cos\theta\,\cos\phi\, E_x + \cos\theta\,\sin\phi\, E_y, \\
\hat{o}\cdot\bar{r}' &= x'\sin\theta\,\cos\phi + y'\sin\theta\,\sin\phi.
\end{aligned} \tag{9-82}$$

We then get

$$\begin{aligned}
E_\theta &= -\frac{jk}{2\pi R}e^{-jkR}\,V_0(-\sin\phi\, g_1 - \cos\phi\, g_2), \\
E_\phi &= +\frac{jk}{2\pi R}e^{-jkR}\,V_0(\cos\theta\,\cos\phi\, g_1 - \cos\theta\,\sin\phi\, g_2),
\end{aligned} \tag{9-83}$$

where

$$V_0 = E_0 t,$$

$$g_1 = [1 - (-1)^n e^{jkb \sin\theta \sin\phi}] \int_0^a \cos\frac{m\pi x'}{a} e^{jkx' \sin\theta \cos\phi}\, dx',$$

$$g_2 = [1 - (-1)^m e^{jka \sin\theta \cos\phi}] \int_0^b \cos\frac{n\pi y'}{b} e^{jky' \sin\theta \sin\phi}\, dy'.$$

Here we assumed that the thickness t of the patch is much smaller than a wavelength, and thus it is assumed that the magnetic current is a narrow line current at the edge of the patch located just above the ground plane.

Microstrip antennas are operated at the lowest resonant frequency with $m = 1$ and $n = 0$, and therefore $k = 2\pi f_r/c = \pi/a\sqrt{\epsilon_r}$. To account for the fringe field, Lo (1979) used the following effective sizes and obtained good agreement between theory and experiment:

$$a_{\text{eff}} = a + \frac{t}{2},$$
$$b_{\text{eff}} = b + \frac{t}{2}. \tag{9-84}$$

Circular patch antennas of radius a can be analyzed similarly by using the following field inside the cavity (Section 4-14):

$$E_z = E_0 J_1(k_{11}\rho) \cos\phi, \tag{9-85}$$

where

$$k_{11} = \frac{1.841}{a}.$$

The magnetic current is then

$$\bar{E}_s \times \hat{n} = \hat{z}E_z \times \hat{\rho} = E_z \hat{\phi}. \tag{9-86}$$

This is then substituted into (9-69). Care should be taken to differentiate the angle ϕ at the observation point (R, θ, ϕ) and the angle ϕ' for the field at the edge of the patch (a, ϕ') [see equations (9-31) and (9-32)]. The far field is then given by

$$E_\theta = -\frac{jk}{2\pi R} e^{-jkR} \int_0^{2\pi} \cos(\phi - \phi') V_0 a \cos\phi' \exp[jka \sin\theta \cos(\phi - \phi')]\, d\phi'$$

$$= -\frac{jk}{R} V_0 a J_1'(ka \sin\theta) \cos\phi, \tag{9-87}$$

$$E_\phi = \frac{jkV_0 a}{R} \frac{J_1(ka \sin\theta)}{ka \sin\theta} \cos\theta \sin\phi,$$

where $V_0 = E_0 t J_1(k_{11} a)$ is the edge voltage at $\rho = a$ and $\phi = 0$, and $k = 1.841/a\sqrt{\epsilon_r}$.

9-7 SELF- AND MUTUAL IMPEDANCES OF WIRE ANTENNAS WITH GIVEN CURRENT DISTRIBUTIONS

Consider a wire antenna excited by a given voltage source (Fig. 9-17). The input impedance Z is given by

$$Z = \frac{V_0}{I(0)}. \tag{9-88}$$

In general, there are two ways of determining the input impedance. One way is that the exact current distribution $I(z)$ for a given voltage can be determined by solving the boundary value problem, then using this current at $z = 0$ to obtain the input impedance. This, however, requires extensive theoretical and numerical work. In many cases we are often not interested in the exact current distribution, but we wish to have a simple method of calculating useful and approximate input impedances. This second approach is shown in this section. It is based on a variational expression for the impedance such that an approximate current distribution is used to obtain the impedance with greater accuracy than the current distribution (Elliott, 1981, Chap. 7).

Let us assume that the wire size a is much smaller than a wavelength and that the current density J_z on the surface of the wire is all directed in the z direction, and therefore, the z component of the electric field E_z is obtained from the following:

$$\bar{E} = \nabla \times \nabla \times \bar{\pi} = \nabla(\nabla \cdot \bar{\pi}) + k^2 \bar{\pi}. \tag{9-89}$$

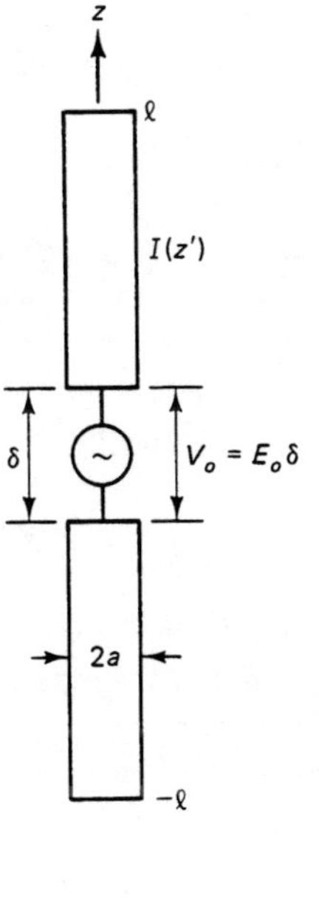

Figure 9-17 Wire antenna excited by a voltage V_0.

Taking the z component and noting that $\bar{\pi} = \pi_z \hat{z}$, we get

$$E_z = \left(\frac{\partial^2}{\partial z^2} + k^2\right)\pi_z,$$
$$\pi_z = \frac{1}{j\omega\epsilon_0}\int_{S'} G_0(\bar{r}, \bar{r}')J_z(\bar{r}')\,ds', \tag{9-90}$$

where S' is the area enclosing the wire. Since $J_z(\bar{r}')$ has no azimuthal variation, we let $ds' = a\,d\phi'\,dz'$, integrate with respect to ϕ', and let $I(z')$ be the total current at z'. We also let the current $I(z')$ be located on the axis of the wire and evaluate E_z on the surface of the wire. Then we get

$$E_z(z) = \int_{-l}^{l} K(z, z')I(z')\,dz', \tag{9-91}$$

where

$$K(z, z') = \frac{1}{4\pi j\omega\epsilon_0}\left(\frac{\partial^2}{\partial z^2} + k^2\right)\frac{e^{-jkr}}{r},$$
$$r = [a^2 + (z - z')^2]^{1/2}.$$

Now the electric field E_z is zero on the surface of the wire except on the gap where it is equal to $-E_0$. Thus we write

$$E_z(z) = \begin{cases} -E_0, & |z| < \dfrac{\delta}{2}, \\ 0, & \dfrac{\delta}{2} < |z| < l. \end{cases} \tag{9-92}$$

If we multiply both sides by $I(z)$ and integrate over $(-l, l)$, we get

$$\int_{-l}^{l} E_z(z)I(z)\,dz = -V_0 I(0). \tag{9-93}$$

From this, we get the input impedance Z:

$$Z = \frac{V_0}{I(0)} = -\frac{\int_{-l}^{l} E_z(z)I(z)\,dz}{I(0)^2}, \tag{9-94}$$

where $E_z(z)$ is the z component of the electric field produced by the current $I(z')$ and is given by (9-91). Thus we can also write

$$Z = -\frac{1}{I(0)^2}\int_{-l}^{l}\int_{-l}^{l} I(z)K(z, z')I(z')\,dz'\,dz. \tag{9-95}$$

Equation (9-94) or (9-95) is a variational expression for Z, and therefore if the current is $I + \delta I$ and the corresponding impedance is $Z + \delta Z$, (9-95) gives $\delta Z = 0$ indicating that for the first-order approximation for I, the impedance has second-order error. Thus the accuracy of the calculated impedance is better than the accuracy of the current distribution used in the equation (see the Appendix to Chapter 7, Section B).

We can now use (9-95) and an approximate current distribution $I(z)$ to obtain the impedance. The current distribution is known to be sinusoidal if $a/\lambda << 1$, and therefore we use

$$I(z) = I_m \sin k(l - |z|). \tag{9-96}$$

Then we get (Elliott, 1981, p. 300)

$$Z = \frac{j60}{\sin^2 kl}\{4\cos^2 kl\, S(kl) - \cos 2kl\, S(2kl) - \sin 2kl[2C(kl) - C(2kl)]\}, \tag{9-97}$$

where

$$C(ky) = \ln\frac{2y}{a} - \frac{1}{2}\,\mathrm{Cin}(2ky) - \frac{j}{2}\,\mathrm{Si}(2ky),$$

$$S(ky) = \frac{1}{2}\,\mathrm{Si}(2ky) - \frac{j}{2}\,\mathrm{Cin}(2ky) - ka.$$

$\mathrm{Si}(x)$ is the sine integral:

$$\mathrm{Si}(x) = \int^x \frac{\sin u}{u}\,du.$$

$\mathrm{Cin}(x)$ is the modified cosine integral:

$$\mathrm{Cin}(x) = \int_0^x \frac{1-\cos u}{u}\,du.$$

It should be noted that the resistive component of the input impedance is related to the radiated power, and therefore it is insensitive to the wire size a. However, the reactive component is related to the reactive power, which represents the stored energy in the near field, and therefore it depends very much on the wire size.

Let us next consider the mutual impedance between antennas 1 and 2 (Fig. 9-18). The voltages V_1 and V_2 and the currents at the terminals $I_1(0)$ and $I_2(0)$ are related by a linear relationship applicable to any linear passive network:

$$\begin{aligned} V_1 &= Z_{11} I_1(0) + Z_{12} I_2(0), \\ V_2 &= Z_{21} I_1(0) + Z_{22} I_2(0), \end{aligned} \tag{9-98}$$

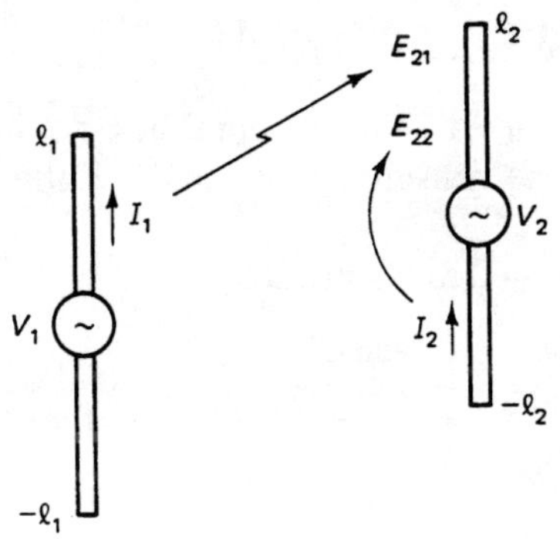

Figure 9-18 Mutual impedance.

where $Z_{12} = Z_{21}$. Z_{11} and Z_{22} are the self-impedances and Z_{12} is the mutual impedance. Now consider the z component of the electric field E_2 on the surface of the wire 2. This consists of the contribution from the currents $I_1(z)$ and $I_2(z)$, and therefore we write

$$E_2(z) = E_{21}(z) + E_{22}(z), \tag{9-99}$$

where $E_{21}(z)$ is the field on the surface of wire 2 produced by I_1 and $E_{22}(z)$ is the field on the surface of wire 2 produced by the current I_2. Now E_2 must be zero on the wire surface except at the gap $|z| < \delta/2$, where the voltage V_2 is applied. Thus we have

$$E_2(z) = \begin{cases} -E_0 & \text{for } |z| < \dfrac{\delta}{2} \\ 0 & \text{for } \dfrac{\delta}{2} < |z| < l_2 \end{cases} \tag{9-100}$$

Multiplying both sides by $I_2(z)$ and integrating over $(-l_2, l_2)$, we get

$$\int_{-l_2}^{l_2} [E_{21}(z)I_2(z) + E_{22}(z)I_2(z)]\,dz = -V_2 I_2(0). \tag{9-101}$$

This can be rewritten as

$$V_2 = Z_{21} I_1(0) + Z_{22} I_2(0), \tag{9-102}$$

where

$$Z_{21} = -\frac{1}{I_1(0)I_2(0)} \int_{-l_2}^{l_2} E_{21}(z)I_2(z)\,dz,$$

$$Z_{22} = -\frac{1}{I_2(0)^2} \int_{-l_2}^{l_2} E_{22}(z)I_2(z)\,dz.$$

The mutual impedance Z_{21} can also be written as

$$Z_{21} = -\frac{1}{I_1(0)I_2(0)} \int_{-l_1}^{l_1} \int_{-l_2}^{l_2} I_2(z)K_{21}(z, z')I_1(z')\,dz\,dz', \tag{9-103}$$

where the field E_{21} due to I_1 is now written as

$$E_{21}(z) = \int_{-l_1}^{l_1} K_{21}(z, z')I_1(z')\,dz'. \tag{9-104}$$

K_{21} is given by the same expression as (9-91) except that a is replaced by the spacing between two wire antennas. The self-impedance Z_{22} has the same form as Z in (9-94), as it should.

If the current $I_1(z')$ is assumed to be sinusoidal,

$$I_1(z') = I_{m1} \sin k(l_1 - |z'|), \tag{9-105}$$

then the integral in (9-104) can be performed (Elliott, 1981, p. 331), giving the following simple and exact expression (Fig. 9-19)

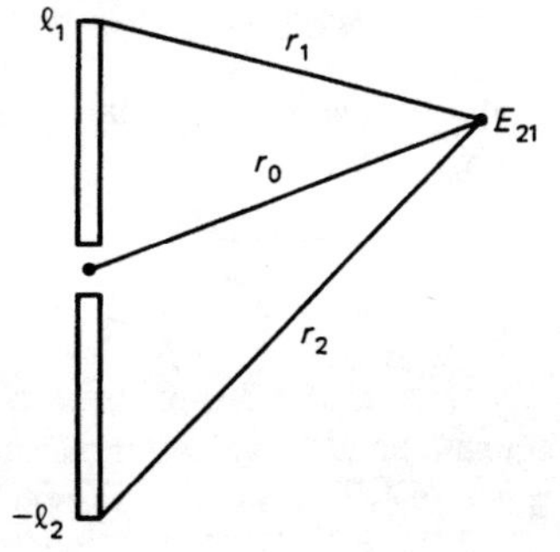

Figure 9-19 The field E_{21} produced by the sinusoidal current I_1.

$$E_{21}(z) = -j30 \cdot I_m \left(\frac{e^{-jkr_1}}{r_1} + \frac{e^{-jkr_2}}{r_2} - 2 \cos kl_1 \frac{e^{-jkr_0}}{r_0} \right). \tag{9-106}$$

This can then be substituted into (9-102) to obtain the mutual impedance.

The input impedance of microstrip antennas can be obtained by using the same formula, (9-94). The field $E_z(z)$ is then the field produced by the current $I(z)$ on the feed in the presence of the microstrip antenna. Here the feedpin size must be considered to obtain the accurate value of the reactive component of the input impedance.

9-8 CURRENT DISTRIBUTION OF A WIRE ANTENNA

In Section 9-7 we used the assumed sinusoidal current distribution to calculate the self- and mutual impedances. These current distributions are good approximations, but the exact distributions must be determined by solving the integral equation as follows: The integral equation is already indicated in (9-92):

$$\int_{-l}^{l} K(z,z')I(z')\,dz' = -E_i(z), \tag{9-107}$$

where $K(z,z')$ is given in (9-91) and $E_i(z)$ is the impressed or incident field.

$$E_i(z) = \begin{cases} E_0 & \text{for } |z| < \dfrac{\delta}{2} \\ 0 & \text{for } \dfrac{\delta}{2} < |z| < l \end{cases}.$$

The integral equation (9-107) is called the *Pocklington integral equation*. The kernel $K(z,z')$ can be further simplified to give the *Richmond form* (Mittra, 1973, p. 13):

$$K(z,z') = \frac{1}{4\pi j\omega\epsilon_0} \frac{e^{-jkr}}{r^5} [(1 + jkr)(2r^2 - 3a^2) + k^2a^2r^2], \tag{9-108}$$

where $r = [a^2 + (z - z')^2]^{1/2}$.

The integral equations above can be solved by using the moment method (see Section 10-14).

PROBLEMS

9-1. The gain G of a large circular aperture antenna is related to the actual aperture area A through the aperture efficiency η_a:

$$G = \eta_a \frac{4\pi A}{\lambda^2}.$$

The half-power beamwidth (in rad) is $\theta_b = \alpha\lambda/D$, where D is the diameter of the aperture. A parabolic antenna of diameter 3 m is used to transmit microwaves at 20 GHz over a distance of 500 km, and an identical antenna is used as a receiving antenna. Assume that $\eta_a = 0.75$ and $\alpha = 1.267$. The transmitting power is 1 W. Find the power flux density at the receiver. Find the gain (in dB) of the antenna and the ratio of the power received to the power transmitted (in dB).

9-2. Consider equation (9-18). If the open-circuit voltage is $V = E_0 l \sin\theta$, show that the current is $V_0/2R_0$ and that the power reradiated due to this current is equal to the received power given in (9-18).

9-3. Consider a linear array of uniformly excited five half-wave wire antennas placed in front of the conducting plane shown in Fig. P9-3. Find the radiation patterns in the x–y plane and in the z–y plane. $h = \lambda/4$ and $d = 3\lambda/4$. Find the first sidelobe level in dB.

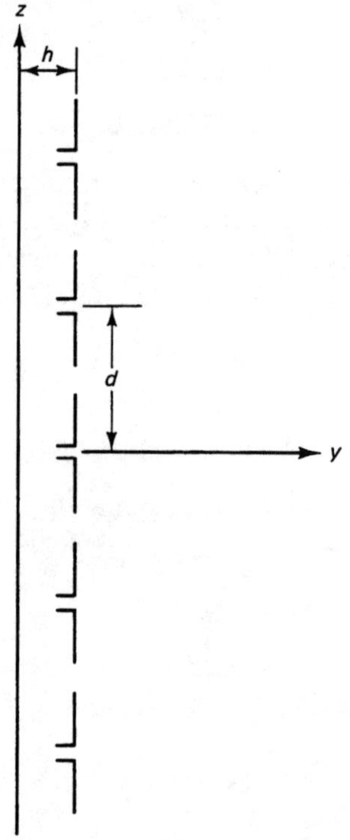

Figure P9-3 Array of half-wave wire antennas.

9-4. Show that the array factor of a circular array of N identical antennas uniformly excited and uniformly distributed on a circle of radius a is given by

$$F(\theta, \phi) = \sum_{m=-\infty}^{\infty} J_{mN}(ka \sin\theta) e^{jmN(\pi/2 - \phi)}.$$

If $ka << N$, this can be approximated by

$$F(\theta, \phi) \approx J_0(ka \sin \theta).$$

[Use Poisson's sum formula (9-61).]

9-5. It is desired to design an array with five elements that would produce a pattern as close to the rectangular pattern shown in Fig. P9-5 as possible. Use the Fourier series method to obtain I_1/I_0 and I_2/I_0. The spacing is $\lambda/2$. Also show the actual pattern.

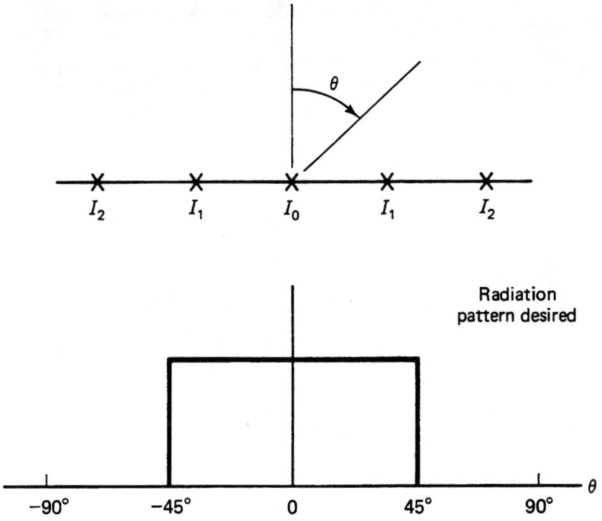

Figure P9-5 Pattern synthesis.

9-6. The field in the rectangular aperture is given by [see (9-70)]

$$E_{sx} = \frac{1}{3} + \frac{2}{3} \cos^2 \frac{\pi x}{a} \qquad \text{for } |x| < \frac{a}{2} \text{ and } |y| < \frac{b}{2}.$$

Find the radiation pattern, first sidelobe level, and half-power beamwidth when $a = 100\lambda$ and $b = 50\lambda$.

9-7. The field in a circular aperture is given by (9-73)

$$E_{sx} = (a^2 - \rho'^2)^n \qquad \text{for } \rho' \leq a,$$

$$a = 50\lambda.$$

Find the radiation pattern, first sidelobe level, and half-power beamwidth when $n = 0$ and $n = 1$.

9-8. Find and plot the radiation pattern of a circular microstrip antenna with diameter of 3 cm, $t = 2$ mm, and $\epsilon_r = 2.5$ in the x–z plane and in the y–z plane.

9-9. Two half-wave wire antennas with radius $a = 0.005\lambda$ are placed side by side with a separation of $\lambda/4$. Find the self- and mutual impedances.

10

Scattering of Waves by Conducting and Dielectric Objects

In radar the radio wave is transmitted toward an object and the scattered wave received by an antenna reveals the characteristics of the object, such as its position and motion. In biomedical applications, microwaves, optical waves, or acoustic waves are propagated through biological media and the scattering from various portions of a body is used to identify the objects for diagnostic purposes. The scattering of waves may be used to probe atmospheric conditions, such as the size, density, and motion of rain, fog, smog, and cloud particles, which in turn gives useful information on the environment and for weather prediction. In microwave and space communication systems, there is an increasing need for higher frequencies and millimeter waves because of their larger channel capacity. However, millimeter waves suffer absorption and scattering by the atmosphere much more than do microwaves, and thus the knowledge of these scattering characteristics is essential to reliable communications.

In this chapter we develop basic formulations of the scattering problem and discuss whenever possible potential applications. We limit ourselves to the scattering by a single object (Kerker, 1969; van de Hulst, 1957). Also in this chapter, we exclude the rigorous boundary value solutions of spherical, cylindrical, and other complex objects as they are covered in Chapters 11, 12, and 13. Some material in Chapter 10 is taken from Chapter 2 of my book *Wave Propagation and Scattering in Random Media* (New York: Academic Press, 1978), with permission.

10-1 CROSS SECTIONS AND SCATTERING AMPLITUDE

When an object is illuminated by a wave, a part of the incident power is scattered out and another part is absorbed by the object. The characteristics of these two phenomena, scattering and absorption, can be expressed most conveniently by assuming an incident plane wave. Let us consider a linearly polarized electromagnetic plane wave propagating in a medium with dielectric constant ϵ_0 and permeability μ_0 with the electric field given by

$$\bar{E}_i(\bar{r}) = \hat{e}_i \exp(-jk\hat{i} \cdot \bar{r}). \tag{10-1}$$

The amplitude $|E_i|$ is chosen to be 1 (volt/m), $k = \omega\sqrt{\mu_0 \epsilon_0} = (2\pi)/\lambda$ is the wave number, λ is a wavelength in the medium, $\hat{i}$ is a unit vector in the direction of wave propagation, and $\hat{e}_i$ is a unit vector in the direction of its polarization.

The object may be a dielectric particle such as a raindrop or ice particle, or a conducting body such as an aircraft (Fig. 10-1). The total field $\bar{E}$ at a distance R from a reference point in the object, in the direction of a unit vector $\hat{o}$, consists of the incident field $\bar{E}_i$ and the field $\bar{E}_s$ scattered by the particle. Within a distance $R < D^2/\lambda$ (where D is a typical dimension of the object such as its diameter), the field $\bar{E}_s$ has complicated amplitude and phase variations because of the interference between contributions from different parts of the object, and the observation point $\bar{r}$ is said to be in the near field of the object. When $R > D^2/\lambda$, however, the scattered field $\bar{E}_s$ behaves as a spherical wave and is given by

$$\bar{E}_s(\bar{r}) = \bar{f}(\hat{o}, \hat{i})\frac{e^{-jkR}}{R} \qquad \text{for } R > \frac{D^2}{\lambda}. \tag{10-2}$$

$\bar{f}(\hat{o}, \hat{i})$ represents the amplitude, phase, and polarization of the scattered wave in the far field in the direction $\hat{o}$ when the object is illuminated by a plane wave propagating in the direction $\hat{i}$ with unit amplitude and is called the *scattering amplitude*. It should be noted that even though the incident wave is linearly polarized, the scattered wave is in general elliptically polarized.

Consider the scattered power flux density S_s at a distance R from the object in

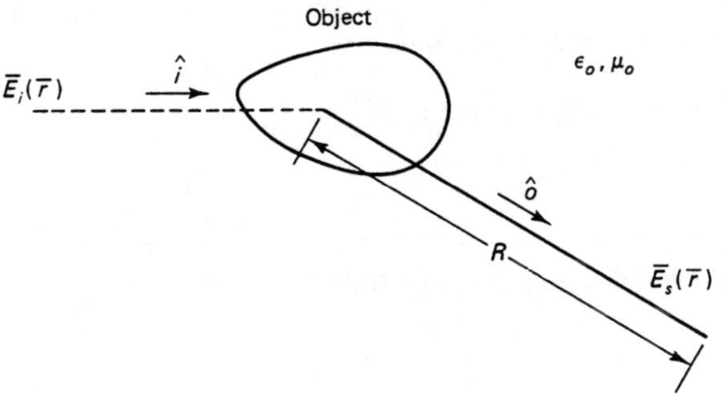

Figure 10-1 A plane wave $\bar{E}_i(\bar{r})$ is incident on an object, and the scattered field $\bar{E}_s(\bar{r})$ is observed in the direction $\hat{o}$ at a distance R.

the direction $\hat{o}$, caused by an incident power flux density S_i. We define the *differential scattering cross section* as follows:

$$\sigma_d(\hat{o},\hat{\imath}) = \lim_{R\to\infty} \frac{R^2 S_s}{S_i} = |f(\hat{o},\hat{\imath})|^2 = \frac{\sigma_t}{4\pi} p(\hat{o},\hat{\imath}), \tag{10-3}$$

where S_i and S_s are the magnitudes of the incident and the scattering power flux density vectors.

$$\bar{S}_i = \frac{1}{2}(\bar{E}_i \times \bar{H}_i^*) = \frac{|E_i|^2}{2\eta_0}\hat{\imath}, \qquad \bar{S}_s = \frac{1}{2}(\bar{E}_s \times \bar{H}_s^*) = \frac{|E_s|^2}{2\eta_0}\hat{o}, \tag{10-4}$$

and $\eta_0 = (\mu_0/\epsilon_0)^{1/2}$ is the characteristic impedance of the medium. We see that σ_d has the dimensions of area/solid angle. It may be defined physically as follows: Suppose that the observed scattered power flux density in the direction $\hat{o}$ is extended uniformly over 1 steradian of solid angle about $\hat{o}$. Then the cross section of an object that would cause just this amount of scattering would be σ_d, so that σ_d varies with $\hat{o}$. The dimensionless quantity $p(\hat{o},\hat{\imath})$ in (10-3) is called the *phase function* and is commonly used in radiative transfer theory. The name "phase function" has its origins in astronomy, where it refers to lunar phases and has no relation to the phase of the wave. σ_t is the *total cross section*, to be defined in (10-9).

In radar applications, the *bistatic radar cross section* σ_{bi} and the *backscattering cross section* σ_b are commonly used. They are related to σ_d through

$$\sigma_{bi}(\hat{o},\hat{\imath}) = 4\pi\sigma_d(\hat{o},\hat{\imath}), \qquad \sigma_b = 4\pi\sigma_d(-\hat{\imath},\hat{\imath}). \tag{10-5}$$

σ_b is also called the *radar cross section*. A physical concept of σ_{bi} may be obtained in a way similar to that used in obtaining σ_d above. Suppose that the observed power flux density in the direction $\hat{o}$ is extended uniformly in all directions from the object over the entire 4π steradians of solid angle. Then the cross section that would cause this would be 4π times σ_d for the direction $\hat{o}$.

Next, let us consider the total observed scattered power *at all angles* surrounding the object. The cross section of an object that would producc this amount of scattering is called the *scattering cross section* σ_s, and is given by

$$\sigma_s = \int_{4\pi} \sigma_d \, d\omega = \int_{4\pi} |\bar{f}(\hat{o},\hat{\imath})|^2 \, d\omega = \frac{\sigma_t}{4\pi}\int_{4\pi} p(\hat{o},\hat{\imath}) \, d\omega, \tag{10-6}$$

where $d\omega$ is the differential solid angle.

Alternatively, σ_s can be written more generally as

$$\sigma_s = \frac{\displaystyle\int_{S_0} \mathrm{Re}(\tfrac{1}{2}\bar{E}_s \times \bar{H}_s^*) \cdot d\bar{a}}{|S_i|}, \tag{10-7}$$

where S_0 is an arbitrary surface enclosing the object and $d\bar{a}$ is a vector representing the differential surface area directed outward (Fig. 10-2).

Next, consider the total power absorbed by the object. The cross section of an object that would correspond to this much power is called σ_a, the absorption cross

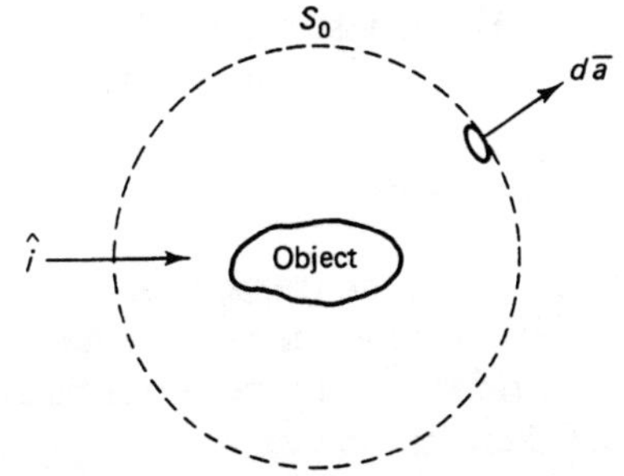

Figure 10-2 Area S_0 surrounding the object.

section. It can be expressed either in terms of the total power flux entering the object or as the volume integral of the loss inside the particle.

$$\sigma_a = \frac{-\int_{S_0} \mathrm{Re}(\frac{1}{2}\bar{E} \times \bar{H}^*) \cdot d\bar{a}}{|S_i|} = \frac{\int_V k\epsilon_r''(\bar{r}')|\bar{E}(\bar{r}')|^2\, dV'}{|E_i|^2}, \tag{10-8}$$

where $\bar{E} = \bar{E}_i + \bar{E}_s$ and $\bar{H} = \bar{H}_i + \bar{H}_s$ are the total fields. Finally, the sum of the scattering and the absorption cross sections is called the *total cross section* σ_t or the *extinction cross section*:

$$\sigma_t = \sigma_s + \sigma_a. \tag{10-9}$$

The ratio W_0 of the scattering cross section to the total cross section is called the *albedo* of an object and is given by

$$W_0 = \frac{\sigma_s}{\sigma_t} = \frac{1}{\sigma_t}\int_{4\pi} |\bar{f}(\hat{o},\hat{\imath})|^2\, d\omega = \frac{1}{4\pi}\int_{4\pi} p(\hat{o},\hat{\imath})\, d\omega. \tag{10-10}$$

The cross sections are also normalized by the geometric cross section and are called the *absorption efficiency* Q_a, the *scattering efficiency* Q_s, and the *extinction efficiency* Q_t.

$$Q_a = \frac{\sigma_a}{\sigma_g}, \qquad Q_s = \frac{\sigma_s}{\sigma_g}, \qquad Q_t = \frac{\sigma_t}{\sigma_g}. \tag{10-11}$$

In the above we assumed that the incident wave is a linearly polarized plane wave. In a more general case, the incident wave should be an elliptically polarized plane wave. The scattering amplitude should then be expressed by a 2×2 scattering amplitude matrix. This is explained in more detail in Section 10-10.

10-2 RADAR EQUATIONS

Let us consider a transmitter Tr illuminating an object at a large distance R_1. Let $G(\hat{\imath})$ be the gain function of the transmitter and P_t be the total power transmitted. The scattered wave is received with a receiver Re at a large distance R_2. We let $A_r(\hat{o})$ be the receiving cross section of the receiver and P_r be the received power. We now wish to find the ratio P_r/P_t (Fig. 10-3). We assume that R_1 and R_2 are large and that the object is in the far field of both antennas. This requires that approximately

$$R_1 > \frac{2D_t^2}{\lambda} \quad \text{and} \quad R_2 > \frac{2D_r^2}{\lambda},$$

where D_t and D_r are the aperture sizes of the transmitter and the receiver.

The gain function $G(\hat{\imath})$ of an antenna is the ratio of the actual power flux $P(\hat{\imath})$ radiated per unit solid angle in the direction $\hat{\imath}$ to the power flux $P_t/4\pi$ of an isotropic radiator per unit solid angle. We therefore have

$$G(\hat{\imath}) = \frac{P(\hat{\imath})}{P_t/4\pi}. \tag{10-12}$$

In terms of the gain $G_t(\hat{\imath})$ of the transmitter, we obtain the incident power flux density S_i at the object

$$S_i = \frac{G_t(\hat{\imath})}{4\pi R_1^2} P_t. \tag{10-13}$$

The power flux density S_r at the receiver is then given by

$$S_r = \frac{\sigma_{bi}(\hat{o}, \hat{\imath}) S_i}{4\pi R_2^2}, \tag{10-14}$$

where σ_{bi} is the *bistatic cross section* given in (10-5). The received power P_r, when a wave is incident on the receiver from a given direction ($\hat{o}$), is given by

$$P_r = A_r(\hat{o}) S_r, \tag{10-15}$$

where $A_r(\hat{o})$ is called the receiving cross section.

It is known that for all matched antennas, the receiving cross section is proportional to the gain function and their ratio is $\lambda^2/4\pi$ (see Section 9-1):

$$A_r(\hat{o}) = \frac{\lambda^2}{4\pi} G_r(-\hat{o}), \tag{10-16}$$

where $G_r(-\hat{o})$ is the gain function in the direction $-\hat{o}$.

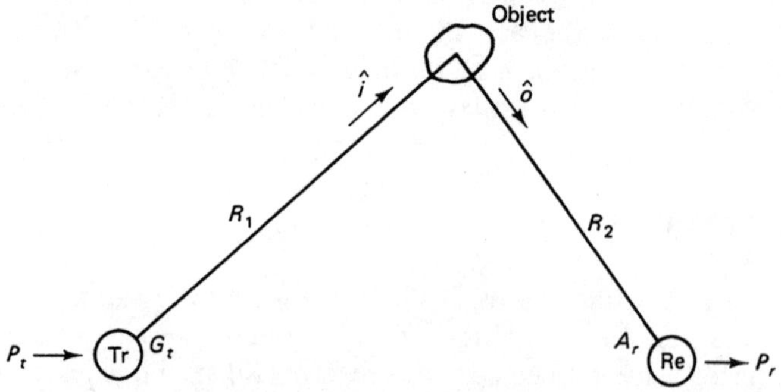

Figure 10-3 Radar equation.

Combining (10-13) to (10-16), we get the ratio of the received to the transmitted power:

$$\frac{P_r}{P_t} = \frac{\lambda^2 G_t(\hat{\imath}) G_r(-\hat{o}) \sigma_{bi}(\hat{o}, \hat{\imath})}{(4\pi)^3 R_1^2 R_2^2}. \tag{10-17}$$

This is the bistatic radar equation. For radar applications, the same antenna is used both as a transmitter and a receiver. Thus $\hat{o} = -\hat{\imath}$ and $R_1 = R_2 = R$. Therefore, for radar, we have

$$\frac{P_r}{P_t} = \frac{\lambda^2 [G_t(\hat{\imath})]^2 \sigma_b(-\hat{\imath}, \hat{\imath})}{(4\pi)^3 R^4}. \tag{10-18}$$

Equations (10-17) and (10-18) give the received power in terms of the antenna gains, the distances, and the cross section. This is applicable when the object is far from both the transmitter and the receiver. It is also required that the receiving antenna be matched to the incoming wave in polarization and impedance. If there is a mismatch, (10-17) and (10-18) must be multiplied by a mismatch factor that is less than unity. In addition, both the transmitter and the receiver must be in the far zone of the object ($R_1 > 2D_o/\lambda^2$ and $R_2 > 2D_o/\lambda^2$; D_o is the object size).

10-3 GENERAL PROPERTIES OF CROSS SECTIONS

Before we discuss the mathematical representations of the various cross sections, it may be worthwhile to present an overall view of how these cross sections are related to the geometric cross section, wavelength, and dielectric constant. If the size of an object is much greater than a wavelength, the total cross section σ_t approaches twice the geometric cross section σ_g of the object as the size increases. To show this, let us consider an incident wave with power flux density S_i (Fig. 10-4). The total flux $S_i\sigma_g$ within the geometric cross section σ_g is either reflected out or absorbed by the object. Behind the object, there should be a shadow region where practically no wave exists. In this shadow region, the scattered wave from the object is exactly equal to the incident wave but 180° out of phase, and this scattered flux is equal to $S_i\sigma_g$ in magnitude. The total scattered and absorbed flux, therefore, approaches $(S_i\sigma_g + S_i\sigma_g)$ and the total cross section σ_t approaches

$$\sigma_t \rightarrow \frac{2S_i\sigma_g}{S_i} = 2\sigma_g. \tag{10-19}$$

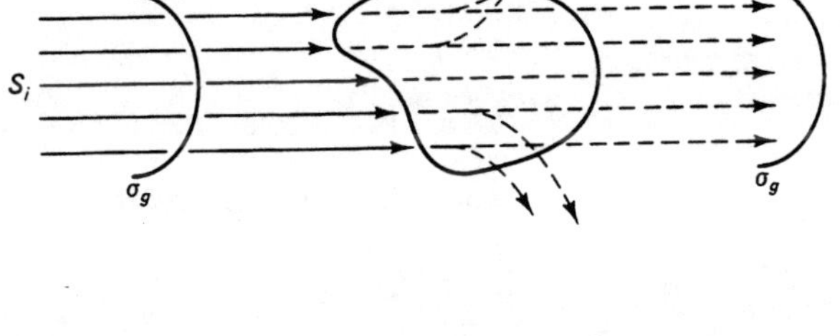

Figure 10-4 Relationship between total cross section and geometric cross section for a large object.

It is also seen that the total absorbed power, when the object is very large, cannot be greater than $S_i \sigma_g$, and thus the absorption cross section σ_a approaches a constant somewhat less than the geometric cross section:

$$\sigma_a \to \sigma_g. \tag{10-20}$$

If the size is much smaller than a wavelength, the scattering cross section σ_s is inversely proportional to the fourth power of the wavelength and proportional to the square of the volume of the object. These characteristics of a small object are generally called *Rayleigh scattering*. The absorption cross section σ_a for a small scatterer is inversely proportional to the wavelength and directly proportional to its volume. Compared with the geometric cross section, we have

$$\frac{\sigma_s}{\sigma_g} \sim \left(\frac{\text{size}}{\lambda}\right)^4 [(\epsilon_r' - 1)^2 + \epsilon_r''^2], \tag{10-21}$$

$$\frac{\sigma_a}{\sigma_g} \sim \frac{\text{size}}{\lambda} \epsilon_r''. \tag{10-22}$$

Curves of the normalized cross section above versus relative size of an object are sketched in Fig. 10-5.

It is also possible to obtain the behavior of the backscattering cross section σ_b for a large object. Consider a point of specular reflection on the surface of the object (Fig. 10-6). An incident wave with power flux density S_i is incident on a small area $\Delta l_1 \, \Delta l_2 = (a_1 \, \Delta\theta_1)(a_2 \, \Delta\theta_2)$. Since the radii of curvature are large, the surface may be considered locally plane, and therefore, using the reflection coefficient for normal incidence on a plane boundary, the reflected power flux density on the surface is given by

$$S_r = \left| \frac{\sqrt{\epsilon_r} - 1}{\sqrt{\epsilon_r} + 1} \right|^2 S_i.$$

At a large distance R from the particle, the flux within this small area $\Delta l_1 \, \Delta l_2$ spreads out over an area $R^2(2\delta\theta_1)(2\delta\theta_2)$, and therefore, the scattered flux density S_s at R is related to S_r through

$$S_s R^2(2\delta\theta_1)(2\delta\theta_2) = S_r(a_1 \, \delta\theta_2)(a_2 \, \delta\theta_2),$$

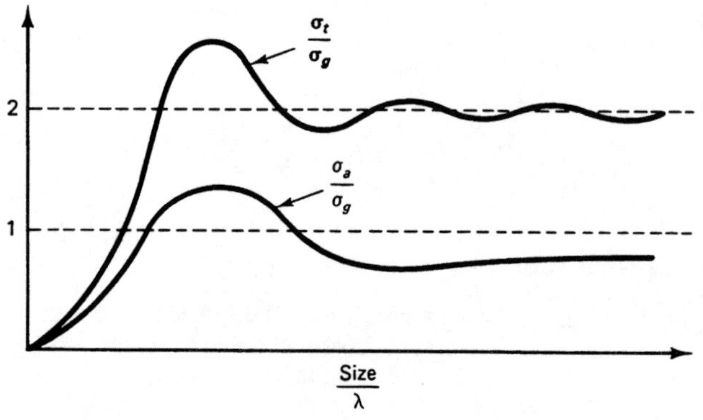

Figure 10-5 Total cross section σ_t and absorption cross section σ_a normalized to geometric cross section σ_g.

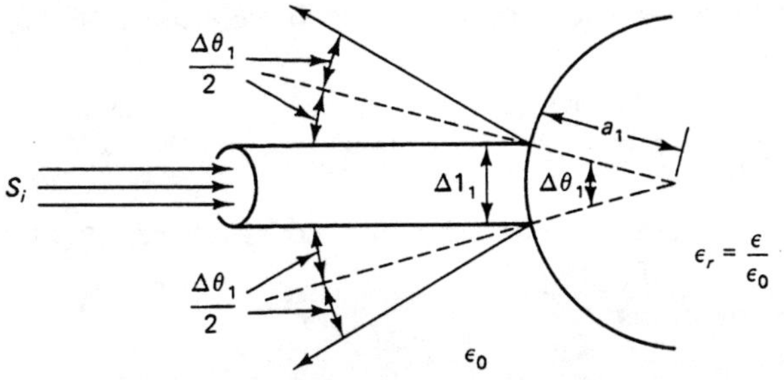

Figure 10-6 Backscattering from a large object.

from which we obtain the backscattering cross section σ_b:

$$\sigma_b = 4\pi\sigma_d(-\hat{\imath}, \hat{\imath}) = \lim_{R\to\infty} \frac{4\pi R^2 S_s}{S_i} = \pi a_1 a_2 \left|\frac{\sqrt{\epsilon_r} - 1}{\sqrt{\epsilon_r} + 1}\right|^2. \tag{10-23}$$

This is the limiting value of σ_b as the object size is increased to infinity, and therefore, for any finite size, σ_b may have a value considerably different from (10-23).

The total cross section σ_t represents the total power loss from the incident wave due to the scattering and absorption of a wave by the object. This loss is closely related to the behavior of the scattered wave in the forward direction, and this general relationship is embodied in the *forward scattering theorem*, also called the *optical theorem*.

The forward scattering theorem states that the total cross section σ_t is related to the imaginary part of the scattering amplitude in the forward direction $\bar{f}(\hat{\imath}, \hat{\imath})$ in the following manner:

$$\sigma_t = -\frac{4\pi}{k} \operatorname{Im}[\bar{f}(\hat{\imath}, \hat{\imath})] \cdot \hat{e}_i, \tag{10-24}$$

where Im denotes the "imaginary part of" and $\hat{e}_i$ is the unit vector in the direction of polarization of the incident wave. The proof is given in the Appendix to Chapter 10, Section A.

10-4 INTEGRAL REPRESENTATIONS OF SCATTERING AMPLITUDE AND ABSORPTION CROSS SECTIONS

Mathematical descriptions of scattering amplitude and absorption cross sections can be made in one of two ways. If the shape of an object is a simple one such as a sphere, it is possible to obtain exact expressions for cross sections and the scattering amplitude. The exact solution for a dielectric sphere, called the Mie solution, is discussed in Chapter 12. In many practical situations, however, the shape of an object is not simple. Therefore, we need a method of determining approximate cross sections for objects with complex shapes. This can be done through general integral representations of the scattering amplitude. The method is also useful for objects with simple shapes because the calculations can be made easily.

Let us consider a dielectric body whose relative dielectric constant is a function of position within the body.

$$\epsilon_r(\bar{r}) = \frac{\epsilon(\bar{r})}{\epsilon_0} = \epsilon_r'(\bar{r}) - j\epsilon_r''(\bar{r}) \qquad \text{in } V. \tag{10-25}$$

The dielectric body occupies the volume V and is surrounded by a medium whose dielectric constant is ϵ_0.

We first write Maxwell's equations:

$$\begin{aligned} \nabla \times \bar{E} &= -j\omega\mu_0 \bar{H}, \\ \nabla \times \bar{H} &= j\omega\epsilon(\bar{r})\bar{E}. \end{aligned} \tag{10-26}$$

Here, we assume that the permeability μ_0 is constant in and outside the dielectric body. If we write the second equation in (10-26) in the following manner:

$$\nabla \times \bar{H} = j\omega\epsilon_0 \bar{E} + \bar{J}_{eq}, \tag{10-27}$$

where

$$\bar{J}_{eq} = \begin{cases} j\omega\epsilon_0[\epsilon_r(\bar{r}) - 1]\bar{E} & \text{in } V \\ 0 & \text{outside,} \end{cases}$$

the term $\bar{J}_{eq}$ may be considered as an equivalent current source which generates the scattered wave. The solution to (10-26) and (10-27) is given by

$$\begin{aligned} \bar{E}(\bar{r}) &= \bar{E}_i(\bar{r}) + \bar{E}_s(\bar{r}), \\ \bar{H}(r) &= \bar{H}_i(\bar{r}) + \bar{H}_s(\bar{r}), \end{aligned} \tag{10-28}$$

where $(\bar{E}_i, \bar{H}_i)$ is the primary (or incident) wave that exists in the absence of the object, and $(\bar{E}_s, \bar{H}_s)$ is the scattered wave originating from it. Using the Hertz vector $\bar{\Pi}_s$, we write

$$\begin{aligned} \bar{E}_s(\bar{r}) &= \nabla \times \nabla \times \bar{\Pi}_s(\bar{r}), \\ \bar{H}_s(\bar{r}) &= j\omega\epsilon_0 \nabla \times \bar{\Pi}_s(\bar{r}), \\ \bar{\Pi}_s(\bar{r}) &= \frac{1}{j\omega\epsilon_0} \int_V G_0(\bar{r}, \bar{r}')\bar{J}_{eq}(\bar{r}')\, dV' \\ &= \int_V [\epsilon_r(\bar{r}') - 1]\bar{E}(\bar{r}')G_0(\bar{r}, \bar{r}')\, dV', \end{aligned} \tag{10-29}$$

where

$$G_0(\bar{r}, \bar{r}') = \frac{\exp(-jk|\bar{r} - \bar{r}'|)}{4\pi|\bar{r} - \bar{r}'|},$$

is the free-space Green's function. Equation (10-29) is valid only for $\bar{r} \neq \bar{r}'$.

To obtain the scattering amplitude, we consider $\bar{E}_s(\bar{r})$ in the far field of the object. Referring to Fig. 10-7, we note that $\bar{r} = R\hat{o}$, and in the far zone the magnitude $1/|\bar{r} - \bar{r}'|$ of Green's function can be approximated by $1/R$. The phase $k|\bar{r} - \bar{r}'|$

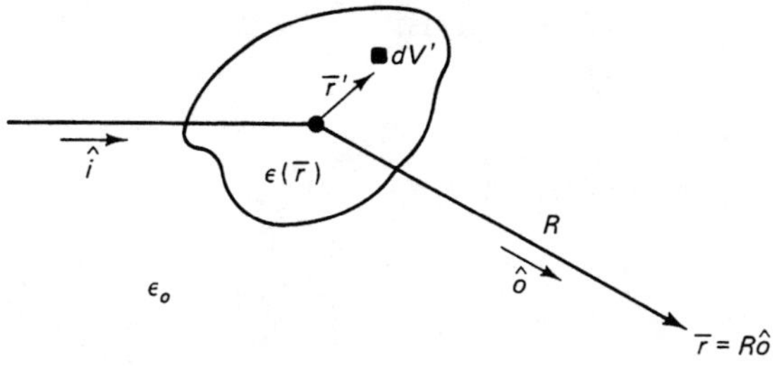

Figure 10-7 Geometry of a point $\bar{r}'$ within the object and observation point $\bar{r}$.

cannot be approximated by kR, however, because the difference may be significant in terms of wavelengths. Expanding $|\bar{r} - \bar{r}'|$ in a binomial series and keeping its first term, we get

$$|\bar{r} - \bar{r}'| = (R^2 + r'^2 - 2R\bar{r}' \cdot \hat{o})^{1/2} \simeq R - \bar{r}' \cdot \hat{o},$$

and Green's function becomes for large R,

$$G_0(\bar{r}, \bar{r}') = \frac{\exp(-jkR + jk\bar{r}' \cdot \hat{o})}{4\pi R}. \tag{10-30}$$

We note also that in the far field,

$$\nabla\left(\frac{e^{-jkR}}{R}\right) \simeq \frac{e^{-jkR}}{R}(-jk\nabla R) = -jk\hat{o}\frac{e^{-jkR}}{R}, \tag{10-31}$$

and thus ∇ is equivalent to $-jk\hat{o}$. Substituting (10-30) and (10-31) into (10-29), we get

$$\bar{E}_s(\bar{r}) = \bar{f}(\hat{o}, \hat{\imath})\frac{\exp(-jkR)}{R},$$

$$\bar{f}(\hat{o}, \hat{\imath}) = \frac{k^2}{4\pi}\int_V [\bar{E} - \hat{o}(\hat{o} \cdot \bar{E})][\epsilon_r(\bar{r}') - 1] \exp(jk\bar{r}' \cdot \hat{o})\, dV', \tag{10-32}$$

where we used $-\hat{o} \times (\hat{o} \times \bar{E}) = \bar{E} - \hat{o}(\hat{o} \cdot \bar{E})$. Note also that $\hat{o}(\hat{o} \cdot \bar{E})$ is the component of $\bar{E}$ along $\hat{o}$, and therefore, $[\bar{E} - \hat{o}(\hat{o} \cdot \bar{E})]$ is the component of $\bar{E}$ perpendicular to $\hat{o}$. This is an exact expression for the scattering amplitude $\bar{f}(\hat{o}, \hat{\imath})$ in terms of the total electric field $\bar{E}(\bar{r}')$ inside the object. This field $\bar{E}(\bar{r}')$ is not known in general, and therefore (10-32) is not a complete description of the scattering amplitude in terms of known quantities. In many practical situations, however, it is possible to approximate $\bar{E}(\bar{r}')$ by some known function and thus obtain a useful approximate expression for $\bar{f}(\hat{o}, \hat{\imath})$. This will be done in the next sections. The absorption cross section σ_a for a dielectric body has been given in (10-8).

We note here that we can develop an alternative integral equation for the magnetic field $\bar{H}(\bar{r})$ rather than $\bar{E}(\bar{r})$. From Maxwell's equations, we obtain the vector wave equation for $\bar{H}$ in the following form:

$$\nabla \times \nabla \times \bar{H}(\bar{r}) - \omega^2 \mu_0 \epsilon_0 \bar{H}(\bar{r}) = \omega^2 \mu_0 \epsilon_0[\epsilon_r(\bar{r}) - 1]\bar{H}(\bar{r}) + j\omega\epsilon_0[\nabla\epsilon_r(\bar{r}) \times \bar{E}(\bar{r})]. \tag{10-33}$$

Therefore, an integral equation for $\overline{H}$ has two terms:

$$\overline{H}(\overline{r}) = \overline{H}_i(\overline{r}) + \overline{H}_s(\overline{r}) = \overline{H}_i(\overline{r}) + \nabla \times \nabla \times \overline{\Pi}_{ms}(\overline{r}),$$

$$\overline{\Pi}_{ms}(\overline{r}) = \int_V [\epsilon_r(\overline{r}') - 1] G_0\left(\frac{\overline{r}}{\overline{r}'}\right) \overline{H}(\overline{r}')\, dV'$$

$$- \frac{1}{\omega\mu_0} \int_V G_0\left(\frac{\overline{r}}{\overline{r}'}\right) \nabla' \epsilon_r(\overline{r}') \times \overline{E}(\overline{r}')\, dV', \qquad (10\text{-}34)$$

where $\nabla' \epsilon_r(\overline{r}')$ is the gradient with respect to $\overline{r}'$. Thus the second term in $\overline{\Pi}_{ms}$ contains the effect of depolarization due to the inhomogeneity of the dielectric constant. For a homogeneous object $\nabla' \epsilon_r(\overline{r})$ gives a delta function on the surface, and thus the second term becomes a surface integral.

10-5 RAYLEIGH SCATTERING FOR A SPHERICAL OBJECT

We have indicated in Section 10-3 the general scattering characteristics of a small object. This is generally known as Rayleigh scattering. In this section we present a detailed analysis for a few simple geometries. Let us consider a dielectric sphere whose size is much smaller than a wavelength. Because of its small size, the impinging electric field within and near the sphere must behave almost as an electrostatic field. It is known in electrostatics that when a constant electric field E_i is applied to a dielectric sphere, the electric field $\overline{E}$ inside the sphere is uniform and given by (Fig. 10-8)

$$\overline{E} = \frac{3}{\epsilon_r + 2} \overline{E}_i, \qquad \overline{E}_i = E_i \hat{e}_i. \qquad (10\text{-}35)$$

We also note in (10-32) that $\exp(jk\overline{r}' \cdot \hat{o}) \approx 1$ because $k\overline{r}' << 1$. Now we can substi-

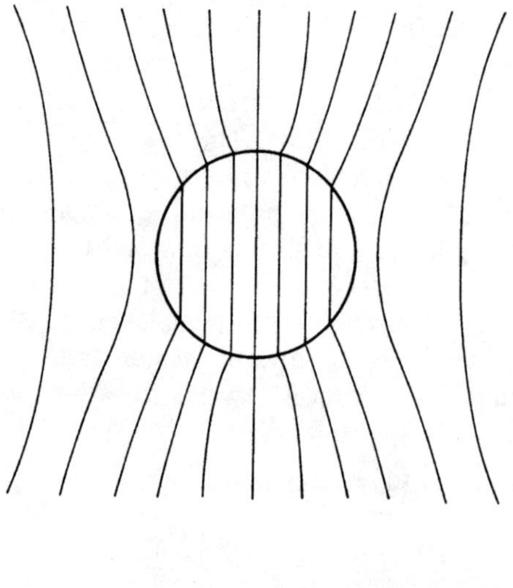

Figure 10-8 Electrostatic field inside a dielectric sphere.

tute (10-35) into (10-32) and obtain the scattering amplitude $\bar{f}$. However, to express $\bar{f}$ in a more convenient form, let us note that the scattering is caused by the equivalent current $\bar{J}_{eq} = j\omega\epsilon_0(\epsilon_r - 1)\bar{E}$ in (10-27). We may use the polarization vector $\bar{P} = \bar{J}_{eq}/j\omega$ and the equivalent dipole moment $\bar{p}$ of the sphere given by the integral of $\bar{P}$ over the volume V of the sphere.

$$\bar{p} = \int_V \bar{P}\, dV' = \int_V \epsilon_0(\epsilon_r - 1)\bar{E}\, dV' = \frac{3(\epsilon_r - 1)}{\epsilon_r + 2}\epsilon_0 V \bar{E}_i. \tag{10-36}$$

We can then rewrite the scattering amplitude for Rayleigh scattering as follows:

$$\bar{f}(\hat{o}, \hat{\imath}) = \frac{k^2}{4\pi\epsilon_0}[\bar{p} - \hat{o}(\hat{o}\cdot\bar{p})], \tag{10-37}$$

where V is the volume of the sphere.

Note that $\bar{p} - \hat{o}(\hat{o}\cdot\bar{p})$ is the component of $\bar{p}$ perpendicular to $\hat{o}$, and therefore its magnitude is equal to $p\ \sin\chi$, where χ is the angle between $\bar{p}$ and $\hat{o}$ (Fig. 10-9). This is to be expected as this represents the radiation pattern of the electric dipole $\bar{p}$. We also note that (10-36) is valid even when the object is lossy and ϵ_r is complex. The differential cross section $\sigma_d(\hat{o}, \hat{\imath})$ is given by

$$\sigma_d(\hat{o}, \hat{\imath}) = \frac{k^4}{(4\pi)^2}\left|\frac{3(\epsilon_r - 1)}{\epsilon_r + 2}\right|^2 V^2 \sin^2\chi, \tag{10-38}$$

where $\sin^2\chi = 1 - (\hat{o}\cdot\hat{e}_i)^2$.

We note that the cross section is inversely proportional to the fourth power of the wavelength and directly proportional to the square of the volume of the scatterer. These two characteristics of a small scatterer were derived by Rayleigh using dimensional analysis and are generally known as Rayleigh scattering. The blue of the sky can be explained by noting that the blue portion of a light spectrum scatters more light than the red portion due to the λ^{-4} dependence. Furthermore, the skylight at right angles to the sun must be linearly polarized, as is evident from Fig. 10-9. These two characteristics, the blue color and the polarization, were a great scientific puzzle in the nineteenth century and were finally explained by Rayleigh (Kerker, 1969). Rayleigh noted that the scatterers need not be water or ice as was commonly believed at that time, but that molecules of air itself can contribute to this scattering. The redness of the sunset is caused by the decrease of the blue portion of the spectrum due to Rayleigh scattering.

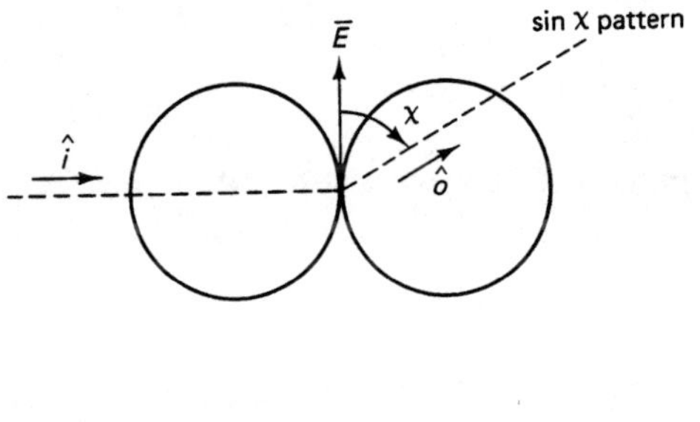

Figure 10-9 Dipole radiation pattern for Rayleigh scattering.

Let us consider the scattering cross section σ_s of a small dielectric sphere.

$$\sigma_s = \int_{4\pi} \sigma_d \, d\omega = \frac{1}{4\pi} \int_{4\pi} \sigma_{bi} \, d\omega$$

$$= \frac{k^4}{(4\pi)^2} \left| \frac{3(\epsilon_r - 1)}{\epsilon_r + 2} \right|^2 V^2 \int_0^\pi \sin\chi \, d\chi \int_0^{2\pi} d\phi \, \sin^2\chi \tag{10-39}$$

$$= \frac{24\pi^3 V^2}{\lambda^4} \left| \frac{\epsilon_r - 1}{\epsilon_r + 2} \right|^2 = \frac{128\pi^5 a^6}{3\lambda^4} \left| \frac{\epsilon_r - 1}{\epsilon_r + 2} \right|^2 .$$

It is often desired to compare the scattering cross section with the actual geometrical cross section πa^2. Thus we obtain the Rayleigh equation,

$$Q_s = \frac{\sigma_s}{\pi a^2} = \frac{8(ka)^4}{3} \left| \frac{\epsilon_r - 1}{\epsilon_r + 2} \right|^2 . \tag{10-40}$$

The Rayleigh equation above is valid only for small ka. The approximate upper limit of the radius of the scatterer is generally taken to be $a = 0.05\lambda$. At this radius, the percent error of the Rayleigh equation (10-40) is less than 4% (Kerker, 1969, p. 85).

The absorption cross section σ_a is obtained using (10-8) with (10-35):

$$\begin{aligned} \sigma_a &= k\epsilon_r'' \left| \frac{3}{\epsilon_r + 2} \right|^2 V, \\ Q_a &= \frac{\sigma_a}{\pi a^2} = ka\epsilon_r'' \left| \frac{3}{\epsilon_r + 2} \right|^2 \frac{4}{3}. \end{aligned} \tag{10-41}$$

The total cross section σ_t is the sum of (10-40) and (10-41). We note that σ_t cannot be obtained by applying the forward scattering theorem to (10-37), since (10-37) gives $\sigma_t = 0$ when $\epsilon_r'' = 0$. In general, for a given approximate value of $\bar{E}(\bar{r}')$ within an object, the scattering cross section as obtained by integrating $|f|^2$ in (10-32) or (10-37) over 4π, plus the absorption cross section in (10-8) or (10-41), gives a much better approximation to the total cross section than is obtained by direct application of the forward scattering theorem to (10-32).

10-6 RAYLEIGH SCATTERING FOR A SMALL ELLIPSOIDAL OBJECT

Many of the particles and objects encountered in practice are not spherical, but they can often be approximated by an ellipsoid whose surface is given by

$$\frac{x^2}{a^2} + \frac{y^2}{b^2} + \frac{z^2}{c^2} = 1. \tag{10-42}$$

If the object size is small compared with a wavelength and the incident field $\bar{E}_i$ has components E_{ix}, E_{iy}, and E_{iz} in the x, y, and z directions, respectively, the components of the field inside the object are given by the following static solution (Stratton, 1941, p. 213; van der Hulst, 1957, p. 71):

$$E_x = \frac{E_{ix}}{1 + (\epsilon_r - 1)L_x},$$

$$L_x = \frac{abc}{2} \int_0^\infty (s + a^2)^{-1} [(s + a^2)(s + b^2)(s + c^2)]^{-1/2}\, ds, \tag{10-43}$$

with an appropriate interchange of a, b, and c for E_y and E_z. It can be easily proved that L_x, L_y, and L_z are functions of the ratio b/a and c/a only and do not depend on the values of a, b, and c. It is also known that

$$L_x + L_y + L_z = 1. \tag{10-44}$$

For a prolate ellipsoid ($a = b < c$),

$$L_z = \frac{1 - e^2}{e^2}\left(-1 + \frac{1}{2e} \ln \frac{1 + e}{1 - e}\right),$$

$$L_x = L_y = \tfrac{1}{2}(1 - L_z), \tag{10-45}$$

$$e^2 = 1 - \left(\frac{a}{c}\right)^2.$$

For an oblate ellipsoid ($a = b > c$)

$$L_z = \frac{1 + f^2}{f^2}\left(1 - \frac{1}{f} \arctan f\right),$$

$$L_x = L_y = \tfrac{1}{2}(1 - L_z), \tag{10-46}$$

$$f^2 = \left(\frac{a}{c}\right)^2 - 1.$$

The scattering amplitude $\bar{f}(\hat{o}, \hat{\imath})$ can then be given by

$$\bar{f}(\hat{o}, \hat{\imath}) = \frac{k^2}{4\pi\epsilon_0}[\bar{p} - \hat{o}(\hat{o} \cdot \bar{p})], \tag{10-47}$$

where

$$V = \tfrac{4}{3}\pi abc,$$

$$\bar{p} = \alpha_x E_{ix}\hat{x} + \alpha_y E_{iy}\hat{y} + \alpha_z E_{iz}\hat{z}$$

$$\alpha_x = \frac{\epsilon_0(\epsilon_r - 1)V}{1 + (\epsilon_r - 1)L_x}.$$

α_y and α_z are obtained by replacing L_x by L_y and L_z, respectively. α_x, α_y, and α_z are the polarizability of the object for the x, y, and z directions.

As an example, consider a plane wave propagating in the direction (θ', ϕ')

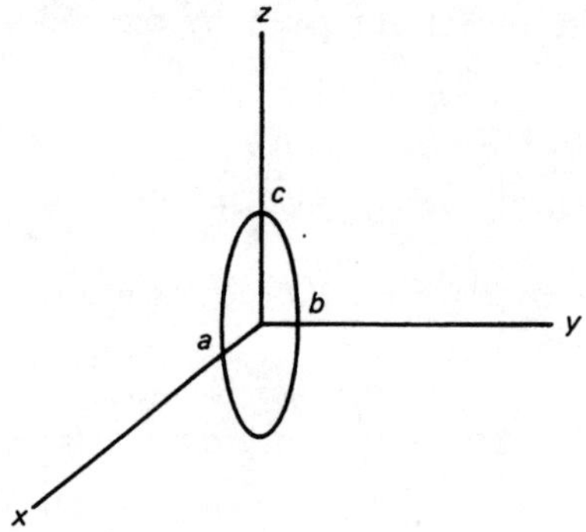

Figure 10-10 Ellipsoid.

with the components E'_θ and E'_ϕ. The object is located at the origin with the principal axes oriented in the coordinate system x_b–y_b–z_b and the surface given by (Fig. 10-10)

$$\frac{x_b^2}{a^2}+\frac{y_b^2}{b^2}+\frac{z_b^2}{c^2}=1. \tag{10-48}$$

We wish to find the scattering amplitude in the direction (θ,ϕ) with the field components E_θ and E_ϕ (Fig. 10-11). We write the scattering amplitude by the following 2×2 matrix $[F]=[f_{ij}]$.

$$\begin{bmatrix}E_\theta\\E_\phi\end{bmatrix}=\frac{e^{-jkR}}{R}\begin{bmatrix}f_{11} & f_{12}\\ f_{21} & f_{22}\end{bmatrix}\begin{bmatrix}E'_\theta\\E'_\phi\end{bmatrix}. \tag{10-49}$$

We can then obtain the scattering amplitude matrix as follows:

$$[F]=\frac{k^2}{4\pi\epsilon_0}[C_1][\alpha][C_2], \tag{10-50}$$

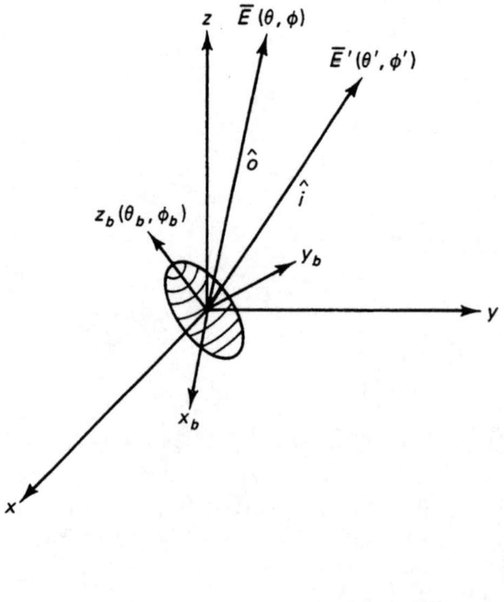

Figure 10-11 Scattering amplitude matrix.

where

$$[C_1] = \begin{bmatrix} \hat{\theta}\cdot\hat{x}_b & \hat{\theta}\cdot\hat{y}_b & \hat{\theta}\cdot\hat{z}_b \\ \hat{\phi}\cdot\hat{x}_b & \hat{\phi}\cdot\hat{y}_b & \hat{\phi}\cdot\hat{z}_b \end{bmatrix},$$

$$[C_2] = \begin{bmatrix} \hat{\theta}'\cdot\hat{x}_b & \hat{\phi}'\cdot\hat{x}_b \\ \hat{\theta}'\cdot\hat{y}_b & \hat{\phi}'\cdot\hat{y}_b \\ \hat{\theta}'\cdot\hat{z}_b & \hat{\phi}'\cdot\hat{z}_b \end{bmatrix},$$

$$[\alpha] = \begin{bmatrix} \alpha_x & 0 & 0 \\ 0 & \alpha_y & 0 \\ 0 & 0 & \alpha_z \end{bmatrix}.$$

The relationship between $(\hat{x}_b, \hat{y}_b, \hat{z}_b)$ and $(\hat{x}, \hat{y}, \hat{z})$ is given by Euler's transformation $[A_e]$:

$$\begin{bmatrix} \hat{x}_b \\ \hat{y}_b \\ \hat{z}_b \end{bmatrix} = [A_e] \begin{bmatrix} \hat{x} \\ \hat{y} \\ \hat{z} \end{bmatrix}. \tag{10-51}$$

There are several representations of Euler's transformation and they are given by Goldstein (1981). For axially symmetric objects with an axis oriented in the (θ_b, ϕ_b) direction (Fig. 10-11), we have $L_x = L_y$ and

$$[A_e] = \begin{bmatrix} \cos\theta_b \cos\phi_b & \cos\theta_b \sin\phi_b & -\sin\theta_b \\ -\sin\phi_b & \cos\phi_b & 0 \\ \sin\theta_b \cos\phi_b & \sin\theta_b \sin\phi_b & \cos\theta_b \end{bmatrix}. \tag{10-52}$$

We can then calculate $[C_1]$ and $[C_2]$ in terms of θ, ϕ, θ', ϕ', and θ_b, ϕ_b, noting that

$$\hat{\theta} = \cos\theta \cos\phi\, \hat{x} + \cos\theta \sin\phi\, \hat{y} - \sin\theta\, \hat{z},$$

$$\hat{\phi} = -\sin\phi\, \hat{x} + \cos\phi\, \hat{y};$$

and $\hat{\theta}'$ and $\hat{\phi}'$ with θ' and ϕ' replacing θ and ϕ.

Let us consider the scattering and absorption cross sections for an axially symmetric object oriented in the x–z plane inclined with angle θ_b from the z axis which is illuminated by the incident wave propagating in the z direction (Fig. 10-12). The scattering cross section must be calculated by integrating $|\bar{f}|^2$ over all the solid angle [see (10-6)] and the absorption cross section is obtained by (10-8). For non-spherical objects, the scattering and absorption cross sections depend on the polarization of the incident wave. If the incident wave is polarized in the x direction ($\overline{E}_i = E_{ix}\hat{x}$), we can perform the calculations above and obtain

$$\sigma_{sx} = \left(\frac{k^2}{4\pi\epsilon_0}\right)^2 \frac{8\pi}{3} \left(|\alpha_x \cos^2\theta_b + \alpha_z \sin^2\theta_b|^2 + |\alpha_x - \alpha_z|^2 \sin^2\theta_b \cos^2\theta_b\right),$$

$$\sigma_{ax} = k\epsilon_r'' \frac{|\alpha_x|^2 \cos^2\theta_b + |\alpha_z|^2 \sin^2\theta_b}{\epsilon_0^2 V |\epsilon_r - 1|^2}. \tag{10-53}$$

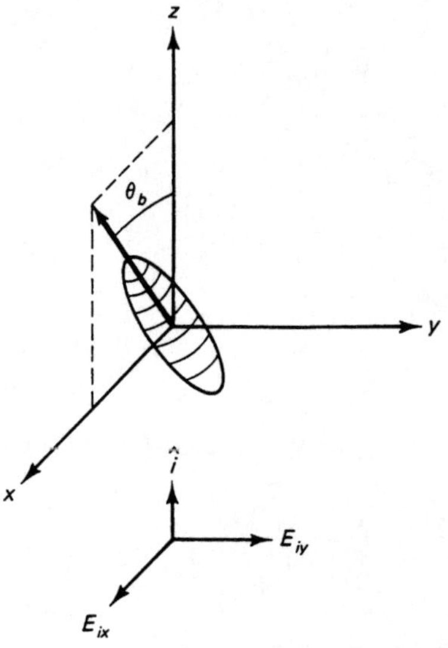

Figure 10-12 Axially symmetric object illuminated by a plane wave.

If the incident wave is polarized in the y direction ($\bar{E}_i = E_{iy}\hat{y}$), we get

$$\sigma_{sy} = \left(\frac{k^2}{4\pi\epsilon_0}\right)^2 |\alpha_y|^2 \frac{8\pi}{3},$$

$$\sigma_{ay} = k\epsilon_r'' \frac{|\alpha_y|^2}{|\epsilon_r - 1|^2 \epsilon_0^2} \frac{1}{V}. \tag{10-54}$$

10-7 RAYLEIGH–DEBYE SCATTERING (BORN APPROXIMATION)

We now consider the scattering characteristics of a scatterer whose relative dielectric constant ϵ_r is close to unity. In this case the field inside the scatterer may be approximated by the incident field.

$$\bar{E}(\bar{r}) = \bar{E}_i(\bar{r}) = \hat{e}_i \exp(-jk\bar{r}\cdot\hat{\imath}). \tag{10-55}$$

Substituting this into (10-32), we get

$$\bar{f}(\hat{o},\hat{\imath}) = \frac{k^2}{4\pi}[-\hat{o}\times(\hat{o}\times\hat{e}_i)]VS(\bar{k}_s), \tag{10-56}$$

$$S(\bar{k}_s) = \frac{1}{V}\int_V [\epsilon_r(\bar{r}') - 1]\exp(-j\bar{k}_s\cdot\bar{r}')\,dV', \tag{10-57}$$

where

$$\bar{k}_s = k\bar{i}_s = k(\hat{\imath} - \hat{o}), \qquad |\bar{i}_s| = 2\sin\frac{\theta}{2},$$

and θ is the angle between $\hat{\imath}$ and $\hat{o}$ (Fig. 10-13).

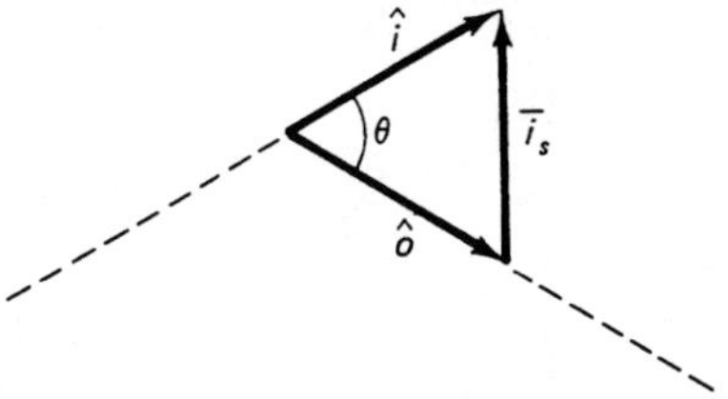

Figure 10-13 Relationship showing $\bar{i}_s$ and θ.

This approximation is valid when

$$(\epsilon_r - 1)kD \ll 1, \tag{10-58}$$

where D is a typical dimension of the object such as its diameter. We note that (10-57) is a Fourier transform of $[\epsilon_r(r') - 1]$ in the direction of $\bar{i}_s$. Therefore, the scattering amplitude $f(\hat{o}, \hat{\imath})$ is proportional to the Fourier transform of $[\epsilon_r(r') - 1]$ evaluated at the wave number $\bar{k}_s$. In general, if $[\epsilon_r(r') - 1]$ is concentrated in a region small compared with a wavelength, the cross section is spread out in $\bar{k}_s$ and thus in angle θ, and the scattering is almost isotropic. If the object size is large compared to a wavelength, the scattering is concentrated in a small forward angular region $\theta \approx 0$. This situation is similar to the relationship between a time function and its frequency spectrum. If a function is limited in time within T, its spectrum is spread out over a frequency range $1/T$.

The Rayleigh–Debye absorption cross section is obtained from (10-8):

$$\sigma_a = k \int_V \epsilon_r''(\bar{r})\, dV. \tag{10-59}$$

Let us take a few examples.

Scattering by a Homogeneous Sphere of Radius *a*

In this case, because of the spherical symmetry, we choose the z' axis in the direction of $\bar{i}_s$ (Fig. 10-14). We then write

$$\bar{f}(\hat{o}, \hat{\imath}) = \frac{k^2}{4\pi}[-\hat{o} \times (\hat{o} \times \hat{e}_i)](\epsilon_r - 1)VF(\theta), \tag{10-60}$$

$$\begin{aligned} F(\theta) &= \frac{1}{V}\int_V \exp(-jk\bar{i}_s \cdot \bar{r}')\, dV' \\ &= \frac{1}{V}\int_0^{2\pi} d\phi' \int_0^{\pi} \sin\theta'\, d\theta' \int_0^a r'^2\, dr' \exp(-jk_s r' \cos\theta') \\ &= \frac{3}{k_s^3 a^3}(\sin k_s a - k_s a \cos k_s a), \end{aligned} \tag{10-61}$$

with $k_s = 2k \sin(\theta/2)$. The plot of $|F(\theta)|^2$ is shown in Fig. 10-15.

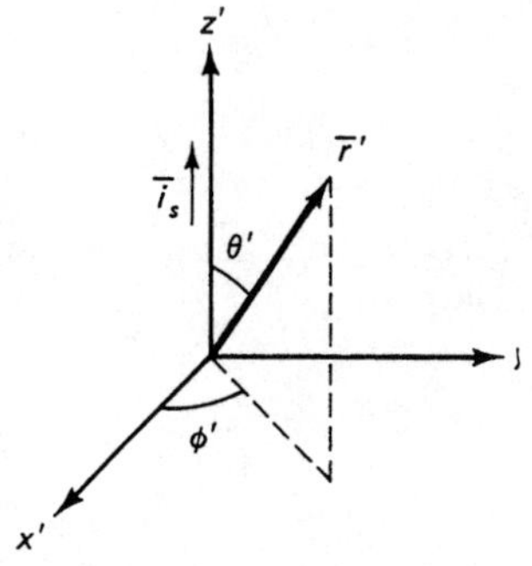

Figure 10-14 Coordinate axes for evaluation of (10-61).

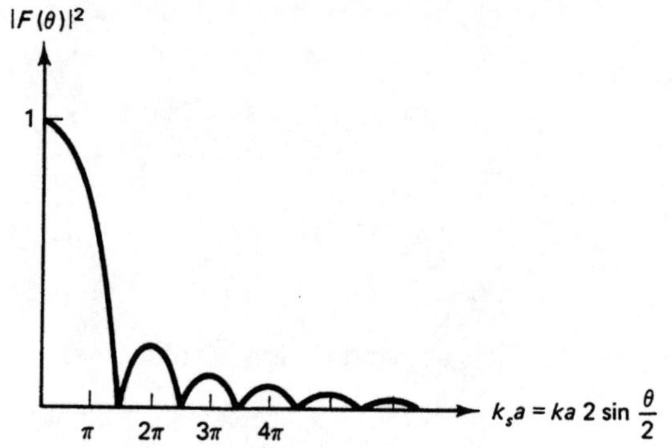

Figure 10-15 Scattering pattern (10-61) for a homogeneous sphere of radius a.

Scattering by an Ellipsoidal Object

Let us consider an ellipsoid as shown in Fig. 10-16. The incident wave is propagating in the direction (θ_i, ϕ_i) and the scattered wave is observed in the direction (θ_0, ϕ_0). The surface of the ellipsoid is given by

$$\frac{x^2}{a^2} + \frac{y^2}{b^2} + \frac{z^2}{c^2} = 1. \tag{10-62}$$

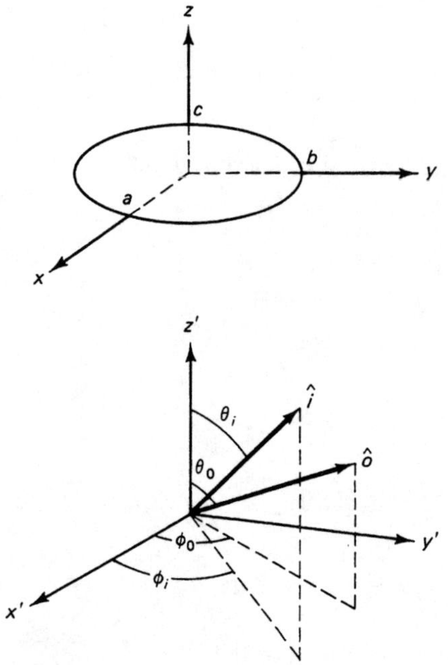

Figure 10-16 Ellipsoidal object and the directions $\hat{\imath}$ and $\hat{o}$ in (10-63).

We express the direction of the incident wave $\hat{i}$ and the direction of the observation point $\hat{o}$ in the spherical coordinate system,

$$\begin{aligned} \bar{k}_s &= k_{s1}\hat{x} + k_{s2}\hat{y} + k_{s3}\hat{z}, \\ k_{s1} &= k(\sin\theta_i\cos\phi_i - \sin\theta_0\cos\phi_0), \\ k_{s2} &= k(\sin\theta_i\sin\phi_i - \sin\theta_0\sin\phi_0), \\ k_{s3} &= k(\cos\theta_i - \cos\theta_0). \end{aligned} \tag{10-63}$$

We then get

$$F = \frac{1}{V}\int_V \exp[-j(k_{s1}x + k_{s2}y + k_{s3}z)]\,dx\,dy\,dz. \tag{10-64}$$

The integration is over the ellipsoidal volume, but if we use the following normalized coordinate, the integration can be performed over a sphere of unit radius.

$$x' = \frac{x}{a}, \qquad y' = \frac{y}{b}, \qquad z' = \frac{z}{c}. \tag{10-65}$$

We then get

$$\begin{aligned} F &= \frac{3}{4\pi}\int \exp(-j\bar{K}\cdot\bar{r})\,dx'\,dy'\,dz' \\ &= \frac{3}{K^3}(\sin K - K\cos K), \end{aligned} \tag{10-66}$$

where

$$K = [(k_{s1}a)^2 + (k_{s2}b)^2 + (k_{s3}c)^2]^{1/2}.$$

Scattering From a Randomly Oriented Object With Axial Symmetry

We note that in (10-63), $\sqrt{k_{s1}^2 + k_{s2}^2}$ and k_{s3} are the components of $k\bar{i}_s = k(\hat{i} - \hat{o})$ in the directions perpendicular and parallel to the z axis. Letting β be the angle between $\bar{i}_s$ and the z axis, we write

$$\sqrt{k_{s1}^2 + k_{s2}^2} = k_s\sin\beta \qquad \text{and} \qquad k_{s3} = k_s\cos\beta.$$

For random orientation, we average the scattered intensity over all possible orientations of the object. Here, due to the randomness, the intensity rather than the field must be averaged.

$$|F|^2_{\text{ave}} = \begin{cases} \dfrac{1}{4\pi}\displaystyle\int |F|^2\,d\omega, & d\omega = \sin\beta\,d\beta\,d\phi, \\ \dfrac{1}{2}\displaystyle\int_{-1}^{1} |F|^2\,d\mu, & \mu = \cos\beta. \end{cases} \tag{10-67}$$

10-8 ELLIPTIC POLARIZATION AND STOKES PARAMETERS

In the preceding sections we considered a linearly polarized incident wave. In general, however, it is necessary to consider an incident wave with elliptic polarization. Let us examine a plane wave propagating in the z direction, whose electric field components as functions of time are given by

$$\begin{aligned} E_x &= \mathrm{Re}(E_1 e^{j\omega t}) = \mathrm{Re}[a_1 \exp(j\omega t - jkz + j\delta_1)] = a_1 \cos(\tau + \delta_1), \\ E_y &= \mathrm{Re}(E_2 e^{j\omega t}) = \mathrm{Re}[a_2 \exp(j\omega t - jkz + j\delta_2)] = a_2 \cos(\tau + \delta_2), \\ E_z &= 0, \end{aligned} \tag{10-68a}$$

where $\tau = \omega t - kz$, and E_1 and E_2 are the phasors for E_x and E_y.

In the above, we used the IEEE convention $\exp(j\omega t)$. In many studies using Stokes' parameters, it is more common to use the $\exp(-i\omega t)$ convention. In this case, (10-68a) should be written as

$$\begin{aligned} E_x &= \mathrm{Re}(E_1 e^{-i\omega t}) = \mathrm{Re}[a_1 \exp(-i\omega t + ikz - i\delta_1)] = a_1 \cos(\tau + \delta_1), \\ E_y &= \mathrm{Re}(E_2 e^{-i\omega t}) = \mathrm{Re}[a_2 \exp(-i\omega t + ikz - i\delta_2)] = a_2 \cos(\tau + \delta_2). \end{aligned} \tag{10-68b}$$

Now consider a general elliptically polarized wave. The endpoint of the electric field vector $\bar{E} = E_x \hat{x} + E_y \hat{y}$ traces an ellipse. The equation for this ellipse is obtained by eliminating τ from (10-68a) or (10-68b).

$$\left(\frac{E_x}{a_1}\right)^2 + \left(\frac{E_y}{a_2}\right)^2 - \frac{2E_x E_y}{a_1 a_2} \cos\delta = \sin^2\delta, \tag{10-69}$$

where $\delta = \delta_2 - \delta_1$ is the phase difference.

To describe the elliptically polarized wave given in (10-68), three independent parameters are needed. For example, they can be a_1, a_2, and δ. It is, however, more convenient to use parameters of the same dimension. In 1852, G. G. Stokes introduced what are now called the *Stokes parameters*. They are

$$\begin{aligned} I &= a_1^2 + a_2^2 = |E_1|^2 + |E_2|^2, \\ Q &= a_1^2 - a_2^2 = |E_1|^2 - |E_2|^2, \\ U &= 2a_1 a_2 \cos\delta = 2\,\mathrm{Re}(E_1 E_2^*), \\ V &= \mp 2a_1 a_2 \sin\delta = 2\,\mathrm{Im}(E_1 E_2^*), \end{aligned} \tag{10-70}$$

where E_1 and E_2 are the phasor representations of the electric field components E_x and E_y given by

$$\begin{aligned} E_1 &= a_1 \exp(j\delta_1 - jkz) = a_1 \exp(-i\delta_1 + ikz), \\ E_2 &= a_2 \exp(j\delta_2 - jkz) = a_2 \exp(-i\delta_2 + ikz). \end{aligned}$$

Note that the upper and lower signs for V are for $\exp(j\omega t)$ and $\exp(-i\omega t)$ dependence, respectively. Among these four parameters, there exists the relationship, obtained from (10-70),

$$I^2 = Q^2 + U^2 + V^2. \tag{10-71}$$

Equations (10-70) and (10-71) together provide three independent quantities that describe an elliptically polarized wave.

As an example, for a wave linearly polarized in the direction ψ_0 with respect to the x axis, we have $a_1 = E_0 \cos\psi_0$, $a_2 = E_0 \sin\psi_0$ and $\delta = 0$, and the Stokes parameters are

$$I = E_0^2, \qquad Q = E_0^2 \cos 2\psi_0, \qquad U = E_0^2 \sin 2\psi_0, \qquad V = 0. \tag{10-72}$$

For a right-handed circularly polarized wave, we have $a_1 = a_2 = E_0$, $\delta = -\pi/2$, and

$$I = 2E_0^2, \qquad Q = 0, \qquad U = 0, \qquad V = \pm 2E_0^2. \tag{10-73}$$

Here the upper and lower signs for V are for $\exp(j\omega t)$ and $\exp(-i\omega t)$ dependence, respectively. It is also common to use the modified Stokes parameters given by

$$I_1 = |E_1|^2, \qquad I_2 = |E_2|^2, \qquad U = 2\,\mathrm{Re}(E_1 E_2^*), \qquad V = 2\,\mathrm{Im}(E_1 E_2^*). \tag{10-74}$$

Alternatively, it is possible to describe the ellipse in Fig. 10-17 in terms of the semimajor (a) and the semiminor (b) axes of the ellipse and the orientation angle (ψ). Using I, b/a, and ψ, the Stokes parameters become, for $\exp(-i\omega t)$ dependence,

$$\begin{aligned} Q &= I \cos 2\chi \cos 2\psi, \\ U &= I \cos 2\chi \sin 2\psi, \\ V &= I \sin 2\chi, \end{aligned} \tag{10-75}$$

where $\tan\chi = \pm b/a$, with a plus sign for left-handed polarization and a minus sign for right-handed polarization. The polarization is defined as right-handed when the electric field rotates as a right-handed screw advancing in the direction of propagation.

From (10-75) it is seen that I and V depend on the total intensity and the ellipticity angle χ and are not affected by the orientation angle ψ of the ellipse, but Q and U vary according to the choice of coordinates.

Equation (10-75) may be compared with the Cartesian coordinates (X, Y, Z) of a point (r, θ, ϕ) on a sphere with radius $r = I$, $\theta = (\pi/2) - 2\chi$ and $\phi = 2\psi$, $X = r \sin\theta \cos\phi$, $Y = r \sin\theta \sin\phi$, and $Z = r \cos\theta$. This sphere is called the *Poincaré sphere*, and its north and south poles represent left- and right-handed circular

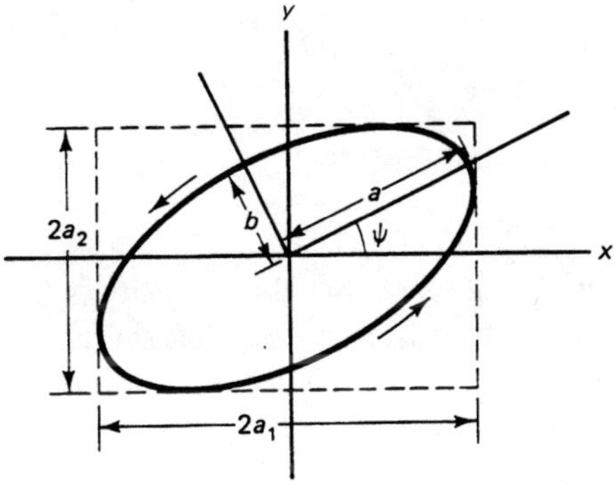

Figure 10-17 Right-handed elliptical polarization.

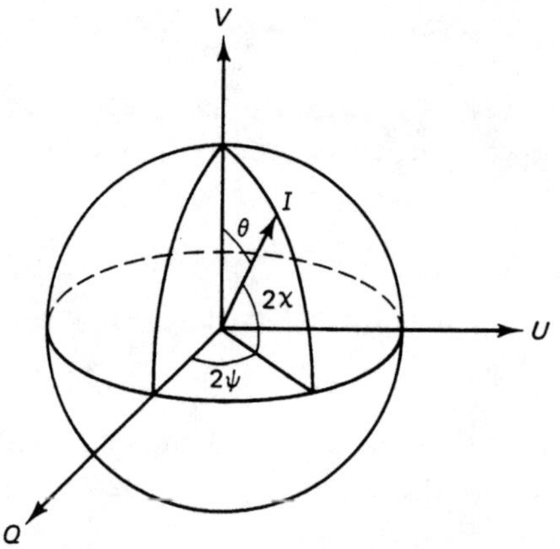

Figure 10-18 Poincaré sphere.

polarizations, respectively. The northern and southern hemispheres represent left- and right-handed elliptic polarizations, and the equator represents linear polarization (Fig. 10-18).

10-9 PARTIAL POLARIZATION AND NATURAL LIGHT

In the elliptic polarization discussed in Section 10-8, the ratio of the amplitudes a_1 and a_2, and the phase difference $\delta = \delta_2 - \delta_1$ are absolute constants. This happens when a wave is purely monochromatic (single-frequency). In the more general case of a polychormatic wave with a certain bandwidth $\Delta\omega$, the amplitudes and the phase difference undergo continuous variations with various frequencies within $\Delta\omega$, and therefore a_1, a_2, and δ are slowly varying random functions of time. In general, therefore, the Stokes parameters should be expressed by the averages. Denoting the time average by angular brackets, $\langle\cdot\rangle$, we have, for $\exp(-i\omega t)$ dependence,

$$
\begin{aligned}
I &= \langle a_1^2\rangle + \langle a_2^2\rangle = \langle|E_1|^2\rangle + \langle|E_2|^2\rangle,\\
Q &= \langle a_1^2\rangle - \langle a_2^2\rangle = \langle|E_1|^2\rangle - \langle|E_2|^2\rangle,\\
U &= 2\langle a_1 a_2 \cos\delta\rangle = 2\,\mathrm{Re}\langle E_1 E_2^*\rangle,\\
V &= 2\langle a_1 a_2 \sin\delta\rangle = 2\,\mathrm{Im}\langle E_1 E_2^*\rangle.
\end{aligned}
\tag{10-76}
$$

For modified Stokes parameters (I_1, I_2, U, V), we have $I_1 = \langle|E_1|^2\rangle$ and $I_2 = \langle|E_2|^2\rangle$. In this case the condition in (10-71) must be replaced by

$$I^2 \geq Q^2 + U^2 + V^2. \tag{10-77}$$

Natural light is characterized by the fact that the intensity is the same in any direction perpendicular to the direction of the ray and that there is no correlation between rectangular components of the field. Therefore, the necessary and sufficient conditions for light to be natural are

$$I = 2\langle |E|^2 \rangle$$

and

$$Q = U = V = 0. \tag{10-78}$$

In general, a wave may be partially polarized. The degree of polarization m is defined by the ratio

$$m = \frac{(Q^2 + U^2 + V^2)^{1/2}}{I}; \tag{10-79}$$

$m = 1$ for elliptic polarization, $0 < m < 1$ for partial polarization, and $m = 0$ for an unpolarized wave (natural light).

It is clear that the Stokes parameters $[I]$ can always be expressed as the sum of the elliptically polarized wave $[I_p]$ and the unpolarized wave $[I_u]$.

$$[I] = [I_p] + [I_u],$$

$$[I] = \begin{bmatrix} I \\ Q \\ U \\ V \end{bmatrix}, \quad [I_p] = \begin{bmatrix} mI \\ Q \\ U \\ V \end{bmatrix}, \quad [I_u] = \begin{bmatrix} (1-m)I \\ 0 \\ 0 \\ 0 \end{bmatrix}. \tag{10-80}$$

10-10 SCATTERING AMPLITUDE FUNCTIONS f_{11}, f_{12}, f_{21}, AND f_{22} AND THE STOKES MATRIX

In Section 10-1 the scattering amplitude $\bar{f}(\hat{o}, \hat{\imath})$ is defined by

$$\bar{E}_s(\bar{r}) = \bar{f}(\hat{o}, \hat{\imath}) \frac{e^{ikR}}{R}, \tag{10-81a}$$

for a linearly polarized incident wave given by

$$\bar{E}_i(\bar{r}) = \hat{e}_i e^{ik\hat{\imath} \cdot \bar{r}}. \tag{10-81b}$$

To generalize the description of the scattered wave to include elliptic, partially polarized, and unpolarized waves, we may choose the following coordinate system (van de Hulst, 1957). We use $\exp(-i\omega t)$ in this section. We choose the z axis to be the direction of the incident wave, and the y–z plane to be the *plane of scattering*, defined as the plane that includes the direction of the incident wave $\hat{\imath}$ and the observation $\hat{o}$ (Fig. 10-19). The incident wave has two components, $E_{ix} = E_{i\perp}$ and $E_{iy} = E_{i\parallel}$, in the directions perpendicular and parallel, respectively, to the plane of scattering. The scattered wave in the direction $\hat{o}$ has two components, $E_{sX} = E_{s\perp}$ and $E_{sY} = E_{s\parallel}$, perpendicular and parallel, respectively, to the plane of scattering. It is clear that $E_{s\perp}$ and $E_{s\parallel}$ are linearly related to $E_{i\perp}$ and $E_{i\parallel}$, and therefore we write

$$\begin{bmatrix} E_{s\perp} \\ E_{s\parallel} \end{bmatrix} = \frac{e^{ikR}}{R} \begin{bmatrix} f_{11} & f_{12} \\ f_{21} & f_{22} \end{bmatrix} \begin{bmatrix} E_{i\perp} \\ E_{i\parallel} \end{bmatrix}. \tag{10-82}$$

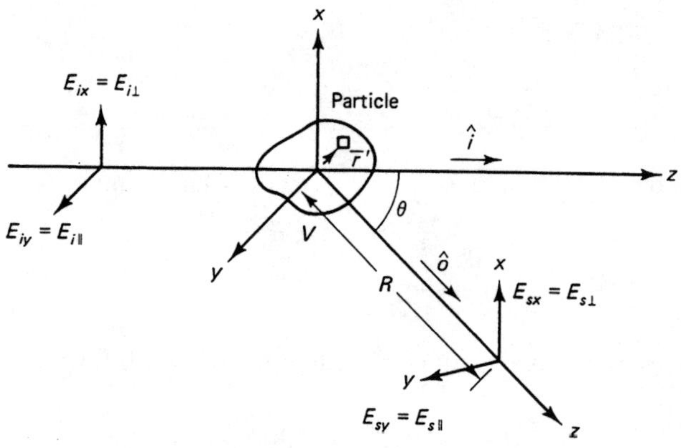

Figure 10-19 Geometry for defining scattering amplitude. The y–z and Y–Z planes are the plane of scattering.

$E_{i\perp}$ and $E_{i\parallel}$ are evaluated at the origin $x = y = z = 0$, and $E_{s\perp}$ and $E_{s\parallel}$ are at a large distance R from the origin. f_{11}, f_{12}, f_{21}, and f_{22} are functions of θ, and they are related to the scattering functions S_1, S_2, S_3, and S_4 used by van de Hulst and in the Mie solution for a sphere (see Chapter 11).

$$f_{11} = \frac{i}{k} S_1, \quad f_{12} = \frac{i}{k} S_4, \quad f_{21} = \frac{i}{k} S_3, \quad f_{22} = \frac{i}{k} S_2. \tag{10-83}$$

Alternatively, we often use the spherical system shown in Fig. 10-11. The scattered wave (E_θ, E_ϕ) is related to the incident wave (E'_θ, E'_ϕ), and f_{ij} is a function of $(\theta, \phi, \theta'$, and $\phi')$:

$$\begin{bmatrix} E_\theta \\ E_\phi \end{bmatrix} = \frac{e^{ikR}}{R} \begin{bmatrix} f_{11} & f_{12} \\ f_{21} & f_{22} \end{bmatrix} \begin{bmatrix} E'_\theta \\ E'_\phi \end{bmatrix}. \tag{10-84}$$

If the scattering functions are known, and if the incident wave has an arbitrary state of polarization and its Stokes parameters are given by I_{1i}, I_{2i}, U_i, and V_i, what are the Stokes parameters I_{1s}, I_{2s}, U_s, and U_s of the scattered wave? This relationship can be obtained using (10-74) and (10-82) in terms of the following 4×4 Mueller matrix $\bar{\bar{S}}$:

$$I_s = \frac{1}{R^2} \bar{\bar{S}} I_i, \tag{10-85}$$

where I_s and I_i are 4×1 column matrices and $\bar{\bar{S}}$ is a 4×4 matrix.

$$I_s = \begin{bmatrix} I_{1s} \\ I_{2s} \\ U_s \\ V_s \end{bmatrix}, \qquad I_i = \begin{bmatrix} I_{1i} \\ I_{2i} \\ U_i \\ V_i \end{bmatrix}, \tag{10-86}$$

$$\bar{\bar{S}} = \begin{bmatrix} |f_{11}|^2 & |f_{12}|^2 & \mathrm{Re}(f_{11} f_{12}^*) & -\mathrm{Im}(f_{11} f_{12}^*) \\ |f_{21}|^2 & |f_{22}|^2 & \mathrm{Re}(f_{21} f_{22}^*) & -\mathrm{Im}(f_{21} f_{22}^*) \\ 2\,\mathrm{Re}(f_{11} f_{21}^*) & 2\,\mathrm{Re}(f_{12} f_{22}^*) & \mathrm{Re}(f_{11} f_{22}^* + f_{12} f_{21}^*) & -\mathrm{Im}(f_{11} f_{22}^* - f_{12} f_{21}^*) \\ 2\,\mathrm{Im}(f_{11} f_{21}^*) & 2\,\mathrm{Im}(f_{12} f_{22}^*) & \mathrm{Im}(f_{11} f_{22}^* + f_{12} f_{21}^*) & \mathrm{Re}(f_{11} f_{22}^* - f_{12} f_{21}^*) \end{bmatrix}.$$

The matrix representations above are used in the formulation of vector radiative transfer theory.

10-11 ACOUSTIC SCATTERING

In this section we describe the absorption and scattering characteristics of an object when illuminated by an incident acoustic wave of unit amplitude (Fig. 10-20):

$$P_i(\bar{r}) = \exp(-jk\hat{\imath} \cdot \bar{r}). \tag{10-87}$$

The scattering amplitude $f(\hat{o}, \hat{\imath})$ is a scalar quantity and the scattered acoustic field is given by

$$P_s(\bar{r}) = f(\hat{o}, \hat{\imath})\frac{e^{-jkR}}{R} \qquad \text{for } R > \frac{D^2}{\lambda}. \tag{10-88}$$

The incident and the scattered power flux are given by

$$\bar{S}_i = \frac{|p_i|^2}{2\eta_0}\hat{\imath}, \qquad \bar{S}_s = \frac{|P_s|^2}{2\eta_0}\hat{o}, \tag{10-89}$$

where $\eta_0 = \rho_0 c_0$ is the characteristic impedance and ρ_0 and c_0 are the equilibrium density of the medium and the velocity of the wave propagation in the medium, respectively. The differential scattering cross section σ_d, the scattering cross section σ_s, the absorption cross section σ_a, and the total cross section σ_t are defined by exactly the same formulas as shown in Section 10-1. The forward scattering theorem is

$$\sigma_t = -\frac{4\pi}{k}\,\mathrm{Im} f(\hat{\imath}, \hat{\imath}). \tag{10-90}$$

Integral representation of the scattering amplitude $f(\hat{o}, \hat{\imath})$ is somewhat different from that for an electromagnetic wave because of the factor $1/\rho_0$ inside the divergence operation in (2-131).

For a time-harmonic case with $\exp(j\omega t)$, (2-131) becomes

$$\nabla^2 p + k^2 p = -k^2\gamma_\kappa p + \mathrm{div}[\gamma_\rho\, \mathrm{grad}\, p], \tag{10-91}$$

where $\gamma_\kappa = (\kappa_e - \kappa)/\kappa$, $\gamma_\rho = (\rho_e - \rho)/\rho_e$, $k^2 = \omega^2/c^2 = \omega^2\kappa\rho$ and κ and ρ are the compressibility and the density, respectively, of the medium surrounding the object, and κ_e and ρ_e are those of the object.

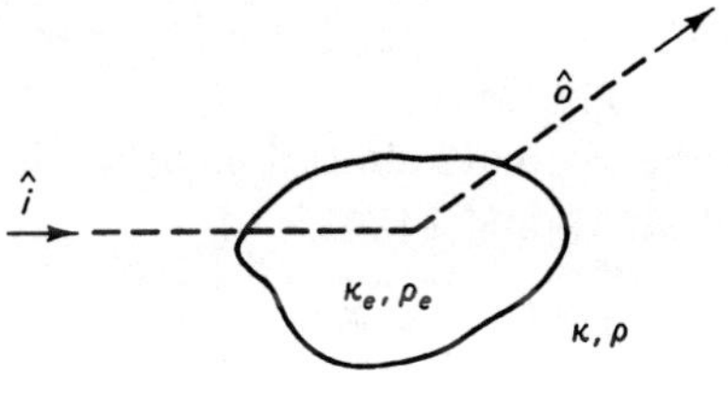

Figure 10-20 Acoustic scattering by an object.

The right side of (10-91) generates the scattered wave, and the scattering amplitude $f(\hat{o},\hat{\imath})$ is given by

$$f(\hat{o},\hat{\imath}) = \frac{k^2}{4\pi}\int_V \left(\gamma_\kappa p + j\gamma_\rho \frac{\hat{o}}{k}\cdot\nabla' p\right) e^{+jk\hat{o}\cdot\bar{r}'}\,dV', \tag{10-92}$$

where we used the divergence theorem to convert the integral involving the second term of the right side of (10-91) to the second term of the integral of (10-92).

Using (10-92), we can obtain the following Born approximation:

$$f(\hat{o},\hat{\imath}) = \frac{k^2}{4\pi}\int_V (\gamma_\kappa + \gamma_\rho \cos\theta)\exp(-j\bar{k}_s\cdot\bar{r}')\,dV', \tag{10-93a}$$

where $\bar{k}_s = k(\hat{\imath}-\hat{o})$, $|\bar{k}_s| = 2k\sin(\theta/2)$, and $\cos\theta = \hat{\imath}\cdot\hat{o}$. This is applicable to the case

$$\left(\frac{\kappa_e\rho_e}{\kappa\rho} - 1\right)kD \ll 1, \tag{10-93b}$$

where D is a typical size of the object.

It should be interesting to note that for a small object, the first term inside the integral in (10-93) gives isotropic scattering similar to the electromagnetic case, but the second term gives scattering proportional to $\hat{o}\cdot\hat{\imath} = \cos\theta$.

For Rayleigh scattering by a small sphere, the incident pressure p_i with magnitude p_0 and the pressure p_e inside the sphere are given by

$$\begin{aligned} p_i &= p_0 e^{-jkx} \approx p_0(1-jkx), \\ p_e &\approx p_0\left(1 - \frac{jkx\cdot 3\rho_e}{\rho + 2\rho_e}\right). \end{aligned} \tag{10-94}$$

Therefore, we get

$$f(\hat{o},\hat{\imath}) = \frac{k^2a^3}{3}\left(\frac{\kappa_e-\kappa}{\kappa} + \frac{3(\rho_e-\rho)}{2\rho_e+\rho}\cos\theta\right),$$

$$\frac{\sigma_s}{\pi a^2} = \frac{4(ka)^4}{9}\left(\left|\frac{\kappa_e-\kappa}{\kappa}\right|^2 + 3\left|\frac{\rho_e-\rho}{2\rho_e+\rho}\right|^2\right).$$

10-12 SCATTERING CROSS SECTION OF A CONDUCTING BODY

Many objects used in aerospace applications have conducting surfaces, examples being airplanes, rockets, spacecraft, and missiles. It is, therefore, important to study the scattering characteristics of a conducting body, particularly the back-scattering cross section for radar applications.

Let us first obtain the general formulation of this problem. Consider a conducting surface S illuminated by an incident wave

$$\bar{E}_i(r) = E_i e^{-jk\hat{\imath}\cdot\bar{r}}\hat{e}_i. \tag{10-95}$$

The scattered field $\bar{E}_s(r)$ is given by (see Section 10-3)

$$\begin{aligned}\bar{E}_s(r) &= \nabla \times \nabla \times \bar{\pi}_s(\bar{r}),\\ \bar{H}_s(r) &= j\omega\epsilon_0 \nabla \times \bar{\pi}_s(\bar{r}),\end{aligned} \tag{10-96}$$

where

$$\bar{\pi}_s(\bar{r}) = \frac{1}{j\omega\epsilon_0}\int_S G_0(\bar{r}, \bar{r}')\bar{J}_s(\bar{r}')\, da,$$

and $\bar{J}_s$ is the surface current on the surface of the conductor. Therefore, we get the scattered field at a large distance from the object:

$$\begin{aligned}\bar{E}_s(\bar{r}) &= \hat{e}_s f(\hat{o}, \hat{\imath})\frac{e^{-jkR}}{R}\\ &= -jk\eta_0\frac{e^{-jkR}}{4\pi R}\int_S [-\hat{o} \times (\hat{o} \times \hat{\jmath})J_s(r')]e^{jk\hat{o}\cdot\bar{r}'}\, da,\end{aligned} \tag{10-97}$$

where $\bar{J}_s(\bar{r}') = J_s(\bar{r}')\hat{\jmath}$ and $\eta_0 = 120\pi$ ohms is the free-space characteristic impedance.

The bistatic cross section is then given by

$$\begin{aligned}\sigma_{bi}(o, i) &= \frac{4\pi|f(\hat{o}, \hat{\imath})|^2}{|E_i|^2}\\ &= \frac{k^2\eta_0^2}{4\pi}\left|\int_S [-\hat{o} \times (\hat{o} \times \hat{\jmath})]\frac{J_s(\bar{r}')}{|E_i|}e^{jk\hat{o}\cdot\bar{r}'}\, da\right|^2.\end{aligned} \tag{10-98}$$

If the exact surface current $\bar{J}_s(\bar{r}')$ is known, the formulas above give the exact scattering characteristics.

Let us next consider the radar cross section σ_b. As is usually the case, the antenna receives the component of the scattered wave along the direction of the polarization of the incident wave $\hat{e}_i$. Thus, in this case, we use

$$\begin{aligned}\sigma_b &= \lim_{R\to\infty}\frac{4\pi R^2|\hat{e}_i\cdot\bar{E}_s|^2}{|E_i|^2}\\ &= \frac{4\pi|\hat{e}_i\cdot\hat{e}_s f(-\hat{\imath}, \hat{\imath})|^2}{|E_i|^2}\\ &= \frac{k^2\eta_0^2}{4\pi}\left|\int_S \frac{\hat{e}_i\cdot\bar{J}_s(\bar{r}')}{|E_i|}e^{-jk\hat{\imath}\cdot\bar{r}'}\, da\right|^2\\ &= \frac{k^2\eta_0^2}{4\pi}\frac{\left|\int_S \bar{E}_i\cdot\bar{J}_s(\bar{r})\, da\right|^2}{|E_i|^4}.\end{aligned} \tag{10-99}$$

The surface current $\bar{J}_s(\bar{r}')$ is still unknown in the formulation above. In the next section, a useful approximation called physical optics is discussed.

10-13 PHYSICAL OPTICS APPROXIMATION

If the object is large compared with a wavelength and the surface is smooth (radius of curvature is much greater than a wavelength), the surface current $\bar{J}_s(\bar{r})$ may be well approximated by the current that would exist if the surface were a conducting plane tangential to the surface at the point $\bar{r}$. Thus the surface is regarded to be locally planar. In this case, in the illuminated region $\bar{J}_s$ is twice the tangential component of the incident magnetic field:

$$\bar{J}_s(\bar{r}) = \begin{cases} 2(\hat{n} \times \bar{H}_i) & \text{in the illuminated region} \\ 0 & \text{in the shadow,} \end{cases} \tag{10-100}$$

where $\hat{n}$ is a unit vector normal to the surface (Fig. 10-21). Using this approximation and noting

$$\bar{E}_i \cdot \bar{J}_s = 2\bar{E}_i \cdot (\hat{n} \times \bar{H}_i) = 2\hat{n} \cdot \bar{E}_i \times \bar{H}_i,$$

we get

$$\sigma_b = \frac{k^2}{\pi} \left| \int_{S_1} \hat{n} \cdot \hat{i} e^{-j2k\hat{i} \cdot \bar{r}'} \, da \right|^2. \tag{10-101}$$

As an example, consider a thin rectangular conducting plate illuminated by a plane wave propagating from the direction (θ, ϕ) (Fig. 10-22). In this case, we note that

$$-\hat{i} = \sin\theta \cos\phi \, \hat{x} + \sin\theta \sin\phi \, \hat{y} + \cos\theta \, \hat{z}$$

$$\hat{n} = \hat{z}.$$

Therefore, we get

$$\sigma_b = \frac{4\pi}{\lambda^2}(A^2 \cos^2\theta) \left[\frac{\sin(2ka \sin\theta \cos\phi)}{(2ka \sin\theta \cos\phi)} \frac{\sin(2kb \sin\theta \sin\phi)}{(2kb \sin\theta \sin\phi)} \right]^2, \tag{10-102}$$

where A is the area of the plate. For a circular plate of radius a, we obtain

$$\sigma_b = \frac{4\pi}{\lambda^2}(a^2 \cos^2\theta) \left[\frac{J_1(2ka \sin\theta)}{ka \sin\theta} \right]^2, \tag{10-103}$$

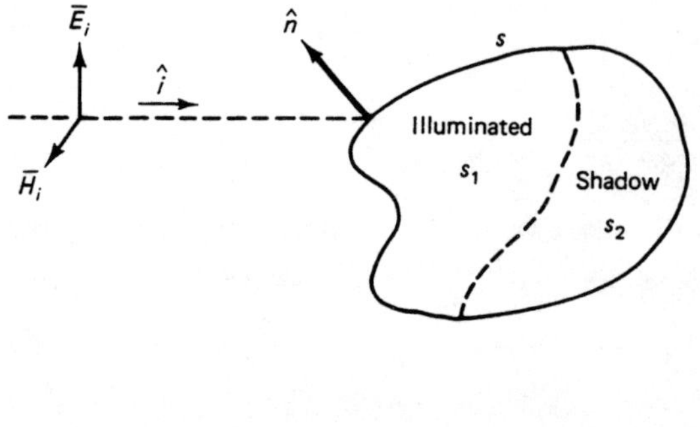

Figure 10-21 Physical optics approximation.

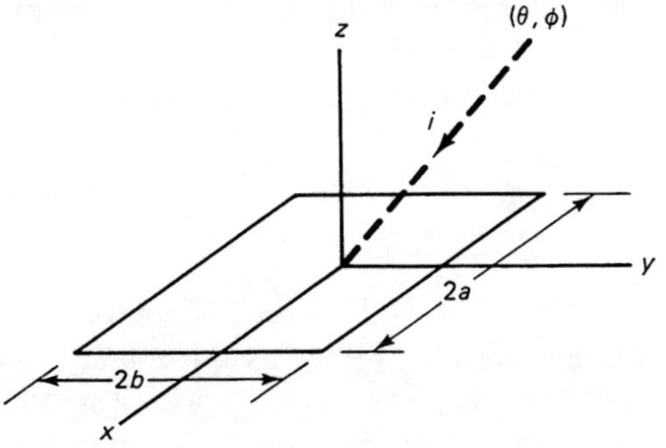

Figure 10-22 Rectangular conducting plate illuminated by a plane wave.

where we made use of the following:

$$\int_0^a r\,dr \int_0^{2\pi} d\phi\, e^{j2kr\sin\theta\sin\phi} = 2\pi \int_0^a r\,dr\, J_0(2kr\sin\theta),$$

and $\int x\,dx\,J_0(x) = xJ_1(x)$. Next consider a wave incident on an infinite cone along its axis (Fig. 10-23). We note that $-\hat{n}\cdot\hat{\imath} = \sin\theta_0$, and thus we get

$$\sigma_b = \frac{4\pi}{\lambda^2}\left|\int_0^{2\pi} d\phi \int_0^{\infty} r\,dr\,\sin^2\theta_0\, e^{j2kr\cos\theta_0}\right|^2$$

$$= \frac{\pi}{4k^2}\tan^4\theta_0. \qquad (10\text{-}104)$$

Physical optics approximations give convenient and simple expressions for the scattering cross section and therefore are widely used. The validity of this approximation is limited to a large conducting object with a smooth surface. Unlike the geometric optical approximation to be discussed later, the physical optics approximation contains wavelength dependence and the results are often in good agree-

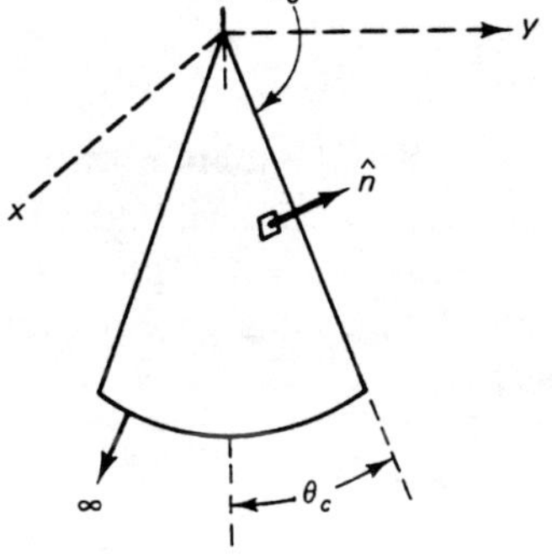

Figure 10-23 Infinite conducting cone.

ment with experimental data, even though it is difficult to establish exactly how valid physical optics is for a general case. Physical optics is equivalent to the Kirchhoff approximation used in the aperture problems discussed in Sections 6-3 and 6-11.

10-14 MOMENT METHOD: COMPUTER APPLICATIONS

The physical optics approximation described in Section 10-13 gives a convenient expression for the surface current, and thus it is widely used for many applications. However, the approximation is valid only for a large object with a smooth surface. For a small object, we need a different technique which can give a reasonable approximation for the surface current. In this section we start with a general formulation for the integral equation for the surface current and apply the moment method to solve for the surface current.

Let us first note that the total field $\bar{E}(\bar{r})$ consists of the incident field $\bar{E}_i(\bar{r})$ and the field $\bar{E}_s$ scattered by the object.

$$\bar{E}(r) = \bar{E}_i(r) + \bar{E}_s(r), \tag{10-105}$$

where the scattered field $\bar{E}_s(r)$ is given by

$$\begin{aligned} \bar{E}_s(r) &= \nabla \times \nabla \times \bar{\pi}(r) \\ &= \frac{1}{j\omega\epsilon_0} \nabla \times \nabla \times \int_S G_0(\bar{r}, \bar{r}')\bar{J}_s(r')\, da, \end{aligned} \tag{10-106}$$

and $G_0(\bar{r}, \bar{r}')$ is a free-space Green's function. Now the boundary condition requires that the tangential component of $\bar{E}(r)$ vanish on the surface of the conductor. Therefore, we write

$$\bar{E}(r)|_{\tan} = \bar{E}_i(r)|_{\tan} + \bar{E}_s(r)|_{\tan} = 0, \tag{10-107}$$

on the surface S.

We write (10-107) in the following form:

$$L(\bar{J}_s) = \bar{E}_i|_{\tan}, \tag{10-108}$$

where

$$\begin{aligned} \bar{E}_s(r)|_{\tan} &= -L(\bar{J}_s) \\ &= \left[\frac{1}{j\omega\epsilon_0} \nabla \times \nabla \times \int_S G_0(r, r')\bar{J}_s(r')\, da \right]_{\tan}. \end{aligned}$$

Equation (10-108) is an integrodifferential equation for the surface current $\bar{J}_s$. We note that in this equation, $\bar{L}$ and $\bar{E}_i|_{\tan}$ are known and $\bar{J}_s$ is unknown. Let us solve (10-108) by means of the *moment method* (see Chapter 18).

We expand the unknown current $\bar{J}_s(r)$ in a series of given current distributions $\bar{J}_n(r)$ with unknown coefficients I_n:

$$\bar{J}_s(r) = \sum_n I_n \bar{J}_n(r), \tag{10-109}$$

where $\bar{J}_n(r)$ is called the *basis function*. Substituting (10-109) into (10-108), we get

$$\sum_n I_n L(\bar{J}_n(r)) = \bar{E}_i|_{\tan}. \tag{10-110}$$

Now we choose a set of testing functions $\overline{W}_1, \overline{W}_2, \ldots$, which are tangential vectors on S. We form the inner product of both sides of (10-110) with $\overline{W}_m$:

$$\sum_n I_n \langle \overline{W}_m, L(\bar{J}_n) \rangle = \langle \overline{W}_m, \bar{E}_i \rangle, \tag{10-111}$$

where the inner product is defined by

$$\langle \overline{W}_m, \bar{E} \rangle = \int_S \overline{W}_m \cdot \bar{E}\, da. \tag{10-112}$$

Now defining matrices

$$I = (I_n) = \begin{bmatrix} I_1 \\ I_2 \\ \vdots \end{bmatrix}, \qquad V = (V_m) = \begin{bmatrix} \langle \overline{W}_1, \bar{E}_i \rangle \\ \langle \overline{W}_2, \bar{E}_i \rangle \\ \vdots \end{bmatrix}$$
$$Z = (Z_{mn}) = \begin{bmatrix} \langle W_1, L(J_1) \rangle & \langle W_1, L(J_2) \rangle & \cdots \\ \langle W_2, L(J_1) \rangle & \langle W_2, L(J_2) \rangle & \cdots \end{bmatrix}, \tag{10-113}$$

we can write (10-111) in the following matrix form:

$$ZI = V. \tag{10-114}$$

We note here that once we choose the basis function $\bar{J}_n(r)$ and the testing functions $\overline{W}_m(r)$, the matrices Z and V are known and I is unknown. Thus, inverting the matrix Z, we obtain the solution

$$I = Z^{-1} V. \tag{10-115}$$

The surface current $\bar{J}(r)$ is then given by

$$\bar{J}(r) = \sum_n I_n \bar{J}_n(r). \tag{10-116}$$

As an example, let us consider the scattering by a thin wire of length l and diameter $2a$ ($l >> a$) (Fig. 10-24). Noting that $\bar{J}_s(r')$ has only the z component, we write

$$\bar{J}_s(r')\, da = I_z(z')\, dz'\, \hat{z}. \tag{10-117}$$

Here we assume that the current $I_z(z')$ is a line current on the axis of the wire but that the boundary condition is to be satisfied on the surface of the wire (Fig. 10-25). This convenient approximation avoids the singularity of Green's function at $\bar{r} = \bar{r}'$.

Using (10-117), the integral equation (10-111) becomes

$$-\frac{1}{j\omega\epsilon_0} \int_{-l/2}^{l/2} \left(\frac{\partial^2}{\partial z^2} + k^2 \right) G(r, r') I_z(z')\, dz' = E_{iz}(z'). \tag{10-118}$$

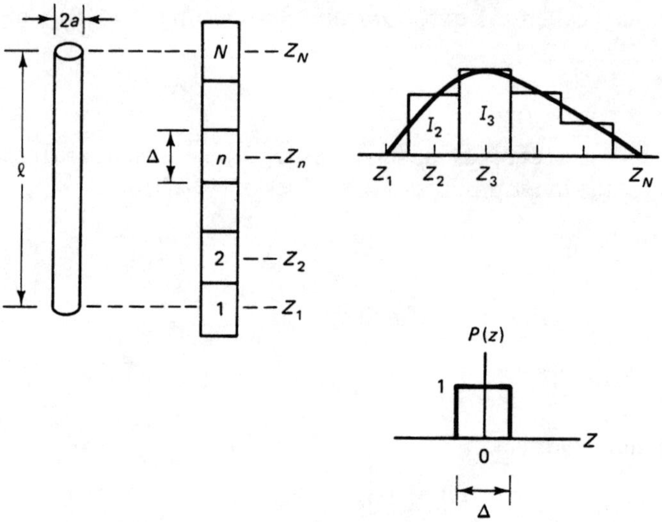

Figure 10-24 Moment method applied to scattering by a thin wire.

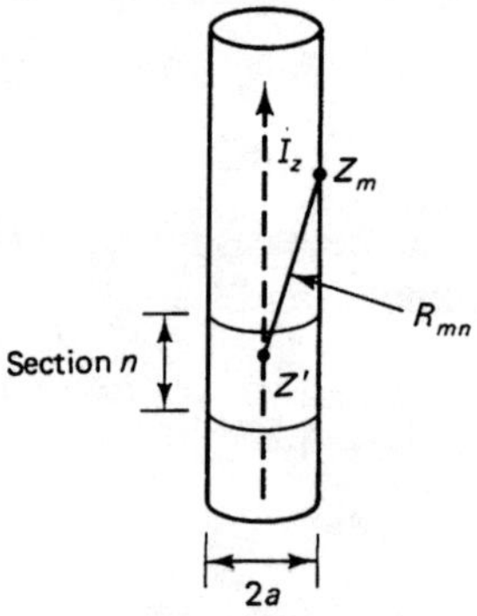

Figure 10-25 The current $I_z(z')$ is on the axis and the boundary condition is satisfied on the surface at z_m.

This equation is often called the *Pocklington equation* (Mittra, 1973). For a straight wire, noting $R = |\bar{r} - \bar{r}'| = [(z - z')^2 + a^2]^{1/2}$, we can also write (10-118)

$$\int_{-l/2}^{l/2} K(z, z')I_z(z')\,dz' = E_{iz}(z'), \tag{10-119}$$

where

$$K(z, z') = -\frac{e^{-jkR}}{j4\pi\omega\epsilon_0}[(1 + jkR)(2R^2 - 3a^2) + k^2a^2R^2]R^{-5}.$$

We now divide the length l into $N - 1$ sections and choose the basis function $\bar{J}_n(r)$ to be a rectangular current $P(z - z_n)$ in the nth section of the wire. Thus $I(z)$ is approximated by a series of steps with the value at the midpoint in each section (Fig. 10-24). Note that section 1 and section N are extended beyond the wire by $\Delta/2$. We can then let $I_1 = I_N = 0$, which is the condition required at the end of the wire. We write

$$I(z) = \sum_{n=1}^{N} I_n P(z - z_n), \tag{10-120}$$

where $P(z - z_n)$ is the basis function. We now choose $\delta(z - z_m)$ as the testing function:

$$W_m = \delta(z - z_m). \tag{10-121}$$

Thus the inner product (10-112) gives the values of (10-119) at those discrete points z_m. Then (10-119) becomes

$$\sum_{n=1}^{N} Z_{mn} I_n = V_m, \tag{10-122}$$

where

$$V_m = E_{iz}(z_m)\Delta,$$

$$Z_{mn} = \Delta \int_{z_n - \Delta/2}^{z_n + \Delta/2} K(z, z')\, dz'.$$

Once Z_{mn} is evaluated, the solution is easily obtained by solving the matrix equation (10-114) by inversion of Z and the current distribution is given by (10-116). Note that $I_1 = I_N = 0$, and therefore $[Z]$ is a $(N-2) \times (N-2)$ matrix.

The backscattered cross section σ_b is then given by (10-99). In matrix form, we get

$$\sigma_b = \frac{k^2 \eta_0^2}{4\pi} \left| \tilde{V} I \right|^2 = \frac{k^2 \eta_0^2}{4\pi} \left| \tilde{V} Z^{-1} V \right|^2, \tag{10-123}$$

where the incident electric field $\bar{E}_i$ is normalized so that $|\bar{E}_i| = 1$.

PROBLEMS

10-1. The surface of the moon is rough and the radar cross section is about 4×10^{-4} of its geometric cross section. Suppose that the moon is illuminated by a radar transmitter with a diameter of 142 ft and an aperture efficiency of 60%. The peak power is 130 kW and the frequency is 400 MHz. Calculate the power received.

10-2. At $\lambda = 5$ cm, the refractive index of water at 20°C is $8.670 - j1.202$. The median diameter (in mm) of raindrops is given by

$$D_m = 1.238 p^{0.182},$$

where p (in mm/h) is the precipitation rate and the terminal velocity (in m/s) is given by

$$v = 200.8 a^{1/2},$$

where a (in m) is the radius of the droplet ($a = D_m/2$). Assuming that the Rayleigh formula is applicable, calculate the scattering and the absorption cross sections of a rain droplet. Also find the number of droplets per m^3. Assume that $p = 12.5$ mm/h. Find the attenuation of the wave in dB/km.

10-3. A lossy prolate ellipsoidal particle ($a = b < c$) is placed at the origin as shown in Fig. 10-11, and $\theta_b = 0$. Find the polarizability in the x, y, and z directions when $\epsilon_r = 1 - j15$, $a = b = 10$ μm, $c = 1$ mm, and $\lambda = 3$ cm. Find the scattering amplitude matrix and the scattering and absorption cross sections when $\theta' = \phi' = 0$ and when $\theta' = \pi/2$ and $\phi' = 0$.

10-4. Using the Rayleigh–Debye approximation, calculate the backscattering cross section of a spherical object of radius a_1 with a spherical core of radius a_2 at $\lambda_1 = 0.6$ μm (Fig. P10-4).

$$n_1 = 1.01, \qquad a_1 = 2\ \mu\text{m},$$
$$n_2 = 1.02, \qquad a_2 = 0.5\ \mu\text{m},$$
$$d = 1\ \mu\text{m}.$$

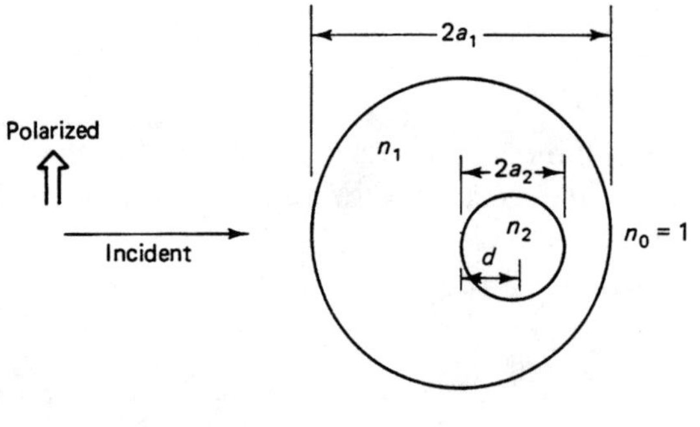

Figure P10-4 Rayleigh–Debye scattering.

10-5. If the Stokes parameters of a wave are

$$I = 3, \qquad U = 2,$$
$$Q = 1, \qquad V = -2,$$

find E_x and E_y and draw the locus similar to Fig. 10-17.

10-6. Consider a wave whose components are given by

$$E_x = 2\cos\left(\omega t + \frac{\pi}{8}\right),$$
$$E_y = 3\cos\left(\omega t + \frac{3\pi}{2}\right).$$

Find the Stokes parameters and Poincaré representation of this wave. Show its location on the Poincaré sphere.

10-7. Calculate the acoustic scattering cross section of a red blood cell in plasma at 5 MHz. Assume that it is a sphere of radius 2.75 μm and that $\kappa_l = 34.1 \times 10^{-12}$ cm^2/dyne and $\rho_l = 1.092$ g/cm^3. The plasma surrounding the blood cell has $\kappa = 40.9 \times 10^{-12}$ cm^2/dyn and $\rho = 1.021$ g/cm^3.

10-8. Using the physical optics approximation, find the backscattering cross section of the finite conducting cylinder shown in Fig. P10-8.

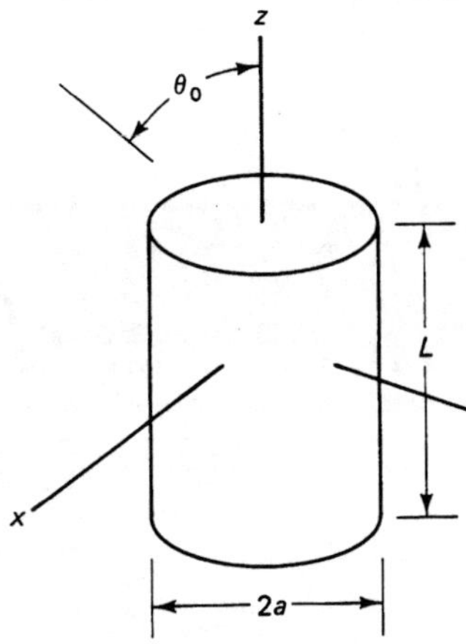

Figure P10-8 Finite conducting cylinder.

10-9. Consider a short wire of length L illuminated by an incident wave polarized in a plane containing the dipole. Find the backscattering cross section of this short wire.

10-10. Calculate the backscattering cross section σ_b/λ^2 of a wire with radius a and length l when a wave polarized in the direction of the wire is incident at broadside. $ka = 0.0314$ and $\frac{1}{2}kl = 1.5$.

11

Waves in Cylindrical Structures, Spheres, and Wedges

Many bodies of practical interest, such as biological media, rockets, and portions of aircraft, may be closely approximated by cylindrical structures, spheres, and wedges. These shapes are well defined and coincide with one or more of the coordinates for which wave equations are separable. The exact solutions can be obtained in closed form for most problems. These bodies may have radiators such as slot antennas and dipole antennas on/or close to their surface, and their radiation characteristics are greatly affected by the geometry of the body. Also, the scattering and absorption characteristics of these bodies, when illuminated from outside, are important in many practical problems, such as radar cross-section studies and microwave hazards. In this chapter we investigate the scattering of waves from these objects and the effects of these structures on the radiation characteristics of antennas.

In analyzing various problems in this chapter, we present a number of powerful analytical techniques, including the Fourier transform, saddle-point technique, Watson transform, residue series representation, and geometric optical solutions. These techniques are useful not only for these problems, but are also important mathematical tools applicable to a large number of other problems. [See Bowman et al. (1969) for detailed treatments of these topics.]

11-1 PLANE WAVE INCIDENT ON A CONDUCTING CYLINDER

The determination of a radar cross section is one of the most important and practical problems. It provides information about the object, and it is useful for the design of a vehicle with a specified radar cross section. Since many objects and vehicles are composed of cylindrical layers, it is important to devise a systematic technique to find scattered waves from such an object when a plane wave is incident on it. Let us consider first an incident plane wave propagating in the direction (θ_0, ϕ_0). We may consider two cases: TM and TE. The TM wave is polarized in a plane parallel to the z axis and therefore $H_z = 0$. The TE wave is polarized perpendicular to the z axis and thus $E_z = 0$. We let E_0 be the magnitude of the incident electric field (Fig. 11-1).

Let us consider the TM incident wave. Note that the TM wave has the following electric field components in the cylindrical coordinates (ρ, ϕ, z).

$$\begin{aligned} E_{zi} &= E_0 \sin\theta_0 e^{-j\bar{k}\cdot\bar{r}}, \\ E_{\rho i} &= -E_0 \cos\theta_0 \cos(\phi - \phi_0) e^{-j\bar{k}\cdot\bar{r}}, \\ E_{\phi i} &= E_0 \cos\theta_0 \sin(\phi - \phi_0) e^{-j\bar{k}\cdot\bar{r}}. \end{aligned} \tag{11-1}$$

where

$$\begin{aligned} \bar{k} &= k(\sin\theta_0 \cos\phi_0 \hat{x} + \sin\theta_0 \sin\phi_0 \hat{y} + \cos\theta_0 \hat{z}), \\ \bar{r} &= x\hat{x} + y\hat{y} + z\hat{z} \\ &= \rho\cos\phi\hat{x} + \rho\sin\phi\hat{y} + z\hat{z}, \\ \bar{k}\cdot\bar{r} &= kz\cos\theta_0 + k\rho\sin\theta_0\cos(\phi - \phi_0). \end{aligned}$$

Now the TM wave is generated by

$$\Pi_{zi} = A_i e^{-j\bar{k}\cdot\bar{r}} = A_i e^{-jkz\cos\theta_0 - jk\rho\sin\theta_0\cos(\phi - \phi_0)}, \tag{11-2}$$

and E_z and Π_z are related by

$$\begin{aligned} E_{zi} &= \left(\frac{\partial^2}{\partial z^2} + k^2\right)\Pi_{zi} \\ &= A_i k^2 \sin^2\theta_0 e^{-j\bar{k}\cdot\bar{r}}. \end{aligned} \tag{11-3}$$

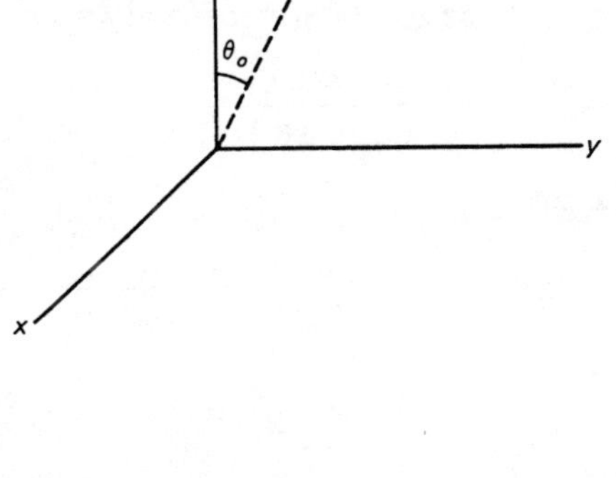

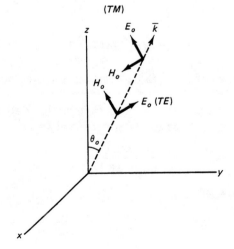

Figure 11-1 Plane wave in the direction (θ_0, ϕ_0): TE, $\bar{E} = E_0\hat{\phi}$ and $\bar{H} = -H_0\hat{\theta}$; TM, $\bar{E} = -E_0\hat{\theta}$ and $\bar{H} = -H_0\hat{\phi}$.

Comparing (11-3) with (11-1), we get

$$A_i = \frac{E_0}{k^2 \sin\theta_0}. \tag{11-4}$$

We will observe shortly that the incident wave (11-2) is not convenient to satisfy the boundary condition in order to determine the scattered wave, and it is necessary to expand the incident wave in a Fourier series in ϕ. Let us expand (11-2) in a Fourier series.

$$\Pi_{zi} = \sum_{n=-\infty}^{\infty} a_n(z, \rho) e^{-jn(\phi - \phi_0)}. \tag{11-5}$$

The coefficient a_n is obtained by

$$a_n = \frac{1}{2\pi}\int_0^{2\pi} \Pi_{zi} e^{jn(\phi - \phi_0)}\, d(\phi - \phi_0). \tag{11-6}$$

Now we make use of the following integral representation of the Bessel function:

$$\begin{aligned} J_n(Z) &= \frac{1}{2\pi}\int_0^{2\pi} e^{jZ\cos\phi + jn(\phi - \pi/2)}\, d\phi, \\ J_{-n}(Z) &= (-1)^n J_n(Z) = J_n(-Z). \end{aligned} \tag{11-7}$$

We then get

$$a_n = A_i e^{-jkz\cos\theta_0} J_n(k\rho \sin\theta_0) e^{-jn(\pi/2)}. \tag{11-8}$$

Thus Π_{zi} in (11-2) is expressed in the following form:

$$\Pi_{zi} = \sum_{n=-\infty}^{\infty} A_i e^{-jkz\cos\theta_0} J_n(k\rho \sin\theta_0) e^{-jn(\phi - \phi_0 + \pi/2)}. \tag{11-9}$$

Let us consider the scattering by a conducting cylinder of radius a when the TM wave (11-9) is incident. We write the scattered wave in terms of the Hertz potential Π_{zs}. Considering that they should satisfy the wave equation and the radiation condition, we write

$$\Pi_{zs} = \sum_{n=-\infty}^{\infty} A_{ns} e^{-jkz\cos\theta_0} H_n^{(2)}(k\rho \sin\theta_0) e^{-jn[\phi - \phi_0 + (\pi/2)]}, \tag{11-10}$$

where A_{ns} are the unknown coefficients to be determined by the boundary condition.

The boundary conditions are that E_z and E_ϕ must be zero at $\rho = a$. This requires that the total $\Pi_z = \Pi_{zi} + \Pi_{zs}$ be zero at $\rho = a$. Thus we get

$$A_{ns} = -\frac{J_n(ka \sin\theta_0)}{H_n^{(2)}(ka \sin\theta_0)} A_i. \tag{11-11}$$

The total Hertz potential Π_z is given by

$$\Pi_z = \sum_{n=-\infty}^{\infty} A_i e^{-jkz\cos\theta_0}\left[J_n(k\rho \sin\theta_0) - \frac{J_n(ka \sin\theta_0) H_n^{(2)}(k\rho \sin\theta_0)}{H_n^{(2)}(ka \sin\theta_0)}\right] e^{-jn[\phi - \phi_0 + (\pi/2)]}. \tag{11-12}$$

The field components including both TM(Π_z) and TE(Π_{mz}) waves are then given by

$$E_z = \left(\frac{\partial^2}{\partial z^2} + k^2\right)\Pi_z, \qquad H_z = \left(\frac{\partial^2}{\partial z^2} + k^2\right)\Pi_{mz},$$

$$E_\rho = \frac{\partial^2}{\partial\rho\,\partial z}\Pi_z - j\omega\mu\frac{1}{\rho}\frac{\partial}{\partial\phi}\Pi_{mz},$$

$$E_\phi = \frac{1}{\rho}\frac{\partial^2}{\partial\phi\,\partial z}\Pi_z + j\omega\mu\frac{\partial}{\partial\rho}\Pi_{mz}, \tag{11-13}$$

$$H_\rho = j\omega\epsilon\frac{1}{\rho}\frac{\partial}{\partial\phi}\Pi_z + \frac{\partial^2}{\partial\rho\,\partial z}\Pi_{mz},$$

$$H_\phi = -j\omega\epsilon\frac{\partial}{\partial\rho}\Pi_z + \frac{1}{\rho}\frac{\partial^2}{\partial\phi\,\partial z}\Pi_{mz}.$$

For example, the current density on the conducting cylinder $\bar{J}$ is given by

$$\begin{aligned} \bar{J} &= J_\phi\hat{\phi} + J_z\hat{z} \\ &= -H_z\hat{\phi} + H_\phi\hat{z} \qquad \text{at } \rho = a. \end{aligned} \tag{11-14}$$

Let us next consider the TE wave incident on the cylinder. We use the magnetic Hertz potential Π_{mz} and write the incident wave as

$$\Pi_{mzi} = B_i e^{-j\bar{k}\cdot\bar{r}} = B_i e^{-jkz\cos\theta_0 - jk\rho\sin\theta_0\cos(\phi-\phi_0)}, \tag{11-15}$$

where $H_0 = E_0/\eta = k_0^2 \sin\theta_0 B_i$, $\eta = (\mu/\epsilon)^{1/2}$. This can be expressed in a Fourier series

$$\Pi_{mzi} = \sum_{n=-\infty}^{\infty} B_i e^{-jkz\cos\theta_0} J_n(k\rho\sin\theta_0) e^{-jn(\phi-\phi_0+\pi/2)}. \tag{11-16}$$

We write the scattered wave in the following form with the unknown coefficients B_{ns}:

$$\Pi_{mzs} = \sum_{n=-\infty}^{\infty} B_{ns} e^{-jkz\cos\theta_0} H_n^{(2)}(k\rho\sin\theta_0) e^{-jn(\phi-\phi_0+\pi/2)}. \tag{11-17}$$

The boundary condition at $\rho = a$ is that $(\partial/\partial\rho)\Pi_{mz} = 0$, which determines B_{ns}:

$$B_{ns} = -\frac{J_n'(ka\sin\theta_0)}{H_n^{(2)\prime}(ka\sin\theta_0)} B_i. \tag{11-18}$$

The final solution is then given by

$$\Pi_{mz} = \sum_{n=-\infty}^{\infty} B_i e^{-jkz\cos\theta_0}\left[J_n(k\rho\sin\theta_0) - \frac{J_n'(ka\sin\theta_0) H_n^{(2)}(k\rho\sin\theta_0)}{H_n^{(2)\prime}(ka\sin\theta_0)}\right] e^{-jn(\phi-\phi_0+\pi/2)}. \tag{11-19}$$

Equations (11-12) and (11-19) are the total fields due to the incident TM and TE waves, respectively. All the electric and magnetic field components are then easily obtained from them. Note that the incident TM wave A_i produces the scattered TM wave only, and the incident TE wave B_i produces the TE wave only, and that there is no coupling between the TM and TE modes. This is true only for certain special cases such as a conducting cylinder or normal incidence on a dielectric cylinder. In general, as can be seen in the next section, the incident TM wave can produce both TM and TE scattered waves.

11-2 PLANE WAVE INCIDENT ON A DIELECTRIC CYLINDER

Let us consider the scattering of a plane wave by a dielectric cylinder of radius a and a relative dielectric constant ϵ_r (Fig. 11-2). We let k and $k_1 = k\sqrt{\epsilon_r}$ be the wave number outside and inside the cylinder, respectively. In general, the plane incident wave consists of TM and TE waves. The Hertz potentials for the incident TM and TE waves are given in (11-9) and (11-16):

$$\begin{aligned}\Pi_{zi} &= \sum_{n=-\infty}^{\infty} A_i e^{-jkz\cos\theta_0} J_n(k\rho\sin\theta_0) e^{-jn(\phi-\phi_0+\pi/2)},\\ \Pi_{mzi} &= \sum_{n=-\infty}^{\infty} B_i e^{-jkz\cos\theta_0} J_n(k\rho\sin\theta_0) e^{-jn(\phi-\phi_0+\pi/2)}.\end{aligned} \tag{11-20}$$

We express the scattered field outside the cylinder using the unknown coefficients A_{sn} and B_{sn}:

$$\begin{aligned}\Pi_{zs} &= \sum_{n=-\infty}^{\infty} A_{sn} e^{-jkz\cos\theta_0} H_n^{(2)}(k\rho\sin\theta_0) e^{-jn(\phi-\phi_0+\pi/2)},\\ \Pi_{mzs} &= \sum_{n=-\infty}^{\infty} B_{sn} e^{-jkz\cos\theta_0} H_n^{(2)}(k\rho\sin\theta_0) e^{-jn(\phi-\phi_0+\pi/2)}.\end{aligned} \tag{11-21}$$

Here we need both TM and TE modes as they are generally coupled.

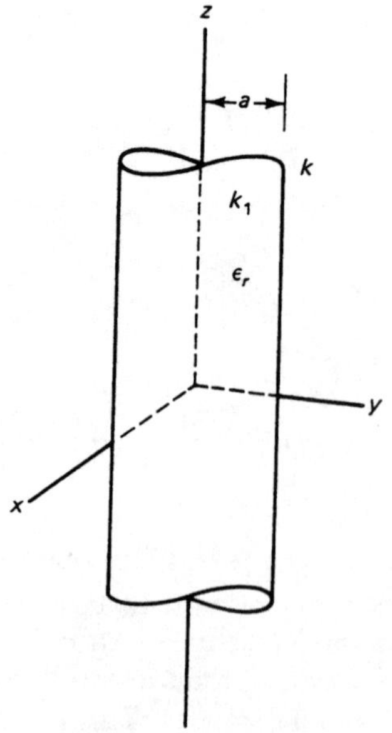

Figure 11-2 Dielectric cylinder.

Inside the dielectric cylinder, we write using the unknown coefficients A_{en} and B_{en},

$$\Pi_{ze} = \sum_{n=-\infty}^{\infty} A_{en}\, e^{-jkz\cos\theta_0} J_n(k_1\rho\sin\theta_1) e^{-jn(\phi-\phi_0+\pi/2)},$$
$$\Pi_{mze} = \sum_{n=-\infty}^{\infty} B_{en}\, e^{-jkz\cos\theta_0} J_n(k_1\rho\sin\theta_1) e^{-jn(\phi-\phi_0+\pi/2)}. \tag{11-22}$$

Here the z dependence $\exp(-jkz\cos\theta_0)$ is the same as that of the incident field because the boundary conditions must be satisfied at all z. This requires that

$$k\cos\theta_0 = k_1\cos\theta_1,$$

and

$$k_1\sin\theta_1 = (k_1^2 - k_1^2\cos^2\theta_1)^{1/2} = (k_1^2 - k^2\cos^2\theta_0)^{1/2}. \tag{11-23}$$

The boundary conditions are that E_z, E_ϕ, H_z, and H_ϕ, are continuous at $\rho = a$. Since $E_z = \left(\frac{\partial^2}{\partial z^2} + k^2\right)\Pi_z$, the continuity of E_z and the continuity of H_z give the following:

$$(k^2\sin^2\theta_0)[A_i J_n(ka\sin\theta_0) + A_{sn} H_n^{(2)}(ka\sin\theta_0)] = k_1^2\sin^2\theta_1 A_{en} J_n(k_1 a\sin\theta_1),$$
$$(k^2\sin^2\theta_0)[B_i J_n(ka\sin\theta_0) + B_{sn} H_n^{(2)}(ka\sin\theta_0)] = k_1^2\sin^2\theta_1 B_{en} J_n(k_1 a\sin\theta_1). \tag{11-24}$$

Next consider the continuity of E_ϕ. Noting that

$$E_\phi = \frac{1}{\rho}\frac{\partial^2}{\partial\phi\,\partial z}\Pi_z + j\omega\mu\frac{\partial}{\partial\rho}\Pi_{mz}, \tag{11-25}$$

we get

$$-\frac{kn\cos\theta_0}{a}[A_i J_n(ka\sin\theta_0) + A_{sn} H_n^{(2)}(ka\sin\theta_0)]$$
$$+ j\omega\mu k\sin\theta_0[B_i J_n'(ka\sin\theta_0) + B_{sn} H_n^{(2)\prime}(ka\sin\theta_0)] \tag{11-26}$$
$$= -\frac{kn\cos\theta_0}{a}[A_{en} J_n(k_1 a\sin\theta_1)] + j\omega\mu k_1\sin\theta_1 B_{en} J_n'(k_1 a\sin\theta_1).$$

Similarly noting that

$$H_\phi = -j\omega\epsilon\frac{\partial}{\partial\rho}\Pi_z + \frac{1}{\rho}\frac{\partial^2}{\partial\phi\,\partial z}\Pi_{mz},$$

we get

$$-j\omega\epsilon k\sin\theta_0[A_i J_n'(ka\sin\theta_0)$$
$$+ A_{sn} H_n^{(2)\prime}(ka\sin\theta_0)] - \frac{kn\cos\theta_0}{a}[B_i J_n(ka\sin\theta_0) + B_{sn} H_n^{(2)}(ka\sin\theta_0)]$$
$$= -j\omega\epsilon_1 k_1\sin\theta_1 A_{en} J_n'(k_1 a\sin\theta_1) - \frac{kn\cos\theta_0}{a} B_{en} J_n(k_1 a\sin\theta_1). \tag{11-27}$$

The equations (11-24), (11-25), (11-26), and (11-27) can be solved for the four unknown coefficients A_{sn}, B_{sn}, A_{en}, and B_{en}. They are then substituted into (11-21) and (11-22) to obtain the final solution for Π and Π_m and the fields are then given by (11-13) with the appropriate k and ϵ in each region.

A similar procedure can be used for a layered dielectric cylinder. Let ϵ_m and k_m be the relative dielectric constant and the wave number of the cylindrical layer with radius from a_{m-1} to a_m. Then the field $(E_z, H_z, E_\phi, H_\phi)$ at a_{m-1} is related to the field at a_m by a 4×4 matrix. We can then apply the boundary condition and obtain the complete solution.

11-3 AXIAL DIPOLE NEAR A CONDUCTING CYLINDER

Dipole and loop antennas are often used in the vicinity of cylindrical structures. In this section we investigate the radiation from an axial dipole located near the conducting cylinder (Fig. 11-3). We first consider the field produced by a dipole in free space. We take an axial electric dipole located at r', (ρ', ϕ', z'). The rectangular components of the Hertz potential satisfy the scalar wave equation:

$$(\nabla^2 + k^2)\Pi_z = -\frac{J_z}{j\omega\epsilon}. \tag{11-28}$$

The axial electric dipole of current I and length L is represented by

$$J_z = IL\,\delta(r - r') \tag{11-29}$$

and thus

$$\Pi_z = \frac{IL}{j\omega\epsilon} G(r/r'), \tag{11-30}$$

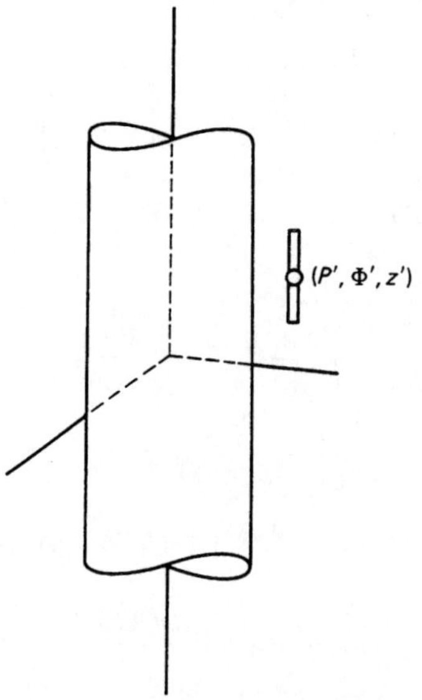

Figure 11-3 Axial dipole near a cylinder.

where Green's function $G(r/r')$ in cylindrical coordinates satisfies

$$\left(\frac{\partial^2}{\partial\rho^2}+\frac{1}{\rho^2}\frac{\partial}{\partial\rho}+\frac{1}{\rho^2}\frac{\partial^2}{\partial\phi^2}+\frac{\partial^2}{\partial z^2}+k^2\right)G(r/r')=-\frac{\delta(\rho-\rho')\delta(\phi-\phi')\delta(z-z')}{\rho}. \tag{11-31}$$

We write the free-space Green's function $G(r/r')$ in a Fourier integral in the z direction and in Fourier series in the ϕ direction (see Section 5-8):

$$G(r/r')=\frac{1}{2\pi}\int_c\sum_{n=-\infty}^{\infty}G_n(h,\rho,\rho')e^{-jn(\phi-\phi')-jh(z-z')}\,dh, \tag{11-32}$$

where

$$G_n(h,\rho,\rho')=\begin{cases}-j\frac{1}{4}J_n(\lambda\rho)H_n^{(2)}(\lambda\rho'), & \text{for } \rho<\rho'\\ -j\frac{1}{4}J_n(\lambda\rho')H_n^{(2)}(\lambda\rho) & \text{for } \rho'<\rho,\end{cases} \tag{11-33}$$

$$\lambda^2=k^2-h^2.$$

We now consider the field outside the cylinder with radius a when excited by an axial dipole at (ρ',ϕ',z'). Then we write the scattered field $G_s(r)$ as follows:

$$G_s(r)=\frac{1}{2\pi}\int_c\sum_{n=-\infty}^{\infty}A_nH_n^{(2)}(\lambda\rho)e^{-jn(\phi-\phi')-jh(z-z')}\,dh. \tag{11-34}$$

The boundary condition is that

$$G(r/r')+G_s(r)=0 \qquad \text{at } \rho=a.$$

Therefore, we obtain

$$A_n=-j\frac{1}{4}\frac{J_n(\lambda a)H_n^{(2)}(\lambda\rho')}{H_n^{(2)}(\lambda a)}. \tag{11-35}$$

Equation (11-34) with (11-35) gives the complete expression of the scattered field due to the conducting cylinder of radius a excited by a dipole at (ρ',ϕ',z'). They are expressed as an inverse Fourier transform with the contour c of integration in the complex h plane.

Note first that the integrand contains λ, and therefore we need to consider whether we should take $\lambda=+(k^2-h^2)^{1/2}$ or $\lambda=-(k^2-h^2)^{1/2}$. This requires study of the branch points and Riemann surfaces discussed in the Appendix to Chapter 11, Section A. In general, the integrand is not single valued because of $\lambda=\pm(k^2-h^2)^{1/2}$. To assure the single value for the integrand, we draw branch cuts from the branch points $h=\pm k$ such as shown in Fig. 11-4 or 11-5. As long as the

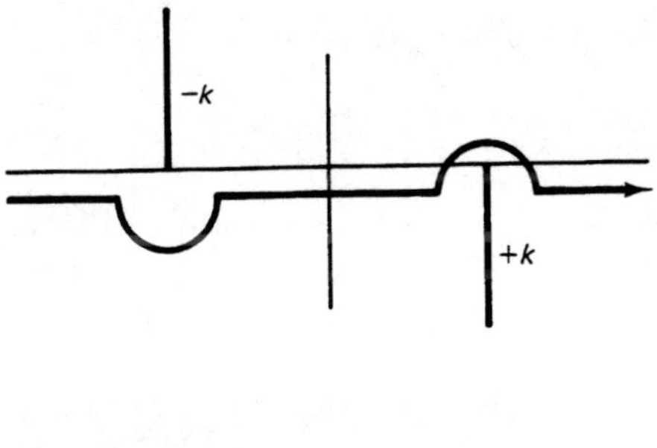

Figure 11-4 Branch cut Re $h=k$.

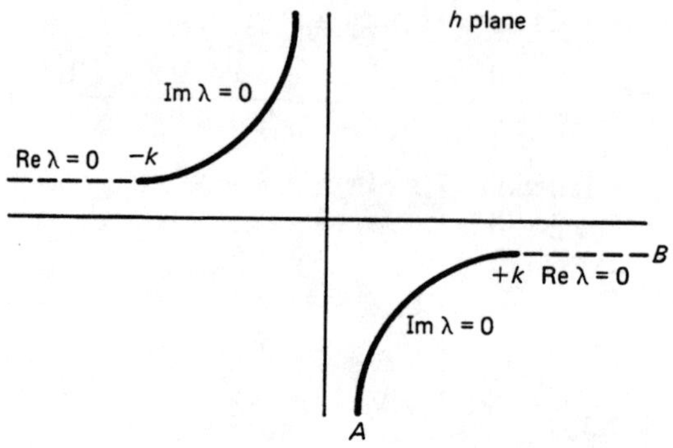

Figure 11-5 Branch cut Im $\lambda = 0$.

contour does not cross the branch cuts, the integrand is single valued (see the Appendix to Chapter 11, Section B).

11-4 RADIATION FIELD

One of the important characteristics of antennas is the radiation pattern, the behavior of the field at a large distance. To represent the radiation pattern, it is convenient to employ the spherical coordinate system and consider the components E_θ, E_ϕ and H_θ, H_ϕ. We note first that in the far field, the electric and magnetic fields are perpendicular to each other and both are transverse to the direction of propagation. Their magnitudes are related by the free-space characteristic impedance η.

$$E_\theta = \eta H_\phi \quad \text{and} \quad E_\phi = -\eta H_\theta, \qquad \eta = \left(\frac{\mu_0}{\epsilon_0}\right)^{1/2}, \tag{11-36}$$

and therefore the Poynting vector is given by

$$\begin{aligned} \bar{S} &= \tfrac{1}{2}\,\mathrm{Re}(\bar{E} \times \bar{H}^*) = S_r \hat{r}, \\ S_r &= \frac{1}{2\eta}(|E_\theta|^2 + |E_\phi|^2) = \frac{1}{2\eta}[\eta^2|H_\phi|^2 + |E_\phi|^2]. \end{aligned} \tag{11-37}$$

The last expression in (11-37) is convenient for the radiation field from cylindrical structures because E_ϕ and H_ϕ are the same in both cylindrical and spherical coordinates. E_ϕ and H_ϕ are calculated from the Hertz potential.

Let us consider the radiation field of an axial dipole near a conducting cylinder, which was discussed in detail in Section 11-3. E_ϕ and H_ϕ are then obtained from (11-30), (11-32), and (11-34). For $\rho > \rho'$ we get

$$\begin{aligned} E_\phi &= \frac{1}{\rho}\frac{\partial^2}{\partial\phi\,\partial z}\Pi_z \\ &= \sum_{n=-\infty}^{\infty} \int_c C_n(h) H_n^{(2)}(\lambda\rho) e^{-jh(z-z')}\,dh, \end{aligned} \tag{11-38}$$

$$H_\phi = -j\omega\epsilon \frac{\partial}{\partial\rho}\Pi_z$$
$$= \sum_{n=-\infty}^{\infty} D_n(h) H_n^{(2)\prime}(\lambda\rho) e^{-jh(z-z')}\, dh, \tag{11-39}$$

where

$$C_n(h) = \frac{IL}{j\omega\epsilon}\frac{jnh}{8\pi\rho}\frac{J_n(\lambda\rho')H_n^{(2)}(\lambda a) - J_n(\lambda a)H_n^{(2)}(\lambda\rho')}{H_n^{(2)}(\lambda a)} e^{-jn(\phi-\phi')},$$
$$D_n(h) = \frac{IL}{j\omega\epsilon}\frac{-j\omega\epsilon\lambda}{2\pi}\frac{J_n(\lambda\rho')H_n^{(2)}(\lambda a) - J_n(\lambda a)H_n^{(2)}(\lambda\rho')}{H_n^{(2)}(\lambda a)} e^{-jn(\phi-\phi')},$$
$$\lambda = \sqrt{k^2 - h^2}.$$

The radiation field is obtained by evaluating (11-38) and (11-39) for the large distance from the antenna. This is done by the saddle-point technique, discussed in the next section.

11-5 SADDLE-POINT TECHNIQUE

Let us consider the integral given in (11-38):

$$I_1 = \int_{-\infty}^{\infty} C_n(h) H_n^{(2)}(\sqrt{k^2-h^2}\,\rho) e^{-jhz}\, dh, \tag{11-40}$$

for a large distance R from the origin. First, we approximate the Hankel function by its asymptotic form

$$H_n^{(2)}(x) = \sqrt{\frac{2}{\pi x}} e^{-jx + j(2n+1)(\pi/4)},$$

which is valid for $|x| >> |n|$. Thus we write (11-40) as

$$I_1 = \int_{-\infty}^{\infty} A(h) e^{-j\sqrt{k^2-h^2}\,\rho - jhz}\, dh, \tag{11-41}$$

where

$$A(h) = C_n(h)\sqrt{\frac{2}{\pi\rho}} \frac{e^{j(2n+1)(\pi/4)}}{(k^2-h^2)^{1/4}}.$$

Let us express (11-41) in spherical coordinates. For convenience, we use $\theta_c = (\pi/2) - \theta$:

$$z = R\cos\theta = R\sin\theta_c,$$
$$\rho = R\sin\theta = R\cos\theta_c, \tag{11-42}$$

We also transform h into the α plane:

$$h = k\sin\alpha. \tag{11-43}$$

Then we get

$$
\begin{aligned}
I_1 &= \int_c A(k \sin\alpha) e^{-jkR\cos(\alpha-\theta_c)} k \cos\alpha \, d\alpha \\
&= \int_c F(k \sin\alpha) e^{-jkR\cos(\alpha-\theta_c)} \, d\alpha.
\end{aligned}
\tag{11-44}
$$

The contours in the h and α planes are shown in Fig. 11-6.

We now wish to evaluate this integral for an observation point far from the origin (i.e., kR large). Let us first recognize that at both ends of the contour, the integrand vanishes. In the example we are considering,

$$
\begin{aligned}
\cos(\alpha-\theta_c) &= \cos(x+jy) \\
&= \cos x \cosh y - j \sin x \sinh y,
\end{aligned}
\tag{11-45}
$$

and the absolute value becomes

$$
|e^{-jkR\cos(\alpha-\theta_c)}| = e^{-kR\sin x \sinh y},
$$

which vanishes for $0 < x < \pi$ and $y \to +\infty$ and for $-\pi < x < 0$ and $y \to -\infty$. In Fig. 11-7, the component $(-kR \sin x \sinh y)$ is shown. We note that the original contour starts from a point in the valley where the exponent is $-\infty$ and thus the magnitude is zero, along path C the magnitude increases, and finally, the magnitude decreases to zero in another valley region. Along this original contour C, both real and imaginary parts of the exponent vary. We therefore deform the path into the steepest descent contour (SDC) along which the imaginary part of the exponent is constant (Figs. 11-7 and 11-8) and evaluate the integral for large kR (see the Appendix to Chapter 11, Section C).

We have

$$
f(\alpha) = -j \cos(\alpha - \theta_c), \tag{11-46}
$$

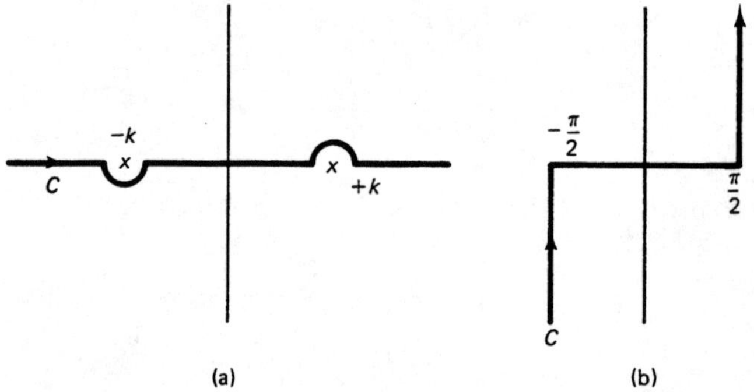

Figure 11-6 Contour C: (a) in the h plane; (b) in the α plane.

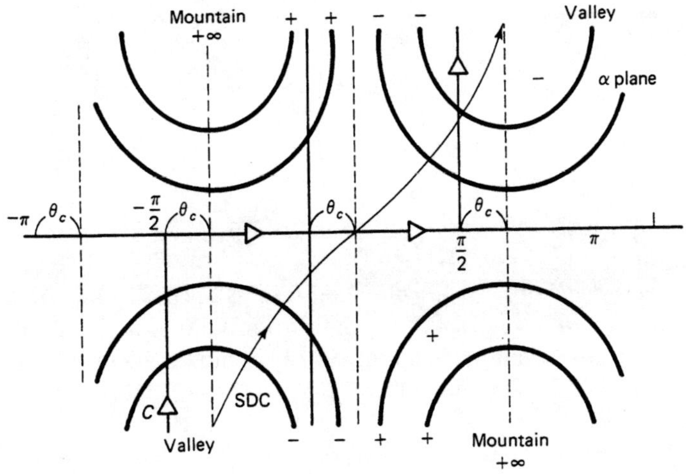

Figure 11-7 Real part of the exponent $-jkR\cos(\alpha - \theta_c)$.

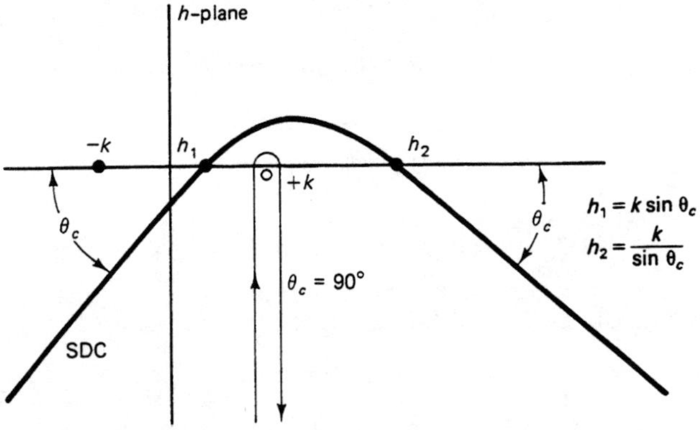

Figure 11-8 Steepest descent contour in the h plane.

and thus the saddle point is at

$$\frac{df}{d\alpha} = j\sin(\alpha - \theta_c) = 0, \tag{11-47}$$

which gives $\alpha_s = \theta_c$.

Now we let $\alpha - \alpha_s = se^{j\gamma}$, and noting that

$$f(\alpha_s) = -j,$$

$$f''(\alpha_s) = j,$$

we expand $f(\alpha)$ about $\alpha = \alpha_s$:

$$f(\alpha) = -j + j\frac{s^2 e^{j2\gamma}}{2}.$$

We must choose γ such that the second term is real and negative. Thus we require that

$$2\gamma + \frac{\pi}{2} = \pm\pi,$$

which gives $\gamma = \pi/4$ or $-3\pi/4$. $\gamma = \pi/4$ represents the path going from the third to the first quadrant and $\gamma = -3\pi/4$ is the path from the first to the third quadrant. Both are the steepest descent path, but obviously in our problem, the choice must be

$$\gamma = \frac{\pi}{4},$$

and

$$f(\alpha) = -j - \frac{s^2}{2}.$$

Using the above γ, we obtain for large kR:

$$\begin{aligned} I &= \int_c A(k \sin\alpha) e^{-jkR\cos(\alpha - \theta_c)} k \cos\alpha \, d\alpha \\ &\simeq F(\theta)\sqrt{\frac{2\pi}{kR}} e^{-jkR + j(\pi/4)} \qquad (11\text{-}48) \\ &= A(k \sin\theta_c) k \cos\theta_c \sqrt{\frac{2\pi}{kR}} e^{-jkR + j(\pi/4)}. \end{aligned}$$

We therefore obtain the following approximate evaluations of the integrals:

$$\begin{aligned} I_1 &= \int_{-\infty}^{\infty} C_n(h) H_n^{(2)}(\sqrt{k^2 - h^2}\rho) e^{-jhz} \, dh, \qquad (11\text{-}49) \\ &\simeq C_n(k\cos\theta_c) \frac{2}{R} e^{-jkR + j(n+1)(\pi/2)} \quad \text{for large } kR. \end{aligned}$$

11-6 RADIATION FROM A DIPOLE AND PARSEVAL'S THEOREM

We now go back to (11-38) and (11-39) and obtain the radiation field. Using (11-49), we first note that E_ϕ has R^{-2} dependence while H_ϕ has R^{-1} dependence for large kR. This means that E_ϕ diminishes faster than H_ϕ, and therefore, H_ϕ and E_θ are the only components of the radiation field. We thus get, using (11-49),

$$H_\phi = (IL) \frac{e^{-jkR}}{R} \sum_{n=-\infty}^{\infty} f_n(\theta) e^{-jn(\phi - \phi')}, \qquad (11\text{-}50)$$

where

$$f_n(\theta) = \frac{k \sin\theta}{\pi} \frac{J_n(k\rho' \sin\theta)H^{(2)}(ka \sin\theta) - J_n(ka \sin\theta)H_n^{(2)}(k\rho' \sin\theta)}{H_n^{(2)}(ka \sin\theta)} e^{jn(\pi/2)}.$$

The Poynting vector S_r in (11-37) is then given by

$$S_r = \tfrac{1}{2}\eta|H_\phi|^2. \tag{11-51}$$

The radiation resistance R_{rad} is defined by

$$\frac{1}{2}I^2 R_{\text{rad}} = P_t = \text{total radiated power},$$
$$P_t = \int_0^\pi \sin\theta\, d\theta \int_0^{2\pi} d\phi\, S_r(\theta, \phi)R^2. \tag{11-52}$$

We therefore get the radiation resistance:

$$\frac{R_{\text{rad}}}{\eta} = L^2 \int_0^\pi \sin\theta\, d\theta \int_0^{2\pi} d\phi \left| \sum_{n=-\infty}^{\infty} f_n(\theta)e^{-jn\phi} \right|^2. \tag{11-53}$$

Let us next consider the integration in (11-53) with respect to ϕ. We get

$$\begin{aligned} \int_0^{2\pi} d\phi \left| \sum_{n=-\infty}^{\infty} f_n e^{-jn\phi} \right|^2 &= \int_0^{2\pi} d\phi \sum_n \sum_{n'} f_n f_{n'}^* e^{-j(n-n')\phi} \\ &= 2\pi \sum_{n=-\infty}^{\infty} |f_n|^2. \end{aligned} \tag{11-54}$$

This relationship (11-54) states that the integral of the square of the magnitude of a periodic function is the sum of the square of the magnitude of each harmonic component, and this is called *Parseval's theorem*. An example is the total power delivered to a resistor by a periodic current. There is no coupling between different frequency components, and the total power is equal to the sum of the power for each frequency component.

Using (11-54), the radiation resistance is given by

$$\frac{R_{\text{rad}}}{\eta} = 2\pi L^2 \int_0^\pi \sin\theta\, d\theta \sum_{n=-\infty}^{\infty} |f_n(\theta)|^2. \tag{11-55}$$

Equation (11-54) is Parseval's theorem for a periodic function. The equivalent Parseval's theorem for a continuous function can be stated as follows:

$$\int_{-\infty}^{\infty} dz \left| \frac{1}{2\pi} \int_{-\infty}^{\infty} f(h)e^{-jhz}\, dh \right|^2 = \frac{1}{2\pi} \int_{-\infty}^{\infty} |f(h)|^2\, dh. \tag{11-56}$$

This can be proved by using the following:

$$\int_{-\infty}^{\infty} e^{-j(h-h')z}\, dz = 2\pi\delta(h - h'). \tag{11-57}$$

11-7 LARGE CYLINDERS AND THE WATSON TRANSFORM

In previous sections the waves in cylindrical structures were given in terms of the Fourier series in ϕ and the Fourier transform in the z direction. These formal solutions are called *harmonic series* representations. There are two considerations associated with harmonic series representations which make it difficult to use them in many practical problems. One is that even if a solution is obtained in a harmonic series form, evaluation of the actual field quantities requires the truncation of an infinite series and thus knowledge of its convergence. It is therefore desirable to have alternative representations which may have different convergence characteristics. This is particularly important because it is often time consuming and expensive to obtain sufficient accuracy in evaluating various Bessel functions of large order by computers.

Second, the harmonic series representation can be obtained only for a small number of structures, whose surfaces coincide with the eleven coordinate systems where the wave equation is separable (rectangular, cylindrical, elliptic cylinder, parabolic cylinder, spherical, conical, parabolic, prolate spheroidal, oblate spheroidal, ellipsoidal, and paraboloidal; see Morse and Feshback, p. 656). Thus it is important to develop an alternative representation that may be useful in describing waves for more practical and complex shapes. The Watson transform technique described in this section offers this alternative representation which may be used for more complex problems (Chapter 13).

Let us first illustrate the problem associated with large cylinders by taking radiation from a dipole shown in (11-50). We recognize that in calculating the radiation pattern, we need to deal with the infinite series

$$S = \sum_{n=-\infty}^{\infty} f(n)e^{-jn\phi}. \tag{11-58}$$

The calculation of this series can be done by summing a finite number of terms instead of infinite terms and if the series is reasonably convergent, the finite sum should give a good solution. The important question then is: How many terms are required to represent the sum adequately? Problems involving cylinders and spheres actually contain the Bessel functions with the argument of order of magnitude of ka, where a is the radius or size of the object. Bessel functions behave quite differently depending on whether the argument is much less than, approximately equal to, or much greater than the order. Because of this, in general, series of the type (11-58) require at least *two ka terms* to represent the sum within a few percent accuracy. For a large cylinder, not only must the large number of terms be summed, but these terms contain Bessel functions of a large argument and large order. In many practical problems, this presents a formidable computational problem.

It would be extremely useful if series of the type (11-58) can be converted into a fast convergent series for large ka, so that only a few terms are required for obtaining numerical results. This is done by the Watson transform technique described in this section. Historically, shortwave radio-wave propagation over the earth was one of the central practical problems in the early twentieth century. The calculation was particularly difficult because of the extremely large *ka* value for the earth. In 1919, G. N. Watson successfully devised this technique, which converted the slow convergent series to a fast convergent series (Bremmer, 1949, p. 6).

First, let us consider the integral

$$I = \int_{C_1+C_2} \frac{A(\nu)}{\sin \nu\pi}\, d\nu. \tag{11-59}$$

The integrand has poles at $\nu = n$ integer, and thus the integral may be evaluated by taking the residues at each pole. The contour $C_1 + C_2$ encloses $2N + 1$ poles. We then let $N \to \infty$ (Fig. 11-9). We get

$$\begin{aligned} I &= 2\pi j \sum_{n=-\infty}^{\infty} \frac{A(\nu)}{\dfrac{\partial}{\partial \nu}(\sin \nu\pi)}\Bigg|_{\nu=n} \\ &= 2j \sum_{-\infty}^{\infty} A(n) e^{-jn\pi}. \end{aligned} \tag{11-60}$$

Comparing (11-60) with (11-58) we note that the series (11-58) can be expressed by the following complex integral (Wait, 1959, Chaps. 8 and 9).

$$S = \sum_{n=-\infty}^{\infty} f(n) e^{-jn\phi} = \frac{1}{2j} \int_{C_1+C_2} \frac{f(\nu) e^{-j\nu(\phi-\pi)}}{\sin \nu\pi}\, d\nu. \tag{11-61}$$

Many series representations for problems involving cylindrical and spherical structures can be transformed into the complex integral of this form. This transformation (11-61) is called the *Watson transform*.

The original series (11-58) is called the harmonic series and is obviously adequate for small cylinders. For large cylinders, however, the convergence of the harmonic series is too slow and the integral (11-61) is more useful. There are three ways to proceed from the integral form of (11-61). Each provides a good representation in different regions of space. Three approaches are:

1. *Residue series representation*: The integral (11-61) may be evaluated at the poles of $f(\nu)$, yielding a series of residues. This representation is highly convergent, requiring only a few terms in the *shadow region*. In this region the wave creeps along the surface and is called the *creeping wave*.

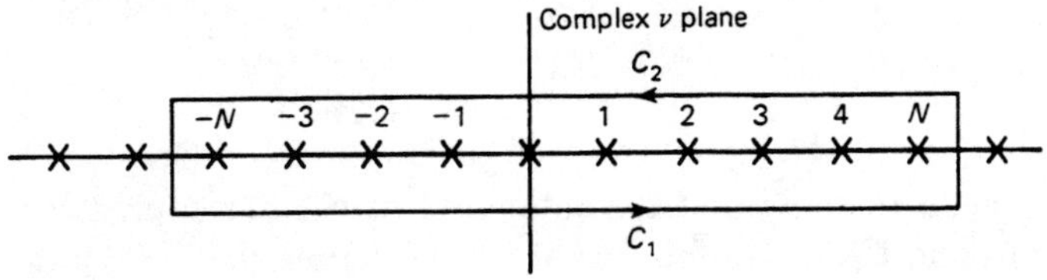

Figure 11-9 Contour integration for the Watson transform.

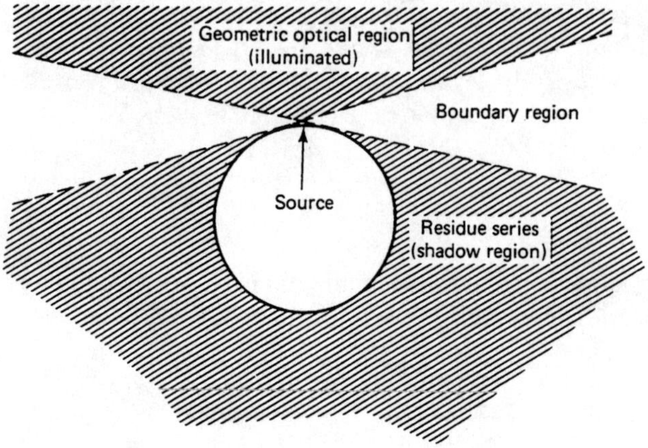

Figure 11-10 Wave representations in three regions.

2. *Geometric optical representation*: The integral (11-61) may be evaluated by means of the saddle-point technique, which gives a simple and useful representation in the *illuminated region*.
3. *Fock function representation*: In the *boundary region* between the illuminated and shadow region, the two preceding techniques are not applicable and the technique developed by Fock must be used.

The three regions discussed above are shown in Fig. 11-10. We now examine each of the three techniques.

11-8 RESIDUE SERIES REPRESENTATION AND CREEPING WAVES

The evaluation of the integral (11-61) can be made by deforming the contour C_1 to C_1' and C_2 to C_2' and taking the residues at the poles of $f(\nu)$. We note that (Fig. 11-11) (see the Appendix to Chapter 11, Section D)

$$\int_{C_1} d\nu = \int_{C_1'} d\nu - 2\pi j \sum_{m=1}^{\infty} (\text{residue at } \nu_m)$$

and

$$\int_{C_2} d\nu = \int_{C_2'} d\nu - 2\pi j \sum_{m=1}^{\infty} (\text{residue at } -\nu_m). \tag{11-62}$$

Next we need to show that the integrand of (11-61) approaches zero along the contour C_1' and C_2'. To do this, we need to examine the behaviors of the integrand for $|\nu| \to \infty$. The detailed explanations are given in Wait (1959) and therefore omitted here. Noting that the integrals in (11-62) along C_1' and C_2' are zero, we get

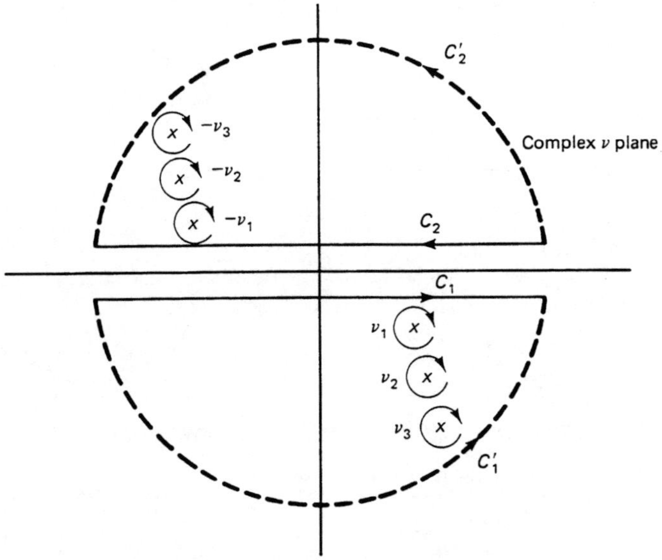

Figure 11-11 Contours for residue series.

$$
\begin{aligned}
S &= \sum_{n=-\infty}^{\infty} f(n)e^{-jn\phi} \\
&= \frac{1}{2j}\int_{C_1+C_2} \frac{f(\nu)e^{-j\nu(\phi-\pi)}}{\sin \nu\pi}\, d\nu \qquad (11\text{-}63) \\
&= -\pi \sum_{m=1}^{\infty} \mathrm{Re}(\nu_m) - \pi \sum_{m=1}^{\infty} \mathrm{Re}(-\nu_m),
\end{aligned}
$$

where

$$\mathrm{Re}(\nu_m) = \text{residue of } \frac{f(\nu)e^{-j\nu(\phi-\pi)}}{\sin \nu\pi}$$

at $\nu = \nu_m$, the poles of $f(\nu)$. The last expression in (11-63) is called the *residue series*.

Let us first examine the locations of the poles ν_m of $f(\nu)$. Note that the denominator of $f(\nu)$ is $H_\nu^{(2)}(ka \sin\theta)$ in (11-50). Therefore, the poles are given in

$$H_{\nu_m}^{(2)}(ka \sin\theta) = 0. \qquad (11\text{-}64)$$

It can be shown that the zeros of $H_\nu^{(2)}(z)$ when $|z| >> 1$ are given approximately by

$$\nu_m = z + \left(\frac{z}{2}\right)^{1/3}\left[\frac{3}{2}\left(m - \frac{1}{4}\right)\right]^{2/3} e^{-j(\pi/3)}, \qquad m = 1, 2, \ldots . \qquad (11\text{-}65)$$

Note that ν_m is complex and its imaginary part is negative (Fig. 11-12).

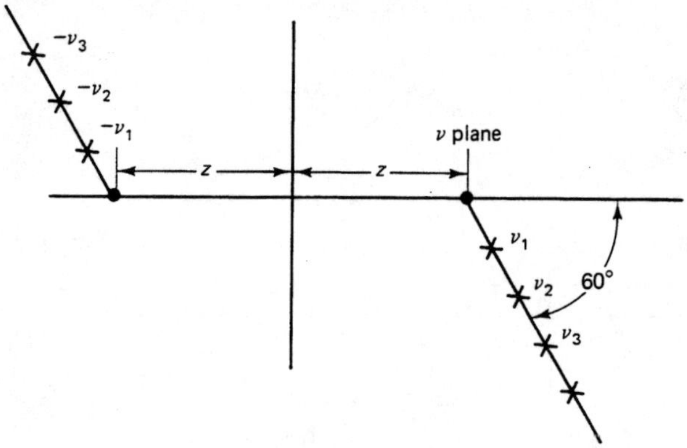

Figure 11-12 Locations of poles for residue series.

Let us rewrite the residue series in the following form:

$$S = \sum_{m=1}^{\infty} \frac{N^+(\nu_m)}{\left(\frac{\partial}{\partial \nu} D\right)\nu_m} e^{-j\nu_m \phi} + \sum_{m=1}^{\infty} \frac{N^-(\nu_m)}{\left(\frac{\partial}{\partial \nu} D\right)\nu_m} e^{-j\nu_m(2\pi - \phi)}, \tag{11-66}$$

where

$$f(\nu) = \frac{N(\nu)}{D(\nu)},$$

$$D(\nu) = H_\nu^{(2)}(ka \sin\theta),$$

$$N^+(\nu) = \frac{(-\pi)N(\nu)e^{j\nu\pi}}{\sin \nu\pi},$$

$$N^-(\nu) = \frac{(-\pi)N(\nu)e^{-j\nu\pi}}{\sin \nu\pi}.$$

The residue of a function $N^+(\nu)/D(\nu)$ at $\nu = \nu_m$ is given by

$$\begin{aligned} \text{residue} &= \lim_{\nu \to \nu_m} (\nu - \nu_m) \frac{N^+(\nu)}{D(\nu)} \\ &= \left. \frac{N^+(\nu)}{\frac{\partial}{\partial \nu} D(\nu)} \right|_{\nu = \nu_m}. \end{aligned} \tag{11-67}$$

Let us examine the residue series (11-66). The first series due to the poles at ν_m has the angular dependence $e^{-j\nu_m\phi}$. As shown in (11-65), ν_m has the real part somewhat greater than $z = ka \sin\theta$ and the negative imaginary part. We therefore have

$$\begin{aligned} e^{-j\nu_m \phi} &= e^{-j\nu_{mr}\phi - \nu_{mi}\phi}, \\ \nu_m &= \nu_{mr} - j\nu_{mi}. \end{aligned} \tag{11-68}$$

When $\theta = \pi/2$, $\nu_{mr} > ka$, and therefore, $\nu_{mr}\phi > ka\phi$. Since $a\phi$ is the distance along the surface, letting $\nu_{mr} = \beta a$, where β is the phase constant for the wave propagating along the surface, we get $\beta > k$, indicating that the phase velocity along the surface is less than the velocity of light. Therefore, the first series represents the wave that propagates with phase velocity slower than the velocity of light on the surface with attenuation. The second series in (11-66), corresponding to $-\nu_m$, represents the wave propagating in the opposite direction with the same propagation constant. These waves creep along the surface and are called the *creeping waves* (Fig. 11-13).

We may also note that the original harmonic series representation has the form

$$\sum_{n=-\infty}^{\infty} (\cdots)e^{-jn\phi},$$

which is a Fourier series representation in the interval $0 \le \phi \le 2\pi$. We may write this as a series of $\cos n\phi$ and $\sin n\phi$. Then we recognize that the harmonic series is in essence the representation of the wave in a series of standing waves ($\sin n\phi$ and $\cos n\phi$) around the cylinder. It is obvious, then, that for a small cylinder, since the wave has little ϕ variation, a small number of terms of the harmonic series should give a reasonable representation of the field. On the other hand, for a large cylinder, the wave radiated from a slot propagates along the cylinder surface and attenuates as it radiates the energy due to the curvature, and thus the wave is essentially a traveling wave along the surface. It is then clear that the suitable representation is the residue series because each term

$$e^{-j\nu_m\phi}$$

is a traveling-wave representation, and it requires only a few terms to describe the wave adequately.

11-9 POISSON'S SUM FORMULA, GEOMETRIC OPTICAL REGION, AND FOCK REPRESENTATION

Before we discuss the wave in the geometric optical region, let us discuss an alternative representation of the series (11-58), which we write as

$$S = \sum_{n=-\infty}^{\infty} F(n). \tag{11-69}$$

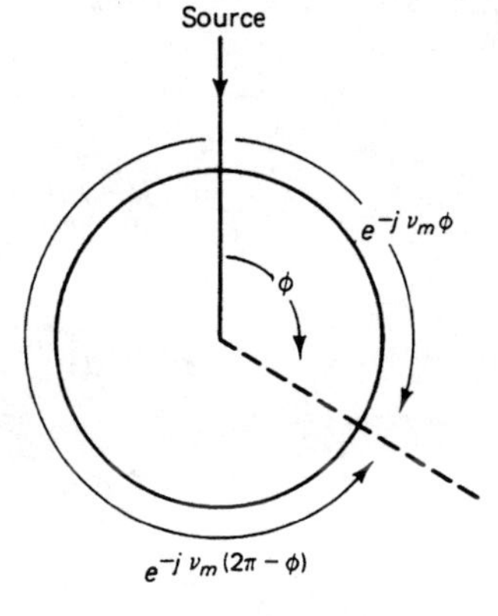

Figure 11-13 Creeping waves.

We will show in this section that the sum (11-69) can be represented as a sum of the Fourier transform $G(m)$ of $F(n)$ in the following form:

$$\sum_{n=-\infty}^{\infty} F(n) = \sum_{m=-\infty}^{\infty} G(m), \tag{11-70}$$

where

$$G(m) = \int_{-\infty}^{\infty} F(\nu)e^{-j2m\pi\nu}\,d\nu.$$

This is called *Poisson's sum formula*. It is useful when the original series is slowly convergent, but the series of the transform is fast convergent.

To prove this, we start with the Watson transform

$$S = \sum_{n=-\infty}^{\infty} F(n).$$

$$= \frac{1}{2j}\int_{C_1+C_2} \frac{F(\nu)e^{j\nu\pi}}{\sin \nu\pi}\,d\nu. \tag{11-71}$$

We write this in a somewhat different way. Along C_1, we note that $\operatorname{Im}\nu < 0$. Thus $|e^{-j\nu\pi}| < 1$. Therefore, we write

$$\frac{1}{\sin \nu\pi} = \frac{2j}{e^{j\nu\pi}(1 - e^{-j2\nu\pi})}$$

$$= 2je^{-j\nu\pi} \sum_{m=0}^{\infty} e^{-j2\nu\pi m}.$$

Thus we get

$$\frac{1}{2j}\int_{C_1} \frac{F(\nu)e^{j\nu\pi}}{\sin \nu\pi}\,d\nu = \sum_{m=0}^{\infty} G(m),$$

where

$$G(m) = \int_{C_1} F(\nu)e^{-j2\nu\pi m}\,d\nu.$$

Along C_2, $\operatorname{Im}\nu > 0$. Thus $|e^{j\nu\pi}| < 1$. Therefore,

$$\frac{1}{\sin \nu\pi} = -2je^{j\nu\pi} \sum_{m=0}^{\infty} e^{j2\nu\pi m}.$$

Thus we get

$$\frac{1}{2j}\int_{C_2} \frac{F(\nu)e^{j\nu\pi}}{\sin \nu\pi}\,d\nu = \sum_{m=1}^{\infty} G(-m),$$

where

$$G(-m) = -\int_{C_2} F(\nu)e^{j2\nu\pi m}\,d\nu$$

$$= \int_{-C_2} F(\nu)e^{j2\nu\pi m}\,d\nu.$$

Noting that $F(\nu)$ along C_1 is the same as $F(\nu)$ along $(-C_2)$, we obtain Poisson's sum formula:

$$\begin{aligned} S &= \sum_{n=-\infty}^{\infty} F(n) \\ &= \frac{1}{2j}\int_{C_1+C_2} \frac{F(\nu)e^{j\nu\pi}}{\sin \nu\pi}\,d\nu \\ &= \sum_{m=-\infty}^{\infty} G(m), \end{aligned} \tag{11-72}$$

where

$$G(m) = \int_{-\infty}^{\infty} F(\nu)e^{-j2\nu\pi m}\,d\nu.$$

Let us apply Poisson's sum formula to our problem (11-58):

$$\begin{aligned} S &= \sum_{n=-\infty}^{\infty} f(n)e^{-jn\phi} \\ &= \sum_{m=-\infty}^{\infty} S_m, \end{aligned} \tag{11-73}$$

$$S_m = \int_{-\infty}^{\infty} f(\nu)e^{-j\nu(\phi+2m\pi)}\,d\nu.$$

Consider S_m for $m \geq 0$. The difference between S_m and S_0 is that S_m has the additional $2m\pi$ to the angle ϕ, and thus S_m represents the wave encircling the cylinder m times (Fig. 11-14a). On the other hand, when $m < 0$, letting $\nu = -\nu'$ and $m = -m'$, we get

$$S_m = S_{-m'} = \int_{-\infty}^{\infty} f(-\nu')e^{j\nu'(\phi-2m'\pi)}\,d\nu'. \tag{11-74}$$

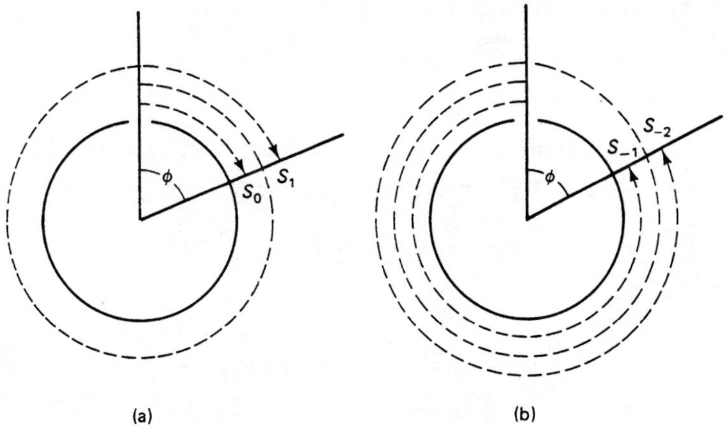

Figure 11-14 Waves encircling the cylinder.

This is recognized as the wave encircling the cylinder in the opposite direction (Fig. 11-14b).

In the illuminated region (Fig. 11-10), it is clear that the major contribution to the field comes from the S_0 term. It is then possible to evaluate S_0 by means of the saddle-point technique, and this leads to the geometric optical solution.

It is clear from previous sections that the wave in the boundary between the shadow and the illuminated region cannot be adequately represented by the residue series or geometric optical representations. In this region it is important to consider the order ν of the Hankel function close to its argument z. Making use of the Hankel approximation, it is possible to represent the field in this region in terms of a function proposed by Fock. The details are given in Wait's book (1959, pp. 64–68).

11-10 MIE SCATTERING BY A DIELECTRIC SPHERE

The exact solution of the scattering of a plane electromagnetic wave by an isotropic, homogeneous dielectric sphere of arbitrary size is usually referred to as *Mie theory*, even though Lorenz gave essentially the same results before Mie's work (Kerker, 1969, Sec. 3.4). In this chapter we employ a technique using the radial components of electric and magnetic hertz vectors. This technique differs from the vector wave equation formulation by Stratton, but it has the definite advantage of working with the scalar wave equations.

In spherical coordinates, it is possible to express the complete electromagnetic field in terms of two scalar functions π_1 and π_2. They are the radial components of the electric and magnetic Hertz vectors (see Section 4-15):

$$\overline{\pi}_e = \pi_1 \hat{r} \qquad \text{and} \qquad \overline{\pi}_m = \pi_2 \hat{r}. \tag{11-75}$$

π_1 produces all the TM modes with $H_r = 0$ and π_2 produces all the TE modes with $E_r = 0$. π_1 and π_2 satisfy the scalar wave equation.

$$\begin{aligned} (\nabla^2 + k^2)\pi_1 &= 0 \\ (\nabla^2 + k^2)\pi_2 &= 0, \end{aligned} \tag{11-76}$$

and the electric and magnetic fields are derived from these two scalar functions.

$$\begin{aligned} \overline{E} &= \nabla \times \nabla \times (r\pi_1 \hat{r}) - j\omega\mu \nabla \times (r\pi_2 \hat{r}), \\ \overline{H} &= j\omega\epsilon \nabla \times (r\pi_1 \hat{r}) + \nabla \times \nabla \times (r\pi_2 \hat{r}). \end{aligned} \tag{11-77}$$

In spherical coordinates, we write

$$E_r = \frac{\partial^2}{\partial r^2}(r\pi_1) + k^2 r\pi_1,$$

$$E_\theta = \frac{1}{r}\frac{\partial^2}{\partial r\,\partial\theta}(r\pi_1) - j\omega\mu \frac{1}{\sin\theta}\frac{\partial}{\partial\phi}\pi_2,$$

$$E_\phi = \frac{1}{r \sin\theta} \frac{\partial^2}{\partial r\, \partial\phi}(r\pi_1) + j\omega\mu \frac{\partial}{\partial\theta}\pi_2,$$

$$H_r = \frac{\partial^2}{\partial r^2}(r\pi_2) + k^2 r\pi_2, \tag{11-78}$$

$$H_\theta = j\omega\epsilon \frac{1}{\sin\theta} \frac{\partial}{\partial\phi}\pi_1 + \frac{1}{r} \frac{\partial^2}{\partial r\, \partial\theta}(r\pi_2),$$

$$H_\phi = -j\omega\epsilon \frac{\partial}{\partial\phi}\pi_1 + \frac{1}{r \sin\theta} \frac{\partial^2}{\partial r\, \partial\phi}(r\pi_2).$$

We now consider a sphere with a complex dielectric constant ϵ_1 and a permeability μ_1 immersed in a medium with ϵ_2 and μ_2 (Fig. 11-15). The incident wave is polarized in the x direction and is given by

$$\bar{E}_{\text{inc}}(z) = e^{-jk_2 z}\hat{x}. \tag{11-79}$$

Let us consider the boundary conditions. At $r = a$, the tangential electric and magnetic field must be continuous, and therefore, designating the field inside by $\bar{E}_1$ and $\bar{H}_1$ and the field outside by $\bar{E}_2$ and $\bar{H}_2$, we write the boundary conditions at $r = a$.

$$E_{1\theta} = E_{2\theta}, \qquad E_{1\phi} = E_{2\phi}, \qquad H_{1\theta} = H_{2\theta}, \qquad H_{1\phi} = H_{2\phi}. \tag{11-80}$$

As can be seen from (11-78), the boundary conditions above contain a mixture of π_1 and π_2. Therefore, it is convenient to reduce them into the boundary conditions on π_1 alone and π_2 alone. To do this, we note that if we take a linear combination of E_θ and E_ϕ in the following manner, the terms containing π_2 drop out and the condition becomes the continuity of

$$\frac{\partial}{\partial\theta}(\sin\theta\, E_\theta) + \frac{\partial}{\partial\phi}E_\phi = \left[\frac{\partial}{\partial\theta}\left(\sin\theta \frac{\partial}{\partial\theta}\right) + \frac{1}{\sin\theta}\frac{\partial^2}{\partial\phi^2}\right]\frac{1}{r}\frac{\partial}{\partial r}(r\pi_1),$$

across the boundary $r = a$. Since this must hold for any θ and ϕ, we require that $\partial(r\pi_1)/\partial r$ be continuous at $r = a$.

Similarly, considering $\partial E_\theta/\partial\phi - \partial(\sin\theta\, E_\phi)/\partial\theta$, we obtain the boundary condition that $\mu\pi_2$ be continuous at $r = a$. Similar considerations for H_θ and H_ϕ give the boundary condition that $\partial(r\pi_2)/\partial r$ and $\epsilon\pi_1$ be continuous at $r = a$.

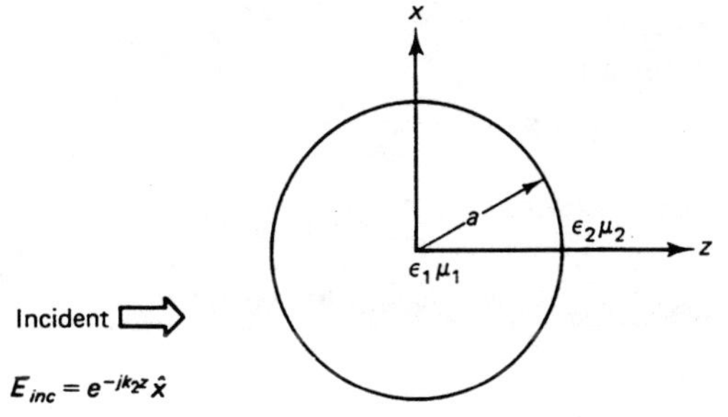

Figure 11-15 Mie scattering.

Let us now express the incident field $\bar{E}_{inc}$ in terms of spherical harmonics. We observe that $\bar{E}_{inc}$ can be derived from two scalar functions π_1^i and π_2^i that satisfy the wave equation. General expressions for π_1^i and π_2^i are

$$\pi_1^i = \sum_{n=0}^{\infty}\sum_{m=0}^{n} j_n(k_2 r)P_n^m(\cos\theta)[A_{mn}^{(1)}\cos m\phi + B_{mn}^{(1)}\sin m\phi],$$
$$\pi_2^i = \sum_{n=0}^{\infty}\sum_{m=0}^{n} j_n(k_2 r)P_n^m(\cos\theta)[A_{mn}^{(2)}\cos m\phi + B_{mn}^{(2)}\sin m\phi]. \tag{11-81}$$

Here we used j_n as it is finite at $r = 0$, and P_n^m as it is finite at $\theta = 0$ and π.

In order to determine the constants $A_{mn}^{(1)}$, $A_{mn}^{(2)}$, $B_{mn}^{(1)}$, and $B_{mn}^{(2)}$, we examine the radial component of $\bar{E}_i$ and $\bar{H}_i$:

$$E_{ir} = \bar{E}_{inc}\cdot\hat{r} = e^{-jk_2 r\cos\theta}\sin\theta\cos\phi. \tag{11-82}$$

We expand E_{ir} in spherical harmonics and equate it to E_{ir} derived from (11-81).

$$E_{ir} = \frac{\partial^2}{\partial r^2}(r\pi_1^i) + k_2^2 r\pi_1^i$$
$$= \sum_{n=0}^{\infty}\sum_{m=0}^{n}\frac{n(n+1)}{r} j_n(k_2 r)P_n^m(\cos\theta)[A_{mn}^{(1)}\cos m\phi + B_{mn}^{(1)}\sin m\phi]. \tag{11-83}$$

Here we used the following:

$$\left[\frac{d^2}{dr^2} + k^2 - \frac{n(n+1)}{r^2}\right][rz_n(kr)] = 0, \tag{11-84}$$

where $z_n(kr)$ is any spherical Bessel function.

The expansion of E_{ir} in (11-82) can be done in the following manner. First, we note that

$$e^{-jkr\cos\theta} = \sum_{n=0}^{\infty}(-j)^n(2n+1)j_n(kr)P_n(\cos\theta). \tag{11-85}$$

This is obtained by expanding $\exp(-jkr\cos\theta)$ in spherical harmonics,

$$\exp(-jkr\cos\theta) = \sum_{n=0}^{\infty} a_n(r)P_n(\cos\theta),$$

and noting that $P_n(\cos\theta)$ are orthogonal and using the relationship

$$\int_0^{\pi} e^{-jkr\cos\theta}P_n(\cos\theta)\sin\theta\, d\theta = 2(-j)^n j_n(kr),$$
$$\int_0^{\pi}[P_n(\cos\theta)]^2\sin\theta\, d\theta = \frac{2}{2n+1}.$$

We then obtain the expansion of E_{ir} from (11-85) by noting that

$$E_{ir} = \frac{1}{jkr}\frac{\partial}{\partial\theta}\exp(-jkr\cos\theta)\cos\phi,$$

$$\frac{\partial}{\partial\theta}P_n(\cos\theta) = -P_n^1(\cos\theta).$$

We thus obtain the following expansion of E_{ir}:

$$E_{ir} = \sum_{n=0}^{\infty} \frac{(-j)^{n-1}(2n+1)}{k_2 r} j_n(k_2 r) P_n^1(\cos\theta)\cos\phi. \tag{11-86}$$

Comparing (11-83) with (11-86), we get

$$B_{mn}^{(1)} = 0, \qquad A_{mn}^{(1)} = 0, \quad m \neq 1, \qquad A_{1n}^{(1)} = \frac{(-j)^{n-1}(2n+1)}{n(n+1)k_2}.$$

Substituting these into (11-81), we get the spherical harmonic representation of the incident Hertz potential π_1^i:

$$r\pi_1^i = \frac{1}{k_2^2}\sum_{n=1}^{\infty} \frac{(-j)^{n-1}(2n+1)}{n(n+1)} \Psi_n(k_2 r) P_n^1(\cos\theta)\cos\phi. \tag{11-87}$$

Similarly, we get

$$r\pi_2^i = \frac{1}{(\mu_2/\epsilon_2)^{1/2} k_2^2}\sum_{n=1}^{\infty} \frac{(-j)^{n-1}(2n+1)}{n(n+1)} \Psi_n(k_2 r) P_n^1(\cos\theta)\sin\phi, \tag{11-88}$$

where $\Psi_n(x) = x j_n(x) = \sqrt{\pi x/2}\, J_{n+1/2}(x)$.

We now have the expressions for the incident Hertz potentials, (11-87) and (11-88). Next we write the general expressions for the scattered fields and satisfy the boundary conditions. We note that the boundary conditions are such that π_1 outside the sphere couples only to π_1 inside the sphere. Thus since π_1^i has $\cos\phi$ dependence, we expect that the scattered field and the field inside the sphere have the same dependence on ϕ. Similarly, all π_2 should have $\sin\phi$ dependence.

Therefore, the general expressions for the scattered fields should be

$$\begin{aligned} r\pi_1^s &= \frac{(-1)}{k_2^2}\sum_{n=1}^{\infty} \frac{(-j)^{n-1}(2n+1)}{n(n+1)} a_n \zeta_n(k_2 r) P_n^1(\cos\theta)\cos\phi, \\ r\pi_2^s &= \frac{(-1)}{\sqrt{\mu_2/\epsilon_2}\, k_2^2}\sum_{n=1}^{\infty} \frac{(-j)^{n-1}(2n+1)}{n(n+1)} b_n \zeta_n(k_2 r) P_n^1(\cos\theta)\sin\phi, \end{aligned} \tag{11-89}$$

where $\zeta_n(x) = x h_n^{(2)}(x) = \sqrt{\pi x/2}\, H_{n+1/2}^{(2)}(x)$. Inside the sphere, we have

$$\begin{aligned} r\pi_1^r &= \frac{1}{k_1^2}\sum_{n=1}^{\infty} \frac{(-j)^{n-1}(2n+1)}{n(n+1)} c_n \Psi_n(k_1 r) P_n^1(\cos\theta)\cos\phi, \\ r\pi_2^r &= \frac{1}{\sqrt{\mu_1/\epsilon_1}\, k_1^2}\sum_{n=1}^{\infty} \frac{(-j)^{n-1}(2n+1)}{n(n+1)} d_n \Psi_n(k_1 r) P_n^1(\cos\theta)\sin\phi, \end{aligned} \tag{11-90}$$

where a_n, b_n, c_n, and d_n are the constants to be determined by the boundary conditions.

Now we apply the boundary condition on π_1 at $r = a$.

$$\begin{aligned} \frac{\partial}{\partial r}[r(\pi_1^i + \pi_1^s)] &= \frac{\partial}{\partial r}[r\pi_1^r], \\ \epsilon_2(\pi_1^i + \pi_1^s) &= \epsilon_1 \pi_1^r. \end{aligned} \tag{11-91}$$

Substituting (11-87), (11-89), and (11-90) into the above, we get

$$m[\Psi_n'(k_2 a) - a_n \zeta_n'(k_2 a)] = c_n \Psi_n'(k_1 a),$$

$$\frac{1}{\mu_2}[\Psi_n(k_2 a) - a_n \zeta_n(k_2 a)] = \frac{1}{\mu_1} c_n \Psi_n(k_1 a).$$

From these two, we obtain

$$a_n = \frac{\mu_1 \Psi_n(\alpha)\Psi_n'(\beta) - \mu_2 m \Psi_n(\beta)\Psi_n'(\alpha)}{\mu_1 \zeta_n(\alpha)\Psi_n'(\beta) - \mu_2 m \Psi_n(\beta)\zeta_n'(\alpha)}, \tag{11-92}$$

where $m = k_1/k_2 = \sqrt{\mu_1 \epsilon_1/\mu_2 \epsilon_2}$, $\alpha = k_2 a$, $\beta = k_1 a$.

To determine b_n for π_2, we use the boundary condition

$$\begin{aligned} \frac{\partial}{\partial r}[r(\pi_2^i + \pi_2^s)] &= \frac{\partial}{\partial r}[r\pi_2^r], \\ \mu_2(\pi_2^i + \pi_2^s) &= \mu_1 \pi_2^r, \end{aligned} \tag{11-93}$$

and obtain

$$b_n = \frac{\mu_2 m \Psi_n(\alpha)\Psi_n'(\beta) - \mu_1 \Psi_n(\beta)\Psi_n'(\alpha)}{\mu_2 m \zeta_n(\alpha)\Psi_n'(\beta) - \mu_1 \Psi_n(\beta)\zeta_n'(\alpha)}. \tag{11-94}$$

The constants c_n and d_n are then obtained from (11-91) and (11-93):

$$\begin{aligned} c_n &= \frac{jm\mu_1}{\mu_1 \zeta_n(\alpha)\Psi_n'(\beta) - \mu_2 m \Psi_n(\beta)J_n'(\alpha)}, \\ d_n &= \frac{jm\mu_1}{\mu_2 m \zeta_n(\alpha)\Psi_n'(\beta) - \mu_1 \Psi_n(\beta)J_n'(\alpha)}. \end{aligned} \tag{11-95}$$

With these constants a_n, b_n, c_n, and d_n (11-89) and (11-90) constitute the complete Mie solution for a dielectric sphere.

We now consider the far field. Noting that for $r \to \infty$,

$$\zeta_n(k_2 r) \to j^{n+1} e^{-jk_2 r},$$

we get

$$\begin{aligned} r\pi_1^s &\to \frac{e^{-jk_2 r}}{k_2^2} \sum_{n=1}^{\infty} \frac{2n+1}{n(n+1)} a_n P_n^1(\cos\theta)\cos\phi, \\ r\pi_2^s &\to \frac{e^{-jk_2 r}}{\sqrt{\frac{\mu_2}{\epsilon_2}}\, k_2^2} \sum_{n=1}^{\infty} \frac{2n+1}{n(n+1)} b_n P_n^1(\cos\theta)\sin\phi. \end{aligned} \tag{11-96}$$

The far fields E_θ and E_ϕ can be obtained from (11-78).

Noting that

$$\frac{\partial}{\partial r}(r\pi_1^s) = -jk_2 r\pi_1^s,$$

$$\frac{\partial}{\partial r}(r\pi_2^s) = -jk_2 r\pi_2^s,$$

we get

$$
\begin{aligned}
E_\theta &= f_\theta(\theta, \phi)\frac{e^{-jk_2 r}}{r}, \\
E_\phi &= f_\phi(\theta, \phi)\frac{e^{-jk_2 r}}{r}, \\
f_\theta &= -\frac{j\cos\phi\, S_2(\theta)}{k_2}, \\
f_\phi &= \frac{j\sin\phi\, S_1(\theta)}{k_2}, \\
S_1(\theta) &= \sum_{n=1}^{\infty}\frac{2n+1}{n(n+1)}[a_n\pi_n(\cos\theta) + b_n\tau_n(\cos\theta)], \\
S_2(\theta) &= \sum_{n=1}^{\infty}\frac{2n+1}{n(n+1)}[a_n\tau_n(\cos\theta) + b_n\pi_n(\cos\theta)],
\end{aligned}
\tag{11-97}
$$

where

$$\pi_n(\cos\theta) = \frac{P_n^1(\cos\theta)}{\sin\theta} \qquad \text{and} \qquad \tau_n = \frac{d}{d\theta}P_n^1(\cos\theta).$$

Equations (11-97) are the expressions for the scattered field in the far zone of a dielectric particle when the incident wave is polarized in the x direction.

Let us calculate the total cross section by means of the forward scattering theorem:

$$\sigma_t = -\frac{4\pi}{k_2}\hat{e}_i \cdot \hat{e}_s I_m f(\hat{\imath}, \hat{\imath}). \tag{11-98}$$

Noting that $\hat{e}_i = \hat{x}$ and

$$\pi_n(\cos\theta)|_{\theta=0} = \tau_n(\cos\theta)|_{\theta=0} = \frac{n(n+1)}{2}, \tag{11-99}$$

we get

$$\sigma_t = \frac{4\pi}{k_2^2}\operatorname{Re} S_1(0) = \frac{4\pi}{k_2^2}\operatorname{Re} S_2(0). \tag{11-100}$$

Thus the normalized total cross section with respect to the geometric cross section πa^2 is given by

$$\frac{\sigma_t}{\pi a^2} = \frac{2}{\alpha^2}\sum_{n=1}^{\infty}(2n+1)[\operatorname{Re}(a_n + b_n)], \qquad \alpha = k_2 a. \tag{11-101}$$

The backscattering cross section σ_b is given by

$$\sigma_b = 4\pi|f|^2{}_{\substack{\theta=\pi,\\ \phi=0}}$$

and since

$$\pi_n(\cos\theta)|_{\theta=\pi} = -\tau_n(\cos\theta)|_{\theta=\pi} = -(-1)^n\frac{n(n+1)}{2},$$

we get

$$\frac{\sigma_b}{\pi a^2} = \frac{1}{\alpha^2}\left|\sum_{n=1}^{\infty}(2n+1)(-1)^n(a_n - b_n)\right|^2 = \frac{|S_2(\pi)|^2}{\alpha^2} \tag{11-102}$$

The scattering cross section σ_s is obtained from

$$\begin{aligned}\sigma_s &= \int_{4\pi} |f(\theta,\phi)|^2 \, d\Omega \\ &= \int_0^{2\pi} d\phi \int d\theta \sin\theta \left[\left|\frac{\cos\phi\, S_2(\theta)}{k_2}\right|^2 + \left|\frac{\cos\phi\, S_1(\theta)}{k_2}\right|^2\right] \\ &= \frac{\pi}{k_2^2}\int_0^{\pi} d\theta \sin\theta [|S_2(\theta)|^2 + |S_1(\theta)|^2].\end{aligned} \tag{11-103}$$

Now from (11-97), we get

$$|S_2(\theta)|^2 = \sum_{n=1}^{\infty}\sum_{m=1}^{\infty}\frac{(2n+1)(2m+1)}{n(n+1)m(m+1)}[a_n a_m^* \pi_n \pi_m + b_n b_n^* \tau_n \tau_m + a_n b_m^* \pi_n \tau_m + a_m^* b_n \pi_m \tau_n],$$

$$|S_1(\theta)|^2 = \sum_{n=1}^{\infty}\sum_{m=1}^{\infty}\frac{(2n+1)(2m+1)}{n(n+1)m(m+1)}[a_n a_m^* \tau_n \tau_m + b_n b_m^* \pi_n \pi_m + a_n b_m^* \tau_n \pi_m + a_m^* b_n \tau_m \pi_n].$$

Adding these two and noting the orthogonality of Legendre functions,

$$\begin{aligned}\int_0^{\pi}(\pi_n \pi_m + \tau_n \tau_m)\sin\theta\, d\theta &= 0 \qquad \text{if } n \neq m \\ &= \frac{2}{2n+1}\frac{(n+1)!}{(n-1)!}n(n+1) \qquad \text{if } n = m,\end{aligned}$$

$$\int_0^{\pi}(\pi_n \tau_m + \tau_n \pi_m)\sin\theta\, d\theta = 0,$$

we get

$$\frac{\sigma_s}{\pi a^2} = \frac{2}{\alpha^2}\sum_{n=1}^{\infty}(2n+1)(|a_n|^2 + |b_n|^2). \tag{11-104}$$

Equations (11-101), (11-102), and (11-104) are the final expressions for the total cross section, the backscattering cross section, and the scattering cross section of a dielectric sphere. In this section we discussed scattering by a homogeneous dielectric sphere. We can, however, extend the foregoing analysis to scattering by a sphere consisting of many concentric dielectric layers by expressing π_1 and π_2 using linear combinations of two spherical Bessel functions with two unknowns in each layer instead of one spherical function shown in (11-90).

11-11 AXIAL DIPOLE IN THE VICINITY OF A CONDUCTING WEDGE

Dipoles located in the vicinity of a wedge (Fig. 11-16) have the number of practical applications. Examples are antennas located near a sharp edge such as aircraft wings or fins and antennas with a corner reflector. In this section we study the

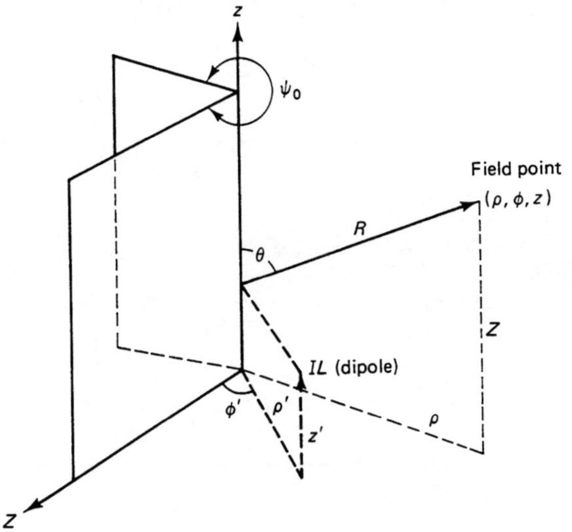

Figure 11-16 Electric dipole in the vicinity of a wedge.

radiation characteristics of an axial dipole affected by a conducting wedge of given angle.

The differential equation governing this case is given by

$$\pi_z = \frac{IL}{j\omega\epsilon} G(r/r'),$$

$$\left(\frac{\partial^2}{\partial\rho^2} + \frac{1}{\rho}\frac{\partial}{\partial\rho} + \frac{1}{\rho^2}\frac{\partial^2}{\partial\phi^2} \frac{\partial^2}{\partial z^2} + k^2\right) G(r/r') = \frac{\delta(\rho-\rho')\delta(\phi-\phi')\delta(z-z')}{\rho}, \tag{11-105}$$

and the boundary condition at the conducting surface is

$$G = 0 \quad \text{at } \phi = 0 \quad \text{and} \quad \phi = \Psi_0. \tag{11-106}$$

We expand $G(r/r')$ in a series of normalized eigenfunctions $\Phi_m(\phi)$:

$$G(r/r') = \sum_{m=1}^{\infty} G_m(\rho, z)\Phi_m(\phi)\Phi_m(\phi'), \tag{11-107}$$

where the eigenfunction is given by

$$\Phi_m(\phi) = \sqrt{\frac{2}{\Psi_0}} \sin \frac{m\pi}{\Psi_0}\phi, \tag{11-108}$$

with the eigenvalue $\nu_m = m\pi/\Psi_0, m = 1, 2, 3, \ldots, \infty$ and

$$\int_0^{\Psi_0} [\phi_m(\phi)]^2 \, d\phi = 1.$$

We note that on the right side of (11-105),

$$\delta(\phi - \phi') = \sum_{m=1}^{\infty} \Phi_m(\phi)\Phi_m(\phi'). \tag{11-109}$$

Therefore, substituting (11-107) and (11-109) into (11-105), we get

$$\left(\frac{\partial^2}{\partial \rho^2} + \frac{1}{\rho}\frac{\partial}{\partial \rho} - \frac{\nu_m^2}{\rho^2} + \frac{\partial^2}{\partial z^2} + k^2\right) G_m = -\frac{\delta(\rho - \rho')\delta(z - z')}{\rho}. \tag{11-110}$$

We now take the Fourier transform in the z direction, and following the procedure given in Section 11-3, obtain

$$G(r/r') = \frac{1}{2\pi}\int_c \sum_{m=1}^{\infty} G_m(\rho, h)\Phi_m(\phi)\Phi_m(\phi')e^{-jh(z-z')}\,dh,$$

$$G_m(\rho, h) = -j\frac{\pi}{2}J_{\nu_m}(\sqrt{k^2 - h^2}\rho')H_{\nu_m}^{(2)}(\sqrt{k^2 - h^2}\rho), \qquad \rho' < \rho \tag{11-111}$$

$$= -j\frac{\pi}{2}J_{\nu_m}(\sqrt{k^2 - h^2}\rho)H_{\nu_m}^{(2)}(\sqrt{k^2 - h^2}\rho'), \qquad \rho' > \rho$$

This gives the Hertz potential produced by an axial dipole. The electric and magnetic fields are obtained by differentiation. The far field is obtained by using the saddle-point technique.

The field due to an axial magnetic dipole can be expressed similarly using the eigenfunction $\Psi_m(\phi), m = 0, 1, 2, \ldots$, which satisfies Neumann's boundary condition:

$$\Psi_m(\phi) = \left(\frac{2}{\Psi_0}\right)^{1/2} \cos\left(\frac{m\pi}{\Psi_0}\phi\right), \qquad m = 1, 2, \ldots$$

$$= \left(\frac{1}{\Psi_0}\right)^{1/2}, \qquad m = 0. \tag{11-112}$$

11-12 LINE SOURCE AND PLANE WAVE INCIDENT ON A WEDGE

Using the results given in Section 11-1, we can obtain the exact solution for a wave excited by a line source and for a wave excited by a plane wave in the presence of a wedge. This solution is important in the study of the geometric theory of diffraction (GTD), discussed in Chapter 13. Let us consider a conducting wedge excited by an electric line current I_0 located at $\bar{r}' = (\rho, \theta_i)$ (Fig. 11-17). The electric field E_z at $\bar{r} = (s, \theta_s)$ satisfies the following equation and the boundary condition:

$$E_z = -j\omega\mu I_0 G_1,$$

$$(\nabla_t^2 + k^2)G_1 = -\delta(\bar{r} - \bar{r}'), \tag{11-113}$$

$$G_1 = 0 \quad \text{at } \theta = 0 \qquad \text{and} \qquad \theta = \Psi_0.$$

Using the results (11-111), we get, for $s > \rho$,

$$G_1 = \left(-j\frac{\pi}{\Psi_0}\right) \sum_{m=1}^{\infty} J_{\nu_m}(k\rho)H_{\nu_m}^{(2)}(ks) \sin \nu_m \theta_i \sin \nu_m \theta_s, \tag{11-114}$$

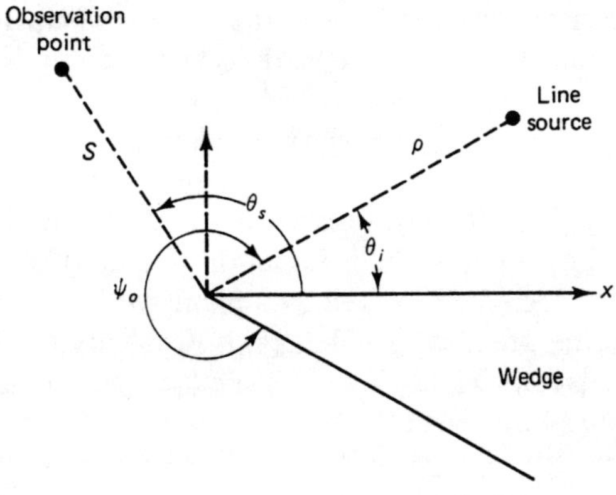

Figure 11-17 Wedge excited by a line source.

where $\nu_m = m\pi/\Psi_0$. For $s < \rho$, ρ and s above should be interchanged:

$$G_1 = \left(-j\frac{\pi}{\Psi_0}\right) \sum_{m=1}^{\infty} J_{\nu_m}(ks) H_{\nu_m}^{(2)}(k\rho) \sin \nu_m \theta_i \sin \nu_m \theta_s. \tag{11-115}$$

For a magnetic line source I_m, we have

$$\begin{aligned} H_z &= -j\omega\epsilon I_m G_2, \\ (\nabla_t^2 + k^2)G_2 &= -\delta(\bar{r} - \bar{r}') \\ \frac{\partial G_2}{\partial n} &= 0 \quad \text{at } \theta = 0 \quad \text{and} \quad \theta = \Psi_0. \end{aligned} \tag{11-116}$$

Green's function G_2 is then given by

$$G_2 = \left(-j\frac{\pi}{\Psi_0}\right) \sum_{m=0}^{\infty} \frac{\epsilon_m}{2} J_{\nu_m}(ks) H_{\nu_m}^{(2)}(k\rho) \cos \nu_m \theta_i \cos \nu_m \theta_s \qquad \text{for } s < \rho, \tag{11-117}$$

where $\epsilon_m = 1$ for $m = 0$, $\epsilon_m = 2$ for $m \neq 0$, and ρ and s are interchanged for $s > \rho$.

Let us next consider the case of a plane wave incident on a wedge. For a plane wave polarized in the z direction incident from the direction θ_i, we let $\rho \to \infty$ and use the asymptotic form of $H_\nu^{(2)}$:

$$H_\nu^{(2)}(k\rho) \sim \left(\frac{2}{\pi k \rho}\right)^{1/2} e^{-j[k\rho - (\nu\pi/2) - (\pi/4)]}. \tag{11-118}$$

Substituting this in (11-115) and noting that the incident field at the origin is

$$\begin{aligned} E_0 &= -j\omega\mu I_0 G_0, \\ G_0 &= -\frac{j}{4} H_0^{(2)}(k\rho) \sim -\frac{j}{4}\left[\frac{2}{\pi k \rho}\right]^{1/2} e^{-j[k\rho - (\pi/4)]}, \end{aligned} \tag{11-119}$$

we get the field due to the plane wave with the incident electric field E_0 at the origin,

$$E_z = E_0 \frac{4\pi}{\Psi_0} \sum_{m=1}^{\infty} e^{j\nu_m(\pi/2)} J_{\nu_m}(ks) \sin \nu_m \theta_i \sin \nu_m \theta_s. \tag{11-120}$$

Similarly, when the field is polarized perpendicular to the z axis, the magnetic field due to the plane wave with the incident magnetic field H_0 at the origin is given by

$$H_z = H_0 \frac{4\pi}{\Psi_0} \sum_{m=0}^{\infty} \frac{\epsilon_m}{2} e^{j\nu_m(\pi/2)} J_{\nu_m}(ks) \cos \nu_m \theta_i \cos \nu_m \theta_s. \tag{11-121}$$

Equations (11-113) to (11-117) give the exact solution for the total field for a conducting wedge when excited by a line source. Equations (11-120) and (11-121) give the exact total field for a wedge when excited by a plane wave. These are the exact series solutions, but in some practical problems it is necessary to obtain the simpler closed-form solutions. A convenient closed-form exact solution is possible only for a plane wave incident on a knife edge (half-plane) $\Psi_0 = 2\pi$. For other cases, such as a plane wave incident on a wedge $\Psi_0 \neq 2\pi$ and a line source excitation, we can obtain some convenient approximate expressions.

11-13 HALF-PLANE EXCITED BY A PLANE WAVE

For a half-plane (knife edge), we let $\Psi_0 = 2\pi$ in (11-120) and (11-121) and obtain the exact solution. However, this half-plane diffraction problem was first solved by Sommerfeld in 1896 and the solution was given in terms of the Fresnel integral. The problem has also been solved by the Wiener–Hopf technique (see James, 1976; Jull, 1981; Noble, 1958). Here we give the final expressions and refer readers to the references above for the details.

For a plane wave polarized in the z direction incident on a half-plane from the direction θ_i (Fig. 11-18) the electric field is given by

$$E_{zi} = E_0 e^{jk\rho \cos(\theta - \theta_i)}. \tag{11-122}$$

The total electric field satisfies the Dirichlet boundary condition on a half-plane and the field at (s, θ_s) is given by

$$E_z = \frac{E_0 e^{j(\pi/4)}}{\sqrt{\pi}} [e^{jks \cos(\theta_s - \theta_i)} F(a_1) - e^{jks \cos(\theta_s + \theta_i)} F(a_2)], \tag{11-123}$$

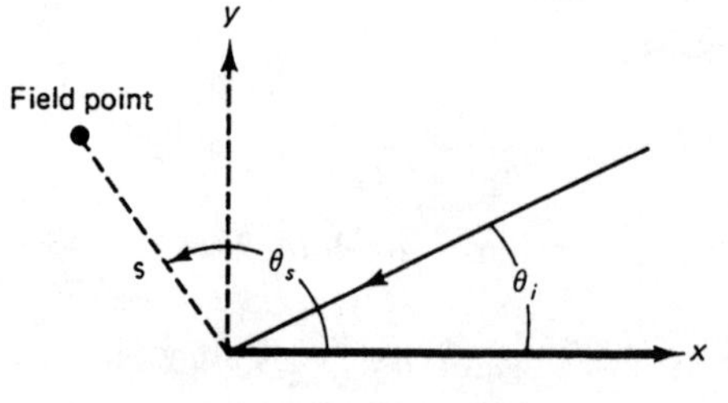

Figure 11-18 Plane wave incident on a half-plane ($x > 0, y = 0$).

where

$$F(a) = \int_a^{\infty} e^{-j\tau^2}\,d\tau \text{ is the complex Fresnel integral,}$$

$$a_1 = -(2ks)^{1/2}\cos\frac{\theta_s - \theta_i}{2},$$

$$a_2 = -(2ks)^{1/2}\cos\frac{\theta_s + \theta_i}{2}.$$

Also note that

$$F(a) + F(-a) = \int_{-\infty}^{\infty} e^{-j\tau^2}\,d\tau = \sqrt{\pi}e^{-j(\pi/4)}. \tag{11-124}$$

For a plane wave polarized perpendicular to the z axis, the incident magnetic field is given by

$$H_{zi} = H_0 e^{jk\rho\cos(\theta - \theta_i)}. \tag{11-125}$$

The total field satisfies the Neumann boundary condition on a half-plane and the field is given by

$$H_z = \frac{H_0 e^{j(\pi/4)}}{\sqrt{\pi}}[e^{jks\cos(\theta_s - \theta_i)}F(a_1) + e^{jks\cos(\theta_s + \theta_i)}F(a_2)]. \tag{11-126}$$

Note that the only difference between (11-123) and (11-126) is the minus or plus sign in front of the second term. We will make use of these results in Chapter 13.

PROBLEMS

11-1. A plane wave is normally incident on a conducting cylinder of radius a. Find the total current (integral over ϕ) on the cylinder for the TM case and for the TE case. $a = 1$ cm and $f = 1$ GHz.

11-2. A plane wave is normally incident on a dielectric cylinder of radius a with relative dielectric constant ϵ_r. Find the electric field on the axis of the cylinder when the incident field E_0 is polarized parallel to the axis (TM case). Find the electric field on the axis if the field is polarized perpendicular to the axis (TE case). $a = 10$ cm, $f = 1$ GHz, the real part of $\epsilon_r = 60$, and the loss tangent $= 0.5$.

11-3. A plane sound wave p_0 is incident obliquely from a medium with C_0 and ρ_0 on a cylinder with C_1 and ρ_1. Find the pressure p inside and outside the cylinder.

11-4. Find the field radiated from a slot on a conducting cylinder. The electric field on the slot is as given in Fig. P11-4. To do this problem, first write general expressions for Π_z and Π_{mz} and find expressions for E_ϕ and E_z. Then let $\rho = a$ and equate E_ϕ and E_z to the slot field given.

11-5. Find the radiation pattern of an annular slot antenna on a conducting cylinder (Fig. P11-5).

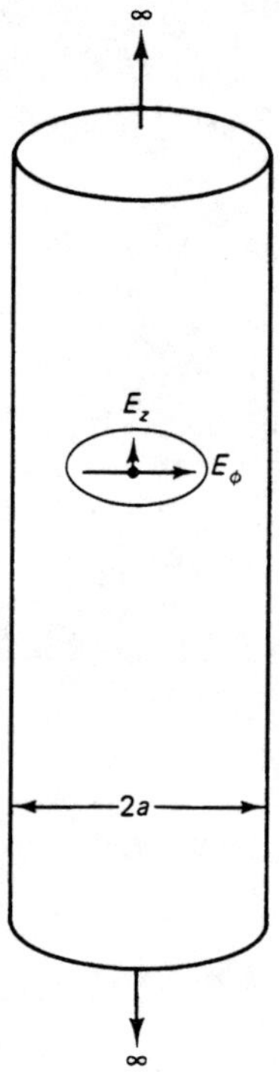

Figure P11-4 Slot on a conducting cylinder.

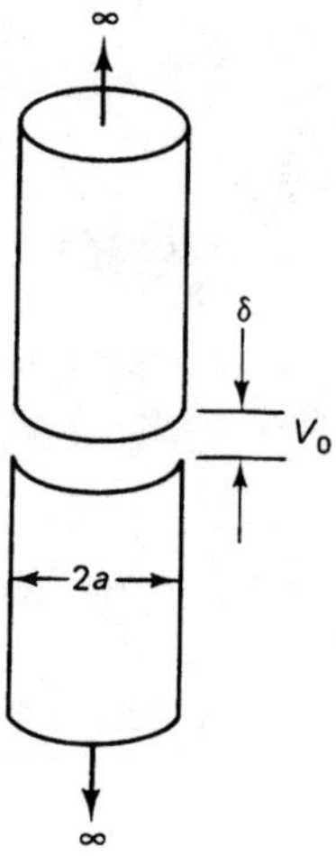

Figure P11-5 Annular slot.

11-6. The gamma (factorial) function is given in the following integral form:

$$\Gamma(z+1) = z! = \int_0^\infty t^z e^{-t}\, dt.$$

For large z, $\Gamma(z)$ is given approximately by Stirling's formula:

$$\Gamma(z) \sim e^{-z} z^{z-1/2} (2\pi)^{1/2}$$

Derive this formula using the saddle-point technique. (*Hint*: Let $t = zv$.)

11-7. Show that the Airy integral $Ai(x)$,

$$Ai(x) = \frac{x^{1/3}}{2\pi} \int_{-\infty}^{\infty} e^{jx[(t^3/3)+t]}\, dt,$$

is given by

$$Ai(x) \sim \frac{x^{1/3}}{2\pi} e^{-2x/3} \sqrt{\frac{\pi}{x}}$$

for large x. Show the saddle points, original contour, and SDC in the complex t plane.

11-8. The integral representation of the Legendre function $P_n(\cos\theta)$ is given by

$$P_n(\cos\theta) = \frac{1}{2\pi} \int_0^{2\pi} (\cos\theta + i \sin\theta \cos\phi)^n\, d\phi.$$

Show that for large n,

$$P_n(\cos\theta) \sim \left(\frac{2}{n\pi \sin\theta}\right)^{1/2} \cos\left[\left(n + \frac{1}{2}\right)\theta - \frac{\pi}{4}\right].$$

11-9. Evaluate the Bessel function $J_0(z)$ for large z using the integral representation

$$J_0(z) = \frac{1}{\pi}\int_0^\pi \exp(jz\,\cos\phi)\,d\phi.$$

11-10. Evaluate the Legendre function $P_n^m(\cos\theta)$ for large n using the following definition:

$$P_n^m(\cos\theta) = \frac{c}{2\pi j}\oint e^{nf(w)}\,dw,$$

where the integral is to be taken over the unit circle in the counterclockwise sense, and

$$f(w) = \ln\left[\cos\theta + \frac{j}{2}\sin\theta\left(w + \frac{1}{w}\right)\right] - \frac{m+1}{n}\ln w,$$

$$c = \frac{(n+m)!}{n!}e^{-jm\pi/2}.$$

11-11. Find the integral I for large k (Fig. P11-11):

$$I = \int_{-\infty}^{\infty} \frac{\exp[-jk(r_1 + r_2)]}{r_1 r_2}\,dy,$$

where $r_1 = \sqrt{y^2 + a^2}$ and $r_2 = \sqrt{(y-c)^2 + b^2}$.

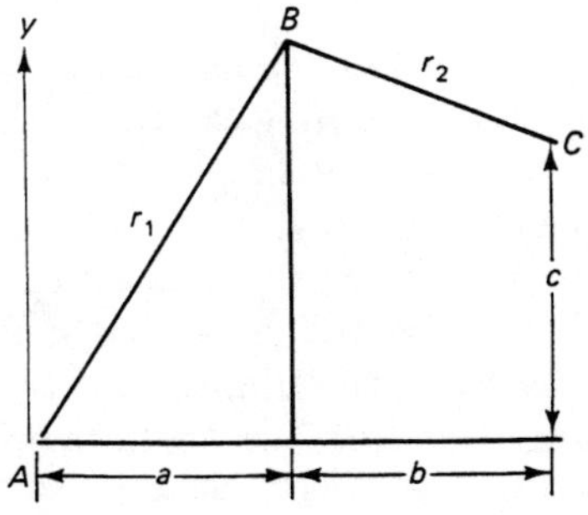

Figure P11-11 Radiation from A to B to C.

11-12. A plane electromagnetic wave is incident on a conducting sphere of radius a. Find the total cross section, backscattering cross section, and scattering cross section of the sphere.

11-13. A plane sound wave p_0 is incident from a medium with c_0 and ρ_0 on a sphere of radius a with c_1 and ρ_1. Find the total cross section, backscattering cross section, and scattering cross section of the sphere.

12

Scattering by Complex Objects

In Chapter 10 we discussed some of the approximate methods of calculating scattering by simple objects such as Rayleigh scattering and Rayleigh–Debye scattering. In Chapter 11 we discussed exact solutions for simple objects such as cylinders and spheres. In many practical situations, the objects often have a complex shape and the size and the dielectric constant are such that the simple solutions discussed in Chapters 10 and 11 are not applicable. Several methods that have been developed to deal with these problems will be discussed. In this chapter we start with integral equation formulations for scalar fields for Dirichlet, Neumann, and two-media problems. EFIE and MFIE are discussed next, followed by the T-matrix method. We then discuss inhomogeneous dielectric media, including Green's dyadic. We conclude this chapter with aperture diffraction problems, small apertures, and Babinet's principle. For radar cross sections, see Ruck et al. (1970). Aperture problems are discussed in Maanders and Mittra (1977), Rahmat-Samii and Mittra (1977), Arvas and Harrington (1983), and Butler, Rahmat-Samii, and Mittra (1978). Low-frequency scattering (Kleinman, 1978; van Bladel, 1968; Stevenson, 1953; Senior, 1984) and narrow strips and slots (Butler and Wilton, 1980; Senior, 1979) are also discussed in this chapter. For dyadic Green's functions, see Tai (1971), Yaghjian (1980), and Livesay and Chen (1974). Also see Lewin et al. (1977) for curved structures.

12-1 SCALAR SURFACE INTEGRAL EQUATIONS FOR SOFT AND HARD SURFACES

In Section 6-1 we used Huygens' principle and obtained the expressions for the field in terms of the incident wave and the contribution from the surface field. In this section we make use of these relationships to obtain the surface integral equations when the object is a soft body or a hard body. A penetrable homogeneous body is discussed in the next section. There are two surface equations. One relates the field to the surface integral and the other relates the normal derivative of the field to the surface integral.

Integral Equation for a Soft Body

If the object is a soft body, the boundary condition is the Dirichlet type.

$$\psi(\bar{r}) = 0 \qquad \text{on the surface.} \tag{12-1}$$

Now we make use of (6-11) (Fig. 12-1):

$$\psi_i(\bar{r}) + \int_S \left[\psi(\bar{r}')\frac{\partial G(\bar{r},\bar{r}')}{\partial n'} - G(\bar{r},\bar{r}')\frac{\partial \psi(\bar{r}')}{\partial n'}\right] dS' = \frac{1}{2}\psi(\bar{r}) \qquad \text{if } \bar{r} \text{ is on the surface } S. \tag{12-2}$$

We then use the boundary condition (12-1), and we get

$$\psi_i(\bar{r}) = \int_{S'} G_0(\bar{r},\bar{r}')\frac{\partial \psi(\bar{r}')}{\partial n'}\, dS', \tag{12-3}$$

where

$$G_0(\bar{r},\bar{r}') = \frac{\exp(-jk|\bar{r}-\bar{r}'|)}{4\pi|\bar{r}-\bar{r}'|} \qquad \text{and } \bar{r} \text{ and } \bar{r}' \text{ are on the surface } S.$$

Note that as $\bar{r}'$ approaches $\bar{r}$, G_0 becomes infinite, and therefore this is an improper integral. However, as is shown in Appendix 12A, the integral is convergent as G_0 is proportional to R^{-1}, where $R = |\bar{r} - \bar{r}'|$ and the surface integral is convergent if the integrand is $R^{-\alpha}$ and $0 < \alpha < 2$. Thus this is a weakly singular integral equation.

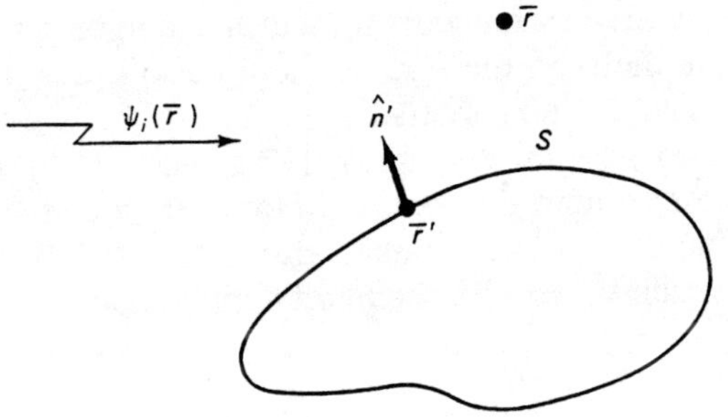

Figure 12-1 Surface integral equation.

Equation (12-3) is the Fredholm integral equation of the first kind for the unknown function $\partial\psi/\partial n'$ (see the Appendix to Chapter 12, Section B).

Alternatively, we can obtain the Fredholm equation of the second kind. To do this, we take the normal derivative of (6-9) and let $\bar{r}$ approach the surface S. As shown in the Appendix to Chapter 12, Section A, we get

$$\frac{\partial}{\partial n}\int_S G_0(\bar{r},\bar{r}')\frac{\partial\psi(\bar{r}')}{\partial n'}dS' = -\frac{1}{2}\frac{\partial\psi(\bar{r})}{\partial n} + \fint \frac{\partial G_0(\bar{r},\bar{r}')}{\partial n}\frac{\partial\psi(\bar{r}')}{\partial n'}dS'. \tag{12-4}$$

We then obtain

$$\frac{\partial\psi_i(\bar{r})}{\partial n} = \frac{1}{2}\frac{\partial\psi(\bar{r})}{\partial n} + \fint \frac{\partial G_0(\bar{r},\bar{r}')}{\partial n}\frac{\partial\psi(\bar{r}')}{\partial n'}dS', \tag{12-5}$$

where both $\bar{r}$ and $\bar{r}'$ are on the surface S. This is the desired Fredholm integral equation of the second kind. It can be proved that the integral is convergent (see the Appendix to Chapter 12, Section A).

Integral Equations for a Hard Body

The boundary condition at the surface of a hard body is the Neumann type:

$$\frac{\partial\psi}{\partial n} = 0. \tag{12-6}$$

From (6-11) we get the Fredholm integral equation of the second kind for $\psi(\bar{r})$:

$$\psi_i(\bar{r}) + \fint_S \psi(\bar{r}')\frac{\partial G_0(\bar{r},\bar{r}')}{\partial n'}dS' = \frac{1}{2}\psi(\bar{r}). \tag{12-7}$$

The integral is convergent (see the Appendix to Chapter 12, Section A).

We can also obtain the Fredholm integral equation of the first kind:

$$\frac{\partial\psi_i(\bar{r})}{\partial n} + \frac{\partial}{\partial n}\int \psi(\bar{r}')\frac{\partial G_0(\bar{r},\bar{r}')}{\partial n'}dS' = 0. \tag{12-8}$$

The kernel in this case is singular (see the Appendix to Chapter 12, Section A).

We note that in the formulations above, the scattered field satisfies the radiation condition because we used Green's function G_0, which is outgoing and satisfies the radiation condition. The field on the surface should also satisfy the edge condition if the surface contains edges and corners.

In the above, we have two integral equations, (12-3) and (12-5), for soft objects. We can use either one to obtain the solution. However, the second kind (12-5) is often more stable in numerical calculations. Also note that if the surface integral in (12-5) is negligibly small, we get the Kirchhoff approximation,

$$\frac{\partial\psi}{\partial n} \simeq 2\frac{\partial\psi_i}{\partial n}. \tag{12-9}$$

Similarly, for the hard surface, (12-7) is more convenient and the Kirchhoff approximation is obtained immediately by neglecting the surface integral:

$$\psi(\bar{r}) \simeq 2\psi_i(\bar{r}). \tag{12-10}$$

12-2 SCALAR SURFACE INTEGRAL EQUATIONS FOR A PENETRABLE HOMOGENEOUS BODY

Let us now consider the scattering of a scalar acoustic wave incident from the media with the density ρ_0 and the sound velocity c_0 on a homogeneous body with ρ_1 and c_1. This is sometimes called the *two-media problem* (Fig. 12-2). We use the field ψ to designate the velocity potential. Then we have

$$\begin{aligned} (\nabla^2 + k_0^2)\psi_0 &= 0 \qquad \text{outside the body} \\ (\nabla^2 + k_1^2)\psi_1 &= 0 \qquad \text{inside the body.} \end{aligned} \tag{12-11}$$

The boundary conditions on S are

$$\rho_0\psi_0 = \rho_1\psi_1 \qquad \text{and} \qquad \frac{\partial\psi_0}{\partial n} = \frac{\partial\psi_1}{\partial n}. \tag{12-12}$$

On the surface S, there are two unknowns: ψ_0 and $\partial\psi_0/\partial n$ (or ψ_1 and $\partial\psi_1/\partial n$). We will therefore develop two coupled surface integral equations for ψ_0 and $\partial\psi_0/\partial n$.

First, consider the region outside the body. As the observation point $\bar{r}$ approaches the surface S from outside, we get (see 6-11)

$$\psi_i(\bar{r}) + \int_S \left(\psi_0(\bar{r}')\frac{\partial G_0(\bar{r},\bar{r}')}{\partial n'} - G_0(\bar{r},\bar{r}')\frac{\partial\psi_0(\bar{r}')}{\partial n'}\right) dS' = \frac{1}{2}\psi_0(\bar{r}), \tag{12-13}$$

where

$$G_0(\bar{r},\bar{r}') = \frac{\exp(-jk_0|\bar{r}-\bar{r}'|)}{4\pi|\bar{r}-\bar{r}'|}.$$

Next we use the same formula (12-11) for the region inside the body and let $\bar{r}$ approach the surface S from inside.

$$\int_S \left(\psi_1(\bar{r}')\frac{\partial G_1(\bar{r},\bar{r}')}{\partial n'} - G_1(\bar{r},\bar{r}')\frac{\partial\psi_1(\bar{r}')}{\partial n'}\right) dS' = -\frac{1}{2}\psi_1(\bar{r}), \tag{12-14}$$

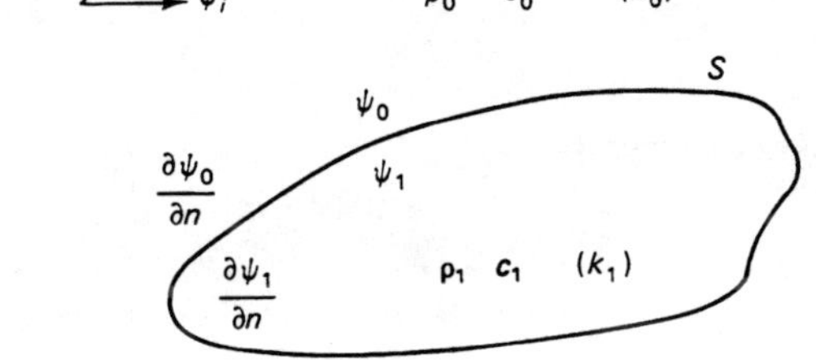

Figure 12-2 Two-media problem.

where

$$G_1(\bar{r}, \bar{r}') = \frac{\exp(-jk_1|\bar{r} - \bar{r}'|)}{4\pi|\bar{r} - \bar{r}'|}.$$

Note that the right side is $-\frac{1}{2}\psi_1$ rather than $+\frac{1}{2}\psi_1$ because $\partial/\partial n'$ is now a normal derivative in the direction toward the outside. Now using the boundary conditions, we can rewrite (12-14) as follows:

$$\oint_S \left[\left(\frac{\rho_0}{\rho_1}\right)\psi_0(\bar{r}')\frac{\partial G_1(\bar{r}, \bar{r}')}{\partial n'} - G_1(\bar{r}, \bar{r}')\frac{\partial \psi_0(\bar{r}')}{\partial n'}\right] dS' \\ = -\frac{1}{2}\left(\frac{\rho_0}{\rho_1}\right)\psi_0(\bar{r}). \tag{12-15}$$

Equations (12-13) and (12-15) give two integral equations for two unknowns $\psi_0(\bar{r}')$ and $\partial\psi_0(\bar{r}')/\partial n'$. Then two equations can be solved numerically.

Alternatively, we take the normal derivative of (6-9) and let $\bar{r}$ approach the surface S. We then get

$$\frac{\partial \psi_i(\bar{r})}{\partial n} + \frac{\partial}{\partial n}\int_S \psi_0(\bar{r}')\frac{\partial G_0(\bar{r}, \bar{r}')}{\partial n'} dS' - \oint \frac{\partial G_0(\bar{r}, \bar{r}')}{\partial n}\frac{\partial \psi_0(\bar{r}')}{\partial n'} dS' = \frac{1}{2}\frac{\partial \psi_0(\bar{r})}{\partial n}. \tag{12-16}$$

Now we use (6-9) for the region inside the body and let $\bar{r}$ approach the surface S from inside. We then use the boundary conditions and obtain

$$\frac{\rho_0}{\rho_1}\frac{\partial}{\partial n}\int_S \psi_0(\bar{r}')\frac{\partial G_1(\bar{r}, \bar{r}')}{\partial n'} dS' - \oint \frac{\partial G_1(\bar{r}, \bar{r}')}{\partial n}\frac{\partial \psi_0(\bar{r}')}{\partial n'} dS' = -\frac{1}{2}\frac{\partial \psi_0(\bar{r})}{\partial n}. \tag{12-17}$$

We note that both (12-16) and (12-17) contain singular kernels, due to the first integral term of (12-16) and the first term of (12-17).

Note also that we have four equations, (12-13), (12-15), (12-16), and (12-17). The first two relate the field outside and inside the surface to the surface integral and the last two relate the normal derivatives of the field outside and inside the surface to the surface integral. We can make different combinations of these four equations to obtain different integral equations with various degrees of computational advantages.

12-3 EFIE AND MFIE

Let us consider an electromagnetic wave $(\bar{E}_i, \bar{H}_i)$ incident on a perfectly conducting body with surface S (Fig. 12-3). There are two surface integral equations. The equation in terms of the electric field is called the *electric field integral equation* (EFIE) and the equation in terms of the magnetic field is called the *magnetic field integral equation* (MFIE).

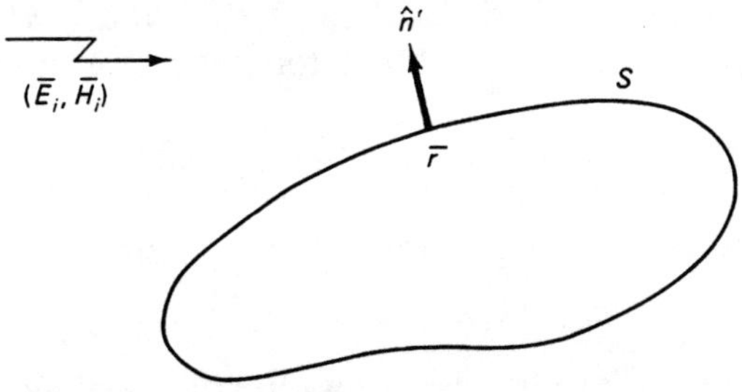

Figure 12-3 EFIE and MFIE.

The EFIE is obtained by using the surface integral representation (6-111) and imposing the boundary condition.

$$\bar{E}_i(\bar{r}) + \int_S \bar{E}_s \, dS' = \tfrac{1}{2}\bar{E}(\bar{r}). \tag{12-18}$$

The boundary condition on S is

$$\hat{n}' \times \bar{E}(\bar{r}) = 0. \tag{12-19}$$

Therefore, we get

$$\hat{n}' \times \bar{E}_i(\bar{r}) + \hat{n}' \times \int \bar{E}_s \, dS' = 0, \tag{12-20}$$

where

$$\bar{E}_s = -[j\omega\mu G(\hat{n}' \times \bar{H}) - (\hat{n}' \times \bar{E}) \times \nabla' G - (\hat{n}' \cdot \bar{E})\nabla' G].$$

Noting that $\hat{n}' \times \bar{H} = \bar{J}_s$ = surface current, $\hat{n}' \cdot \bar{E} = \rho_s/\epsilon$, ρ_s = surface charge, and using the continuity condition

$$\nabla_s \cdot \bar{J}_s + j\omega\rho_s = 0,$$

where $\nabla_s \cdot$ is the surface divergence, we get the following EFIE for the surface current $\bar{J}_s$:

$$\hat{n}' \times \bar{E}_i(\bar{r}) = \frac{1}{j\omega\epsilon}\hat{n}' \times \int_S [-k^2 G(\bar{r}, \bar{r}')\, \bar{J}_s(\bar{r}') + (\nabla_s' \cdot \bar{J}_s(\bar{r}'))\nabla' G(\bar{r}, \bar{r}')]\, dS',$$

where

$$G = G(\bar{r}, \bar{r}') = \frac{e^{-jk|\bar{r} - \bar{r}'|}}{4\pi|\bar{r} - \bar{r}'|}. \tag{12-21}$$

Next, we consider the MFIE. We start with (6-111):

$$\bar{H}_i + \int_S \bar{H}_s \, dS' = \tfrac{1}{2}\bar{H}, \tag{12-22}$$

where

$$\bar{H}_s = j\omega\epsilon G\hat{n}' \times \bar{E} + (\hat{n}' \times \bar{H}) \times \nabla' G + (\hat{n}' \cdot \bar{H})\nabla' G.$$

However, the boundary conditions on the surface are $\hat{n}' \times \bar{E} = 0$ and $\hat{n}' \cdot \bar{H} = 0$. Letting $\hat{n}' \times \bar{H} = \bar{J}_s$ = surface current, we get the MFIE for the surface current $\bar{J}_s$:

$$\hat{n}' \times \bar{H}_i(\bar{r}) + \int_S [\hat{n}' \times \bar{J}_s(\bar{r}') \times \nabla' G(\bar{r}, \bar{r}')]\, dS' = \frac{1}{2} \bar{J}_s(\bar{r}). \qquad (12\text{-}23)$$

Note that for MFIE, the first term gives the physical optics approximation:

$$\bar{J}_s = 2\hat{n}' \times \bar{H}_i. \qquad (12\text{-}24)$$

Therefore, the MFIE should be useful for a large object with a smooth surface whose radius of curvature is large compared with a wavelength. Then physical optics is a good approximation and the surface integral is a small correction term. On the other hand, the EFIE should be more useful for a thin object such as a wire.

12-4 *T*-MATRIX METHOD (EXTENDED BOUNDARY CONDITION METHOD)

Up to this point in this chapter we have discussed the formulation of the scattering problem in terms of the surface integral equations for the unknown surface fields. In this section we present a different technique, originated by P. C. Waterman (1969), which makes use of the extinction theorem (Sections 6-1 and 6-9). According to the extinction theorem, the surface field produces a field contribution inside the object which exactly cancels or extinguishes the incident field throughout the interior of the object. First, we will show that by using this extinction theorem as the *extended boundary condition*, we can get an integral equation for the surface field which holds throughout the interior of the object. Second, we note that this integral equation need not be satisfied throughout the interior; it need hold only in any portion of the interior volume because of the *analytic continuity*. These two are the key elements of the T-matrix method. In actual applications, these two ideas are used to derive the transition matrix (T-matrix) which relates the scattered wave to the incident wave. We illustrate this technique and explain its meaning by using Dirichlet's problem as an example.

Let us consider an incident wave ψ_i impinging on an object with surface S where Dirichlet's boundary condition holds (Fig. 12-4). This is a two-dimensional problem, and we first express the incident field ψ_i in a series of appropriate basis functions. For a two-dimensional problem, the natural choice is cylindrical harmonics. Then we write

$$\psi_i(\bar{r}) = \sum_{n=-\infty}^{\infty} a_n J_n(kr) e^{jn\phi}. \qquad (12\text{-}25)$$

Note that the incident field ψ_i is finite everywhere, including the origin, and therefore $J_n(kr)$ must be used. For a given incident wave, a_n is the known coefficient. For example, for a plane wave $\psi_i(\bar{r}) = \exp(-jk\hat{i} \cdot \bar{r})$ propagating in the direction ϕ_i, we have

$$\hat{i} = \cos\phi_i \hat{x} + \sin\phi_i \hat{y},$$
$$\bar{r} = r\cos\phi\hat{x} + r\sin\phi\hat{y}, \tag{12-26}$$
$$a_n = e^{-jn(\phi_i + \pi/2)}.$$

Next, we write the scattered wave $\psi_s(\bar{r})$ in region I outside the circumscribed cylinder S_c in the same cylindrical harmonics.

$$\psi_s(\bar{r}) = \sum_{n=-\infty}^{\infty} b_n H_n^{(2)}(kr)e^{jn\phi}. \tag{12-27}$$

In this region the waves are outgoing, and we used the Hankel function of the second kind to satisfy the radiation condition.

The transition matrix or the T-matrix $[T]$ is defined by

$$[b] = [T][a], \tag{12-28}$$

where $[b] = [b_n]$ and $[a] = [a_n]$ are the column matrices and $[T]$ is the square matrix. Here these matrices are truncated and finite for numerical calculations. The T-matrix is then obtained systematically by use of the extinction theorem, as discussed below.

To obtain the T-matrix, let us first write the extinction theorem (6-12):

$$\psi_i(\bar{r}) + \int_S \left[\psi(\bar{r}')\frac{\partial G(\bar{r},\bar{r}')}{\partial n'} - G(\bar{r},\bar{r}')\frac{\partial \psi(\bar{r}')}{\partial n'}\right] dS' = 0 \qquad \text{if } \bar{r} \text{ is inside } S. \tag{12-29}$$

This theorem applies to regions II and II′ in Fig. 12-4. However, let us apply this only to region II inside the inscribed cylinder S_i. Then $|\bar{r}| < |\bar{r}'|$, and therefore Green's function $G(\bar{r},\bar{r}')$, can be written for $r < r'$:

$$G(\bar{r},\bar{r}') = -\frac{j}{4}\sum_{n=-\infty}^{\infty} J_n(kr)H_n^{(2)}(kr')e^{jn(\phi-\phi')}. \tag{12-30}$$

Now we substitute this and (12-25) into (12-29). We then note that ψ_i in (12-29) is a series involving $J_n(kr)\exp(jn\phi)$, and the integral term in (12-29) is also a series in-

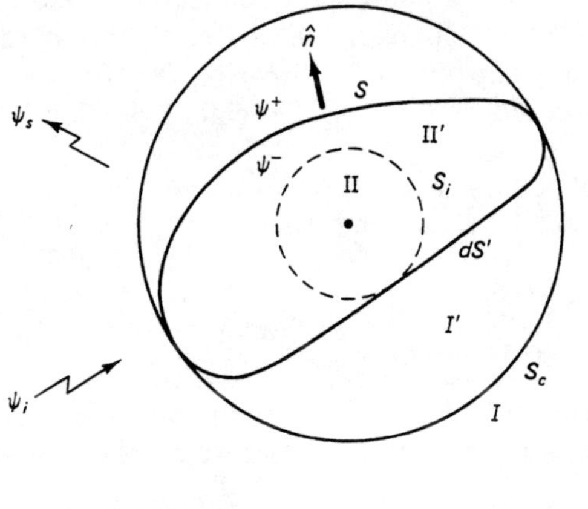

Figure 12-4 T-matrix method.

volving $J_n(kr)\exp(jn\phi)$, which appears in $G(\bar{r},\bar{r}')$ of (12-30). Since $J_n(kr)\exp(jn\phi)$ is orthogonal, we can equate each coefficient of $J_n\exp(jn\phi)$ to zero and obtain

$$a_n = \frac{j}{4}\int_S\left[\psi(\bar{r}')\frac{\partial}{\partial n'}(H_n^{(2)}(kr')e^{-jn\phi'} - \frac{\partial\psi(\bar{r}')}{\partial n'}H_n^{(2)}(kr')e^{-jn\phi'}\right]dS', \qquad (12\text{-}31)$$

where $dS' = [r'^2 + (dr'/d\phi')^2]^{1/2}\,d\phi'$ is the elementary distance along the surface S, where $r' = r'(\phi')$ and

$$\frac{\partial}{\partial n'} = \hat{n}\cdot\nabla = \left[r'^2 + \left(\frac{dr'}{d\phi'}\right)^2\right]^{-1/2}\left(r'\frac{\partial}{\partial r'} - \frac{1}{r'}\frac{\partial r'}{\partial\phi'}\frac{\partial}{\partial\phi'}\right).$$

This equation relates the coefficients a_n of the incident wave to the surface field $\psi(\bar{r}')$ and $\partial\psi(\bar{r}')/\partial n'$ and is obtained by applying the extinction theorem to region II only. Therefore, the surface field obtained from (12-31) when substituted into the extinction theorem (12-29) should cancel the incident field completely in region II. However, the left side of the extinction theorem (12-29) is a regular solution of the wave equations in both II′ and II, and therefore if it is null in II, it should also be null in II′ by analytic continuation. Thus the surface field determined from (12-31) should be the true surface field satisfying (12-29) for all interior regions II and II′.

Next we obtain the scattered field ψ_s in region I using the following:

$$\begin{aligned}\psi_s(\bar{r}) &= \int_S\left[\psi(\bar{r}')\frac{\partial G(\bar{r},\bar{r}')}{\partial n'} - G(\bar{r},\bar{r}')\frac{\partial\psi(\bar{r}')}{\partial n'}\right]dS',\\ G(\bar{r},\bar{r}') &= -\frac{j}{4}\sum_{n=-\infty}^{\infty}J_n(kr')H_n^{(2)}(kr)e^{jn(\phi-\phi')} \qquad \text{for } r > r'.\end{aligned} \qquad (12\text{-}32)$$

Noting the expansion of ψ_s in (12-27) and using the orthogonality of $H_n^{(2)}(kr)e^{jn\phi}$, we obtain

$$b_n = -\frac{j}{4}\int_S\left[\psi(\bar{r}')\frac{\partial}{\partial n'}(J_n(kr')e^{-jn\phi'}) - \frac{\partial\psi(\bar{r}')}{\partial n'}J_n(kr')e^{-jn\phi'}\right]dS'. \qquad (12\text{-}33)$$

We now have a_n in (12-31) and b_n in (12-33) expressed in terms of the surface field. We next use the boundary condition and express the surface field in terms of a complete set of functions. We then eliminate the surface field from the two equations (12-31) and (12-33) and finally obtain the T matrix.

Let us illustrate this procedure by using Dirichlet's problem ($\psi = 0$ on S). Then we get

$$\begin{aligned}a_n &= -\frac{j}{4}\int_S\frac{\partial\psi(\bar{r}')}{\partial n'}H_n^{(2)}(kr')e^{-jn\phi'}\,dS',\\ b_n &= \frac{j}{4}\int_S\frac{\partial\psi(\bar{r}')}{\partial n'}J_n(kr')e^{-jn\phi'}\,dS'.\end{aligned} \qquad (12\text{-}34)$$

We now expand the unknown surface function $\partial\psi/\partial n'$ in a series of a complete set of functions. The choice of the function is arbitrary, but we use the following:

$$\frac{\partial\psi(\bar{r}')}{\partial n'} = \sum_{n=-\infty}^{\infty} \alpha_n \frac{\partial}{\partial n'} [J_n(kr')e^{jn\phi'}], \tag{12-35}$$

where α_n is the unknown coefficient and we assume that $(\partial/\partial n')[J_n \exp(jn\phi')]$ is a complete set. Substituting (12-35) into (12-34), we get

$$\begin{aligned} [a] &= [Q^-][\alpha], \\ [b] &= -[Q^+][\alpha], \end{aligned} \tag{12-36}$$

where $[a]$, $[b]$, and $[\alpha]$ are column matrices and

$$\begin{aligned} Q^-_{mn} &= -\frac{j}{4}\int_S \psi_m \phi'_n \, dS', \\ Q^+_{mn} &= -\frac{j}{4}\int_S \psi_{rm} \phi'_n \, dS', \\ \psi_m &= H^{(2)}_m(kr')e^{-jm\phi'}, \\ \psi_{rm} &= J_m(kr')e^{-jm\phi'}, \\ \phi'_n &= \frac{\partial}{\partial n'}[J_n(kr')e^{jn\phi'}]. \end{aligned}$$

Finally, we eliminate the surface field $[\alpha]$ from (12-36) and get the T matrix

$$\begin{aligned} [b] &= [T][a], \\ [T] &= -[Q^+][Q^-]^{-1}. \end{aligned} \tag{12-37}$$

Let us next consider Neumann's problem for which the boundary condition on S is $(\partial/\partial n')\psi(\bar{r}') = 0$. From (12-31) and (12-33) we get a_n and b_n in terms of $\psi(\bar{r}')$ on the surface. We expand $\psi(\bar{r}')$ in a complete set of functions $J_n(kr') \exp(jn\phi')$, and then we get

$$\psi(\bar{r}') = \sum_{n=-\infty}^{\infty} \alpha_n J_n(kr')e^{jn\phi'}, \tag{12-38}$$

where α_n are the unknown coefficients. Substituting this into (12-31) and (12-33), we get

$$\begin{aligned} [a] &= [Q^-][\alpha], \\ [b] &= -[Q^+][\alpha] = [T][a], \\ [T] &= -[Q^+][Q^-]^{-1}, \end{aligned} \tag{12-39}$$

where

$$\begin{aligned} Q^-_{mn} &= \frac{j}{4}\int_S \psi'_m \phi_n \, dS', \\ Q^+_{mn} &= \frac{j}{4}\int_S \psi'_{rm} \phi_n \, dS', \end{aligned} \tag{12-40}$$

and

$$\psi_m' = \frac{\partial}{\partial n'}[H_m^{(2)}(kr')e^{-jm\phi'}] = \frac{\partial}{\partial n'}\psi_m,$$

$$\psi_{rm}' = \frac{\partial}{\partial n'}[J_m(kr')e^{-jm\phi'}] = \frac{\partial}{\partial n'}\psi_{rm}, \tag{12-41}$$

$$\phi_n = J_n(kr')e^{jn\phi'}.$$

For a two-media problem, where the wave numbers and the densities outside the surface S are k_0 and ρ_0 and those inside S are k_1 and ρ_1, we let ψ^+ and ψ^- denote the surface field just outside and inside the surface S (Fig. 12-4). The field inside S satisfies the wave equation with the wave number k_1 and is regular, and therefore it can be expanded in the following cylindrical harmonics:

$$\psi(\bar{r}) = \sum_{n=-\infty}^{\infty} \beta_n J_n(k_1 r)e^{jn\phi}, \tag{12-42}$$

where β_n is the unknown coefficient. Using this, we can write the field just inside S as follows:

$$\psi^-(\bar{r}') = \sum_{n=-\infty}^{\infty} \beta_n J_n(k_1 r')e^{jn\phi'},$$

$$\frac{\partial \psi^-(\bar{r}')}{\partial n'} = \sum_{n=-\infty}^{\infty} \beta_n \frac{\partial}{\partial n'}[J_n(k_1 r')e^{jn\phi'}]. \tag{12-43}$$

Now the boundary conditions are:

$$\psi^+(\bar{r}') = \frac{\rho_1}{\rho_0}\psi^-(\bar{r}'),$$

$$\frac{\partial \psi^+(\bar{r}')}{\partial n'} = \frac{\partial \psi^-(\bar{r}')}{\partial n'}.$$

The expressions for a_n in (12-31) and b_n in (12-33) contain the field just outside the surface S, and they are related to the field inside given by β_n.

$$[a] = [Q^-][\beta],$$

$$[b] = -[Q^+][\beta] = [T][a], \tag{12-44}$$

$$[T] = -[Q^+][Q^-]^{-1},$$

where

$$Q_{mn}^- = \frac{j}{4}\int_S \left(\frac{\rho_1}{\rho_0}\psi_m'\phi_n - \psi_m\phi_n'\right) dS',$$

$$Q_{mn}^+ = \frac{j}{4}\int_S \left(\frac{\rho_1}{\rho_0}\psi_{rm}'\phi_n - \psi_{rm}\phi_n'\right) dS',$$

$$\psi_m = H_m^{(2)}(k_0 r')e^{-jm\phi'},$$

$$\psi'_m = \frac{\partial}{\partial n'}\psi_m,$$

$$\psi_{rm} = J_m(k_0 r')e^{-jm\phi'},$$

$$\psi'_{rm} = \frac{\partial}{\partial n'}\psi_{rm},$$

$$\phi_n = J_n(k_1 r')e^{jn\phi'},$$

$$\phi'_n = \frac{\partial}{\partial n'}\phi_n.$$

The T-matrix method has been used successfully for scattering from a body of complex shapes such as particles with irregular shapes. Since it makes use of the expansion of the field in cylindrical harmonics (spherical harmonics for three-dimensional objects and Fourier expansion for periodic structures), the matrix may become ill-conditioned if the axial ratio (ratio of the major to minor axes) is much greater than about 5 or if the object has corners.

In this section we illustrated the T-matrix method with a two-dimensional problem and cylindrical harmonics. The T-matrix method can be equally applicable to three-dimensional problems using spherical harmonics. For electromagnetic problems, the complete cylindrical or spherical expansion of the vector field must be used (Waterman, 1969).

12-5 SYMMETRY AND UNITARITY OF THE *T* MATRIX AND THE SCATTERING MATRIX

The T matrix satisfies a certain symmetric relationship because of the reciprocity principle. If a delta function source is located at $\bar{r}_1(r_1, \phi_1)$ and the scattered field $\psi_s(\bar{r}_2)$ is observed at $\bar{r}_2(r_2, \phi_2)$, and next a delta function source is located at $\bar{r}_2$ and the scattered field $\psi_s(\bar{r}_1)$ is observed at $\bar{r}_1$, then according to reciprocity,

$$\psi_s(\bar{r}_1) = \psi_s(\bar{r}_2). \tag{12-45}$$

For the first case, the incident wave is

$$\psi_i(\bar{r}) = -\frac{j}{4}\sum_n J_n(kr)H_n^{(2)}(kr_1)e^{+jn(\phi-\phi_1)}. \tag{12-46}$$

Therefore, we have

$$a_n = -\frac{j}{4}H_n^{(2)}(kr_1)e^{-jn\phi_1}. \tag{12-47}$$

The scattered field $\psi_s(\bar{r}_2)$ is then given by

$$\begin{aligned}\psi_s(\bar{r}_2) &= \sum_n b_n H_n^{(2)}(kr_2)e^{jn\phi_2} \\ &= \left(-\frac{j}{4}\right)\sum_n\sum_m T_{nm} H_m^{(2)}(kr_1)H_n^{(2)}(kr_2)e^{jn\phi_2 - jm\phi_1}.\end{aligned} \tag{12-48}$$

If we switch $\bar{r}_1$ and $\bar{r}_2$, we get

$$\psi_s(\bar{r}_1) = \left(-\frac{j}{4}\right)\sum_n\sum_m T_{nm} H_m^{(2)}(kr_2)H_n^{(2)}(kr_1)e^{jn\phi_1 - jm\phi_2}. \tag{12-49}$$

Now noting that $\psi_s(\bar{r}_1) = \psi_s(\bar{r}_2)$ and letting $m = -n'$ and $n = -m'$ in (12-49), and using $H_{-n}^{(2)} = (-1)^n H_n^{(2)}$, we get the following:

$$T_{mn} = (-1)^{m+n} T_{-n,-m}. \tag{12-50}$$

This is the symmetry relationship satisfied by the T matrix.

Let us consider the scattering matrix S. We note first that the incident wave is regular at the origin and is expressed in a series of regular cylindrical harmonics $J_n \exp(jn\phi)$. If we rewrite the incident wave using the identity

$$J_n(z) = \frac{1}{2}(H_n^{(1)}(z) + H_n^{(2)}(z)),$$

the part containing $H_n^{(2)}$ is the outgoing wave, while the part containing $H_n^{(1)}$ is the incoming wave:

$$\begin{aligned}\psi_i(\bar{r}) &= \sum_n a_n J_n(kr)e^{jn\phi} \\ &= \psi_i^-(\bar{r}) + \psi_i^+(\bar{r}), \\ \psi_i^-(\bar{r}) &= \sum_n a_n \tfrac{1}{2} H_n^{(1)}(kr)e^{jn\phi}, \\ \psi_i^+(\bar{r}) &= \sum_n a_n \tfrac{1}{2} H_n^{(2)}(kr)e^{jn\phi}.\end{aligned} \tag{12-51}$$

Also we note that ψ_s is outgoing:

$$\psi_s(\bar{r}) = \sum_n b_n H_n^{(2)}(kr)e^{jn\phi}. \tag{12-52}$$

We can now regroup the wave as consisting of the incoming and the outgoing waves:

$$\begin{aligned}\psi_i^-(\bar{r}) &= \sum_n \frac{a_n}{2} H_n^{(1)}(kr)e^{jn\phi}, && \text{incoming} \\ \psi_i^+(\bar{r}) + \psi_s(\bar{r}) &= \sum_n \left(\frac{a_n}{2} + b_n\right) H_n^{(2)}(kr)e^{jn\phi} && \text{outgoing.}\end{aligned} \tag{12-53}$$

The scattering matrix S is defined by

$$\frac{a_n}{2} + b_n = \sum_m S_{nm} \frac{a_m}{2}$$

or

$$\tfrac{1}{2}[a] + [b] = [S]\tfrac{1}{2}[a]. \tag{12-54}$$

Since $[b] = [T][a]$, clearly $[S]$ is related to $[T]$:

$$[U] + 2[T] = [S], \tag{12-55}$$

where $[U]$ is a unit matrix.

Next, let us consider the conservation of power for a lossless object. If the object is lossless, the total incoming power should be equal to the total outgoing power. The total incoming power is given by

$$P_i = \int_0^{2\pi} d\phi \, |\psi_i^-(\bar{r})|^2. \tag{12-56}$$

If we evaluate this total power using the orthogonality of $\exp(jn\phi)$, we get

$$P_i = \frac{\pi}{2} \sum_n |A_n|^2, \tag{12-57}$$

where $A_n = a_n H_n^{(1)}(kr)$. Similarly, for the outgoing power, we get

$$\begin{aligned} P_0 &= \int_0^{2\pi} d\phi \, |\psi_i^+(\bar{r}) + \psi_s(\bar{r})|^2 \\ &= \frac{\pi}{2} \sum_n |A_n + 2B_n|^2, \end{aligned} \tag{12-58}$$

where $B_n = b_n H_n^{(1)}(kr)$ and we used the relation $|H_n^{(1)}(kr)| = |H_n^{(2)}(kr)|$.

We now express (12-57) and (12-58) in matrix form. We let A be the column matrix $[A_n]$. Then $P_i = (\pi/2)A^+ A$, where A^+ is the complex conjugate of the transpose of A, called the adjoint matrix (see the Appendix to Chapter 8, Section A). Similarly, (12-58) becomes

$$P_0 = \frac{\pi}{2}(A + 2TA)^+(A + 2TA).$$

Equating P_0 to P_i and writing them in matrix form, we get

$$A^+ A = A^+ S^+ S A, \tag{12-59}$$

where we used (12-55). From (12-59) it is clear that the scattering matrix S is unitary for a lossless object:

$$S^+ S = U. \tag{12-60}$$

12-6 *T*-MATRIX SOLUTION FOR SCATTERING FROM PERIODIC SINUSOIDAL SURFACES

In Section 7-5 we discussed the scattering of waves from periodic surfaces using the Rayleigh hypothesis. Even though that technique was limited to the case where the slope is less than 0.448, the solution is simple and has been used extensively for many problems in gratings and corrugated walls. More rigorous solutions can be obtained by using the T-matrix method.

Let us consider Dirichlet's problem for the surface defined by

$$\zeta = -h \cos\frac{2\pi x}{L}. \tag{12-61}$$

The incident wave ψ_i is given by

$$\psi_i = A_0 e^{+jqz - j\beta z}, \tag{12-62}$$

where $\beta = k \sin\theta_i$ and $q = k \cos\theta_i$ and θ_i is the angle of incidence. The scattered field ψ_s in the region $z > h$ is given by the space harmonics

$$\psi_s = \sum_n B_n e^{-jq_n z - j\beta_n x}, \tag{12-63}$$

where

$$\beta_n = \beta + \frac{2n\pi}{L},$$

$$q_n = \begin{cases} (k^2 - \beta_n^2)^{1/2} & \text{if } k > |\beta_n| \\ -j(\beta_n^2 - k^2)^{1/2} & \text{if } k < |\beta_n| \end{cases}.$$

Now we use the extinction theorem (12-29) for the region $z < -h$ and use the periodic Green's function

$$G(\bar{r}, \bar{r}') = \sum_{n=-\infty}^{\infty} \frac{1}{2jq_n L} e^{-jq_n(z'-z) - j\beta_n(x-x')} \quad \text{for } z < z'. \tag{12-64}$$

We also let

$$\psi_i(\bar{r}) = \sum_n a_n e^{jq_n z - j\beta_n x}. \tag{12-65}$$

For our problem with (12-62), $a_0 = A_0$ and $a_n = 0$ for $n \neq 0$. However, we can formulate the problem for the more general case given by (12-65). We substitute (12-64) and (12-65) into (12-29) and note that $\psi(\bar{r}') = 0$ on the surface. We also express the surface field $(\partial/\partial n')\psi(\bar{r}')$ by the following orthogonal space harmonics:

$$\frac{\partial}{\partial n'}\psi(\bar{r}')\, dS' = \sum_n \alpha_n e^{-j\beta_n x'}\, dx'. \tag{12-66}$$

Note that

$$\frac{\partial}{\partial n'} dS' = \left(\frac{\partial}{\partial z'} - \frac{\partial \zeta}{\partial x'}\frac{\partial}{\partial x'}\right) dx', \tag{12-67}$$

and therefore we used the expansion (12-66) rather than expanding $(\partial/\partial n')\psi(\bar{r}')$ in space harmonics.

Substituting (12-66) in (12-29), and letting $z' = \zeta = -h \cos(2\pi x/L)$, we get

$$[a] = [Q^-][\alpha], \tag{12-68}$$

where

$$Q^-_{mn} = \frac{1}{2jq_m} J_{|m-n|}(q_m h) e^{j(\pi/2)|m-n|}.$$

Next we use (12-63) in (12-32) with $\psi(\bar{r}') = 0$ on the surface, and the periodic Green's function

$$G(\bar{r}, \bar{r}') = \sum_n \frac{1}{2jq_n L} e^{-jq_n(z-z') - j\beta_n(x-x')} \qquad \text{for } z > z'. \tag{12-69}$$

We then get

$$[b] = -[Q^+][\alpha], \tag{12-70}$$

where

$$Q^+_{mn} = \frac{1}{2jq_m} J_{|m-n|}(q_m h) e^{-j(\pi/2)|m-n|}.$$

Combining (12-68) and (12-70), we get

$$\begin{aligned} [b] &= [T][a], \\ [T] &= -[Q^+][Q^-]^{-1}. \end{aligned} \tag{12-71}$$

The T-matrix solution in (12-71) should be exact and applicable for the periodic surface given by (12-61) even when the slope is higher than 0.448. In actual numerical calculations, however, the matrices become ill-conditioned if the slope becomes too high.

12-7 VOLUME INTEGRAL EQUATIONS FOR INHOMOGENEOUS BODIES: TM CASE

In the preceding sections we discussed two techniques of solving the scattering problem for complex objects: the surface integral equation method and the T-matrix method. Both methods are applicable to Dirichlet's, Neumann's, and two-media problems. However, these two methods cannot be applied to the problem of scattering by inhomogeneous dielectric bodies or anisotropic bodies. The volume integral equations discussed in this section can be applied to these problems. It should be noted, however, that the surface integral equations and the T-matrix method deal with the integral equations for the unknown "surface" field distribution, while the volume integral equations deal with the "volume" distribution of the unknown field, which requires larger matrices than the surface field for an object of similar size.

Let us consider a two-dimensional dielectric object (Fig. 12-5). The dielectric constant $\epsilon(x, y)$ is a function of position. From Maxwell's equations, we note that the problem of dielectric material with $\epsilon = \epsilon_0 \epsilon_r$ is equivalent to having the equivalent current source $\bar{J}_{\text{eq}}$.

$$
\begin{aligned}
\nabla \times \bar{E} &= -j\omega\mu_0 \bar{H}, \\
\nabla \times \bar{H} &= j\omega\epsilon\bar{E} = j\omega\epsilon_0 \bar{E} + \bar{J}_{\text{eq}}, \\
\bar{J}_{\text{eq}} &= j\omega\epsilon_0(\epsilon_r - 1)\bar{E}.
\end{aligned}
\tag{12-72}
$$

The total field $\bar{E}$ consists of the incident field $\bar{E}_i$ and the field $\bar{E}_s$ produced by the equivalent current $\bar{J}_{\text{eq}}$. Using the Hertz vector $\bar{\pi}$, we get (see Section 12-9 for three-dimensional objects)

$$
\begin{aligned}
\bar{E} &= \bar{E}_i + \bar{E}_s, \\
\bar{E}_s &= \nabla\nabla \cdot \bar{\pi} + k_0^2 \bar{\pi}, \\
\bar{\pi} &= \int G(\bar{r}, \bar{r}') \frac{\bar{J}_{\text{eq}}(\bar{r}')}{j\omega\epsilon_0} dV'.
\end{aligned}
\tag{12-73}
$$

Let us consider the two-dimensional TM waves ($\partial/\partial z = 0$ and $\bar{H}$ is transverse to the z axis) with the field components E_z, H_x, and H_y. The total field $\bar{E}$ consists of the incident field $\bar{E}_i$ and the scattered field $\bar{E}_s$ produced by $\bar{J}_{\text{eq}}$. For the TM wave, the

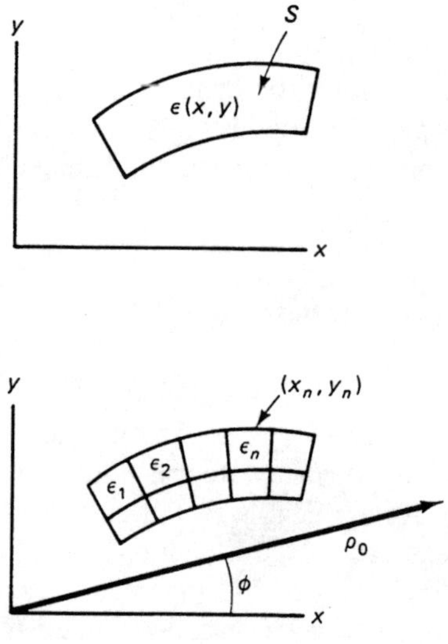

Figure 12-5 Scattering by an inhomogeneous body.

Hertz vector $\bar{\pi}$ and the electric field $\bar{E}$ have only the z component, and therefore we get from (12-73)

$$E_z(\bar{r}) = E_{zi}(\bar{r}) + E_{zs}(\bar{r}),$$
$$E_{zs}(\bar{r}) = -j\omega\mu_0 \int_S G(\bar{r}, \bar{r}')\bar{J}_{eq}(\bar{r}')\, dS', \tag{12-74}$$

where

$$G(\bar{r}, \bar{r}') = -j\tfrac{1}{4} H_0^{(2)}(k_0 \rho), \qquad \rho = |\bar{r} - \bar{r}'| \tag{12-75}$$

and $dS' = dx'\,dy'$ and the integral is over the cross section S. Rewriting (12-74), we get

$$E_z(\bar{r}) + \frac{jk_0^2}{4} \int_S [\epsilon_r(\bar{r}') - 1] E_z(\bar{r}') H_0^{(2)}(k_0 \rho)\, dS' = E_{zi}(\bar{r}). \tag{12-76}$$

Now we divide the cross section S into sufficiently small cells so that the electric field and the dielectric constant can be assumed to be constant over each cell (Fig. 12-25b). We denote by E_n and ϵ_n the electric field and the dielectric constant of the nth cell. Evaluating (12-76) at the center of the mth cell, we get

$$\sum_{n=1}^{N} C_{mn} E_n = E_{mi}, \qquad m = 1, 2, \ldots, N \tag{12-77}$$

where

$$C_{mn} = \delta_{mn} + \frac{jk_0^2}{4}(\epsilon_n - 1) \int_{\text{cell } n} H_0^{(2)}(k_0 \rho)\, dS'.$$

δ_{mn} is called the *Kronecker delta* and is defined by

$$\delta_{mn} = \begin{cases} 1 & \text{if } m = n \\ 0 & \text{if } m \neq n, \end{cases}$$

$$\rho = [(x_m - x')^2 + (y_m - y')^2]^{1/2},$$

(x_m, y_m) is the center of the cell m,

E_{mi} is the incident field at (x_m, y_m).

The integration over each cell in C_{mn} can be performed numerically. However, since each cell may have a different shape, the calculation may become tedious. A simple approximate method of calculating C_{mn} was proposed by Richmond. For a sufficiently small cell, we can replace the cell by a circular cell with the same cross-sectional area. The radius of the nth equivalent circular cell is a_n. It is then possible to obtain a simple analytical expression for C_{mn}. For $m \neq n$, we use the following expansion of $H_0^{(2)}(k_0 \rho)$ valid for $\rho_{mn} > \rho'$ (Fig. 12-6):

$$H_0^{(2)}(k_0 \rho) = \sum_{n=-\infty}^{\infty} e^{-jn(\phi - \phi')} J_n(k_0 \rho') H_n^{(2)}(k_0 \rho_{mn}). \tag{12-78}$$

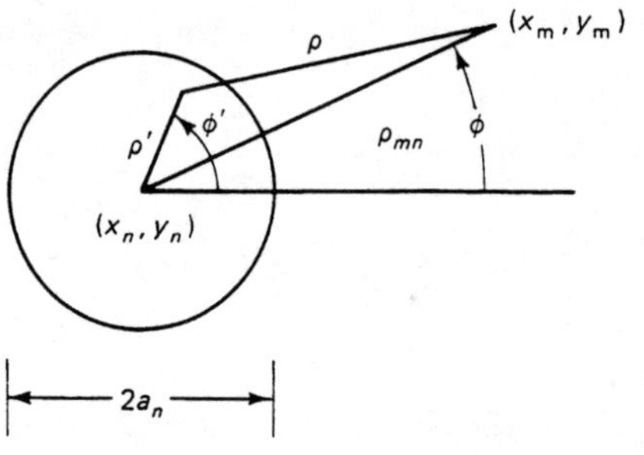

Figure 12-6 Equivalent circular cell.

Noting that $dS' = \rho'\, d\rho'\, d\phi'$ and

$$\int J_0(x)x\, dx = xJ_1(x), \tag{12-79}$$

we get

$$C_{mn} = \frac{j\pi k_0 a_n}{2}(\epsilon_n - 1)\, J_1(k_0 a_n)H_0^{(2)}(k_0 \rho_{mn}), \tag{12-80}$$

where a_n is the radius of the equivalent circular cell. For $m = n$, we use

$$\int H_0^{(2)}(x)x\, dx = xH_1^{(2)}(x),$$

$$H_1^{(2)}(x) \approx j\frac{2}{\pi x} \qquad \text{as } x \to 0.$$

We then get

$$C_{nn} = 1 + \frac{j\pi}{2}(\epsilon_n - 1)\left[k_0 a_n H_0^{(2)}(k_0 a_n) - \frac{j2}{\pi}\right]. \tag{12-81}$$

Equations (12-80) and (12-81) give convenient approximate expressions for C_{mn}.

Rewriting (12-77) in matrix form and inverting the matrix, we get

$$[E] = [C]^{-1}[E_i], \tag{12-82}$$

where $[E] = [E_n]$ and $[E_i] = [E_{ni}]$ are $1 \times N$ column matrices and $[C] = [C_{mn}]$ is an $N \times N$ square matrix.

Let us next consider the scattered field at the observation point (x, y) far from the object. The scattered field E_s from the cell n is given by $-C_{mn}E_n$ as (x_m, y_m) is moved to (x, y). Thus we get

$$E_s(x, y) = -\sum_{n=1}^{N} C_{mn}E_n, \tag{12-83}$$

where $x_m = x$ and $y_m = y$, and C_{mn} is given by (12-80). In the far-field zone of the object, we get (Fig. 12-6)

$$\rho_{mn} = [(x - x_n)^2 + (y - y_n)^2]^{1/2}$$
$$\to \rho_0 - x_n \cos\phi - y_n \sin\phi. \tag{12-84}$$
$$H_0^{(2)}(k_0 \rho_{mn}) \to \left(\frac{2j}{\pi k_0 \rho_0}\right)^{1/2} e^{-jk_0\rho_{mn}},$$

where ρ_{mn} in the amplitude is replaced by ρ_0.

The scattering pattern of the object when illuminated by a plane wave can be expressed in terms of the *echo width* $W(\phi)$ (Harrington, 1961):

$$W(\phi) = \lim_{\rho_0 \to \infty} 2\pi\rho_0 \left|\frac{E_s}{E_i}\right|^2. \tag{12-85}$$

Using (12-84), we get the echo width in units of wavelength:

$$\frac{W(\phi)}{\lambda} = \left(\frac{\pi}{2}\right) \left|\sum_{n=1}^{N} (\epsilon_n - 1) \frac{E_n}{|E_i|} k_0 a_n J_1(k_0 a_n) e^{jk_0(x_n \cos\phi + y_n \sin\phi)}\right|^2. \tag{12-86}$$

The cell size a_n in the foregoing method should not exceed $0.06/\sqrt{\epsilon_r}$ wavelengths for accurate results, and this cell size and the computation time determine the size of the object that can be handled by this volume integral method.

12-8 VOLUME INTEGRAL EQUATIONS FOR INHOMOGENEOUS BODIES: TE CASE

For the TM case we assumed that the medium is inhomogeneous, but isotropic, and therefore there is no coupling between E_z and the other components of the electric field. If the medium is anisotropic, the coupling among E_x, E_y, and E_z needs to be included. In this section we discuss the TE case for an inhomogeneous and isotropic body.

The analysis for the TM case discussed in Section 12-7 can be extended to the TE case. For the TE case, however, there are couplings between E_x and E_y. Let the incident wave be

$$\bar{E}_i = \hat{x}E_{ix} + \hat{y}E_{iy}. \tag{12-87}$$

The equivalent current $\bar{J}_{eq}$ is then given by

$$\bar{J}_{eq} = j\omega\epsilon_0(\epsilon_r - 1)\bar{E}$$
$$= \hat{x}J_x + \hat{y}J_y. \tag{12-88}$$

From (12-73), we get

$$\bar{E}_s = \hat{x}E_{sx} + \hat{y}E_{sy},$$

$$E_{sx} = \left(\frac{\partial^2}{\partial x^2} + k_0^2\right)\pi_x + \frac{\partial^2}{\partial x\,\partial y}\pi_y,$$

$$E_{sy} = \frac{\partial^2}{\partial x\, \partial y}\pi_x + \left(\frac{\partial^2}{\partial y^2} + k_0^2\right)\pi_y. \tag{12-89}$$

$$\pi_x(\bar{r}) = \int G(\bar{r}, \bar{r}')\frac{J_x}{j\omega\epsilon_0}\,dS',$$

$$\pi_y(\bar{r}) = \int G(\bar{r}, \bar{r}')\frac{J_y}{j\omega\epsilon_0}\,dS'.$$

Now we follow the procedure for the TM case and obtain the integral equations for E_x and E_y:

$$\begin{aligned} E_x(\bar{r}) - E_{sx}(\bar{r}) &= E_{ix}(\bar{r}), \\ E_y(\bar{r}) - E_{sy}(\bar{r}) &= E_{iy}(\bar{r}), \end{aligned} \tag{12-90}$$

where E_{sx} and E_{sy} are given in (12-89) as integrals involving J_x and J_y which are related to E_x and E_y through (12-88).

Now we divide the cross section into N small cells and then approximate each cell by a circular cell of radius a_n with the same cross-sectional area as that of the original cell. We then assume that the field and the dielectric constant are constant over each cell. In the nth cell, we let E_{xn} and E_{yn} denote the electric fields E_x and E_y, respectively. The relative dielectric constant for the nth cell is ϵ_n and the incident field at the center $\bar{r}_n$ of the nth cell has the components E_{ixn} and E_{iyn}. We can then convert (12-90) into the following $2N$ linear equations:

$$\begin{aligned} \sum_{n=1}^{N} (A_{mn} E_{xn} + B_{mn} E_{yn}) &= E_{ixm}, \\ \sum_{n=1}^{N} (C_{mn} E_{xn} + D_{mn} E_{yn}) &= E_{iym}. \end{aligned} \tag{12-91}$$

To calculate these coefficients A_{mn}, B_{mn}, C_{mn}, and D_{mn}, we need to calculate E_s at the mth cell due to the constant current $\bar{J}$ in the nth cell. First let us examine E_{xm} due to the current J_{xn} and J_{yn} in the nth cell ($m \neq n$):

$$\begin{aligned} J_{xn} &= j\omega\epsilon_0(\epsilon_n - 1)E_{xn}, \\ J_{yn} &= j\omega\epsilon_0(\epsilon_n - 1)E_{yn}. \end{aligned} \tag{12-92}$$

We make use of the expansion (12-78) and (12-89) and obtain

$$\begin{bmatrix} E_{sxm} \\ E_{sym} \end{bmatrix} = K \begin{bmatrix} h_{11} & h_{12} \\ h_{21} & h_{22} \end{bmatrix} \begin{bmatrix} J_{xn} \\ J_{yn} \end{bmatrix}, \tag{12-93}$$

where

$$h_{11} = [k_0 \rho y^2 H_0^{(2)}(k_0\rho) + (x^2 - y^2)H_1^{(2)}(k_0\rho)],$$

$$h_{12} = h_{21} = xy[2H_1^{(2)}(k_0\rho) - k_0\rho H_0^{(2)}(k_0\rho)],$$

$$h_{22} = [k_0 \rho x^2 H_0^{(2)}(k_0 \rho) + (y^2 - x^2) H_1^{(2)}(k_0 \rho)].$$

$$K = -\frac{\pi a_n J_1(k_0 a_n)}{2\omega\epsilon_0 \rho^3}.$$

Here we used

$$\rho = [(x_m - x_n)^2 + (y_m - y_n)^2]^{1/2},$$

$$x = x_m - x_n,$$

$$y = y_m - y_n.$$

To obtain E_{sx} and E_{sy} at the nth cell due to J_x and J_y in the same nth cell ($m = n$), we note that ρ is the distance between $\bar{r}$ and $\bar{r}'$ inside the same circular cell. Using (12-78) and (12-89) and performing integration inside the cell, we get

$$E_{sxn} = h_0 J_{xn},$$

$$E_{syn} = h_0 J_{yn}, \tag{12-94}$$

$$h_0 = -\frac{1}{4\omega\epsilon_0}[\pi k_0 a_n H_1^{(2)}(k_0 a_n) - 4j].$$

Using (12-93) and (12-94), we finally get the coefficients in (12-91). For $m \neq n$,

$$A_{mn} = K' h_{11},$$

$$B_{mn} = C_{mn} = K' h_{12}, \tag{12-95}$$

$$D_{mn} = K' h_{22},$$

where

$$K' = Kj\omega\epsilon_0(\epsilon_n - 1) = \frac{j\pi a_n J_1(k_0 a_n)(\epsilon_n - 1)}{2\rho^3};$$

h_{11}, h_{12}, h_{21}, and h_{22} are given in (12-93). For $m = n$, we have

$$A_{nn} = D_{nn} = 1 - h_0 j\omega\epsilon_0(\epsilon_n - 1),$$

$$B_{nn} = C_{nn} = 0, \tag{12-96}$$

where h_0 is given in (12-94).

The scattered field for a plane incident wave can be obtained by following the procedure for the TM case. At a distance far from the object, the scattered field has only a ϕ component, which is produced by the ϕ component of the electric field in the body:

$$E_\phi = E_y \cos\phi - E_x \sin\phi. \tag{12-97}$$

We therefore get the echo width in units of wavelength:

$$\frac{W(\phi)}{\lambda} = \left(\frac{\pi}{2}\right)\left|\sum_{n=1}^{N} (\epsilon_n - 1) k_0 a_n J_1(k_0 a_n) \frac{E_{\phi n}}{|E_i|} e^{j\psi}\right|^2, \tag{12-98}$$

where

$$E_{\phi n} = E_{yn} \cos\phi - E_{xn} \sin\phi,$$

$$\psi = k(x_n \cos\phi + y_n \sin\phi).$$

12-9 THREE-DIMENSIONAL DIELECTRIC BODIES

In the two preceding sections, we discussed the scattering by two-dimensional dielectric objects and formulated the integral equations for the electric field inside the body. This technique can be generalized to three-dimensional objects. However, careful attention must be paid to the singularity of the Green's function, which has been studied extensively (Yaghjian, 1980; van Bladel, 1964; Tai, 1971).

Let us start with (12-72) and (12-73). For three-dimensional objects, these equations are valid if the observation point $\bar{r}$ is outside the medium. However, to construct an integral equation for $\bar{E}(\bar{r})$, the observation point $\bar{r}$ must be inside the medium. At $\bar{r}$ inside the medium, the total field $\bar{E}(\bar{r})$ is equal to

$$\bar{E}(\bar{r}) = \bar{E}_i(\bar{r}) + \bar{E}_s(\bar{r}), \tag{12-99}$$

where $\bar{E}_i$ is the incident field and $\bar{E}_s$ is the scattered field. The scattered field $\bar{E}_s(\bar{r})$ is produced by the equivalent current $\bar{J}_{eq}(\bar{r}')$ and the point $\bar{r}'$ can coincide with $\bar{r}$. To investigate the case where $\bar{r} \neq \bar{r}'$ and the case where $\bar{r}$ can coincide with $\bar{r}'$, we divide the volume of the dielectric medium into a small spherical volume V_δ centered at $\bar{r}$ and the remaining volume $V - V_\delta$. In the volume $V - V_\delta$, $\bar{r}'$ does not coincide with $\bar{r}$ and therefore we can use (12-73). In the volume V_δ, it has been shown that the electric field is equal to $-\bar{J}_{eq}/3j\omega\epsilon_0$. Thus we write

$$\bar{E}_s(\bar{r}) = (-j\omega\mu_0) \lim_{\delta\to 0} \int_{V-V_\delta} \bar{\bar{G}}(\bar{r}, \bar{r}')\bar{J}_{eq}(\bar{r}')\, dv' - \frac{\bar{J}_{eq}(\bar{r})}{3j\omega\epsilon_0}, \tag{12-100}$$

where $\bar{\bar{G}}$ is the free-space electric dyadic Green's function given by

$$\begin{aligned} \bar{\bar{G}}(\bar{r}, \bar{r}') &= \frac{1}{k^2}\nabla \times \nabla \times (G_0\bar{\bar{I}}) \\ &= \left[\bar{\bar{I}} + \frac{\nabla\nabla}{k^2}\right] G_0(\bar{r}, \bar{r}'), \\ G_0(\bar{r}, \bar{r}') &= \frac{\exp[-jk|\bar{r} - \bar{r}'|]}{4\pi|\bar{r} - \bar{r}'|} \\ \bar{\bar{I}} &= \text{unit dyadic.} \end{aligned} \tag{12-101}$$

In Cartesian coordinates, $\bar{\bar{G}}$ is given by

$$\overline{G}(\bar{r},\bar{r}') = \begin{bmatrix} 1+\frac{1}{k^2}\frac{\partial^2}{\partial x^2} & \frac{1}{k^2}\frac{\partial^2}{\partial x\,\partial y} & \frac{1}{k^2}\frac{\partial^2}{\partial x\,\partial z} \\ \frac{1}{k^2}\frac{\partial^2}{\partial x\,\partial y} & 1+\frac{1}{k^2}\frac{\partial^2}{\partial y^2} & \frac{1}{k^2}\frac{\partial^2}{\partial y\,\partial z} \\ \frac{1}{k^2}\frac{\partial^2}{\partial x\,\partial z} & \frac{1}{k^2}\frac{\partial^2}{\partial y\,\partial z} & 1+\frac{1}{k^2}\frac{\partial^2}{\partial^2 z} \end{bmatrix} G_0(\bar{r},\bar{r}') \tag{12-102}$$

Substituting (12-100) into (12-99), we get the integral equation for $\overline{E}(\bar{r})$:

$$\left[1+\frac{\epsilon_r(\bar{r})-1}{3}\right]\overline{E}(\bar{r}) - k^2 \fint \overline{\overline{G}}(\bar{r},\bar{r}')[\epsilon_r(\bar{r}')-1]\overline{E}(\bar{r}')\,dv' = \overline{E}_i(\bar{r}), \tag{12-103}$$

where $\fint$ means $\lim_{\delta\to 0}\int_{V-V_\delta}$, and this can be approximated by taking a finite small volume V_δ.

In the above, we took a small spherical volume for V_δ. The formulations are also valid if V_δ is a cubic (Livesay and Chen, 1974) (see Fig. 12-7). However, if V_δ has other shapes, the electric field at $\bar{r}$ is different and depends on the shape. The electric field due to the volume V_δ with ellipsoidal, right circular cylinder, rectangular parallelepiped, or pillbox has been calculated (Yaghjian, 1980).

12-10 ELECTROMAGNETIC APERTURE INTEGRAL EQUATIONS FOR A CONDUCTING SCREEN

In the preceding sections we discussed scattering by conducting and dielectric bodies. In this section we consider scattering and wave transmission through apertures on conducting screens. An example is electromagnetic penetration through gaps in and apertures on an enclosure, which affect the performance of electronic systems inside.

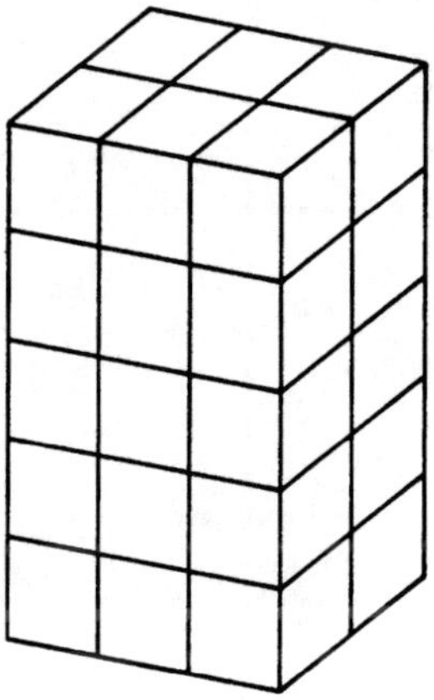

Figure 12-7 Dielectric body divided into many cubic cells.

Let us consider a conducting screen S with the aperture A illuminated from the left. According to the equivalence theorem, this problem can be replaced by a closed screen with the equivalent magnetic current sources (Fig. 12-8).

$$\bar{J}_{ms2} = \bar{E} \times \hat{n},$$
$$\bar{J}_{ms1} = \bar{E} \times (-\hat{n}), \tag{12-104}$$

where $\bar{E}$ is the actual field on the aperture.

The magnetic field on the left side of the screen is equal to the incident field $\bar{H}_i$, the field reflected from the screen with the aperture closed $\bar{H}_r$, and the contributions from $\bar{J}_{ms1}$. The contribution from $\bar{J}_{ms1}$ in front of the screen is the same as the contribution from $\bar{J}_{ms1}$ and that of its image in the absence of the screen. The image of $\bar{J}_{ms1}$ is in the same direction as $\bar{J}_{ms1}$ and $\bar{J}_{ms1}$ and its image is equal to $2\bar{J}_{ms1}$. Therefore, using the Franz formula (Section 6-9) or the magnetic Hertz vector in Section 2-9, we get

$$\bar{H}_1(\bar{r}) = \bar{H}_i(\bar{r}) + \bar{H}_r(\bar{r}) + [\nabla\nabla\cdot + k^2]\bar{\pi}_{m1}(\bar{r}),$$

where

$$\bar{\pi}_{m1}(\bar{r}) = \int_A G(\bar{r} - \bar{r}')\frac{2\bar{J}_{ms1}(\bar{r}')}{j\omega\mu}\,dS',$$
$$G(\bar{r}, \bar{r}') = \frac{e^{-jk|\bar{r}-\bar{r}'|}}{4\pi|\bar{r}-\bar{r}'|}. \tag{12-105}$$

The field on the right side of the screen is similarly given by

$$\bar{H}_2(\bar{r}) = [\nabla\nabla\cdot + k^2]\bar{\pi}_{m2}, \tag{12-106}$$

where

$$\bar{\pi}_{m2}(\bar{r}) = \int_A G(\bar{r}, \bar{r}')\frac{2\bar{J}_{ms2}}{j\omega\mu}\,dS'.$$

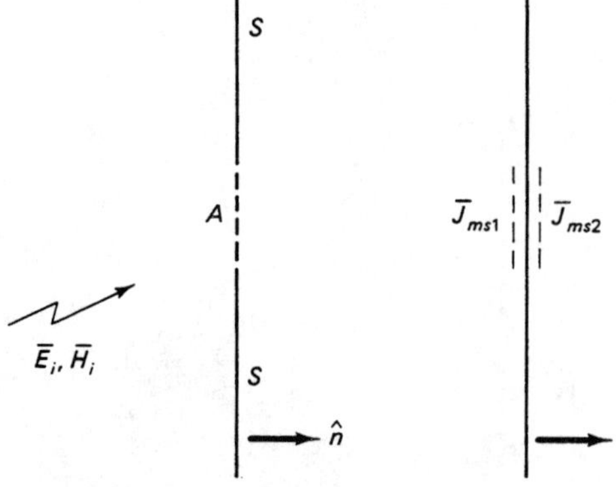

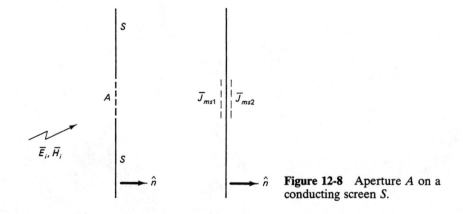

Figure 12-8 Aperture A on a conducting screen S.

Now the boundary condition on the aperture is the continuity of the tangential electric field and the tangential magnetic field. The continuity of the tangential electric field was used in (12-104). The continuity of the magnetic field is

$$\hat{n} \times \bar{H}_1 = \hat{n} \times \bar{H}_2 \qquad \text{on } A. \tag{12-107}$$

Substituting (12-105) and (12-106) into (12-107), we get the desired surface integral equation for $\bar{J}_{ms1}$:

$$\hat{n} \times [(\nabla\nabla\cdot + k^2)\bar{\pi}_m(\bar{r}) + \bar{H}_i(\bar{r})] = 0 \qquad \text{on } A, \tag{12-108}$$

where

$$\bar{\pi}_m(\bar{r}) = \int_A G(\bar{r}, \bar{r}')\frac{2\bar{J}_{ms1}(\bar{r}')}{j\omega\mu}\,dS',$$

and $\hat{n} \times [\bar{H}_i + \bar{H}_r] = 2\hat{n} \times \bar{H}_i$ on A. This is the basic integral equation for the unknown magnetic current $\bar{J}_{ms1}$ in the aperture, and in general it must be solved numerically using the moment method or other numerical techniques. The magnetic current $\bar{J}_{ms1} = \bar{E} \times (-\hat{n})$ must also satisfy the edge condition on the rim of the aperture.

On the aperture, the tangential magnetic field and the normal electric field satisfy the following simple relationships.

1. The tangential magnetic field is identical to the tangential incident field:

$$\hat{n} \times \bar{H}_1 = \hat{n} \times \bar{H}_2 = \hat{n} \times \bar{H}_i \qquad \text{on } A. \tag{12-109}$$

To derive these, we use (12-108) and (12-109) and write

$$\begin{aligned} \hat{n} \times \bar{H}_1(\bar{r}) &= 2\hat{n} \times \bar{H}_i(\bar{r}) - \hat{n} \times \bar{H}_i(\bar{r}) \\ &= \hat{n} \times \bar{H}_i(\bar{r}) \qquad \text{on } A. \end{aligned} \tag{12-110}$$

2. The normal component of the electric field is the same as that of the incident field:

$$\hat{n} \cdot \bar{E}_1 = \hat{n} \cdot \bar{E}_2 = \hat{n} \cdot \bar{E}_i \qquad \text{on } A. \tag{12-111}$$

To show this, consider the electric field. The electric field $\bar{E}_1$ on the left side of the screen is given by

$$\bar{E}_1(\bar{r}) = \bar{E}_i(\bar{r}) + \bar{E}_r(\bar{r}) - j\omega\mu\nabla \times \bar{\pi}_{m1}(\bar{r}). \tag{12-112}$$

Similarly, the electric field $\bar{E}_2$ on the right side is

$$\bar{E}_2(\bar{r}) = -j\omega\mu\nabla \times \bar{\pi}_{m2}(\bar{r}). \tag{12-113}$$

Noting that $\bar{\pi}_{m1} = -\bar{\pi}_{m2}$, the normal components of $\bar{E}_1$ and $\bar{E}_2$ are given by

$$\begin{aligned} \hat{n} \cdot \bar{E}_1 &= \hat{n} \cdot (\bar{E}_i + \bar{E}_r) + \hat{n} \cdot (-j\omega\mu\nabla \times \bar{\pi}_m) \\ \hat{n} \cdot \bar{E}_2 &= \hat{n} \cdot (j\omega\mu\nabla \times \pi_m). \end{aligned} \tag{12-114}$$

The normal components of $\bar{E}$ must also be continuous on A, and therefore we get

$$\hat{n} \cdot \bar{E}_1 = \hat{n} \cdot \bar{E}_2 = \frac{1}{2}\hat{n} \cdot (\bar{E}_i + \bar{E}_r). \tag{12-115}$$

However, the normal component of $\bar{E}_i$ and $\bar{E}_r$ when the aperture is closed is twice the normal component of $\bar{E}_i$, proving (12-111).

12-11 SMALL APERTURES

If the aperture size is small compared with a wavelength, we can express the effects of the aperture field in terms of the equivalent magnetic and electric dipoles (Butler et al., 1978). Let us consider the electric field $\bar{E}_2$ (12-113) at the distance $R >> \lambda$ in the direction of a unit vector $\hat{o}$ in the region $z > 0$ (Fig. 12-9). Green's function $G(\bar{r}, \bar{r}')$ can then be approximated by

$$G(\bar{r}, \bar{r}') = \frac{1}{4\pi R} \exp(-jkR + jk\bar{r}' \cdot \hat{o}), \tag{12-116}$$

where $|\bar{r} - \bar{r}'| \simeq R - \bar{r}' \cdot \hat{o}$. Using this in (12-113) and noting that $\nabla = -jk\hat{o}$, we get

$$\bar{E}_2(\bar{r}) = j2k\frac{e^{-jkR}}{4\pi R}\hat{o} \times \int_A \bar{J}_{ms2}(\bar{r}')e^{jk\bar{r}' \cdot \hat{o}}\, dS'. \tag{12-117}$$

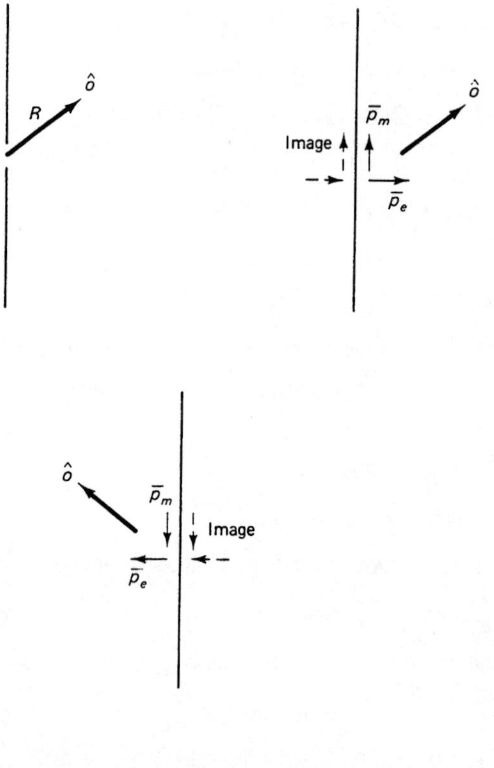

Figure 12-9 Small apertures.

Since the aperture size is much smaller than a wavelength, we expand $\exp(jk\bar{r}' \cdot \hat{o})$ in Taylor's series and write

$$\bar{E}_2(\bar{r}) = \sum_{n=0}^{\infty} \bar{E}_{2n}(\bar{r}). \tag{12-118}$$

It is possible to show that the term for $n = 0$ is identical to the field created by the magnetic dipole $\bar{p}_m$ in the presence of the screen, which is the sum of the contribution by $\bar{p}_m$ and its image $\bar{p}_m$ (Fig. 12-9).

$$\bar{E}_{20}(\bar{r}) = -\omega\mu k \frac{e^{-jkR}}{4\pi R} \hat{o} \times (2\bar{p}_m), \tag{12-119}$$

where

$$2\bar{p}_m = \frac{2}{j\omega\mu} \int \bar{J}_{ms2}(\bar{r}')\, dS'.$$

It can also be shown that the magnetic field and $\bar{p}_m$ have dimensions of (volume) × (magnetic field). The magnetic moment $\bar{p}_m$ is proportional to the tangential component of the magnetic field $\bar{H}_{sc}$ at the aperture when the aperture is closed (short circuited) and can be expressed by

$$\bar{p}_m = -\bar{\bar{\alpha}}_m \cdot \bar{H}_{sc}, \qquad \text{for the region } z > 0 \tag{12-120}$$

or in matrix form,

$$\begin{bmatrix} p_{mx} \\ p_{my} \end{bmatrix} = -\begin{bmatrix} \alpha_{mxx} & \alpha_{mxy} \\ \alpha_{myx} & \alpha_{myy} \end{bmatrix} \begin{bmatrix} H_{scx} \\ H_{scy} \end{bmatrix},$$

where $\bar{\bar{\alpha}}_m$ is called the *magnetic polarizability*. The value of $\bar{\bar{\alpha}}_m$ for a circular aperture of radius a is given by

$$\begin{aligned} \alpha_{mxx} &= \alpha_{myy} = \tfrac{4}{3}a^3, \\ \alpha_{mxy} &= \alpha_{myx} = 0. \end{aligned} \tag{12-121}$$

Similarly, noting (12-112) and (12-113), for the region $z < 0$, the equivalent magnetic moment $\bar{p}_m$ is given by

$$\bar{p}_m = +\bar{\bar{\alpha}}_m \cdot \bar{H}_{sc}. \tag{12-122}$$

Detailed derivations of the results above using the magnetic polarizability for elliptic apertures are given in Butler et al. (1978).

The next term for $n = 1$ in (12-118) gives the field due to an equivalent electric dipole and quadrupole moments:

$$\begin{aligned} \bar{E}_{21}(\bar{r}) &= \bar{E}_{21d}(\bar{r}) + \bar{E}_{21q}(\bar{r}), \\ \bar{E}_{21d} &= -\frac{k^2}{\epsilon} \frac{e^{-jkR}}{4\pi R} \hat{o} \times [\hat{o} \times (2\bar{p}_e)], \\ \bar{E}_{21q} &= \text{quadrupole}, \end{aligned} \tag{12-123}$$

where

$$\bar{p}_e = -\frac{\epsilon}{2}\int \bar{r}' \times \bar{J}_{ms2}(\bar{r}')\, dS'.$$

$\bar{E}_{21d}$ above is identical to the field created by the electric dipole moment $\bar{p}_e$ in the presence of the screen, which is equal to the sum of the contributions by $\bar{p}_e$ and its image $\bar{p}_e$ (Fig. 12-9). The dipole moment $\bar{p}_e$ is pointed in the z direction and is proportional to the normal component of the electric field E_{scz} at the aperture when the aperture is closed.

$$\bar{p}_e = p_e \hat{z} = \epsilon\alpha_e E_{scz}\, \hat{z}, \tag{12-124}$$

where α_e is called the *electric polarizability*. The value of α_e for a circular aperture of radius a is given by

$$\alpha_e = \tfrac{2}{3}a^3. \tag{12-125}$$

For the region $z < 0$, the dipole moment $\bar{p}_e$ is given by

$$\bar{p}_e = -\epsilon\alpha_e E_{scz}\, \hat{z}. \tag{12-126}$$

Limiting ourselves to the lowest orders of scattering, we can conclude that a small aperture is equivalent to the magnetic and electric dipoles given in (12-120), (12-122), (12-124), and (12-126).

In the above, we discussed the small aperture on a plane screen. This can be generalized to the coupling of electromagnetic waves between waveguides or cavities through a small aperture from region 1 to region 2. Let $(\bar{E}_s, \bar{H}_s)$ be the field in region 1 when the aperture is closed (short-circuited), and let E_{sz} be the normal component of $\bar{E}_s$ at the aperture and $\bar{H}_{st}$ be the tangential component of $\bar{H}_s$ at the aperture (Fig. 12-10). Then when the aperture is open, the field in region 1 is the sum of $(\bar{E}_s, \bar{H}_s)$ and the field produced by $\bar{p}_e$ and $\bar{p}_m$ placed on the closed aperture in region 1 (Fig. 12-10):

$$\begin{aligned} \bar{p}_e &= -\epsilon\alpha_e E_{sz}\, \hat{z}, \\ \bar{p}_m &= \bar{\bar{\alpha}}_m \cdot \bar{H}_{st}. \end{aligned} \tag{12-127}$$

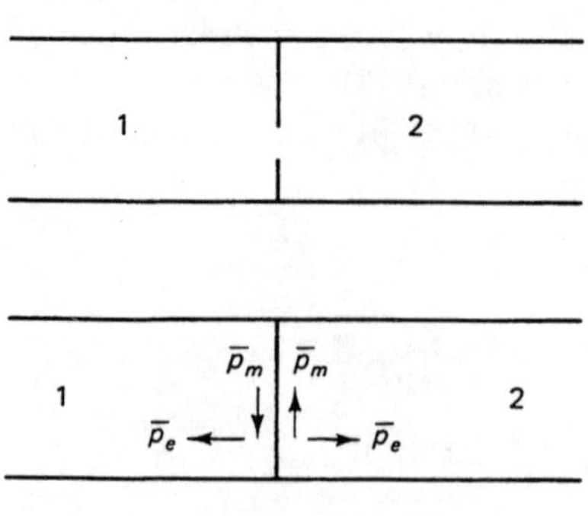

Figure 12-10 Coupling through a small hole.

The field in region 2 is then created by the following $\bar{p}_e$ and $\bar{p}_m$ placed on the closed aperture in region 2:

$$\begin{aligned} \bar{p}_e &= \epsilon\alpha_e E_{sz}\hat{z}, \\ \bar{p}_m &= -\bar{\bar{\alpha}}_m \cdot \bar{H}_{st}. \end{aligned} \tag{12-128}$$

12-12 BABINET'S PRINCIPLE AND SLOT AND WIRE ANTENNAS

Consider a slot on a conducting screen and its complementary problem where the slot is replaced by a conducting piece and the screen becomes an aperture. Because of the symmetry of these two complementary problems, we expect that the solution to one can be used to solve the other. This equivalent relationship is called *Babinet's principle*.

Consider the two problems shown in Fig. 12-11. In Fig. 12-11a, the conducting screen with a slot is illuminated by an electric current $\bar{J}$, and $(\bar{E}_1, \bar{H}_1)$ is the field diffracted by the slot. Now consider the complementary problem shown in Fig. 12-11b, where the metallic wire is excited by the source current $(\epsilon/\mu)^{1/2}\bar{J}_{mc}$. The field scattered by the wire is $(\bar{E}_{cs}, \bar{H}_{cs})$. Then Babinet's principle states that

$$\begin{aligned} \bar{E}_1 &= -\sqrt{\frac{\mu}{\epsilon}}\bar{H}_{cs}, \\ \bar{H}_1 &= \sqrt{\frac{\epsilon}{\mu}}\bar{E}_{cs}. \end{aligned} \tag{12-129}$$

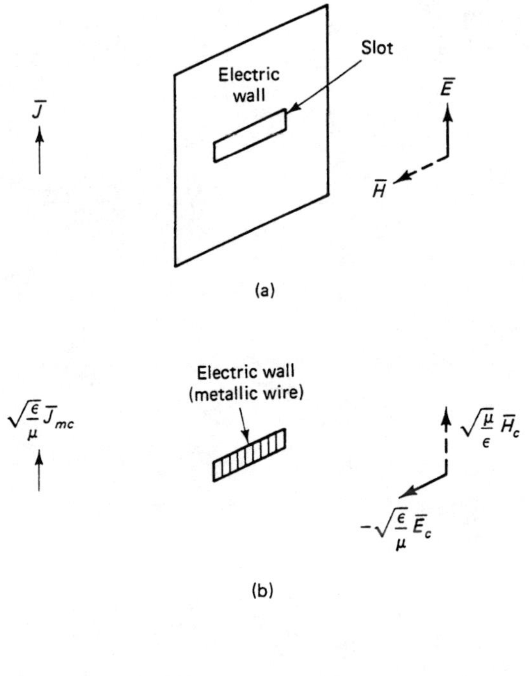

Figure 12-11 Babinet's principle applied to (a) a slot and (b) a wire.

The proof of (12-129) can be done in two stages. First we can consider three cases shown in Fig. 12-12. In part (a) we have a source $\bar{J}$ and a perfectly conducting screen (electric wall on which the tangential electric field is zero) with a slot and let the fields behind the screen be $\bar{E}_1$ and $\bar{H}_1$ (Fig. 12-12a). Next we consider a complementary problem where the screen and the slot are interchanged and the electric wall is replaced by a magnetic wall (tangential magnetic field is zero). Let the fields at the same position in this case be $\bar{E}_2$ and $\bar{H}_2$ (Fig. 12-12b).

Third, consider the fields in the absence of the screen and let the fields at the same position in this case be $\bar{E}_T$ and $\bar{H}_T$ (Fig. 12-12c). Then Babinet's principle states that

$$\begin{aligned} \bar{E}_1 + \bar{E}_2 &= \bar{E}_T, \\ \bar{H}_1 + \bar{H}_2 &= \bar{H}_T. \end{aligned} \tag{12-130}$$

To prove this, we note from (12-109) that the tangential magnetic field on the slot in part (a) is equal to the tangential magnetic field of the incident field. The tangential electric field E_1 is zero on S_1. Thus for part (a), the total tangential electric and magnetic fields (E_1, H_1) are

$$\begin{aligned} E_1 = E_T &= 0 \qquad \text{on screen } (S_1), \\ H_1 &= H_T \qquad \text{on slot } (S_2). \end{aligned} \tag{12-131}$$

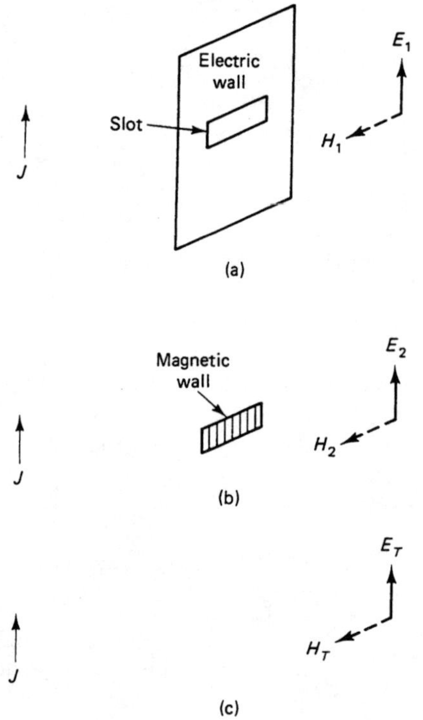

Figure 12-12 Three cases showing Babinet's principle (12-130).

For the magnetic screen of Fig. 12-12b,

$$\begin{aligned} E_2 &= E_T && \text{on aperture } (S_1), \\ H_2 &= H_T = 0 && \text{on magnetic wall } (S_2). \end{aligned} \tag{12-132}$$

Adding the above two fields, we get

$$\begin{aligned} E_1 + E_2 &= E_T && \text{on } S_1, \\ H_1 + H_2 &= H_T && \text{on } S_2. \end{aligned} \tag{12-133}$$

We now note that from the uniqueness theorem E_1 on S_1 and H_1 on S_2 uniquely determine all fields everywhere. Similarly, E_2 on S_1 and H_2 on S_2 uniquely determine all fields everywhere. But the superposition of (E_1, H_1) and (E_2, H_2) gives the field E_T on S_1 and H_T on S_2, which by the uniqueness theorem should for $z > 0$ give fields everywhere identical to the incident field. Thus this proves Babinet's principle (12-130).

In the above, we stated Babinet's principle using a conducting screen with an aperture and a disk with a magnetic wall. More practical and useful is Babinet's principle applied to a slot on a conducting screen and a wire. This can be obtained by noting the duality of Maxwell's equations and their invariance when all the quantities are replaced by the following with the subscript c.

$$\begin{aligned} E &\to \sqrt{\frac{\mu}{\epsilon}} H_c, \\ H &\to -\sqrt{\frac{\epsilon}{\mu}} E_c, \\ J_m &\to -\sqrt{\frac{\mu}{\epsilon}} J_c, \\ J &\to \sqrt{\frac{\epsilon}{\mu}} J_{mc}, \\ \rho_m &\to -\sqrt{\frac{\mu}{\epsilon}} \rho_c, \\ \rho &\to \sqrt{\frac{\epsilon}{\mu}} \rho_{mc}. \end{aligned} \tag{12-134}$$

Furthermore, if the electric and magnetic walls are interchanged, then we have identical boundary conditions, and the solution should be the same. For example, E and H in Fig. 12-13a are identical to $\sqrt{\mu/\epsilon}H_c$ and $-\sqrt{\epsilon/\mu}E_c$ in Fig. 12-13b. Using this duality, we see that the situation in Fig. 12-11a is identical to that in Fig. 12-11b. Thus (12-133) can be written as

$$\begin{aligned} E_1 + \sqrt{\frac{\mu}{\epsilon}} H_c &= E_T, \\ H_1 - \sqrt{\frac{\epsilon}{\mu}} E_c &= H_T. \end{aligned} \tag{12-135}$$

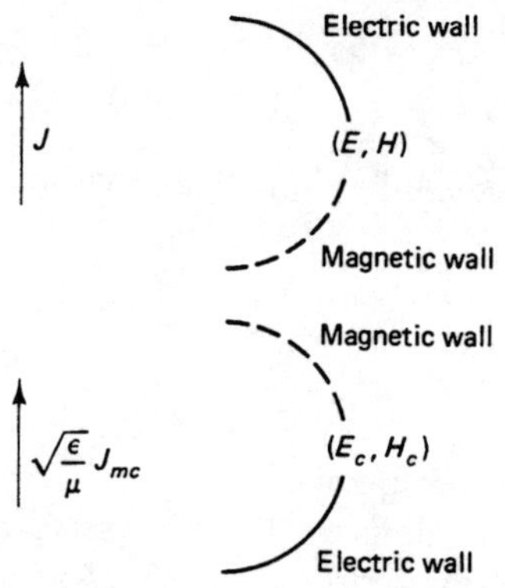

Figure 12-13 Duality principle.

If we write $\overline{H}_c$ and $\overline{E}_c$ as the sum of the incident and scattered waves, we have

$$\begin{aligned}\sqrt{\frac{\mu}{\epsilon}} H_c &= \sqrt{\frac{\mu}{\epsilon}} H_{cT} + \sqrt{\frac{\mu}{\epsilon}} H_{cs} \\ &= E_T + \sqrt{\frac{\mu}{\epsilon}} H_{cs}, \\ -\sqrt{\frac{\epsilon}{\mu}} E_c &= -\sqrt{\frac{\epsilon}{\mu}} E_{cT} - \sqrt{\frac{\epsilon}{\mu}} E_{cs} \\ &= H_T - \sqrt{\frac{\epsilon}{\mu}} E_{cs},\end{aligned} \tag{12-136}$$

where H_{cs} and E_{cs} are the scattered fields produced by the electric current on the wire. Then we can rewrite (12-135) as (12-129), proving Babinet's principle.

Using Babinet's principle, we can show the relationship between the impedance of a slot antenna and the impedance of its complementary wire antenna. Consider the slot and its complementary wire antenna shown in Fig. 12-14. The fields for the slot and for the wire antenna are $E_1 H_1$ and E_{cs} and H_{cs}, respectively, and are related by (12-129). For the slot antenna, the voltage V_s and the current I_s are given by

$$\begin{aligned}V_s &= \int_a^b E_1 \cdot dl, \\ I_s &= 2\int_c^d H_1 \cdot dl.\end{aligned} \tag{12-137}$$

The current I_s is given by $\int_c^d H_1 \cdot dl$ on one side of the screen plus $\int_d^c H_1 \cdot dl$ on the other side. For the wire antenna, the voltage V_w and the current I_w are given by

$$\begin{aligned}V_w &= \int_c^d E_{cs} \cdot dl, \\ I_w &= 2\int_b^a H_{cs} \cdot dl.\end{aligned} \tag{12-138}$$

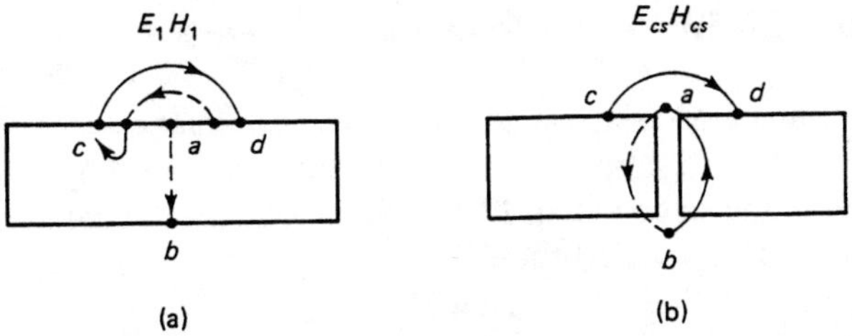

Figure 12-14 (a) Slot and (b) wire antennas.

Points c and d above are taken infinitesimally close to a. Using (12-129) and (12-137), we write (12-138) as

$$V_w = \eta \int_c^d H_1 \cdot dl = \frac{\eta I_s}{2},$$
$$I_w = \frac{2}{\eta} \int_a^b E_1 \cdot dl = \frac{2}{\eta} V_s, \tag{12-139}$$

and thus the wire impedance $Z_w = (V_w/I_w)$ and the slot impedance $Z_s = (V_s/I_s)$ are related by

$$Z_w Z_s = \frac{\eta^2}{4}, \qquad \eta = \sqrt{\mu/\epsilon}. \tag{12-140}$$

This important relationship is called *Booker's relation* and it shows that if the slot impedance (or wire antenna impedance) is known, the wire antenna impedance (or slot impedance) can be found from (12-140).

Mushiake pointed out that as a consequence of Booker's relation, if the slot and the wire have the same shape (Fig. 12-15), the antenna is self-complementary, and the impedance is invariant and equal to $\eta/2$ independent of the shape of the antenna and frequency

$$Z = \frac{\eta}{2}. \tag{12-141}$$

This is called the *Mushiake relation* and is obviously related to the concept of frequency-independent antennas (Rumsey, 1966).

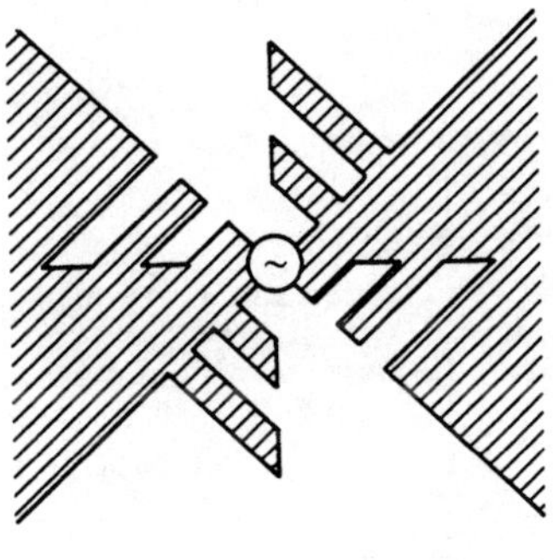

Figure 12-15 Self-complementary antenna.

12-13 ELECTROMAGNETIC DIFFRACTION BY SLITS AND RIBBONS

As examples of the electromagnetic scattering by apertures on a conducting screen discussed in Section 12-10, we consider a simpler two-dimensional problem $\partial/\partial y = 0$ of diffraction by a slit on a conducting screen (Fig. 12-16). We start with the general integral equation (12-108). Let us consider the TE ($\bar{E} \perp \hat{y}$) case. The $\bar{J}_{ms1} = \bar{E} \times (-\hat{z}) = E_x \hat{y}$. The incident wave with $\bar{E}_i = E_i \hat{x}$ and $\bar{H}_i = H_i \hat{y}$ is incident on the screen. Then (12-108) becomes

$$2j\omega\epsilon_0 \int_{-a}^{a} G(x,x')E_x(x')\,dx' = H_i, \tag{12-142}$$

where $G(x,x') = -(j/4)H_0^{(2)}(k|x - x'|)$.

This can be solved numerically using the method of moments. If the wave is normally incident, $H_i = A\exp(-jkz)$ and the solution has been obtained for a narrow slit:

$$j\omega\epsilon_0 E_x(x') = C_1\left[1 - \left(\frac{x'}{a}\right)^2\right]^{-1/2}, \tag{12-143}$$

$$C_1 = \frac{-A}{a(\ln ka + \ln(\gamma/4) + j\pi/2)}, \tag{12-144}$$

where $\gamma = 1.78107$ (Euler's constant) and $\ln\gamma = 0.5772$. Note that E_x satisfies the edge condition (see the Appendix to Chapter 7, Section C).

The far field on the right side of the screen is given by

$$H_y = -\frac{j}{2}a\pi C_1\left(\frac{2}{\pi kr}\right)^{1/2} e^{-jkr + j\pi/4}. \tag{12-145}$$

For TM($\bar{H} \perp \hat{y}$), $\bar{J}_{ms1} = -E_y\hat{x}$ and we get

$$\left(\frac{\partial^2}{\partial x^2} + k^2\right)\int_{-a}^{a} G(x,x')\frac{2E_y(x')}{j\omega\mu}\,dx' = H_{ix}. \tag{12-146}$$

This can be solved numerically. However, for a narrow slit, when the wave is normally incident, $E_{iy} = A\exp(-jkz)$ and the aperture field is given by

$$E_y = C_2\left[1 - \left(\frac{x'}{a}\right)^2\right]^{1/2}, \qquad C_2 = jkaA. \tag{12-147}$$

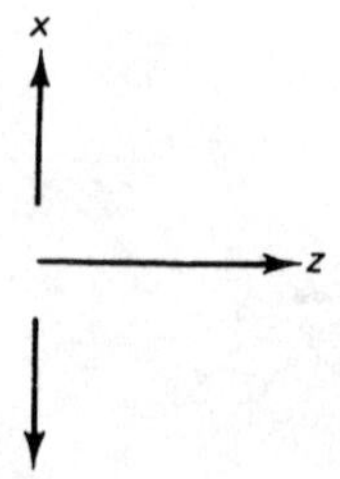

Figure 12-16 Diffraction by a slit.

The far field in the right side of the screen is given by

$$E_y = \frac{\pi k a C_2}{4}\left(\frac{2}{\pi k r}\right)^{1/2}\frac{z}{r}e^{-jkr+j(\pi/4)}. \tag{12-148}$$

The results above can be used to find the diffraction by ribbons using Babinet's principle.

12-14 RELATED PROBLEMS

Transient phenomena can be treated in the frequency domain and then their Fourier transform yields the time-domain solution. Alternatively, the transient can be investigated in the time domain using time-stepping (Felsen, 1976). The finite-difference time-domain method (FDTD) is a useful numerical technique in a time-domain approach (Yee, 1966). Time-domain solutions can be expressed in a series of complex exponentials that correspond to the singularities in the Laplace transform. This was proposed by Baum in 1971 and is called the singularity expansion method (SEM) (see Baum, 1976; Tesche, 1973; Felsen, 1976; Uslenghi, 1978).

There are other numerical techniques to deal with scattering by complex bodies. Yasuura's method (Ikuno and Yasuura, 1978; Yasuura and Okuno, 1982), introduced in late 1960, makes use of the smoothing process on the mode-matching method. This is shown to be a powerful numerical technique with high accuracy and efficiency.

PROBLEMS

12-1. Consider a two-dimensional Dirichlet's problem for the sinusoidal surface defined by (12-61) illuminated by a plane wave (12-62). Find an integral equation similar to (12-3) and (12-5). Find the scattered wave using the Kirchhoff approximation.

12-2. For the sinusoidal surface of Problem 12-1, find integral equations for the two-media problem.

12-3. Consider a knife edge illuminated by a plane wave as shown in Fig. 11-18. For TM wave with the incident wave E_{zi} given in (11-122), find EFIE and MFIE for the surface current. Also for TE wave with (11-125), find EFIE and MFIE for the surface current.

12-4. Apply the T-matrix method to find the scattered wave from a conducting ellipsoidal cylinder whose surface is given by

$$\frac{x^2}{a^2}+\frac{y^2}{b^2}=1.$$

The incident wave is a plane wave polarized in the z direction.

12-5. Consider the two-dimensional problem shown in Fig. 12-5. If the object has a square cross section of $a \times a$, use four cells as shown in Fig. P12-5, and obtain the echo width.

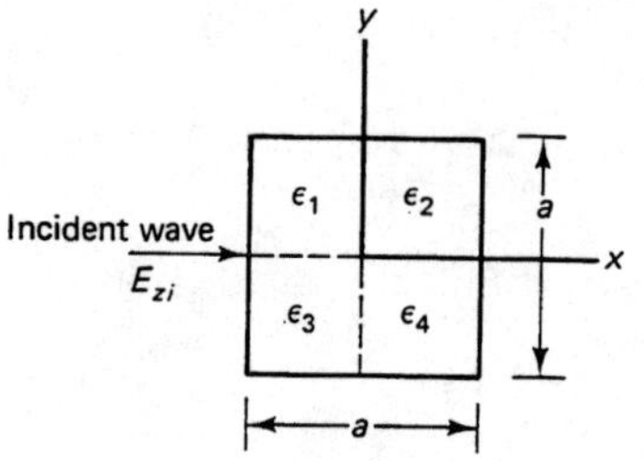

Figure P12-5 Scattering by a dielectric object. $a = 0.1\lambda_0$, $\varepsilon_1 = 1.1$, $\varepsilon_2 = 1.2$, $\varepsilon_3 = 1.3$, and $\varepsilon_4 = 1.4$.

12-6. A linearly polarized electromagnetic wave with power density P_0 at 10 GHz is normally incident on a small circular aperture of radius 2 mm in a conducting screen. Find the total power transmitted through the aperture.

12-7. A conducting screen has a slit of width 1 mm. The screen is normally illuminated by a TE electromagnetic wave with power density P_0 at 10 GHz. Find the power transmitted through a unit length of the slit. Do this with a TM incident wave.

13

Geometric Theory of Diffraction and Low-Frequency Techniques

In Chapter 12 we discussed the integral equation method for dealing with scattering and diffraction from objects. It should be noted that even though the integral equation itself is exact, the actual solution usually requires extensive numerical and matrix calculations, such as the moment method and the T-matrix method. It is clear, then, that the size of the matrix depends on the size of the object. In fact, in many applications, approximately 10 points per wavelength may often be needed to keep the error within a few percent; therefore, the exact integral equation method becomes impractical for a large object. On the other hand, if the object size is much smaller than a wavelength, the solution should approach a static case. In this chapter we examine the high-frequency technique, which is applicable to objects much greater than a wavelength, and the low-frequency technique, which is applicable to object sizes much smaller than a wavelength. For object size close to a wavelength, called the *resonance region*, the method of moment and other numerical techniques can be used effectively.

Extensive literature is available for the geometric theory of diffraction (GTD) and its related subjects. Important papers on GTD have been collected in the IEEE Press Reprint Series (Hansen, 1981). It is also discussed extensively in recent books by James (1976) and Jull (1981). UTD is covered in Kouyoumjian and Pathak (1974), and UAT is discussed in Deschamps et al. (1984) and Lee (1977).

13-1 GEOMETRICAL THEORY OF DIFFRACTION

As the wavelength λ approaches zero, the field can be described by *geometric optics*, but it contains no diffraction effect. An important extension of geometric optics to include diffraction was proposed by J. Keller and it is called the *geometrical theory of diffraction* or GTD (Keller, 1962). It introduces *diffracted rays* in addition to the usual geometric optical rays. Unlike geometric optics, the diffracted rays can enter the shadow regions. The GTD deals with the diffracted ray originating from edges, corners (vertices), and curved surfaces and is based on the following postulates:

1. Fermat's principle can be generalized and is applicable to diffracted rays, and thus the diffracted ray follows a curve that has a stationary optical path among all the paths between two points.
2. Diffraction is a local phenomena at high frequencies, and thus the magnitude of the diffracted ray depends on the nature of the incident wave and on the boundary in the neighborhood of the point of diffraction.
3. The phase of the diffracted ray is proportional to the optical length of the ray, and the amplitude varies to conserve the power in a narrow tube of rays.

According to the postulates above, the diffracted rays are proportional to the product of the incident wave and the *diffraction coefficient*, in analogy to the geometric optical rays reflected from the surface, where the reflected ray is proportional to the incident wave and the *reflection coefficient*. In general, the diffraction coefficient is proportional to $\lambda^{1/2}$ for edges, to λ for vertices, and decreases exponentially with λ^{-1} for surfaces. Therefore, the edge diffraction is strongest, the corner diffraction is weaker, and the surface diffraction is weakest. The diffraction coefficient is obtained by considering the asymptotic form of the exact solutions of the simpler *canonical* problem. For example, the edge diffraction coefficient is determined by the asymptotic form of the exact solution for an infinite wedge.

Let us illustrate the edge diffraction by considering a two-dimensional perfectly conducting knife edge (Fig. 13-1). We first consider the TM case (E_z, H_x, H_y).

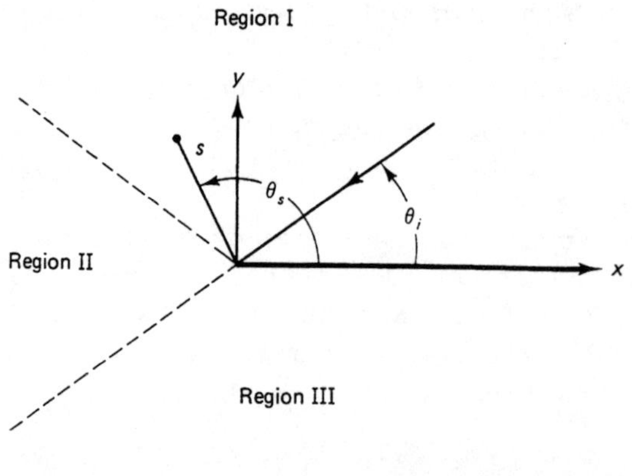

Figure 13-1 Edge diffraction.

The magnetic field is transverse to the z axis. The field component E_z satisfies the wave equation and Dirichlet's boundary condition (soft surface) ($E_z = 0$) on the conductor ($x > 0, y = 0$). This is equivalent to a scalar problem for the scalar field $\psi = E_z$ satisfying the wave equation and Dirichlet's condition $\psi = 0$ on the surface.

Let A_i be the incident wave ψ_i at the edge. Then according to the GTD, the diffracted ray ψ_d at (s, θ_s) is given by

$$\psi_d(s, \theta_s) = A_i D(\theta_s, \theta_i)\frac{e^{-jks}}{\sqrt{s}}, \tag{13-1}$$

where $D(\theta_s, \theta_i)$ is the diffraction coefficient, and $\exp(-jks)/s^{1/2}$ represents the cylindrical wave.

The determination of the diffraction coefficient $D(\theta_s, \theta_i)$ in (13-1) will be made by considering a canonical problem of plane wave diffraction by a knife edge. Since the exact solution of the knife edge diffraction is well known, the diffraction coefficient is obtained by comparing the asymptotic form of the exact solution with (13-1).

Let us consider a plane incident wave given by

$$\psi_i = A_i e^{jks\cos(\theta_s - \theta_i)}. \tag{13-2}$$

The exact solution to this knife-edge problem has already been discussed (Section 11-13). The exact total field at (s, θ_s) is given by

$$\psi(s, \theta_s) = \frac{A_i e^{j(\pi/4)}}{\sqrt{\pi}}[e^{jks\cos(\theta_s - \theta_i)} F(a_1) - e^{jks\cos(\theta_s + \theta_i)} F(a_2)], \tag{13-3}$$

where

$$F(a) = \int_a^\infty e^{-j\tau^2}\, d\tau \text{ is the Fresnel integral,}$$

$$a_1 = -(2ks)^{1/2}\cos\frac{\theta_s - \theta_i}{2},$$

$$a_2 = -(2ks)^{1/2}\cos\frac{\theta_s + \theta_i}{2}.$$

The diffraction coefficient D in (13-1) is then determined by examining the exact field (13-3) for a large distance s from the edge and comparing this asymptotic form with (13-1).

Let us first examine the total field ψ at a large distance from the edge in each of regions I, II, and III in Fig. 13-1. In region I we should have

$$\begin{aligned} \psi &= \psi_i + \psi_r + \psi_d, \\ \psi_r &= -A_i e^{jks\cos(\theta_s + \theta_i)}, \end{aligned} \tag{13-4}$$

where ψ_i is the incident field given in (13-2), ψ_r is the reflected field, and ψ_d is the diffracted field given in (13-1).

In region II we do not have the reflected wave, and thus

$$\psi = \psi_i + \psi_d. \tag{13-5}$$

In region III there is no incident or reflected wave, and thus

$$\psi = \psi_d. \tag{13-6}$$

Note that these asymptotic forms show discontinuous behavior at the boundaries between regions I, II, and III, although the exact field is continuous. Since (13-3) is the exact solution, it should reduce to (13-4), (13-5), and (13-6) for large kr. This can be shown by first noting that the Fresnel integral has different asymptotic forms; depending on the sign of a. If $a > 0$, we have

$$F(a) \approx \frac{e^{-ja^2}}{j2a}, \qquad a > \sqrt{10}. \tag{13-7}$$

However, if $a < 0$, we use the following identity:

$$F(a) + F(-a) = \sqrt{\pi}e^{-j(\pi/4)}, \tag{13-8}$$

and get

$$F(a) = \frac{e^{-ja^2}}{j2a} + \sqrt{\pi}e^{-j(\pi/4)}, \qquad a < 0 \quad \text{and} \quad |a| > \sqrt{10}. \tag{13-9}$$

The additional constant $\sqrt{\pi}e^{-j(\pi/4)}$ will be shown to correspond to the discontinuous behavior of the asymptotic forms (13-4), (13-5), and (13-6).

Now let us examine (13-3) in region I. Here we get $a_1 < 0$ and $a_2 < 0$, and thus using (13-9), we get

$$\psi = \psi_i + \psi_r + \psi_d, \tag{13-10}$$

$$\psi_d = A_i D(\theta_s, \theta_i) \frac{e^{-jks}}{\sqrt{s}}, \tag{13-11}$$

$$D(\theta_s, \theta_i) = -\frac{e^{-j(\pi/4)}}{2(2\pi k)^{1/2}} \left\{ \frac{1}{\cos[(\theta_s - \theta_i)/2]} - \frac{1}{\cos[(\theta_s + \theta_i)/2]} \right\}.$$

In region II we get $a_1 < 0$ and $a_2 > 0$, and therefore we have

$$\psi = \psi_i + \psi_d. \tag{13-12}$$

In region III, $a_1 > 0$ and $a_2 > 0$, and therefore

$$\psi = \psi_d, \tag{13-13}$$

where ψ_d in the above is given in (13-11). We conclude, therefore, that the asymptotic form of the diffracted field is given by (13-11) for plane-wave incidence.

We now generalize this to state that the diffracted ray from an edge is also given by (13-11) when an arbitrary incident field at the edge is given by A_i. Making use of this generalization, we are now in a position to construct a GTD solution for more complex problems. It should be noted, however, that the diffraction coefficient given in (13-11) becomes infinite at the *reflection boundary* between regions I and II where $\theta_s + \theta_i = \pi$, and at the *shadow boundary* between regions II and III where $\theta_s - \theta_i = \pi$. This is to be expected because a_1 or a_2 becomes zero at these boundaries and the asymptotic forms (13-7) and (13-9) cannot be valid in the *transition region* near these boundaries. The failure of GTD in the transition regions can be overcome by the *uniform geometric theory of diffraction* (UTD) or the *uniform asymptotic theory of edge diffraction* (UAT). These techniques will be discussed later.

For the TE case (H_z, E_x, E_y), the field component H_z satisfies the wave equation and Neumann's boundary condition on the conductor ($x > 0, y = 0$). In this case we use H_z in place of E_z. Then the exact solution is also given by (13-3) except that the minus sign for the second term is replaced by a plus sign. Therefore, we can summarize both the TM and TE cases as follows: Let ψ be E_z for TM and H_z for TE. Then if the incident wave ψ_i at the edge is denoted by A_i, the two-dimensional diffracted field at (s, θ_s) in the GTD approximation is given by

$$\psi_d(s, \theta_s) = A_i D(\theta_s, \theta_i) \frac{e^{-jks}}{\sqrt{s}}, \tag{13-14}$$

where D is the diffraction coefficient given by

$$D(\theta_s, \theta_i) = -\frac{e^{-j(\pi/4)}}{2(2\pi k)^{1/2}} \left\{ \frac{1}{\cos[(\theta_s - \theta_i)/2]} \mp \frac{1}{\cos[(\theta_s + \theta_i)/2]} \right\}. \tag{13-15}$$

The upper sign should be used for the TM (Dirichlet's) problem and the lower sign should be used for the TE (Neumann's) problem. Whenever convenient, we use D_s for Dirichlet's problem (soft screen) and D_h for Neumann's problem (hard screen). Note also that (13-14) is valid only in the region $|a_1| > \sqrt{10}$ and $|a_2| > \sqrt{10}$. If $|a_1| < \sqrt{10}$ or $|a_2| < \sqrt{10}$, UTD (Section 13-4) or other high-frequency techniques (Section 13-9) should be used.

13-2 DIFFRACTION BY A SLIT FOR DIRICHLET'S PROBLEM

Let us consider the two-dimensional problem of finding the field diffracted by a slit of width $2a$ on a conducting plane (Fig. 13-2). First, we consider Dirichlet's (TM)

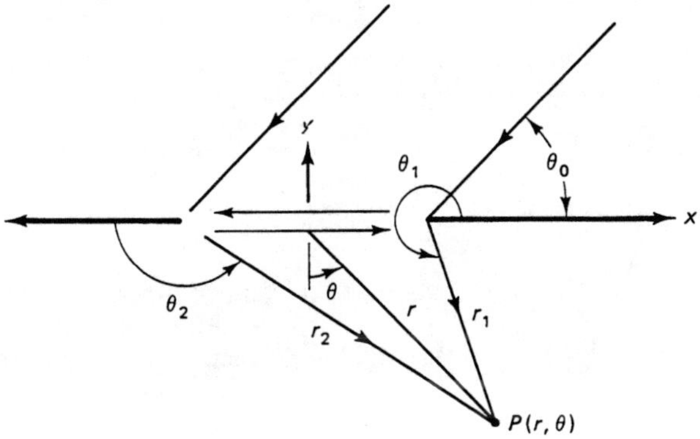

Figure 13-2 Diffraction by a slit of width $2a$.

problem, where $\psi = E_z$. We now obtain the GTD solution according to the formulation in Section 13-1. At the observation point P, the field diffracted once at the edge ($x = a, y = 0$) is given by

$$\psi_{s1} = A_1 D(\theta_1, \theta_0) \frac{e^{-jkr_1}}{\sqrt{r_1}}, \tag{13-16}$$

where $D(\theta_1, \theta_0)$ is given in (13-15).

Letting $D(r_1, \theta_1, \theta_0) = D(\theta_1, \theta_0) e^{-jkr_1}/\sqrt{r_1}$, we get the single diffraction (Fig. 13-3a)

$$\psi_{s1} = A_1 D(r_1, \theta_1, \theta_0), \tag{13-17}$$

where A_1 is the incident field at the edge.

If the incident wave ψ_i is a plane wave given by

$$\psi_i = A_0 e^{jk(x \cos\theta_0 + y \sin\theta_0)}, \tag{13-18}$$

then A_1 is given by

$$A_1 = A_0 e^{jka \cos\theta_0}. \tag{13-19}$$

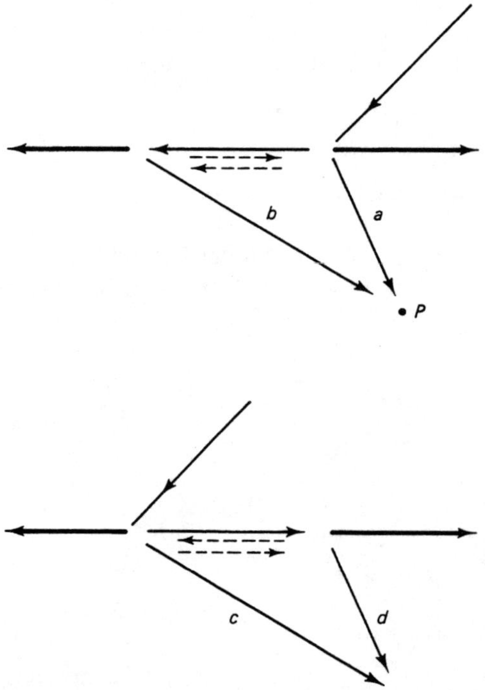

Figure 13-3 Four diffracted rays: a and c are single diffraction; b and d are double diffraction. Dashed lines represent multiple diffraction.

Similarly, the single diffraction by the edge at $(x = -a, y = 0)$ is given by (Fig. 13-3c)

$$\psi_{s2} = A_2 D(r_2, \theta_2, \pi + \theta_0), \tag{13-20}$$

where A_2 is the incident wave at the edge, and for a plane incident wave (13-18), it is given by

$$A_2 = A_0 e^{-jka \cos\theta_0}. \tag{13-21}$$

Let us next consider the double diffraction shown in Fig. 13-3b and d.

$$\begin{aligned} \psi_{d1} &= A_1 D(2a, \pi, \theta_0) D(r_2, \theta_2, \pi), \\ \psi_{d2} &= A_2 D(2a, \pi, \pi + \theta_0) D(r_1, \theta_1, \pi). \end{aligned} \tag{13-22}$$

The single and double diffraction terms are dominant and usually give good accuracy for $ka > 1.5$. Triple and multiple diffractions can also be included. We can add to ψ_{s1} the ray which traveled from the edge at $x = a$ to the edge at $x = -a$ and back to the edge at $x = a$ and then diffracted to P. This triply diffracted field is given by

$$\psi_{t1} = A_1 D(2a, \pi, \theta_0) D(2a, \pi, \pi) D(r_1, \theta_1, \pi). \tag{13-23}$$

The ray that traveled from $x = a$ to $x = -a$ n times and then diffracted to P is given by

$$[D(2a, \pi, \pi)]^{2n} \psi_{t1}.$$

Therefore, adding all these multiple diffracted rays shown in Fig. 13-3a, we get

$$\psi_{t1} \sum_{n=0}^{\infty} [D(2a, \pi, \pi)]^{2n} = \psi_{t1}[1 - D(2a, \pi, \pi)^2]^{-1}. \tag{13-24}$$

For the ray b, we get

$$\psi_{d1} \sum_{n=0}^{\infty} [D(2a, \pi, \pi)]^{2n} = \psi_{d1}[1 - D(2a, \pi, \pi)^2]^{-1}, \tag{13-25}$$

where $\psi_{d1} = A_1 D(2a, \pi, \theta_0) D(r_2, \theta_2, \pi)$. Similarly, we get all the rays for c and d in Fig. 13-3. The complete diffracted field ψ_d is therefore given by

$$\psi_d = \psi_{s1} + \psi_{s2} + (\psi_{d1} + \psi_{d2} + \psi_{t1} + \psi_{t2})[1 - D(2a, \pi, \pi)^2]^{-1}, \tag{13-26}$$

where ψ_{s1} and ψ_{s2} are given in (13-17) and (13-20), ψ_{d1} and ψ_{d2} are given in (13-22), ψ_{t1} is given in (13-23), and

$$\psi_{t2} = A_2 D(2a, \pi, \pi + \theta_0) D(2a, \pi, \pi) D(r_2, \theta_2, \pi).$$

Let us examine the diffracted field ψ_d in (13-26) at a large distance ($kr >> 1$) from the slit when the wave is normally incident on the slit. We then get (Fig. 13-2)

$$
\begin{aligned}
r_1 &= r - a \sin\theta, \\
r_2 &= r + a \sin\theta, \\
\theta_0 &= \frac{\pi}{2}, \\
\theta_1 &= \frac{3\pi}{2} + \theta, \\
\theta_2 &= \frac{\pi}{2} + \theta, \\
A_1 &= A_2 = A_0.
\end{aligned}
\tag{13-27}
$$

For the far field, the diffracted wave is cylindrical, and therefore we write

$$\psi_d = A_1 f_d(\theta)\left(\frac{k}{2\pi r}\right)^{1/2} e^{-jkr - j(\pi/4)}, \tag{13-28}$$

where $f_d(\theta)$ is the scattering amplitude. We now examine the single and double diffractions. For the far field, the scattering amplitudes for single diffraction are derived from (13-17) and (13-20):

$$
\begin{aligned}
f_{s1} &= \frac{e^{jka\sin\theta}}{2k}\left[\frac{1}{\sin(\theta/2)} - \frac{1}{\cos(\theta/2)}\right], \\
f_{s2} &= \frac{e^{-jka\sin\theta}}{2k}\left[\frac{-1}{\sin(\theta/2)} + \frac{-1}{\cos(\theta/2)}\right].
\end{aligned}
\tag{13-29}
$$

Note that $\theta = 0$ is the boundary between the illuminated and shadow regions for each edge, and therefore both ψ_{s1} and ψ_{s2} become infinite as $\theta \to 0$. However, in the far field, these two singularities cancel each other, producing a finite diffracted field:

$$f_{s1} + f_{s2} = \frac{1}{k}\left[\frac{j\sin(ka\sin\theta)}{\sin(\theta/2)} - \frac{\cos(ka\sin\theta)}{\cos(\theta/2)}\right]. \tag{13-30}$$

For a wide slit, the multiple diffractions ψ_{d1}, ψ_{d2}, ψ_{t1}, and ψ_{t2} can be neglected and (13-30) gives a good approximation. This can be shown to be consistent with the Kirchhoff diffraction theory for large ka (Section 6-3).

The scattering amplitudes of the single diffraction terms f_{s1} and f_{s2} are proportional to k^{-1}, as seen in (13-30). The scattering amplitudes of the double diffraction terms are f_{d1} and f_{d2}, and they are proportional to $k^{-3/2}$. The terms f_{t1} and f_{t2} are proportional to k^{-2}, and the multiple diffraction terms $[1 - D(2a, \pi, \pi)^2]^{-1}$ give all the higher negative powers of k. These terms can be obtained easily from the general formulations (13-26) with (13-27). For example, the scattering amplitudes in the forward direction $\theta = 0$ are given by

$$
\begin{aligned}
f_{s1}(0) + f_{s2}(0) &= j2a - \frac{1}{k}, \\
f_{d1}(0) = f_{d2}(0) &= \frac{e^{-jk2a - j(\pi/4)}}{\sqrt{\pi}(ka)^{1/2}k}, \\
f_{t1}(0) = f_{t2}(0) &= \frac{e^{-jk4a + j(\pi/2)}}{2\pi k^2 a}, \\
D^2(2a, \pi, \pi) &= \frac{e^{-jk4a - j(\pi/2)}}{2\pi ka}.
\end{aligned}
\tag{13-31}
$$

The transmission cross section σ of an aperture on a screen is defined as the ratio of the total power transmitted through the aperture to the incident power flux density when a plane wave is incident. According to the forward scattering theorem, the transmission cross section σ of the two-dimensional slit is given by the imaginary part of the forward scattering amplitude:

$$\sigma = \mathrm{Im}\, f_d(\theta = 0), \tag{13-32}$$

where f_d is as defined in (13-28) and σ is defined by the total transmitted power per unit power flux density per unit slit length. If we consider only the single diffraction f_{s1} and f_{s2}, we get

$$\frac{\sigma}{2a} = 1. \tag{13-33}$$

The transmission cross section is equal to the geometric cross section $\sigma_g = 2a$.

If we include the double diffraction terms f_{d1} and f_{d2}, we get

$$\frac{\sigma}{2a} = 1 - \frac{\cos(2ka - \pi/4)}{\sqrt{\pi}(ka)^{3/2}}. \tag{13-34}$$

This gives good agreement with the exact solution for ka greater than about 2 and fairly good agreement even for $ka > 1$.

13-3 DIFFRACTION BY A SLIT FOR NEUMANN'S PROBLEM AND SLOPE DIFFRACTION

The two-dimensional diffraction coefficient for a knife edge with Neumann's condition is given in (13-15) with $\psi = H_z$. The single diffraction ψ_{s1} and ψ_{s2} is given by

$$
\begin{aligned}
\psi_{s1} &= A_1 D(r_1, \theta_1, \theta_0), \\
\psi_{s2} &= A_2 D(r_2, \theta_2, \pi + \theta_0),
\end{aligned}
\tag{13-35}
$$

where

$$D(r_s, \theta_s, \theta_i) = \frac{e^{-jkr_s}}{\sqrt{r_s}} D(\theta_s, \theta_i),$$

$$D(\theta_s, \theta_i) = -\frac{e^{-j(\pi/4)}}{2(2\pi k)^{1/2}} \left[\frac{1}{\cos[(\theta_s - \theta_i)/2]} + \frac{1}{\cos[(\theta_s + \theta_i)/2]} \right].$$

Next consider the double refraction

$$\psi_{d1} = A_1 D(2a, \pi, \theta_0) D(r_2, \theta_2, \pi). \tag{13-36}$$

For Dirichlet's problem, this was finite and nonzero. However, for Neumann's problem, as seen in (13-15), $D(\theta_s, \theta_i) = 0$ if either $\theta_s = \pi$ or $\theta_i = \pi$ (Fig. 13-4). Therefore, the wave incident on the edge is proportional to $D(2a, \pi, \theta_0)$ and is zero. Since the diffracted field is obviously not zero, this means that we need to consider the higher-order term (Keller, 1962; Jull, 1981). The incident field H_z is zero at the edge, but its derivative, which is proportional to E_x, is not zero and contributes to the higher-order term. Consider a wave that is zero at the edge but is nonuniform in the neighborhood of the edge. The simplest such wave is

$$\psi_i = y A_0 e^{-jkx}. \tag{13-37}$$

Its normal derivative in the direction of $-y$ at the edge is

$$\frac{\partial \psi_i}{\partial n} = -\frac{\partial \psi_i}{\partial y} = -A_0. \tag{13-38}$$

This incident wave can also be described as the derivative of a plane wave with respect to θ_i.

$$\psi_i = -\frac{A_0}{jk} \frac{\partial}{\partial \alpha} [e^{jk(x \cos\alpha + y \sin\alpha)}]_{\alpha = \pi}. \tag{13-39}$$

The diffracted wave corresponding to this incident wave has been obtained by (Keller, 1962)

$$\psi_d(r_s, \theta_s) = \frac{\partial \psi_i}{\partial n} D'(\theta_s, \theta_i) \frac{e^{-jkr_s}}{\sqrt{r_s}}, \tag{13-40}$$

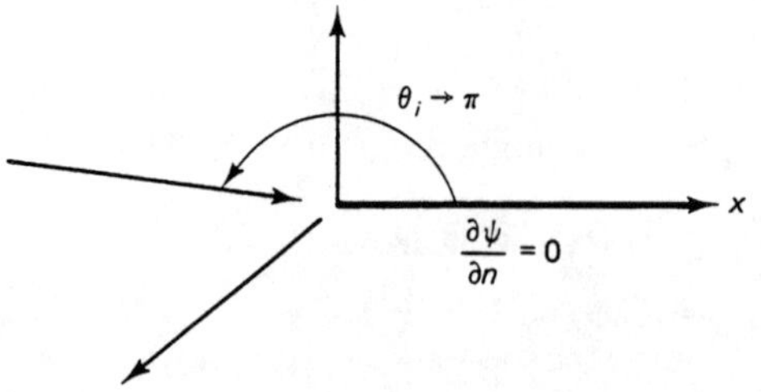

Figure 13-4 Diffraction by a hard half-plane when the incident angle is π.

where

$$D'(\theta_s, \theta_i) = \frac{1}{jk}\left[\frac{\partial}{\partial \alpha} D(\theta_s, \alpha)\right]_{\alpha=\pi}.$$

Using (13-40), we get the doubly refracted field ψ_{d1}.

$$\psi_{d1} = A_1 \frac{\partial}{\partial n} D(2a, \pi, \theta_0) D'(r_2, \theta_2, \pi), \tag{13-41}$$

where

$$\begin{aligned}
\frac{\partial}{\partial n} D(2a, \pi, \theta_0) &= \left[\frac{-1}{r}\frac{\partial}{\partial \alpha} D(r, \alpha, \theta_0)\right]_{\substack{r=2a \\ \alpha=\pi}} \\
&= \frac{\exp(-jk2a - j\pi/4)}{8(\pi k a)^{1/2} a} \frac{\cos(\theta_0/2)}{\sin^2(\theta_0/2)}, \\
D'(r_2, \theta_2, \pi) &= \left[\frac{1}{jk}\frac{\partial}{\partial \alpha} D(r_2, \theta_2, \alpha)\right]_{\alpha=\pi} \\
&= \frac{\exp(-jkr_2 + j\pi/4)}{2k(2\pi k r_2)^{1/2}} \frac{\cos(\theta_2/2)}{\sin^2(\theta_2/2)}.
\end{aligned}$$

Use of these higher-order terms involving the normal derivative is called *slope diffraction*.

For a far field, using the scattering amplitude, we write the diffracted field as

$$\psi_d = A_0 f_d(\theta)\left(\frac{k}{2\pi r}\right)^{1/2} e^{-jkr - j(\pi/4)}. \tag{13-42}$$

For normal incidence, we get the scattering amplitude for single diffraction:

$$\begin{aligned}
f_{s1} &= \frac{e^{jka\sin\theta}}{2k}\left[\frac{1}{\sin(\theta/2)} + \frac{1}{\cos(\theta/2)}\right], \\
f_{s2} &= \frac{e^{-jka\sin\theta}}{2k}\left[\frac{-1}{\sin(\theta/2)} + \frac{1}{\cos(\theta/2)}\right], \qquad (13\text{-}43) \\
f_{s1} + f_{s2} &= \frac{1}{k}\left[\frac{j\sin(ka\sin\theta)}{\sin(\theta/2)} + \frac{\cos(ka\sin\theta)}{\cos(\theta/2)}\right].
\end{aligned}$$

The transmission cross section using the single and double diffractions can be obtained by using the forward scattering theorem. We get

$$\frac{\sigma}{2a} = 1 - \frac{\sin(2ka - \pi/4)}{8\sqrt{\pi}(ka)^{5/2}}. \tag{13-44}$$

Note that the double diffraction term is proportional to $(ka)^{-5/2}$ for Neumann's problem, whereas it is $(ka)^{-3/2}$ for Dirichlet's problem, as shown in (13-34); therefore, the double diffraction is weaker for Neumann's problem than for Dirichlet's problem. The expression (13-44) can be used for $ka > 1.5$ without much error.

13-4 UNIFORM GEOMETRICAL THEORY OF DIFFRACTION FOR AN EDGE

We have already noted in Section 3-1 that the diffraction coefficients D_s and D_h in (13-15) become singular at $\theta_i \pm \theta_s = \pi$ and therefore cannot be used in the transition regions where $|a_1| < \sqrt{10}$ or $|a_2| < \sqrt{10}$ and a_1 and a_2 are given in (13-3). Kouyoumjian (1974) and others extended the GTD so that the diffraction coefficients remain valid in the transition regions. We will illustrate this technique called UTD using the two-dimensional knife edge problems discussed in Section 13-1 (Fig. 13-5).

Consider a line source located at (ρ, θ_i) and the observation point at (s, θ_s). The total field is given by (Fig. 13-5)

$$\begin{aligned} \psi &= \psi_i + \psi_r + \psi_d && \text{in region I,} \\ \psi &= \psi_i + \psi_d && \text{in region II,} \\ \psi &= \psi_d && \text{in region III,} \end{aligned} \tag{13-45}$$

where ψ_i, ψ_r, and ψ_d are the incident, reflected, and diffracted fields, respectively. If we use the GTD solution (13-14) for ψ_d, it becomes singular in the transition region. We can, however, use the Fresnel integral representation (13-3). This is continuous across the transition region, but it is the solution to a plane-wave incidence, not a cylindrical-wave incidence from a line source. The exact solution for a half-plane excited by a line source is available, but it is not in a closed form with known functions. We can, however, modify (13-3) for a line source located far from the edge ($k\rho >> 1$).

The approximate edge-diffracted field excited by a line source according to UTD (Kouyoumjian and Pathak, 1974) is then given by

$$\psi_d = A_i D(\theta_s, \theta_i) \frac{e^{-jks}}{\sqrt{s}}, \tag{13-46}$$

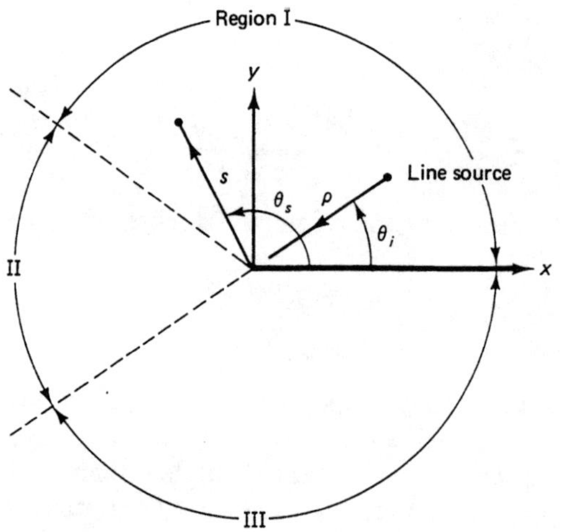

Figure 13-5 UTD for knife-edge diffraction.

where the incident cylindrical wave A_i at the edge is given by

$$A_i = A_0 \frac{e^{-jk\rho}}{\sqrt{\rho}}.$$

The UTD diffraction coefficient $D(\theta_s, \theta_i)$ is given by

$$D(\theta_s, \theta_i) = D(\theta_s - \theta_i) \mp D(\theta_s + \theta_i), \tag{13-47}$$

where the upper sign is for a soft (Dirichlet) surface and the lower sign is for a hard (Neumann) surface, and $D(\beta)$ is given by

$$\begin{aligned} D(\beta) &= -\frac{e^{-j(\pi/4)}}{2(2\pi k)^{1/2}} \frac{2j\sqrt{X}e^{jX}F(\sqrt{X})}{\cos(\beta/2)}, \\ X &= 2kL\cos^2\frac{\beta}{2}, \\ \sqrt{X} &= (2kL)^{1/2}\left|\cos\frac{\beta}{2}\right|, \\ F(a) &= \int_a^\infty e^{-j\tau^2}\,d\tau, \\ L &= \text{distance parameter} = \frac{s\rho}{s+\rho}. \end{aligned} \tag{13-48}$$

Now noting that $\sqrt{X}/\cos(\beta/2) = \pm(2kL)^{1/2}$, depending on the sign of $\cos(\beta/2)$, we can rewrite (13-48) as follows:

$$D(\beta) = e^{j(\pi/4)}(L/\pi)^{1/2} e^{jX} F(\sqrt{X}) \operatorname{sgn}(\beta - \pi), \tag{13-49}$$

where

$$\operatorname{sgn}(\beta - \pi) = \begin{cases} 1 & \text{if } \beta - \pi > 0, \\ -1 & \text{if } \beta - \pi < 0. \end{cases}$$

The function $D(\beta)$ is therefore discontinuous across $\beta = \pi$, which is the transition region. This discontinuity exactly cancels the discontinuity of ψ_i and ψ_r and yields the continuous total field everywhere.

To show this, consider the transition region between regions I and II where $\theta_s + \theta_i = \pi$. In this case, noting that $F(0) = (\sqrt{\pi})/2e^{-j(\pi/4)}$, we get, as θ_s approaches $\pi - \theta_i$,

$$D(\theta_s + \theta_i) = \begin{cases} -\dfrac{\sqrt{L}}{2} & \text{if } \theta_s < \pi - \theta_i \\ +\dfrac{\sqrt{L}}{2} & \text{if } \theta_s > \pi - \theta_i \end{cases}. \tag{13-50}$$

Therefore, for a soft surface, we have

$$\psi_d = A_i D(\theta_s - \theta_i)\frac{e^{-jks}}{\sqrt{s}} \pm A_0 \frac{e^{-jk(\rho+s)}}{\sqrt{\rho+s}}\frac{1}{2}, \tag{13-51}$$

where the upper sign is for $\theta_s < \pi - \theta_i$ and the lower sign is for $\theta_s > \pi - \theta_i$. Now the incident wave ψ_i is continuous across the transition angle, but ψ_r is discontinuous. For a soft surface, we have

$$\psi_r = \begin{cases} -A_0 \dfrac{e^{-jk(\rho+s)}}{\sqrt{\rho+s}} & \text{if } \theta_s < \pi - \theta_i \\ 0 & \text{if } \theta_s > \pi - \theta_i. \end{cases} \tag{13-52}$$

Adding (13-51) and (13-52), we have the continuous total field. See Figure 13-6 for a pictorial explanation. Summarizing this section, the total field is given by (13-45), (13-46), (13-47), and (13-49). We now have two choices. The ordinary GTD discussed in Sections 13-1, 13-2, and 13-3 is simple and useful outside the transition regions. The UTD discussed in this section makes use of Fresnel integrals and therefore requires more numerical work than the GTD. However, it can be used in all regions and gives continuous and useful results even in the transition regions.

13-5 EDGE DIFFRACTION FOR A POINT SOURCE

Up to this point, we have considered only two-dimensional problems of a knife edge excited by a plane wave or a line source. If a knife edge is illuminated by a point source, we need to take into account the angle of incidence (Fig. 13-7). According to Fermat's principle, the diffracted rays should be along the optical length, which is stationary among all paths from the source to the observation point. From this we conclude that the incident ray and the diffracted rays make equal angles with the edge. This angle is denoted by β_0 in Fig. 13-7.

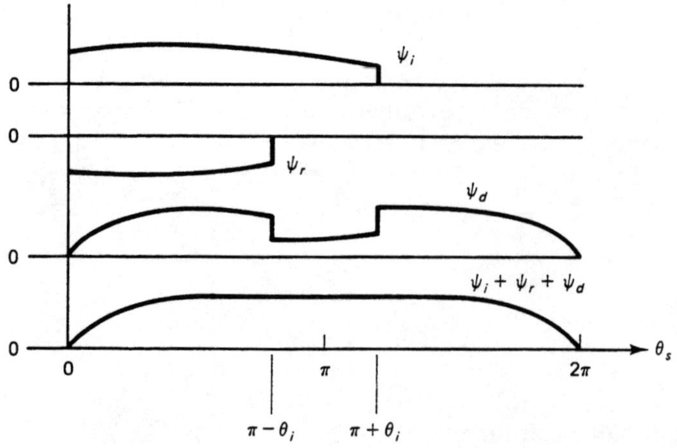

Figure 13-6 Pictorial explanation of UTD and cancellation of discontinuities of ψ_i, ψ_r, and ψ_d at the reflection ($\pi - \theta_i$) and the shadow ($\pi + \theta_i$) boundaries.

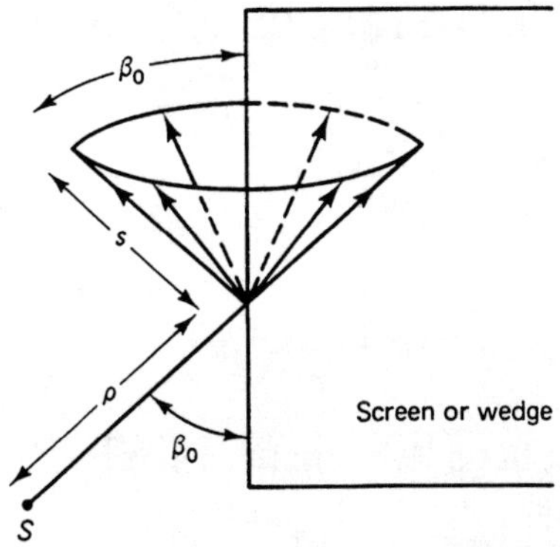

Figure 13-7 Cone of diffracted rays from an edge illuminated by a point source S.

Now we write the GTD solution for the diffracted field due to a point source. The scalar diffracted field ψ_0 at the observation point is given by

$$\psi_d = A_i D(\theta_s, \theta_i)\left[\frac{\rho}{s(s+\rho)}\right]^{1/2} e^{-jks}, \tag{13-53}$$

where $A_i = A_0 e^{-jk\rho}/\rho$ is the incident wave at the edge.

The *divergence factor* $[\rho/s(s+\rho)]^{1/2}$ is obtained by observing the conservation of power in a tube of geometric optical rays. In Fig. 13-8 the total power through the small cross-sectional area dA is conserved.

$$|\psi_d^2|\, dA = |\psi_0|^2\, dA_0. \tag{13-54}$$

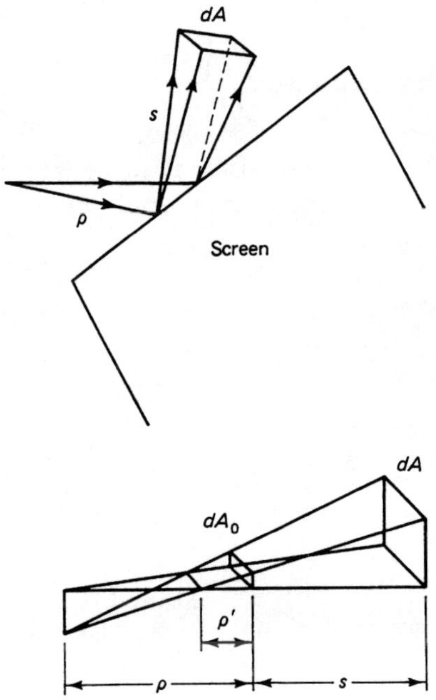

Figure 13-8 Conservation of power.

Note the ratio of the cross-sectional area is

$$\frac{dA}{dA_0} = \frac{(\rho + s)(\rho' + s)}{\rho\rho'}. \tag{13-55}$$

Therefore, we get

$$|\psi_d|^2 = \frac{\rho\rho'}{(\rho + s)(\rho' + s)}|\psi_0|^2. \tag{13-56}$$

As $\rho' \to 0$, $|\psi_0|$ becomes infinite, but $\rho'|\psi_0|^2$ is finite. Therefore, we conclude that ψ_d is proportional to $[\rho/s(\rho + s)]^{1/2}$ as $\rho' \to 0$.

The diffraction coefficient $D(\theta_s, \theta_i)$ for the region not close to the transition region can be obtained by examining the exact solution for the field when a plane wave is obliquely incident on the edge. Let the incident field be (Fig. 13-9)

$$\psi_i = A_i \exp[jkr \sin\beta_0 \cos(\theta - \theta_i) + jkz \cos\beta_0]. \tag{13-57}$$

Then the exact total field at (r, θ, z) is given by

$$\psi = \frac{A_i e^{jkz\cos\beta_0 + j(\pi/4)}}{\sqrt{\pi}}[e^{jkr\sin\beta_0\cos(\theta-\theta_i)}F(a_1) \mp e^{jkr\sin\beta_0\cos(\theta+\theta_i)}F(a_2)], \tag{13-58}$$

where

$$a_1 = -(2kr \sin\beta_0)^{1/2} \cos\frac{\theta - \theta_i}{2},$$

$$a_2 = -(2kr \sin\beta_0) \cos\frac{\theta + \theta_i}{2},$$

and the minus sign in front of the second term is for Dirichlet's problem and the plus sign is for Neumann's problem. If we consider the far-field approximation of (13-58)

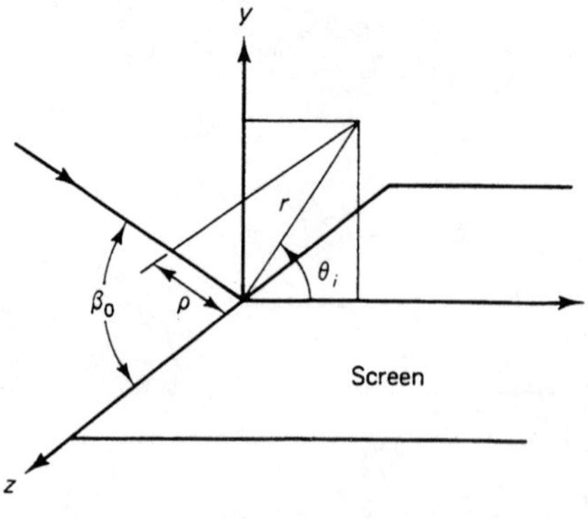

Figure 13-9 Plane wave incident on an edge from the direction (β_o, θ_i).

as we did in Section 13-1, and note that $s = r \sin \beta_0 - z \cos \beta_0$, we get the diffraction coefficient:

$$D(\theta_s, \theta_i) = -\frac{e^{-j(\pi/4)}}{2(2\pi k)^{1/2} \sin \beta_0} \left\{ \frac{1}{\cos[(\theta_s - \theta_i)/2]} \mp \frac{1}{\cos[(\theta_s + \theta_i)/2]} \right\}. \tag{13-59}$$

The UTD solution applicable to all angles, including the transition regions, is therefore given by

$$\begin{aligned} \psi &= \psi_i + \psi_r + \psi_d && \text{in region I} \\ &= \psi_i + \psi_d && \text{in region II} \\ &= \psi_d && \text{in region III,} \end{aligned}$$

$$\psi_d = A_i D(\theta_s, \theta_i) \left[\frac{\rho}{s(s+\rho)} \right]^{1/2} e^{-jks}, \tag{13-60}$$

$$A_i = A_0 \frac{e^{-jk\rho}}{\rho},$$

$$D(\theta_s, \theta_i) = D(\theta_s - \theta_i) \mp D(\theta_s + \theta_i),$$

where the upper sign is for Dirichlet's surface and the lower sign is for Neumann's surface.

$$D(\beta) = \frac{e^{j(\pi/4)}}{\sin \beta_0} \left(\frac{L}{\pi} \right)^{1/2} e^{jX} F(\sqrt{X}) \operatorname{sgn}(\beta - \pi),$$

$$X = 2kL \cos^2 \frac{\beta}{2},$$

$$\sqrt{X} = (2kL)^{1/2} \left| \cos \frac{\beta}{2} \right|, \tag{13-61}$$

$$F(a) = \int_a^\infty e^{-j\tau^2} \, d\tau,$$

$$L = \frac{s\rho \sin^2 \beta_0}{\rho + s}.$$

Equation (13-59) gives the GTD solution applicable to the regions outside the transition region and (13-60) is the UTD solution applicable to all angles. Both apply to the region not too close to the edge, $kL > 1.0$.

The scalar solutions above can easily be extended to the electromagnetic problem. We note that the components of the electric field parallel and perpendicular to the plane of incidence $(E^i_{\beta 0}, E^i_\phi)$ are proportional to E^i_z and H^i_z, respectively, and therefore we get the diffracted field $(E^d_{\beta 0}, E^d_\phi)$ using the diffraction coefficients for soft D_s and hard D_h surfaces (Fig. 13-10).

$$\begin{bmatrix} E^d_{\beta 0} \\ E^d_\phi \end{bmatrix} = \begin{bmatrix} -D_s & 0 \\ 0 & -D_h \end{bmatrix} \begin{bmatrix} E^i_{\beta 0} \\ E^i_\phi \end{bmatrix} \left[\frac{\rho}{s(\rho + s)} \right]^{1/2} e^{-jks}. \tag{13-62}$$

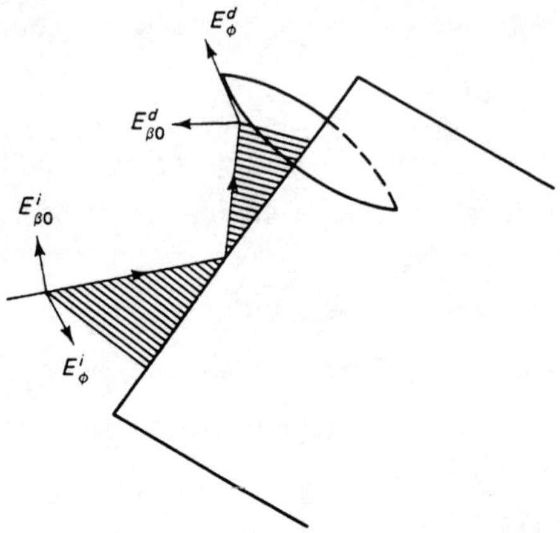

Figure 13-10 Electromagnetic diffraction by an edge.

13-6 WEDGE DIFFRACTION FOR A POINT SOURCE

Up to this point we have discussed GTD and UTD for a knife edge. We now consider GTD and UTD for a wedge with the angle $(2-n)\pi$ (Fig. 13-11). Exact solutions for a wedge illuminated by a point source, a line source, and a plane wave are available in the form of infinite series and Fourier integrals. However, they are not in a convenient closed form. For the field point far from the wedge, the exact solution can be evaluated to give an asymptotic closed-form expression. In this section we summarize the results for GTD and UTD.

The GTD solution for the diffracted field for a wedge is applicable to the region far from the edge and excluding the transition region. For a two-dimensional problem, the diffracted field is given by

$$\psi_d = A_i D(\theta_s, \theta_i)\frac{e^{-jks}}{\sqrt{s}}, \tag{13-63}$$

where A_i is the incident wave at the edge, and for a line source it is given by

$$A_i = A_0\frac{e^{-jk\rho}}{\sqrt{\rho}}.$$

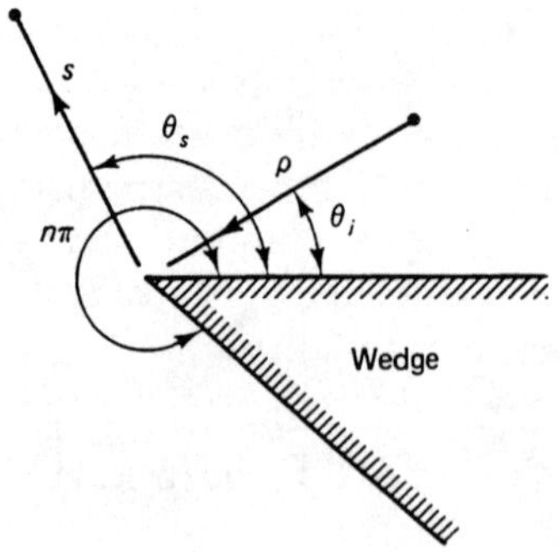

Figure 13-11 Wedge diffraction.

For a wedge illuminated by a point source, the diffracted field is given by (Fig. 13-7)

$$\psi_d = A_i D(\theta_s, \theta_i)\left[\frac{\rho}{s(s+\rho)}\right]^{1/2} e^{-jks}, \tag{13-64}$$

where $A_i = A_0 e^{-jk\rho}/\rho$. The GTD diffraction coefficient is given by

$$D(\theta_s, \theta_i) = D(\theta_s - \theta_i) \mp D(\theta_s + \theta_i), \tag{13-65}$$

where the upper (lower) sign is for a soft (hard) surface wedge, and

$$D(\beta) = \frac{e^{-j(\pi/4)} \sin(\pi/n)}{n(2\pi k)^{1/2} \sin\beta} \frac{1}{\cos(\pi/n) - \cos(\beta/n)}. \tag{13-66}$$

Note that for a knife edge, $n = 2$ and $D(\beta)$ reduces to the knife-edge form. Note also that $D(\theta_s, \theta_i)$ is singular at the reflection and shadow boundaries.

The UTD diffraction coefficient is valid for all regions and is given by Kouyoumjian and Pathak (1974, p. 1453).

$$D(\theta_s, \theta_i) = D_+(\theta_s - \theta_i) + D_-(\theta_s - \theta_i) \mp [D_+(\theta_s + \theta_i) + D_-(\theta_s + \theta_i)], \tag{13-67}$$

where the upper (lower) sign is for a soft (hard) surface,

$$D_\pm(\beta) = -\frac{e^{-j(\pi/4)}}{2n(2\pi k)^{1/2} \sin\beta_0} \cos\frac{\pi \pm \beta}{2n} F(X_\pm).$$

$$X_\pm = 2kL \cos^2\frac{2n\pi N^\pm - \beta}{2},$$

$N^\pm$ are integers that most nearly satisfy $2\pi n N^\pm - \beta = \pm\pi$,

$$F(X) = 2j\sqrt{|X|}\, e^{jX} \int_{\sqrt{|X|}}^{\infty} e^{-j\tau^2}\, d\tau.$$

13-7 SLOPE DIFFRACTION AND GRAZING INCIDENCE

We have discussed in Section 13-3 that if the incident wave is zero at the edge, we need to include the derivative of the incident wave. In general, if the incident wave is not slowly varying, the derivative or higher-order terms need to be included. We therefore write for a soft surface:

$$\psi_d = \left(\psi_i D_s + \frac{\partial \psi_i}{\partial n} d_s\right)\left[\frac{\rho}{s(\rho+s)}\right]^{1/2} e^{-jks}, \tag{13-68}$$

where

$$d_s = \frac{1}{jk \sin\beta_0} \frac{\partial}{\partial \theta_i} D_s(\theta_s, \theta_i),$$

and $D_s(\theta_s, \theta_i)$ is given in (13-67). For example, if the wave is incident on a soft wedge at the grazing angle, $\theta_i = 0$, the incident wave consists of the direct wave and the reflected wave. They are canceled on the surface and therefore the incident wave is zero. In this case, the slope term is dominant and given by

$$\psi_d = \frac{1}{2}\frac{\partial \psi_i}{\partial n}\frac{1}{jk\ \sin\beta_0}\frac{\partial}{\partial\theta_i}D_s\left[\frac{\rho}{s(\rho+s)}\right]^{1/2} e^{-jks}. \tag{13-69}$$

The factor $\frac{1}{2}$ is needed because only half of the incident wave is the direct wave incident on the edge.

If the wedge is hard and the wave is incident at the grazing angle, we have

$$\psi_d = \frac{1}{2}\psi_i D_h\left[\frac{\rho}{s(s+\rho)}\right]^{1/2} e^{-jks}. \tag{13-70}$$

Here the factor of $\frac{1}{2}$ is also needed.

13-8 CURVED WEDGE

If the wedge has a curved edge with a radius of curvature a, the divergence factor $[\rho/s(s+\rho)]^{1/2}$ must be modified to include the effects of the radius of curvature. The diffracted ray should appear as originating at a distance ρ' rather than ρ. To derive ρ', first consider the case when the incident and diffracted rays are in the plane formed by the edge and the normal $\hat{n}$ (or the radius of curvature) (Fig. 13-12). We can then easily see that

$$\Delta\theta = \Delta\theta_1 + 2\Delta\theta_2. \tag{13-71}$$

Since $\Delta\theta = \Delta l\ \cos\phi/\rho'$, $\Delta\theta_1 = \Delta l\ \cos\phi/\rho$, and $\Delta\theta_2 = \Delta l/a$, we get

$$\frac{1}{\rho'} = \frac{1}{\rho} + \frac{2}{a\ \cos\phi}. \tag{13-72}$$

If we consider the incident and diffracted rays that are not in this plane, we need to consider the projection of these rays on this plane. Thus we get

$$\frac{1}{\rho'} = \frac{1}{\rho} - \frac{\hat{n}\cdot(\hat{\rho}-\hat{s})}{a\ \sin^2\beta_0}. \tag{13-73}$$

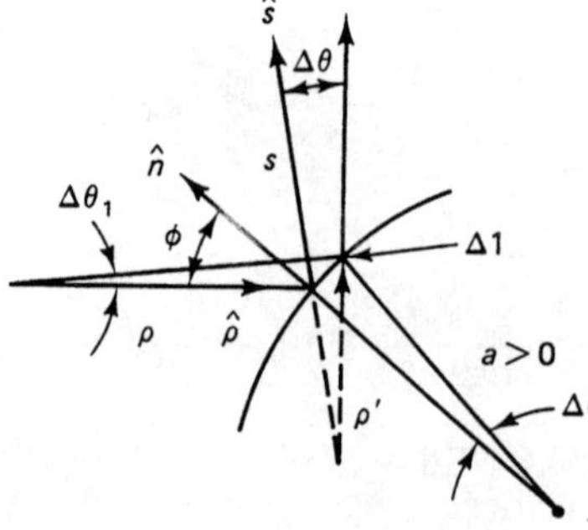

Figure 13-12 Divergence factor for a curved wedge.

The divergence factor should then be given by

$$\left[\frac{\rho'}{s(s+\rho')}\right]^{1/2}. \tag{13-74}$$

ρ' is therefore the distance between the edge and the caustic (focal point) of the diffracted ray.

For example, if a plane wave is normally incident on a circular aperture of radius a, $\rho \to \infty$, $\beta_0 = \pi/2$, $\hat{n} \cdot \hat{\rho} = 0$, and therefore (Fig. 13-13)

$$\frac{1}{\rho'} = \frac{\hat{n} \cdot \hat{s}}{a} = -\frac{\sin\theta_1}{a}. \tag{13-75}$$

The divergence factor is therefore

$$\left[\frac{\rho'}{s(s+\rho')}\right]^{1/2} = \frac{1}{\{s[1-(s\sin\theta_1)/a]\}^{1/2}}. \tag{13-76}$$

The diffraction coefficients for the curved wedge are the same as those for the straight wedge, since it is not affected by the radius of curvature.

13-9 OTHER HIGH-FREQUENCY TECHNIQUES

We have discussed the geometrical theory of diffraction (GTD) developed by Keller and the uniform geometric theory of diffraction (UTD) developed by Kouyoumjian and Pathak. In addition to UTD, the uniform asymptotic theory of edge diffraction (UAT) has been developed by Lewis, Boersma, Ahluwalia, Lee, and Deschamps (Lee, 1977, 1978). In this section we outline some basic ideas of UAT, but we will omit the detailed description.

We start with Keller's GTD and write it as

$$\psi(\text{Keller}) = \psi_g + \psi_d, \tag{13-77}$$

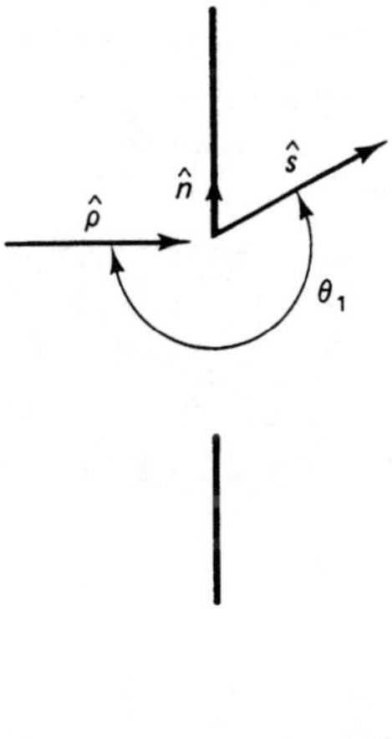

Figure 13-13 Wave incident on a circular aperture.

where ψ_g is the geometric optical solution and ψ_d is Keller's diffracted field. As we discussed in Section 13-1 and 13-4, ψ_d is singular at the shadow and reflection boundaries. The UTD can be written as

$$\psi(\text{UTD}) = \psi_g + \psi(\text{K-P}), \tag{13-78}$$

where $\psi(\text{K-P})$ is the diffracted field developed by Kouyoumjian and Pathak and is continuous in all regions. The UAT can be written as

$$\psi(\text{UAT}) = \psi(A) + \psi_d, \tag{13-79}$$

where ψ_d is Keller's GTD diffracted field and $\psi(A)$ is the new asymptotic term. $\psi(\text{UAT})$ is then continuous in all regions, and it may also be applicable in the near field of the edge, although its mathematical expressions may be somewhat complex.

Ufimtsev (1975) developed a technique called the physical theory of diffraction (PTD) in 1962. This is an extension of physical optics to include the field produced by the fringe current. For a perfectly conducting body, the PTD can be written as

$$\psi(\text{PTD}) = \psi(\text{PO}) + \psi(\text{fringe}), \tag{13-80}$$

where $\psi(\text{PO})$ is the physical optics field produced by the current $2\hat{n} \times \overline{H}_i$ ($\overline{H}_i$ is the incident magnetic field and $\hat{n}$ is the unit vector normal to the surface) and $\psi(\text{fringe})$ is the fringe field produced by the fringe current. Far from the edge, $\psi(\text{PO})$ approaches ψ_g and $\psi(\text{fringe})$ approaches ψ_d (Knott and Senior, 1974).

13-10 VERTEX AND SURFACE DIFFRACTION

We have shown that the GTD diffracted field from an edge or wedge is given by

$$\psi_d = \psi_i D \left[\frac{\rho}{s(s+\rho)}\right]^{1/2} e^{-jks}, \tag{13-81}$$

where ψ_i is the incident field at the edge, D the diffraction coefficient, $[\rho/s(s+\rho)]^{1/2}$ the divergence factor, s the distance from the edge to the observation point, and ρ the distance between the edge and the caustic of the diffracted ray [see (13-53)].

For the diffracted wave from a vertex, we should have a spherical wave,

$$\psi_d = \psi_i D \frac{e^{-jks}}{s}. \tag{13-82}$$

The diffraction coefficient D can be determined from the study of the appropriate canonical problems, such as the corner of an edge or the corner of a cube. Some studies have been done on this difficult problem (Bowman et al., 1969).

The GTD diffraction field for a smooth convex surface is obtained in the following manner. Consider the wave originating at P and observed at Q (Fig. 13-14). We assume that the diffracted ray obeys Fermat's principle that the total optical path length between P and Q is minimum. Thus the diffracted ray follows a

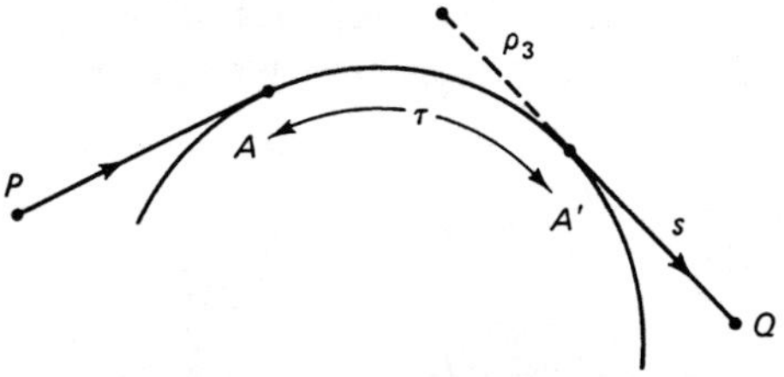

Figure 13-14 Diffracted ray for a smooth convex surface.

geodesic path from P to A and A' and Q. The wave originating at P is tangentially incident on the suface at A and is a creeping wave from A to A' behaving at each point as if it were a creeping wave on a circular cylinder with the radius equal to the radius of curvature of the actual surface. The wave then leaves the surface at A' and propagates to Q. The diffracted field at Q is therefore given by

$$\psi_d = \psi_i(A) D \left[\frac{\rho_3}{(\rho_3 + s)s} \right]^{1/2} e^{-jk(\tau + s)}, \tag{13-83}$$

where $\psi_i(A)$ is the incident field at A; the diffraction coefficient includes the attenuation constant $\alpha(\tau')$ of the creeping wave from A to $A'[\exp(-\int_0^\tau \alpha(\tau')\,d\tau')]$. The radius of curvature ρ_3 in (13-83) represents the curvature in the plane perpendicular to the plane containing the ray P–A–A'–Q. For a detailed discussion on this, see James (1976, Chap. 6).

13-11 LOW-FREQUENCY SCATTERING

Up to this point in this chapter, we have considered the GTD and some of its variations. These are an improvement over geometric optics. In general, the high-frequency approximations of the field (ψ, $\bar{E}$, and $\bar{H}$) may be obtained by expanding the field in inverse powers of $k = \omega/c$.

$$\bar{E} = \sum_{n=0}^{\infty} \frac{\bar{E}_n}{(-jk)^n}. \tag{13-84}$$

The first term ($n = 0$) represents geometric optics. In contrast, the low-frequency approximations are obtained by expanding the field in powers of $k = \omega/c$.

Let us consider the scattered field from a dielectric object whose size is small compared with a wavelength. We write Maxwell's equations as

$$\begin{aligned}
\nabla \times \bar{E} &= (-jk)\eta_0 \bar{H}, \\
\nabla \times \bar{H} &= (-jk)\left(-\frac{\epsilon_r}{\eta_0}\right)\bar{E}, \\
\nabla \cdot (\epsilon_r \bar{E}) &= 0, \\
\nabla \cdot \bar{H} &= 0,
\end{aligned} \tag{13-85}$$

where $k = \omega/c$ is the free-space wave number, $\eta_0 = \mu_0/\epsilon_0$ is the free-space characteristic impedance, and ϵ_r is the relative dielectric constant. We now expand the field in powers of k.

$$\bar{E} = \sum_{n=0}^{\infty} (-jk)^n \bar{E}_n,$$
$$\bar{H} = \sum_{n=0}^{\infty} (-jk)^n \bar{H}_n. \tag{13-86}$$

Substituting (13-86) into (13-85) and equating the like powers of $(-jk)$, we get the zeroeth-order equations:

$$\begin{aligned} \nabla \times \bar{E}_0 &= 0, \\ \nabla \times \bar{H}_0 &= 0, \\ \nabla \cdot (\epsilon_r \bar{E}_0) &= 0, \\ \nabla \cdot \bar{H}_0 &= 0. \end{aligned} \tag{13-87}$$

Note that the electric and magnetic fields are not coupled, and they are identical to the equations for the static case.

The first-order equations are

$$\begin{aligned} \nabla \times \bar{E}_1 &= \eta_0 \bar{H}_0, \\ \nabla \times \bar{H}_1 &= -\frac{\epsilon_r}{\eta_0} \bar{E}_0, \\ \nabla \cdot (\epsilon_r \bar{E}_1) &= 0, \\ \nabla \cdot \bar{H}_1 &= 0. \end{aligned} \tag{13-88}$$

Here $\bar{E}_1$ and $\bar{H}_1$ are generated by $\bar{H}_0$ and $\bar{E}_0$, respectively.

As an example, consider a plane wave incident on a dielectric ellipsoid. The incident wave $(\bar{E}_i, \bar{H}_i)$ is expanded in powers of k.

$$\bar{E}_i = \sum_{n=0}^{\infty} (-jk)^n \bar{E}_n,$$
$$\bar{H}_i = \sum_{n=0}^{\infty} (-jk)^n \bar{H}_n. \tag{13-89}$$

The zeroeth-order solution for $\bar{E}_0$ is identical to the electrostatic case (Stratton, 1941, p. 211) and is given in Section 10-6. The zeroeth-order solution for $\bar{H}_0$ is obtained by (13-87). The incident field is $\bar{H}_i = E_0/\eta_0$, and these equations are the same as those in free space and therefore $\bar{H}_0 = \bar{H}_i = \bar{E}_0/\eta_0$. Higher-order solutions are given in van Bladel (1964, p. 279).

Note that for the zeroeth-order electric field $\bar{E}_0$, we used the dielectric constant ϵ_r, which is in general complex. For a lossy medium with conductivity σ, we have $\epsilon_r = \epsilon_r' - j\epsilon_r'' = \epsilon_r' - j\sigma/\omega\epsilon_0$. If the frequency is low but not zero, we should use the complex ϵ_r rather than separating ϵ_r into ϵ_r' and σ. If dc fields are involved, we should rewrite Maxwell's equations as

$$
\begin{aligned}
\nabla \times \bar{E} &= (-jk)\eta_0 \bar{H}, \\
\nabla \times \bar{H} &= (-jk)\left(-\frac{\epsilon_r'}{\eta_0}\right)\bar{E} + \sigma\bar{E}, \\
\nabla \cdot \bar{H} &= 0, \\
\nabla \cdot \left[\sigma\bar{E} - (-jk)\frac{\epsilon_r'}{\eta_0}\bar{E}\right] &= 0.
\end{aligned}
\tag{13-90}
$$

We can then expand $\bar{E}$ and $\bar{H}$ in powers of k and obtain the zeroeth-order equations.

$$
\begin{aligned}
\nabla \times \bar{E}_0 &= 0, \\
\nabla \times \bar{H}_0 &= \sigma\bar{E}_0, \\
\nabla \cdot (\sigma\bar{E}_0) &= 0, \\
\nabla \cdot \bar{H}_0 &= 0.
\end{aligned}
\tag{13-91}
$$

This is the equation for magnetostatics. Solutions for an ellipsoidal conductor are given by Stratton (1941, p. 207).

Let us next consider the dc current distribution in a conducting body. Since $\nabla \times \bar{E}_0 = 0$ from (13-91), we get

$$\bar{E}_0 = -\nabla V. \tag{13-92}$$

Substituting this into the divergence equation in (13-91), we get

$$\nabla \cdot (\sigma \nabla V) = 0. \tag{13-93}$$

The current density $\bar{J}$ is given by

$$\bar{J} = \sigma\bar{E}_0 = -\sigma\nabla V. \tag{13-94}$$

As an example, consider the conducting body shown in Fig. 13-15. At the surface S_1, the voltage V_0 is applied and at S_2, the voltage is zero. The voltage V then satisfies (13-93). The boundary condition at the surface S is that the current normal to the surface is zero. From (13-94), this means that

$$\frac{\partial V}{\partial n} = 0 \qquad \text{on } S.$$

The total current I_0 is then given by

$$I_0 = \int_{S_1} \bar{J} \cdot \hat{n}\, ds = -\int_{S_1} \sigma \frac{\partial V}{\partial n}\, ds. \tag{13-95}$$

Numerical techniques such as the finite-element method are often employed to solve (13-93) with appropriate boundary conditions.

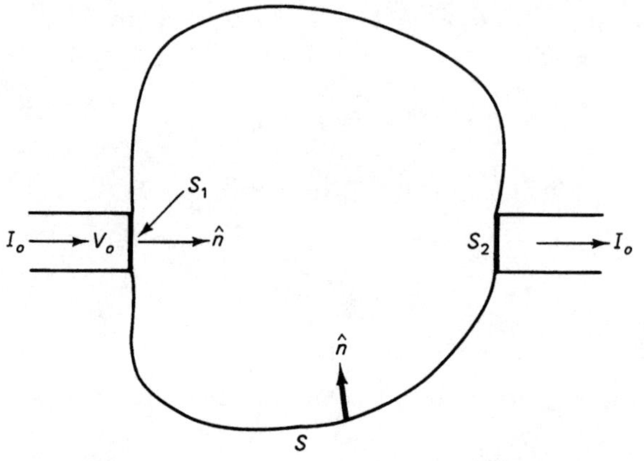

Figure 13-15 DC current distribution in a conducting body.

PROBLEMS

13-1. A conducting knife edge is illuminated by an electric line source I_z as shown in Fig. P13-1. Find and plot the field $E_z(x)$ at $y = 0, h = d = 5\lambda$ as a function of x. $\lambda = 300$ m.

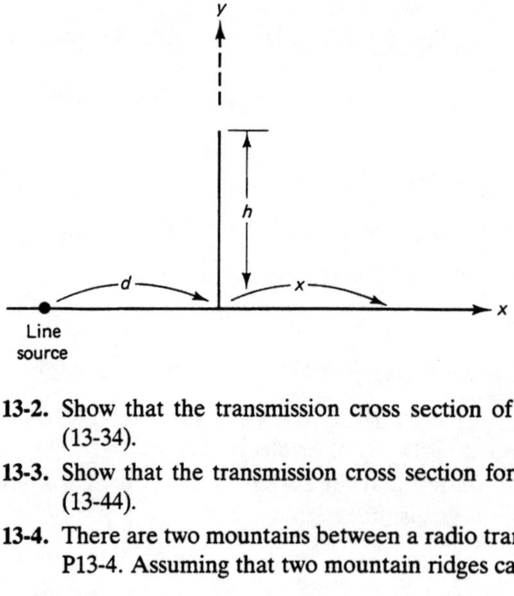

Figure P13-1 Diffraction by a knife edge.

13-2. Show that the transmission cross section of a slit for Dirichlet screen is given by (13-34).

13-3. Show that the transmission cross section for a slit on Neumann screen is given by (13-44).

13-4. There are two mountains between a radio transmitter and a receiver as shown in Fig. P13-4. Assuming that two mountain ridges can be approximated by two knife edges,

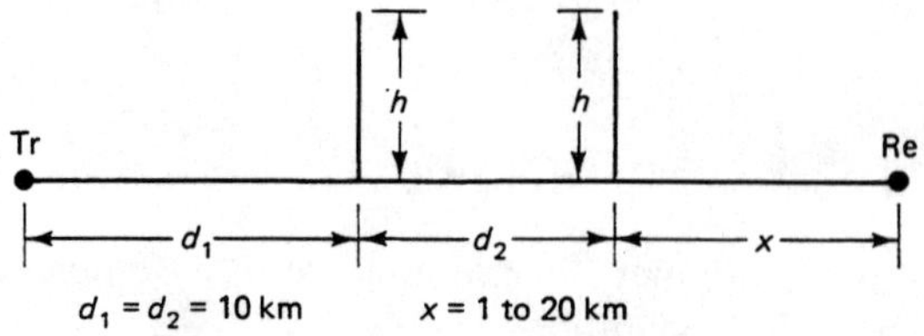

Figure P13-4 Double ridge diffraction.

calculate the received field normalized to the direct field in free space as a function of the distance x. Assume that the edges are soft.

13-5. A point source is located near a 90° wedge as shown in Fig. P13-5. Find the diffracted field $\psi(x)$. $h = d = 5\lambda$.

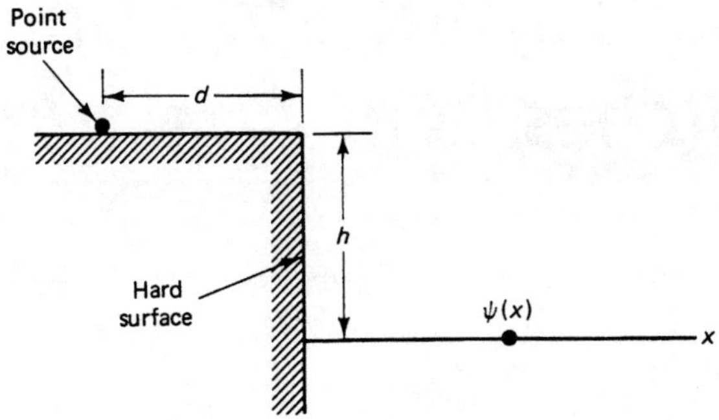

Figure P13-5 90° hard wedge.

13-6. Find the total resistance of the square plate shown in Fig. P13-6. Assume that the two terminal wires A and B are perfect conductors.

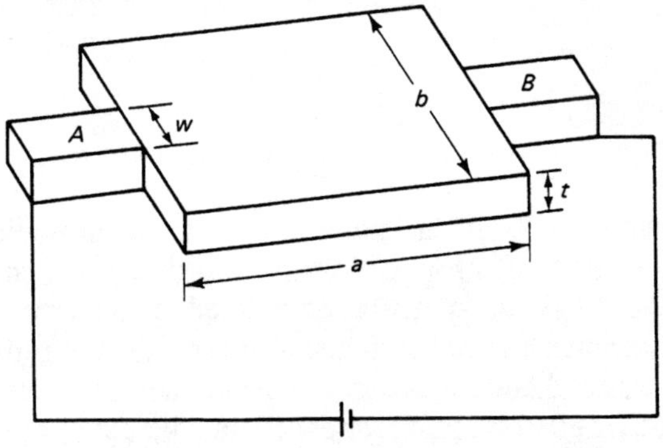

Figure P13-6 Square conducting plate with conductivity σ.

14

Planar Layers, Strip Lines, Patches, and Apertures

Planar dielectric layers and slabs are found in many applications. Examples are dielectric coating on conducting surface, propagation in the atmosphere and the ocean, strip lines, and periodic patches and apertures embedded in dielectric layers. In this chapter we first discuss excitation of waves in planar layers. We then consider strip lines, patches, and apertures in dielectric layers which are useful in microwave and millimeter wave applications. See Itoh (1987) for excellent collections of papers on planar structures, and Unger (1977) for planar waveguides. The spectral domain method is discussed in Yamashita and Mittra (1968), Itoh (1980), and Scott (1989).

14-1 EXCITATION OF WAVES IN A DIELECTRIC SLAB

Let us consider the excitation of TM waves on a dielectric slab placed on a conducting plane. This is a two-dimensional problem, and the field components are E_x, E_z, and H_y. Let us assume that the wave is excited by a small two-dimensional slot shown in Fig. 14-1. At the slot ($x = 0$), $E_z = E_0$ is given. Now according to the equivalence theorem (Section 6-10), the field excited by the aperture field $\bar{E}_a$ on an aperture in a conducting surface is the same as the field excited by the following magnetic surface current density $\bar{K}_s$ placed on the conducting surface:

$$\bar{K}_s = \bar{E}_a \times \hat{n}, \tag{14-1}$$

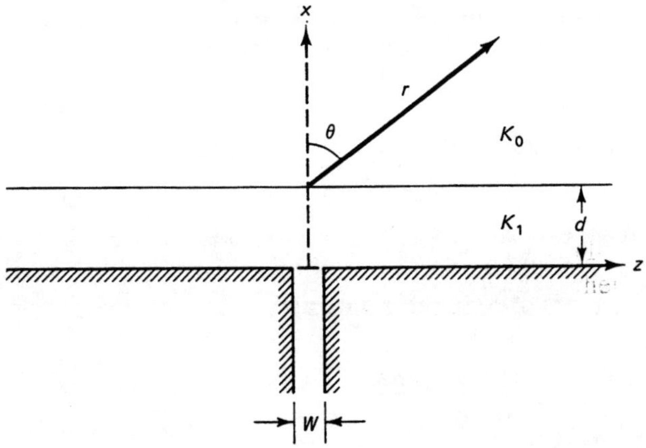

Figure 14-1 Dielectric slab on a conducting plane excited by a slot.

where $\hat{n}$ is the unit vector directed into the observation point (Fig. 14-2). Therefore, the field is the same as that excited by a magnetic surface current density $\bar{K}_s = E_0 \hat{y}$.

We assume that the slot width W is small compared with a wavelength, and therefore we can approximate the magnetic source current density $\bar{J}_{ms}$ by

$$\bar{J}_{ms} = \bar{K}_s \delta(x) = E_0 W \delta(x)\delta(z)\hat{y}. \tag{14-2}$$

Note that the unit for $\bar{J}_{ms}$ is volt/m^2, and the unit for $\bar{K}_s$ is volt/m.

Now from Maxwell's equations,

$$\begin{aligned} \nabla \times \bar{E} &= -j\omega\mu\bar{H} - \bar{J}_{ms}, \\ \nabla \times \bar{H} &= j\omega\epsilon\bar{E}, \end{aligned} \tag{14-3}$$

and letting $\partial/\partial y = 0$, we get

$$\left(\frac{\partial^2}{\partial x^2} + \frac{\partial^2}{\partial z^2} + k^2\right) H_y = j\omega\epsilon E_0 W \delta(x)\delta(z), \tag{14-4}$$

where $k = k_0$ in air and $k = k_1$ in dielectric. The boundary conditions are $E_z = 0$ at $x = 0$ and the continuity of E_z and H_y at $x = d$. In terms of H_y, the boundary conditions are

$$\begin{aligned} &\frac{\partial}{\partial x} H_y = 0 \qquad \text{at } x = 0. \\ &\frac{1}{\epsilon}\frac{\partial}{\partial x} H_y \text{ and } H_y \text{ are continuous at } x = d. \end{aligned} \tag{14-5}$$

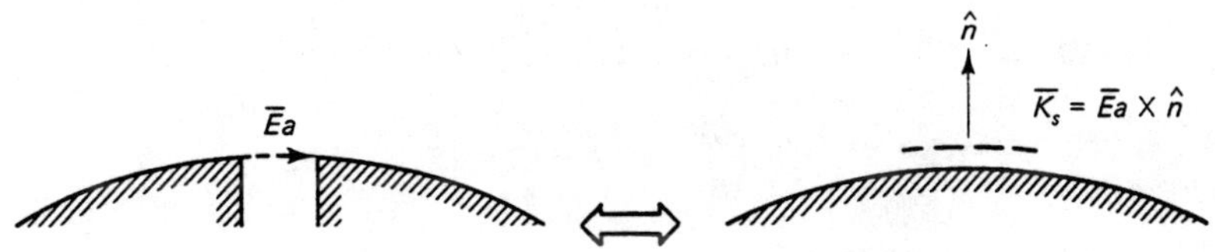

Figure 14-2 Equivalence of the aperture field and magnetic current.

Let us start with the delta functions in (14-4) located at an arbitrary point (x', z') in the slab and then let x' and z' be zero later. We then have $H_y = -j\omega\epsilon_1 E_0 W G$ and

$$\left(\frac{\partial^2}{\partial x^2} + \frac{\partial^2}{\partial z^2} + k_1^2\right)G = -\delta(x - x')\delta(z - z') \qquad \text{in slab}$$
$$\left(\frac{\partial^2}{\partial x^2} + \frac{\partial^2}{\partial z^2} + k_0^2\right)G = 0 \qquad \text{outside.} \tag{14-6}$$

We now take the Fourier transform of (14-6) in the z direction.

$$g(x, \beta) = \int_{-\infty}^{\infty} G(x, z)e^{j\beta z}\,dz,$$
$$G(x, z) = \frac{1}{2\pi}\int g(x, \beta)e^{-j\beta z}\,d\beta. \tag{14-7}$$

Inside the slab, we get

$$\left(\frac{d^2}{dx^2} + k_1^2 - \beta^2\right)g(x, \beta) = -\delta(x - x')e^{j\beta z'}. \tag{14-8}$$

The solution is then given by a sum of the primary field g_p and the secondary field g_s. The primary field is the field that is excited by the source in the infinite homogeneous space in the absence of any boundaries, and the secondary field represents all the effects of the boundaries. The primary field may also be called the incident field and is the particular solution of the inhomogeneous differential equation (14-8). The secondary field can be called the scattered field and is the complementary solution of (14-8).

The primary field is easily obtained following Section 5-5.

$$g_p(x, \beta) = \frac{\exp[-jq_1|x - x'|]}{2jq_1}, \tag{14-9}$$

where

$$q_1 = \begin{cases} (k_1^2 - \beta^2)^{1/2} & \text{for } |\beta| < |k_1| \\ -j(\beta^2 - k_1^2)^{1/2} & \text{for } |\beta| > |k_1| \end{cases}.$$

The secondary field, g_s, is given by two waves traveling in the $+x$ and $-x$ directions.

$$g_s(x, \beta) = B\,\exp(-jq_1 x) + C\,\exp(+jq_1 x). \tag{14-10}$$

In the air $x > d$, we have

$$\left(\frac{d^2}{dx^2} + k_0^2 - \beta^2\right)g(x, \beta) = 0. \tag{14-11}$$

Therefore, the solution is

$$g(x, \beta) = A\,\exp[-jq_0(x - d)], \tag{14-12}$$

where

$$q_0 = \begin{cases} (k_0^2 - \beta^2)^{1/2} & \text{for } |\beta| < |k| \\ -j(\beta^2 - k_0^2)^{1/2} & \text{for } |\beta| > |k| \end{cases}.$$

Note that, in general, (14-12) should consist of both the outgoing wave satisfying the radiation condition and the incoming wave $\exp(+jq_0 x)$, but since there is no incoming wave, this term is zero.

We now satisfy the boundary conditions. At $x = 0$, we have $(\partial/\partial x)G = 0$. Thus we get

$$\frac{\exp(-jq_1 x')}{2jq_1} - B + C = 0. \tag{14-13}$$

At $x = d$, we have G and $(1/\epsilon)(\partial/\partial x)G$ being continuous. Thus we get

$$\begin{aligned} &\frac{\exp[-jq_1(d - x')]}{2jq_1} + B \exp(-jq_1 d) + C \exp(+jq_1 d) = A, \\ &\frac{q_1}{\epsilon_1}\left\{\frac{\exp[-jq_1(d - x')]}{2jq_1} + B \exp(-jq_1 d) - C \exp(+jq_1 d)\right\} = \frac{q_0}{\epsilon_0} A. \end{aligned} \tag{14-14}$$

Equations (14-13) and (14-14) are easily solved for three unknowns A, B, and C. Here we write the field H_y in air ($x > d$) when the slot is at $x = 0$ ($x' = 0$):

$$\begin{aligned} H_y(x, z) &= -j\omega\epsilon_1 E_0 W G, \\ G &= \frac{1}{2\pi}\int_c A(\beta) \exp[-jq_0(x - d) - j\beta z]\, d\beta. \\ A(\beta) &= \frac{e^{-jq_1 d}}{jq_1} \frac{T}{1 - R \exp(-j2q_1 d)}, \\ R &= \frac{(q_1/\epsilon_1) - (q_0/\epsilon_0)}{(q_1/\epsilon_1) + (q_0/\epsilon_0)}, \qquad T = 1 + R. \end{aligned} \tag{14-15}$$

The physical interpretation of this solution is that $\exp(-jq_1 d)/jq_1$ is twice the direct primary wave $\exp(-jq_1 d)/(2jq_1)$ and represents a sum of the waves from the magnetic current source and from its image. The images for electric and magnetic current sources can easily be seen noting that the electric field tangential to the conducting surface is zero (Fig. 14-3). T is the transmission coefficient from dielectric to air, and R is the reflection coefficient. If we expand $[1 - R \exp(-j2q_1 d)]^{-1}$ in a series, we get

$$A(\beta) = \frac{e^{-jq_1 d}}{jq_1} \sum_{n=0}^{\infty} TR^n \exp(-j2q_1 nd). \tag{14-16}$$

Each term can be identified as the multiple reflected wave (Fig. 14-4).

Let us consider the integral in (14-15). We note first that there are poles located at the roots $\beta = \beta_s$ of the denominator:

$$1 - R \exp(-j2q_1 d) = 0. \tag{14-17}$$

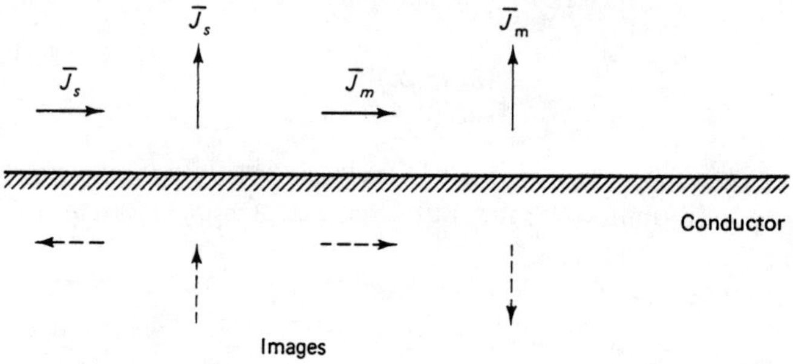

Figure 14-3 Images for a conducting plane.

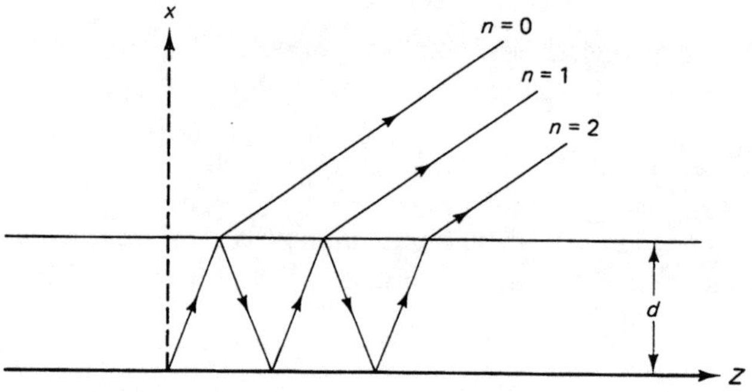

Figure 14-4 Multiple reflected waves.

In addition, there are branch points at $\beta = \pm k_0$ due to $q_0 = (k_0^2 - \beta^2)^{1/2}$. There is no branch point, however, at $\beta = \pm k_1$ even if $q_1 = (k_1^2 - \beta^2)^{1/2}$ appears in the integral. This is because q_1 represents the propagation constant in the x direction within the slab, and since the slab is bounded at $x = 0$ and d, changing $+q_1$ to $-q_2$ simply interchanges the positive-going and negative-going waves inside the slab and the results are unchanged. Of course, this can be verified mathematically by changing q_1 to $-q_1$ in $A(\beta)$.

Let us next examine the poles given in (14-17). This equation can easily be converted to

$$\frac{q_1}{\epsilon_1} \tan q_1 d = \frac{jq_0}{\epsilon_0}. \tag{14-18}$$

This is identical to the eigenvalue equation (3-84) for the trapped surface wave.

The integrand in (14-15) therefore contains a finite number of trapped surface wave poles β_p and these are located on the real axis (Fig. 14-5). It is often convenient to use the transformation from β to α.

$$\beta = k_0 \sin \alpha. \tag{14-19}$$

We also let

$$x - d = r \cos \theta, \qquad z = r \sin \theta. \tag{14-20}$$

Then (14-15) becomes

$$G = \int_c F(\alpha) \exp[k_0 r f(\alpha)]\, d\alpha, \tag{14-21}$$

where

$$F(\alpha) = \frac{A(\alpha) k_0 \cos \alpha}{2\pi},$$

$$f(\alpha) = -j \cos(\alpha - \theta).$$

$F(\alpha)$ has poles at $\alpha_p(\beta_p = k_0 \sin \alpha_p)$.

Let us first consider the radiation field far from the source and the surface. The radiation field in the form of (14-21) has been evaluated in Section 11-5. According to (11-48), which is obtained by the saddle-point technique, the radiation field is given by

$$G_r = \frac{A(\theta) k_0 \cos \theta}{(2\pi k_0 r)^{1/2}} e^{-jk_0 r + j(\pi/4)}. \tag{14-22}$$

This is valid when $k_0 r >> 1$ and $\theta \neq \pi/2$. The saddle point α_s is located at $\alpha_s = \theta$ in the complex α plane (Fig. 14-5).

Let us next consider the field on the surface where $\theta = \pi/2$ (Fig. 14-1). In the complex α plane, the steepest descent contour (SDC) now goes through the saddle point $\alpha_s = \pi/2$. Then the original contour c is deformed as shown in Fig. 14-6. Therefore, we have

$$\int_C g(\alpha)\, d\alpha = \int_{\text{SDC}} g(\alpha)\, d\alpha - 2\pi j \sum \text{residue at } \alpha_p \tag{14-23}$$

where $g(\alpha)$ is the integrand in (14-21). The first term is (14-22) which vanishes on the surface $\theta = \pi/2$. The residue terms can be evaluated from (14-15). The surface wave poles are given by (14-18). Let β_s be the propagation constant of the surface wave and $q_0 = -j\alpha_t$. Then the residue term in (14-23) is given by

$$-2\pi j \sum \text{residue} = -j \sum \frac{N(\beta_s)}{D'(\beta_s)} e^{-\alpha_t(x-d) - j\beta_s z}, \tag{14-24}$$

where $A(\beta) = N(\beta)/D(\beta)$ (see the Appendix to Chapter 11, Section D).

In summary, the radiation field for $\theta \neq \pi/2$ is given by (14-22) and the surface wave at $\theta = \pi/2$ is given by (14-24). In Fig. 14-7 we also show a leaky wave pole at $\alpha = \alpha_l$, which can be found in some structures. If leaky wave poles exist, then as the

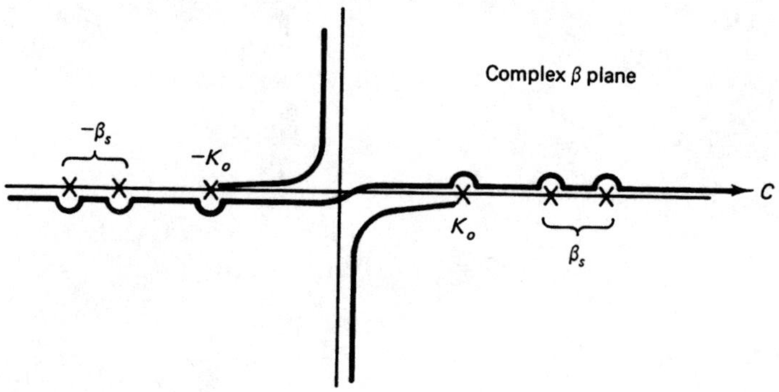

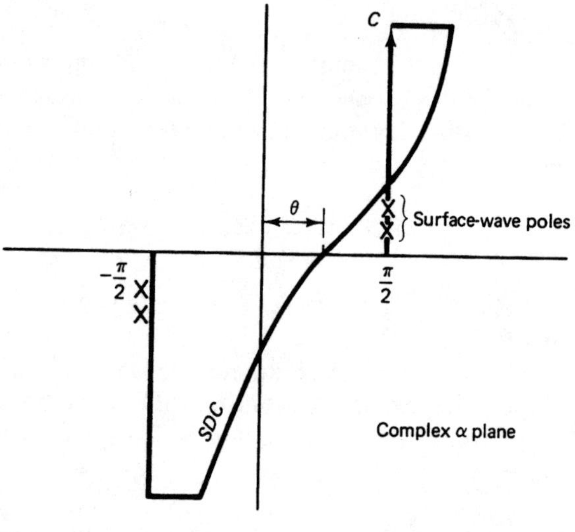

Figure 14-5 Original contour C, SDC, and surface-wave poles.

observation angle θ increases beyond θ_l, the leaky wave contribution appears. It is seen in Section 3-10 that the leaky wave is an improper wave and cannot exist by itself. However, as seen in the above, leaky waves can exist in a part of the space $\theta > \theta_l$.

In the above, the leaky wave appears discontinuously at $\theta = \theta_l$. This is, of course, not physical, and is caused by our separate evaluation of the saddle-point and the surface-wave contributions. These two contributions should be evaluated together to obtain a continuous total field. This is the modified saddle-point technique to be discussed in Chapter 15.

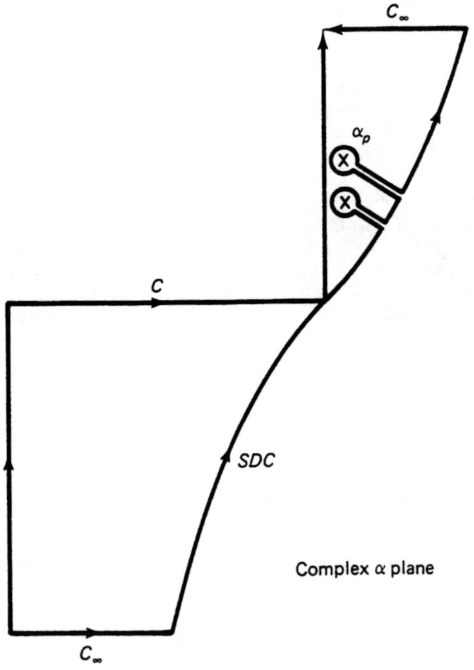

Figure 14-6 Original contour C is deformed to SCD.

14-2 EXCITATION OF WAVES IN A VERTICALLY INHOMOGENEOUS MEDIUM

Let us consider the excitation of waves by a point source located in an inhomogeneous medium whose refractive index $n(z)$ varies as a function of z only (Fig. 14-8). An example is acoustic excitation in the ocean. We formulate this problem using the WKB approximation.

Let us consider the scalar Green's function $G(\bar{r},\bar{r}')$, $\bar{r}=(x,y,z)$ and $\bar{r}'=(x,y',z')$ in an inhomogeneous medium whose refractive index n is a function of z only.

$$[\nabla^2+k_0^2n^2(z)]G(\bar{r},\bar{r}')=-\delta(\bar{r}-\bar{r}'). \qquad (14\text{-}25)$$

First, we take the Fourier transform in the x and y directions.

$$g(q_1,q_2,z)=\iint G(\bar{r},\bar{r}')\exp(jq_1x+jq_2y)\,dx\,dy. \qquad (14\text{-}26)$$

Equation (14-25) is then transformed to

$$\left[\frac{d^2}{dz^2}+q^2(z)\right]g(q_1,q_2,z)=-\delta(z-z')e^{jq_1x'+jq_2y'}, \qquad (14\text{-}27)$$

where $q^2(z)=k_0^2n^2(z)-q_1^2-q_2^2$.

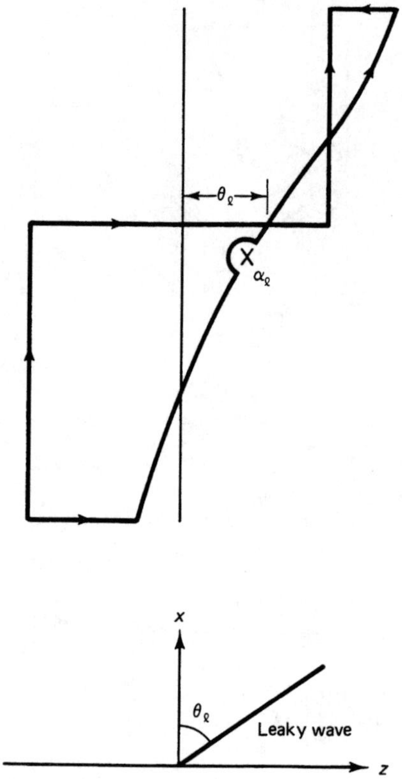

Figure 14-7 Leaky wave contributions.

The WKB solution to (14-27) is given by (Sections (3-14) and 5-5)

$$g(q_1, q_2, z) = \begin{cases} \dfrac{y_1(z) y_2(z') e^{jq_1 x' + jq_2 y'}}{\Delta} & \text{if } z > z' \\ \dfrac{y_1(z') y_2(z') e^{jq_1 x' + jq_2 y'}}{\Delta} & \text{if } z < z' \end{cases}, \tag{14-28}$$

where

$$\Delta = y_1 y_2' - y_1' y_2 = 2j,$$

$$y_1(z) = q^{-1/2} \exp\left(-j \int_{z'}^{z} q \, dz \right),$$

$$y_2(z) = q^{-1/2} \exp\left(+j \int_{z'}^{z} q \, dz \right).$$

Note that $y_1(z)$ and $y_2(z)$ represent the outgoing waves in the $+z$ and $-z$ directions satisfying the radiation condition as $z \to +\infty$ and $z \to -\infty$, respectively, and Δ is the

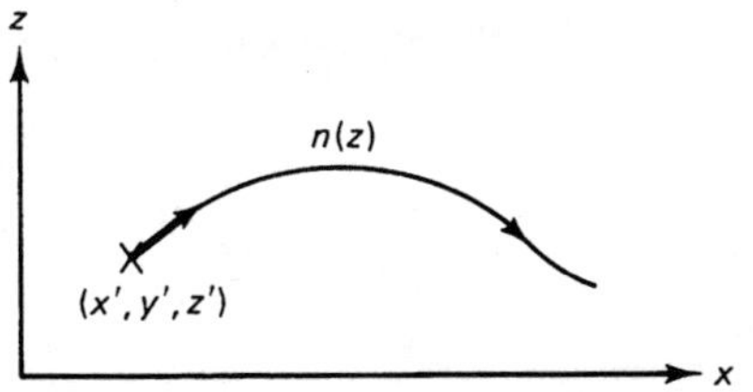

Figure 14-8 Wave excitation in a vertically inhomogeneous medium.

Wronskian of y_1 and y_2. Green's function $G(\bar{r}, \bar{r}')$ is then given by the inverse Fourier transform of $g(q_1, q_2, z)$:

$$G(\bar{r}, \bar{r}') = \frac{1}{(2\pi)^2} \iint \frac{\exp[-jf(z, q_1, q_2)]}{2jq(z)^{1/2} q(z')^{1/2}} dq_1 \, dq_2 \qquad \text{for } z > z', \tag{14-29}$$

where $f = \int_{z'}^{z} q \, dz + q_1(x - x') + q_2(y - y')$.

Green's function for $z < z'$ is given by the same expression except that f is replaced by

$$f = -\int_{z'}^{z} q \, dz + q_1(x - x') + q_2(y - y'). \tag{14-30}$$

Equation (14-29) represents the WKB solution to Green's function. It is applicable to the case away from the turning point.

Let us evaluate (14-29) approximately using the saddle-point method (stationary phase). The saddle points $q_1 = q_{1s}$ and $q_2 = q_{2s}$ are given by

$$\frac{\partial f}{\partial q_1} = 0 \quad \text{and} \quad \frac{\partial f}{\partial q_2} = 0. \tag{14-31}$$

Green's function is then given by (14A-9) in the Appendix to Chapter 14, Section A:

$$G = \frac{1}{(2\pi)^2} \frac{e^{-jf_s}}{2jq(z)^{1/2} q(z')^{1/2}} \frac{(2\pi) e^{j(\pi/2)}}{(f_{11} f_{22} - f_{12}^2)^{1/2}}, \tag{14-32}$$

where

$$f_s = f(z, q_{1s}, q_{2s}), \qquad f_{11} = \frac{\partial^2}{\partial q_1^2} f, \qquad f_{22} = \frac{\partial^2}{\partial q_2^2} f, \qquad f_{12} = \frac{\partial^2}{\partial q_1 \partial q_2} f,$$

and f_{11}, f_{22}, and f_{12} are evaluated at q_{1s} and q_{2s}. Equation (14-32) gives Green's function at (x, y, z) when the source is located at (x', y', z') in the medium with the refractive index $n(z)$. This requires that the saddle points q_{1s} and q_{2s} be found from (14-31). It is easier, however, to assume q_{1s} and q_{2s} first and then to determine (x, y, z) for these saddle points, as will be shown shortly. The saddle points give the direction of the ray at the source point, and the ray equations below can be used to determine the ray path $x = x(z)$ and $y = y(z)$.

Using the definition of f in (14-29), the saddle points q_{1s} and q_{2s} in (14-31) are given by

$$x - x' = \int_{z'}^{z} \frac{q_{1s}\, dz}{[k_0^2 n^2(z) - q_{1s}^2 - q_{2s}^2]^{1/2}},$$
$$y - y' = \int_{z'}^{z} \frac{q_{2s}\, dz}{[k_0^2 n^2(z) - q_{1s}^2 - q_{2s}^2]^{1/2}}. \tag{14-33}$$

These two equations will be shown to be identical to the ray equation of geometric optics (Chapter 15). We can also determine the physical meaning of the saddle points q_{1s} and q_{2s} as follows: Consider the ray for a given q_{1s} and q_{2s}. From (14-33), a small distance ds along this ray is given by

$$ds^2 = dx^2 + dy^2 + dz^2, \tag{14-34}$$

where

$$dx = \frac{q_{1s}\, dz}{q_s}, \qquad q_s = [k_0^2 n^2(z) - q_{1s}^2 - q_{2s}^2]^{1/2}$$
$$dy = \frac{q_{2s}\, dz}{q_s}.$$

Therefore, we have

$$ds^2 = \frac{k_0^2 n^2\, dz^2}{q_s^2} = \frac{k_0^2 n^2\, dx^2}{q_{1s}^2} = \frac{k_0^2 n^2\, dy^2}{q_{2s}^2}.$$

If we let

$$\frac{dx}{ds} = \sin\theta\, \cos\phi, \qquad \frac{dy}{ds} = \sin\theta\, \sin\phi, \tag{14-35}$$

then we get

$$q_{1s} = k_0 n(z) \sin\theta\, \cos\phi,$$
$$q_{2s} = k_0 n(z) \sin\theta\, \sin\phi. \tag{14-36}$$

Equation (14-36) shows that for given saddle points q_{1s} and q_{2s}, $n(z) \sin\theta \cos\phi$ and $n(z) \sin\theta \sin\phi$ are constant. Since $q_{2s}/q_{1s} = \tan\phi = \text{constant}$, ϕ is constant along the ray, and we have

$$n(z) \sin\theta(z) = \text{constant along the ray.} \tag{14-37}$$

This is identical to Snell's law. The stationary phase solution is therefore equivalent to the geometric optics solution.

The procedure for obtaining Green's function is as follows. Suppose that the ray is started at (x', y', z') in the direction (θ_0, ϕ_0). Then q_{1s} and q_{2s} are given by

$$q_{1s} = k_0 n(z') \sin\theta_0\, \cos\phi_0,$$
$$q_{2s} = k_0 n(z') \sin\theta_0\, \sin\phi_0. \tag{14-38}$$

These q_{1s} and q_{2s} are used in the ray equation (14-33) to determine (x, y) at z. These (x, y, z) and q_{1s} and q_{2s} are then used in (14-32) to obtain Green's function.

As an example, consider the parabolic refractive index profile:

$$n^2(z) = n_0^2\left[1 - \left(\frac{z}{z_0}\right)^2\right]. \tag{14-39}$$

The source is located at $x' = 0$ and $z' = 0$, and the ray is launched in the direction θ_0 and $\phi_0 = 0$. Then the ray equation (14-33) can be integrated to give

$$z = z_0 \cos\theta_0 \sin\left[\frac{x}{z_0 \sin\theta_0}\right]. \tag{14-40}$$

Green's function is then given by

$$G = \frac{e^{-jfs}}{4\pi} \frac{q_{1s} q_0^{1/2}}{[x^2 q_s q_0^2 + xzq_{1s}^3]^{1/2}}, \tag{14-41}$$

where $q_0 = q(z') = q(0)$ and q_{1s} and q_s are given in (14-38) and (14-34), respectively. If $n(z) = 1$, (14-41) reduces to $\exp(-jkr)/(4\pi r)$, as it should.

14-3 STRIP LINES

We now turn to a different topic. Strip lines have many attractive features. They are compact, inexpensive, and can easily be produced as printed circuits. The most common types have a strip of conductor on a grounded dielectric slab (Fig. 14-9). Even though this is a two-conductor line, this structure cannot support TEM modes

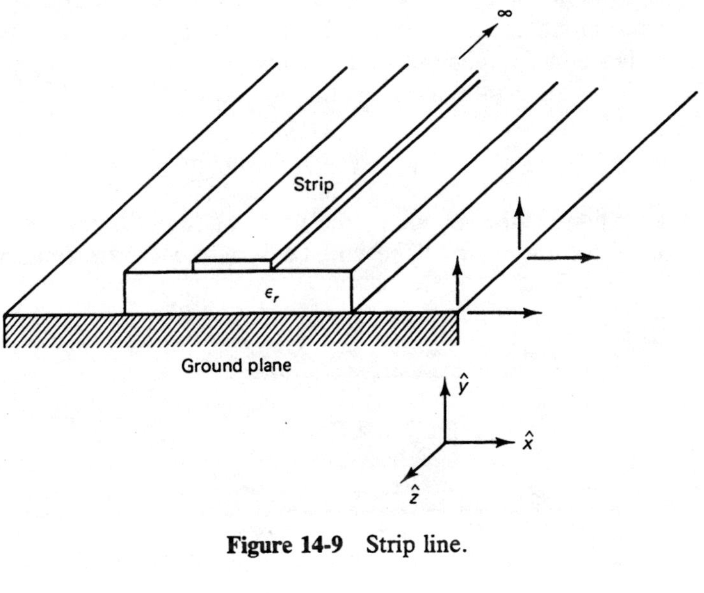

Figure 14-9 Strip line.

because the dielectric occupies only a part of the cross section (see Section 4-9). Thus, in general, both E_z and H_z are nonzero, and we need to consider hybrid modes consisting of both TE and TM modes.

Even though the hybrid mode analysis is required for exact analysis of strip lines, simpler techniques can be used for approximate analysis of strip lines. One useful method is the quasi-static approximation, in which E_z and H_z are neglected and the field is assumed to be TEM. The error introduced by this approximation is usually a fraction of a percent. One drawback of the quasi-static approximation is that since this is a TEM solution, the propagation constant is independent of frequency (no modal dispersion) except for the frequency dependence of the di-electric material itself. In this section we start with the TEM solution and then discuss the quasi-TEM approximation and the exact hybrid solution. Important papers on this topic are compiled by Itoh (1987).

TEM Solution

We have discussed in Section 4-9 that if we have a transmission line consisting of two conductors and a homogeneous dielectric medium, this line can support a TEM mode. An example is that of boxed strip lines (Fig. 14-10). As shown in Section 4-9, TEM modes are given by the following:

$$\begin{aligned}\bar{E}_t &= -(\nabla_t V)e^{-jkz},\\ \bar{H}_t &= \frac{1}{Z}\hat{z}\times\bar{E}_t,\end{aligned} \tag{14-42}$$

where $\bar{E}_t$ and $\bar{H}_t$ are the electric and magnetic fields transverse to the direction of wave propagation $\hat{z}$ and are given by $k = \omega/v$, $v = (\mu\epsilon)^{-1/2}$, and $Z = (\mu/\epsilon)^{1/2}$, which are the wave number, the velocity of the electromagnetic wave, and the characteristic impedance of the medium, respectively. The electrostatic potential $V(x, y)$ in (14-12) satisfies the two-dimensional Laplace equation:

$$\left(\frac{\partial^2}{\partial x^2}+\frac{\partial^2}{\partial y^2}\right)V(x,y) = 0. \tag{14-43}$$

We next define the voltage V_0, the current I_0, and the characteristic impedance Z_0 of the transmission line. The voltage V_0 is the electrostatic potential at one

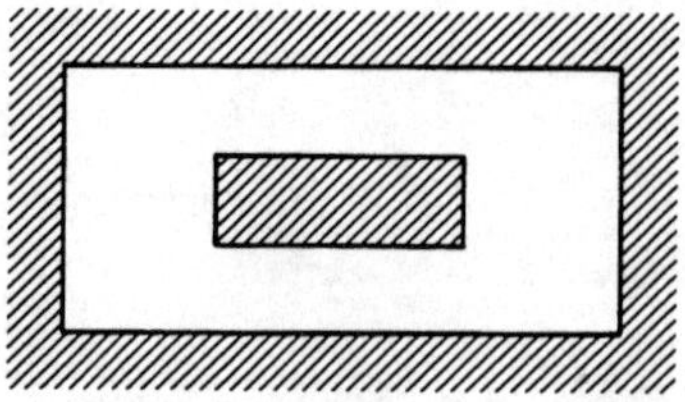

Figure 14-10 Boxed strip line.

conductor with respect to the other and is given by the line integral of the electric field (Fig. 14-11).

$$V_0 = -\int_{s_1}^{s_2} \bar{E}_t \cdot d\bar{l} = \int_{s_1}^{s_2} \nabla_t V \cdot d\bar{l}. \tag{14-44}$$

The current is given by the integral of the surface current density J_s around one conductor,

$$I_0 = \oint |\bar{J}_s|\, dl. \tag{14-45}$$

The current density $\bar{J}_s$ is related to $\bar{H}_t$ by $\bar{J}_s = \hat{n} \times \bar{H}_t$, where $\hat{n}$ is the unit vector normal to the surface of the conductor, and $\bar{H}_t$ is related to $\bar{E}_t$. Thus we have

$$\begin{aligned} \bar{J}_s &= \hat{n} \times \bar{H}_t = \frac{1}{Z}\hat{n} \times (\hat{z} \times \bar{E}_t) \\ &= \frac{\hat{z}}{Z}(\hat{n} \cdot \bar{E}) = \frac{\hat{z}}{Z}\frac{\rho_s}{\epsilon} = \hat{z} v \rho_s, \end{aligned} \tag{14-46}$$

where ρ_s is the surface charge density. The current I_0 is therefore related to the total charge Q,

$$I_0 = \oint v \rho_s\, dl = vQ. \tag{14-47}$$

The characteristic impedance Z_c of the transmission line is then given by

$$Z_c = \frac{V_0}{I_0} = \frac{V_0}{vQ} = \frac{1}{vC_e}, \tag{14-48}$$

where C_e is the electrostatic capacitance per unit length of the line. It is, therefore, clear that once the electrostatic capacitance is found, the characteristic impedance is given by (14-48) and that the solution to the TEM line problem is reduced to finding the capacitance C_e per unit length of the line.

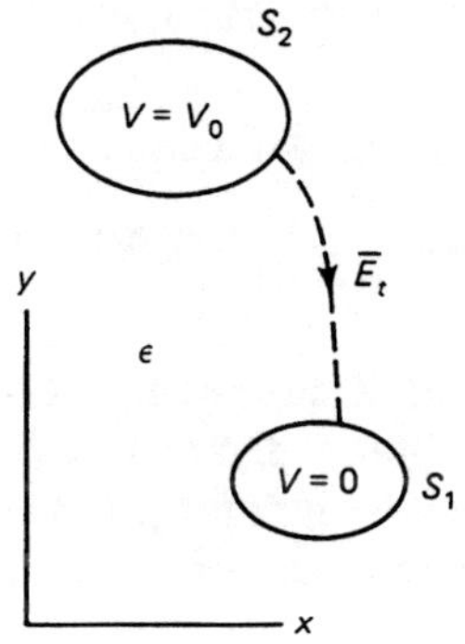

Figure 14-11 Cross-sectional view of a two-conductor line with a homogeneous medium.

The capacitance C_e can be expressed using the time-averaged electric stored energy W_e:

$$W_e = \frac{1}{4} C_e V_0^2 = \frac{1}{4} \frac{Q^2}{C_e}. \tag{14-49}$$

Now W_e can be expressed either in terms of the energy density in the medium or the charge on the conductor:

$$W_e = \frac{1}{4} \int_a \epsilon |\bar{E}|^2 \, dx \, dy = \frac{1}{4} \int_a \epsilon |\nabla_t V|^2 \, dx \, dy, \tag{14-50}$$

$$W_e = \frac{1}{4} \oint_{s_2} \rho_s V \, dl, \tag{14-51}$$

where a is the area between two conductors and (14-51) is the integral around the conductor s_2.

Using (14-49) and (14-50), we write

$$\begin{aligned} C_e &= \frac{\int_a \epsilon |\nabla_t V|^2 \, dx \, dy}{V_0^2} \\ &= \frac{\int_a \epsilon |\nabla_t V|^2 \, dx \, dy}{\left| \int_{s_1}^{s_2} \nabla_t V \cdot d\bar{l} \right|^2}. \end{aligned} \tag{14-52}$$

If the electrostatic potential $V(x, y)$ is found, the formula above can be used to calculate the capacitance C_e. We can also use (14-49) and (14-51) to obtain

$$\begin{aligned} \frac{1}{C_e} &= \frac{\oint \rho V \, dl}{Q^2} \\ &= \frac{\oint dl \oint dl' \, \rho_s(x, y) G(x, y; x', y') \rho_s(x', y')}{\left| \oint \rho_s(x, y) \, dl \right|^2}, \end{aligned} \tag{14-53}$$

where G is Green's function satisfying the equation

$$\nabla^2 G(\bar{r}, \bar{r}') = -\frac{\delta(\bar{r}, \bar{r}')}{\epsilon}. \tag{14-54}$$

It has been shown (Collin, 1966, Chap. 4) that (14-52) and (14-53) are the variational expressions for the capacitance C_e and that (14-52) gives the upper bound on C_e while (14-53) gives the lower bound on C_e. Once the solution for $V(x, y)$ or for $\rho_s(x, y)$ is obtained, C_e is obtained by either (14-52) or (14-53).

Let us also note that the time-averaged magnetic stored energy W_m is equal to

$$W_m = \frac{L_e I_0^2}{4} = W_e. \tag{14-55}$$

Therefore, the charactierstic impedance Z_c, the propagation constant k, and the total transmitted power P can also be given by

$$Z_c = \left(\frac{L_e}{C_e}\right)^{1/2},$$

$$k = \frac{\omega}{v} = \omega(\mu\epsilon)^{1/2} = \omega(L_e C_e)^{1/2}, \tag{14-56}$$

$$P = \frac{V_0 I_0}{2} = \frac{Z_0 I_0^2}{2}.$$

There are several techniques to obtain the solutions for $V(x, y)$ or $\rho_s(x, y)$ or the capacitance C_e. Analytical techniques include conformal mapping and the Fourier transform. The variational technique is also used with numerical techniques and Fourier transforms.

Quasistatic Approximation

The TEM solutions discussed above are applicable only to two-conductor lines with a homogeneous medium. If the medium is inhomogeneous and consists of different dielectric slabs as shown in Fig. 14-9, the TEM wave cannot exist and both TE and TM modes are needed. However, the z components of electric and magnetic fields are often negligibly small, and the characteristics of the transmission line with an inhomogeneous dielectric medium can be approximated by the quasi-static approximation to be described below. The error introduced may be less than a fraction of a percent for most applications.

In the quasi-static approximation, we assume that the transverse electric and magnetic field distributions are approximately equal to that of the static electric and magnetic field distributions. The time-averaged electric and magnetic stored energies are assumed to be

$$W_e = W_m = \tfrac{1}{4} C_e V_0^2 = \tfrac{1}{4} L_e I_0^2. \tag{14-57}$$

The characteristic impedance Z_c is given approximately by $(L_e/C_e)^{1/2}$. However, since the inductance L_e does not depend on the dielectric material, it is the same as that of free space. Therefore, we write

$$Z_c = Z_0 \left(\frac{C_0}{C_e}\right)^{1/2}, \tag{14-58}$$

where C_e is the actual capacitance per unit length of the line, $Z_0 = (L_e/C_0)^{1/2}$ is the characteristic impedance of the transmission line, and C_0 is the capacitance per unit length of the line if the medium is free space. We also approximate the propagation constant k by

$$k = \omega(L_e C_e)^{1/2} = k_0\left(\frac{C_e}{C_0}\right)^{1/2}, \tag{14-59}$$

where $k_0 = 2\pi/\lambda_0$ is the free-space wave number.

Under the quasi-static approximation, we only need to calculate C_e and C_0 in (14-58) and (14-59). To find C_e in an inhomogeneous medium, we first note that $\nabla_t \times \bar{E} = 0$, where $\nabla_t = \hat{x}(\partial/\partial x) + \hat{y}(\partial/\partial y)$, and $\bar{E}$ is given by the electrostatic potential V.

$$\bar{E} = -\nabla_t V. \tag{14-60}$$

Substituting this into $\nabla_t \cdot \bar{D} = \nabla_t \cdot (\epsilon \bar{E}) = 0$, we get $\nabla_t \cdot (\epsilon \nabla_t V) = 0$, which we write as

$$\left[\frac{\partial}{\partial x}\left(\epsilon_r \frac{\partial}{\partial x}\right) + \frac{\partial}{\partial y}\left(\epsilon_r \frac{\partial}{\partial y}\right)\right] V(x, y) = 0, \tag{14-61}$$

where $\epsilon_r(x, y)$ is the relative dielectric constant of the medium.

Once (14-61) is solved for $V(x, y)$ satisfying the boundary condition that $V = V_1 =$ constant on one conductor and $V = V_2 =$ constant on the other conductor, the capacitance C_e is obtained by (14-52). We can also use the other variational form (14-53) if we use

$$\nabla_t \cdot (\epsilon \nabla_t) G(\bar{r}, \bar{r}') = -\delta(\bar{r} - \bar{r}'), \tag{14-62}$$

in place of (14-54). This is of course the same as solving (14-61) with the delta function as the source term.

Exact Hybrid Solution

Both the TEM solution and the quasi-static solution depend only on the electrostatic capacitance, and therefore are nondispersive as long as the frequency dependence of the dielectric constant is negligible. However, these TEM solutions are approximate, and the exact solutions are not TEM but a combination of TE and TM modes. Only this exact hybrid solution exhibits the complete dispersion characteristics of a strip line.

Let us start with the exact formulation of the problem. We need both TM and TE modes as shown in (4-11) and (4-16). Thus we write the fields in air ($i = 1$) and in dielectric ($i = 2$)

$$\begin{aligned}
E_{zi} &= k_{ci} \phi_i(x, y) e^{-j\beta z}, \\
H_{zi} &= k_{ci} \psi_i(x, y) e^{-j\beta z}, \\
\bar{E}_{ti} &= [-j\beta \nabla_t \phi_i(x, y) + j\omega\mu_i \hat{z} \times \nabla_t \psi_i(x, y)] e^{-j\beta z}, \\
\bar{H}_{ti} &= [-j\omega\epsilon_i \hat{z} \times \nabla_t \phi_i(x, y) - j\beta \nabla_t \psi_i(x, y)] e^{-j\beta z},
\end{aligned} \tag{14-63}$$

where

$$(\nabla_t^2 + k_{ci}^2)\phi_i = 0,$$
$$(\nabla_t^2 + k_{ci}^2)\psi_i = 0,$$
$$k_{ci}^2 = k_i^2 - \beta^2.$$

The boundary conditions are that the tangential components of the electric and magnetic fields are continuous across the air–dielectric boundary, the tangential electric field is zero on the conductor, and the field satisfies the radiation condition at infinity. Several methods have been proposed to solve this exact hybrid problem. The spectral domain approach has been used successfully to solve this problem. We outline this approach in a later section. A numerical technique such as the finite-element method has also been used to solve this problem (*IEEE Trans. MTT*, October 1985).

14-4 WAVES EXCITED BY ELECTRIC AND MAGNETIC CURRENTS PERPENDICULAR TO DIELECTRIC LAYERS

In Section 14-1 we considered the two-dimensional problem of wave excitation in dielectric layers. In this section we present formulations for a general three-dimensional problem. Consider the dielectric layers shown in Fig. 14-12. It can be shown that the vertical electric dipole in the presence of dielectric layers excites only the TM waves, which are generated by the electric Hertz vector $\Pi'\hat{z}$. Also the vertical magnetic dipole excites only the TE waves, which are generated by the magnetic Hertz vector $\Pi''\hat{z}$.

Let us first consider the waves excited by the electric current density $J_z(x', y', z')$ pointed in the z direction. In layer 1, the electric Hertz vector $\Pi'\hat{z}$ satisfies the scalar wave equation:

$$(\nabla^2 + k_1^2)\Pi' = -\frac{J_z}{j\omega\epsilon_1}. \tag{14-64}$$

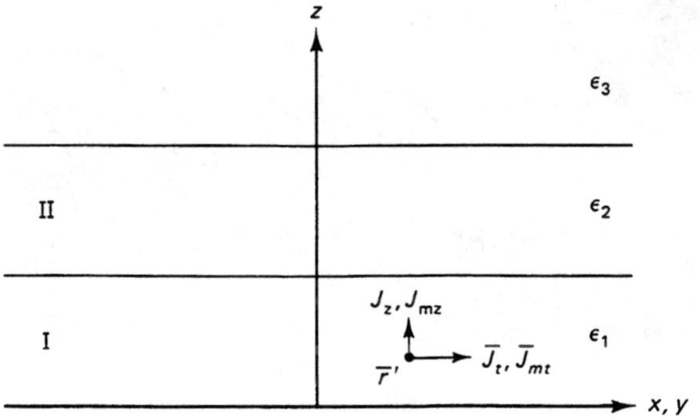

Figure 14-12 Dielectric layers excited by $\bar{J}$ and $\bar{J}_m$ located at $\bar{r}'(x', y', z')$.

Let us take Fourier transform in the x and y directions:

$$\tilde{\Pi}'(\alpha,\beta,z)=\int \Pi'(x,y,z)e^{j\alpha x+j\beta y}\,dx\,dy,$$
$$\tilde{J}_z(\alpha,\beta,z)=\int J_z(x,y,z)e^{j\alpha x+j\beta y}\,dx\,dy. \tag{14-65}$$

Equation (14-64) can then be transformed to

$$\left(\frac{d^2}{dz^2}+\gamma_1^2\right)\tilde{\Pi}'=-\frac{\tilde{J}_z}{j\omega\epsilon_1}, \tag{14-66}$$

where

$$\gamma_1=\begin{cases}(k_1^2-\alpha^2-\beta^2)^{1/2} & \text{if } k_1^2>\alpha^2+\beta^2\\ -j(\alpha^2+\beta^2-k_1^2)^{1/2} & \text{if } k_1^2<\alpha^2+\beta^2.\end{cases}$$

Now the tangential electric and magnetic fields are given by

$$\bar{E}_t=\nabla_t\frac{\partial}{\partial z}\Pi',$$
$$\bar{H}_t=j\omega\epsilon(\nabla_t\Pi'\times\hat{z}). \tag{14-67}$$

Fourier transforms of $\bar{E}_t$ and $\bar{H}_t$ are obtained from (14-67) by letting $\nabla_t=-j\alpha\hat{x}-j\beta\hat{y}=-j\bar{\alpha}$. We also note that while $\bar{H}_t$ is proportional to Π', $\bar{E}_t$ is proportional to $\partial\Pi'/\partial z$. This means that when a delta function source is located at z', $\bar{E}_t$ changes its sign depending on whether $z>z'$ or $z<z'$. Thus it is more convenient to express $\bar{E}_t$ by the current, which changes its sign in the same manner. We introduce the magnetic voltage V'_m, which is proportional to $\bar{H}_t$, and the magnetic current I'_m, which is proportional to $\bar{E}_t$, and then write the Fourier transforms of $\bar{E}_t$ and $\bar{H}_t$ as follows:

$$\bar{e}=\int \bar{E}_t e^{j\alpha x+j\beta y}\,dx\,dy,$$
$$\bar{h}=\int \bar{H}_t e^{j\alpha x+j\beta y}\,dx\,dy,$$
$$\bar{e}=-j\bar{\alpha}(-I'_m)\frac{\tilde{J}_z}{j\omega\epsilon_1},$$
$$\bar{h}=-j\hat{z}\times\bar{\alpha}(-V'_m)\frac{\tilde{J}_z}{j\omega\epsilon_1}, \tag{14-68}$$

where

$$\bar{\alpha}=\alpha\hat{x}+\beta\hat{y},$$
$$V'_m=j\omega\epsilon_i G,$$
$$Z'_m=\frac{\omega\epsilon_i}{\gamma_i}.$$

G is Green's function given by

$$\left(\frac{d^2}{dz^2}+\gamma_1^2\right)G=-\delta(z-z'). \tag{14-69}$$

The boundary conditions that the tangential electric and magnetic fields are continuous are equivalent to the continuity of the voltage V_m' and the current I_m' at the junctions. Thus V_m' and I_m' can be formed by using the equivalent transmission line shown in Fig. 14-13. Note that at perfectly conducting plane, $\bar{e}=0$ and thus $I_m'=0$. Therefore, the perfectly conducting plane is represented by an open terminal in the transmission line. Note also that the choice of V_m' and I_m' in (14-68) are made so that the current injected at the source point is $I_m'=1$ when $\bar{J}_z=1$, as shown in Fig. 14-13.

The procedure of finding $\bar{E}_t$ and $\bar{H}_t$ is as follows: First we solve the transmission line problem in Fig. 14-13 with the unit current $I_m=1$ at the source point z'. Each layer has the characteristic impedance $Z_{mi}'=\omega\epsilon_i/\gamma_i$ $(i=1,2,\ldots)$, and the propagation constant γ_i. Once we find the voltage and current distribution V_m' and I_m', $\bar{e}$ and $\bar{h}$ are obtained in (14-68). We then take the inverse Fourier transform to obtain $\bar{E}_t$:

$$\bar{E}_t=\frac{1}{(2\pi)^2}\int \bar{e}\,e^{-j\alpha x-j\beta y}\,d\alpha\,d\beta.$$

Similarly, we get $\bar{H}_t$ from $\bar{h}$. The Hertz vector $\Pi'\hat{z}$ is given by

$$\Pi'=\frac{1}{(2\pi)^2}\int \frac{G\bar{J}_z}{j\omega\epsilon_1}e^{-j\alpha x-j\beta y}\,d\alpha\,d\beta. \tag{14-70}$$

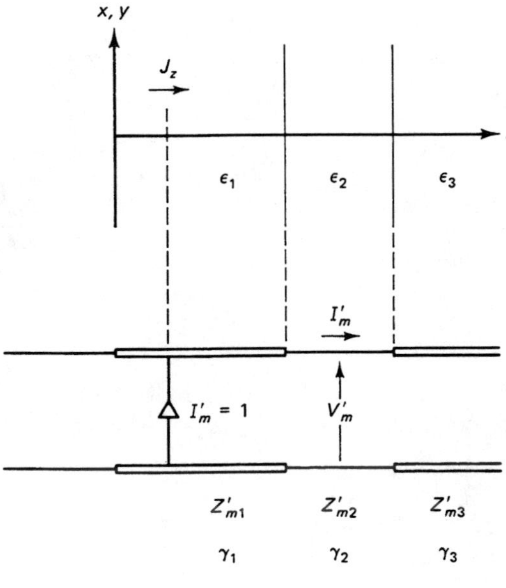

Figure 14-13 Equivalent transmission line for J_z. Perfect conducting plane is "open terminal."

As an example, consider an electric dipole $J_z = I_0 L_0 \delta(\bar{r} - \bar{r}')$ located in air $\epsilon = \epsilon_0$ at $\bar{r}' = (0, 0, h)$ above the ground with dielectric constant ϵ_2. From Fig. 14-13 the voltage incident on the boundary $z = 0$ is

$$V_{in} = \frac{Z_1'}{2} e^{-j\gamma_1 h}.$$

Then the reflected wave is

$$V_{\text{ref}} = \left[\frac{Z_2' - Z_1'}{Z_2' + Z_1'}\right] \frac{Z_1'}{2} e^{-j\gamma_1 h - j\gamma_1 z}.$$

The reflected Hertz vector Π_r' in air is then given by (14-70)

$$\Pi_r' = \frac{1}{(2\pi)^2} \int \frac{V_{\text{ref}}}{j\omega\epsilon_1} \left[\frac{I_0 L_0}{j\omega\epsilon_1}\right] e^{-j\alpha x - j\beta y} \, d\alpha \, d\beta.$$

When converted to a cylindrical system, this is identical to the expression (15-23) for the Sommerfeld dipole problem.

Next consider the magnetic current source J_{mz}. We use the duality principle and find the following:

$$\begin{aligned} \bar{E}_t &= j\omega\mu(\hat{z} \times \nabla_t \Pi''), \\ \bar{H}_t &= \nabla_t \frac{\partial}{\partial z} \Pi''. \end{aligned} \tag{14-71}$$

The Fourier transforms $\bar{e}$ and $\bar{h}$ are therefore given by

$$\begin{aligned} \bar{e} &= -j\bar{\alpha} \times \hat{z}(-V'') \frac{\tilde{J}_{mz}}{j\omega\mu}, \\ \bar{h} &= -j\bar{\alpha}(-I'') \frac{\tilde{J}_{mz}}{j\omega\mu}, \\ \bar{\alpha} &= \alpha\hat{z} + \beta\hat{y}, \\ V'' &= j\omega\mu G, \\ Z_i'' &= \frac{\omega\mu}{\gamma_i}. \end{aligned} \tag{14-72}$$

$\tilde{J}_{mz}$ is the Fourier transform of J_{mz}. The voltage V'' and current I'' are obtained by solving the transmission-line problem (Fig. 14-14). The magnetic Hertz vector $\Pi''\hat{z}$ is given by

$$\Pi'' = \frac{1}{(2\pi)^2} \int \frac{G\tilde{J}_{mz}}{j\omega\mu} e^{-j\alpha x - j\beta y} \, d\alpha \, d\beta. \tag{14-73}$$

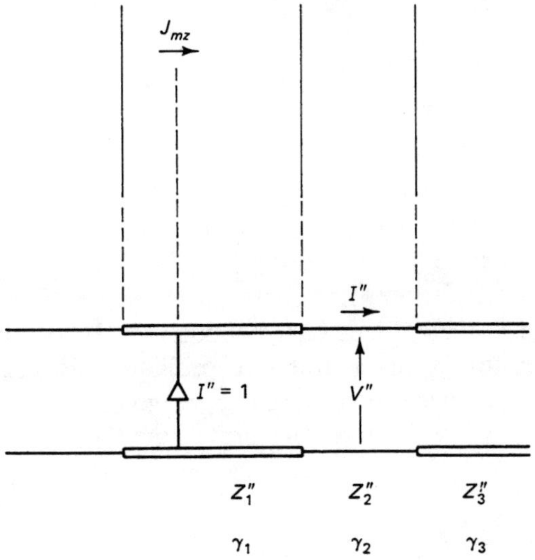

Figure 14-14 Transmission line for J_{mz}. Perfect conducting plane is a short circuit.

14-5 WAVES EXCITED BY TRANSVERSE ELECTRIC AND MAGNETIC CURRENTS IN DIELECTRIC LAYERS

Microstrip lines carry electric currents in the x–y plane embedded in layers. Frequency-selective surfaces are composed of periodic patches and apertures carrying electric and magnetic currents in the x–y plane in dielectric layers. It is therefore necessary to formulate the problem for waves excited by the currents $\bar{J}_t$ and $\bar{J}_{mt}$ which are transverse to the z direction.

First we note that Π_x and Π_y are simply related to J_x and J_y. However, Π_x and Π_y are coupled at the dielectric interface and the formulation will become quite involved. A simpler technique is to use $\Pi'\hat{z}$ and $\Pi''\hat{z}$, because there is no coupling between Π' and Π'' at the dielectric interface. This formulation was proposed by Itoh (1980) and by Felsen and Marcuvitz (1973). Both give the same results. We will follow Felsen–Marcuvitz formulation. First we note that $\bar{E}$ and $\bar{H}$ are given by

$$\bar{E} = \nabla \times \nabla \times (\Pi'\hat{z}) - j\omega\mu\nabla \times (\Pi''\hat{z}),$$
$$\bar{H} = j\omega\epsilon\nabla \times (\Pi'\hat{z}) + \nabla \times \nabla \times (\Pi''\hat{z}). \tag{14-74}$$

The Hertz potentials Π' and Π'' are produced by $\bar{J}$ and $\bar{J}_m$.

$$\Pi' = g_0'\frac{J_z}{j\omega\epsilon} + \left(\frac{1}{j\omega\epsilon}\bar{J}_t\frac{\partial}{\partial z'} + \bar{J}_{mt} \times \hat{z}\right) \cdot \nabla_t' g',$$

$$\Pi'' = g_0''\frac{J_{mz}}{j\omega\mu} + \left(\hat{z} \times \bar{J}_t + \frac{1}{j\omega\mu}\bar{J}_{mt}\frac{\partial}{\partial z'}\right) \cdot \nabla_t' g'',$$

$$
\begin{aligned}
(\nabla^2 + k^2)g_0' &= -\delta(\bar{r} - \bar{r}'),\\
(\nabla^2 + k^2)g_0'' &= -\delta(\bar{r} - \bar{r}'),\\
-\nabla_t^2 g' &= g_0',\\
-\nabla_t^2 g'' &= g_0'',\\
\bar{J} &= J_z \hat{z} + \bar{J}_t,\\
\bar{J}_m &= J_{mz} \hat{z} + \bar{J}_{mt}.
\end{aligned}
\qquad (14\text{-}75)
$$

The first terms with J_z and J_{mz} were discussed in Section 14.4. In the above, ∇_t' is the transverse gradient operator on the source $\bar{r}'$ coordinate and $\partial/\partial z'$ is the derivative with respect to z'. The derivation of (14-75) is given by Felsen and Marcuvitz (1973, p. 445). Now we take Fourier transform of the transverse electric and magnetic field:

$$
\begin{aligned}
\bar{e}_t &= \int \bar{E}_t e^{j\alpha x + j\beta y}\, dx\, dy\\
&= \bar{\bar{e}} \cdot \bar{J},\\
\bar{h}_t &= \int \bar{H}_t e^{j\alpha x + j\beta y}\, dx\, dy\\
&= \bar{\bar{h}} \cdot \bar{J},
\end{aligned}
\qquad (14\text{-}76)
$$

where $\bar{J}$ is the Fourier transform of $\bar{J}_t$:

$$
\bar{J} = \int \bar{J}_t e^{j\alpha x + j\beta y}\, dx\, dy
$$

and $\bar{\bar{e}}$ and $\bar{\bar{h}}$ are Green's dyadic. Once we find $\bar{e}_t$ and $\bar{h}_t$, we get $\bar{E}_t$ and $\bar{H}_t$ by taking the inverse Fourier transform:

$$
\begin{aligned}
\bar{E}_t &= \frac{1}{(2\pi)^2} \int \bar{e}_t e^{-j\alpha x - j\beta y}\, dx\, d\beta,\\
\bar{H}_t &= \frac{1}{(2\pi)^2} \int \bar{h}_t e^{-j\alpha x - j\beta y}\, d\alpha\, d\beta.
\end{aligned}
\qquad (14\text{-}77)
$$

Green's dyadic $[\bar{\bar{e}}]$ and $[\bar{\bar{h}}]$ can be obtained based on (14-75). We give the final results for $\bar{\bar{e}}$ and $\bar{\bar{h}}$ using the equivalent transmission-line voltages and currents shown in Fig. 14-15.

$$
\begin{aligned}
[\bar{\bar{e}}] &= \begin{bmatrix} \alpha^2 & \alpha\beta \\ \alpha\beta & \beta^2 \end{bmatrix} \frac{-V'}{\alpha^2 + \beta^2} + \begin{bmatrix} \beta^2 & -\alpha\beta \\ -\alpha\beta & \alpha^2 \end{bmatrix} \frac{-V''}{\alpha^2 + \beta^2},\\
[\bar{\bar{h}}] &= \begin{bmatrix} \alpha\beta & \alpha^2 \\ -\alpha^2 & -\alpha\beta \end{bmatrix} \frac{I'}{\alpha^2 + \beta^2} + \begin{bmatrix} \alpha\beta & -\alpha^2 \\ \beta^2 & -\alpha\beta \end{bmatrix} \frac{-I''}{\alpha^2 + \beta^2},
\end{aligned}
\qquad (14\text{-}78)
$$

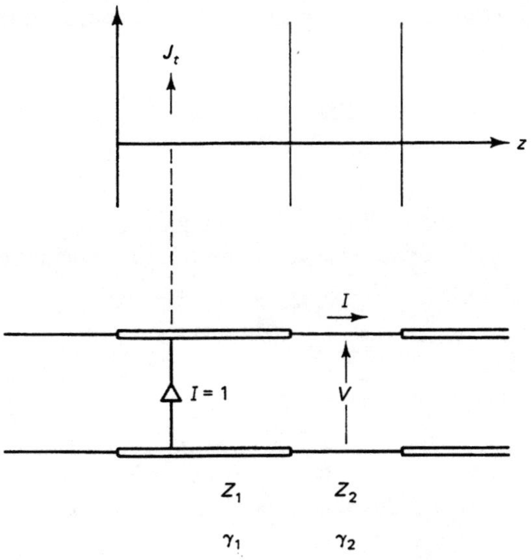

Figure 14-15 Equivalent transmission line excited by transverse current source J_t. Use γ_i and Z_i' to find V' and I', and γ_i and Z_i'' to find V'' and I''.

where

$$V' = -\frac{\partial}{\partial z}\left(\frac{1}{j\omega\epsilon_1}\frac{\partial}{\partial z'}g'\right),$$

$$I' = \frac{\epsilon}{\epsilon_1}\frac{\partial}{\partial z'}g',$$

$$V'' = j\omega\mu g'',$$

$$I'' = -\frac{\partial}{\partial z}g''.$$

These definitions of V', V'', I', and I'' are used to derive $[\bar{\bar{e}}]$ and $[\bar{\bar{h}}]$. They are defined so that the boundary conditions at dielectric interfaces that the tangential electric and magnetic fields are continuous, are automatically satisfied.

In actual calculation, we solve the transmission line in Fig. 14-15 with the unit current source as shown. We get V' and I' if we use the characteristic impedance Z', and we get V'' and I'' if we use the characteristic impedance Z''. These are then substituted in (14-78) to obtain the Green's dyadic $[\bar{\bar{e}}]$ and $[\bar{\bar{h}}]$.

The characteristic impedance and the propagation constant for the ith region is given by

$$Z_i' = \frac{\gamma_i}{\omega\epsilon_i},$$

$$Z_i'' = \frac{\omega\mu}{\gamma_i}, \tag{14-79}$$

$$\gamma_i = (k_i^2 - \alpha^2 - \beta^2)^{1/2} \qquad \text{if } k_i^2 > \alpha^2 + \beta^2$$

$$= -j(\alpha^2 + \beta^2 - k_i^2)^{1/2} \qquad \text{if } k_i^2 < \alpha^2 + \beta.$$

At conducting surfaces, V' and V'' are zero (short-circuited); the Fourier transform e_z of E_z is obtained by using $\nabla \cdot \bar{E} = 0$:

$$e_z = -\frac{\alpha e_x + \beta e_y}{\gamma}, \tag{14-80}$$

where $\bar{e}_t = \hat{x}e_x + \hat{y}e_y$.

Let us next consider the excitation by apertures embedded in dielectric layers (Fig. 14-16). The Fourier transforms of the transverse electric and magnetic fields are given by

$$\begin{aligned} \bar{e}_t &= \bar{\bar{e}}_m \cdot \bar{e}_a, \\ \bar{h}_t &= \bar{\bar{h}}_m \cdot \bar{e}_a. \end{aligned} \tag{14-81}$$

where $\bar{e}_a$ is the Fourier transform of the aperture field $\bar{E}_a$. See Fig. 14-16.

Using the equivalent transmission line in Fig. 14-16, we get V'_m and I'_m using Z'_{mi} and get V''_m and I''_m using Z''_{mi}. The Green's dyadic $\bar{\bar{e}}_m$ and $\bar{\bar{h}}_m$ in matrix form are given by

$$\begin{aligned} [\bar{\bar{e}}_m] &= \begin{bmatrix} \alpha^2 & \alpha\beta \\ \alpha\beta & \beta^2 \end{bmatrix} \frac{I'_m}{\alpha^2+\beta^2} + \begin{bmatrix} \beta^2 & -\alpha\beta \\ -\alpha\beta & \alpha^2 \end{bmatrix} \frac{I''_m}{\alpha^2+\beta^2}, \\ [\bar{\bar{h}}_m] &= \begin{bmatrix} \alpha\beta & \alpha^2 \\ -\alpha^2 & -\alpha\beta \end{bmatrix} \frac{-V'_m}{\alpha^2+\beta^2} + \begin{bmatrix} \alpha\beta & -\alpha^2 \\ \beta^2 & -\alpha\beta \end{bmatrix} \frac{V''_m}{\alpha^2+\beta^2}, \\ Z'_m &= \frac{\omega\epsilon}{\gamma_i} \qquad Z''_m = \frac{\gamma_i}{\omega\mu}. \end{aligned} \tag{14-82}$$

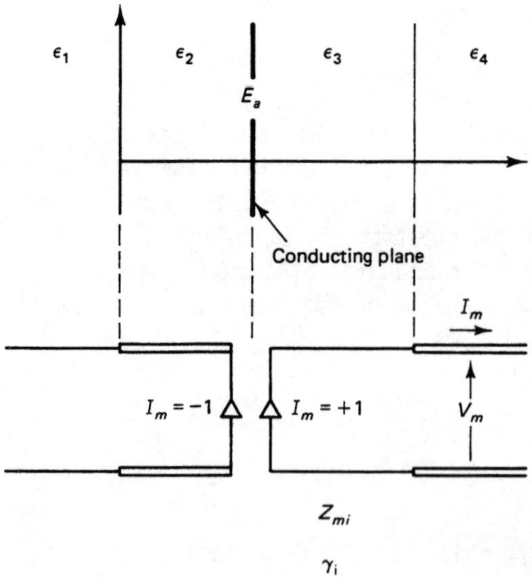

Figure 14-16 Equivalent transmission line for J_{mt} (the aperture field $\bar{E}_a$). Use γ_i and Z'_{mi} to find V'_m and I'_m, and γ_i and Z''_{mi} to find V''_m and I''_m.

At conducting surfaces, I'_m and I''_m are zero (open-circuited). The voltages and currents are defined by

$$\begin{aligned} V'_m &= j\omega\epsilon g', \\ I'_m &= -\frac{\partial}{\partial z} g', \\ V''_m &= -\frac{1}{j\omega\mu}\frac{\partial}{\partial z}\frac{\partial}{\partial z'} g'', \\ I''_m &= \frac{\partial}{\partial z'} g''. \end{aligned} \tag{14-83}$$

These definitions are needed to derive (14-82), but they are not needed to find Green's dyadic.

14-6 STRIP LINES EMBEDDED IN DIELECTRIC LAYERS

Making use of the formulaton given in Section 14-5, we now discuss the method to calculate the propagation constant of a strip line (Fig. 14-17). At $z = 0$, the tangential electric field is zero on the strip. We also note that the unknown in the problem is β, the propagation constant in the y direction. Thus we only have the integral with respect to α. We then have

$$\bar{E}_t = \frac{1}{2\pi}\int \bar{\bar{e}} \cdot \bar{J} e^{-j\alpha x}\, d\alpha = 0 \tag{14-84}$$

at $z = 0$ and $|x| < w/2$.

Now we use the moment method. We express $\bar{J}$, the Fourier transform of $\bar{J}_t$ in a series of N basis functions $\bar{J}_n$ and N unknown coefficients C_n:

$$\bar{J} = \sum_{n=1}^{N} \bar{J}_n C_n, \tag{14-85}$$

where

$$\bar{J}_n = \begin{bmatrix} J_{nx} \\ J_{ny} \end{bmatrix},$$

$$\bar{C}_n = \begin{bmatrix} C_1 \\ C_2 \\ \vdots \\ C_N \end{bmatrix};$$

and $\bar{J}_n$ is the Fourier transform of the basis function $\bar{J}_{tn}$:

$$\bar{J}_n = \int \bar{J}_{tn} e^{+j\alpha x}\, dx. \tag{14-86}$$

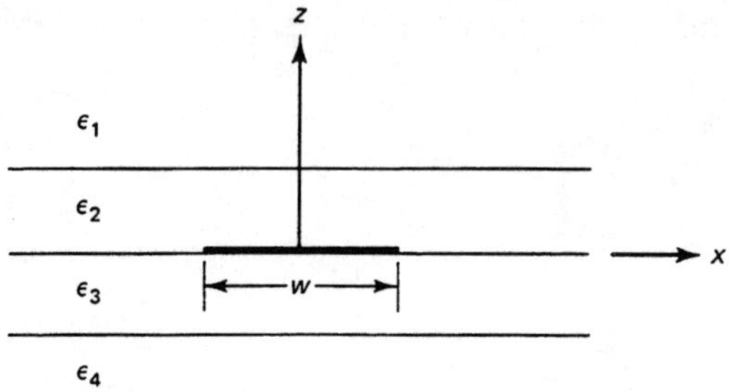

Figure 14-17 Strip lines. Wave propagates in the y direction.

We then multiply (14-84) from the left by the transpose of $\bar{J}_{tm}$ and integrate over the strip with respect to x. This gives the transpose of the conjugate of $\bar{J}_m, m = 1, 2, \ldots, N$. Thus we get the following $N \times N$ matrix equation:

$$[K_{mn}][C_n] = 0, \tag{14-87}$$

where

$$K_{mn} = \int [J^*_{mx} J^*_{my}][\bar{\bar{e}}]\begin{bmatrix} J_{nx} \\ J_{ny} \end{bmatrix} d\alpha.$$

In order to have nonzero solution for $[C_n]$, the determinant of $[K]$ must be zero:

$$|K_{mn}| = 0. \tag{14-88}$$

The solution of (14-88) gives the propagation constant β. The foregoing technique of using the basis function as the weighting function is called *Galerkin's method*.

The basis functions satisfying the edge condition have been proposed [Itoh, 1980]. For example, we can use for $n = 1, 2, \ldots, N$,

$$\begin{aligned} J_{tnx} &= \frac{\sin(2n\pi x/w)}{[1-(2x/w)^2]^{1/2}}, \\ J_{tny} &= 0, \end{aligned} \tag{14-89}$$

and for $n = N+1, N+2, \ldots, 2N$,

$$\begin{aligned} J_{tnx} &= 0, \\ J_{tny} &= \frac{\cos[2(n-1)\pi x/w]}{[1-(2x/w)^2]^{1/2}}. \end{aligned} \tag{14-90}$$

The corresponding Fourier transforms are

$$\begin{aligned} J_{nx} &= \frac{\pi w}{4j}\left[J_0\left(\left|\frac{w\alpha}{2} + n\pi\right|\right) - J_0\left(\left|\frac{w\alpha}{2} - n\pi\right|\right)\right], \\ J_{ny} &= \frac{\pi w}{4j}\left[J_0\left(\left|\frac{w\alpha}{2} + (n-1)\pi\right|\right) + J_0\left(\left|\frac{w\alpha}{2} - (n-1)\pi\right|\right)\right]. \end{aligned} \tag{14-91}$$

14-7 PERIODIC PATCHES AND APERTURES EMBEDDED IN DIELECTRIC LAYERS

Let us consider periodically placed patches and apertures embedded in dielectric layers. These structures are often used as frequency-selective surfaces. For periodic structures, we can use all the formulations developed in the preceding sections if we replace Fourier integrals with space harmonic Floquet representations (see Chapter 7). Specifically we make the following changes:

$$\frac{1}{(2\pi)^2}\iint f(\alpha,\beta)e^{-j\alpha x - j\beta y}\,d\alpha\,d\beta \rightarrow \sum_m \sum_n \frac{f(\alpha_m,\beta_n)}{l_x l_y} e^{-j\alpha_m x - j\beta_n y}, \tag{14-92}$$

where

$$\alpha_m = \alpha_0 + \frac{2m\pi}{l_x},$$

$$\beta_n = \beta_0 + \frac{2n\pi}{l_y},$$

$$\alpha_0 = k \sin\theta_0 \cos\phi_0,$$

$$\beta_0 = k \sin\theta_0 \sin\phi_0.$$

The periods in the x and the y directions are l_x and l_y, respectively, and (θ_0, ϕ_0) is the direction of the incident wave.

For skewed periodic structure (Fig. 14-18), noting that it is periodic in x' and y', we have

$$\begin{aligned} y' \sin\Omega &= y, \\ x' + y' \cos\Omega &= x. \end{aligned} \tag{14-93}$$

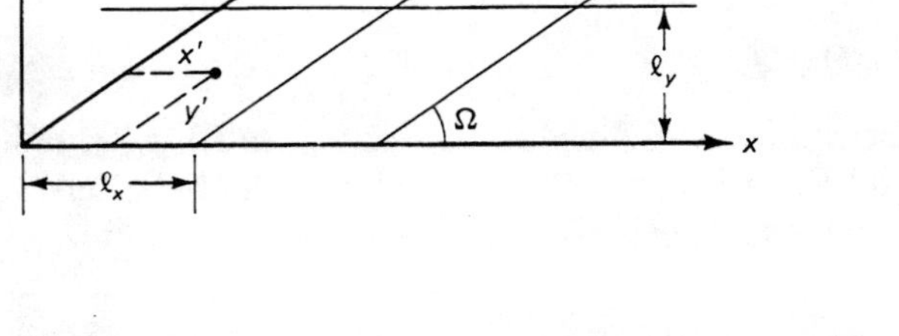

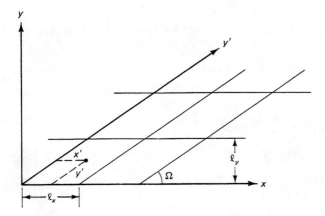

Figure 14-18 Skewed periodic structure.

From this we get $y' = y/\sin\Omega$ and $x' = x - y\cot\Omega$. The Floquet modes, which are periodic in x' and y', are given by

$$\exp\left[-j\alpha_0 x - j\frac{2\pi m}{l_x}x' - j\beta_0 y - j\frac{2\pi n \sin\Omega}{l_y}y'\right]$$
$$= \exp\left[-j\left(\alpha_0 + \frac{2\pi m}{l_x}\right)x - j\left(\beta_0 + \frac{2\pi n}{l_y} - \frac{2m\pi}{l_x}\cot\Omega\right)y\right].$$

Therefore, we use for skew periodic structure,

$$\alpha_m = \alpha_0 + \frac{2m\pi}{l_x},$$
$$\beta_{mn} = \beta_0 + \frac{2n\pi}{l_y} - \frac{2m\pi}{l_x}\cot\Omega. \tag{14-94}$$

Let us consider the current distribution on the patch in periodic structure. The total tangential electric field is the sum of incident and the scattered waves, and this must be zero on the patch. Therefore, we write

$$\bar{E}_{\text{inc}} + \bar{E}_t = 0 \qquad \text{on patch} \tag{14-95}$$

In Floquet mode representations, we have

$$\bar{E}_{\text{inc}} + \frac{1}{l_x l_y}\sum_m \sum_n \bar{\bar{e}}\bar{J}e^{-j\alpha_m x - j\beta_n y} = 0. \tag{14-96}$$

We use the moment method and express the current $\bar{J}_t$ in a series of the basis functions $\bar{J}_{tj}$ and its Fourier transform $\bar{J}_j$:

$$\bar{J}_t = \sum_j \bar{J}_{tj} C_j,$$
$$\bar{J} = \sum_j \bar{J}_j C_j. \tag{14-97}$$

We substitute (14-97) into (14-96), multiply (14-96) by the transpose of $\bar{J}_{ti}$, and integrate over the patch as we did in Section 14.6. We then get

$$[K_{ij}][C_j] + [E_i] = 0, \tag{14-98}$$

where

$$K_{ij} = \frac{1}{l_x l_y}\sum_m \sum_n [J_{ix}^* J_{iy}^*][\bar{\bar{e}}]\begin{bmatrix} J_{jx} \\ J_{jy} \end{bmatrix},$$
$$[E_i] = [J_{i0}^*]^t[\bar{E}_0],$$

where $[J_{i0}^*]^t$ is the transpose of the conjugate of J_i evaluated at $\alpha = \alpha_0$ and $\beta = \beta_0$, and the incident wave $\bar{E}_{\text{inc}}$ is given by $\bar{E}_{\text{inc}} = \bar{E}_0 e^{-j\alpha_0 x - j\beta_0 y}$. The solution of (14-98) gives the coefficient C_j and the current distribution is given by (14-97).

Next we consider the field of aperture embedded in dielectric layers. On one side of the aperture, the tangential magnetic field is given by $\overline{H}_{\text{inc}} + \overline{H}_1$, where $\overline{H}_1$ is the contribution from the aperture field $\overline{E}_a$. On the other side we have the magnetic field $\overline{H}_2$ contributed by the same aperture field $\overline{E}_a$. These two magnetic fields must be equal at the aperture and therefore we write

$$\overline{H}_{\text{inc}} + \overline{H}_1 = \overline{H}_2 \qquad \text{on aperture} \tag{14-99}$$

Making use of the formulations given in (14-81) we express $\overline{H}_1$ and $\overline{H}_2$ in terms of the aperture field $\overline{E}_a$:

$$\overline{H}_1 = \frac{1}{l_x l_y} \sum_m \sum_n \overline{\overline{h}}_m \cdot \bar{e}_a e^{-j\alpha_m x - j\beta_n y}. \tag{14-100}$$

$\overline{H}_2$ is given in the same form except that $\overline{\overline{h}}_m$ is chosen for the other side of the aperture.

Next, the aperture field $\overline{E}_a$ and its Fourier transform $\bar{e}_a$ are expressed in series of basis functions $\overline{E}_j$ and its Fourier transform $\bar{e}_j$ with unknown coefficients C_j:

$$\begin{aligned} \overline{E}_a &= \sum_j \overline{E}_j C_j, \\ \bar{e}_a &= \sum_j \bar{e}_j C_j. \end{aligned} \tag{14-101}$$

We then take the vector product of the aperture basis functions $\overline{E}_i$ and (14-99), and integrate over the aperture:

$$\int dx\, dy\, \overline{E}_i \times (\overline{H}_{\text{inc}} + \overline{H}_1) = \int dx\, dy\, \overline{E}_i \times \overline{H}_2. \tag{14-102}$$

The above is the equivalent of multiplying (14-99) by $\hat{z} \times \overline{E}_i$ and integrating over the aperture. Thus we obtain the following matrix equation:

$$[H_i] + [L_{ij}][C_j] = [M_{ij}][C_j], \tag{14-103}$$

where

$$L_{ij} = \frac{1}{l_x l_y} \sum_m \sum_n [-e_{iy}^* \; e_{ix}^*][\overline{\overline{h}}] \begin{bmatrix} e_{jx} \\ e_{jy} \end{bmatrix}.$$

M_{ij} is the same as L_{ij} except that $[\overline{\overline{h}}]$ is on the other side of the aperture.

$$\begin{aligned} [H_i] &= [-e_{iy0}^* \; e_{ix0}^*][\overline{H}_0], \\ \overline{H}_{\text{inc}} &= \overline{H}_0 e^{-j\alpha_0 x - j\beta_0 y}. \end{aligned} \tag{14-104}$$

The matrix equation (14-103) is now solved for the unknown coefficients $[C_j]$ and the final expression for the aperture field is given by (14-101).

PROBLEMS

14-1. Find the field E_y where an electric line source I_0 is located as shown in Fig. P14-1. Also find the radiation field.

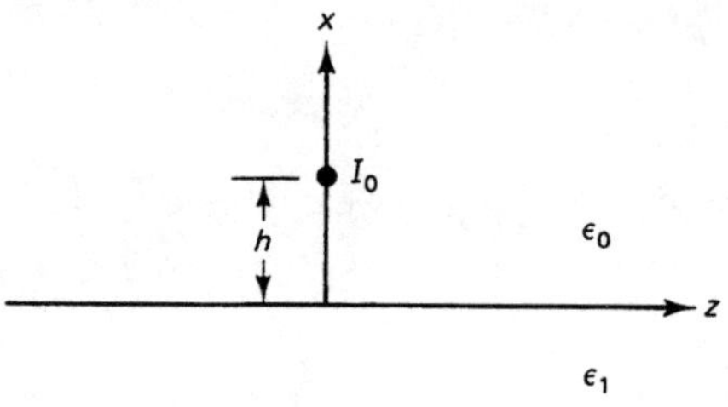

Figure P14-1 Line source above a dielectric half-space.

14-2. A magnetic line source I_m is located above the impedance plane where $E_z/H_y = Z$ is given (Fig. P14-2). Find the field for $x > 0$ and the radiation field, where $Z = R_s + jX_s = [j\omega(\mu_0/\sigma)]^{1/2}$. Do this when $Z = jX$.

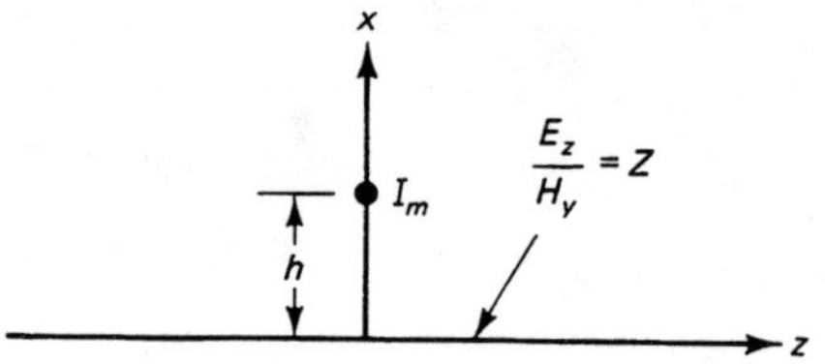

Figure P14-2 Impedance surface.

14-3. Consider the Green's function for the parabolic profile shown in (14-39). If $\theta_0 = 45°$, $\phi_0 = 0$, $z_0 = 1$, and $n_0 = 2$, trace the ray and calculate the magnitude of the Green's function.

14-4. Find the characteristic impedance of the strip line shown in Fig. P14-4. First find the Green's function and use (14-53). The charge density ρ_s satisfies the edge condition and is given by

$$\rho_s = \frac{\rho_0 \delta(y)}{[1 - (2x/W_0)^2]^{1/2}}.$$

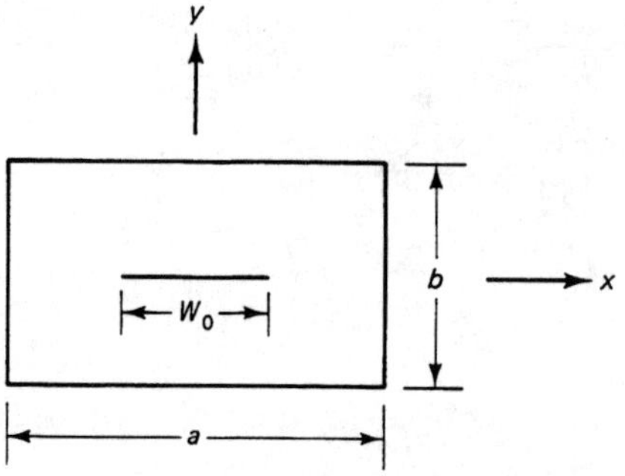

Figure P14-4 Strip line.

14-5. Use the quasi-static approximation to find the normalized phase velocity and the characteristic impedance of the coaxial line filled with two concentric dielectric materials as shown in Fig. P14-5, where a is the radius of the outer conductor and b is the radius of the inner conductor. $a = 0.5$ cm, $b = 0.2$ cm, $c = 0.3$ cm, $\epsilon_1/\epsilon_0 = 1.5$, and $\epsilon_2/\epsilon_0 = 2$.

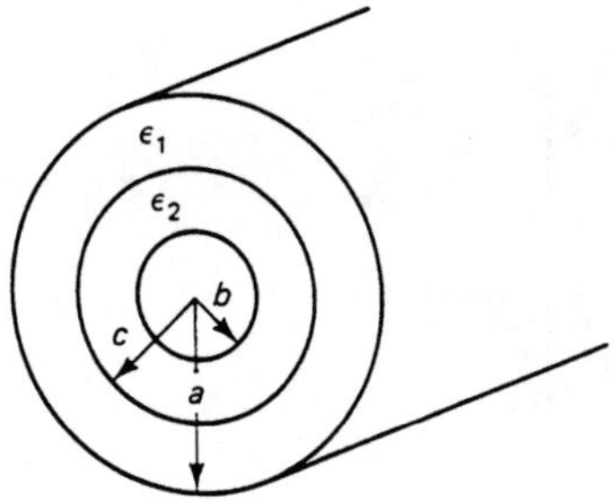

Figure P14-5 Coaxial line.

14-6. A vertical electric dipole is located in dielectric layers as shown in Fig. P14-6. Find the radiation field in air.

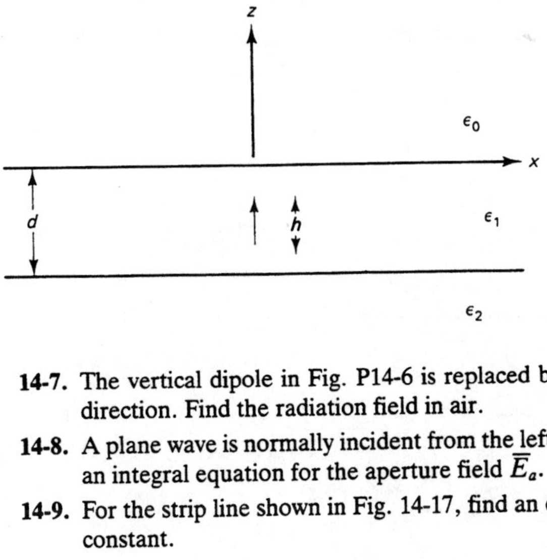

Figure P14-6 Electric dipole in a dielectric layer.

14-7. The vertical dipole in Fig. P14-6 is replaced by a horizontal dipole pointed in the x direction. Find the radiation field in air.

14-8. A plane wave is normally incident from the left on the layer shown in Fig. 14-16. Find an integral equation for the aperture field $\overline{E}_a$.

14-9. For the strip line shown in Fig. 14-17, find an equation to determine the propagation constant.

14-10. Find an equation to determine the current on the periodic patches shown in Fig. P14-10.

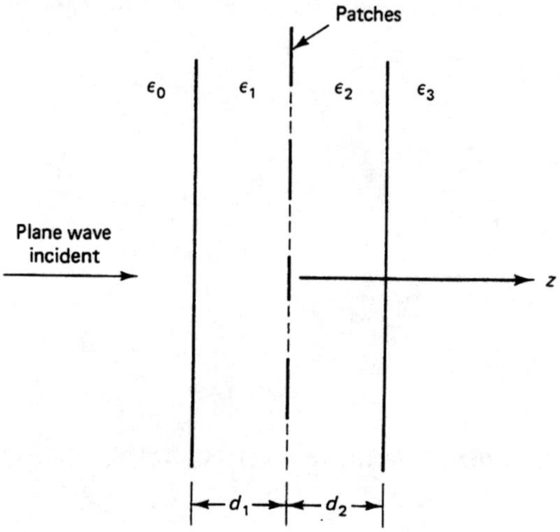

Figure P14-10 Periodic patches.

14-11. Find an equation to determine the aperture field on the periodic aperture shown in Fig. P14-11.

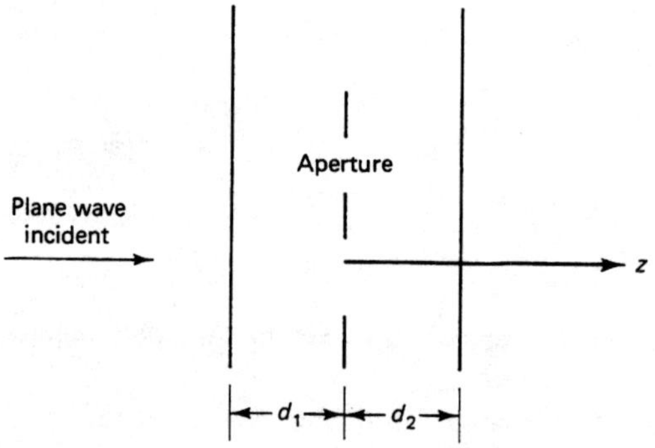

Figure P14-11 Periodic apertures.

15

Radiation from a Dipole on the Conducting Earth

15-1 SOMMERFELD DIPOLE PROBLEM

One of the important practical problems in radio is the determination of the characteristics of radio-wave propagation over the earth. The theoretical investigation of this problem, however, involved some subtle mathematical analyses that attracted attention from a great number of mathematicians and scientists over the past several decades. Historically, the problem dates back to the work by Zenneck in 1907, in which he investigated the characteristics of the wave propagating over the earth's surface, now called the *Zenneck wave*. Sommerfeld in 1909 investigated the excitation of the Zenneck wave by a dipole source. One portion of this solution had all the characteristics of the Zenneck wave on the surface and thus was called the *surface wave*.

The work of Sommerfeld, however, initiated a widely known controversy which was not completely resolved until the works of Ott, Van der Waerden, and Banos and Wesley in the late 1940s. The controversy ranged over the existence of the Zenneck surface wave, the definition of the surface wave, the choice of the branch cuts, and the poles in right or wrong Riemann surfaces. In addition, the 1909 paper by Sommerfeld contained an error in a sign, adding considerable confusion, even though the 1926 paper by Sommerfeld was correct (see Banos, 1966; Brekhovskikh, 1960; Sommerfeld, 1949; Wait, 1962, 1981; see also related topics in

Bremmer, 1949, and Wait, 1982). In this chapter we present a systematic study of this problem, and whenever appropriate, we indicate the source of the historical controversies.

15-2 VERTICAL ELECTRIC DIPOLE LOCATED ABOVE THE EARTH

Let us consider the radiation from a vertical dipole above the earth (Fig. 15-1). The earth is characterized by the relative dielectric constant ϵ_r and the conductivity σ or, equivalently, the complex relative dielectric constant ϵ/ϵ_0 or complex index of refraction n:

$$n^2 = \frac{\epsilon}{\epsilon_0} = \epsilon_r - j\frac{\sigma}{\omega\epsilon_0}. \tag{15-1}$$

The dipole is located at $z = h$ in air, where the wave number is k; the wave number within the earth is k_e. In this example k is real and k_e is complex, but the formulation is equally applicable to cases where k is complex (such as in the ionosphere) and k_e is real (air), or both k and k_e are complex (ice and water or ground).

Let us first consider the field in air ($z > 0$). We recognize that the field can be described by the Hertz vector $\overline{\pi}$, whose rectangular components satisfy a scalar wave equation,

$$(\nabla^2 + k^2)\pi_z = -\frac{J_z}{j\omega\epsilon_0}, \tag{15-2}$$

and the electric dipole with the equivalent length L carrying current I located at r' is given by

$$J_z = IL\,\delta(r - r'). \tag{15-3}$$

For convenience, we let

$$\frac{IL}{j\omega\epsilon_0} = 1, \tag{15-4}$$

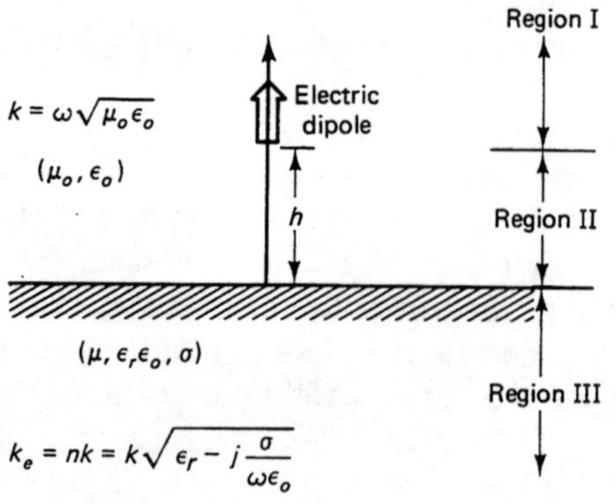

Figure 15-1 Vertical dipole above the ground.

and thus in air we have

$$(\nabla^2 + k^2)\pi_z = -\delta(r - r'). \tag{15-5}$$

Within the ground ($z < 0$) we have

$$(\nabla^2 + k_e^2)\pi_z = 0. \tag{15-6}$$

The boundary condition at $z = 0$ is that both tangential electric and magnetic fields are continuous across the boundary. Equations (15-5) and (15-6) together with the boundary conditions and the radiation condition give a complete mathematical description of the problem. We write the solution of (15-5) as a sum of the primary wave and the secondary wave. The primary wave is the wave radiated from the dipole in an infinite space in the absence of the boundary and has the correct singularity at the location of the antenna. The secondary wave represents the effects of the boundary but has no singularity at the antenna location.

Let us consider the primary wave π_p. This is obviously given by

$$\pi_p = \frac{e^{-jk|r-r'|}}{4\pi|r - r'|}. \tag{15-7}$$

To satisfy the boundary conditions, we need to express (15-7) in terms of cylindrical waves that have the same radial wave number in air and within the ground. This is done by the Fourier–Bessel transform shown below.

Let us write (15-5) in a cylindrical coordinate system

$$\left\{\frac{1}{\rho}\frac{\partial}{\partial\rho}\left(\rho\frac{\partial}{\partial\rho}\right) + \frac{1}{\rho^2}\frac{\partial^2}{\partial\phi^2} + \frac{\partial^2}{\partial z^2} + k^2\right\}\pi_p = -\frac{\delta(\rho - \rho')\delta(\phi - \phi')\delta(z - z')}{\rho}. \tag{15-8}$$

First, we expand π_p in a Fourier series in ϕ. Noting that π_p is a function of $\phi - \phi'$, we write

$$\pi_p = \sum_{m=-\infty}^{\infty} \pi_m(\rho, z)e^{-jm(\phi - \phi')}, \tag{15-9}$$

where

$$\pi_m(\rho, z) = \frac{1}{2\pi}\int_0^{2\pi} \pi_p e^{jm(\phi - \phi')}\, d\phi.$$

We obtain

$$\left[\frac{1}{\rho}\frac{\partial}{\partial\rho}\left(\rho\frac{\partial}{\partial\rho}\right) - \frac{m^2}{\rho^2} + \frac{\partial^2}{\partial z^2} + k^2\right]\pi_m = -\frac{\delta(\rho - \rho')\delta(z - z')}{2\pi\rho}. \tag{15-10}$$

We now express π_m in a Fourier–Bessel transform

$$\pi_m(\rho, z) = \int_0^{\infty} g_m(\lambda, z)J_m(\lambda\rho)\lambda\, d\lambda, \tag{15-11}$$

where

$$g_m(\lambda, z) = \int_0^{\infty} \pi_m(\rho, z)J_m(\lambda\rho)\rho\, d\rho.$$

Equation (15-11) is the Fourier–Bessel transform, which is the cylindrical equivalent of the usual Fourier transform. Note that λ is a complex variable commonly used in Fourier–Bessel transforms and is not a wavelength here.

Applying the Fourier–Bessel transform to both sides of (15-10), we obtain

$$\left(\frac{d^2}{dz^2} + k^2 - \lambda^2\right) g_m(\lambda, z) = -\frac{\delta(z - z')J_m(\lambda\rho')}{2\pi}. \tag{15-12}$$

To obtain (15-12) from (15-10), we use integration by parts and note the behaviors of π_m as $\rho \to 0$ and $\rho \to \infty$. Equation (15-12) is easily solved to give

$$g_m(\lambda, z) = \frac{e^{-jq|z-z'|}}{2jq} \frac{J_m(\lambda\rho')}{2\pi}, \tag{15-12a}$$

where $\lambda^2 + q^2 = k^2$.

Substituting (15-12a) into (15-11) and then into (15-9), we obtain

$$\pi_p = \frac{1}{4\pi} \sum_{m=-\infty}^{\infty} e^{-jm(\phi - \phi')} \int_0^\infty J_m(\lambda\rho) J_m(\lambda\rho') e^{-jq|z-z'|} \frac{\lambda\, d\lambda}{jq}. \tag{15-13}$$

Equation (15-13) is, of course, equal to (15-7), but it is written in terms of the cylindrical wave with the propagation constant λ [i.e., $J_m(\lambda\rho)$]. By expanding the field above and below the boundary in terms of the same wave number λ, we can satisfy the boundary conditions at any ρ. In contrast, when the boundary is in parallel to the z axis, such as in a dielectric cylinder, we use the Fourier transform in the z direction and express the field inside and outside the cylinder in terms of the same wave number in the z direction, e^{-jhz}.

In particular, when the antenna is located at $\rho' = 0$ and $z' = h$, (15-13) becomes

$$\pi_p(\rho, z) = \frac{1}{4\pi} \int_0^\infty J_0(\lambda\rho) e^{-jq|z-h|} \frac{\lambda\, d\lambda}{jq}, \tag{15-14}$$

where $\lambda^2 + q^2 = k^2$.

Let us now examine the problem pictured in Fig. 15-1. In air (regions I and II), π_z satisfies the differential equation

$$(\nabla^2 + k^2)\pi_z = -\delta(r - r'), \tag{15-15}$$

where r' is at $\rho = 0$ and $z = h$. We write π_z as a sum of the primary wave π_p and the scattered wave π_s:

$$\pi_z = \pi_p + \pi_s. \tag{15-16}$$

We write the primary wave π_p in region I as

$$\pi_p = \frac{1}{4\pi} \int_0^\infty J_0(\lambda\rho) e^{-jq(z-h)} \frac{\lambda\, d\lambda}{jq}, \tag{15-17a}$$

and in region II as

$$\pi_p = \frac{1}{4\pi}\int_0^\infty J_0(\lambda\rho)e^{-jq(h-z)}\frac{\lambda\, d\lambda}{jq}, \tag{15-17b}$$

where $\lambda^2 + q^2 = k^2$.

The difference in exponents $(z-h)$ and $(h-z)$ in (15-17a) and (15-17b) represents the singularity at $z=h$. The scattered wave π_s has no singularity at $z=h$ and it satisfies the homogeneous wave equation. Therefore, for both regions I and II, we write

$$\pi_s = \frac{1}{4\pi}\int_0^\infty R(\lambda)J_0(\lambda\rho)e^{-jq(z+h)}\frac{\lambda\, d\lambda}{jq}. \tag{15-18}$$

In region III there is no primary wave, and thus the scattered wave π_s satisfying the wave equation

$$(\nabla^2 + k_e^2)\pi_s = 0, \tag{15-19}$$

can be written as

$$\pi_s = \frac{1}{4\pi}\int_0^\infty T(\lambda)J_0(\lambda\rho)e^{+jq_e z - jqh}\frac{\lambda\, d\lambda}{jq}, \tag{15-20}$$

where $\lambda^2 + q_e^2 = k_e^2$. The choice of $(-jq)$ instead of $(+jq)$ for (15-17) and (15-18) is made so that this represents the outgoing wave in the $+z$ direction satisfying the "radiation condition" when q is in the fourth quadrant. Similarly, $(+jq_e z)$ is chosen for (15-20) to represent the outgoing wave in the $-z$ direction when q_e is in the fourth quadrant. Other commonly used notations are μ or γ in place of jq. We choose jq because q represents the wave number in the z direction in the same sense that λ is the wave number in the ρ direction.

Equations (15-16), (15-17a), (15-17b), (15-18), and (15-20) represent the complete expressions of the fields, which are expressed in terms of two unknown functions, $R(\lambda)$ and $T(\lambda)$. These two functions are now determined by applying the boundary conditions at $z=0$. The conditions are that the tangential electric and tangential magnetic fields are continuous across the boundary. Because of the symmetry of the problem, the only tangential electric field is E_ρ, and the only magnetic field is H_ϕ. Thus, noting that

$$E_\rho = \frac{\partial^2}{\partial\rho\,\partial z}\pi_z \quad \text{and} \quad H_\phi = -j\omega\epsilon\frac{\partial}{\partial\rho}\pi_z,$$

the boundary conditions are given by

$$\left.\begin{aligned} \frac{\partial}{\partial z}\pi_z^{(2)} &= \frac{\partial}{\partial z}\pi_z^{(3)} \\ \pi_z^{(2)} &= n^2\pi_z^{(3)} \end{aligned}\right\} \quad \text{at } z=0, \tag{15-21}$$

where $\pi_z^{(2)}$ and $\pi_z^{(3)}$ are π_z in regions II and III, respectively.

Applying (15-21) to (15-17), (15-18), and (15-20), we get

$$R(\lambda) = \frac{n^2 q - q_e}{n^2 q + q_e},$$
$$T(\lambda) = \frac{2q}{n^2 q + q_e}. \tag{15-22}$$

Therefore, the solution to the original problem pictured in Fig. 15-1 is

$$\pi_z = \left(\frac{IL}{j\omega\epsilon_0}\right)\left[\frac{1}{4\pi}\int_0^\infty J_0(\lambda\rho)e^{-jq|z-h|}\frac{\lambda\,d\lambda}{jq} + \frac{1}{4\pi}\int_0^\infty \frac{n^2 q - q_e}{n^2 q + q_e} J_0(\lambda\rho)e^{-jq(z+h)}\frac{\lambda\,d\lambda}{jq}\right] \quad \text{for } z > 0 \tag{15-23}$$

and

$$\pi_z = \left(\frac{IL}{j\omega\epsilon_0}\right)\frac{1}{4\pi}\int_0^\infty \frac{2}{n^2 q + q_e} J_0(\lambda\rho)e^{jq_e z - jqh}\frac{\lambda\,d\lambda}{j} \quad \text{for } z < 0 \tag{15-24}$$

where $\lambda^2 + q^2 = k^2$ and $\lambda^2 + q_e^2 = k_e^2$ and q and q_e are chosen to be in the fourth quadrant, and $IL/j\omega\epsilon_0$ is restored in this expression. Equations (15-23) and (15-24) are the general expression of the Hertz potential, and the other field components are obtained by differentiation:

$$E_z = \left(\frac{\partial^2}{\partial z^2} + k^2\right)\pi_z,$$
$$E_\rho = \frac{\partial^2}{\partial\rho\,\partial z}\pi_z, \tag{15-25}$$
$$H_\phi = -j\omega\epsilon\frac{\partial}{\partial\rho}\pi_z.$$

15-3 REFLECTED WAVES IN AIR

Let us consider (15-23) without $IL/j\omega\epsilon_0$. We write

$$\pi_z = \pi_p + \pi_s \tag{15-26}$$

and

$$\pi_p = \frac{e^{-jk|r-r'|}}{4\pi|r-r'|},$$
$$\pi_s = \frac{1}{4\pi}\int_0^\infty \frac{n^2 q - q_e}{n^2 q + q_e} J_0(\lambda\rho)e^{-jq(z+h)}\frac{\lambda\,d\lambda}{jq}.$$

π_p represents the direct wave from the antenna to the observation point, and π_s represents the reflected wave as pictured in Fig. 15-2.

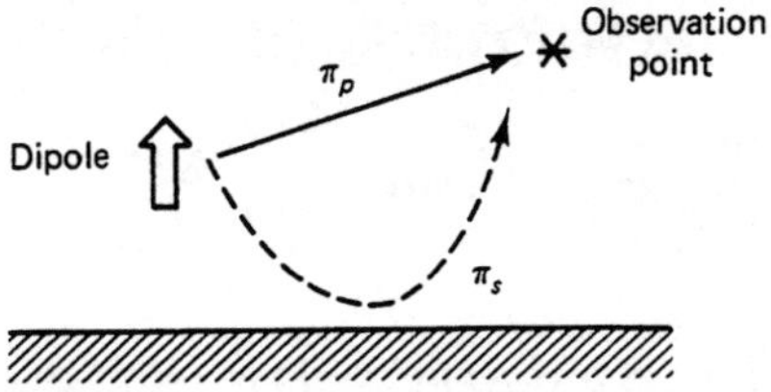

Figure 15-2 Direct and reflected waves.

To evaluate the integral, it is convenient to convert the range of the integral from $(0 \rightarrow \infty)$ to $(-\infty \rightarrow +\infty)$. To this end we use the identity

$$J_0(\lambda\rho) = \tfrac{1}{2}[H_0^{(1)}(\lambda\rho) + H_0^{(2)}(\lambda\rho)] \tag{15-27}$$

and write the integral as

$$\int_{W_1} f(\lambda)J_0(\lambda\rho)\lambda\, d\lambda = \tfrac{1}{2}\int_{W_1} f(\lambda)H_0^{(1)}(\lambda\rho)\lambda\, d\lambda + \tfrac{1}{2}\int_{W_1} f(\lambda)H_0^{(2)}(\lambda\rho)\lambda\, d\lambda, \tag{15-28}$$

where the contour W_1 is shown in Fig. 15-3. We now convert the first integral into the contour W_2, which is symmetric to W_1 about the origin. Thus writing

$$\lambda = -\lambda',$$

the first integral becomes

$$\tfrac{1}{2}\int_{W_1} f(\lambda)H_0^{(1)}(\lambda\rho)\lambda\, d\lambda = \tfrac{1}{2}\int_{W_2} f(-\lambda')H_0^{(1)}(-\lambda'\rho)\lambda'\, d\lambda'. \tag{15-29}$$

Noting the identity

$$H_\nu^{(1)}(e^{\pi j} Z) = -e^{-\nu\pi j} H_\nu^{(2)}(Z),$$

(15-29) becomes

$$-\tfrac{1}{2}\int_{W_2} f(-\lambda')H_0^{(2)}(\lambda'\rho)\lambda'\, d\lambda'.$$

We reverse the path W_2 to $-W_2$, and noting that in our problem $f(\lambda)$ is an even function of λ, we finally obtain

$$\begin{aligned}\int_{W_1} f(\lambda)J_0(\lambda\rho)\lambda\, d\lambda &= \tfrac{1}{2}\int_{-W_2} f(-\lambda)H_0^{(2)}(\lambda\rho)\lambda\, d\lambda + \tfrac{1}{2}\int_{W_1} f(\lambda)H_0^{(2)}(\lambda\rho)\lambda\, d\lambda \\ &= \tfrac{1}{2}\int_{W} f(\lambda)H_0^{(2)}(\lambda\rho)\lambda\, d\lambda.\end{aligned} \tag{15-30}$$

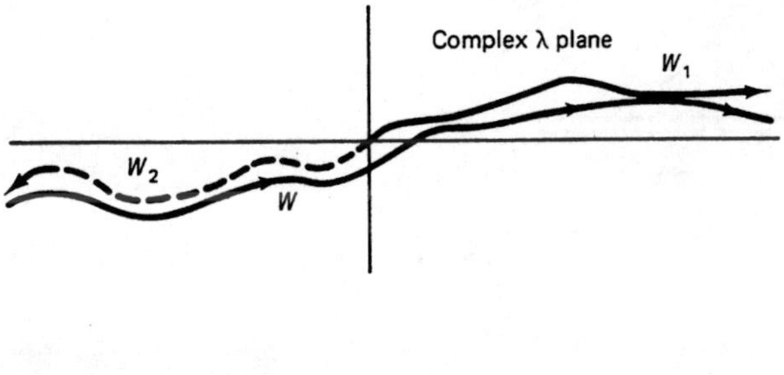

Figure 15-3 The integral from 0 to ∞ is converted to the integral from $-\infty$ to $+\infty$.

Using (15-30), we write the reflected wave equation (15-26) as

$$\pi_s = \frac{1}{8\pi}\int_W \frac{n^2 q - q_e}{n^2 q + q_e} H_0^{(2)}(\lambda\rho)e^{-jq(z+h)}\frac{\lambda\, d\lambda}{jq}. \tag{15-31}$$

The contour W is shown in Fig. 15-4.

It is not possible to evaluate the integral (15-31) in terms of elementary functions in a closed form. Thus it is necessary to evaluate the integral by approximate techniques. The approximate evaluation of the integral (15-31) may be classified into the following three cases:

1. *Radiation field using the saddle-point technique*. We may evaluate (15-31) for a large distance from the image point using the saddle-point technique. This yields a simple expression for the radiation pattern of the reflected wave.
2. *The field along the surface*. The evaluation of the field on the earth surface is the central point of this Sommerfeld problem. This is accomplished by the saddle-point technique modified by the presence of the pole (Sommerfeld pole).
3. *Lateral wave*. If the index of refraction n is smaller than unity, the integration along the branch cut must be considered. This results in the lateral wave.

Mathematically, the three cases above result from the *saddle-point* evaluation, the effect of the *pole*, and the *branch point*. The consideration of these three points is most important when evaluating a complex integral of this type. In addition to the three analytical methods above, we can also carry out the integration *numerically*. A significant improvement in efficiency results when the integration path is deformed into the steepest descent path. This is discussed in Section 15-7.

15-4 RADIATION FIELD: SADDLE-POINT TECHNIQUE

When the observation point is far from the antenna, it is possible to evaluate the integral in (15-31) and obtain a simple expression for the radiation field. This far-field solution is not valid for the region close to the surface, where it is necessary to employ the modified saddle-point technique described in the next section.

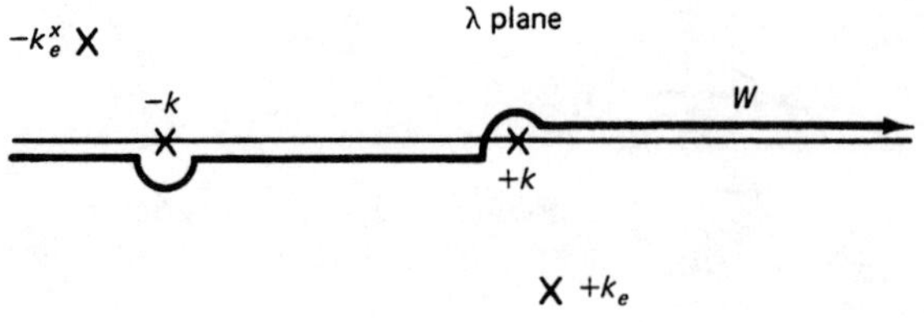

Figure 15-4 Contour W for (15-31).

Let us evaluate (15-31) by means of the saddle-point technique. We first approximate the Hankel function by its asymptotic form

$$H_0^{(2)}(\lambda\rho) = \left(\frac{2}{\pi\lambda\rho}\right)^{1/2} e^{-j\lambda\rho + j(\pi/4)}. \tag{15-32}$$

This is valid only for $|\lambda\rho| >> 1$. But λ ranges from $-\infty$ to $+\infty$. Note, however, that the main contribution to the integral comes from the neighborhood of $\lambda = k \sin\theta$ and thus $|\lambda\rho| \sim kR_2 \sin^2\theta$. Therefore, (15-32) can be used as long as θ is not too small (the radiation close to the vertical axis) (see Banos, 1966, p. 76). If θ is small, the double-saddle-point method should be used. Then the first term of the rigorous asymptotic solution is identical to that obtained by (15-32). Using (15-32), we get

$$\pi_s = \frac{1}{8\pi}\int_W \frac{n^2 q - q_e}{n^2 q + q_e}\left(\frac{2}{\pi\lambda\rho}\right)^{1/2} \frac{e^{j(\pi/4)}}{jq} e^{-j\lambda\rho - jq(z+h)}\lambda\, d\lambda. \tag{15-33}$$

Using the saddle-point evaluation of the integral

$$\int_W f(\lambda)e^{-j\lambda\rho - jq(z+h)}\, d\lambda \approx f(k \sin\theta)\sqrt{\frac{2\pi}{kR_2}}\, k \cos\theta e^{-jkR_2 + j(\pi/4)}, \tag{15-34}$$

where

$$R_2 = \sqrt{\rho^2 + (z+h)^2}, \qquad z + h = R_2 \cos\theta, \qquad \rho = R_2 \sin\theta,$$

we get

$$\pi_s = \left(\frac{n^2 \cos\theta - \sqrt{n^2 - \sin^2\theta}}{n^2 \cos\theta + \sqrt{n^2 - \sin^2\theta}}\right)\frac{e^{-jkR_2}}{4\pi R_2}, \tag{15-35}$$

where $\lambda = k \sin\theta$ is substituted in $q = \sqrt{k^2 - \lambda^2}$ and $q_e = \sqrt{k^2 n^2 - \lambda^2}$, and R_2 is the distance from the image point (Fig. 15-5).

Equation (15-35) is the spherical wave originating at the image point, but the magnitude is multiplied by the reflection coefficient $R(\theta)$.

$$R(\theta) = \frac{n^2 \cos\theta - \sqrt{n^2 - \sin^2\theta}}{n^2 \cos\theta + \sqrt{n^2 - \sin^2\theta}}. \tag{15-36}$$

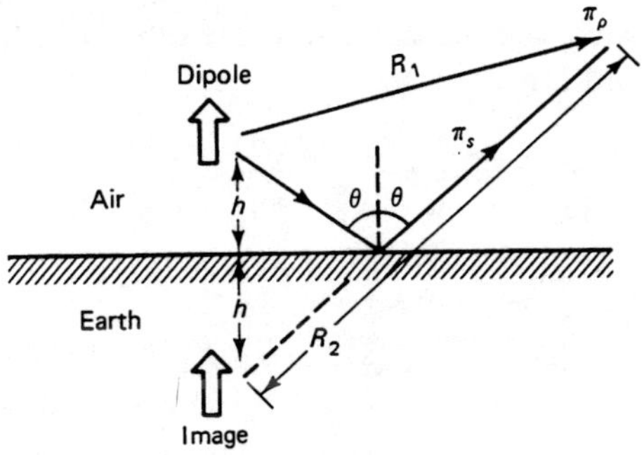

Figure 15-5 Radiation field.

This reflection coefficient is exactly the same as the plane-wave reflection coefficient when the incident angle is θ. This is to be expected because the saddle-point technique is valid for the field far from the source and thus the wave is expected to behave locally as a plane wave.

The total field is then given by

$$\begin{aligned}\pi &= \pi_p + \pi_s \\ &= \frac{e^{-jkR_1}}{4\pi R_1} + R(\theta)\frac{e^{-jkR_2}}{4\pi R_2}.\end{aligned} \tag{15-37}$$

We note, however, that on the surface of the earth, $R_1 = R_2 = R$ and at a large distance, θ approaches $\pi/2$, and thus $R(\theta)$ approaches -1. Therefore the expression (15-37) reduces to zero on the surface far from the source. This is to be expected because far from the source the wave should behave essentially as a spherical wave, but on the surface the air and the earth have two different wave numbers, k and k_e. Thus it is impossible to have two different spherical waves in air and within the earth having the same phase relationship to satisfy boundary conditions.

In general, a spherical wave of the type e^{-jkR}/R should vanish along the interface between two different media. This of course does not imply that other wave types, such as $1/R^2$, vanish, and in fact, the Sommerfeld problem in the next section is essentially that of finding the field on the surface.

15-5 FIELD ALONG THE SURFACE AND THE SINGULARITIES OF THE INTEGRAND

The evaluation of the field on the surface due to a dipole source is central to the Sommerfeld problem. This is an important practical problem of radio-wave propagation. It also presents a rather involved mathematical subtlety; therefore, a thorough understanding of the basic techniques is very important.

Let us first write the total field as a sum of the field when the earth is perfectly conducting and the term representing the finiteness of the conductivity of the earth:

$$\pi = \frac{e^{-jkR_1}}{4\pi R_1} + \frac{e^{-jkR_2}}{4\pi R_2} - 2P. \tag{15-38}$$

The first term is the incident wave, the second term is the reflected wave when the earth is perfectly conducting, and P is given by

$$P = \frac{1}{8\pi}\int_w \frac{q_e}{n^2 q + q_e} H_0^{(2)}(\lambda\rho) e^{-jq(z+h)} \frac{\lambda\, d\lambda}{jq}, \tag{15-39a}$$

$$\approx \frac{1}{8\pi}\int_w \frac{q_e}{n^2 q + q_e}\left(\frac{2}{\pi\lambda\rho}\right)^{1/2} \frac{e^{j(\pi/4)}}{jq} e^{-j\lambda\rho - jq(z+h)}\lambda\, d\lambda, \tag{15-39b}$$

where the approximation (15-32) has been used in (15-39b).

We can work with this form (15-39b), but as will be shown shortly, this form contains branch points at $\lambda = \pm k$ and $\lambda = \pm k_e$. It is possible to eliminate the branch points at $\lambda = \pm k$ by the following transformation of the variable of integration from λ to α:

$$\lambda = k \sin\alpha. \tag{15-40}$$

Furthermore, using R_2 and θ instead of ρ and z through the tranformation

$$\begin{aligned} \rho &= R_2 \sin\theta, \\ z + h &= R_2 \cos\theta, \end{aligned} \tag{15-41}$$

equation (15-39b) becomes

$$P = \int_c F(\alpha) e^{-jkR_2 \cos(\alpha - \theta)}\, d\alpha, \tag{15-42}$$

where

$$F(\alpha) = \frac{ke^{-j(\pi/4)}}{4\pi}\left(\frac{1}{2\pi k R_2}\frac{\sin\alpha}{\sin\theta}\right)^{1/2}\frac{\sqrt{n^2 - \sin^2\alpha}}{n^2\cos\alpha + \sqrt{n^2 - \sin^2\alpha}}.$$

Equations (15-39b) and (15-42) are the basic integrals that must be evaluated. The actual evaluation in this section is done in the α plane using (15-42) because the procedure is simpler and clearer in the α plane than in the λ plane. However, the corresponding discussion in the λ plane is given in the Appendix, Chapter 15, Section B, to clarify the situation, particularly with respect to the historical controversies of this Sommerfeld problem.

In evaluating complex integrals, it is essential first to examine all the singularities of the integrand. In general, there are three kinds of singularities: poles, essential singularities, and branch points. The singularities of a function $f(\lambda)$ are the points where $f(\lambda)$ is not analytic.

1. *Pole (isolated singularity)*. In the neighborhood of the pole, $f(\lambda)$ can be expanded in a Laurent series with *finite* negative powers.
2. *Essential singularities*. $f(\lambda)$ is expressed in an *infinite* series of negative powers. For example, $e^{1/\lambda} = 1 + 1/\lambda + 1/2!\,\lambda^2 + \cdots + 1/n!\,\lambda^n + \cdots$.
3. *Branch points*. The function $f(\lambda)$ has more than one value at a given λ, representing more than one branch. The points where these branches meet are the branch points. For example, $f(\lambda) = \lambda^{1/2}, f(\lambda) = \ln\lambda$.

In our problem there are poles where the denominator of the integrand vanishes:

$$n^2 q + q_e = 0 \quad \text{in the } \lambda \text{ plane,} \tag{15-43a}$$

$$n^2\cos\alpha + \sqrt{n^2 - \sin^2\alpha} = 0 \quad \text{in the } \alpha \text{ plane,} \tag{15-43b}$$

and branch points

$$\text{at } \lambda = \pm k \text{ and } \lambda = \pm k_e \quad \text{in the } \lambda \text{ plane,} \tag{15-44a}$$

and

$$\text{at } \sin\alpha = \pm n \qquad \text{in the } \alpha \text{ plane.} \tag{15-44b}$$

In addition, there is a branch point at $\lambda = 0$ due to the Hankel function $H_0^{(2)}(\lambda\rho)$. However, since λ represents the propagation constant in the radial direction, the wave corresponding to this branch point $\lambda = 0$ does not propagate on the surface, and thus this branch point has practically no effect on the field.

The integration in (15-42) will be carried out using the saddle-point technique. However, the pole in the integrand is located close to the saddle point and thus the modified saddle-point technique must be used. It is therefore important to determine the location of the pole, particularly with respect to the different branches of the complex plane resulting from the branch points.

Let us consider the branch points in (15-42). Because of the square root in $\sqrt{n^2 - \sin^2\alpha}$, there are two Riemann surfaces. We draw the branch cut in the α plane along

$$\operatorname{Im}\sqrt{n^2 - \sin^2\alpha} = 0. \tag{15-44c}$$

We note that

$$q_e = k\sqrt{n^2 - \sin^2\alpha}$$

is the wave number in the $-z$ direction inside the earth and thus $\operatorname{Im} q_e$ must be negative because

$$|e^{jq_e z}| = e^{-(\operatorname{Im} q_e)z} \qquad \text{for } z < 0$$

vanishes as $z \to -\infty$. Therefore,

$$\operatorname{Im}\sqrt{n^2 - \sin^2\alpha} < 0 \tag{15-45a}$$

corresponds to the wave attenuating in the $-z$ direction satisfying the radiation condition, and thus this may be called the *proper Riemann surface*. On the other hand,

$$\operatorname{Im}\sqrt{n^2 - \sin^2\alpha} > 0 \tag{15-45b}$$

corresponds to the wave exponentially increasing toward $z \to -\infty$, and thus, this may be called the *improper Riemann surface*. Obviously, the original contour c in (15-42) is located in the proper Riemann surface. This is shown in Fig. 15-6 (see the Appendix to Chapter 15, Section B).

15-6 SOMMERFELD POLE AND ZENNECK WAVE

Let us consider the pole given by (15-43a):

$$n^2 q + q_e = 0. \tag{15-46}$$

This is identical to the equation used to determine the propagation constant for the Zenneck wave. To show this, consider the Zenneck wave propagating in the x direction along the surface at $z = 0$. Then for $z > 0$, we have

$$\pi_1 = Ae^{-j\lambda x - jqz}, \tag{15-47}$$

and for $z < 0$, we write

$$\pi_2 = Be^{-j\lambda x + jq_e z}. \tag{15-48}$$

Now satisfying the boundary condition:

$$n^2 \pi = \text{continuous} \qquad \text{and} \qquad \frac{\partial \pi}{\partial z} = \text{continuous}$$

we get

$$A = n^2 B \qquad \text{and} \qquad -jqA = jq_e B.$$

From this, we get (15-46).

Let us find the solution λ_p from (15-46). Substituting $q^2 = k^2 - \lambda^2$ and $q_e^2 = k^2 n^2 - \lambda^2$ into (15-46) and squaring both sides of $n^2 q = -q_e$, we get $n^4(k^2 - \lambda_p^2) = k^2 n^2 - \lambda_p^2$. From this we get

$$\frac{1}{\lambda_p^2} = \frac{1}{k^2} + \frac{1}{k^2 n^2}. \tag{15-49}$$

We note here that by squaring $q_e = -n^2 q$, we introduced an additional solution corresponding to $q_e = +n^2 q$. This is the solution with $\exp(+jqz)$ for $z > 0$, and therefore it is in the improper Riemann surface. To see this more clearly, we obtain q using (15-49):

$$q = \pm \frac{k}{(n^2 + 1)^{1/2}}. \tag{15-50}$$

Now we need to choose the sign for q such that it satisfies (15-46). Noting that $-n^2 q = q_e$ and that $\text{Im}\, q_e$ must be negative, we require that $\text{Im}(-n^2 q)$ be negative.

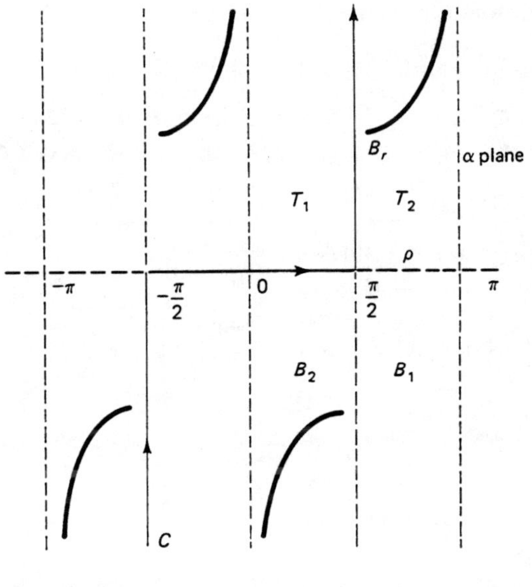

Figure 15-6 Riemann surface corresponding to $\text{Im}\sqrt{n^2 - \sin^2\alpha} < 0$. The branch points are at $\sin\alpha = \pm n$ and the branch cuts are along IM $\sqrt{n^2 - \sin^2\alpha} = 0$.

However, considering that $\mathrm{Im}(n^2)$ is negative, we see that $\mathrm{Im}[n^2/(n^2+1)^{1/2}]$ is also negative. Thus, to make $\mathrm{Im}(-n^2 q)$ negative, we must choose the minus sign in (15-50).

$$q = -\frac{k}{(n^2+1)^{1/2}},$$
$$\lambda_p = \frac{nk}{(n^2+1)^{1/2}}, \tag{15-51}$$
$$q_e = \frac{n^2 k}{(n^2+1)^{1/2}}.$$

For detailed behavior of the locations of the Sommerfeld poles, see the Appendix to Chapter 15, Section B.

For conducting earth, $|n|$ is normally much greater than unity, and therefore we can get the following approximate solution from (15-51):

$$\lambda_p \approx k\left(1 - \frac{1}{2n^2}\right). \tag{15-52}$$

Next we examine the Sommerfeld pole in the complex α plane:

$$\lambda_p = k \sin \alpha_p. \tag{15-53}$$

Following the procedure indicated above for q, we get

$$\cos \alpha_p = -\frac{1}{(n^2+1)^{1/2}},$$
$$\sin \alpha_p = \frac{n}{(n^2+1)^{1/2}}. \tag{15-54}$$

Here $\cos \alpha_p$ is in the third quadrant:

$$\mathrm{Re}(\cos \alpha_p) < 0 \quad \text{and} \quad \mathrm{Im}(\cos \alpha_p) < 0.$$

Therefore, α_p must be located in T_2 of Fig. 15-6. Considering that $|n| >> 1$, we see that α_p is located very close to $\pi/2$ and is slightly below the 45° line, as can be seen below and in Fig. 15-7.

$$\cos \alpha_p = \cos\left(\frac{\pi}{2} + \Delta\right) = -\sin \Delta = -\frac{1}{(n^2+1)^{1/2}},$$
$$\Delta = \sin^{-1}\frac{1}{(n^2+1)^{1/2}} = \left(\frac{\omega\epsilon_0}{\sigma}\right)^{1/2} e^{j(\pi/4)},$$
$$n^2 = \epsilon_r - j\frac{\sigma}{\omega\epsilon_0} \approx -j\frac{\sigma}{\omega\epsilon_0}, \tag{15-55}$$
$$\alpha_p = \frac{\pi}{2} + \sin^{-1}\frac{1}{(n^2+1)^{1/2}} \approx \frac{\pi}{2} + \frac{1}{n}.$$

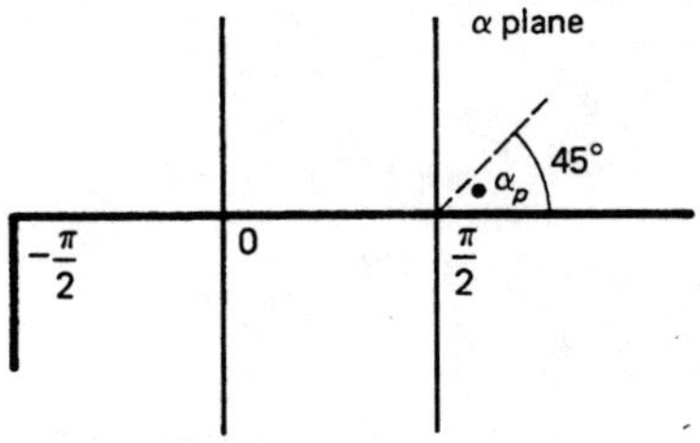

Figure 15-7 Location of Sommerfeld pole α_p.

15-7 SOLUTION TO THE SOMMERFELD PROBLEM

We now evaluate the integral given in (15-42):

$$P = \int_c F(\alpha) e^{-jkR_2 \cos(\alpha - \theta)} \, d\alpha,$$

where $F(\alpha)$ has the pole at $\alpha = \alpha_p$ given in Section 15-6. The pole is located below the original contour c, and therefore, we can make use of the modified saddle-point technique discussed in the Appendix to Chapter 15, Section C.

$$\begin{aligned} I_0 &= \int_c F(\alpha) e^{zf(\alpha)} \, d\alpha \\ &= I_1 + I_2, \end{aligned} \tag{15-56}$$

where

$$I_1 = e^{zf(\alpha_s)} \int_{-\infty}^{\infty} \left[F(s) \frac{d\alpha}{ds} - \frac{R_1(s_p)}{s - s_p} \right] \exp\left(-z \frac{s^2}{2}\right) ds,$$

$$I_2 = -j\pi R(s_p) \operatorname{erfc}(j\sqrt{z/2}\, s_p),$$

$$\begin{aligned} R(s_p) &= R_1(s_p) \exp[sf(\alpha_p)] \\ &= \text{residue of } F(\alpha) \exp[zf(\alpha)] \text{ at the pole } \alpha = \alpha_p, \end{aligned}$$

$$f(\alpha) - f(\alpha_s) = -\frac{s^2}{2}, \qquad \operatorname{erfc}(z) = 1 - \operatorname{erf}(z), \qquad \operatorname{erf}(z) = \frac{2}{\sqrt{\pi}} \int_0^z e^{-t^2} dt.$$

The saddle point is located at $s = s_s$, and the pole is located at $s = s_p$.

For our problem (15-42), we have

$$\begin{aligned} f(\alpha) &= -j \cos(\alpha - \theta), \\ z &= kR_2. \end{aligned} \tag{15-57}$$

Therefore, the saddle point is at $\alpha_s = \theta$, and

$$f(\alpha) - f(\alpha_s) = j[\cos(\alpha - \theta) - 1] = -\frac{s^2}{2}. \tag{15-58}$$

From this we get

$$s = 2e^{-j(\pi/4)} \sin\frac{\alpha - \theta}{2},$$

$$\frac{d\alpha}{ds} = \left(1 - j\frac{s^2}{4}\right)^{-1/2} e^{j(\pi/4)}, \tag{15-59}$$

$$\alpha = \theta + 2\sin^{-1}\left(\frac{s}{2}e^{j(\pi/4)}\right).$$

The path of integration for I_1 is on the steepest descent contour (SDC) and is along the real axis of s. The original contour of c and SDC in the α plane and the s plane are pictured in Fig. 15-8. Because of (15-59), two branch points (B_{r1} and B_{r2}) appear at $s = \pm 2e^{-j(\pi/4)}$ in the s plane, but these branch points are far from saddle point and have little effect on the integral. Note that the integrand for I_1 has no pole, and therefore I_1 can be evaluated numerically by expressing the integrand in (15-56) as a function of s using (15-59). We can use the following integration formula (Abramowitz and Stegun, 1964, p. 890):

$$\int_{-\infty}^{\infty} e^{-x^2} f(x)\,dx = \sum_{i=1}^{n} W_i f(x_i) + R_n, \tag{15-60}$$

where x_i is the ith zero of the Hermite polynomial $H_n(x)$, W_i the weighting coefficient, and R_n the remainder. The table for x_i and W_i is available in Abramowitz and Stegun (1964, p. 924).

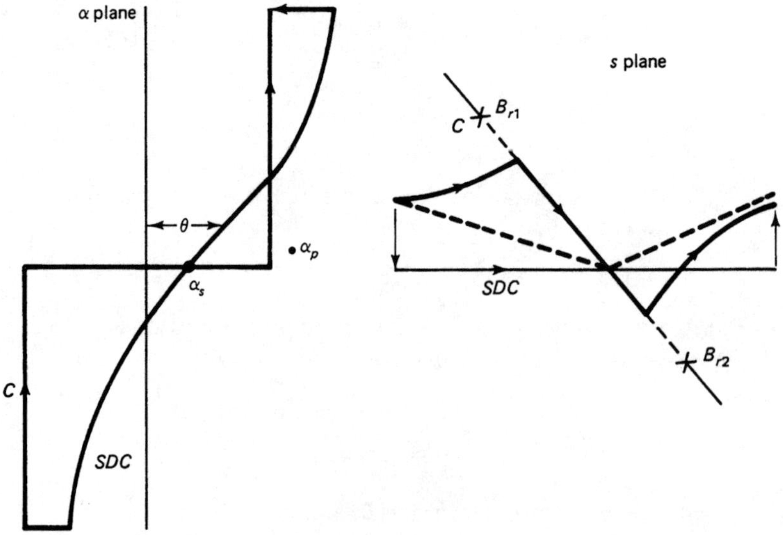

Figure 15-8 Original contour C and the steepest descent contour (SDC) in the α plane and s plane.

Alternatively, we can use the series expansion (15C-31) in the Appendix to Chapter 15, Section C. The series expansion is given by

$$I_1 = e^{-jkR_2} \sum_{n=0} B_{2n}\left(\frac{2}{kR_2}\right)^{n+1/2} \Gamma\left(n+\frac{1}{2}\right), \tag{15-61}$$

where

$$B_0 = \left[F(s)\frac{d\alpha}{ds}\right]_{s=0} + \frac{R_1(s_p)}{s_p}$$

$$= F(\alpha_s)e^{j(\pi/4)} + \frac{R_1(s_p)}{s_p}.$$

However, this term B_0 can be shown to be vanishingly small as $|n|$ becomes large, and $\theta \to \pi/2$. Therefore, the series (15-61) starts with the term $n = 1$, and thus for large R_2, I_1 is proportional to $1/R_2^2$. This term decays much faster than I_p and I_1 is negligible compared with I_p. Therefore, P is given approximately by I_2.

$$P \approx -j\pi R(s_p)\ \mathrm{erfc}\left(j\sqrt{\frac{z}{2}}s_p\right). \tag{15-62a}$$

For a large refractive index $|n|$ and $\theta \to \pi/2$, this can be simplified to

$$P = \frac{e^{-jkR_2}}{4\pi R_2}[\sqrt{\pi}j\sqrt{p_1}e^{-p_1}\ \mathrm{erfc}(j\sqrt{p_1})], \tag{15-62b}$$

where

$$p_1 = kR_2\frac{s_p^2}{2}.$$

Substituting this into (15-38), we finally obtain for the field on the surface,

$$\pi = 2\frac{e^{-jkR}}{4\pi R}[1 - \sqrt{\pi}j\sqrt{p_1}e^{-p_1}\ \mathrm{erfc}(j\sqrt{p_1})]. \tag{15-63}$$

The quantity

$$F = 1 - \sqrt{\pi}j\sqrt{p_1}e^{-p_1}\ \mathrm{erfc}(j\sqrt{p_1}) \tag{15-64}$$

represents the attenuation of the wave from the value for a perfectly conducting surface and is called the *attenuation function*.

Let us examine the attenuation function F. F depends only upon p_1. p_1 is given in (62b), which we write as

$$p_1 = jkR\cos(\alpha_p - \theta) - jkR. \tag{15-65}$$

The first term, $jkR\cos(\alpha_p - \theta)$, is the total complex phase from the origin to the observation point for the Zenneck wave, because

$$e^{-jkR\cos(\alpha_p - \theta)} = [e^{-j\lambda\rho - jq(z+h)}]_{\lambda=\lambda_p},$$

where λ_p is the radial propagation constant for the Zenneck wave given in (15-51). Therefore, p_1 represents the difference between the propagation constant for the Zenneck wave and the free-space wave.

It is clear, then, that the characteristics of the total wave do not depend on the distance (kR) alone, but depend on how large or small p_1 is. The magnitude of p_1 is called the *numerical distance* by Sommerfeld.

For small $|p_1|$ we have

$$F = 1 - j\sqrt{\pi}\sqrt{p_1}\,e^{-p_1} + \cdots, \tag{15-66}$$

which approaches 1 as $p_1 \to 0$.

For large $|p_1|$ ($|p_1| > 10$), we use the asymptotic expansion

$$\operatorname{erfc}(Z) = \frac{e^{-Z^2}}{\sqrt{\pi}Z}\left[1 - \frac{1}{2Z^2} + \frac{1\cdot 3}{(2Z^2)^2} - \cdots\right], \tag{15-67}$$

valid for Re $Z > 0$, and we get

$$F = -\frac{1}{2p_1} + \cdots. \tag{15-68}$$

Note also that for large n,

$$j\sqrt{p_1} = j\sqrt{\frac{kR}{2}}\,\frac{e^{-j(\pi/4)}}{n}, \tag{15-69}$$

and thus the angle of $j\sqrt{p_1}$ is close to but slightly less than $+90°$.

Equation (15-63) may be written in the following form:

$$\pi = 2\frac{e^{-jkR}}{4\pi R}\left(1 - j\sqrt{p_1\pi}\,e^{-p_1} - 2\sqrt{p_1}\,e^{-p_1}\int_0^{\sqrt{p_1}} e^{\alpha^2}\,d\alpha\right). \tag{15-70}$$

This is the form given in Sommerfeld's 1926 paper. (See the Appendix to Chapter 15, Section A, for the controversies over the sign error.) Equation (15-70) is valid for large kR and moderate numerical distance. It is not valid for small kR, where the quasi-static approach must be used. It is also not valid for extremely large kR and large numerical distances in which case the effect of the pole is negligible and the conventional saddle-point technique is applicable. Here the field on the surface behaves as

$$\frac{e^{-jkR}}{R^2} \tag{15-71}$$

and there is no term exhibiting the Zenneck wave characteristics.

15-8 LATERAL WAVES: BRANCH CUT INTEGRATION

As noted in Section 15-3, we need to consider three points in the complex integration: the saddle point, poles, and branch points. As may be noted from Section 15-7 for the Sommerfeld problem of wave propagation over the earth, the pole is very close to the saddle point, but the branch points are far from the saddle point. Thus, only the effect of the pole needs to be considered for the Sommerfeld problem for dipoles on earth.

However, for other problems, where the index of refraction n is smaller than unity, the pole is far from the saddle point, but the branch point is very close to the saddle point. Thus the effect of the pole can be neglected, but the effect of the branch points needs to be taken into account. We first examine the mathematical technique of handling the branch point and later discuss the physical significance.

Let us examine the reflected wave given in (15-26).

$$\pi_z = \pi_p + \pi_s,$$
$$\pi_p = \frac{e^{-jk|r-r'|}}{4\pi|r-r'|}, \tag{15-72}$$
$$\pi_s = \frac{1}{8\pi}\int_W \frac{n^2 q - q_e}{n^2 q + q_e} H_0^{(2)}(\lambda\rho) e^{-jq(z+h)} \frac{\lambda\, d\lambda}{jq}.$$

We use the approximation (15-32) and the transformation (15-40) from λ to α and write

$$\pi_s = \frac{1}{4\pi}\left(\frac{k}{2\pi\rho}\right)^{1/2} e^{-j(\pi/4)} \int_C R(\alpha) e^{-jkR_2 \cos(\alpha-\theta)} \sqrt{\sin\alpha}\, d\alpha, \tag{15-73}$$

where

$$R(\alpha) = \frac{n^2 \cos\alpha - \sqrt{n^2 - \sin^2\alpha}}{n^2 \cos\alpha + \sqrt{n^2 - \sin^2\alpha}},$$

is the reflection coefficient.

As is evident from (15-73), the branch point is located at

$$\sin\alpha_b = n. \tag{15-74}$$

If $|n| > 1$ as in the case of the earth, α_b is located far from the location of the saddle point $\alpha_s = \theta$ $(0 \le \theta \le \pi/2)$, and this branch point α_b has little effect on the wave propagation over the earth.

However, if n is real and $0 < n < 1$, then α_b is real and $0 \le \alpha_b \le \pi/2$, and therefore the branch cut integration for α_b must be taken into account when the saddle-point integration is performed. For example, consider an antenna located in air below the ionosphere or an antenna buried underground. In these cases the index of refraction n_2 of the medium on the other side of the boundary from the antenna is less than that of the medium n_1 where the antenna is located, and thus $|n| = |n_2/n_1| < 1$ (Fig. 15-9).

We evaluate (15-73) by means of the saddle-point technique. The saddle-point contour is shown in Fig. 15-10. We note, first, that if the observation angle θ is smaller than α_b,

$$\theta < \alpha_b, \qquad \alpha_b = \sin^{-1} n,$$

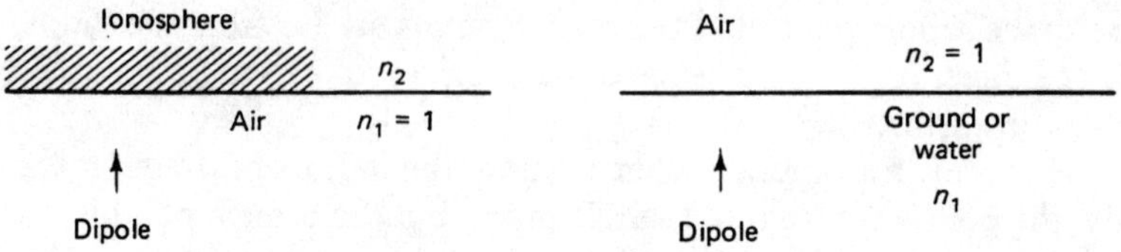

Figure 15-9 Dipole is located in the medium n_1 and $|n| = |n_2/n_1| < 1$.

then the branch point α_b has practically no effect and thus

$$\pi_s = \int_C = \int_{SDC} = R(\theta)\frac{e^{-jkR_2}}{4\pi R_2}, \tag{15-75}$$

which is the reflected wave previously obtained.

On the other hand, if the observation angle θ is greater than α_b,

$$\theta > \alpha_b,$$

the integral along C becomes (Fig. 15-11)

$$\int_C = \int_{C_1}.$$

However, the contour C_1 can be deformed to the branch cut integration (A–B–C) and SDC (D–E) shown in Fig. 15-12. Note that a portion of the SDC is in the bottom Riemann surface, as shown by the dotted line:

$$\int_C = \int_{C_1} = \int_{B_r} + \int_{SDC}.$$

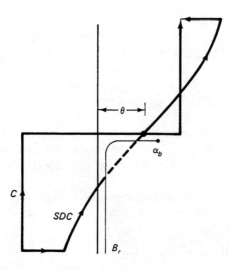

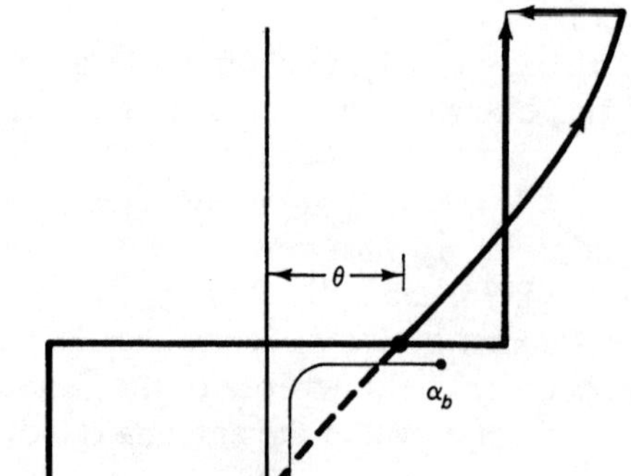

Figure 15-10 Branch cut B_r is drawn along IM $\sqrt{n^2 - \sin^2\alpha} = 0$.

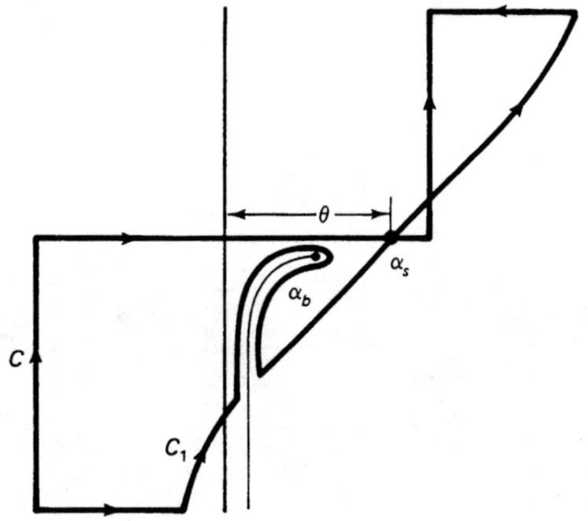

Figure 15-11 Path of integration C_1.

Therefore, for $\theta > \alpha_b$,

$$\pi_s = R(\theta)\frac{e^{-jkR_2}}{4\pi R_2} + \pi_l, \tag{15-76}$$

where π_l is the branch cut integration:

$$\pi_l = \frac{1}{4\pi}\left(\frac{k}{2\pi\rho}\right)^{1/2} e^{-j(\pi/4)} \int_{B_r} R(\alpha)e^{-jkR_2\cos(\alpha-\theta)}\sqrt{\sin\alpha}\, d\alpha. \tag{15-77}$$

The physical significance of the angle α_b is shown in Fig. 15-13. When the observation point is within region A ($\theta < \alpha_b$), the wave consists of the primary wave and the reflected wave. Note that the transmitted wave makes an angle θ_t with the vertical axis as given by Snell's law.

$$\sin\theta = n\sin\theta_t. \tag{15-78}$$

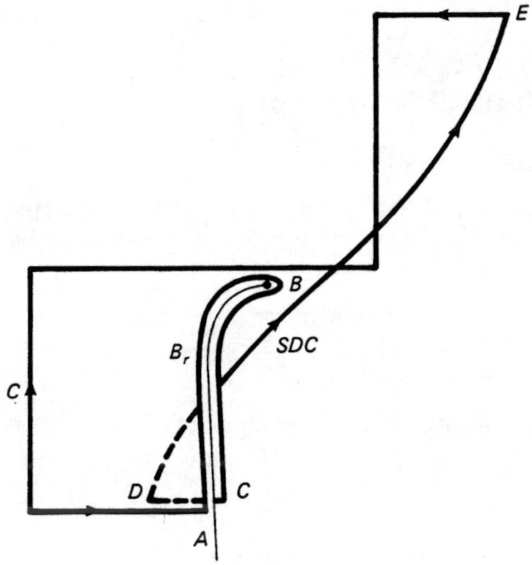

Figure 15-12 The integral along C_1 is equal to the branch cut integration along B_r and the steepest descent contour SDC.

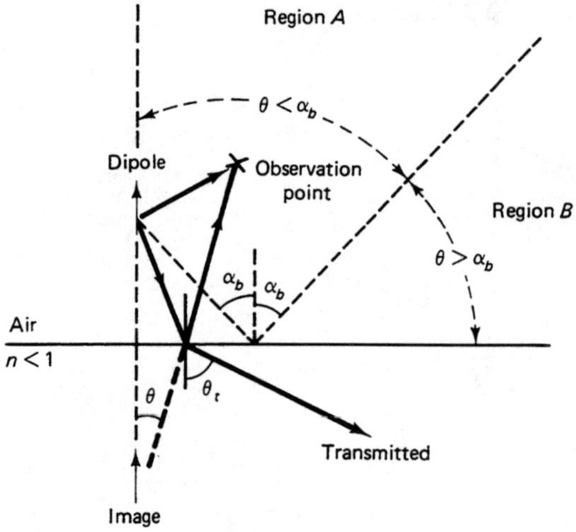

Figure 15-13 The critical angle α_b for the total reflection.

When θ exceeds the angle that makes $\theta_t = \pi/2$, total reflection occurs. This critical angle is α_b:

$$\sin \alpha_b = n. \tag{15-79}$$

Therefore, regions A and B correspond to the cases where the angle of incidence θ is less than or greater than the critical angle for the total reflection.

Let us evaluate (15-77) for large kR_2:

$$\begin{aligned} \pi_l &= \frac{1}{4\pi}\left(\frac{k}{2\pi\rho}\right)^{1/2} e^{-j(\pi/4)} I, \\ I &= \int_{B_r} F(\alpha) e^{-f(\alpha)}\, d\alpha, \end{aligned} \tag{15-80}$$

where

$$f(\alpha) = jkR_2 \cos(\alpha - \theta),$$
$$F(\alpha) = R(\alpha)\sqrt{\sin \alpha}.$$

The integral is along the paths A–B and B–C shown in Fig. 15-12. We then note that the value of $F(\alpha)$ along A–B is different from that along B–C because $F(\alpha)$ contains $\sqrt{n^2 - \sin^2 \alpha}$. Note that along A–B,

$$F_1(\alpha) = \frac{n^2 \cos \alpha - \sqrt{n^2 - \sin^2 \alpha}}{n^2 \cos \alpha + \sqrt{n^2 - \sin^2 \alpha}} \sqrt{\sin \alpha}, \tag{15-81}$$

where $\operatorname{Re} \sqrt{n^2 - \sin^2 \alpha} > 0$, but along B–C, the sign of $\sqrt{n^2 - \sin^2 \alpha}$ must be changed as it is on the other side of the branch cut. Thus we should write

$$F_2(\alpha) = \frac{n^2 \cos \alpha + \sqrt{n^2 - \sin^2 \alpha}}{n^2 \cos \alpha - \sqrt{n^2 - \sin^2 \alpha}} \sqrt{\sin \alpha}. \tag{15-82}$$

Using (15-81) and (15-82), we get for (15-80),

$$I = \int_A^B [F_1(\alpha) - F_2(\alpha)] e^{-f(\alpha)}\, d\alpha. \tag{15-83}$$

To evaluate this integral, let us first expand $f(\alpha)$ about $\alpha = \alpha_b$ and keep the first term. As will be shown shortly, this is justified because the major contribution to the integral comes from the immediate neighborhood of α_b.

$$\begin{aligned} f(\alpha) &= f(\alpha_b) + (\alpha - \alpha_b)\left(\frac{\partial f}{\partial \alpha}\right)_{\alpha_b} \\ &= jkR_2 \cos(\theta - \alpha_b) + [jkR_2 \sin(\theta - \alpha_b)](\alpha - \alpha_b). \end{aligned} \tag{15-84}$$

Next we deform the contour A–B such that the exponent in (15-84) decays exponentially in a steepest descent path from the branch point α_b. To this end we write the exponent as

$$e^{-jkR_2 \sin(\theta - \alpha_b)(\alpha - \alpha_b)} = e^{-[kR_2 \sin(\theta - \alpha_b)]s},$$

where s is real representing the distance from the branch point and obviously

$$\alpha - \alpha_b = -js. \tag{15-85}$$

We then get

$$I = e^{-jkR_2 \cos(\theta - \alpha_b)} \int_0^\infty [-F_1(\alpha) + F_2(\alpha)] e^{-[kR_2 \sin(\theta - \alpha_b)]s} (-j)\, ds, \tag{15-86}$$

$$[-F_1(\alpha) + F_2(\alpha)] = \frac{4n^2 \cos\alpha \sqrt{n^2 - \sin^2\alpha}}{n^4 \cos^2\alpha - (n^2 - \sin^2\alpha)} \sqrt{\sin\alpha}. \tag{15-87}$$

Because the integrand decays exponentially in (15-86), most of the contribution comes from the neighborhood of $s = 0$. It is therefore necessary to examine the behavior of (15-87) in the neighborhood of $s = 0$.

Near $s = 0$, $(\alpha = \alpha_b)$ we expand $(n^2 - \sin^2\alpha)$ in a Taylor's series about $\alpha = \alpha_b$ and keep the first term.

$$\begin{aligned} n^2 - \sin^2\alpha &\simeq -2 \sin\alpha_b \cos\alpha_b (\alpha - \alpha_b) \\ &= [2jn \cos\alpha_b] s. \\ \sqrt{n^2 - \sin^2\alpha} &= (2n \cos\alpha_b) e^{j(\pi/4)} \sqrt{s}. \end{aligned} \tag{15-88}$$

Thus, near $s = 0$, (15-87) may be approximated by

$$[-F_1(\alpha) + F_2(\alpha)] \simeq \frac{4\sqrt{2}}{n(\cos\alpha_b)^{1/2}} e^{j(\pi/4)} \sqrt{s}. \tag{15-89}$$

The integral (15-86) then becomes

$$I = e^{-jkR_2 \cos(\theta - \alpha_b)} \frac{4\sqrt{2} e^{j(\pi/4)}(-j)}{n(\cos\alpha_b)^{1/2}} \int_0^\infty \sqrt{s}\, e^{-[kR_2 \sin(\theta - \alpha_b)]s}\, ds. \tag{15-90}$$

Using the integral

$$\int_0^\infty \sqrt{s}\, e^{-as}\, ds = \frac{\sqrt{2\pi}}{(2a)^{3/2}},$$

we finally get

$$I = e^{-jkR_2 \cos(\theta - \alpha_b)} \frac{2\sqrt{2\pi}\, e^{j(\pi/4)}(-j)}{n(\cos \alpha_b)^{1/2}} \frac{1}{[kR_2 \sin(\theta - \alpha_b)]^{3/2}}. \tag{15-91}$$

Substituting this in (15-80), we get the expression for the branch cut contribution

$$\pi_l = \frac{(-j2)}{4\pi kn(\rho \cos \alpha_b)^{1/2}} \frac{e^{-jkR_2 \cos(\theta - \alpha_b)}}{[R_2 \sin(\theta - \alpha_b)]^{3/2}}. \tag{15-92}$$

We now explain the physical meaning of π_l by rewriting this in terms of the distances pictured in Fig. 15-14. We note that

$$kR_2 \cos(\theta - \alpha_b) = k(L_0 + L_2) + knL_1,$$

and

$$R_2 \sin(\theta - \alpha_b) = L_1 \cos \alpha_b,$$

and therefore

$$\pi_l = \frac{(-j2)}{4\pi kn(1-n^2)} \frac{1}{\rho^{1/2} L_1^{3/2}} e^{-jk(L_0 + L_2) - jknL_1}. \tag{15-93}$$

Let us examine the physical meaning of (15-93). First, we note that the phase front is given by

$$kL_0 + knL_1 + kL_2 = \text{constant}.$$

This shows that this wave first propagates over the distance L_0 with free-space propagation constant k and is incident on the surface at the critical angle α_b. Then the wave propagates just below the surface with the propagation constant of the

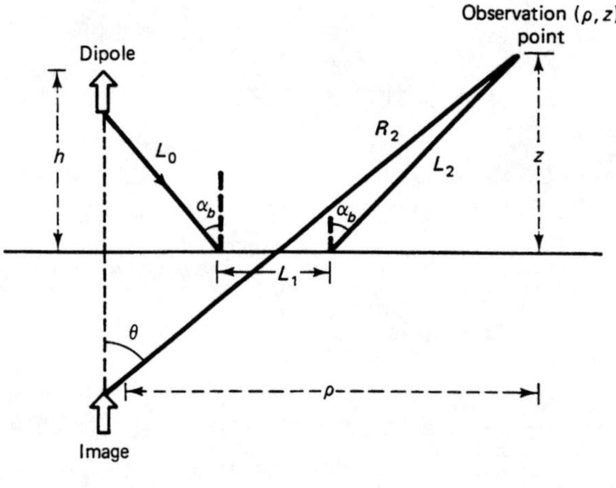

Figure 15-14 Lateral wave contribution.

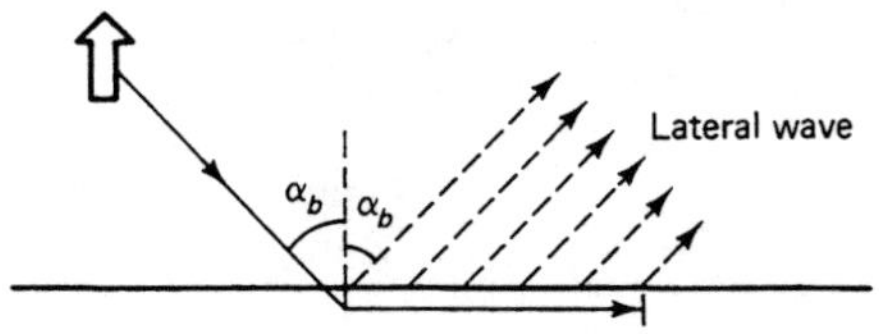

Figure 15-15 Lateral wave.

lower medium kn, and at the same time, radiating into the upper medium in the direction of the critical angle. This is pictured in Fig. 15-15. This particular wave is called the *lateral wave*, *headwave* (*kopfwelle*), or *flank wave (flankenwelle)*. The lateral wave is related to the following seismic phenomenon. Let the seismic impulse (the earthquake) originate at a point on the surface at $t = 0$. The wave propagates faster in the ground than in the air. [v(air) < v(ground) and thus $n = v(\text{air})/v(\text{ground}) < 1$.] Thus the wave fronts at t in the air and in the ground are different on the surface (Fig. 15-16). But there should be no discontinuity in the wavefront, and therefore there is a wave whose wavefront is tangential from the wavefront in the ground to the wavefront in the air. This particular wavefront propagates in exactly the same direction as the lateral wave. Moreover, on the ground this wave arrives first and thus is called the head wave.

In addition to this particular phase characteristic, the lateral wave has a distinct amplitude characteristic given by

$$\frac{1}{\rho^{1/2}}\frac{1}{L_1^{3/2}}.$$

Because of this dependence, π_l becomes infinite as $L_1 \to 0$. This is because of the approximation used to obtain (15-93). In fact, as $L_1 \to 0$, the saddle point approaches the branch point ($\theta \to \alpha_b$), and thus the saddle point and the branch cut integration cannot be performed separately. The field in the neighborhood of $L_1 \to 0$ (or $\theta \to \alpha_b$) is examined in Brekhovskikh (1960) showing the "caustic" behavior.

The lateral wave described above occurs when $0 < n < 1$. An example is the ionosphere, in which $n = \sqrt{1 - (\omega_p/\omega)^2}$. Another important case where the lateral wave plays a dominant role is the problem of communication between two points in the absorbing medium such as the ocean or the earth (Fig. 15-17). In this case, k is complex and its magnitude is large, but kn is the free-space wave number, and thus $|n| < 1$. Due to absorption, the direct wave and the reflected wave are almost

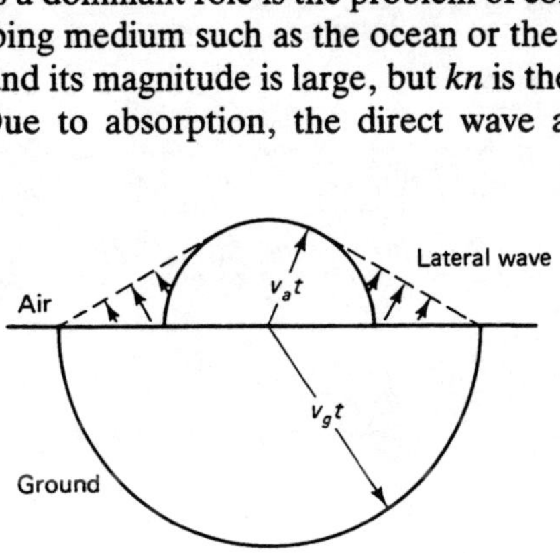

Figure 15-16 Lateral wave and seismic wave.

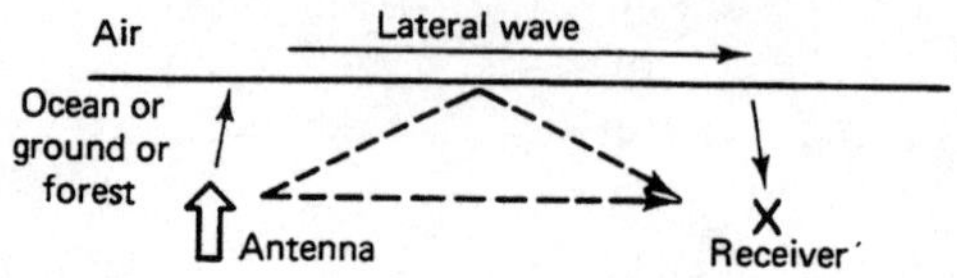

Figure 15-17 Communication by lateral wave.

completely attenuated in the medium. But the lateral wave propagates from the antenna to the surface suffering some attenuation, then propagates in the air without attenuation over a long distance and then arrives at the receiver. Thus the communication is only through the lateral wave in this case. The communication between two points in the forest is often done by the lateral wave.

15-9 REFRACTED WAVE

We now return to (15-24) and examine the field in the lower medium ($z < 0$).

$$\pi_z = \frac{1}{4\pi}\int_0^\infty \frac{2}{n^2 q + q_e} J_0(\lambda\rho) e^{jq_e z - jqh} \frac{\lambda\, d\lambda}{j}. \tag{15-94}$$

We use (15-30) to obtain the integral along $W(-\infty$ to $+\infty)$ and use the asymptotic form (15-32) for $H_0^{(2)}(\lambda\rho)$. We thus have

$$\pi_z = \frac{1}{8\pi}\int_W \frac{2}{n^2 q + q_e}\left(\frac{2}{\pi\lambda\rho}\right)^{1/2} \frac{e^{-j(\pi/4)}\lambda}{j} e^{-j\lambda\rho - jqh + jq_e z}\, d\lambda. \tag{15-95}$$

We evaluate this by means of the saddle-point technique. We use the first term of the asymptotic series given by

$$\int_W F(\lambda) e^{-f(\lambda)}\, d\lambda \simeq F(\lambda_s) e^{-f(\lambda_s)} \sqrt{\frac{2\pi}{f''(\lambda_s)}}, \tag{15-96}$$

where λ_s is the saddle point given by

$$f'(\lambda_s) = \left.\frac{\partial f}{\partial \lambda}\right|_{\lambda=\lambda_s} = 0.$$

The saddle point for (15-95) is given by

$$\begin{aligned}\frac{\partial f}{\partial \lambda} &= \frac{\lambda}{\partial \lambda}(j\lambda\rho + jqh - jq_e z)\\ &= j\left(\rho + \frac{\partial q}{\partial \lambda}h - \frac{\partial q_e}{\partial \lambda}z\right)\\ &= 0.\end{aligned}$$

But

$$\frac{\partial q}{\partial \lambda} = -\frac{\lambda}{q} \quad \text{and} \quad \frac{\partial q_e}{\partial \lambda} = -\frac{\lambda}{q_e}.$$

Thus the saddle point λ_s is given by

$$\rho - \frac{\lambda_s}{q_s}h + \frac{\lambda_s}{q_{es}}z = 0, \qquad (15\text{-}97)$$

where q_s and q_{es} are q and q_e evaluated at λ_s.

The physical meaning of (15-97) is clear when we use the transformation

$$\lambda = k \sin\alpha.$$

Then (15-97) becomes (see Fig. 15-18)

$$\rho - (\tan\alpha_1)h + (\tan\alpha_2)z = 0, \qquad (15\text{-}98)$$

where α_1 and α_2 satisfy Snell's law:

$$\sin\alpha_1 = n \sin\alpha_2.$$

Thus the wave from the dipole arrives at the surface and refracts according to Snell's law and propagates to the observation point. The total phase is then

$$\begin{aligned} f(\lambda_s) &= j(\lambda\rho + qh - q_e z)_{\lambda_s} \\ &= j(kL_0 + knL_t). \end{aligned} \qquad (15\text{-}99)$$

Therefore, the complete refracted field is given by using (15-96).

$$\pi_z = \frac{1}{4\pi} \frac{T(\alpha_1)(\sin\alpha_1)^{1/2} e^{-j(kL_0 + knL_t)}}{\rho^{1/2} \cos\alpha_1[h/\cos^3\alpha_1 + (-z)/n \cos^3\alpha_2]^{1/2}}, \qquad (15\text{-}100)$$

where $T(\alpha_1)$ is the transmission coefficient for the incident angle α_1:

$$T(\alpha_1) = \frac{2\cos\alpha_1}{n^2\cos\alpha_1 + \sqrt{n^2 - \sin^2\alpha_1}}. \qquad (15\text{-}101)$$

Equation (15-100) is based on the saddle-point technique, and therefore it is valid when the dipole and the observation point are sufficiently far from the surface. It can be shown that the expression (15-100) is identical to that obtained by the application of geometric optical techniques. In general, use of the saddle-point technique leads to the geometric optical solution.

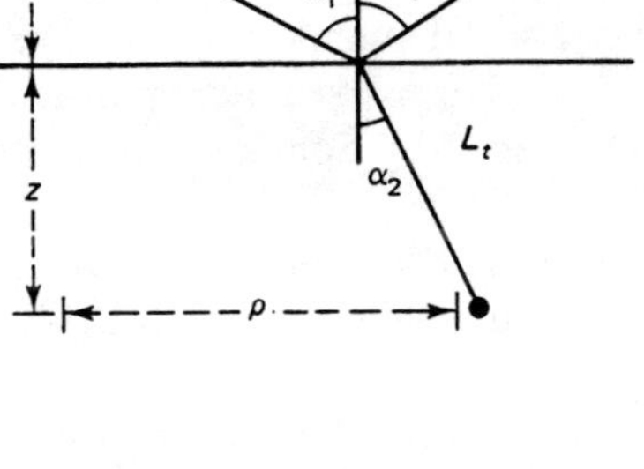

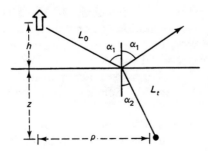

Figure 15-18 Refracted wave.

15-10 RADIATION FROM A HORIZONTAL DIPOLE

In contrast with the azimuthally uniform radiation from a vertical dipole, the radiation from a horizontal dipole is directional. Furthermore, on the surface of the ground most radiation is in the direction of the dipole axis. This is in contrast with the radiation in free space, where the radiation is broadside (in the direction perpendicular to the axis). One important practical case where a horizontal dipole is essential is the radiation from a buried antenna (antenna buried underground, in ice, or submerged underwater). In this case it can be shown that the horizontal dipole is most effective and practical and that a vertical dipole is an ineffective radiator.

Let us now consider a horizontal dipole located at $(0, 0, h)$ and oriented in the x direction (Fig. 15-19). The primary field can be easily obtained from the x component of the Hertz potential π_x:

$$(\nabla^2 + k^2)\pi_x = -\frac{I_x L}{j\omega\epsilon_0}\delta(r - r'). \tag{15-102}$$

The primary field π_{xp} is then given by

$$\begin{aligned}\pi_{xp} &= \frac{e^{-jk|r-r'|}}{4\pi|r-r'|} \\ &= \frac{1}{4\pi}\int_0^\infty J_0(\lambda\rho)e^{-jq|z-h|}\frac{\lambda\, d\lambda}{jq} \\ &= \frac{1}{8\pi}\int_W H_0^{(2)}(\lambda\rho)e^{-jq|z-h|}\frac{\lambda\, d\lambda}{jq}.\end{aligned} \tag{15-103}$$

Here we omit $I_x L/j\omega\epsilon_0$ for convenience. The final results must be multiplied by $I_x L/j\omega\epsilon_0$ to obtain the true field.

Now we write the secondary fields in the air and the ground

$$\begin{aligned}\pi_{xs1} &= \frac{1}{4\pi}\int_0^\infty R(\lambda)J_0(\lambda\rho)e^{-jq|z+h|}\frac{\lambda\, d\lambda}{jq} \qquad \text{for } z > 0, \\ \pi_{xs2} &= \frac{1}{4\pi}\int_0^\infty T(\lambda)J_0(\lambda\rho)e^{+q_e z - jqh}\frac{\lambda\, d\lambda}{jq} \qquad \text{for } z < 0.\end{aligned} \tag{15-104}$$

At this point it may appear that these two functions $R(\lambda)$ and $T(\lambda)$ may be determined by applying the boundary conditions as was done for a vertical dipole. However, this is not possible. Note that, in general, a complete description of the electromagnetic field requires two scalar functions (such as π_z and π_z^* for TM and TE modes). Therefore, we need another scalar function in addition to π_{xs1} and π_{xs2}. Convenient choice, which was first used by Sommerfeld, is the z component of π, π_{zs1}, and π_{zs2}.

Let us now examine the boundary conditions at $z = 0$. At $z = 0$, the tangential electric fields E_x and E_y and the tangential magnetic field, H_x and H_y must be

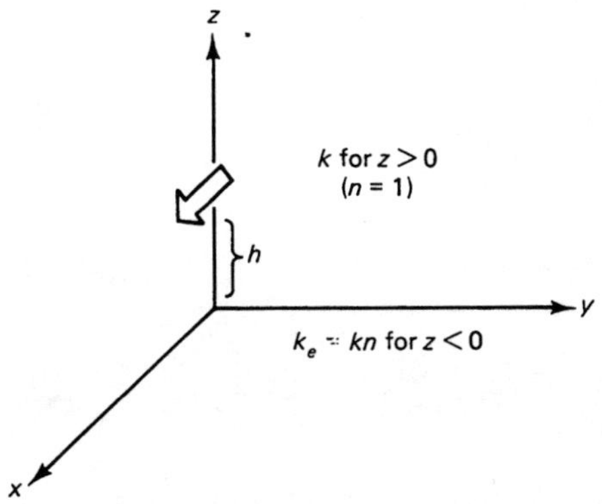

Figure 15-19 Horizontal dipole.

continuous. We write these conditions in terms of π_x and π_z, noting that $\bar{E} = \nabla(\nabla \cdot \bar{\pi}) + k^2\bar{\pi}$ and $\bar{H} = j\omega\epsilon\nabla \times \bar{\pi}$.

$$\text{Continuity of } E_x: \quad \pi_{x1} = n^2\pi_{x2}, \tag{15-105a}$$

$$\text{Continuity of } E_y: \quad \frac{\partial}{\partial x}\pi_{x1} + \frac{\partial}{\partial z}\pi_{z1} = \frac{\partial}{\partial x}\pi_{x2} + \frac{\partial}{\partial z}\pi_{z2}, \tag{15-105b}$$

$$\text{Continuity of } H_x: \quad \pi_{z1} = n^2\pi_{z2}, \tag{15-105c}$$

$$\text{Continuity of } H_y: \quad \frac{\partial}{\partial z}\pi_{x1} = n^2\frac{\partial}{\partial z}\pi_{x2}. \tag{15-105d}$$

We use (15-105a) and (15-105d) to determine $R(\lambda)$ and $T(\lambda)$ for π_x in (15-104).

$$R(\lambda) = \frac{q - q_e}{q + q_e},$$
$$T(\lambda) = \frac{1}{n^2}\frac{2q}{q + q_e}. \tag{15-106}$$

Therefore, we write

$$\pi_{x1} = \frac{1}{4\pi}\int_0^\infty J_0(\lambda\rho)e^{-jq|z-h|}\frac{\lambda\, d\lambda}{jq}$$
$$+ \frac{1}{4\pi}\int_0^\infty \frac{q - q_e}{q + q_e}J_0(\lambda\rho)e^{-jq|z+h|}\frac{\lambda\, d\lambda}{jq}, \qquad z > 0,$$
$$\pi_{x2} = \frac{1}{4\pi}\int_0^\infty \frac{2q}{n^2(q + q_e)}J_0(\lambda\rho)e^{jq_e z - jqh}\frac{\lambda\, d\lambda}{jq}, \qquad z < 0. \tag{15-107}$$

We note that if the second medium is perfectly conducting, Equation (15-107) reduces to a correct image representation. Now we use (15-105b) and (15-105c) to obtain π_{z1} and π_{z2}. First we note from (15-105b) that

$$\frac{\partial}{\partial z}(\pi_{z1} - \pi_{z2}) = \frac{\partial}{\partial x}(\pi_{x2} - \pi_{x1}). \tag{15-108}$$

Note that

$$\frac{\partial}{\partial x}=\frac{\partial\rho}{\partial x}\frac{\partial}{\partial\rho}+\frac{\partial\phi}{\partial x}\frac{\partial}{\partial\phi}\quad\text{and}\quad\frac{\partial\rho}{\partial x}=\cos\phi\quad\text{and}\quad\frac{\partial}{\partial\phi}=0$$

for our problem.

Therefore, we write the right side of (15-108), using (15-107) and $J_0'(\lambda\rho)=-J_1(\lambda\rho)$,

$$\frac{\partial}{\partial x}(\pi_{x2}-\pi_{x1})=\frac{\cos\phi}{4\pi}\int_0^\infty[T(\lambda)-1-R(\lambda)][-J_1(\lambda\rho)]e^{-jph}\frac{\lambda^2\,d\lambda}{jq}.$$

Since this must be equal to the left side of (15-108), π_z on the left side must have the same form as (15-109):

$$\cos\phi\int_0^\infty J_1(\lambda\rho)\cdots.$$

Thus we write

$$\begin{aligned}\pi_{z1}&=\cos\phi\int_0^\infty A(\lambda)J_1(\lambda\rho)e^{-jq(z+h)}\lambda^2\,d\lambda &&\text{for } z>0,\\ \pi_{z2}&=\cos\phi\int_0^\infty B(\lambda)J_1(\lambda\rho)e^{jq_e z-jqh}\lambda^2\,d\lambda &&\text{for } z<0.\end{aligned}\tag{15-109}$$

Now satisfying (15-108) and (15-105c), we determine $A(\lambda)$ and $B(\lambda)$:

$$\begin{aligned}A(\lambda)&=-\frac{2}{k^2}\frac{q-q_e}{n^2q+q_e},\\ B(\lambda)&=-\frac{2}{n^2k^2}\frac{q-q_e}{n^2q+q_e}.\end{aligned}\tag{15-110}$$

Equations (15-107) and (15-109) constitute the complete expressions for the field of a horizontal dipole.

We note that the radiation from (15-107) is mostly in the direction perpendicular to the axis, as this is similar to the radiation from a horizontal dipole and its image. On the surface, however, this field is extremely small because the radiation from the image tends to cancel the direct radiation.

On the other hand, the field due to π_z as given in (15-109) is directional because of the $\cos\phi$ factor. Also, the coefficients $A(\lambda)$ and $B(\lambda)$ have the same Sommerfeld denominator (n^2q+q_e) as in the vertical dipole. Thus π_x produced by a horizontal dipole does not contribute much to the field on the surface, but it gives rise to π_z, which propagates mostly in the direction of the dipole axis and behaves in a manner similar to that of the vertical dipole.

15-11 RADIATION IN LAYERED MEDIA

In many practical problems we need to consider more than one interface. For example, wave propagation over the earth is greatly affected by the presence of the ionosphere, particularly for VLF and lower frequencies. Also, the thermal ionosphere, which will be discussed later, may be represented by layers. Another example would be the radiation from slot or dipole antennas on the surface of a spacecraft or high-speed vehicle, which may be covered by some protective material (Fig. 15-20). It is therefore important to study the effects of layers on the radiation field.

Let us consider a vertical dipole located within two interfaces, as shown in Fig. 15-21. As was shown in Section 15-3, the field between two boundaries consists of the primary wave π_{p0} and the secondary wave π_{s0}.

$$\pi_{z0} = \pi_{p0} + \pi_{s0}, \qquad 0 < z < h, \tag{15-111}$$

where

$$\pi_{p0} = \frac{1}{4\pi}\int_0^\infty e^{-jq|z-z_0|} J_0(\lambda\rho)\frac{\lambda\, d\lambda}{jq}, \tag{15-112}$$

$$\pi_{s0} = \frac{1}{4\pi}\int_0^\infty [a_0(\lambda)e^{-jqz} + b_0(\lambda)e^{+jqz}] J_0(\lambda\rho)\frac{\lambda\, d\lambda}{jq}, \tag{15-113}$$

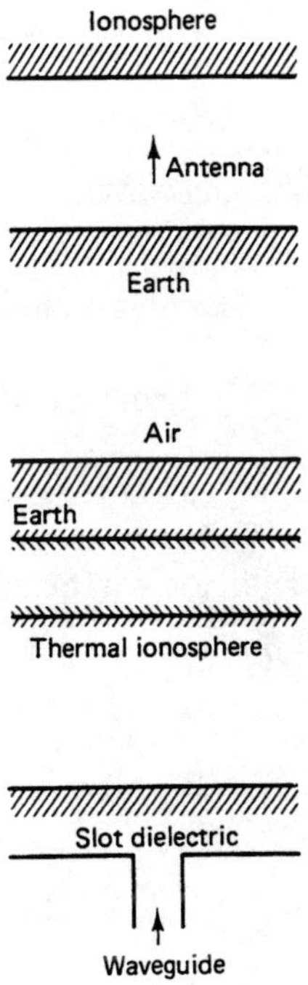

Figure 15-20 Radiation in layered media.

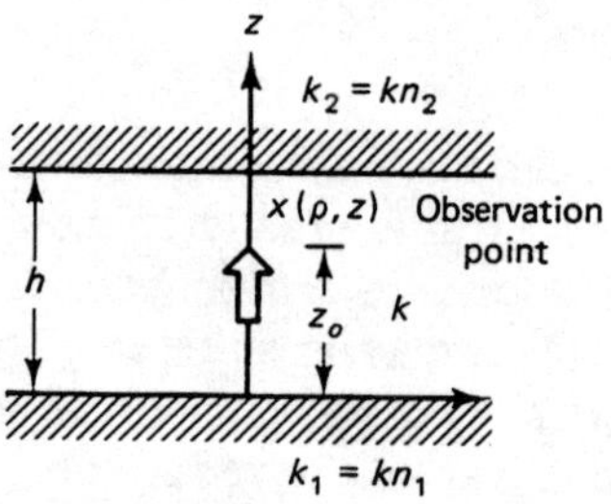

Figure 15-21 Vertical dipole in a layer.

Here we again omit $IL/j\omega\epsilon_0$ for convenience. Note also that a_0 and b_0 in (15-113) represent the upward and downward waves.

The field below the lower boundary is then given by

$$\pi_{s1} = \frac{1}{4\pi}\int_0^\infty [b_1(\lambda)e^{+jq_1 z}]J_0(\lambda\rho)\frac{\lambda\, d\lambda}{jq}, \qquad z < 0, \tag{15-114}$$

and the field above the upper boundary is given by

$$\pi_{s2} = \frac{1}{4\pi}\int_0^\infty [a_2(\lambda)e^{+jq_2 z}]J_0(\lambda\rho)\frac{\lambda\, d\lambda}{jq}, \qquad z > h, \tag{15-115}$$

where

$$\begin{aligned} \lambda^2 + q^2 &= k^2, \\ \lambda^2 + q_1^2 &= k_1^2, \\ \lambda^2 + q_2^2 &= k_2^2. \end{aligned} \tag{15-116}$$

Here b_1 and a_2 represent the outgoing waves.

The boundary conditions at $z = 0$ and $z = h$ are

$$\begin{aligned} n^2\pi &= \text{continuous}, \\ \frac{\partial \pi}{\partial z} &= \text{continuous}. \end{aligned} \tag{15-117}$$

Applying these four conditions, we can determine four constants a_0, b_0, b_1, and a_1. Instead of simply applying (15-117), however, we wish to express these constants in terms of the reflection and transmission coefficients for the upper and lower media:

$$\begin{aligned} R_2 &= \frac{n_2^2 q - q_2}{n_2^2 q + q_2}, \qquad T_2 = \frac{2q}{n_2^2 q + q_2}, \\ R_1 &= \frac{n_1^2 q - q_1}{n_1^2 q + q_1}, \qquad T_1 = \frac{2q}{n_1^2 q + q_1}. \end{aligned} \tag{15-118}$$

To do this, we note that for the upper medium, the incident wave is given by the Fourier–Bessel transform of

$$[e^{-jq(z-z_0)} + a_0 e^{-jqz}],$$

and the reflected wave is given by

$$[b_0 e^{jqz}].$$

The ratio of these two at $z = h$ must be the reflection coefficient R_2. Thus

$$R_2[e^{-jq(h-z_0)} + a_0 e^{-jqh}] = [b_0 e^{jqh}]. \tag{15-119}$$

Similarly, at $z = 0$, we have

$$R_1[e^{-jqz_0} + b_0] = a_0. \tag{15-120}$$

Solving these two equations for a_0 and b_0, we obtain

$$a_0 = \frac{R_1 e^{-jqz_0} + R_1 R_2 e^{-jq(2h-z_0)}}{1 - R_1 R_2 e^{-j2qh}}, \tag{15-121}$$

$$b_0 = \frac{R_2 e^{-jq(h-z_0)} + R_1 R_2 e^{-jq(h+z_0)}}{1 - R_1 R_2 e^{-j2qh}}. \tag{15-122}$$

Similarly, we can obtain $b_1(\lambda)$ and $a_2(\lambda)$ by using T_1 and T_2:

$$b_1 = \frac{T_1 e^{-jqz_0} + T_1 R_2 e^{-jq(2h-z_0)}}{1 - R_1 R_2 e^{-j2qh}}, \tag{15-123}$$

$$a_2 = \frac{T_2 e^{-jqz_0} + T_2 R_1 e^{-jq(h+z_0)}}{1 - R_1 R_2 e^{-j2qh}} e^{jq_2 h}. \tag{15-124}$$

Substituting (15-121) to (15-124) into (15-112) to (15-115), we get a complete expression of the field.

Let us now examine these expressions. We first look for singularities in the integrand. It is clear that there are poles located at the roots of

$$1 - R_1 R_2 e^{-j2qh} = 0. \tag{15-125}$$

In general, there are infinite numbers of roots for (15-125) and thus there is a series of poles in the λ plane. This will be discussed further later.

It is also noted that the integrand contains q, q_1, and q_2, given by

$$q = \sqrt{k^2 - \lambda^2},$$
$$q_1 = \sqrt{k_1^2 - \lambda^2},$$
$$q_2 = \sqrt{k_2^2 - \lambda^2},$$

and thus it *appears* that there are three branch points in the λ plane: at $\lambda = \pm k$, $\lambda = \pm k_1$, and $\lambda = \pm k_2$. However, closer examination reveals that there is no branch point at $\lambda = \pm k$. This can be proved by showing that the integrand is unchanged when q is changed to $-q$. Another point of view is to recognize that in describing the field in (15-113), there is no need to differentiate $+q$ and $-q$. For example, in (15-113), e^{-jqz} represents the upward wave, but by changing it to e^{+jqz}, we simply interchange the role of a_0 (and b_0) from upward (downward) to downward (upward). However, for q_1 and q_2, we do need to differentiate $+q_1$ and $-q_1$, $+q_2$ and $-q_2$

because one represents the outgoing wave satisfying the radiation condition, and the other represents the incoming wave. In general, if there are many layers as shown in Fig. 15-22, even though the integrand contains

$$q_i = \sqrt{k_i^2 - \lambda^2}, \qquad i = 1, 2, \ldots, 6,$$

the only branch points are at

$$\lambda = \pm k_1 \quad \text{and} \quad \lambda = \pm k_6,$$

and the integrands are even functions of q_2, q_3, q_4, and q_5 and have no branch points at k_2, k_3, k_4, and k_5.

15-12 GEOMETRIC OPTICAL REPRESENTATION

Let us now evaluate the integrals in the expressions (15-111) to (15-115). We may proceed along the following two lines:

1. *Geometric optical representation*. This is based on the saddle-point technique and is applicable to the case where the height h is many wavelengths high.
2. *Mode and lateral wave representation*. This is based on the residue series and the branch cut integration and is useful when the height h is small. We now show the details of these two approaches.

As an example, we take the field in the upper medium $z > h$, which is given by (15-115) and (15-124). Noting the two exponential terms in (15-124), we write π_{s2} as follows:

$$\pi_{s2} = \pi'_{s2} + \pi''_{s2}, \tag{15-126}$$

where

$$\pi'_{s2} = \frac{1}{4\pi}\int_0^\infty \left[\frac{T_2 e^{-jq(h-z_0)-jq_2(z-h)}}{1-R_1R_2e^{-j2qh}}\right] J_0(\lambda\rho)\frac{\lambda\, d\lambda}{jq}, \tag{15-127}$$

$$\pi''_{s2} = \frac{1}{4\pi}\int_0^\infty \left[\frac{T_2R_1 e^{-jq(h+z_0)-jq_2(z-h)}}{1-R_1R_2e^{-j2qh}}\right] J_0(\lambda\rho)\frac{\lambda\, d\lambda}{jq}. \tag{15-128}$$

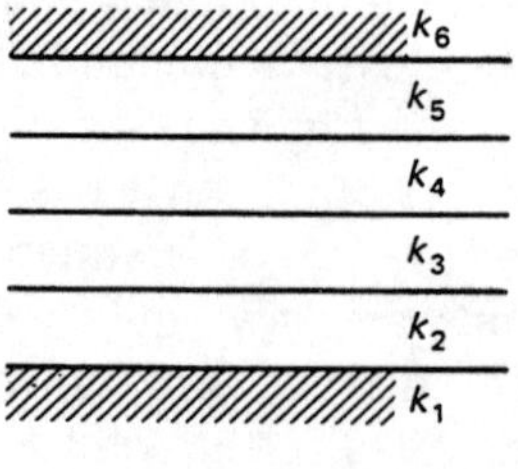

Figure 15-22 There is no branch point at $\lambda = \pm k_2$, $\pm k_3$, $\pm k_4$, and $\pm k_5$. The only branch points are at $\lambda = \pm k_1$ and $\pm k_6$.

Let us first consider π'_{s2} and its z-dependent term in the brackets. First we expand

$$\frac{1}{1 - R_1 R_2 e^{-j2qh}} = \sum_{n=0}^{\infty} (R_1 R_2)^n e^{-j2qnh}, \tag{15-129}$$

and write the quantity in brackets in (15-127) in a series form

$$[\cdot] = \sum_{n=0}^{\infty} u_n, \tag{15-130}$$

where

$$u_n = T_2 e^{-jq(h - z_0) - jq_2(z - h)}(R_1 R_2)^n e^{-j2qnh}.$$

We now show that each of u_n represents the wave successively refracted and reflected at the boundaries.

We note that the first term,

$$u_0 = T_2 e^{-jq(h - z_0) - jq_2(z - h)},$$

represents the wave originating from the dipole, traveling over the distance $(h - z_0)$ with the propagation constant q, reaching the upper surface, traveling over the distance $(z - h)$ with the propagation constant q_2 and arriving at the observation point (Fig. 15-23).

The second term,

$$u_1 = T_2 e^{-jq(h - z_0) - jq_2(z - h)} R_1 R_2 e^{-j2qh},$$

represents the wave propagating over an additional distance of $2h$ with the propagation constant q and reflected by both sides once $(R_1 R_2)$. Similarly, the rest of the terms represent the multiple reflections of the wave between two boundaries.

Similarly, for π''_{s2} we write

$$[\cdot] = \sum_{n=0}^{\infty} v_n, \tag{15-131}$$

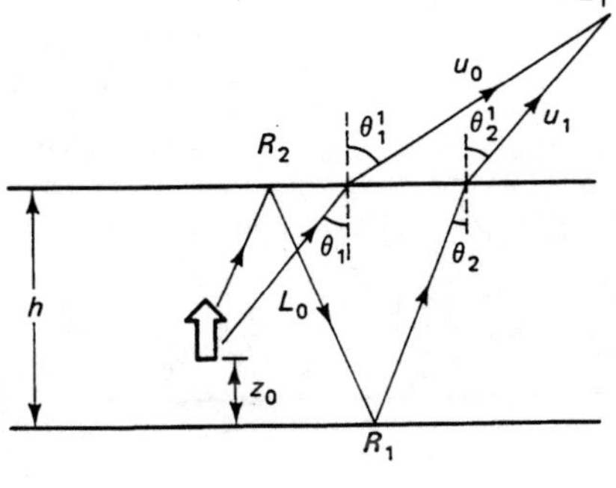

Figure 15-23 Geometric optical representation for u_0 and u_1.

where

$$v_n = T_2 R_1 e^{-jq(h+z_0) - jq_2(z-h)}(R_1 R_2)^n e^{-j2qnh}.$$

Each of v_n represents the wave shown in Fig. 15-24.

Each of the waves above can be evaluated by means of the saddle-point technique shown below:

$$\frac{1}{4\pi}\int_0^\infty A(\lambda)e^{-jqH - jq_2 Z} J_0(\lambda\rho)\frac{\lambda\, d\lambda}{jq} = \frac{1}{8\pi}\int_W A(\lambda)e^{-jqH - jq_2 Z} H_0^{(2)}(\lambda\rho)\frac{\lambda\, d\lambda}{jq}$$
$$\simeq \frac{1}{8\pi}\int_W \left[A(\lambda)\left(\frac{2}{\pi\lambda\rho}\right)^{1/2}\frac{\lambda e^{j(\pi/4)}}{jq}\right]e^{-jf(\lambda)}\, d\lambda, \tag{15-132}$$

where

$$f = qH + q_2 Z + \lambda\rho.$$

Using the saddle-point technique, we get

$$\frac{1}{8\pi}\left\{A(\lambda)\left(\frac{2}{\pi\lambda\rho}\right)^{1/2}\frac{\lambda e^{j(\pi/4)}}{jq}\right\}e^{j(\pi/4)}\sqrt{\frac{2\pi}{-f''(\lambda)}},$$

evaluated at the saddle point given by

$$\frac{\partial f}{\partial \lambda} = 0.$$

Thus we obtain

$$\pi'_{s2} = \sum_{n=0}^{\infty} U_n,$$
$$\pi''_{s2} = \sum_{n=0}^{\infty} V_n, \tag{15-133}$$

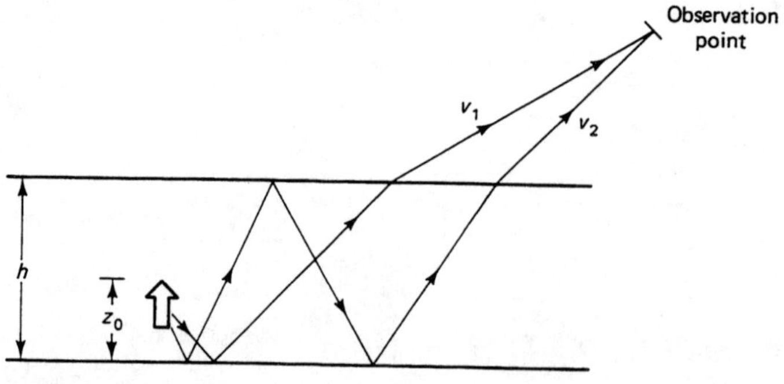

Figure 15-24 Geometric optical representation for v_1 and v_2.

where

$$U_n = \frac{1}{4\pi}\int_0^\infty u_n J_0(\lambda\rho)\frac{\lambda\, d\lambda}{jq},$$

$$V_n = \frac{1}{4\pi}\int_0^\infty v_n J_0(\lambda\rho)\frac{\lambda\, d\lambda}{jq},$$

and the saddle-point evaluations of U_0 and U_1 are

$$U_0 \simeq 4\pi \frac{T_2(\theta_1(\sin\theta_1)^{1/2}}{\rho^{1/2}\cos\theta_1}\frac{e^{-j(kL_0+k_2L_1)}}{\sqrt{L_0/\cos^2\theta_1 + L_1/\cos^2\theta_1'}},$$

$$U_0 \simeq 4\pi \frac{T_2(\theta_1)(\sin\theta_1)^{1/2}}{\rho^{1/2}\cos\theta_1}\frac{e^{-j(kL_0+k_2L_1)}}{\sqrt{L_0/\cos^2\theta_1 + L_1/\cos^2\theta_1'}},$$

$$U_1 \simeq 4\pi \frac{T_2(\theta_2)R_1(\theta_2)R_2(\theta_2)(\sin\theta_2)^{1/2}}{\rho^{1/2}\cos\theta_2}\frac{e^{-j(kL_2+k_2L_3)}}{\sqrt{L_2/\cos\theta_2 + L_3/\cos\theta_2'}}, \tag{15-134}$$

where θ_1, θ_1' and θ_2, θ_2 are the angles corresponding to the saddle point. L_0 and L_2 are the total path lengths from the dipole to the point where the ray leaves the surface, and L_1 and L_2 are the path lengths from the surfaces to the observation point. All U_n and V_n can be expressed in a similar manner.

Since the saddle-point technique is used to obtain (15-134), it is valid only when the distances L_0, L_1, L_2, and L_3 are large. This occurs when the height h is large and the observation point is far from the surface.

On the other hand, if h is small but the observation point is far from the surface, we can apply the saddle-point technique to (15-126) directly. We write (15-127) and (15-128) as follows:

$$\pi_{s2}' = \frac{1}{4\pi}\int_0^\infty (A_1)e^{-jq_2(z-h)}J_0(\lambda\rho)\frac{\lambda\, d\lambda}{jq},$$

$$\pi_{s2}'' = \frac{1}{4\pi}\int_0^\infty (A_2)e^{-jq_2(z-h)}J_0(\lambda\rho)\frac{\lambda\, d\lambda}{jq}, \tag{15-135}$$

assume A_1 and A_2 to be a slowly varying function of λ and obtain the far field in terms of the distance $R = \sqrt{(z-h)^2 + \rho^2}$. This is the usual technique of obtaining the radiation pattern from an antenna covered with a dielectric layer.

15-13 MODE AND LATERAL WAVE REPRESENTATION

Let us consider the field within the layer $0 < z < h$ given in (15-111). We note that π_{z0} can be written in the following manner:

$$\pi_{z0} = \frac{1}{8\pi}\int_{-\infty}^{\infty}\frac{A(\lambda)e^{-jqz} + B(\lambda)e^{+jqz}}{1 - R_1R_2e^{-j2qh}}H_0^{(2)}(\lambda\rho)\frac{\lambda\, d\lambda}{jq}. \tag{15-136}$$

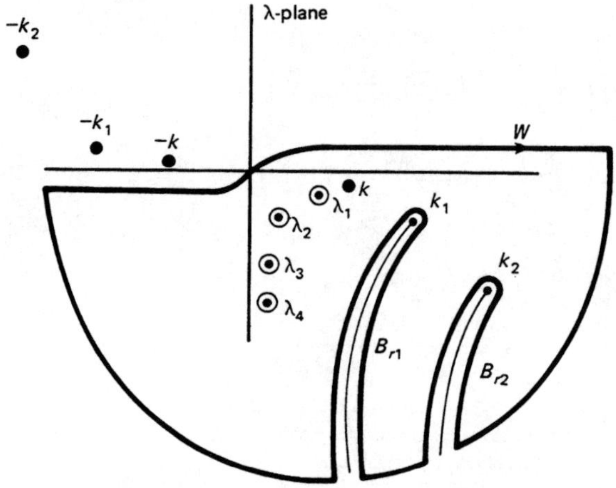

Figure 15-25 Poles at λ_n and branch points at $\pm k_1$ and $\pm k_2$.

The integrand has a series of poles at the roots of

$$1 - R_1 R_2 e^{-j2qh} = 0 \tag{15-137}$$

and two branch points at

$$\lambda = \pm k_1 \qquad \text{and} \qquad \pm k_2.$$

Therefore, the integral can be expressed as

$$\pi_{z0} = -2\pi j \sum_{n=1}^{\infty} \text{residue at poles } \lambda_n + \int_{B_{r1}} + \int_{B_{r2}}, \tag{15-138}$$

where B_{r1} and B_{r2} are the branch cuts for K_1 and K_2 (Fig. 15-25).

In general, the residues have a radial dependence of the form

$$e^{-j\lambda_n \rho} \tag{15-139}$$

and represent waveguide modes. On the other hand, the branch cut integration has the characteristics of the lateral wave as shown in Fig. 15-26. Therefore, if k_1 and k_2 are lossy, the waveguide modes dominate as we would expect in the case of a lossy waveguide. On the other hand, if k_1 and k_2 are lossless, the lateral wave contributes most to the field.

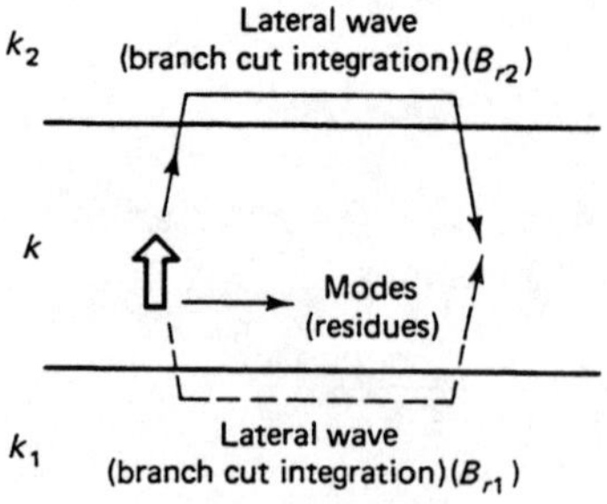

Figure 15-26 Lateral waves and waveguide modes.

It is obvious that if the height is small, there are small numbers of propagating modes and the other modes are almost cut off, implying that the reside series is highly convergent. On the other hand, if h is large, there may be a great number of propagating modes and the residue series is slowly convergent, and thus the geometric optical approach is more useful.

PROBLEMS

15-1. Find expressions for $\bar{E}$ and $\bar{H}$ for the problem shown in Fig. 15-1, and find the radiation field. The relative dielectric constant of the ground is 10 and the conductivity is $\sigma = 5 \times 10^{-3}$. The frequency is 1 MHz.

15-2. A vertical dipole is located on the ground operating at 1 MHz. The ground conductivity is $\sigma = 5 \times 10^{-3}$ (S/m), and the relative dielectric constant (real part) is 10.
- (a) Find the Sommerfeld pole α_p in the α plane.
- (b) Find the numerical distance p at a distance of 10 km on the ground.
- (c) Calculate and plot the attenuation factor as a function of $|p|$ ($10^{-2} \le |p| \le 10^2$).
- (d) Calculate the ratio of $|E_z|$ on the ground to $|E_z|$ in free space at a distance of 10 km.

15-3. A short vertical dipole of length 1 m is carrying a current of 1 A at 50 MHz. The dipole is located on flat ground. The relative dielectric constant of the ground is 15 and the conductivity is 5×10^{-3}.
- (a) Find the propagation constant of the Zenneck wave.
- (b) Find the numerical distance at a distance of 3 km.
- (c) Find the field strength as a function of distance.

15-4. Find the field in the ground when a vertical magnetic dipole is located on the ground.

15-5. Find the solution for the Sommerfeld problem when a magnetic horizontal dipole is located at height h above the ground.

15-6. A vertical dipole is located at height $h = 10$ km above a semi-infinite lossless plasma medium with $0 < n_2 < 1$ as shown in Fig. 15-9. The plasma frequency is 1 MHz and the operating frequency is 2 MHz. Find the field in air.

15-7. A vertical dipole is located above the interface between air $n_1 = 1$ and a lossless medium $n_2 = 2$ as shown in Fig. 15-18. Find the field transmitted in the second medium. $h = 5$ km, and the frequency is 1 MHz.

16

Inverse Scattering

In the usual scattering problem, we specify the object and the incident wave, and then we attempt to find the scattered wave. This is the *direct problem*. In contrast, in the *inverse problem*, we measure the scattered wave for a given incident wave, and then we attempt to determine the properties of the object. Two considerations are important in the inverse problem. First, our measurements are normally limited and we can only measure certain quantities within some ranges. Second, we need an effective inverse method so that we can determine the object characteristics with limited measured data. It is clear, therefore, that the inverse solution may not be unique and that the existence of the solution may not be apparent. It is also common that the inverse solution is unstable, so that a slight error in the measurement may create a large error in the unknown. In this chapter we outline several inversion techniques and their advantages and disadvantages. Also in this chapter, we use the convention $\exp(-i\omega t)$ commonly used in optics and acoustics, rather than $\exp(j\omega t)$ used in electrical engineering and by the IEEE, because there are many references in acoustics and physics on this topic.

16-1 RADON TRANSFORM AND TOMOGRAPHY

In a CT scanner (computed tomography scanner or X-ray tomography), an object is illuminated by an X-ray and the intensity of the transmitted X-ray is recorded for various angles of illumination. These recorded data are then used to reconstruct the image of the object. Let us consider a two-dimensional cross section of an object whose attenuation coefficient is given by $f(x, y)$ (Fig. 16-1). The object is illuminated uniformly in the direction of $\hat{\eta}$ with the intensity I_0. The transmitted power $I_t(\xi)$ is reduced by the total attenuation through the object and is given by

$$I_t(\xi) = I_0 \exp[-\int f(x, y)\, d\eta]. \tag{16-1}$$

Letting $\ln(I_0/I_t) = P_\phi(\xi)$, we have

$$P_\phi(\xi) = \int f(x, y)\, d\eta. \tag{16-2}$$

The function $P_\phi(\xi)$ represents the total attenuation at ξ when the object is illuminated at the angle ϕ and is called the *projection*. The inverse problem is then to find $f(x, y)$ from the measured projection $P_\phi(\xi)$. We can also rewrite (16-2) as follows:

$$P_\phi(\xi) = \int f(x, y)\delta(\xi - \bar{r} \cdot \hat{\xi})\, dx\, dy, \tag{16-3}$$

where $\bar{r} = x\hat{x} + y\hat{y}$ and $\bar{r} \cdot \hat{\xi} = x \cos\phi + y \sin\phi$ and $0 \le \phi < \pi$, $-\infty \le \xi \le +\infty$. Equation (16-3) can be considered a type of transform from $f(x, y)$ to $P_\phi(\xi)$ and is called the *Radon transform*. Therefore, the inverse problem is that of finding the *inverse Radon transform*. This was first studied by Radon in 1917 (Devaney, 1982; Kak, 1979; Herman, 1979).

Let us first consider a one-dimensional Fourier transform of the projection.

$$\bar{P}_\phi(K) = \int_{-\infty}^{\infty} P_\phi(\xi) e^{-jK\xi}\, d\xi. \tag{16-4}$$

Substituting (16-3) and integrating with respect to ξ, we get

$$\bar{P}_\phi(K) = \int f(x, y) e^{-iK\hat{\xi} \cdot \bar{r}}\, dx\, dy. \tag{16-5}$$

This is the two-dimensional Fourier transform of $f(x, y)$ evaluated at $\bar{K} = K\hat{\xi}$:

$$\bar{P}_\phi(K) = F(K\hat{\xi}), \tag{16-6}$$

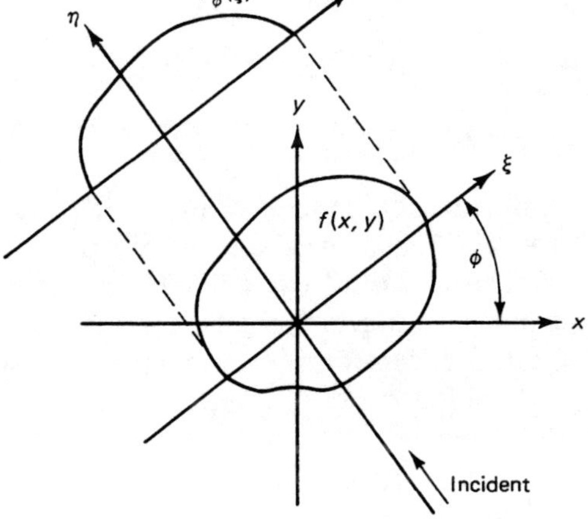

Figure 16-1 Radon transform or projection $P_\phi(\xi)$ of the object $f(x, y)$.

where

$$F(\overline{K}) = \int f(x,y) e^{-i\overline{K}\cdot\overline{r}}\,dx\,dy,$$

and $0 \le \phi < \pi$, $-\infty \le K \le \infty$.

We may think of $\overline{P}_\phi(K)$ as the Fourier transform of $f(x,y)$ along a slice at angle ϕ (Fig. 16-2). Therefore, we state that the one-dimensional Fourier transform of the projection of an object $f(x,y)$ is a slice at angle ϕ of the Fourier transform $F(\overline{K})$ of $f(x,y)$. This is called the *projection slice theorem*.

Now the complete Fourier transform of $f(x,y)$ is obtained by summing the slices over all the K space and $f(x,y)$ is obtained by the inverse Fourier transform. To do this, consider the inverse Fourier transform:

$$f(x,y) = \frac{1}{(2\pi)^2}\int F(\overline{K}) e^{i\overline{K}\cdot\overline{r}}\,d\overline{K}, \tag{16-7}$$

where $F(\overline{K}) = \overline{P}_\phi(K)$ and $d\overline{K} = K\,dK\,d\phi$. In order to use the range $0 \le \phi < \pi$, we write the integral as

$$\int_0^\infty K\,dK \int_0^\pi d\phi + \int_0^\infty K\,dK \int_\pi^{2\pi} d\phi.$$

Then using $\phi' = \phi - \pi$ and $K' = -K$, the second integral becomes

$$-\int_{-\infty}^0 K'\,dK' \int_0^\pi d\phi' = \int_{-\infty}^0 |K'|\,dK' \int_0^\pi d\phi'.$$

We therefore get

$$f(x,y) = \frac{1}{(2\pi)^2}\int_{-\infty}^\infty |K|\,dK \int_0^\pi d\phi\, \overline{P}_\phi(K) e^{iK\hat{\xi}\cdot\overline{r}}. \tag{16-8}$$

We rewrite this in the following form:

$$f(x,y) = \frac{1}{2\pi}\int_0^\pi d\phi\, Q_\phi(t).$$

$$Q_\phi(t) = \frac{1}{2\pi}\int_{-W}^{W} |K|\,dK\,\overline{P}_\phi(K\hat{\xi}) e^{iKt}, \tag{16-9}$$

$$t = \hat{\xi}\cdot\overline{r} = x\cos\phi + y\sin\phi.$$

Note that $Q_\phi(t)$ is the Fourier transform of the product of $\overline{P}_\phi$ and $|K|$ and therefore $|K|$ acts as a filter function $\overline{h}(K) = |K|$. We also used the highest spatial frequency W, because the projection can be measured only over a finite bandwidth $|K| \le W$. $Q_\phi(t)$ is called the *filtered projection*. It is a function of $t = \xi$ and independent of η.

Since $Q_\phi(t)$ is given by the Fourier transform of the product of two Fourier transforms, we can express $Q_\phi(t)$ as the following convolution integral:

$$Q_\phi(t) = \int_{-\infty}^\infty P_\phi(\xi') h(t - \xi')\,d\xi', \tag{16-10}$$

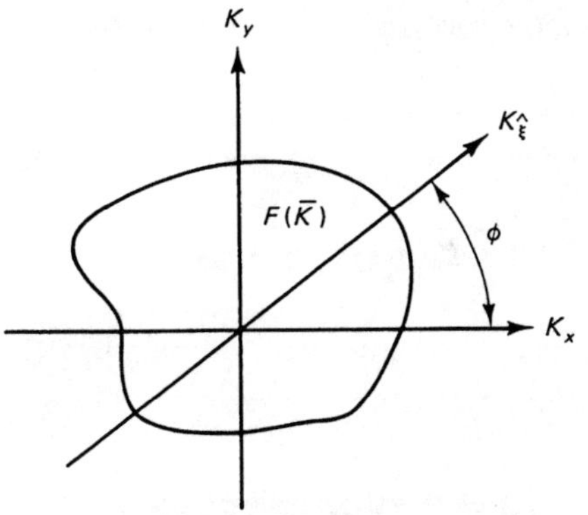

Figure 16-2 Projection slice theorem.

where

$$h(t) = \frac{1}{2\pi}\int_{-W}^{W} |K| e^{iKt}\, dK$$
$$= \frac{W^2}{\pi}\frac{\sin Wt}{Wt} - \frac{W^2}{2\pi}\left[\frac{\sin(Wt/2)}{(Wt/2)}\right]^2, \tag{16-11}$$

where W is the highest spatial frequency.

Summarizing this section, we reconstruct the object $f(x,y)$ from the projection $P_\phi(\xi)$ as follows:

$$f(x,y) = \frac{1}{2\pi}\int_0^{\pi} d\phi\, Q_\phi(t), \qquad t = x\cos\phi + y\sin\phi. \tag{16-12}$$

The filtered projection $Q_\phi(t)$ is given in the following two alternative forms:

$$Q_\phi(t) = \frac{1}{2\pi}\int_{-W}^{W} \bar{h}(K)\bar{P}_\phi(K) e^{iKt}\, dK$$
$$= \int_{-\infty}^{\infty} h(t-\xi')P_\phi(\xi')\, d\xi', \tag{16-13}$$

where

$$h(t) = \frac{1}{2\pi}\int_{-W}^{W} |K| e^{iKt}\, dK,$$
$$\bar{h}(K) = |K|.$$

Note that the filtered projection $Q_\phi(t)$ is obtained from the projection $P_\phi(\xi)$ either by the inverse Fourier transform of $\bar{P}_\phi(K)$ filtered with $\bar{h}(K)$ or by the convolution integral. Once $Q_\phi(t)$ is obtained for a given ϕ, it is summed over all ϕ from 0 to π as shown in (16-12) and the object is reconstructed. This process is called *back projection*. Therefore, we speak of the entire process as the *back projection of the filtered projection*. Equation (16-12) can be regarded as an inverse Radon transform to obtain $f(x,y)$ from the Radon transform $P_\phi(\xi)$. Note that if the object is a delta function $f(x, y) = \delta(x)\delta(y)$, the reconstructed image is $[W/(2\pi r)]J_1(Wr)$, where $r = (x^2 + y^2)^{1/2}$.

In practice, the calculations of (16-13) are made by digital processing (Kak, 1979). For example, the sampling interval of $P_\phi(\xi)$ must be $\tau = 1/(2W)$ and

$$Q_\phi(n\tau) = \tau \sum_m P_\phi(m\tau)h[(n-m)\tau]. \tag{16-14}$$

This can also be done by using FFT in the frequency domain in (16-13).

16-2 ALTERNATIVE INVERSE RADON TRANSFORM IN TERMS OF THE HILBERT TRANSFORM

In Section 16-1 we gave an inverse Radon transform in the form of the back projection of the filtered projection. It is also possible to express the inverse Radon transform using the Hilbert transform as was done by Radon in 1917. This inversion formula, however, contains a derivative and is more sensitive to noise.

Let us first consider the Fourier transform of the projection:

$$\bar{P}_\phi(K) = \int P_\phi(\xi)e^{-iK\xi}\,d\xi. \tag{16-15}$$

Note that we are using $i = \sqrt{-1}$ rather than $j = \sqrt{-1}$ in this chapter.

We integrate this by parts and noting $P_\phi(\xi)$ vanishes as $\xi \to \pm\infty$, we obtain

$$\bar{P}_\phi(K) = \int \frac{1}{iK}\frac{\partial}{\partial\xi}P_\phi(\xi)e^{-iK\xi}\,d\xi. \tag{16-16}$$

Substituting this in (16-13), we get

$$Q_\phi(t) = \frac{1}{2\pi}\int \frac{|K|}{iK}\,dK \int \frac{\partial}{\partial\xi}P_\phi(\xi)e^{iK(t-\xi)}\,d\xi. \tag{16-17}$$

Now the integration with respect to K can be performed using the following (see the Appendix to Chapter 16, Section A):

$$\frac{1}{2\pi}\int i\,\operatorname{sgn}(K)e^{iKt}\,dK = -\frac{1}{\pi t},$$
$$\operatorname{sgn}(K) = \frac{|K|}{K} = \begin{cases} 1 & \text{if } K > 0 \\ -1 & \text{if } K < 0 \end{cases} \tag{16-18}$$

The results can be expressed using the following Hilbert transform (see the Appendix):

$$F_h(t) = \frac{1}{\pi}\int_{-\infty}^{\infty} \frac{f(t')}{t'-t}\,dt', \tag{16-19}$$

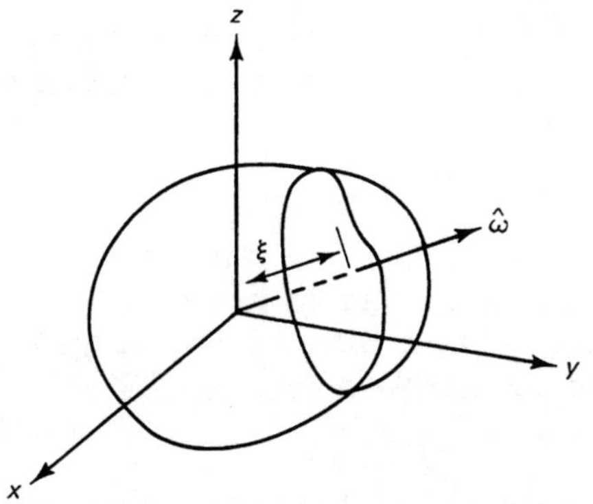

Figure 16-3 Three-dimensional radon transform.

where the integral is the Cauchy principal value. Thus we get

$$f(x, y) = \frac{1}{2\pi}\int_0^{\pi} d\phi \, Q_\phi(t)$$

$$Q_\phi(t) = -\text{Hilbert transform of } \frac{\partial}{\partial \xi} P_\phi(\xi) \tag{16-20}$$

$$= \int_{-\infty}^{\infty} \frac{1}{\pi(t-\xi)} \frac{\partial}{\partial \xi} P_\phi(\xi) \, d\xi.$$

This is an alternative form of the inverse Radon transform.

It is also possible to consider a three-dimensional Radon transform (Fig. 16-3):

$$R(\xi, \hat{\omega}) = \int f(\bar{r}) \delta(\xi - \bar{r} \cdot \hat{\omega}) \, dx \, dy \, dz. \tag{16-21}$$

The inverse Radon transform is then given by

$$f(\bar{r}) = -\frac{1}{8\pi^2} \int \frac{\partial^2 R(\xi, \hat{\omega})}{\partial \xi^2} d\omega, \tag{16-22}$$

where $d\omega$ is the differential element in the solid angle.

16-3 DIFFRACTION TOMOGRAPHY

The X-ray tomography described in the preceding sections deals with the reconstruction of the image of an object from the projection obtained by illuminating the object with uniform X-rays. All rays are assumed to travel in straight lines, and this is true at the limit of zero wavelengths. If acoustic or electromagnetic waves are used to reconstruct the object, diffraction effects due to finite wavelengths cannot be ignored. In general, the formulations of inverse problems, including diffraction, are extremely complicated and general solutions are not yet available. However, if the scattering is weak, it is possible to formulate the inverse problem in a manner similar to X-ray tomography and general solutions can be obtained. This will be presented in this section (see Devaney, 1982).

For a weakly scattering medium, the field U is represented by the Born approximation or the Rytov approximation. Let us assume that the field U satisfies the wave equation

$$[\nabla^2 + k^2 n^2(x,y)]U(x,y) = 0, \tag{16-23}$$

where k is the wave number of the background medium and $n(x,y)$ is the refractive index of the object. For example, the background medium is water in the case of ultrasound imaging of an object in water, and n represents the deviation of the refractive index of the object from water. We rewrite (16-23):

$$(\nabla^2 + k^2)U = -f(x,y)U, \tag{16-24}$$

where

$$f(x,y) = k^2(n^2 - 1).$$

This can be converted to the following integral equation for U:

$$U(\bar{r}) = U_i(\bar{r}) + \int G(\bar{r},\bar{r}')f(\bar{r}')U(\bar{r}')\,d\bar{r}', \tag{16-25}$$

where U_i is the incident wave, and $G(\bar{r},\bar{r}')$ is the Green's function satisfying the equation:

$$(\nabla^2 + k^2)G = -\delta(\bar{r} - \bar{r}').$$

The first Born approximation is obtained by approximating U in the integrand by the incident wave:

$$U(\bar{r}) = U_i(\bar{r}) + \int G(\bar{r},\bar{r}')f(\bar{r}')U_i(\bar{r}')\,d\bar{r}'. \tag{16-26}$$

This is valid when

$$|k(n-1)D| << 1, \tag{16-27}$$

where D is a typical size of the object.

The Rytov approximation is obtained by considering the total complex phase Ψ of U. We let

$$U(x,y) = U_i(\bar{r})e^{\Psi(x,y)}. \tag{16-28}$$

The first Rytov approximation is then given by (see the Appendix to Chapter 16, Section B)

$$\Psi(x,y) = \frac{1}{U_i(\bar{r})}\int G(\bar{r},\bar{r}')f(\bar{r}')U_i(\bar{r}')\,d\bar{r}'. \tag{16-29}$$

Note that if we expand the exponent in (16-28) and keep the first term, we obtain the first Born approximation, and therefore the Rytov approximation contains more scattering terms and is a better approximation than the first Born approximation. In either the Born or Rytov solution, we can use the scattered field U_s:

$$U_s(\bar{r}) = \int G(\bar{r},\bar{r}')f(\bar{r}')U_i(\bar{r}')\,d\bar{r}'. \tag{16-30}$$

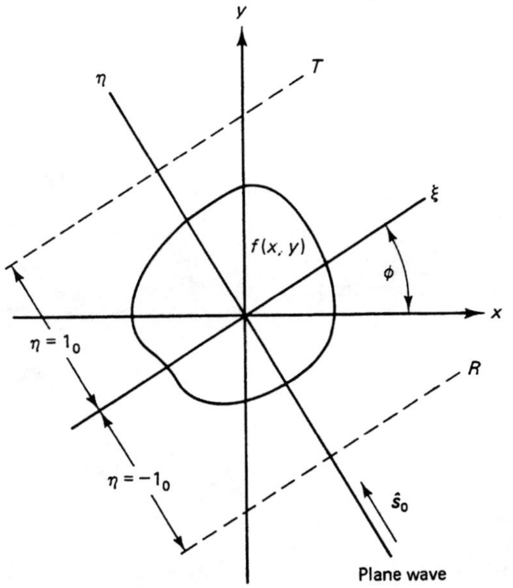

Figure 16-4 Diffraction tomography. Plane wave is incident on the object $f(x, y)$. T is the receiver plane for transmission tomography and R is the receiver plane for reflection tomography.

In the Rytov solution, for a given incident field U_i we measure the total field U and obtain the field U_s by calculating $U_i \ln(U/U_i)$. The function $f(x, y)$ represents the object, and the inverse problem is then reduced to finding the object $f(x, y)$ from U_s.

Let us consider a two-dimensional cross section of the object $f(x, y)$ illuminated by a plane wave propagating in the direction $\hat{s}_0$ (Fig. 16-4).

$$U_i(\bar{r}) = e^{ik\hat{s}_0 \cdot \bar{r}} = e^{ik\eta'}. \tag{16-31}$$

Note that we are using the convention $\exp(-i\omega t)$ in this chapter.

The two-dimensional Green's function is given by

$$\begin{aligned} G(\bar{r}, \bar{r}') &= \frac{i}{4} H_0^{(1)}(k|\bar{r} - \bar{r}'|) \\ &= \frac{1}{2\pi} \int \frac{i}{2K_2} e^{iK_1(\xi - \xi') + iK_2|\eta - \eta'|} \, dK_1, \end{aligned} \tag{16-32}$$

where

$$K_2 = \begin{cases} (k^2 - K_1^2)^{1/2} & \text{if } |K_1| < k \\ i(K_1^2 - k^2)^{1/2} & \text{if } |K_1| > k \end{cases}$$

The spatial frequency K_1 and K_2 are chosen parallel to $\hat{\xi}$ and $\hat{\eta}$ axes, respectively (Fig. 16-5).

Let us first consider U_s for the transmission tomography. Substituting (16-31) and (16-32) into (16-30), and noting that $|\eta - \eta'| = l_0 - \eta'$, we get

$$U_s(\xi, l_0) = \frac{1}{2\pi} \int \frac{i}{2K_2} e^{iK_1\xi + iK_2 l_0} F(K_1, K_2 - k) \, dK_1, \tag{16-33}$$

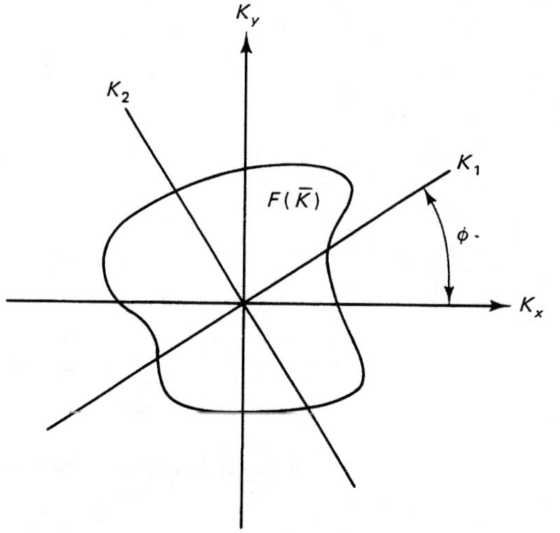

Figure 16-5 Spatial frequency space K.

where F is the Fourier transform of $f(x, y)$.

$$F(K_1, K_2 - k) = \int f(x', y')e^{-iK_1\xi' - i(K_2 - k)\eta'}\, d\xi'\, d\eta'.$$

Note that (16-33) can be seen as a one-dimensional Fourier transform, and therefore, if we take the one-dimensional Fourier transform of the observed data $U_s(\xi, l_0)$ at $\eta = l_0$, we get from (16-33),

$$\begin{aligned} \bar{U}_s(K_1, l_0) &= \int U_s(\xi, l_0)e^{-iK_1\xi}\, d\xi \\ &= \frac{ie^{iK_2 l_0}}{2K_2} F(K_1, K_2 - k). \end{aligned} \tag{16-34a}$$

Note that K_1 and K_2 are related through

$$K_1^2 + K_2^2 = k^2, \tag{16-34b}$$

as given in (16-32), and therefore, if we limit ourselves to the real K_1 and K_2, this represents a circle with radius k in the $\bar{K}$ space. The real K_1 and K_2 means that only the propagating wave is considered and the evanescent wave is neglected. We may rewrite (16-34) using

$$K_1\hat{\xi} + K_2\hat{\eta} = k\hat{s}, \qquad K\hat{\eta} = k\hat{s}_0, \qquad \bar{U}_s(K_1, l_0) = \frac{ie^{iK_2 l_0}}{2K_2} F[k(\hat{s} - \hat{s}_0)]. \tag{16-35}$$

This means that the Fourier transform of the observed data $U_s(\xi, l_0)$ at $\eta = l_0$ is proportional to the Fourier transform $F(\bar{K})$ of the object $f(x, y)$ evaluated at $\bar{K} = k(\hat{s} - \hat{s}_0)$. This is a semicircle in $\bar{K}$ space where $k > K_2 > 0$, shown in Fig. 16-6. This is a generalization of the projection slice theorem for conventional tomography discussed in Section 16-1. Here, instead of a slice at an angle θ, we have a semicircle centered at $\bar{K} = -k\hat{s}_0$. Note that in the high-frequency limit, $k \to \infty$ and the semicircle is stretched to a slice for conventional tomography.

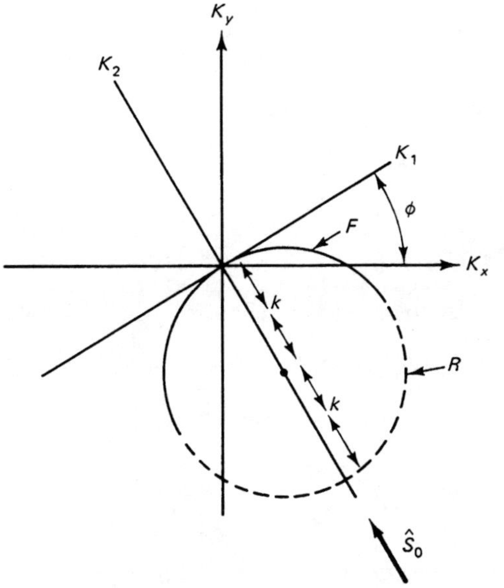

Figure 16-6 Semicircle arc (F) for forward tomography and (R) for reflection tomography.

For reflection tomography, we note that $|\eta - \eta'| = \eta' + l_0$ and the Fourier transform of the observed data $U_s(\xi, -l_0)$ at $\eta = -l_0$ is given by

$$\bar{U}_s(K_1, -l_0) = \frac{ie^{iK_2 l_0}}{2K_2} F(K_1, -K_2 - k). \tag{16-36}$$

This is the lower half of the semicircle as indicated in Fig. 16-6.

From the discussion above, it is clear that the Fourier transform of the observed data for a given incident direction $\hat{s}_0$ is proportional to the Fourier transform of the object evaluated on a semicircle. If we do this process for all angles $0 \le \phi < \pi$, we can cover the K space within the circle of radius $\sqrt{2}k$ for transmission tomography and within the band between the radii from $\sqrt{2}k$ to $2k$ for reflection tomography. We can then invert the Fourier transform to reconstruct the object. This reconstruction process is similar to the back projection for conventional tomography, but since it includes the propagation effect, it is called the back propagation and is explained below. We also note that we can cover a larger K space by varying the frequency $k = \omega/c$.

Let us consider the forward tomography. $U_s(\xi, l_0)$ is measured and its one-dimensional Fourier transform $\bar{U}_s(K_1, l_0)$ is calculated. We then calculate the function

$$F(K_1, K_2 - k) = -i2K_2 e^{iK_2 l_0} \bar{U}_s(K_1, l_0). \tag{16-37}$$

From this we get the object by the inverse Fourier transform:

$$f(x, y) = \frac{1}{(2\pi)^2} \int F(\bar{K}) e^{i\bar{K}\cdot\bar{r}} \, dK_x \, dK_y, \tag{16-38}$$

where $\bar{K} = k(\hat{s} - \hat{s}_0)$.

Now we note that

$$
\begin{aligned}
\bar{K} &= K_1 \hat{\xi} + (K_2 - k)\hat{\eta} \\
\hat{\xi} &= \cos\phi\,\hat{x} + \sin\phi\,\hat{y} \\
\hat{\eta} &= -\sin\phi\,\hat{x} + \cos\phi\,\hat{y}.
\end{aligned}
\tag{16-39}
$$

Therefore, we get

$$
\begin{aligned}
\bar{K} &= K_x \hat{x} + K_y \hat{y} \\
K_x &= K_1 \cos\phi - (K_2 - k)\sin\phi \\
K_y &= K_1 \sin\phi + (K_2 - k)\cos\theta.
\end{aligned}
\tag{16-40}
$$

Changing the variables from (K_x, K_y) to (K_1, ϕ), and noting $K_1^2 + K_2^2 = k^2$, we get

$$
dK_x\,dK_y = \begin{vmatrix} \frac{\partial K_x}{\partial K_1} & \frac{\partial K_x}{\partial \phi} \\ \frac{\partial K_y}{\partial K_1} & \frac{\partial K_y}{\partial \phi} \end{vmatrix} dK_1\,d\phi = \frac{kK_1}{K_2}\,dK_1\,d\phi
\tag{16-41}
$$

Following the procedure in Section 16-1 and using (16-37), we get

$$
\begin{aligned}
f(x, y) &= \frac{1}{2\pi}\int_0^{\pi} d\phi\, Q_\phi(\xi, \eta), \\
Q_\phi(\xi, \eta) &= \frac{1}{2\pi}\int_{-k}^{+k} h(K_1, \eta)\bar{U}_s(K_1, l_0)e^{iK_1\xi}\,dK_1,
\end{aligned}
\tag{16-42}
$$

where $h(K_1, \eta)$ is the filter function given by

$$
h(K_1, \eta) = -i2k|K_1|e^{-ikl_0 + i(K_2 - k)(\eta - l_0)},
$$

where

$$
\begin{aligned}
\xi &= x\cos\phi + y\sin\phi, \\
\eta &= -x\sin\phi + y\cos\phi, \\
K_2 &= (k^2 - K_1^2)^{1/2}.
\end{aligned}
$$

We can conclude here that Q_ϕ is a generalization of the filtered projection and the reconstruction process is called the filtered back propagation. It is also clear that Q_ϕ can also be written as a convolution integral as shown in Section 16-1.

We also note that if the object is a delta function $f(x, y) = \delta(x)\delta(y)$, the reconstructed image is an Airy disk given by

$$
f(x, y) = \frac{k}{\sqrt{2}\pi r} J_1(\sqrt{2}kr).
\tag{16-43}
$$

16-4 PHYSICAL OPTICS INVERSE SCATTERING

Based on the physical optics approximation, it is possible to derive an inversion formula that gives the size and shape of a conducting object from the knowledge of the monostatic scattering for all frequencies and all aspect angles. This inverse scattering formula was first obtained by Bojarski in 1967 and is now known as *Bojarski's identity* (see the survey paper by Bojarski, 1982b, and Lewis, 1969).

Let us consider the scattered far field $\mathbf{E}_s$ from a conducting object:

$$\mathbf{E}_s = -jk\eta_0 \frac{e^{-jkR}}{4\pi R} \int_s [-\hat{o} \times (\hat{o} \times \mathbf{J}_s)] e^{jk\hat{o}\cdot\bar{r}'} \, ds', \tag{16-44}$$

where $\eta_0 = (\mu_0/\epsilon_0)^{1/2}$, R is the distance from the reference point on the object, $\mathbf{J}_s$ is the surface current density, and $\hat{o}$ is the direction of observation.

For monostatic scattering, $\hat{o} = -\hat{\imath}$ where $\hat{\imath}$ is the direction of the incident wave. For physical optics approximation, we have $\mathbf{J}_s = 2(\hat{n} \times \mathbf{H}_i)$ where $\hat{n}$ is the unit vector normal to the surface. We note that

$$\begin{aligned} \hat{\imath} \times (\hat{n} \times \mathbf{H}_i) &= -\mathbf{H}_i(\hat{\imath} \cdot \hat{n}), \\ -\hat{o} \times (\hat{o} \times \mathbf{J}_s) &= \hat{\imath} \times \mathbf{H}_i(\hat{\imath} \cdot \hat{n}). \end{aligned} \tag{16-45}$$

The scattered wave $\mathbf{E}_s$ can then be expressed as

$$\mathbf{E}_s = jk\eta_0 \hat{e}_i \frac{e^{-jkR}}{4\pi R} \int_s 2(\hat{\imath} \cdot \hat{n}) e^{j2k\hat{o}\cdot\mathbf{r}'} \, ds' \tag{16-46}$$

where $\hat{e}_i$ is the unit vector in the direction of the polarization of the incident wave $\mathbf{E}_i = E_i \hat{e}_i$. This formula shows that the monostatic scattered wave is polarized in the direction $\hat{e}_i$, and therefore there is no cross polarization in the monostatic scattered wave based on the physical optics.

Let us now define the normalized complex far-field scattering amplitude $\rho(\mathbf{K})$ when illuminated in the direction $\hat{\imath}$. We also use $\exp(-i\omega t)$ dependence rather than $\exp(j\omega t)$ in this section, as this is commonly used in this work.

$$\rho(\mathbf{K}) = \frac{i}{\sqrt{4\pi}} \int_{\mathbf{K}\cdot\hat{n}>0} e^{-i\mathbf{K}\cdot\mathbf{r}'} \mathbf{K} \cdot d\mathbf{s}', \tag{16-47}$$

where $\mathbf{K} = 2k\hat{o} = -2k\hat{\imath}$, $d\mathbf{s}' = \hat{n}\, ds'$, and $\mathbf{K}\cdot\hat{n} > 0$ means the illuminated surface (Figure 16-7). The normalized complex scattering amplitude $\rho(\mathbf{K})$ is defined so that the back-scattering cross section σ_b is given by [see (10-101)]

$$\sigma_b = \rho\rho^*. \tag{16-48}$$

Now we consider the scattered wave when illuminated from the opposite direction ($\mathbf{K} \to -\mathbf{K}$). We also take the complex conjugate and obtain

$$\rho^*(-\mathbf{K}) = \frac{i}{\sqrt{4\pi}} \int_{\mathbf{K}\cdot\hat{n}<0} e^{-i\mathbf{K}\cdot\mathbf{r}'} \mathbf{K} \cdot d\mathbf{s}'. \tag{16-49}$$

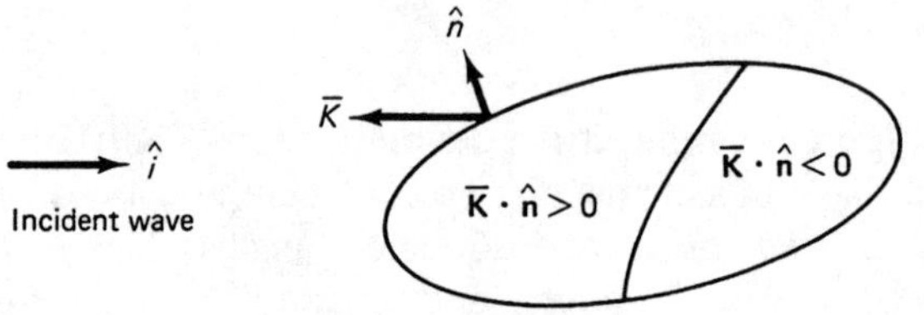

Figure 16-7 Illuminated surface $\mathbf{K}\cdot\hat{\mathbf{n}} > 0$.

Adding $\rho(\mathbf{K})$ and $\rho^*(-\mathbf{K})$, we get

$$\rho(\mathbf{K}) + \rho^*(-\mathbf{K}) = \frac{i}{\sqrt{4\pi}}\int_s e^{-i\mathbf{K}\cdot\mathbf{r}'}\,\mathbf{K}\cdot d\mathbf{s}', \tag{16-50}$$

where s is the complete surface of the object.

By using the divergence theorem, the right side of (16-50) becomes

$$\frac{i}{\sqrt{4\pi}}\int_v \nabla\cdot(e^{-i\mathbf{K}\cdot\mathbf{r}'}\mathbf{K})\,dv' = \frac{2k^2}{\sqrt{\pi}}\int_v e^{-i\mathbf{K}\cdot\mathbf{r}'}\,dv', \tag{16-51}$$

where v is the volume of the object.

We can now define the complex scattering amplitude $\Gamma(\mathbf{K})$ and the characteristic function $\gamma(\mathbf{r}')$ of the scatterer.

$$\Gamma(\mathbf{K}) = \frac{\sqrt{\pi}}{2k^2}[\rho(\mathbf{K}) + \rho^*(-\mathbf{K})],$$
$$\gamma(\mathbf{r}') = \begin{cases} 1 & \text{if } \mathbf{r}' \text{ is inside the scatterer} \\ 0 & \text{if } \mathbf{r}' \text{ is outside.} \end{cases} \tag{16-52}$$

We then get the following three-dimensional Fourier transform relationship:

$$\Gamma(\mathbf{K}) = \int_v \gamma(\mathbf{r}')e^{-i\mathbf{K}\cdot\mathbf{r}'}\,dv'. \tag{16-53}$$

For scatterers of finite volume, we can invert this Fourier transform and obtain

$$\gamma(\mathbf{r}) = \frac{1}{(2\pi)^3}\int \Gamma(\mathbf{K})e^{i\mathbf{K}\cdot\mathbf{r}}\,d\mathbf{K}. \tag{16-54}$$

This is Bojarski's identity and shows that the object shape can be determined by measuring the back-scattered far field over all **K.** This requires that the back scattering be known for all frequencies and all angles, which presents a practical measurement difficulty. Also, the identity is based on the physical optics approximation, which is valid only for high frequencies and should not be used for lower frequencies or resonance regions. This is another source of theoretical difficulty. Therefore, much of the work is directed to obtaining solutions even if the scattering information is incomplete (see Bojarski, 1982b; Lewis, 1969).

If the measurement of the scattered wave can be made only in some portion D of the **K** space, we write

$$A(\mathbf{K}) = \begin{cases} 1 & \text{if } \mathbf{K} \text{ is in } D \\ 0 & \text{if } \mathbf{K} \text{ is outside.} \end{cases} \tag{16-55}$$

Then we can measure $A(\mathbf{K})\Gamma(\mathbf{K})$. The inverse Fourier transform is then given by

$$f(\mathbf{r}) = \frac{1}{(2\pi)^3}\int A(\mathbf{K})\Gamma(\mathbf{K})e^{i\mathbf{K}\cdot\mathbf{r}}\,d\mathbf{K}. \tag{16-56}$$

This is given by

$$\begin{aligned} f(\mathbf{r}) &= \int \mathbf{A}(\mathbf{r}-\mathbf{r}')\gamma(\mathbf{r}')\,dv' \\ &= \int \mathbf{A}(\mathbf{r}')\gamma(\mathbf{r}-\mathbf{r}')\,dv', \end{aligned} \tag{16-57}$$

where

$$\mathbf{A}(\mathbf{r}) = \frac{1}{(2\pi)^3}\int A(\mathbf{K})e^{i\mathbf{K}\cdot\mathbf{r}}\,d\mathbf{K}.$$

In (16-57), $f(\mathbf{r})$ is calculated from the measurement and $\mathbf{A}(\mathbf{r})$ is also known, and therefore this constitutes an integral equation for $\gamma(\mathbf{r})$.

As an example, suppose that the measurement is conducted only from one direction $\hat{z}$. Then D is the line $K_x = K_y = 0$ and therefore

$$A(\mathbf{K}) = \delta(K_x)\delta(K_y). \tag{16-58}$$

We then get

$$\mathbf{A}(\mathbf{r}) = \frac{1}{(2\pi)^3}\int e^{iK_z z}\,dK_z = \frac{1}{(2\pi)^2}\delta(z).$$

Substituting this in (16-57), we get

$$f(\mathbf{r}) = \frac{1}{(2\pi)^2}\iint \gamma(x,y,z)\,dx\,dy. \tag{16-60}$$

This means that $f(\mathbf{r})$ calculated from the measured data gives the cross-sectional area of the target as a function of z.

16-5 HOLOGRAPHIC INVERSE SOURCE PROBLEM

If the field Ψ and its normal derivative $\partial\Psi/\partial n$ produced by a source $\rho(\mathbf{r})$ are measured over a surface S, it should be possible to reconstruct the source distribution $\rho(\mathbf{r})$ from the knowledge of Ψ and $\partial\Psi/\partial n$ on S. This inversion technique has been developed independently by Porter and Bojarski. This is based on the same principle as conventional holography, in which the scattered field is recorded on a photographic film which, when illuminated, reproduces the source distribution (see Bojarski, 1982a, for a review; see also Porter and Devaney, 1982; and Tsang et al. 1987).

Let us first write Green's theorem for the volume enclosed by the surface S (Fig. 16-8).

$$\int_V (u\nabla^2 v - v\nabla^2 u)\,dV = \int_s \left(u\frac{\partial v}{\partial n} - v\frac{\partial u}{\partial n}\right)ds, \tag{16-61}$$

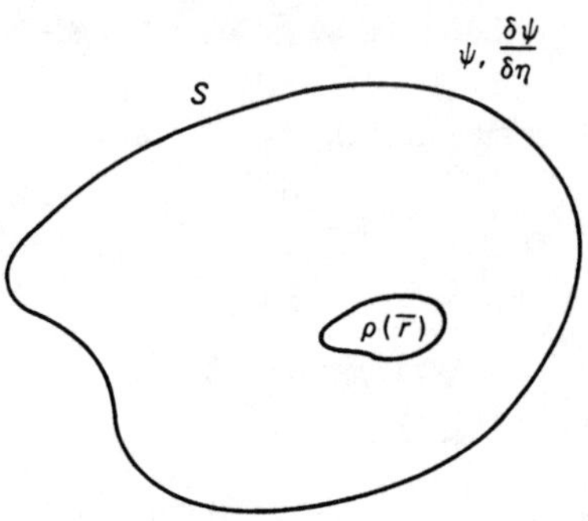

Figure 16-8 Holographic inversion.

where $\partial/\partial n$ is the outward normal derivative. We now let the scalar function u be the field Ψ. The field Ψ is produced by the source ρ and satisfies the following inhomogeneous wave equation:

$$(\nabla^2 + k^2)\Psi = -\rho. \tag{16-62}$$

We then back-propagate the field toward the source. To do this, we use $v = G^*$, where G is Green's function satisfying

$$(\nabla^2 + k^2)G = -\delta(\mathbf{r} - \mathbf{r}'), \qquad G = \frac{\exp(ik|\mathbf{r} - \mathbf{r}'|)}{4\pi|\mathbf{r} - \mathbf{r}'|} \tag{16-63}$$

Substituting (16-62) and (16-63) into (16-61), we get

$$\int_s \left(\Psi\frac{\partial G^*}{\partial n} - G^*\frac{\partial \Psi}{\partial n}\right) ds = -\Psi + \int G^* \rho \, dV. \tag{16-64}$$

Also, if we use $v = G$, we get

$$\int_s \left(\Psi\frac{\partial G}{\partial n} - G\frac{\partial \Psi}{\partial n}\right) ds = -\Psi + \int G\rho \, dV = 0. \tag{16-65}$$

Note that the surface integral vanishes, as it produces no scattered field inside S. Subtracting (16-64) from (16-65), we get the following integral equation for ρ:

$$\Gamma(\mathbf{r}) = \int K(\mathbf{r}, \mathbf{r}')\rho(\mathbf{r}')\, dV', \tag{16-66}$$

where

$$\Gamma(\mathbf{r}) = \int_s \left\{\left[\Psi(\mathbf{r}_s)\frac{\partial g(\mathbf{r}, \mathbf{r}_s)}{\partial n}\right] - \left[g(\mathbf{r}, \mathbf{r}_s)\frac{\partial \Psi(\mathbf{r}_s)}{\partial n}\right]\right\} ds,$$

$$K(\mathbf{r}, \mathbf{r}') = \frac{\sin k|\mathbf{r} - \mathbf{r}'|}{k|\mathbf{r} - \mathbf{r}'|} = G - G^*,$$

$$g(\mathbf{r}, \mathbf{r}_s) = \frac{\sin k|\mathbf{r} - \mathbf{r}_s|}{k|\mathbf{r} - \mathbf{r}_s|} = G - G^*.$$

From knowledge of the surface field $\Psi(\mathbf{r}_s)$ and $\partial\Psi(\mathbf{r}_s)/\partial n$, $\Gamma(\mathbf{r})$ can be calculated, and by solving the integral equation, the source distribution $\rho(\mathbf{r}')$ can be obtained.

An alternative form of the integral equation is possible. We may use the conjugate field Ψ^* and obtain

$$\Delta^*(\mathbf{r}) = \int K(\mathbf{r},\mathbf{r}')\rho^*(r')\,dV'. \tag{16-67}$$

The solutions to the integral equation (16-66) and (16-67) are, however, not unique. The source ρ consists of the nonradiating sources ρ_n and the radiating sources ρ_r. The nonradiating source generates a field that is identically zero outside the source region and does not contribute to the field on S. It has been shown that if we minimize the source energy E,

$$E = \int |\rho|^2\,dV, \tag{16-68}$$

it yields a unique solution. For a detailed discussion on the uniqueness and nonradiating sources, see Porter and Devaney (1982) and Cohen and Bleistein (1979). Also see Devaney and Porter (1985) and Tsang et al. (1987) for inversion of inhomogeneous and attenuating media.

16-6 INVERSE PROBLEMS AND ABEL'S INTEGRAL EQUATION APPLIED TO PROBING OF THE IONOSPHERE

Consider the ionosphere whose electron density profile $N_e(z)$ is known as a function of height z. The refractive index is given by

$$N(\omega) = \left(1 - \frac{\omega_p^2}{\omega^2}\right)^{1/2}, \tag{16-69}$$

$$\omega_p^2 = \frac{e^2 N_e}{m\epsilon_0},$$

where $f_p = \omega_p/2\pi$ is plasma frequency, e and m are the charge and mass of an electron, ϵ_0 is the free-space permittivity, and the loss is neglected.

If we send up a radio wave of angular frequency ω, it reaches the height h where ω_p is equal to ω and returns to the ground. The time $T(\omega)$ for the wave to travel to $z = h$ and return to $z = 0$ is given by

$$T(\omega) = 2\int_0^h \frac{dz}{v_g}, \tag{16-70}$$

where v_g is the group velocity given by

$$\frac{1}{v_g} = \frac{\partial k}{\partial \omega} = \frac{\omega}{c}\frac{1}{(\omega^2 - \omega_p^2)^{1/2}}$$

$$k = \frac{\omega}{c}n(\omega)$$

and the height h at the turning point at a frequency ω is given by

$$\omega = \omega_p(h).$$

For a given profile of $N_e(z)$, and therefore $\omega_p(z)$, we can calculate $T(\omega)$ from (16-70). This is called the *direct problem*.

Now consider the inverse problem. We send up radio waves at various frequencies and measure $T(\omega)$ as a function of ω. From the measured data $T(\omega)$, we attempt to determine the plasma frequency profile $\omega_p(z)$ and the electron density profile $N_e(z)$. This is the *inverse problem* (Fig. 16-9).

Let us rewrite (16-70):

$$g(\omega) = \frac{T(\omega)}{2\omega/c} = \int_0^h \frac{dz}{[\omega^2 - \omega_p^2(z)]^{1/2}}. \tag{16-71}$$

Here $g(\omega)$ is the measured data and $\omega_p(z)$ is the unknown. This is a nonlinear equation for $\omega_p(z)$. To simplify (16-71), we let $\omega^2 = E$ and $\omega_p^2(z) = V(z)$, and write (16-71) in the following form:

$$\begin{aligned} g(E) &= \int_0^h \frac{dz}{[E - V(z)]^{1/2}} \\ &= \int_0^E \frac{1}{(E-V)^{1/2}} \frac{dz}{dV} dV, \end{aligned} \tag{16-72}$$

where the upper limit of integration is determined by $E = V(h)$. We assume that $V(z)$ is a monotonic function of z. Here we used E and V, as this problem is identical to the problem of sliding a particle up a frictionless hill with initial kinetic energy E and measuring the time $T(E)$ required for the particle to return. The time $T(E)$ is measured for different E and the shape of the hill represented by the potential energy $V(h) = mgh$ is to be determined. This is the problem solved by Abel in 1826.

Equation (16-72) is Abel's integral equation,

$$g(E) = \int_0^E \frac{f(V)}{(E-V)^{1/2}} dV, \tag{16-73}$$

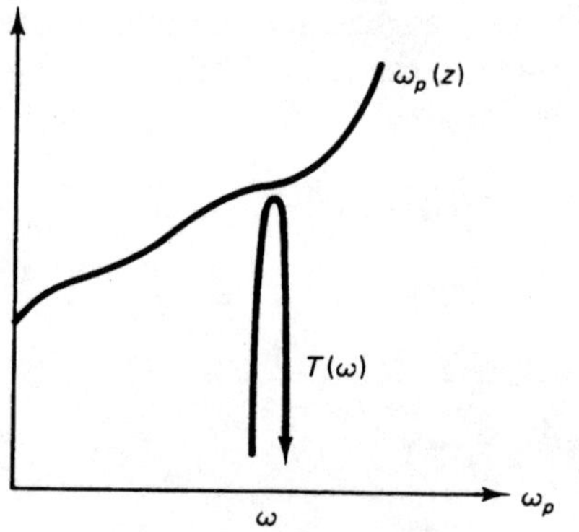

Figure 16-9 Probing of the electron density profile.

where $g(E)$ is known and $f(v)$ is the unknown function. This is a Volterra integral equation of the first kind with the kernel $(E - V)^{-1/2}$. Its solution is given by (see the Appendix to Chapter 16, Section C)

$$z = \int_0^V f(V)\,dV = \frac{1}{\pi}\int_0^V \frac{g(E)}{(V-E)^{1/2}}\,dE. \tag{16-74}$$

Converting $g(E)$, E, and V to $T(\omega)$, ω, and ω_p, we get

$$z = z(\omega_p) = \frac{c}{\pi}\int_0^{\omega_p} \frac{T(\omega)\,d\omega}{(\omega_p^2 - \omega^2)^{1/2}}. \tag{16-75}$$

By measuring $T(\omega)$, we can determine $\omega_p(z)$ from (16-75). For example, if $T(\omega) = (T_0/\omega_0)\omega$, then $z = z(\omega_p) = (cT_0/\pi)(\omega_p/\omega_0)$ and therefore $\omega_p(z) = (\omega_0\pi/cT_0)z$.

16-7 RADAR POLARIMETRY AND RADAR EQUATION

In Section 10-2 we discussed the conventional radar equation in the following form:

$$\frac{P_r}{P_t} = \frac{\lambda^2}{(4\pi)^3}\frac{G_tG_r}{R_1^2R_2^2}\sigma_{bi}m, \tag{16-76}$$

where P_r is the received power; P_t the transmitted power; G_t and G_r the gain of the transmitter and receiver; R_1 and R_2 are distance from transmitter to object and distance from object to receiver, respectively; σ_{bi} the bistatic cross section of the object; and m the mismatch factor. If both impedance and polarizations are matched, $m = 1$, but $0 < m < 1$ otherwise. This conventional radar equation deals with the total power received, but it gives no specific information about the relationships among the polarization charcteristics of the transmitter, the object, and the receiver. However, recent advances in measurement techniques have made possible acquisition of more detailed polarization information and thus have stimulated intensive research on radar polarimetry, which is the utilization of the complete polarization characteristics in radar. The polarimetric techniques are also applicable to the remote sensing of the terrain and the discrimination of signals from clutter, interference, and jamming (Boerner, 1985; Huynen, 1978).

Let us now reexamine the radar equation. Following Section 10-2, we first consider a field incident on the object. We assume that the object is in the far field of both the transmitting and receiving antennas. Since the incident flux density S_i at the object is $S_i = |\overline{E}_t|^2/2\eta$, where $\overline{E}_t$ is the electric field, we write

$$\overline{E}_t = (2\eta S_i)^{1/2}\,\overline{E}_{tn}, \qquad S_i = \frac{G_tP_t}{4\pi R_1^2}, \tag{16-77}$$

where E_{tn} is the normalized transmitted field with $|\overline{E}_{tn}| = 1$. We now express $\overline{E}_t$ in an orthogonal coordinate system:

$$\overline{E}_t = E_{t1}\hat{x}_1 + E_{t2}\hat{x}_2. \tag{16-78}$$

For example, in a spherical system, we have

$$\overline{E}_t = E_{t\theta}\hat{\theta} + E_{t\phi}\hat{\phi}. \tag{16-79}$$

This wave, $\overline{E}_t$, is incident on the target. The scattered wave, $\overline{E}_s$, at the receiver is then given by

$$[E_s] = \frac{e^{ikR_2}}{R_2}[F][E_t], \tag{16-80}$$

where we used matrix notation:

$$[E_s] = \begin{bmatrix} E_{s1} \\ E_{s2} \end{bmatrix}, \qquad [F] = \begin{bmatrix} f_{11} & f_{12} \\ f_{21} & f_{22} \end{bmatrix}, \qquad [E_t] = \begin{bmatrix} E_{t1} \\ E_{t2} \end{bmatrix}. \tag{16-81}$$

Now consider the received power P_r when $\overline{E}_s$ is incident on the receiving antenna. This problem has been investigated in detail. (See Collin and Zucker, 1969, Chap. 4; Lo and Lee, 1988, Chap. 6). First, it has been shown that when the wave $\overline{E}_s$ is incident on the receiving antenna, the open-circuit voltage V_0 is given by

$$V_0 = \overline{h} \cdot \overline{E}_s, \tag{16-82}$$

where $\overline{h}$ is called the *complex effective height* of the antenna (Fig. 16-10). It is given by the radiation field $\overline{E}_r$ when the receiver is used as a transmitter. If the receiver is fed by the current $\overline{I}_r$, the far field is given by (9-26) of Section 9-2:

$$\overline{E}_r = -\frac{j\omega\mu_0}{4\pi R} e^{-jkR}\,\overline{N}. \tag{16-83}$$

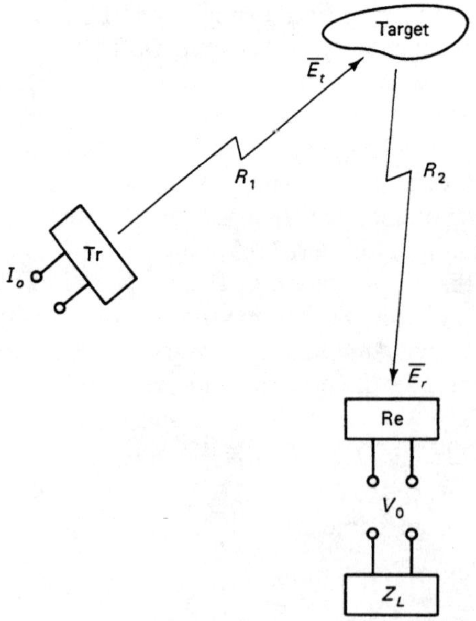

Figure 16-10 Transmitter, target, and receiver.

The complex effective height $\bar{h}$ is defined by

$$\bar{h} = \frac{\bar{N}}{I_r}. \tag{16-84}$$

If the input impedance of the antenna is Z_i and the load impedance is $Z_l = R_l + iX_l$, the received power P_r is given by

$$P_r = \frac{1}{2}\frac{|V_0|^2 R_l}{|Z_i + Z_l|^2}. \tag{16-85}$$

Considering that the maximum received power is obtained when $Z_l = Z_i^*$, we write

$$\frac{R_l}{|Z_i + Z_l|^2} = \frac{q}{4R_i}, \qquad q = 1 - \left|\frac{Z_l - Z_i^*}{Z_l + Z_i}\right|^2 \tag{16-86}$$

We then write

$$P_r = A_r S_r, \qquad S_r = \frac{|E_s|^2}{2\eta}, \tag{16-87}$$

where S_r is the power flux density and the receiving cross section A_r is given by

$$A_r = \frac{\eta}{4R_i}\frac{|\bar{h} \cdot \bar{E}_s|^2}{|\bar{E}_s|^2} q. \tag{16-88}$$

If the impedance is matched, $q = 1$, and if the polarization is matched, $|\bar{h} \cdot \bar{E}_s|^2$ takes its maximum value $|\bar{h}|^2 |\bar{E}_s|^2$.

If both the impedance and polarization are matched, the receiving cross section is equal to $(\lambda^2/4\pi)G_r$, where G_r is the gain of the receiving antenna. Therefore, we write

$$A_r = \frac{\lambda^2}{4\pi} G_r pq, \tag{16-89}$$

where

$$p = \frac{|\bar{h} \cdot \bar{E}_s|^2}{|\bar{h}|^2 |\bar{E}_s|^2} = |\bar{h}_n \cdot \bar{E}_{sn}|^2;$$

$\bar{h}_n$ and E_{sn} are normalized such that $|\bar{h}_n| = 1$ and $|\bar{E}_{sn}| = 1$. Combining (16-77), (16-80), and (16-89), we get

$$\frac{P_r}{P_t} = \frac{\lambda^2 G_r qp}{4\pi}\frac{|\bar{E}_s|^2}{2\eta}, \tag{16-90}$$

where

$$\bar{E}_s = \frac{1}{R^2}\bar{\bar{F}} \cdot \bar{E}_t.$$

This can be rewritten in the following form of radar equation:

$$\frac{P_r}{P_t} = \frac{\lambda^2 G_r G_t q}{(4\pi)^2 R_1^2 R_2^2} |\bar{h}_n \cdot \bar{E}_{sn}|^2 |\bar{\bar{F}} \cdot \bar{E}_{tn}|^2, \tag{16-91}$$

where $\bar{h}_n$, $\bar{E}_{sn}$, and $\bar{E}_{tn}$ are all normalized so that $|\bar{h}_n|^2 = 1$, $|\bar{E}_{sn}|^2 = 1$, and $|\bar{E}_{tn}|^2 = 1$. In matrix notation, we write

$$V_{0n} = \bar{h}_n \cdot \bar{E}_{sn} = [h_n]^t [E_{sn}]$$
$$V = \bar{\bar{F}} \cdot \bar{E}_{tn} = [F][E_{tn}]. \tag{16-92}$$

The sense of polarization used in V_{0n} can be confusing, and therefore it is important to test them for known physical problems. First, if LHC is transmitted to a specular reflector such as a conducting plate, the scattered wave will be RHC propagating toward the receiver. Then the received voltage is null. For this case

$$\bar{h}_n = \frac{1}{\sqrt{2}} (\hat{x} - i\hat{y}),$$
$$\bar{E}_{sn} = \frac{1}{\sqrt{2}} (\hat{x} - i\hat{y}).$$

Therefore, $V_{0n} = 0$.

Another example is two identical helical antennas facing each other. Then if LHC is transmitted, $\bar{E}_{sn}$ at the receiver is

$$\bar{E}_{sn} = \frac{1}{\sqrt{2}} (\hat{x} + i\hat{y}).$$

The complex effective height for the identical antennas is then

$$\bar{h}_n = \frac{1}{\sqrt{2}} (\hat{x} - i\hat{y}).$$

Thus $V_{0n} = 1$ and they are polarization matched.

16-8 OPTIMIZATION OF POLARIZATION

Let us now consider the optimization problem. We attempt to find the polarization of the transmitter and the polarization of the receiver such that the received power is maximum. This will be done in three stages (Kostinski and Boerner, 1986). First, we attempt to find the polarization of the transmitter to maximize $|V|^2$ in (16-92). We have, using matrix notation,

$$|V|^2 = V^*V = [E_{tn}]^+ [F]^+ [F][E_{tn}]$$
$$= [E_{tn}]^+ [G][E_{tn}], \tag{16-93}$$

where + means "adjoint" (complex conjugate of transpose). $[G] = [F]^+[F]$ is called *Graves power matrix* and is Hermitian (see the Appendix to Chapter 8, Section A):

$$[G]^+ = [G]. \tag{16-94}$$

To maximize $|V|^2$, consider the eigenvalue equation

$$[G][X] = \lambda[X] \tag{16-95}$$

where λ is the eigenvalue and $[X]$ is normalized so that $[X]^+[X] = 1$. Now we multiply (16-95) from the left by $[X]^+$. We then get

$$[X]^+[G][X] = \lambda[X]^+[X] = \lambda. \tag{16-96}$$

Therefore, the maximum value of $[X]^+[G][X]$ is given by the maximum eigenvalue λ. Furthermore,

$$\lambda^* = \{[X]^+[G][X]\}^+ = [X]^+[G]^+[X] = [X]^+[G][X] = \lambda. \tag{16-97}$$

Therefore, the eigenvalue λ is real. The eigenvalue λ is easily found from (16-95):

$$\begin{vmatrix} g_{11} - \lambda & g_{12} \\ g_{21} & g_{22} - \lambda \end{vmatrix} = 0. \tag{16-98}$$

The polarization of the transmitted wave is therefore given by (16-95)

$$[E_{tn}] = \frac{1}{[1 + |a|^2]^{1/2}} \begin{bmatrix} 1 \\ a \end{bmatrix}$$
$$a = \frac{\lambda_1 - g_{11}}{g_{12}}, \qquad \lambda_1 > \lambda_2. \tag{16-99}$$

The second stage consists of calculating $[E_s] = [F][E_t]$ using the optimum polarization for $[E_t]$ obtained in (16-99), and in the third stage we adjust the receiver polarization state $\bar{h}_n$ to maximize $|V_{0n}|^2 = |\bar{h}_n \cdot \bar{E}_{sn}|^2$. This optimum polarizations state is given by (Collin and Zucker, 1969, p. 108)

$$\bar{h}_n = \bar{E}_{sn}^*. \tag{16-100}$$

16-9 STOKES VECTOR RADAR EQUATION AND POLARIZATION SIGNATURE

Let us reformulate the radar equation developed in Section 16-7 using Stokes vector' formulation. First we write transmitted flux density $[S_i]$ at the object using the normalized Stokes vector I_{tn}:

$$[S_i] = \frac{G_t P_t}{4\pi R_1^2} [I_{tn}], \tag{16-101}$$

where

$$[I_{tn}] = \begin{bmatrix} I_{tn1} \\ I_{tn2} \\ U_{tn} \\ V_{tn} \end{bmatrix}$$

and $[I_t]$ is normalized so that $I_{tn1} + I_{tn2} = 1$. This is incident on the object and the scattered Stokes vector $[I_s]$ is given by

$$[I_s] = [M][S_i], \tag{16-102}$$

where $[M]$ is 4×4 Mueller matrix (Section 10-10). Now consider the received power P_r. Since this is proportional to $|\bar{h} \cdot \bar{E}_s|^2$, we first express $|\bar{h} \cdot \bar{E}_s|^2$ using the Stokes vector. $\bar{h} = h_1\hat{e}_1 + h_2\hat{e}_2$ is the transmitted wave and $\bar{E}_s = E_{s1}\hat{e}_1 + E_{s2}(-\hat{e}_2)$ is the incoming wave (Fig. 16-11). Note that $\bar{h}$ is directed to $\hat{e}_1 \times \hat{e}_2$, but $\bar{E}_s$ is directed to $\hat{e}_1 \times (-\hat{e}_2)$. Therefore, we get

$$\begin{aligned} |\bar{h} \cdot \bar{E}_s|^2 &= |h_1 E_{s1} - h_2 E_{s2}|^2 \\ &= I_{h1} I_{s1} + I_{h2} I_{s2} - \tfrac{1}{2} U_h U_s + \tfrac{1}{2} V_h V_s, \end{aligned} \tag{16-103}$$

where $[I_h]$ and $[I_s]$ are the Stokes vectors for $\bar{h}$ and $\bar{E}_s$, respectively.

$$[I_h] = \begin{bmatrix} I_{h1} \\ I_{h2} \\ U_h \\ V_h \end{bmatrix}, \qquad [I_s] = \begin{bmatrix} I_{s1} \\ I_{s2} \\ U_s \\ V_s \end{bmatrix}.$$

We therefore get the Stokes vector radar equation:

$$\frac{P_r}{P_t} = \frac{\lambda^2 G_r G_t q}{(4\pi)^2 R_1^2 R_2^2} [\widetilde{h_n}][I_s], \tag{16-104}$$

where $[\widetilde{h_n}]$ is the normalized effective height Stokes vector defined by,

$$[\widetilde{h_n}] = \frac{1}{I_{h1} + I_{h2}} \left(I_{h1}, I_{h2}, -\frac{U_h}{2}, \frac{V_h}{2} \right),$$

$$[I_s] = [M][I_{tn}].$$

If a single antenna is used as both transmitter and receiver, we can express the transmitting Stokes vector using the orientation angle ψ and the ellipticity angle χ (see Section 10-8):

$$\begin{aligned} [I_{tn}] &= \begin{bmatrix} \frac{1}{2}(1 + \cos 2\chi \cos 2\psi) \\ \frac{1}{2}(1 - \cos 2\chi \cos 2\psi) \\ \cos 2\chi \sin 2\psi \\ \sin 2\chi \end{bmatrix}, \\ [h_n] &= \begin{bmatrix} \frac{1}{2}(1 + \cos 2\chi \cos 2\psi) \\ \frac{1}{2}(1 - \cos 2\chi \cos 2\psi) \\ -\frac{1}{2}\cos 2\chi \sin 2\psi \\ \frac{1}{2}\sin 2\chi \end{bmatrix}. \end{aligned} \tag{16-105}$$

The quantity $P_s = [\widetilde{h_n}][M][I_{tn}]$ is called the *polarization signature* and is displayed as a function of the ellipticity angle χ and the orientation angle ψ. It is used

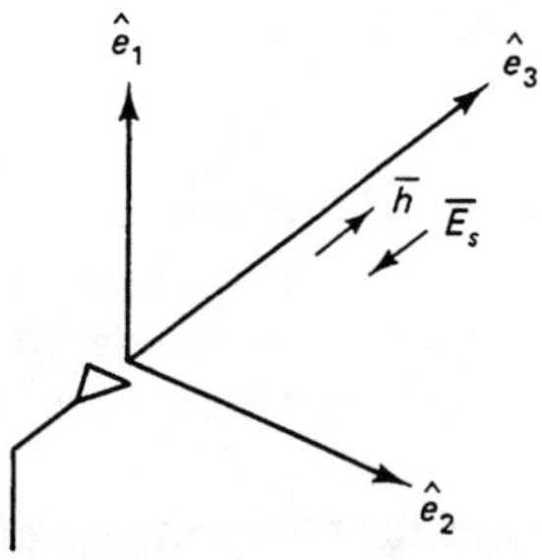

Figure 16-11 $\bar{h}$ is directed to $\hat{e}_1 \times \hat{e}_2$, while $\bar{E}_s$ is directed to $\hat{e}_1 \times (-\hat{e}_2)$.

for identification of scattering mechanisms of the objects, terrain, vegetation, and so on, in imaging radar (Fig. 16-12). As a simple example, consider RHC wave transmitted toward a specular reflector; then

$$[h_n] = \begin{bmatrix} \frac{1}{2} \\ \frac{1}{2} \\ 0 \\ -\frac{1}{2} \end{bmatrix}, \qquad [I_{tn}] = \begin{bmatrix} \frac{1}{2} \\ \frac{1}{2} \\ 0 \\ 1 \end{bmatrix},$$

Therefore, we get $P_s = 0$.

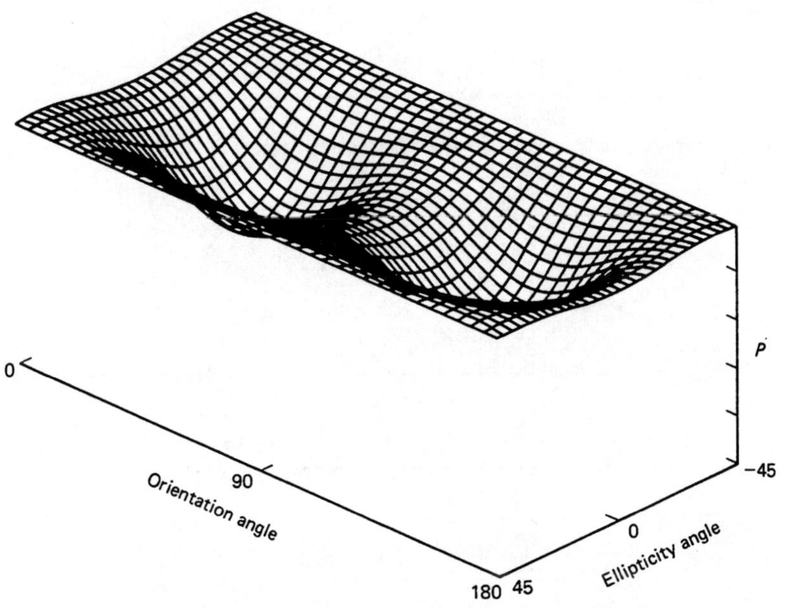

Figure 16-12 Example of polarization signature.

16-10 MEASUREMENT OF STOKES PARAMETER

Stokes parameter (I_1, I_2, U, V) can be obtained by measuring the amplitude and phase of E_1 and E_2 and computing $I_1 = \langle |E_1|^2 \rangle$, $I_2 = \langle |E_2|^2 \rangle$, $U = 2\,\mathrm{Re}\langle E_1 E_2^* \rangle$, and $V = 2\,\mathrm{IM}\langle E_1 E_2^* \rangle$. This is called *coherent* measurement, as it involves the measurement of the phase. When E_1 and E_2 are randomly varying in time, the coherent measurement must be done within a fraction of the coherent time and therefore requires fast and accurate measurement.

Incoherent measurement uses the power measurement and therefore need not be fast; however, it may be more susceptible to noise. I_1 and I_2 can be obtained directly by measuring the power in the x and y components (Fig. 16-13). Next measure the power for the component at 45°. Then we get

$$\begin{aligned} P_{45} &= \langle |E_{x'}|^2 \rangle \\ &= \frac{1}{2}[\langle |E_x|^2 \rangle + \langle |E_y|^2 \rangle + 2\,\mathrm{Re}\langle E_x E_y^* \rangle] \\ &= \frac{1}{2}[I_1 + I_2 + U] \end{aligned} \tag{16-106}$$

where $E_{x'} = 1/\sqrt{2}\,[E_x + E_y]$. Similarly for 135°, we get

$$P_{135} = \tfrac{1}{2}[I_1 + I_2 - U]. \tag{16-107}$$

From these two, we get

$$\frac{U}{I} = \frac{P_{45} - P_{135}}{P_{45} + P_{135}}, \qquad I = I_1 + I_2. \tag{16-108}$$

The denominator is to ensure the normalization.

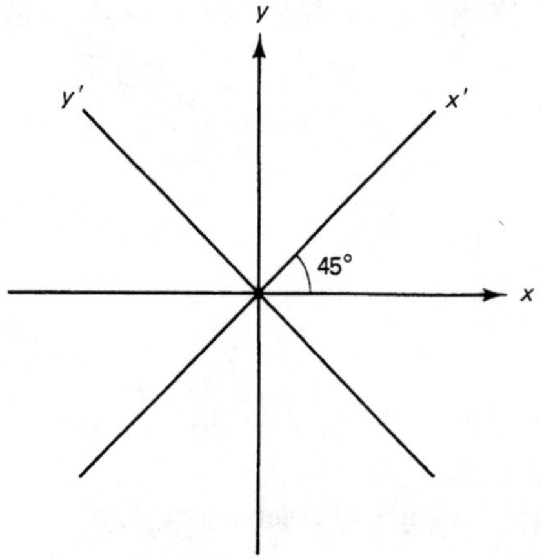

Figure 16-13 Measurement of Stokes parameters.

Next we use the receiver that accepts the RHC wave. This has $\bar{h}_n = 1/\sqrt{2}\,(\hat{x} + i\hat{y})$ and therefore the power measured with this antenna is

$$\begin{aligned} P_R &= \langle |\bar{h}_n \cdot \bar{E}_s|^2 \rangle \\ &= \langle \tfrac{1}{2} |E_x + iE_y|^2 \rangle \\ &= \tfrac{1}{2}[I_1 + I_2 + V]. \end{aligned} \tag{16-109}$$

Similarly for the receiver that accepts the LHC wave, the power measured is

$$P_L = \tfrac{1}{2}[I_1 + I_2 - V]. \tag{16-110}$$

Thus we get

$$\frac{V}{I} = \frac{P_R - P_L}{P_R + P_L}. \tag{16-111}$$

PROBLEMS

16-1. Show that if the object is a delta function $f(x, y) = \delta(x)\delta(y)$, the reconstructed image is $[W/(2\pi r)]J_1(W\gamma)$.

16-2. Prove the three-dimensional inverse Radon transform shown in (16-21) and (16-22).

16-3. Show that if the object is a delta function, the reconstructed image using the diffraction tomography is given by (16-43).

16-4. If the target is a sphere of radius a, find the complex scattering amplitude $\Gamma(\bar{K})$ and the characteristic function $\gamma(\bar{r})$ discussed in Section 16-4.

16-5. Derive the holographic inverse source solution for the one-dimensional source distribution $\rho(x)$.

16-6. Assume that the plasma frequency is given by $\omega_p^2 = A(z - z_0)^2$ for $z > z_0$ and $\omega_p = 0$ for $z < z_0$. Find the transit time $T(\omega)$ in (16-70) and the height h where $\omega = \omega_p(h)$. Show that $T(\omega)$ and the height satisfy (16-25).

16-7. A left-handed helical antenna is transmitting an LHC wave, which is normally incident on a conducting plate. The scattered wave is then received by the same helical antenna. Find $\bar{h}_n$, $\bar{E}_{tn}$, $\bar{E}_{sn}$, V_{0n}, and V given in Section 16-7. If the scattered wave is received by a right-handed helical antenna, find the power received.

16-8. Assume that the scattering matrix $[F]$ is given by

$$[F] = \begin{bmatrix} 2j & \frac{1}{2} \\ \frac{1}{2} & j \end{bmatrix}.$$

Find the eigenvalue λ and the optimum polarization $[E_{tn}]$ of the transmitting wave. Next find the complex effective height of the receiver to maximize the power received.

16-9. Assume that a target is a corner reflector that has the scattering matrix

$$[F] = \begin{bmatrix} 1 & 0 \\ 0 & -1 \end{bmatrix}.$$

Find the Mueller matrix and the polarization signature.

17

Radiometry, Noise Temperature, and Interferometry

In this chapter we first discuss radiometry, which is the passive detection of natural radiation from various media, targets, and objects. Included in this discussion are brightness, antenna temperature, radiative transfer, and emissivity. The effects of the receiving system on the system noise temperature and the minimum detectable temperature are discussed, and the chapter concludes with the use of interferometry for mapping the brightness distribution (see Kraus, 1966; Skolnik, 1970, 1980; Brookner, 1977; King, 1970; Ulaby et al., 1981; and Tsang et al., 1985).

17-1 RADIOMETRY

All natural and man-made objects, terrain, and atmospheric media emit electromagnetic energy. Thermal emission is generally dominant. They also scatter the radiation incident on them. A radiometer is a very sensitive, low-noise receiver that detects natural incoherent radiation from these objects. Typical radiometers monitor broadband continuous radiation and the received power is proportional to the bandwidth of the receiver. Radiometers are used on earth-orbiting satellites and on the ground to probe atmospheric conditions and terrain and to detect targets at microwave, millimeter wave, and infrared frequencies. They are also used for medical applications to probe radiation from biological media.

17-2 BRIGHTNESS AND FLUX DENSITY

The fundamental quantity in radiometry is the brightness B. Consider a small area da and the power flux density incident on this area from the direction $\hat{s}$ within a unit frequency band centered at frequency ν within a unit solid angle (Fig. 17-1). This quantity $B(\hat{s})$ is called the *brightness* and is measured in W m^{-2} Hz^{-1} sr^{-1} (sr = steradian = unit solid angle). The brightness B is also called the *specific intensity* in radiative transfer theory and *radiance* in infrared radiometry. The amount of power dP flowing through the area da within the solid angle $d\Omega$ in a frequency interval $(\nu, \nu + d\nu)$ is therefore given by

$$dP = B \cos\theta \, da \, d\Omega \, d\nu. \tag{17-1}$$

At a given location, the brightness B is a function of the direction $\hat{s}$, and therefore we may call $B(\hat{s})$ the *brightness distribution*. The variation of the brightness B with the frequency is called the *brightness spectrum*.

The brightness B of the solar disk at $\lambda = 0.5$ μm is 1.33×10^{-12} and the brightness of rough ground at $\lambda = 3.9$ cm is 5.4×10^{-24}. Coherent sources such as lasers and radars present a completely different situation. For example, the brightness of an argon–ion laser at $\lambda = 0.5145$ μm is 7.1×10^{3}, and the brightness of a Haystack X-band radar at $\lambda =$ 3.9 cm is 4.8×10^{3} (Skolnik, 1970).

Let us consider a receiving antenna whose receiving cross section is given by $A_r(\theta, \phi)$. The received power P_r in W/Hz is then given by (Fig. 17-2)

$$P_r = \frac{1}{2}\int_{4\pi} B(\theta, \phi) A_r(\theta, \phi) \, d\Omega, \tag{17-2}$$

where the integral is over all solid angles, and the factor $\frac{1}{2}$ is introduced because normally the brightness radiation is incoherent and unpolarized while any antenna receives only one polarization. In general, however, if the brightness is partially polarized, the factor can be between 0 and 1.

If we normalize the receiving cross section A_r to the maximum value A_{rm}, we write

$$A_r(\theta, \phi) = A_{rm} P_n(\theta, \phi). \tag{17-3}$$

We then get

$$P_r = \frac{1}{2} A_{rm} S, \tag{17-4}$$

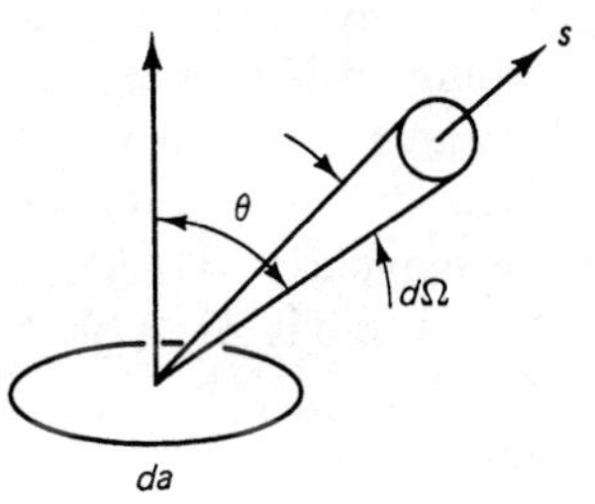

Figure 17-1 Brightness $B(\hat{s})$ and the received power dP.

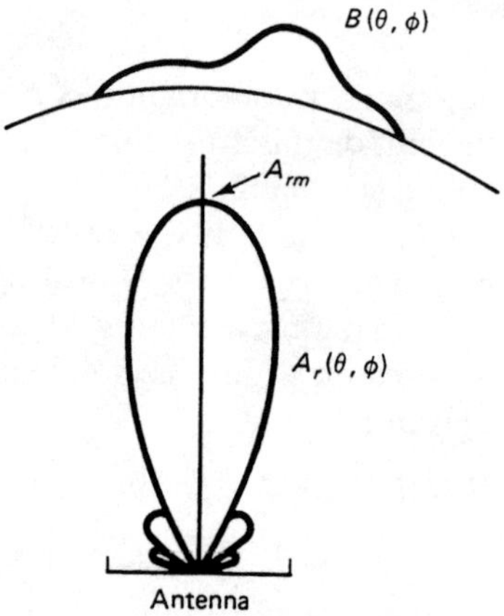

Figure 17-2 Receiving cross section and brightness.

where $S\,(\mathrm{W\,m^{-2}\,Hz^{-1}})$ is called the *observed flux density*.

$$S = \int_{4\pi} B(\theta, \phi)\, P_n(\theta, \phi)\, d\Omega. \tag{17-5}$$

Note also that the maximum receiving cross section A_{rm} and the maximum gain G_m are related by

$$A_{rm} = \frac{\lambda^2}{4\pi} G_m. \tag{17-6}$$

The total flux density from the source is given by

$$S_s = \int_{4\pi} B(\theta, \phi)\, d\Omega, \tag{17-7}$$

and is called the *source flux density*. The unit for flux density is $\mathrm{W\,m^{-2}\,H_z^{-1}}$; in radio astronomy this is called 1 jan $(= 1\ \mathrm{w\,m^{-2}\,H_z^{-1}})$ after the pioneer radio astronomer K. G. Jansky. The flux density of most radio sources in radio astronomy is on the order of 10^{-26} jan.

17-3 BLACKBODY RADIATION AND ANTENNA TEMPERATURE

All objects emit electromagnetic energy. They may also absorb and scatter the energy incident on them. According to Kirchhoff, a good absorber of electromagnetic energy is also a good emitter. A perfect absorber that absorbs electromagnetic energy at all wavelengths is called a *blackbody* and is also a perfect emitter.

The brightness of the electromagnetic radiation from a blackbody depends only on its temperature and frequency and is given by Planck's radiation law:

$$B(\text{blackbody}) = \frac{2h\nu^3}{c^2} \frac{1}{\exp(h\nu/KT) - 1}, \tag{17-8}$$

where h is Planck's constant (6.63×10^{-34} J·s), ν the frequency (Hz), c the velocity of light (3×10^8 m s^{-1}), K is Boltzmann's constant (1.38×10^{-23} J/K) and T is the temperature (K).

For microwave and millimeter waves, $h\nu$ is much smaller than KT; only in infrared and shorter wavelengths does $h\nu$ become comparable to or greater than KT, exhibiting quantum effects. For microwave and millimeter waves, therefore, we can approximate $\exp(h\nu/KT)$ by $1 + h\nu/KT$ and obtain the *Rayleigh–Jeans law:*

$$B(\text{blackbody}) = \frac{2K}{\lambda^2} T. \tag{17-9}$$

Note that the brightness is proportional to the temperature (Fig. 17-3).

For an actual object that is not a blackbody, the brightness is not proportional to its actual temperature as given in (17-9). However, we can define the equivalent blackbody temperature T_s which gives the brightness identical to the actual brightness B:

$$B = \frac{2K}{\lambda^2} T_s. \tag{17-10}$$

This equivalent temperature T_s is called the *source temperature.* Since B is a function of the direction $B(\theta, \phi)$, the source temperature is also a function of the direction $T_s(\theta, \phi)$.

The received power P_r in (17-2) is then given by

$$P_r = \frac{K}{\lambda^2} \int T_s(\theta, \phi) A_r(\theta, \phi)\, d\Omega. \tag{17-11}$$

Nothing that

$$\int A_r(\theta, \phi)\, d\Omega = \lambda^2, \tag{17-12}$$

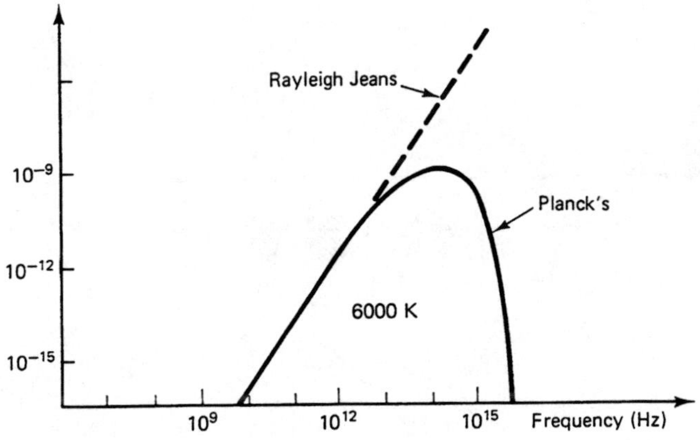

Figure 17-3 Planck's law and Rayleigh–Jean's law.

as shown in Section 9-1, we get

$$P_r = KT_A, \tag{17-13}$$

where K is Boltzmann's constant and

$$T_A = \frac{\int T_s(\theta, \phi)\, P_n(\theta, \phi)\, d\Omega}{\int P_n(\theta, \phi)\, d\Omega}.$$

This quantity T_A is called the *antenna temperature*. As an example, if the source is uniform over a small solid angle Ω_s and the antenna receiving pattern is confined within a solid angle Ω_A, we have

$$T_A = \begin{cases} \dfrac{T_s \Omega_s}{\Omega_A} & \text{if } \Omega_s < \Omega_A \\ T_s & \text{if } \Omega_s > \Omega_A. \end{cases} \tag{17-14}$$

This ratio Ω_s/Ω_A is called the *fill factor*.

The antenna temperature T_A is also equal to the temperature of a resistor that produces the same noise power as the actual power P_r (Fig. 17-4). According to Nyquist's formula, the open-circuit rms noise voltage (Johnson noise or thermal noise) across a resistor R at temperature T in the frequency band $d\nu$ is given by

$$V = (4RKT\, d\nu)^{1/2}. \tag{17-15}$$

The available power W per unit frequency band produced by this resistor when it is transmitted to a matched load is $\frac{1}{2}(V^2/2R)$, and therefore

$$W = KT. \tag{17-16}$$

To show that this is equal to the antenna temperature, note that the antenna temperature is also equal to the temperature of a blackbody enclosure in which the

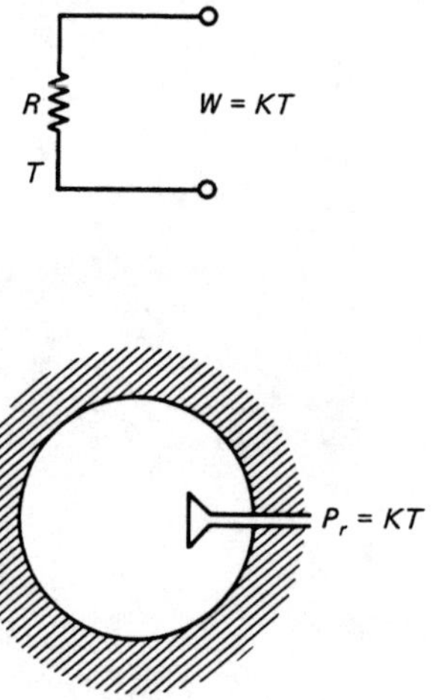

Figure 17-4 Antenna temperature.

antenna is embedded (Fig. 17-4). In this case T_s is constant, and therefore, from (17-13)

$$P_r = KT. \tag{17-17}$$

17-4 EQUATION OF RADIATIVE TRANSFER

In Section 17-3 we discussed the source temperature and the antenna temperature. The source temperature may originate outside the atmosphere, such as the sun. Also, the source temperature may come from the atmosphere itself or the ground. This section and the next section deal with the emission, propagation, absorption, and scattering of the temperature in the atmosphere and the ground.

Let us consider the propagation of the brightness in a medium emitting electromagnetic radiation. An example is the brightness in the atmosphere. Since the brightness B is proportional to the equivalent temperature T, we may use the temperature in place of the brightness. As the brightness B (or T) propagates in the direction $\hat{s}$ through the medium, it is partly absorbed and partly scattered (Fig. 17-5). We therefore write

$$\frac{dT}{ds} = -\gamma T, \tag{17-18}$$

where

$$\begin{aligned} \gamma &= \text{extinction coefficient} \\ &= \gamma_a + \gamma_s, \\ \gamma_a &= \text{absorption coefficient}, \\ \gamma_s &= \text{scattering coefficient}. \end{aligned}$$

The brightness is also increased by the scattering into the direction $\hat{s}$ from the brightness coming from all other directions:

$$\frac{1}{4\pi}\int_{4\pi} p(\hat{s}, \hat{s}')\, T(\hat{s}')\, d\Omega',$$

where $p(\hat{s}, \hat{s}')$ is called the *phase function* (Ishimaru, 1978). The brightness is also increased by the emission from the medium with temperature T_m. The emission is

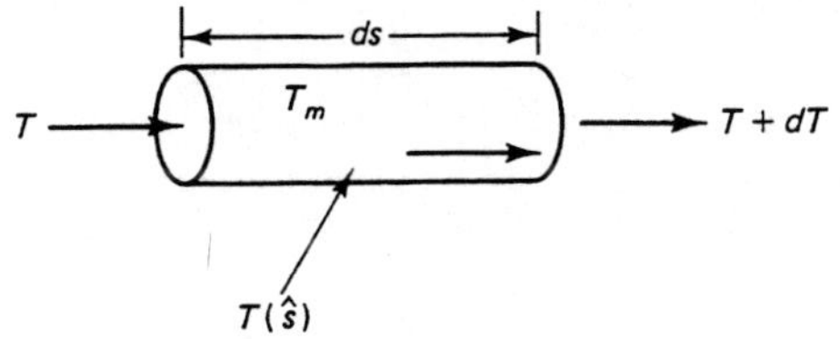

Figure 17-5 Equation of radiative transfer.

equal to the absorption according to Kirchhoff's law; therefore, we have $\gamma_a T_m$. Collecting all these terms, we get the equation of radiative transfer (Fig. 17-5):

$$\frac{dT}{ds} = -\gamma T + \frac{1}{4\pi}\int p(\hat{s},\hat{s}')\,T(\hat{s}')\,d\Omega' + \gamma_a T_m. \tag{17-19}$$

In most microwave radiometry, the scattering effect is negligible compared with the absorption, and therefore we write

$$\frac{dT}{ds} = -\gamma T + \gamma_a T_m, \tag{17-20}$$

where $\gamma \approx \gamma_a$.

Let us consider a ground-based radiometer pointed upward (Fig. 17-6). The absorption coefficient of the atmosphere, clouds, or rain at height z is given by $\gamma_a(z)$ and their temperature by $T_m(z)$. The temperature of the external source such as the sun is T_e. The radiometer measures a temperature given by the solution of (17-20):

$$T = \int_0^\infty \gamma_a(z) T_m(z) \exp\left[-\int_0^z \gamma(z')\sec\theta\, dz'\right]\sec\theta\, dz + T_e \exp\left[-\int_0^\infty \gamma(z')\sec\theta\, dz'\right]. \tag{17-21}$$

Note also that if $T_e = 0$ and T_m, γ_a, and γ are uniform over the height H, we get

$$T = T_m \frac{\gamma_a}{\gamma}[1 - \exp(-\gamma H \sec\theta)]. \tag{17-22}$$

For example, rain may be approximated by $T_m = 273°$, $H \approx 3$ km, and γ_a = absorption coefficient of rain.

The total attenuation τ over a distance l due to the scattering and absorption is called the *optical depth:*

$$\tau = \int_0^l \gamma\, ds. \tag{17-23}$$

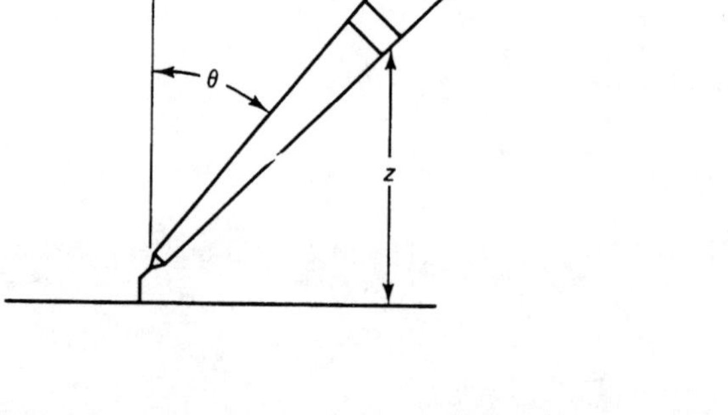

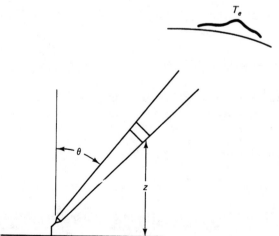

Figure 17-6 Ground-based radiometer.

The path loss L is defined as

$$L = \exp(\tau). \tag{17-24}$$

17-5 SCATTERING CROSS SECTIONS AND ABSORPTIVITY AND EMISSIVITY OF A SURFACE

All surfaces absorb, scatter, and emit electromagnetic radiation. Let us consider the plane wave $\overline{E}^i$ incident on a surface in the direction $\hat{i}$ and the wave $\overline{E}^s$ scattered in the direction $\hat{s}$ at a large distance R from the surface (Fig. 17-7). The scattering cross section per unit area of the surface is given by

$$\sigma^o_{\beta\alpha}(\hat{s}, \hat{i}) = \lim_{R\to\infty} \frac{4\pi|\overline{E}^s_\beta|^2 R^2}{|\overline{E}^i_\alpha|^2 A}, \tag{17-25}$$

where α and β represent the polarization states of the incident and scattered wave respectively, and α and β can be v or h for vertical and horizontal polarizations. Thus we have σ^o_{vv}, σ^o_{hh}, σ^o_{vh}, and σ^o_{hv}.

The scattering coefficient $\gamma_{\beta\alpha}$ is defined in terms of the projected area of the incident wave:

$$\gamma_{\beta\alpha}(\hat{s}, \hat{i}) = \lim_{R\to\infty} \frac{4\pi|\overline{E}^s_\beta|^2 R^2}{\cos\theta_i |\overline{E}^i_\alpha|^2 A}. \tag{17-26}$$

Thus we have

$$\sigma^o_{\beta\alpha}(\hat{s}, \hat{i}) = \cos\theta_i \gamma_{\beta\alpha}(\hat{s}, \hat{i}). \tag{17-27}$$

From the reciprocity theorem, we get

$$\sigma^o_{\beta\alpha}(\hat{s}, \hat{i}) = \sigma^o_{\alpha\beta}(\hat{i}, \hat{s}). \tag{17-28}$$

When a wave with the polarization state α is incident on a surface in the

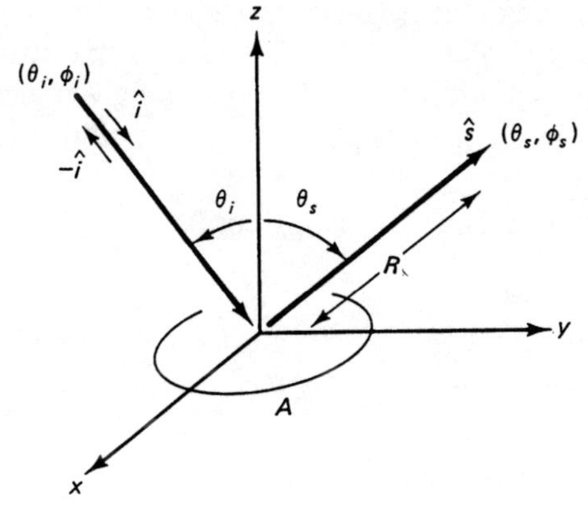

Figure 17-7 Scattering from surface A.

direction $\hat{\imath}$, the total scattered power when a unit power flux density is incident in the projected area $A \cos\theta_i$ is given by (Fig. 17-7)

$$\begin{aligned}\Gamma_\alpha(\hat{\imath}) &= \frac{1}{4\pi}\int_{2\pi}[\gamma_{\beta\alpha}(\hat{s},\hat{\imath}) + \gamma_{\alpha\alpha}(\hat{s},\hat{\imath})]\,d\Omega_s \\ &= \frac{1}{4\pi\cos\theta_i}\int_{2\pi}[\sigma^o_{\beta\alpha}(\hat{s},\hat{\imath}) + \sigma^o_{\alpha\alpha}(\hat{s},\hat{\imath})]\,d\Omega_s,\end{aligned} \tag{17-29}$$

where the scattered power in both α and β polarizations are included. The integration is over the upper hemisphere and 4π comes from the definition (17-26). This fraction $\Gamma_\alpha(\hat{\imath})$, called the *albedo,* is the ratio of the total scattered power to the incident power in the projected area $A\cos\theta_i$. The fractional power absorbed by the surface is therefore

$$a_\alpha(\hat{\imath}) = 1 - \Gamma_\alpha(\hat{\imath}). \tag{17-30}$$

This quantity $a_\alpha(\hat{\imath})$ is called the *absorptivity*.

Let us next consider the brightness or the temperature $T_\alpha(\hat{\imath})$ emitted in the direction $(-\hat{\imath})$ in the polarization state α from a surface kept at the temperature T (Fig. 17-7). If the surface is a blackbody, the source temperature should be T. The ratio $\epsilon_\alpha(\hat{\imath})$ of the actual emission $T_\alpha(\hat{\imath})$ to that of a blackbody at temperature T is called the *emissivity*.

$$T_\alpha(\hat{\imath}) = \epsilon_\alpha(\hat{\imath})\,T. \tag{17-31}$$

According the Kirchhoff's law, if the surface is in thermal equilibrium, the absorption must be equal to the emission, and therefore the emissivity is equal to the absorptivity:

$$\epsilon_\alpha(\hat{\imath}) = a_\alpha(\hat{\imath}). \tag{17-32}$$

A rigorous proof of the Kirchhoff law (17-32) can be made by considering thermodynamic equilibrium when the surface is surrounded by a blackbody. (Tsang et al. 1985).

Equation (17-31) gives the emission from the surface at temperature T, and (17-29), (17-30), and (17-32) relate the emissivity to the scattering characteristics of the surface. For example, if the surface is perfectly rough and scatters the radiation in all directions, it is called a *Lambertian surface.* In this case the scattered power $|\bar{E}^s|^2$ is proportional to the projected surface area $A\cos\theta_s$ for a given incident flux in the projected area $A\cos\theta_i$ and $\gamma_{\beta\alpha}(\hat{s},\hat{\imath})$ is independent of the polarization. Thus we have

$$\sigma^o_{\beta\alpha}(\hat{s},\hat{\imath}) + \sigma^o_{\alpha\alpha}(\hat{s},\hat{\imath}) = \sigma^o_o \cos\theta_s \cos\theta_i. \tag{17-33}$$

Substituting this into (17-29), we get

$$\begin{aligned}\Gamma &= \frac{\sigma^o_o}{4} \\ \epsilon &= a = 1 - \frac{\sigma^o_o}{4}.\end{aligned} \tag{17-34}$$

If the surface is smooth, the scattered wave is in the specular direction and there is no cross polarization. The albedo $\Gamma_\alpha(\hat{i})$ is then given by the Fresnel reflection coefficient $R_\alpha(\hat{i})$:

$$\Gamma_\alpha(\hat{i}) = |R_\alpha(\hat{i})|^2, \tag{17-35}$$

where

$$R_v = \frac{n_1 \cos\theta_2 - n_2 \cos\theta_1}{n_1 \cos\theta_2 + n_2 \cos\theta_1} \quad \text{for vertical polarization,}$$

$$R_h = \frac{n_1 \cos\theta_1 - n_2 \cos\theta_2}{n_1 \cos\theta_1 + n_2 \cos\theta_2} \quad \text{for horizontal polarization,}$$

and the wave is incident from the medium with refractive index n_1 to the medium with n_2 at the angle of incidence θ_1 and $\cos\theta_2 = [1 - (n_1/n_2)^2 \sin^2\theta_1]^{1/2}$. If the incident wave is completely unpolarized, we have

$$\Gamma(\hat{i}) = \tfrac{1}{2}[|R_v|^2 + |R_h|^2]. \tag{17-36}$$

As an example, consider a radiometer looking down on the earth (Fig. 17-8). The temperature T at the radiometer consists of the atmospheric emission T_1, the contribution from the ground T_2, and the scattering by the ground T_3:

$$T = T_1 + T_2 + T_3. \tag{17-37}$$

The atmospheric emission T_1 is given by the emission from the medium with the temperature $T_m(z)$ attenuated through the atmosphere:

$$T_1 = \int_o^h \gamma_a(z) T_m(z) \exp\left[-\int_z^h \gamma(z') \sec\theta \, dz'\right] \sec\theta \, dz. \tag{17-38}$$

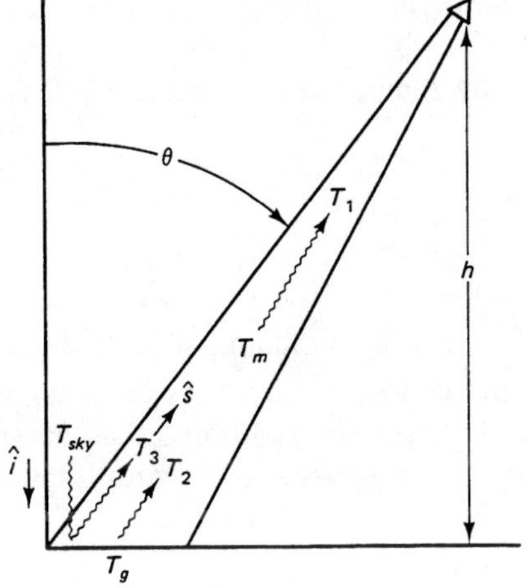

Figure 17-8 Radiometer looking down on the earth.

The contribution T_2 from the ground emission attenuated through the atmosphere is given by

$$T_2 = \epsilon T_g \exp\left[-\int_o^h \gamma(z') \sec\theta \, dz'\right], \tag{17-39}$$

where ϵ is the emissivity of the earth and T_g is the actual temperature of the ground. The emissivity depends on the angle. The approximate values of the emissivity normal to the surface are close to unity for grass, 0.93 for dry soil, 0.68 for moist soil, 0.85 for concrete, 0.45 for water, 0.9 for asphalt, and close to zero for metal.

The scattered component T_3 is the temperature scattered by the ground when the sky temperature T_{sky} is incident on the ground and then attenuated through the atmosphere. Thus we have

$$T_3 = T_{3s} \exp\left[-\int_o^h \gamma(z') \sec\theta \, dz'\right], \tag{17-40}$$

where T_{3s} is the temperature scattered by the ground when the sky temperature T_{sky} is incident. If the ground is a Lambertian surface, we have (Fig. 17-7)

$$\begin{aligned} T_{3s}(\hat{s}) &= \frac{1}{4\pi\cos\theta_s}\int_{2\pi} T_{sky}(\hat{\imath})\,\sigma_o^o \cos\theta_i \cos\theta_s \, d\Omega_i \\ &= \frac{\sigma_o^o}{4\pi}\int_{2\pi} T_{sky}(\hat{\imath}) \cos\theta_i \, d\Omega_i, \end{aligned} \tag{17-41a}$$

where the integration is over the hemisphere. If the sky temperature is uniform, we get

$$T_{3s} = \frac{\sigma_o^o}{4} T_{sky}, \tag{17-41b}$$

where $\sigma_o^o/4$ is the albedo of the surface. More generally, we have

$$T_\alpha(\hat{s}) = \frac{1}{4\pi\cos\theta_s}\int_{2\pi} [\sigma_{\alpha\beta}^o(\hat{s},\hat{\imath})\, T_\beta(\hat{\imath}) + \sigma_{\alpha\alpha}^o(\hat{s},\hat{\imath})\, T_\alpha(\hat{\imath})]\, d\Omega_i, \tag{17-42}$$

where $T_\alpha(\hat{\imath})$ and $T_\beta(\hat{\imath})$ are the sky temperature incident on the surface with the polarization state α and β.

17-6 SYSTEM TEMPERATURE

In the preceding sections we have discussed how the noise temperature is emitted, propagated, absorbed, and scattered, and then reaches the antenna. In this section we discuss what happens in the receiver. In addition to the antenna temperature given in (17-13), the receiver itself contributes to the total noise of the receiving system. Thus we have the system noise temperature T_{sys}.

$$T_{sys} = T_A + T_r, \tag{17-43}$$

where T_r is the receiver noise temperature. If the receiver consists of several cascaded stages with temperature and gain as shown in Fig. 17-9, the gain of the previous stages effectively reduces the contribution of each stage to the overall system temperature. Therefore, the system temperature is given by

$$T_{sys} = T_A + T_r,$$
$$T_r = T_1 + \frac{T_2}{G_1} + \frac{T_3}{G_1 G_2}. \tag{17-44}$$

Note that the total available noise power at the output is given by $T_{out} = G_1 G_2 T_{sys}$. If the first stage is a transmission line with loss L (gain $= \epsilon = L^{-1} < 1$) kept at a physical temperature of T_L, the effective noise temperature at the end of the transmission line is $T_L(1-\epsilon)$. This is seen by noting that the noise generated in dx is equal to the absorbed power $\alpha T_L\, dx$, where α is the attenuation constant. At the end of the line of length l, we have

$$\int_o^l \alpha T_L e^{-\alpha(l-x)}\, dx = T_L(1-\epsilon),$$

where $\epsilon = e^{-\alpha l}$. We therefore have the noise temperature $T_L(1-\epsilon) + T_r$ for the attenuator with loss L (gain $= \epsilon$) followed by the receiver. The total system temperature is then given by

$$\begin{aligned} T_{sys} &= T_A + \frac{1}{\epsilon}[T_L(1-\epsilon) + T_r] \\ &= T_A + T_L(L-1) + LT_r, \end{aligned} \tag{17-45}$$

where T_r is given in (17-44).

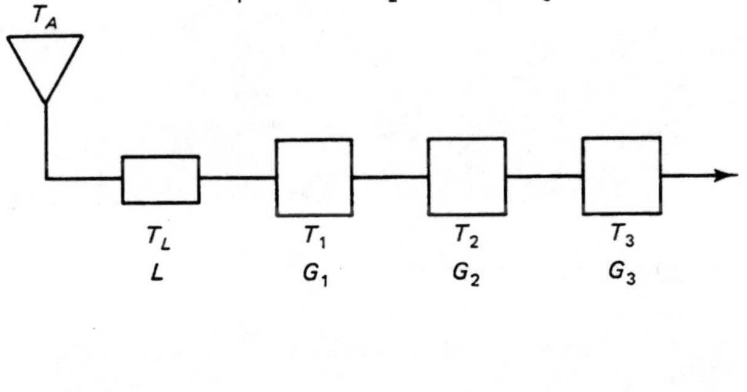

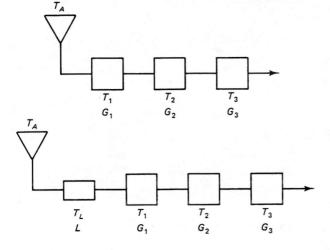

Figure 17-9 System temperature.

17-7 MINIMUM DETECTABLE TEMPERATURE

A receiver that measures the total noise power is called the *"total power receiver"* (Fig. 17-10). The detector is normally a square-law device and the output voltage is proportional to the output noise power. The incident power consists of the broadband system noise temperature T_{sys} and the signal noise temperature ΔT. If the bandwidth of the receiver is B, the output power for the system noise temperature is proportional to $(KT_{sys}B)^2$ and the output power for the signal is proportional to $(K\,\Delta T\,B)^2$. The system noise, however, is B effectively independent noise pulses per second. These pulses are averaged over the integration time τ, and therefore there are $B\tau$ independent pulses. These independent contributions are largely canceled by each other over this period, and thus the system noise contribution is reduced by $B\tau$.

$$W_{sys} \sim \frac{(KT_{sys}\,B)^2}{B\tau}. \tag{17-46}$$

This means that the system temperature is effectively reduced to $T_{sys}/(B\tau)^{1/2}$. The signal output W_s is

$$W_s \sim (K\,\Delta T\,B)^2. \tag{17-47}$$

The *sensitivity* or the *minimum detectable signal* ΔT_{min} is defined as ΔT, which produces the output W_s equal to W_{sys}. Thus we have

$$\Delta T_{min} = \frac{K_s T_{sys}}{\sqrt{B\tau}}, \tag{17-48}$$

where K_s is the sensitivity constant of the order of unity. For example, $K_s = 1$ for the total power receiver and $K_s = \pi/\sqrt{2}$ for a Dick receiver.

The total power receiver is affected by the variation of the system gain $\Delta G/G$. The *Dick receiver* switches at a constant rate between the receiving antenna and a reference load, thus canceling out gain instability (Kraus, 1966, Chap. 7; Skolnik, 1970).

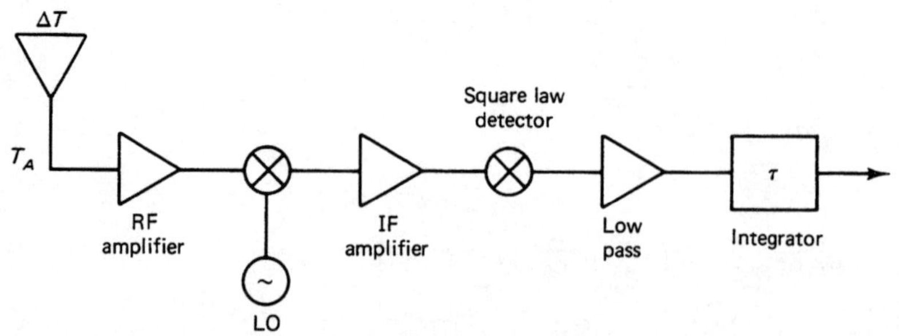

Figure 17-10 Total power receiver.

17-8 RADAR RANGE EQUATION

When the noise temperature is included in a radar equation, there is a maximum range in which a radar can detect a target. Radar equations give the received power P_r for a given transmitted power P_t (see Section 10-2):

$$\frac{P_r}{P_t} = \frac{\lambda^2}{(4\pi)^3} \frac{G^2}{R^4} \sigma. \tag{17-49}$$

The system noise power P_n for the system with temperature T_{sys} and bandwidth B is given by

$$P_n = KT_{sys} B. \tag{17-50}$$

The signal-to-noise ratio is therefore

$$\frac{S}{N} = \frac{P_r}{P_n}. \tag{17-51}$$

We therefore obtain the minimum detectable value of P_r

$$P_r = KT_{sys} B \frac{S}{N}. \tag{17-52}$$

Substituting this into (17-49), we get the maximum range in which the radar can detect a target in terms of S/N, T_{sys}, and B. If the output can be integrated over the time τ, the system temperature is effectively reduced to $T_{sys}/(B\tau)^{1/2}$.

17-9 APERTURE ILLUMINATION AND BRIGHTNESS DISTRIBUTIONS

In visible wavelengths, the atmosphere is transparent, and in wavelengths between 1 cm and 100 m the atmosphere is again transparent. For wavelengths longer than about 10 m, radio waves may not penetrate the ionosphere. Between 1 cm and around 10 μm, there are a considerable number of molecular absorptions. Also for wavelengths shorter than 0.1 μm, there is considerable molecular absorption. Therefore, we have *optical windows* and *radio windows* through the atmosphere. Radio astronomy makes use of this radio window, whereas optical astronomy makes use of the optical window (see Kraus, 1966).

Let us consider a radio telescope pointed in the sky (Fig. 17-11). The aperture field distribution is given by $A(x, y)$ and the radiation pattern $g(\theta, \phi)$ is related to the aperture distribution through a Fourier transform (see Section 9-5):

$$g(\theta, \phi) = \int A(x, y) e^{ik_x x + jk_y y} \, dx \, dy, \tag{17-53}$$

where

$$k_x = k \sin\theta \cos\phi,$$

$$k_y = k \sin\theta \sin\phi.$$

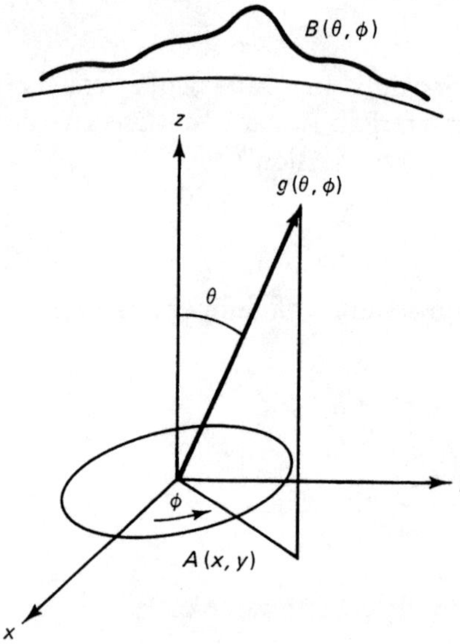

Figure 17-11 Aperture distribution $A(x,y)$ and radiation pattern $g(\theta,\phi)$.

We now consider the antenna beam scanning the sky in the x–z plane ($y=0$, $\phi=0$). Letting $g(\theta,0)=g(\beta)$ and $\int A(x,y)dy = A(x)$, we rewrite (17-53):

$$g(\beta) = \int A(x)e^{j\beta x}\,dx, \tag{17-54}$$

where $\beta = k\sin\theta$.

Next we consider the power pattern $p(\beta)$, which is given by

$$\begin{aligned} p(\beta) &= g(\beta)g^*(\beta) \\ &= \iint A(x)A^*(x')e^{j\beta(x-x')}\,dx\,dx'. \end{aligned} \tag{17-55}$$

The Fourier transform of the power pattern $P(x'')$ is then given by

$$\begin{aligned} P(x'') &= \frac{1}{2\pi}\int p(\beta)e^{-j\beta x''}\,d\beta \\ &= \int A(x)A^*(x-x'')\,dx. \end{aligned} \tag{17-56}$$

This is a convolution integral of the aperture distribution. These relationships are expressed conveniently in Fig. 17-12.

Next we let the antenna beam be pointed to the direction θ_o and let the beam scan the sky with brightness distribution $B(\theta)=B(\beta)$. This scanning is often accomplished by the rotation of the earth. The output $S(\beta_o)$ then depends on the scanning angle θ_o and thus (Fig. 17-13)

$$S(\beta_o) = \int B(\beta)p(\beta_o-\beta)d\beta, \qquad \beta_o = k\sin\theta_o. \tag{17-57}$$

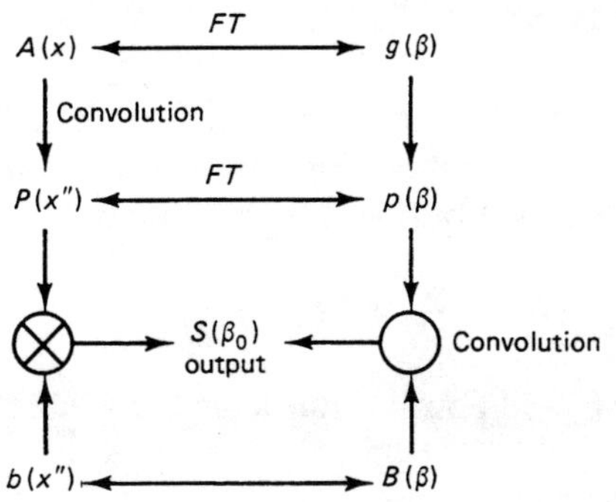

Figure 17-12 Relationships among aperture distribution $A(x)$, radiation pattern $p(\beta)$, brightness distribution $B(\beta)$, and output $S(\beta_o)$.

The Fourier transform of the output $S(\beta_o)$ is then given by the following product:

$$\begin{aligned} S(x) &= \frac{1}{2\pi}\int S(\beta_o)e^{-j\beta_o x}\,d\beta_o \\ &= b(x)P(x), \end{aligned} \tag{17-58}$$

where

$$b(x) = \frac{1}{2\pi}\int B(\beta)e^{-j\beta x}\,d\beta,$$

$$P(x) = \frac{1}{2\pi}\int p(\beta)e^{-j\beta x}\,d\beta.$$

The foregoing relationships among the brightness distribution $B(\beta)$, power pattern $p(\beta)$, and output $S(\beta_o)$ are shown in Fig. 17-12. Equation (17-58) shows that the spectrum of the brightness distribution is filtered by the antenna pattern and produces the output.

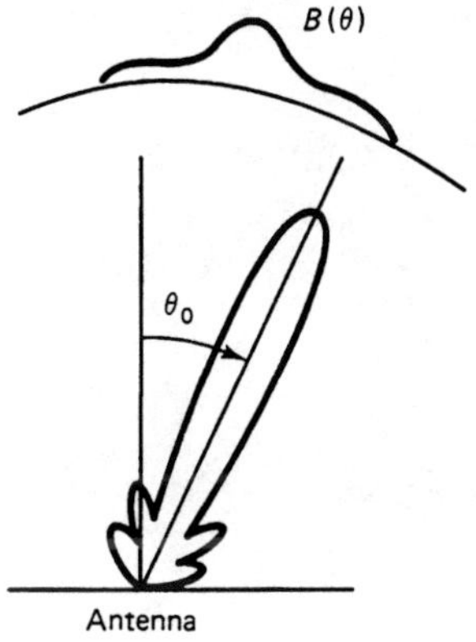

Figure 17-13 Antenna beam scanning the sky.

17-10 TWO-ANTENNA INTERFEROMETER

As an example of the use of the relationships shown in Section 17-9, we consider a two-antenna interferometer. Two point receiving antennas are separated by a distance a (Fig. 17-14). The aperture distribution $A(x)$ is given by

$$A(x) = \frac{1}{2}\delta\left(x - \frac{a}{2}\right) + \frac{1}{2}\delta\left(x + \frac{a}{2}\right). \tag{17-59}$$

The field pattern, power pattern, and Fourier transform are given by

$$\begin{aligned} g(\beta) &= \cos\frac{\beta a}{2} \\ p(\beta) &= \frac{1}{2}(1 + \cos\beta a). \end{aligned} \tag{17-60}$$

The observed output $S(\beta_o)$ as a function of the scan angle $\theta_o (\beta_o = k \sin\theta_o)$ is

$$S(\beta_o) = \frac{S_o}{2}[1 + V(\beta_o, a)], \tag{17-61}$$

where

$$S_o = \int B(\beta)\, d\beta,$$

$$V(\beta_o, a) = \frac{1}{S_o}\int B(\beta)\cos[(\beta_o - \beta)a]\, d\beta.$$

Now we consider the Fourier transform of the brightness distribution $B(\beta)$:

$$V_c(a) = \frac{1}{S_o}\int B(\beta)e^{-j\beta a}\, d\beta. \tag{17-62}$$

This function $V_c(a)$ is called the *complex visibility function* and is the normalized correlation function of the wave at the receiver as a function of the separation a. When the separation a is zero, V_c is unity, and as the separation a increases, the correlation decreases. The function $V_c(a)$ is the normalized *mutual coherence func-*

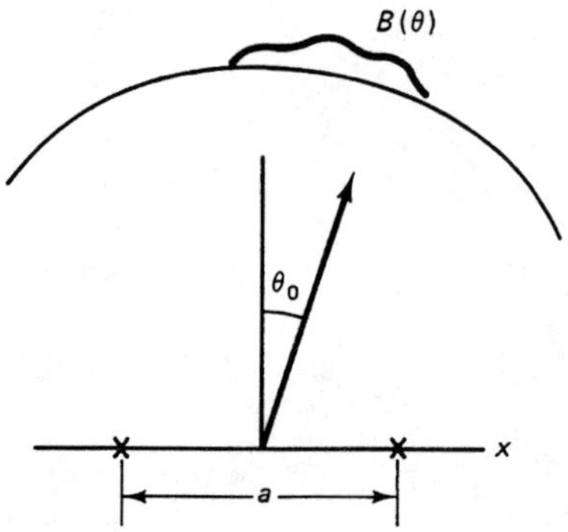

Figure 17-14 Two antenna interferometer.

tion or *degree of coherence*. The Fourier transform relationship (17-62) between the source brightness distribution of the incoherent source and the degree of coherence at the observation point is a special case of the van Cittert–Zernike theorem when the source and the observation point are far from each other (Born and Wolf, 1970, Chap. 10).

The complex visibility function $V_c(a)$ can be expressed as

$$V_c(a) = V_o(a)e^{-j\Delta(a)}. \tag{17-63}$$

Then the observed output $S(\beta_o)$ is given by

$$S(\beta_o) = \frac{S_o}{2}\{1 + V_o(a)\cos[\beta_o a - \Delta(a)]\}, \tag{17-64}$$

where

$$V(\beta_o, a) = \mathrm{Re}[V_c(a)e^{j\beta_o a}].$$

The observed output $S(\beta_o)$ as a function of the scan angle β_o is sketched in Fig. 17-15. In radio astronomy, the scanning is accomplished by the rotation of the earth.

It is now possible to determine the complex visibility function from measurement of the observed output $S(\beta_o)$. The maximum and minimum values of $S(\beta_o)$ are given by

$$\begin{aligned} S_{\max} &= \frac{S_o}{2}[1 + V_o(a)], \\ S_{\min} &= \frac{S_o}{2}[1 - V_o(a)]. \end{aligned} \tag{17-65}$$

From this we get

$$V_o(a) = \frac{S_{\max} - S_{\min}}{S_{\max} + S_{\min}}. \tag{17-66}$$

This magnitude $V_o(a)$ is called the *fringe visibility* or *visibility*. Also at $\beta_o = 0$, we get

$$V(0, a) = V_o(a)\cos\Delta(a). \tag{17-67}$$

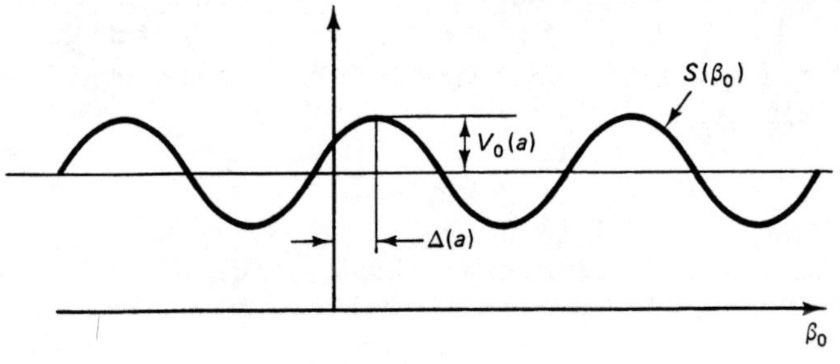

Figure 17-15 Observed output $S(\beta_o)$ and the complex visibility function.

From (17-66) and (17-67) we get $V_o(a)$ and $\Delta(a)$, and thus $V_c(a)$. Once we get $V_c(a)$ for all separations a, it should be possible to obtain the brightness distribution $B(\beta)$ by the inverse Fourier transform of (17-62).

The Fourier transform relationship (17-62) can be generalized to two dimensions as indicated in (17-53). The technique of measuring the complex visibility function and then obtaining the map of the brightness distribution of the radio sources in the sky is called the *aperture synthesis*. To obtain the complex visibility function, the measurement must be made at a sufficient number of separations. The Very Large Array (VLA) constructed in New Mexico can have a separation up to 35 km for frequencies ranging from 1.4 to 24 GHz.

PROBLEMS

17-1. Assume that the sun has an angular size of 1° and is equivalent to a blackbody with 6000 K. An antenna at $f = 10$ GHz has a beamwidth of 0.5° and the receiver bandwidth is 1 MHz. This antenna is directed toward the sun. Calculate:
- **(a)** The gain of the antenna (in dB)
- **(b)** The source flux density
- **(c)** The observed flux density
- **(d)** The source temperature
- **(e)** The antenna temperature

17-2. Do Problem 17-1 with a beamwidth of 2°.

17-3. Assume that raindrops are uniformly distributed from the ground to a height of 3 km and that the rain attenuation is 0.3 dB/km at 10 GHz when the precipitation rate is 12.5 mm/h. The rain temperature is assumed to be 273 K. If an antenna is always pointed toward the sun, find the antenna temperature as a function of angle θ. Assume that the scattering effect is negligible.

17-4. Consider the radiometer shown in Fig. 17-8, located at $h = 3$ km and $\theta = 30°$. The antenna has a beamwidth of 1° at 10 GHz. Assume a rain medium as described in Problem 17-3. Find the rain emission T_1. Find the ground emission contribution T_2 assuming a ground temperature of 283 K for moist soil. Find the scattered component T_3 assuming a uniform sky temperature of 273 K and a Lambertian surface for the ground. Calculate the total temperature $T_1 + T_2 + T_3$. If the ground is covered by a smooth metallic surface, what is the temperature?

17-5. Consider the system shown in Fig. 17-9. Assume that $T_A = 40$ K, $T_L = 290$ K, $L = 0.5$ dB, $T_1 = T_2 = T_3 = 290$ K, and $G_1 = G_2 = G_3 = 20$ dB. Find the system temperature.

17-6. A microwave antenna operating at 10 GHz has a beamwidth of 1° and a bandwidth of 1 MHz. Assuming that the system noise temperature of the antenna is 290 K, the radar cross section of the object is 10 m^2, the required signal-to-noise ratio is at least 10 dB, and the maximum range at which this object should be detected is 100 km, find the transmitting power required.

17-7. Assume that an extraterrestrial civilization (ETC) exists at a distance of R light-years. An antenna 26 m in diameter is used to detect the signal from the ETC. Assume that the ETC is sending a signal using an antenna with a transmitting power of 1 MW and a diameter of 100 m. Assume that the receiver has a system noise of 20 K, and that

$\lambda = 12$ cm, $\Delta f = 1$ Hz and $\tau = 1$ s. Calculate the maximum distance R (light-year) at which the signal from the ETC can be detected.

17-8. Consider an antenna with aperture distribution $A(x)$ pointed to the sun (Fig. P17-8). Assume that the sun has an equivalent angular size of 1° at an equivalent blackbody temperature of 6000 K. The wavelength is 3 cm and the receiver bandwidth is 1 MHz.

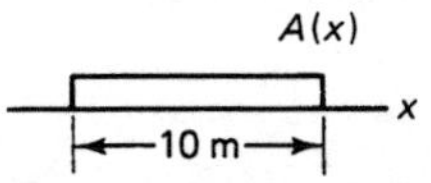

Figure P17-8 Aperture antenna.

Calculate:

(a) The source flux density
(b) The observed flux density
(c) The source temperature
(d) The antenna temperature

Use appropriate approximations, if necessary. Also find:

(e) $g(\lambda)$, $\lambda = k \sin \theta$
(f) $P(\lambda)$
(g) $\bar{P}(x'')$
(h) $B(\lambda)$
(i) $\bar{B}(x'')$
(j) $S(\lambda_0)$

17-9. If a star has the brightness distribution shown in Fig. P17-9, calculate and plot $S(\lambda_0)$ for two-antenna interferometry. The frequency is 1 GHz and $a = 500$ m.

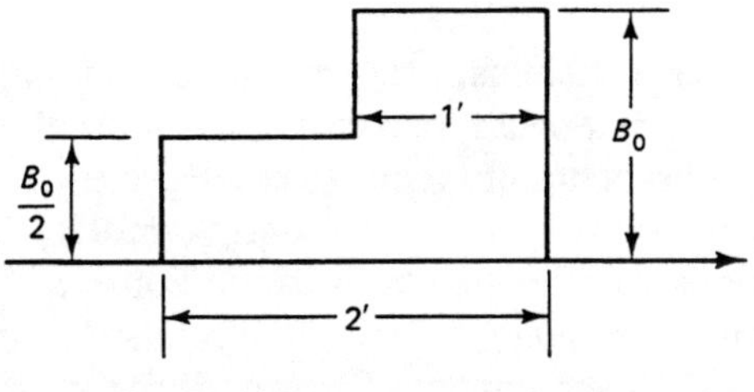

Figure P17-9 Brightness distribution.

17-10. Assume that a certain nebula has an angular size of 1′ at 1 GHz and a source flux density of 1000×10^{-26} jansky. The receiving antenna has a diameter of 10 m and a bandwidth of 5 MHz. Find the antenna temperature. If the integration time is 1 s, what is the maximum receiver noise temperature allowed to detect this nebula?

17-11. For Problem 7-10, two identical receiving antennas are used to determine the angular size of the nebula. Calculate and plot the visibility as a function of the distance between the two antennas.

18

Numerical Techniques

For many practical electromagnetic problems, exact analytical solutions are generally not available, and therefore it is important to consider methods of obtaining approximate numerical solutions. This generally requires extensive use of computers, and it is essential to develop methods to discretize and formulate the problem systematically so that the solutions are obtained most expeditiously. In this chapter we present several fundamental techniques, including the method of moment, the Rayleigh–Ritz method, the finite-element method, and the finite-difference method (see Harrington, 1968; Finlayson, 1972; Zienkiewicz, 1977; and Brebbia et al., 1984. For applications to electromagnetics, see Morgan (1990), Itoh (1989), Mittra (1973, 1975), and Moore and Pizer (1984).

18-1 MOMENT METHOD

The moment method developed by Harrington (1968) has been used successfully to solve many practical electromagnetic problems. In this section we illustrate this method by considering a simple eigenvalue problem with Dirichlet boundary conditions.

Let us consider the eigenvalue problem given by

$$Lf = \lambda f \tag{18-1}$$

in the domain Ω, where L is the operator (differential or integral), f the eigenfunction, and λ the eigenvalue. An example is that of TM modes in waveguides.

$$(\nabla_t^2 + k_p^2)\phi_p = 0, \qquad \phi_p = 0 \quad \text{on the boundary,} \tag{18-2}$$

where $-\nabla_t^2 = L$, $k_p^2 = \lambda$ is the eigenvalue, and $\phi_p = f$ is the eigenfunction. The problem "domain" is the cross section of the waveguide.

Let us solve the eigenvalue problem (18-1) using the moment method (or method of moments, MoM) developed by Harrington. We wish to find an approximate solution f_a of (18-1) in terms of a series of given independent functions N_m:

$$f_a = \sum_{m=1} \alpha_m N_m. \tag{18-3}$$

The function N_m is called the *basis function, expansion function,* or *shape function.* These functions are independent, but are generally not orthogonal. Since f_a is an approximation, it does not really satisfy (18-1), and therefore we define the *error* or the *residual R.*

$$R = Lf_a - \lambda f_a. \tag{18-4}$$

Let us attempt to reduce this residual R by requiring that the integrals of R over the domain Ω weighted in different ways be zero.

$$\int_\Omega W_l R \, d\Omega = 0, \qquad l = 1, 2, \ldots, M, \tag{18-5}$$

where W_l are independent functions and are called the *weighting function* or *testing function.* Substituting (18-3) and (18-4) into (18-5), we get

$$\int_\Omega W_l(L - \lambda) \sum_{m=1}^{M} \alpha_m N_m \, d\Omega = 0, \qquad l = 1, 2, \ldots, M. \tag{18-6}$$

Rewriting this, we get the following matrix equation:

$$\overline{K}\overline{\alpha} = \lambda \overline{M}\overline{\alpha}, \tag{18-7}$$

where

$$\overline{K} = M \times M \text{ matrix} = [K_{lm}],$$

$$\overline{\alpha} = M \times 1 \text{ matrix} = [\alpha_m],$$

$$\overline{M} = M \times M \text{ matrix} = [M_{lm}],$$

$$K_{lm} = \langle W_l, LN_m \rangle,$$

$$M_{lm} = \langle W_l, N_m \rangle.$$

The inner product $\langle W_l, N_m \rangle$ is defined as

$$\langle W_l, N_m \rangle = \int_\Omega W_l N_m \, d\Omega.$$

The original eigenalue problem (18-1) is now converted into the matrix eigenvalue problem (18-7). If (18-7) is satisfied for all l as $M \to \infty$, the residual $R \to 0$ at all points of the domain and the solution f_a in (18-3) converges to f. For finite M, f_a is not equal to f. However, in the following, we may simply use f to represent the approximation f_a whenever no confusion is expected to arise.

This general procedure of obtaining the matrix equation using the basis functions and the weighting function and then solving the matrix equation to obtain the approximation solution is called the method of moments (MoM) or moment method because the process of multiplying R by the different weighting functions is equivalent to taking different moments of R and equating each moment to zero (Kantorovich and Krylov, 1958). This general procedure is also called the *method of weighted residuals* (Finlayson, 1972).

Let us solve the matrix eigenvalue problem (18-7). The eigenvalue λ can be obtained by noting that $(\overline{\overline{K}} - \lambda\overline{\overline{M}})\overline{\alpha} = 0$ has a nonzero solution for $\overline{\alpha}$ only when the determinant of the matrix $\overline{\overline{K}} - \lambda\overline{\overline{M}}$ is zero:

$$\det\{\overline{\overline{K}} - \lambda\overline{\overline{M}}\} = 0. \tag{18-8}$$

From this we should get M eigenvalues, $\lambda_p, p = 1, 2, \ldots, M$. For each eigenvalue λ_p, we use (18-7) to obtain the eigenvector $[\alpha]_p = [\alpha_{pm}]$. The solution f_p for the eigenvalue λ_p is given by

$$f_p = \sum_{m=1}^{M} \alpha_{pm} N_m. \tag{18-9}$$

As an example, consider the problem

$$\begin{gathered} -\frac{d^2}{dx^2} f = \lambda f, \\ f = 0 \quad \text{at } x = 0 \quad \text{and} \quad x = 1. \end{gathered} \tag{18-10}$$

Let us use, following Harrington (1968),

$$\begin{aligned} f_a &= \sum_{m=1}^{M} \alpha_m N_m, \\ N_m(x) &= x(1 - x^m), \\ W_l(x) &= x(1 - x^l). \end{aligned} \tag{18-11}$$

Here we used the basis function $N_m(x)$, which satisfies Dirichlet's boundary condition. This is not necessary, but it is convenient for this problem. We examine other basis functions and boundary conditions in Sections 18-3 and 18-4. The weighting function W_l is chosen so as to be the same form as the basis function. This choice of the weighting function $W_l = N_l$, called *Galerkin's method*, often leads to a simpler symmetric matrix for $\overline{\overline{K}}$.

Using (18-11), we get the matrix equation:

$$\bar{K}\bar{\alpha} = \lambda \bar{M}\bar{\alpha},$$

$$K_{lm} = \int_0^1 W_l L N_m \, dx, \tag{18-12}$$

$$M_{lm} = \int_0^1 W_l N_m \, dx.$$

For example, if $M = 2$, we get

$$\bar{K} = \begin{bmatrix} \frac{1}{3} & \frac{1}{2} \\ \frac{1}{2} & \frac{4}{5} \end{bmatrix}, \qquad \bar{M} = \begin{bmatrix} \frac{1}{30} & \frac{1}{20} \\ \frac{1}{20} & \frac{8}{105} \end{bmatrix}.$$

The eigenvalues λ_p, $p = 1$ and 2, are therefore the roots of the following equation:

$$\begin{vmatrix} \frac{1}{3} - \frac{\lambda}{30} & \frac{1}{2} - \frac{\lambda}{20} \\ \frac{1}{2} - \frac{\lambda}{20} & \frac{4}{5} - \frac{8\lambda}{105} \end{vmatrix} = 0.$$

From this we get

$$\lambda_1 = 10,$$
$$\lambda_2 = 42. \tag{18-13}$$

The eigenvectors $[\alpha]_p$ can be obtained from

$$(\bar{K} - \lambda_p \bar{M})\alpha_p = 0. \tag{18-14}$$

For the eigenvalue $\lambda_1 = 10$, (18-14) becomes

$$\begin{bmatrix} 0 & 0 \\ 0 & \frac{4}{105} \end{bmatrix} \begin{bmatrix} \alpha_{11} \\ \alpha_{12} \end{bmatrix} = 0,$$

from which we get $\alpha_{12} = 0$, and therefore

$$[\alpha]_1 = \begin{bmatrix} \alpha_{11} \\ 0 \end{bmatrix},$$

$$f_1 = \alpha_{11} N_1 = \alpha_{11} x(1 - x).$$

The constant a_{11} is arbitrary, but we may choose it to normalize f_1:

$$\int_0^1 f_1^2 \, dx = 1.$$

Then we get $\alpha_{11} = (30)^{1/2} = 5.4772$. For the eigenvalue $\lambda_2 = 42$, we get

$$\begin{bmatrix} \frac{1}{3} - \frac{42}{30} & \frac{1}{2} - \frac{42}{20} \\ \frac{1}{2} - \frac{42}{20} & \frac{4}{5} - \frac{336}{105} \end{bmatrix} \begin{bmatrix} \alpha_{21} \\ \alpha_{22} \end{bmatrix} = 0,$$

from which we get $\alpha_{22} = -\frac{2}{3}\alpha_{21}$, Therefore,

$$f_2 = \alpha_{21}[x(1-x) - \tfrac{2}{3}x(1-x^2)].$$

If we normalize f_2, α_{21} is 43.474.

Summarizing the results, we have

$$\begin{aligned} f_1(x) &= 5.4772x(1-x), & \lambda_1 &= 10 \\ f_2(x) &= 43.474[x(1-x) - \tfrac{2}{3}x(1-x^2)], & \lambda_2 &= 42, \end{aligned} \tag{18-15}$$

These solutions are close to the following exact solutions of (9-10):

$$\begin{aligned} f_1(x) &= 2^{1/2}\sin \pi x, & \lambda_1 &= \pi^2 = 9.8696 \\ f_2(x) &= 2^{1/2}\sin 2\pi x, & \lambda_2 &= (2\pi)^2 = 39.4784. \end{aligned} \tag{18-16}$$

In this section we have presented a brief outline of the basic formulation of the moment method. However, many questions need to be examined, including basis functions, weighting functions, and boundary conditions. For more extensive discussions, see Harrington (1968) and Moore and Pizer (1984) (modern applications).

18-2 CHOICE OF BASIS FUNCTIONS AND WEIGHTING FUNCTIONS

In Section 18-1 we used a simple whole domain basis function $N_m(x) = x(1-x^m)$ defined in the whole domain $0 \le x \le 1$. The Fourier series expansion is equivalent to using the basis function $\sin m\pi x$ and $\cos m\pi x$ over the entire domain $0 \le x \le 1$. However, in many practical problems, it is often more convenient to use the subdomain basis function defined over a subsection of the domain.

Step Approximation

The simplest choice is the *step approximation* given by the pulse function $P_n(x)$ (Fig. 18-1).

$$N_m(x) = P_m(x) = \begin{cases} 1 & \text{when } x \text{ is in the subdivision } \Delta x_m \\ 0 & \text{when } x \text{ is not in } \Delta x_m. \end{cases} \tag{18-17}$$

$$f(x) = \sum_{m=1}^{M} \alpha_m P_m(x). \tag{18-18}$$

The coefficients α_m are therefore the value of the function at x_m:

$$\alpha_m = f(x_m). \tag{18-19}$$

If the operator L contains the second derivative, as is often the case, the second derivative of $P_m(x)$ contains the derivative of the delta function, and therefore the second derivative operator is usually approximated by the finite-difference operator.

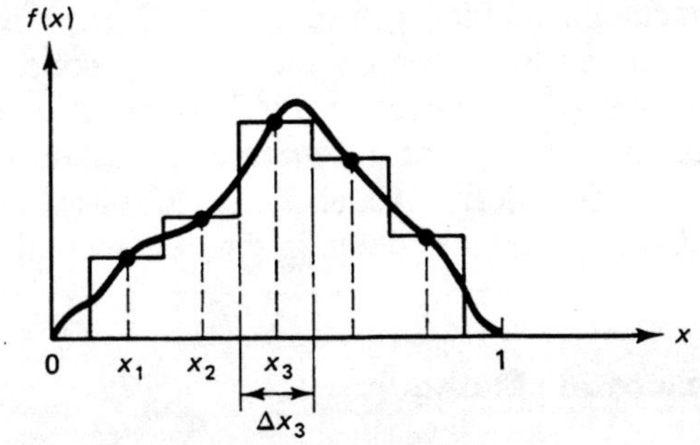

Figure 18-1 Step approximation.

Piecewise Linear Approximation

Another choice is the use of the triangle function, which is commonly used in the finite-element method (Fig. 18-2). This is also called the *piecewise linear approximation.* The function $f(x)$ is approximated by

$$
\begin{aligned}
f(x) &= \sum_{m=1}^{M} \alpha_m N_m(x), \\
\alpha_m &= f(x_m), \\
N_m(x) &= \frac{(x - x_{m-1})}{l_{m-1}}, \qquad x_{m-1} \le x \le x_m \\
&= \frac{l_m - (x - x_m)}{l_m}, \qquad x_m \le x \le x_{m+1}.
\end{aligned}
\tag{18-20}
$$

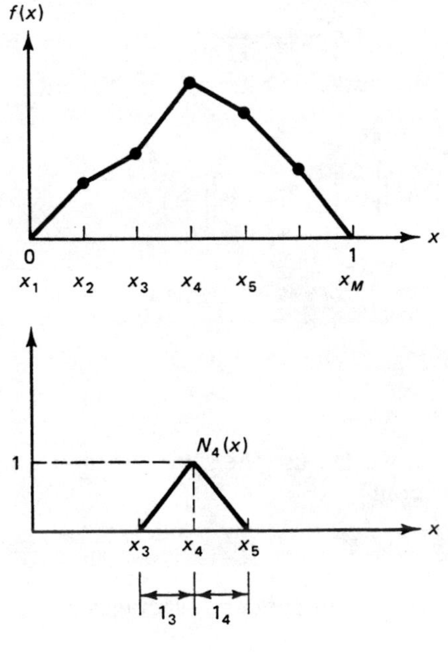

Figure 18-2 Piecewise linear approximation.

The advantage of this basis function is that it is continuous. In general, if the operator has a second derivative, the basis function should have a continuous first derivative, and if the operator has a first derivative, the basis function should be continuous. We will show later that the operator L with a second derivative can be converted to an operator with a first derivative, and thus this basis function can be used conveniently. This is discussed in more detail in the next section.

Point-Matching Method (Collocation Method)

Let us consider the weighting functions W_l. The moment method requires calculation of the inner products shown in (18-7). This calculation is often difficult, but if the weighting function W_l is chosen to be a delta function,

$$W_l = \delta(\bar{r} - \bar{r}_l), \tag{18-21}$$

the inner product can be obtained immediately:

$$K_{lm} = \langle W_l, LN_m \rangle = \int_\Omega \delta(\bar{r} - \bar{r}_l) LN_m \, d\Omega = [LN_m]_{\bar{r} = \bar{r}_l},$$

$$M_{lm} = \langle W_l, N_m \rangle = [N_m]_{\bar{r} = \bar{r}_l}.$$

This is equivalent to satisfying $Lf_a = \lambda f_a$ at the discrete points $\bar{r} = \bar{r}_l (l = 1, 2, \ldots, M)$, and therefore it is called the *point-matching method* or the *collocation method.*

For the example of (18-10), using $N_m(x) = x(1 - x^m)$, we get

$$K_{lm} = (m+1) m x_l^{m-1},$$

$$M_{lm} = x_l(1 - x_l^m).$$

For $N = 2$ we get

$$\begin{bmatrix} 2 & 2 \\ 2 & 4 \end{bmatrix} \begin{bmatrix} \alpha_1 \\ \alpha_2 \end{bmatrix} = \lambda \begin{bmatrix} \frac{2}{9} & \frac{8}{27} \\ \frac{2}{9} & \frac{10}{27} \end{bmatrix} \begin{bmatrix} \alpha_1 \\ \alpha_2 \end{bmatrix}$$

and obtain $\lambda_1 = 9$ and $\lambda_2 = 27$.

Galerkin Method

As we discussed in Section 18-1, we can use the same function for both the basis function and the weighting function:

$$W_l = N_l. \tag{18-22}$$

This is called the *Galerkin method.* We discuss its advantages later.

18-3 WEAK FORMULATION AND EXAMPLES

Previously we indicated that if the operator contains the second derivative, the basis function should have a continuous first derivative, and if the operator contains the first derivative, the basis function should be continuous. For most practical problems, the operator contains the second derivative, and therefore the basis functions such as the step approximations and the piecewise-continuous approximation cannot be used directly and the operator must be approximated by the finite-difference operator. However, it is possible to reduce the order of the derivative in the operator to contain only the first derivative. This is called the *weak form* or *weak formulation.* We will illustrate this formulation using Dirichlet's and Neumann's eigenvalue problems.

Let us consider a one-dimensional eigenvalue problem:

$$-\frac{d^2f}{dx^2} = \lambda f. \tag{18-23}$$

with the Dirichlet boundary condition $f(x) = 0$ at $x = 0$ and $x = 1$. We express the function f in a series of the basis function $N_m(x)$ with unknown coefficients α_m.

$$f(x) = \sum_{m=1}^{M} \alpha_m N_m(x). \tag{18-24}$$

Substituting this into (18-23) and integrating (18-23) with the weighting function $W_l(x)$, we get

$$\int_0^1 W_l(x)\left(\frac{d^2f}{dx^2} + \lambda f\right) dx = 0. \tag{18-25}$$

Note that the integrand contains the second derivative of f. We can lower the order of the derivative by integration by parts. We get

$$-\int_0^1 \frac{dW_l}{dx}\frac{df}{dx}\, dx + W_l \frac{df}{dx}\bigg|_0^1 + \lambda\int_0^1 W_l f\, dx = 0. \tag{18-26}$$

This is the weak form or weak formulation. Here the integrand contains only the first derivative of f, and therefore we can choose the basis function that is continuous but whose first derivative may be discontinuous.

Let us choose the piecewise linear function shown in (18-20) for the basis function.

$$\begin{aligned} N_m(x) &= \frac{x - x_{m-1}}{l_{m-1}} && \text{for } x_{m-1} \le x \le x_m \\ &= \frac{l_m - (x - x_m)}{l_m} && \text{for } x_m \le x \le x_{m+1}. \end{aligned} \tag{18-27}$$

We divide the range (0, 1) into $M-1$ sections as shown in Fig. 18-2. We also use the Galerkin method and let the weighting function be the same as the basis function:

$$W_l(x) = N_l(x). \tag{18-28}$$

We then get, from (18-26),

$$\bar{K}\bar{\alpha} = \lambda \bar{M}\bar{\alpha} + \bar{b}, \tag{18-29}$$

where

$$K_{lm} = \int_0^1 \frac{dN_l(x)}{dx}\frac{dN_m(x)}{dx}\,dx,$$

$$M_{lm} = \int_0^1 N_l(x)N_m(x)\,dx,$$

$$b_1 = -\left.\frac{df}{dx}\right|_{x_1},$$

$$b_M = -\left.\frac{df}{dx}\right|_{x_M},$$

$$b_l = 0 \qquad \text{for } l \neq 1, \qquad l \neq M.$$

The elements K_{lm} and M_{lm} can easily be calculated, and for $M = 5$ we get

$$\bar{K} = \begin{bmatrix} \frac{1}{l_1} & -\frac{1}{l_1} & 0 & 0 & 0 \\ -\frac{1}{l_1} & \frac{1}{l_1}+\frac{1}{l_2} & -\frac{1}{l_2} & 0 & 0 \\ 0 & -\frac{1}{l_2} & \frac{1}{l_2}+\frac{1}{l_3} & -\frac{1}{l_3} & 0 \\ 0 & 0 & -\frac{1}{l_3} & \frac{1}{l_3}+\frac{1}{l_4} & -\frac{1}{l_4} \\ 0 & 0 & 0 & -\frac{1}{l_4} & \frac{1}{l_4} \end{bmatrix},$$

$$\bar{M} = \begin{bmatrix} \frac{l_1}{3} & \frac{l_1}{6} & 0 & 0 & 0 \\ \frac{l_1}{6} & \frac{l_1}{3}+\frac{l_2}{3} & \frac{l_2}{6} & 0 & 0 \\ 0 & \frac{2}{6} & \frac{l_2}{3}+\frac{l_3}{3} & \frac{l_3}{6} & 0 \\ 0 & 0 & \frac{l_3}{6} & \frac{l_3}{3}+\frac{l_4}{3} & \frac{l_4}{6} \\ 0 & 0 & 0 & \frac{l_4}{6} & \frac{l_4}{3} \end{bmatrix}, \tag{18-30}$$

$$\bar{b} = \begin{bmatrix} -\left.\dfrac{df}{dx}\right|_{x_1} \\ 0 \\ 0 \\ 0 \\ \left.\dfrac{df}{dx}\right|_{x_M} \end{bmatrix}, \qquad \bar{\alpha} = \begin{bmatrix} \alpha_1 \\ \alpha_2 \\ \alpha_3 \\ \alpha_4 \\ \alpha_5 \end{bmatrix}.$$

Note that up to this point we have not applied the boundary condition $f(x) = 0$ at $x = 0$ and $x = 1$. The boundary condition in terms of $\bar{\alpha}$ is

$$\alpha_1 = 0 \quad \text{and} \quad \alpha_5 = 0. \tag{18-31}$$

By using this, the matrix equation (118-29) is reduced to the 3×3 matrix. If we use $l_1 = l_2 = l_3 = l_4 = \frac{1}{4}$, we get

$$\begin{bmatrix} 8 & -4 & 0 \\ -4 & 8 & -4 \\ 0 & -4 & 8 \end{bmatrix} \begin{bmatrix} \alpha_2 \\ \alpha_3 \\ \alpha_4 \end{bmatrix} = \lambda \begin{bmatrix} \frac{1}{6} & \frac{1}{24} & 0 \\ \frac{1}{24} & \frac{1}{6} & \frac{1}{24} \\ 0 & \frac{1}{24} & \frac{1}{6} \end{bmatrix} \begin{bmatrix} \alpha_2 \\ \alpha_3 \\ \alpha_4 \end{bmatrix}, \tag{18-32}$$

Letting the determinant of $(\bar{K} - \lambda\bar{M})$ be zero, we get three eigenvalues λ_1, λ_2, and λ_3:

$$\lambda_1 = 10.39, \qquad \lambda_2 = 48, \qquad \lambda_3 = 126.76,$$

which may be compared with the exact values π^2, $(2\pi)^2$, and $(3\pi)^2$. Because the number of terms is only three, λ_1 is reasonably close, but λ_2 and λ_3 are not very close to the exact values.

For each eigenvalue we can find the values of α_2, α_3, α_4. For λ_1, using (18-32) with $\lambda_1 = 10.39$, we get $\alpha_3 = \sqrt{2}\,\alpha_2$ and $\alpha_4 = \alpha_2$. Thus we have

$$f(x) = \alpha_2[N_2(x) + \sqrt{2}N_3(x) + N_4(x)], \tag{18-33}$$

where α_2 is an arbitrary constant.

A plot of $f(x)$ for λ_1 is shown in Fig. 18-3a. α is chosen to be $1/\sqrt{2}$, so that $f(x) = 1$ at $x = 0.5$. Note that $f(x)$ is close to the exact eigenfunction $\sin \pi x$. For $\lambda_2 = 48$, we get $\alpha_3 = 0$ and $\alpha_2 = -\alpha_4$, and thus

$$f(x) = \alpha_2[N_2(x) - N_4(x)]. \tag{18-34}$$

This is shown in Fig. 18-3b.

In the solution above, we did not apply the boundary condition until the last stage in (18-31). This is intentional because the first and the last equations of

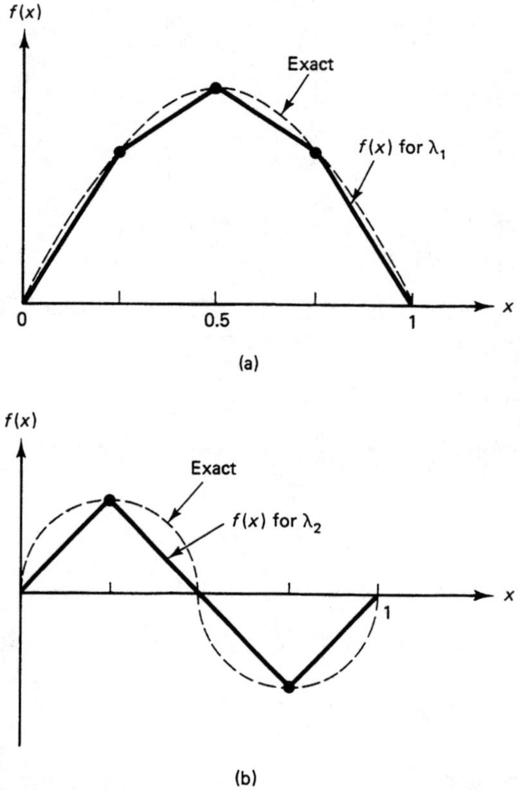

Figure 18-3 Approximation $f(x)$ and the exact solutions $\sin \pi x$ and $\sin 2\pi x$.

(18-29), which were not used to obtain the solution, give extra information about the solution. We write these two equations keeping in mind that $\alpha_1 = \alpha_5 = 0$. We get

$$
\begin{aligned}
-\frac{1}{l_2}\alpha_2 &= \lambda\frac{l_1}{6}\alpha_2 - \left.\frac{df}{dx}\right|_{x_1}, \\
-\frac{1}{l_4}\alpha_4 &= \lambda\frac{l_4}{6}\alpha_4 + \left.\frac{df}{dx}\right|_{x_5},
\end{aligned} \tag{18-35}
$$

From this we get the derivative of the function at x_1 and x_5 for each eigenvalue. The derivatives in (18-35) give a much better approximation than do the derivatives obtained from (18-33) and (18-34). For example, (18-35) gives, with $\alpha_2 = 1/\sqrt{2}$,

$$\left.\frac{df}{dx}\right|_{x_1} = 3.13.$$

This is much closer to the exact value $(d/dx) \sin \pi x|_{x=0} = \pi$ than is the approximation $\alpha_2/l_1 = 2.828$ in (18-33). We may also note that we used the Galerkin method, and therefore, as can be seen from (18-29), K_{lm} is symmetric: $K_{lm} = K_{ml}$. This is one of the advantages of the Galerkin method.

Let us next consider the eigenvalue problem with Neumann's boundary condition:

$$-\frac{d^2}{dx^2}f(x)=\lambda f(x),$$
$$\frac{\partial f}{\partial x}=0 \qquad \text{at } x=0 \quad \text{and} \quad x=1. \tag{18-36}$$

We start with the representation using the basis function $N_m(x)$:

$$f(x)=\sum_{m-1}^{M} N_m(x)\alpha_m. \tag{18-37}$$

Instead of using the weighting function $W_l(x)$ and forming (18-25), we use the weighting function $W_l(x)$ for (18-36) and the weighting function $\overline{W}_l(x)$ for the boundary condition:

$$\int_0^1 W_l(x)\left(\frac{d^2f}{dx^2}+\lambda f\right)dx+\overline{W}_l\frac{df}{dx}\bigg|_0^1=0. \tag{18-38}$$

The reason for the use of $\overline{W}_l$ will be explained shortly. Now we convert (18-38) into the weak form and obtain

$$-\int_0^1 \frac{dW_l}{dx}\frac{df}{dx}dx+W_l\frac{df}{dx}\bigg|_0^1+\lambda\int_0^1 W_l f\,dx+\overline{W}_l\frac{df}{dx}\bigg|_0^1=0. \tag{18-39}$$

Now we see that if we choose $\overline{W}_l$ to be

$$\overline{W}_l=-W_l, \tag{18-40}$$

(18-39) is reduced to

$$-\int_0^1 \frac{dW_l}{dx}\frac{df}{dx}dx+\lambda\int_0^1 W_l f\,dx=0. \tag{18-41}$$

The reason for the use of W_l and $\overline{W}_l$ in (18-38) is now clear. By using $\overline{W}_l$ for the boundary condition and letting $\overline{W}_l=-W_l$, we did not have to use the boundary condition $df/dx=0$ at $x=0$ and 1. This is convenient because it avoids the use of the derivative of the approximation (18-37) at the boundary. The formulation above therefore eliminates the need for the actual calculation of df/dx at the boundary, and the Neumann boundary condition is natural for this problem. For this reason this is called the *natural boundary condition.* In contrast, Dirichlet's condition must be imposed as in (18-31) and is often called the *forced condition* or the *essential boundary condition.*

We now have (18-41), and this can be converted to the matrix equation

$$\overline{\overline{K}}\,\overline{\alpha}=\lambda\overline{\overline{M}}\,\overline{\alpha}, \tag{18-42}$$

where $\overline{\overline{K}}$ and $\overline{\overline{M}}$ are given in (18-29) and (18-30). This is a 5×5 matrix equation rather than the 3×3 matrix equation (9-32) for Dirichlet's problem.

18-4 VARIATIONAL EXPRESSIONS AND THE RAYLEIGH–RITZ METHOD

In general, we can obtain analytic closed-form solutions only for objects with simple geometric shapes such as a sphere and a cylinder. In many practical problems, however, the shapes are complex and we need to resort to approximate techniques. The variational method gives a powerful technique to obtain an approximate solution.

Let us consider a simple example of finding the eigenvalue for a waveguide with a complex cross section. The eigenvalue $\lambda = k_c^2$ is given by (4-55):

$$\lambda = \frac{\int \phi \nabla^2 \phi \, da}{\int \phi^2 \, da}. \tag{18-43}$$

This is exact if the exact eigenfunction ϕ is known. Consider a function ϕ_a that differs from the exact function ϕ by a small amount $\delta\phi$, called a *variation*. Then we can show that the eigenvalue as calculated from (18-43) is stationary with respect to the variation $\delta\phi$. This means that even if we use an approximate function ϕ_a in (18-43), the resulting eigenvalue λ as calculated by (18-43) is a much closer approximation to the true value than is the approximation ϕ_a.

To prove these stationary characteristics, we show that $\delta\lambda$ as calculated from (18-43) is zero. We rewrite (18-43) as

$$\lambda \int \phi^2 \, da + \int \phi \nabla^2 \phi \, da = 0. \tag{18-44}$$

We then take the variation of this equation:

$$\delta\lambda \int \phi^2 \, da + \lambda \int \delta\phi^2 \, da + \int \delta\phi \, \nabla^2 \phi \, da + \int \phi \delta \nabla^2 \phi \, da = 0. \tag{18-45}$$

The second term becomes

$$\lambda \int 2\phi \, \delta\phi \, da.$$

The third and the fourth terms are equal because the operator ∇^2 is self-adjoint:

$$\int \delta\phi \nabla^2 \phi \, da = \int \phi \nabla^2 \delta\phi \, da. \tag{18-46}$$

To prove this, use Green's theorem:

$$\int (u \nabla^2 v - v \nabla^2 u) \, da = \int \left(u \frac{\partial v}{\partial n} - v \frac{\partial u}{\partial n} \right) dl.$$

By letting $u = \delta\phi$ and $v = \phi$, and noting that the right side vanishes if ϕ and $\delta\phi$ satisfy the Dirichlet or Neumann boundary condition, we get (18-46). Using (18-46) in (18-45), we get

$$\delta\lambda \int \phi^2 \, da + 2 \int (\nabla^2 \phi + \lambda \phi) \, \delta\phi \, da = 0.$$

Since $(\nabla^2 + \lambda)\,\phi = 0$ when ϕ is exact, we get $\delta\lambda = 0$, proving that (18-43) is stationary with respect to the variation $\delta\phi$.

For more general eigenvalue problems, we write (18-43) using the operator L and the function f:

$$\lambda = \frac{\int \phi L \phi \, da}{\int \phi^2 \, da} = \frac{\langle f, Lf \rangle}{\langle f, f \rangle}. \tag{18-47}$$

This is a variational expression for λ. Now we wish to find the function f such that the eigenvalue λ is stationary with respect to the variation in f. This is called the *Rayleigh–Ritz method.* We expand f in a series of basis functions N_m:

$$f = \sum_{m=1}^{M} \alpha_m N_m. \tag{18-48}$$

We substitute this into (18-47) and attempt to find α_m such that λ is stationary with respect to the change in α_m. We get

$$\begin{aligned} \lambda &= \frac{N}{D}, \\ N &= \sum_l \sum_m \alpha_l K_{lm} \alpha_m, \\ D &= \sum_l \sum_m \alpha_l M_{lm} \alpha_m, \end{aligned} \tag{18-49}$$

where

$$K_{lm} = \langle N_l, LN_m \rangle,$$
$$M_{lm} = \langle N_l, N_m \rangle.$$

Now to make λ stationary with respect to the change in α_m, we let

$$\frac{\partial \lambda}{\partial \alpha_l} = 0. \tag{18-50}$$

This gives for each l,

$$\frac{1}{D^2}\left(D \sum_m K_{lm}\alpha_m - N \sum M_{lm}\alpha_m \right) = 0. \tag{18-51}$$

Rewriting this, we get

$$\bar{\bar{K}}\bar{\alpha} = \lambda \bar{\bar{M}} \bar{\alpha}. \tag{18-52}$$

This is identical to the moment method solution with $W_l = N_l$ and this is Galerkin's method. Thus we have showed that the Rayleigh–Ritz method is identical to Galerkin's method when applied to the eigenvalue problem.

18-5 FINITE-ELEMENT METHOD

The finite-element method was developed originally for problems in structure mechanics. However, in recent years it has been applied to many other physical problems and its relations with other numerical techniques have been clarified. In this section we discuss the basic concept using some simple examples.

Let us consider a solution to a two-dimensional Laplace equation. An example is the determination of the quasi-static characteristic impedance of strip lines (see Section 14-3). The potential $V(x, y)$ satisfies the Laplace equation with the boundary condition that $V = V_1$ at the surface of one conductor and $V = V_2$ at the surface of the other conductor.

$$\nabla \cdot (\epsilon \nabla V) = 0, \qquad (18\text{-}53)$$

$$V = V_1 \text{ on } C_1,$$

$$V = V_2 \text{ on } C_2.$$

The dielectric constant ϵ may be different in different regions (Fig. 18-4). To solve this problem using the finite-element method, we consider the electrostatic energy W:

$$W = \iint \frac{\epsilon}{2} |E|^2 \, dx\, dy = \iint \frac{\epsilon}{2} |\nabla V|^2 \, dx\, dy. \qquad (18\text{-}54)$$

First we show that if V satisfies the Laplace equation, the energy W is stationary with respect to the variation δV. To prove this, we take a variation of (18-54):

$$\delta W = \iint \epsilon \nabla V \cdot \nabla(\delta V) \, dx\, dy. \qquad (18\text{-}55)$$

Now we make use of the divergence theorem:

$$\int_s \nabla \cdot \bar{A} \, dx\, dy = \oint \bar{A} \cdot \hat{n} \, dl$$

and let $\bar{A} = (\delta V)\, \epsilon \nabla V$. Then noting that $\nabla \cdot \bar{A} = \epsilon \nabla V \cdot \nabla(\delta V) + \delta V \nabla \cdot (\epsilon \nabla V)$, we get

$$\delta W = -\iint \nabla \cdot (\epsilon V)\, \delta V \, dx\, dy + \oint \frac{\partial(\epsilon V)}{\partial n} \delta V \, dl. \qquad (18\text{-}56)$$

Now $\delta V = 0$ on the boundary if we choose V satisfying the boundary condition $V = \text{constant}$. We therefore conclude from (18-56) that if $\nabla \cdot (\epsilon V) = 0$, then $\delta W = 0$.

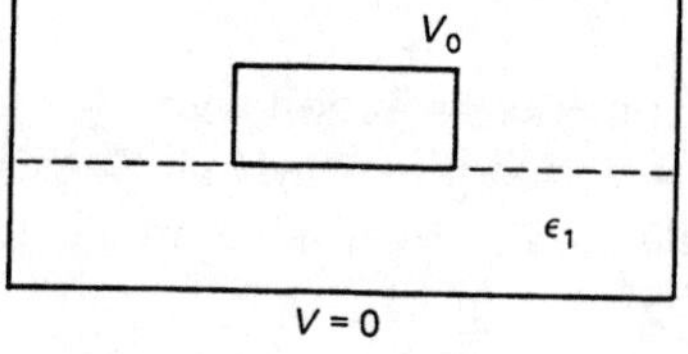

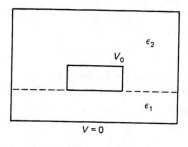

Figure 18-4 Cross section of a strip line.

Conversely, if we can find V such that $\delta W = 0$, V satisfies the equation $\nabla \cdot (\epsilon V) = 0$. The finite-element solution consists of finding the approximate solution V such that the energy W is stationary $\delta W = 0$.

We can generalize the above to show that if a function u satisfies

$$\nabla \cdot (p \nabla u) + k^2 u = g, \tag{18-57a}$$

then the variation $\delta W = 0$, where

$$W = \tfrac{1}{2} \int [p |\nabla u|^2 - k^2 u^2 + 2gu] \, dx \, dy, \tag{18-57b}$$

provided that u satisfies Dirichlet's condition on the boundary.

Let us now apply the finite-element method to this problem. We first divide the domain under consideration into small cells, called *elements*. The elements are connected at specified points called *nodes*. The unknown function $V(x, y)$ is then expressed in terms of the values of the function at these nodes, and between the nodes the function is approximated linearly. We write

$$V(x, y) = \sum_{m=1}^{M} N_m(x, y) \alpha_m, \tag{18-58}$$

where the basis function $N_m(x, y)$, called the *shape function,* is unity at (x_m, y_m) and decreases linearly to zero at the neighboring nodes. Therefore, it is a pyramid function. $N_3(x, y)$ and $N_5(x, y)$ are shown in Fig. 18-5. The nodes can be chosen at any convenient point, and in particular, nodes can be placed closer to each other where the field is expected to vary rapidly and far apart where the field may vary slowly (Fig. 18-6). This flexibility is an important advantage of the finite-element method.

The shape function $N_m(x, y)$ and the potential $V(x, y)$ in each element can be expressed in terms of the locations of three nodes and the values of the potential at these nodes. Consider the function $V(x, y)$ inside the triangular element with nodes at (x_1, y_1), (x_2, y_2), and (x_3, y_3) (Fig. 18-7). Since the function is linear, we let

$$V(x, y) = a + bx + cy. \tag{18-59}$$

The values of $V(x, y)$ at these three nodes are V_1, V_2, and V_3. Therefore, we get

$$\begin{aligned} V_1 &= a + bx_1 + cy_1, \\ V_2 &= a + bx_2 + cy_2, \\ V_3 &= a + bx_3 + cy_3. \end{aligned} \tag{18-60}$$

Writing this in matrix form, we get

$$\begin{bmatrix} V_1 \\ V_2 \\ V_3 \end{bmatrix} = \begin{bmatrix} 1 & x_1 & y_1 \\ 1 & x_2 & y_2 \\ 1 & x_3 & y_3 \end{bmatrix} \begin{bmatrix} a \\ b \\ c \end{bmatrix}. \tag{18-61}$$

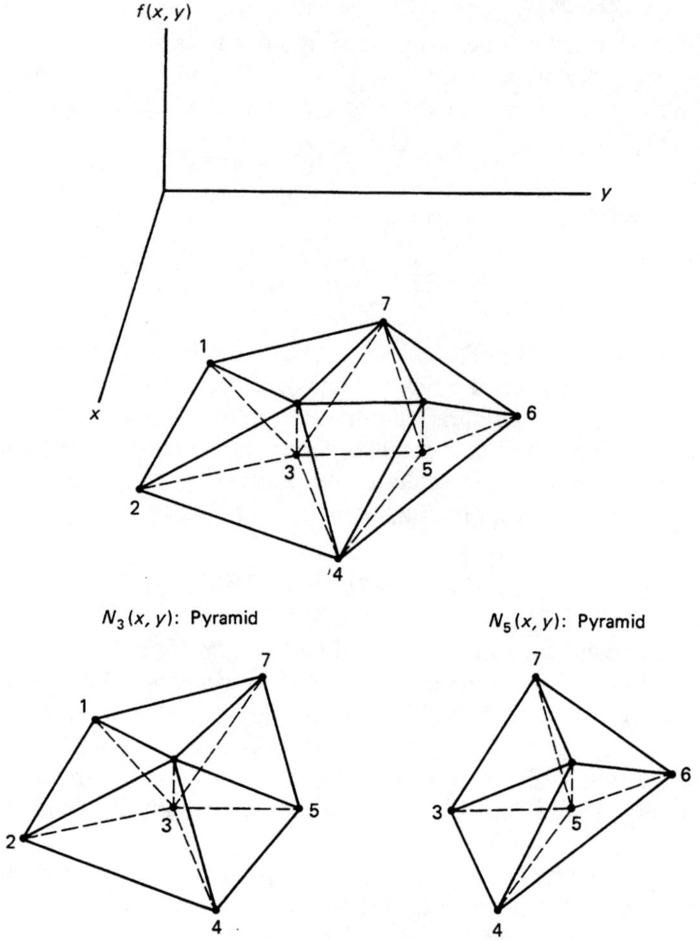

Figure 18-5 Basis function.

From this we easily get a, b, and c. The function $V(x, y)$ is, therefore, given by

$$V(x,y) = [1 \quad x \quad y]\begin{bmatrix} 1 & x_1 & y_1 \\ 1 & x_2 & y_2 \\ 1 & x_3 & y_3 \end{bmatrix}^{-1}\begin{bmatrix} V_1 \\ V_2 \\ V_3 \end{bmatrix}$$

$$= V_1N_1(x,y) + V_2N_2(x,y) + V_3N_3(x,y),$$

where

$$N_1 = \frac{1}{2A}[x_2y_3 - x_3y_2) + (y_2 - y_3)x + (x_3 - x_2)y].$$

$$A = \text{area of triangle} = \frac{1}{2}\begin{vmatrix} 1 & x_1 & y_1 \\ 1 & x_2 & y_2 \\ 1 & x_3 & y_3 \end{vmatrix}.$$

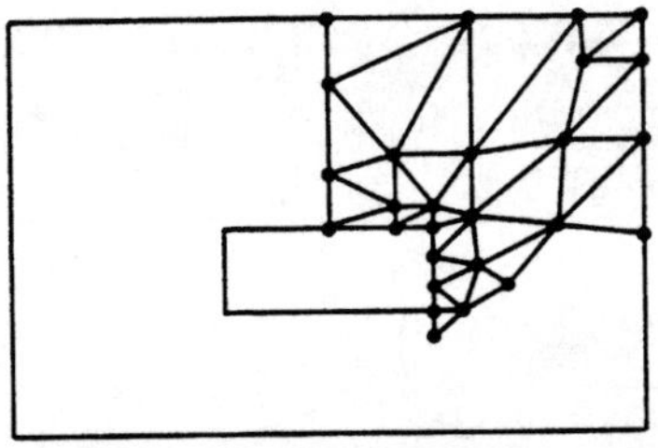

Figure 18-6 Finite elements and nodes.

N_2 and N_3 are obtained by cyclic interchange of the subscripts (1→2, 2→3, 3→1). Note that N_1, N_2, and N_3 are the shape function given in (18-58). The gradient ∇V is then given by

$$\nabla V = V_1 \nabla N_1 + V_2 \nabla N_2 + V_3 \nabla N_3. \tag{18-63}$$

Summarizing this, we write for the element (e),

$$\begin{aligned} V(x, y) &= \sum_{i=1}^{3} V_i N_i, \\ \nabla V(x, y) &= \sum_{i=1}^{3} V_i \nabla N_i, \end{aligned} \tag{18-64}$$

where N_i is given in (18-62).

Now we substitute (18-64) into the energy equation (18-54). For the element (e), we have

$$\begin{aligned} W^e &= \frac{1}{2} \sum_{i=1}^{3} \sum_{j=1}^{3} V_i V_j \iint \varepsilon \nabla N_i \cdot \nabla N_j \, dx \, dy \\ &= \frac{1}{2} \tilde{V} S V, \end{aligned} \tag{18-65}$$

where

$$V = \begin{bmatrix} V_1 \\ V_2 \\ V_3 \end{bmatrix}, \qquad S = 3 \times 3 \text{ symmetric matrix} = [S_{ij}],$$

$$S_{ij} = \iint \varepsilon \nabla N_i \cdot \nabla N_j \, dx \, dy,$$

$$S_{11} = \frac{\varepsilon}{4A} [(y_2 - y_3)^2 + (x_3 - x_2)^2],$$

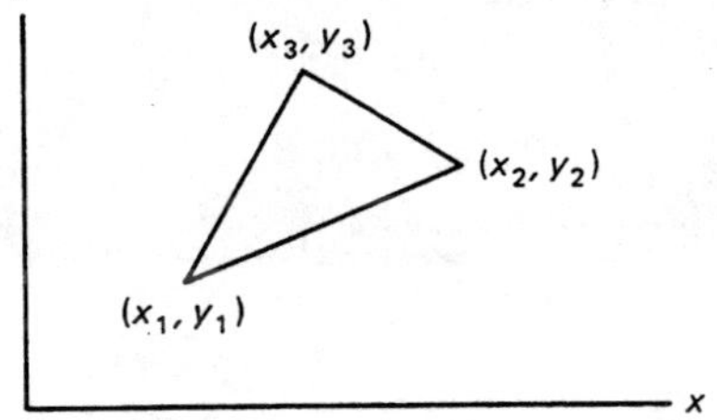

Figure 18-7 Triangular element with nodes at (x_1, y_1), (x_2, y_2), and (x_3, y_3).

$$S_{22} = \frac{\varepsilon}{4A}[(y_3 - y_1)^2 + (x_1 - x_3)^2],$$

$$S_{33} = \frac{\varepsilon}{4A}[(y_1 - y_2)^2 + (x_2 - x_1)^2],$$

$$S_{12} = \frac{\varepsilon}{4A}[(y_2 - y_3)(y_3 - y_1) + (x_3 - x_2)(x_1 - x_3)],$$

$$S_{13} = \frac{\varepsilon}{4A}[(y_2 - y_3)(y_1 - y_2) + (x_3 - x_2)(x_2 - x_1)],$$

$$S_{23} = \frac{\varepsilon}{4A}[(y_3 - y_1)(y_1 - y_2) + (x_1 - x_3)(x_2 - x_1)].$$

It is clear from the expression for S_{ij} that the matrix S is symmetric, $\tilde{S} = S$.

We now assemble the contributions from all elements:

$$W = \sum_e W^e, \tag{18-66}$$

To facilitate this assembly, first consider the "disjoint" elements (1) and (2) (Fig. 18-8). We can write the total energy (18-66) using $V_1, V_2, \ldots, V_6$.

$$W = \tfrac{1}{2}\tilde{V}_d S_d V_d, \tag{18-67}$$

where

$$V_d = \begin{bmatrix} V_1 \\ V_2 \\ \vdots \\ V_6 \end{bmatrix}$$

$$S_d = \begin{bmatrix} S_d^{(1)} & 0 \\ 0 & S_d^{(2)} \end{bmatrix}$$

$$\begin{aligned} S_d^{(1)} &= 3 \times 3 \text{ matrix for the element (1)} \\ &= [S_{ij}^{(1)}], \qquad i, j = 1, 2, 3 \\ S_d^{(2)} &= 3 \times 3 \text{ matrix for the element (2)} \\ &= [S_{ij}^{(2)}], \qquad i, j = 4, 5, 6 \end{aligned}$$

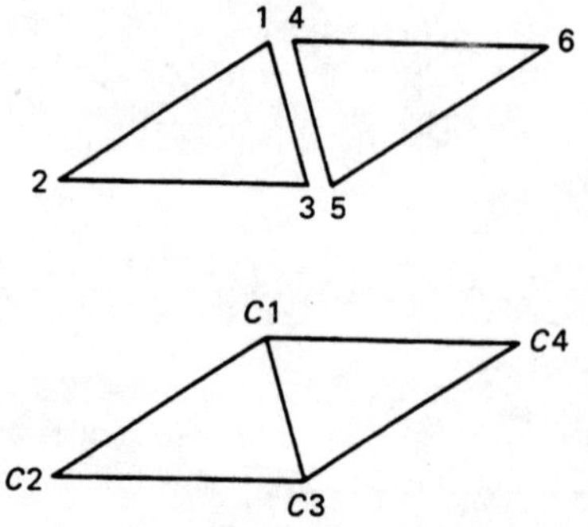

Figure 18-8 Disjoint and conjoint elements.

Note that $S_d^{(1)}$ and $S_d^{(2)}$ are symmetric, and therefore S_d is symmetric. After assembling the elements, $V_1 = V_4$, $V_3 = V_5$, and therefore for the connected problem, we have only V_{c1}, V_{c2}, V_{c3}, and V_{c4}. These are related to the disjoint set V_d by the following:

$$V_d = \begin{bmatrix} 1 & 0 & 0 & 0 \\ 0 & 1 & 0 & 0 \\ 0 & 0 & 1 & 0 \\ 1 & 0 & 0 & 0 \\ 0 & 0 & 1 & 0 \\ 0 & 0 & 0 & 1 \end{bmatrix} \begin{bmatrix} V_{c1} \\ V_{c2} \\ V_{c3} \\ V_{c4} \end{bmatrix}. \tag{18-68}$$

We write this as

$$V_d = CV_c, \tag{18-69}$$

where V_c means the potentials for the connected elements. The total energy is then given by

$$W = \tfrac{1}{2}\tilde{V}_c S V_c, \tag{18-70}$$

where $S = \tilde{C}S_dC$. Since S_d is symmetric, S is also symmetric; $\tilde{S} = S$.

Now we have the total energy (18-70) for the connected problem. We next find V_c such that W is stationary with respect to the variations of V_c. To do this, we first divide V_c into the potentials V_f at N nodes where the potentials are "free" to vary, and the potentials V_p at P nodes where the potentials are "prescribed." For example, the potentials at the conducting surface are prescribed while the potentials between the conductors are free to vary. Thus we write

$$W = \tfrac{1}{2}[\tilde{V}_f \quad \tilde{V}_p] \begin{bmatrix} S_{ff} & S_{fp} \\ S_{pf} & S_{pp} \end{bmatrix} \begin{bmatrix} V_f \\ V_p \end{bmatrix}, \tag{18-71}$$

where

$$V_f = N \times 1 \text{ matrix},$$

$$V_p = P \times 1 \text{ matrix},$$

$$S_{ff} = N \times N \text{ matrix},$$

$$S_{fp} = \tilde{S}_{pf} = N \times P \text{ matrix},$$

$$S_{pp} = P \times P \text{ matrix}.$$

If we let

$$\tilde{V}_f = [V_1, V_2, \ldots, V_k, \ldots, V_N],$$

then W is stationary if

$$\frac{\partial W}{\partial V_K} = 0, \qquad k = 1, \ldots, N. \tag{18-72}$$

The differentiation in (18-72) is limited to those with respect to the N free potentials V_f since the prescribed potentials V_p cannot vary. Performing the differentiation (18-72), and noting that S is symmetric and therefore $\tilde{S}_{pf} = S_{fp}$, we get

$$[S_{ff} \quad S_{fp}]\begin{bmatrix} V_f \\ V_p \end{bmatrix} = 0. \tag{18-73}$$

The unknown potential V_f is then given by the following matrix equation:

$$[S_{ff}][V_f] = -[S_{fp}][V_p]. \tag{18-74}$$

The solution $[V_f]$ is therefore given by

$$[V_f] = -[S_{ff}]^{-1}[S_{fp}][V_p]. \tag{18-75}$$

This, together with the prescribed potential $[V_p]$, gives the complete potential distribution at all the nodes. Once the potential at each node is found, the potential at any point is given by (18-62).

In this section we outlined the basic ideas of the finite-element method using a single two-dimensional electrostatic problem. The finite-element method has been applied to many electromagnetic problems. In this section we considerd the two-dimensional problem and therefore used pyramid functions for the shape function $N_m(x, y)$. For three-dimensional problems, however, we need to use tetrahedral functions for the shape function that is unity at (x_i, y_i, z_i) and decreases linearly to zero at (x_j, y_j, z_j), (x_m, y_m, z_m), and (x_p, y_p, z_p).

$$N(x,y,z) = \frac{a_i + b_i x + c_i y + d_i z}{6V}, \tag{18-76}$$

$$6V = \begin{vmatrix} 1 & x_i & y_i & z_i \\ 1 & x_j & y_j & z_j \\ 1 & x_m & y_m & z_m \\ 1 & x_p & y_p & z_p \end{vmatrix},$$

$$a_i = \begin{vmatrix} x_j & y_j & z_j \\ x_m & y_m & z_m \\ x_p & y_p & z_p \end{vmatrix},$$

$$b_i = -\begin{vmatrix} 1 & y_j & z_j \\ 1 & y_m & z_m \\ 1 & y_p & z_p \end{vmatrix},$$

$$c_i = -\begin{vmatrix} x_j & 1 & z_j \\ x_m & 1 & z_m \\ x_p & 1 & z_p \end{vmatrix},$$

$$d_i = -\begin{vmatrix} x_j & y_j & 1 \\ x_m & y_m & 1 \\ x_p & y_p & 1 \end{vmatrix}.$$

Up to this point we have used pyramid and tetrahedral functions for the shape function N_n. They are all characterized by *linear* dependence on x, y, or z. However, it is possible and sometimes more convenient to use polynomials of a higher order. For example, quadratic (x, y, xy, x^2, and y^2), cubic, or quartic shape functions can be used.

18-6 FINITE-DIFFERENCE METHOD

One of the simplest numerical methods for solving differential equations is to use the finite difference approximation of derivatives. In this section we outline the method using some simple examples.

Consider an electrostatic problem. The electric field $\overline{E}$ satisfies the following equations:

$$\begin{aligned} \nabla \times \overline{E} &= 0, \\ \nabla \cdot (\varepsilon \overline{E}) &= \rho. \end{aligned} \tag{18-77}$$

Since $\nabla \times \overline{E} = 0$, we can use the potential V.

$$\overline{E} = -\nabla V. \tag{18-78}$$

Substituting this into the second equation of (18-77), we get

$$\nabla \cdot (\varepsilon \nabla V) = -\rho. \tag{18-79}$$

The dielectric constant ε can be a function of position. Let us first consider a one-dimensional problem:

$$\frac{\partial}{\partial x}\left(\varepsilon \frac{\partial}{\partial x} V\right) = -\rho. \tag{18-80}$$

We assume that the charge density $\rho = \rho(x)$ is known and we impose the Dirichlet boundary condition:

$$\begin{aligned} V &= V_0 \quad \text{at } x_0, \\ V &= V_l \quad \text{at } x_l. \end{aligned}$$

We now convert the differential equation into a difference equation. We subdivide the range $x_0 \le x \le x_l$ into N equal intervals Δx. The derivative at $x = x_i$ is then approximated by

$$\frac{d}{dx} V \approx \frac{V_{i+1} - V_{i-1}}{2\Delta x}, \tag{18-81}$$

where $V_i = V(x_i)$. Here we used the *central difference approximation*. Its error is of the order of $(\Delta x)^2$. In contrast, the forward $(V_{i+1} - V_i)/\Delta x$ and the backward

$(V_i - V_{i-1})/\Delta x$ difference approximations have the error of the order of Δx. For the second derivative in (18-80), we use (18-81):

$$\frac{d}{dx}\left(\varepsilon\frac{dV}{dx}\right) = \frac{\varepsilon \left.\frac{dV}{dx}\right|_{i+1/2} - \varepsilon \left.\frac{dV}{dx}\right|_{i-1/2}}{\Delta x}. \qquad (18\text{-}82)$$

But

$$\left.\frac{dV}{dx}\right|_{i+1/2} = \frac{V_{i+1} - V_i}{\Delta x},$$

$$\left.\frac{dV}{dx}\right|_{i-1/2} = \frac{V_i - V_{i-1}}{\Delta x}.$$

We therefore obtain

$$\frac{d}{dx}\left(\varepsilon\frac{dV}{dx}\right) \approx \frac{\varepsilon_{i+1/2}V_{i+1} - (\varepsilon_{i+1/2} + \varepsilon_{i-1/2})V_i + \varepsilon_{i-1/2}V_{i-1}}{\Delta x^2}. \qquad (18\text{-}83)$$

Substituting this into (18-80), we get the following matrix equation for $V_1, V_2, \ldots, V_{N-1}$:

$$\bar{\bar{K}}\bar{V} = \bar{f}, \qquad (18\text{-}84)$$

where the matrix elements are shown for $N = 4$ using $\varepsilon_{i+} = \varepsilon_{i+1/2}$ and $\varepsilon_{i-} = \varepsilon_{i-1/2}$ (Fig. 18-9):

$$\bar{\bar{K}} = \begin{bmatrix} \varepsilon_{1-} + \varepsilon_{1+} & -\varepsilon_{1+} & 0 & 0 \\ -\varepsilon_{1+} & \varepsilon_{2-} + \varepsilon_{2+} & -\varepsilon_{2+} & 0 \\ 0 & -\varepsilon_{2+} & \varepsilon_{3-} + \varepsilon_{3+} & -\varepsilon_{3+} \\ 0 & 0 & -\varepsilon_{3+} & \varepsilon_{4-} + \varepsilon_{4+} \end{bmatrix},$$

$$\bar{V} = \begin{bmatrix} V_1 \\ V_2 \\ V_3 \\ V_4 \end{bmatrix},$$

$$\bar{f} = \begin{bmatrix} \rho_1\Delta x^2 + \varepsilon_{1-}V_0 \\ \rho_2\Delta x^2 \\ \rho_3\Delta x^2 \\ \rho_4\Delta x^2 + \varepsilon_{4+}V_l \end{bmatrix}.$$

Note that $\bar{\bar{K}}$ is a symmetric matrix. From (18-84) we obtain the solution

$$\bar{V} = \bar{\bar{K}}^{-1}f. \qquad (18\text{-}85)$$

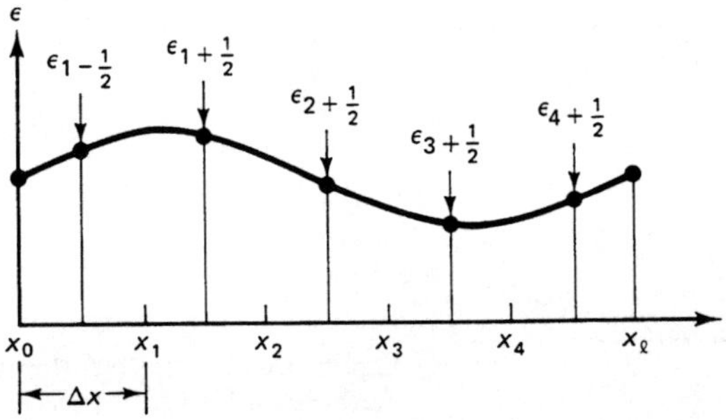

Figure 18-9 Dielectric constant as a function of x.

Let us next consider Neumann's boundary condition. At $x = x_l$, we assume that

$$\frac{dV}{dx} = -E_0. \tag{18-86}$$

where E_0 is a given constant. We then have the additional unknown voltage V_l. The matrix equation is therefore given by

$$\bar{\bar{K}}\bar{V} = \bar{f}, \tag{18-87}$$

where

$$\bar{\bar{K}} = \begin{bmatrix} \varepsilon_{1-} + \varepsilon_{1+} & -\varepsilon_{1+} & 0 & 0 & 0 \\ -\varepsilon_{1+} & \varepsilon_{2-} + \varepsilon_{2+} & -\varepsilon_{2+} & 0 & 0 \\ 0 & -\varepsilon_{2+} & \varepsilon_{3-} + \varepsilon_{3+} & -\varepsilon_{3+} & 0 \\ 0 & 0 & -\varepsilon_{3+} & \varepsilon_{4-} + \varepsilon_{4+} & -\varepsilon_{4+} \end{bmatrix},$$

$$\bar{V} = \begin{bmatrix} V_1 \\ V_2 \\ V_3 \\ V_4 \\ V_l \end{bmatrix},$$

$$\bar{f} = \begin{bmatrix} \rho_1 \Delta x^2 + \varepsilon_{1-} V_0 \\ \rho_2 \Delta x^2 \\ \rho_3 \Delta x^2 \\ \rho_4 \Delta x^2 \end{bmatrix}.$$

This is a set of four equations for five unknowns. The additional equation is provided by the boundary condition:

$$\frac{V_l - V_4}{\Delta x} = -E_0. \tag{18-88}$$

These two equations (18-87) and (18-88) give five equations for five unknowns, V_1, V_2, V_3, V_4, and V_l.

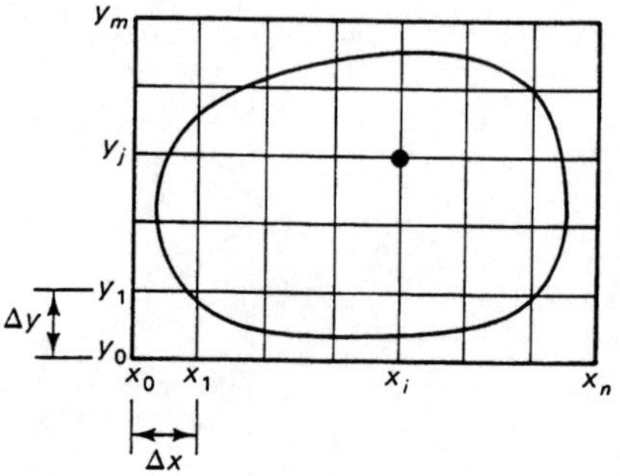

Figure 18-10 Finite difference method applied to a two-dimensional problem.

Let us next consider the two-dimensional problem:

$$\frac{\partial}{\partial x}\left(\varepsilon\frac{\partial V}{\partial x}\right)+\frac{\partial}{\partial y}\left(\varepsilon\frac{\partial V}{\partial y}\right)=-\rho. \tag{18-89}$$

We divide the region into an equally spaced grid (Fig. 18-10). At the point (x_i, y_j), the derivatives of the potential $V_{ij} = V(x_i, y_j)$ can be expressed, following the same procedure as shown for the one-dimensional problem. We convert (9-89) into the finite-difference equation by using the following:

$$\frac{d}{dx}\left(\varepsilon\frac{dV}{dx}\right)\approx\frac{\varepsilon_{i+,j}V_{i+1,j}-(\varepsilon_{i+,j}+\varepsilon_{i-,j})V_{i,j}+\varepsilon_{i-,j}V_{i-1,j}}{\Delta x^2},$$

$$\frac{d}{dy}\left(\varepsilon\frac{dV}{dy}\right)\approx\frac{\varepsilon_{i,j+}V_{i,j+1}-(\varepsilon_{i,j+}+\varepsilon_{i,j-})V_{i,j}+\varepsilon_{i,j-}V_{i-1,j}}{\Delta y^2}, \tag{18-90}$$

$$\rho\approx\rho_{ij},$$

where $\varepsilon_{1+,j} = \varepsilon_{i+1/2,j}$. We can then apply the boundary condition and obtain the matrix equation in a manner similar to the one-dimensional problem.

PROBLEMS

18-1. Solve the following eigenvalue problem numerically, and find the first two equivalent λ_1 and λ_2 values.

$$\left[\frac{d^2}{dx^2}+\lambda(1-x^2)\right]f=0,$$

$$f=0 \quad \text{for } x=0 \text{ and } x=1.$$

18-2. Solve the following problem using the finite-element method, and compare with the exact solution

$$\frac{d^2}{dx^2}f=-1-x,$$

$$f=0 \quad \text{at } x=0 \text{ and } x=1.$$

Appendix to Chapter 2

A. MATHEMATICAL FORMULAS

2A-1 VECTOR FORMULAS AND THEOREMS

$$\nabla \cdot (\phi \overline{A}) = \phi \nabla \cdot \overline{A} + \overline{A} \cdot \nabla \phi$$

$$\nabla \cdot (\overline{A} \times \overline{B}) = \overline{B} \cdot \nabla \times \overline{A} - \overline{A} \cdot \nabla \times \overline{B}$$

$$\nabla \times (\phi \overline{A}) = \nabla \phi \times \overline{A} + \phi \nabla \times \overline{A}$$

$$\nabla \times (\overline{A} \times \overline{B}) = \overline{A} \nabla \cdot \overline{B} - \overline{B} \nabla \cdot \overline{A} + (\overline{B} \cdot \nabla)\overline{A} - (\overline{A} \cdot \nabla)\overline{B}$$

$$\nabla \cdot \nabla \times \overline{A} = 0$$

$$\nabla \times \nabla \phi = 0$$

$$\overline{A} \cdot \overline{B} \times \overline{C} = \overline{B} \cdot \overline{C} \times \overline{A} = \overline{C} \cdot \overline{A} \times \overline{B}$$

$$\overline{A} \times (\overline{B} \times \overline{C}) = \overline{B}(\overline{A} \cdot \overline{C}) - \overline{C}(\overline{A} \cdot \overline{B})$$

Divergence Theorem

$$\int_v \nabla \cdot \bar{A}\, dv = \int_s \bar{A} \cdot d\bar{a}.$$

Stokes' Theorem

$$\int_a \nabla \times \bar{A} \cdot d\bar{a} = \oint_l \bar{A} \cdot d\bar{l}.$$

2A-2 GRADIENT, DIVERGENCE, CURL, AND LAPLACIAN

Cartesian System

$$\nabla f = \left(\frac{\partial}{\partial x}\hat{x} + \frac{\partial}{\partial y}\hat{y} + \frac{\partial}{\partial z}\hat{z}\right) f$$

$$\nabla \cdot A = \frac{\partial}{\partial x}A_x + \frac{\partial}{\partial y}A_y + \frac{\partial}{\partial z}A_z$$

$$\nabla \times A = \begin{vmatrix} \hat{x} & \hat{y} & \hat{z} \\ \dfrac{\partial}{\partial x} & \dfrac{\partial}{\partial y} & \dfrac{\partial}{\partial z} \\ A_x & A_y & A_z \end{vmatrix}$$

$$\nabla^2 f = \left(\frac{\partial^2}{\partial x^2} + \frac{\partial^2}{\partial y^2} + \frac{\partial^2}{\partial z^2}\right) f$$

Cylindrical System

$$\nabla f = \left(\frac{\partial}{\partial \rho}\hat{\rho} + \frac{1}{\rho}\frac{\partial}{\partial \phi}\hat{\phi} + \frac{\partial}{\partial z}\hat{z}\right) f$$

$$\nabla \cdot \bar{A} = \frac{1}{\rho}\frac{\partial}{\partial \rho}(\rho A_\rho) + \frac{1}{\rho}\frac{\partial}{\partial \phi}A_\phi + \frac{\partial}{\partial z}A_z$$

$$\nabla \times \bar{A} = \frac{1}{\rho}\begin{vmatrix} \hat{\rho} & \rho\hat{\phi} & \hat{z} \\ \dfrac{\partial}{\partial \rho} & \dfrac{\partial}{\partial \phi} & \dfrac{\partial}{\partial z} \\ A_\rho & \rho A_\phi & A_z \end{vmatrix}$$

$$\nabla^2 f = \left[\frac{1}{\rho}\frac{\partial}{\partial \rho}\left(\rho\frac{\partial}{\partial \rho}\right) + \frac{1}{\rho^2}\frac{\partial^2}{\partial \phi^2} + \frac{\partial^2}{\partial z^2}\right] f$$

Spherical System

$$\nabla f = \left(\hat{r}\frac{\partial}{\partial r} + \hat{\theta}\frac{\partial}{r\,\partial\theta} + \hat{\phi}\frac{1}{r\sin\theta}\frac{\partial}{\partial\phi}\right)f$$

$$\nabla\cdot\overline{A} = \frac{1}{r^2}\frac{\partial}{\partial r}(r^2 A_r) + \frac{1}{r\sin\theta}\frac{\partial}{\partial\theta}(\sin\theta\, A_\theta) + \frac{1}{r\sin\theta}\frac{\partial}{\partial\phi}A_\phi$$

$$\nabla\times\overline{A} = \frac{1}{r^2\sin\theta}\begin{vmatrix} \hat{r} & r\hat{\theta} & r\sin\theta\hat{\phi} \\ \dfrac{\partial}{\partial r} & \dfrac{\partial}{\partial\theta} & \dfrac{\partial}{\partial\phi} \\ A_r & rA_\theta & r\sin\theta\, A_\phi \end{vmatrix}$$

$$\nabla^2 f = \left[\frac{1}{r^2}\frac{\partial}{\partial r}\left(r^2\frac{\partial}{\partial r}\right) + \frac{1}{r^2\sin\theta}\frac{\partial}{\partial\theta}\left(\sin\theta\frac{\partial}{\partial\theta}\right) + \frac{1}{r^2\sin^2\theta}\frac{\partial^2}{\partial\phi^2}\right]f$$

Appendix to Chapter 3

A. THE FIELD NEAR THE TURNING POINT

Consider the differential equation

$$\left[\frac{d^2}{dz^2} + q^2(z)\right]u(z) = 0, \tag{3A-1}$$

with $q^2(z)$ varying as shown in Fig. 3-26. We will discuss the field in the neighborhood of the turning point z (region II). Here the WKB approximation fails, and we need a different approach. In this region, let us expand $q^2(z)$ about the turning point in a Taylor's series and retain its first term:

$$q^2(z) = -a(z - z_0), \tag{3A-2}$$

where a is the slope at z_0:

$$a = -\left.\frac{d(q^2)}{dz}\right|_{z=z_0}$$

Then the differential equation becomes

$$\left[\frac{d^2}{dz^2} - a(z - z_0)\right]u(z) = 0. \tag{3A-3}$$

We now convert this into the Stokes differential equation by using

$$t = a^{1/3}(z - z_0) = -a^{-2/3} q^2(z), \tag{3A-4}$$

and obtain

$$\left(\frac{d^2}{dt^2} - t\right) u(t) = 0. \tag{3A-5}$$

This is Stokes equation, and the solution satisfying the radiation condition as $t \to \infty$ is given by (see Appendix 3B)

$$u_a(t) = D_0 A_i(t), \tag{3A-6}$$

where D_0 is constant and $A_i(t)$ is an Airy integral.

Let us examine the behavior of (3A-6) toward region I. In this region, t is negative and large as seen from (3A-4), and therefore, asymptotically,

$$\begin{aligned} u_a(t) &= D_0 \frac{1}{\sqrt{\pi}(-t)^{1/4}} \sin\left[\frac{2}{3}(-t)^{3/2} + \frac{\pi}{4}\right] \\ &= \frac{D_0}{\sqrt{\pi}(-t)^{1/4}} \frac{e^{+j[2/3(-t)^{3/2} + \pi/4]} - e^{-j[2/3(-t)^{3/2} + \pi/4]}}{2j}. \end{aligned} \tag{3A-7}$$

Now we note from (3A-2) that

$$-t = a^{-2/3} q^2(z) = -a^{1/3}(z - z_0),$$

and

$$\begin{aligned} \frac{2}{3}(-t)^{3/2} &= \int_0^{(-t)} (-t)^{1/3}\, d(-t) \\ &= \int_{z_0}^{z} q(z)\, dz. \end{aligned} \tag{3A-8}$$

Therefore, we get

$$u_a(z) = \frac{D_0 e^{+j(\pi/4)}}{2j\sqrt{\pi}\, a^{-1/6} q^{1/2}} \left[e^{-j\int_{z_0}^{z} q(z)dz} - e^{+j\int_{z_0}^{z} q(z)dz - j(\pi/2)}\right]. \tag{3A-9}$$

Now, we see that (3A-9) can be smoothly connected to the WKB solution in region I:

$$u_i(z) = \frac{A_0}{q^{1/2}} e^{-j\int_{z_0}^{z} q dz}. \tag{3A-10}$$

The reflected WKB wave is

$$u_r(z) = \frac{B_0}{q^{1/2}} e^{+j\int_{z_0}^{z} q dz}. \tag{3A-11}$$

Comparing this with the second term of (3A-9), we get

$$B_0 = A_0 e^{-j\int_{z}^{z_0} 2q dz + j(\pi/2)}. \tag{3A-12}$$

The transmitted wave is obtained by examining $u_a(t)$ of (3A-6) in region III where $t > 0$ and is large:

$$u_a(t) = D_0 \frac{1}{2\sqrt{\pi}\, t^{1/4}} e^{-(2/3)t^{3/2}}, \tag{3A-13}$$

which may be written as

$$u_a(t) = \frac{D_0 e^{-j(\pi/4)}}{2\sqrt{\pi}\, a^{-1/6} q^{1/2}} e^{-j\int_{z_0}^{z} q\,dz}. \tag{3A-14}$$

Comparing this with the transmitted WKB solution,

$$u_t(z) = \frac{C_0}{q^{1/2}} e^{-j\int_{z_0}^{z} q\,dz}, \tag{3A-15}$$

we get

$$C_0 = A_0 e^{-j\int_0^{z_0} q\,dz}. \tag{3A-16}$$

B. STOKES DIFFERENTIAL EQUATION AND AIRY INTEGRAL

Consider the Stokes differential equation

$$\left(\frac{d^2}{dt^2} - t\right) u(t) = 0. \tag{3B-1}$$

The two independent solutions can be written in the following form:

$$u(t) = w_1(t) = \frac{1}{\sqrt{\pi}} \int_{\Gamma_1} e^{t\lambda - \lambda^3/3}\, d\lambda, \tag{3B-2}$$

$$u(t) = w_2(t) = \frac{1}{\sqrt{\pi}} \int_{\Gamma_2} e^{t\lambda - \lambda^3/3}\, d\lambda, \tag{3B-3}$$

where Γ_1 and Γ_2 are two independent paths originating in one of the shaded regions and ending in another region in the λ plane where the integrand vanishes (Fig. 3B-1).

To show that $w_1(t)$ and $w_2(t)$ are solutions of the Stokes equation, we substitute them into (3B-1). Then we get

$$\begin{aligned}\left(\frac{d^2}{dt^2} - t\right) w_1(t) &= \frac{1}{\sqrt{\pi}} \int_{\Gamma_2} (\lambda^2 - t) e^{t\lambda - \lambda^3/3}\, d\lambda, \\ &= \frac{-1}{\sqrt{\pi}} \int_{\Gamma_2} d(e^{t\lambda - \lambda^3/3}) \\ &= \frac{-1}{\sqrt{\pi}} e^{t\lambda - \lambda^3/3} \Big|_A^B = 0.\end{aligned} \tag{3B-4}$$

Similarly, $w_2(t)$ also satisfies (3B-1).

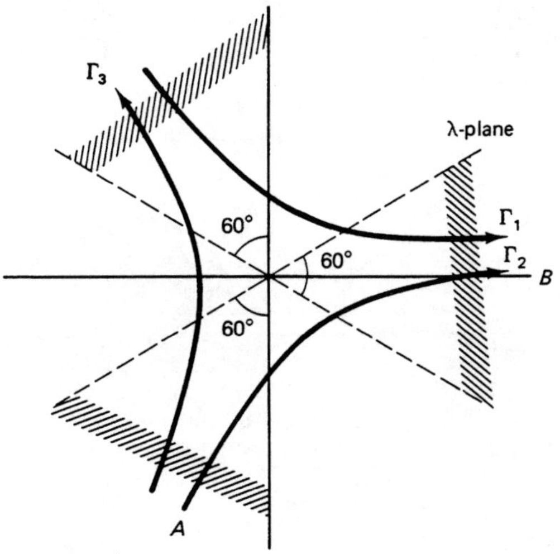

Figure 3B-1 Complex plane for Airy integrals.

It is also possible to obtain two independent solutions by linear combinations of (3B-2) and (3B-3). Two commonly used functions are $A_i(t)$ and $B_i(t)$, defined by

$$A_i(t) = \frac{1}{2\sqrt{\pi}}[w_2(t) - w_1(t)]$$
$$= \frac{1}{\pi}\int_0^\infty \cos\left(\frac{x^3}{3} + tx\right) dx, \tag{3B-5}$$

$$B_i(t) = \frac{1}{2\sqrt{\pi}}[w_2(t) + w_1(t)]$$
$$= \frac{1}{\pi}\int_0^\infty \left[e^{tx - x^3/3} + \sin\left(\frac{x^3}{3} + tx\right)\right] dx. \tag{3B-6}$$

To show (3B-5), we note that

$$w_2(t) - w_1(t) = \frac{1}{\sqrt{\pi}}\int_{\Gamma_3} e^{t\lambda - \lambda^3/3}\, d\lambda,$$

where Γ_3 is as shown in Fig. 3B-1. We now choose the path along the imaginary axis by letting $\lambda = jx$ and

$$\int_{\Gamma_3} d\lambda = \int_{-\infty}^{\infty} j\, dx$$

and obtain (3B-5). To obtain (3B-6), we note that

$$\int_{\Gamma_2} d\lambda = j\int_{-\infty}^{0} e^{j(tx + x^3/3)}\, dx + \int_0^\infty e^{tx - x^3/3}\, dx,$$

$$\int_{\Gamma_1} d\lambda = j\int_{\infty}^{0} e^{j(tx + x^3/3)}\, dx + \int_0^\infty e^{tx - x^3/3}\, dx,$$

and combine these two.

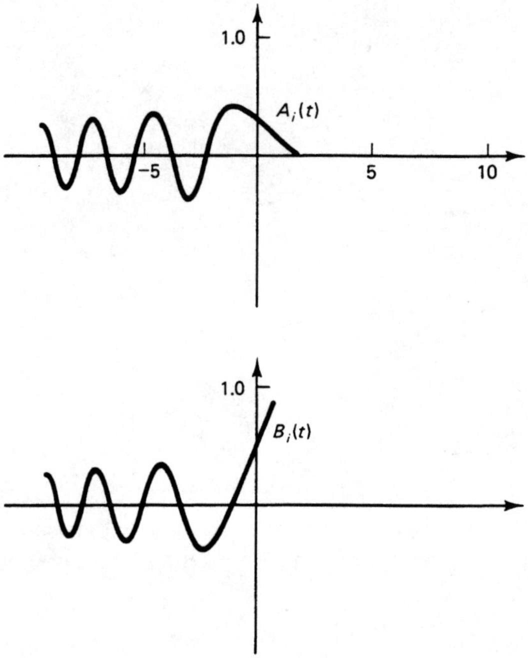

Figure 3B-2 Airy functions.

Along the real axis of t, $A_i(t)$ and $B_i(t)$ behave as shown in Fig. 3B-2. The asymptotic form of $A_i(t)$ and $B_i(t)$ is given below.

For $|t| \to \infty$ and $|\arg t| < \pi/3$,

$$A_i(t) \sim \frac{1}{2\sqrt{\pi}\, t^{1/4}} e^{-(2/3)t^{3/2}},$$

$$B_i(t) \sim \frac{1}{\sqrt{\pi}\, t^{1/4}} e^{(2/3)t^{3/2}}.$$

For $|t| \to \infty$ and $|\arg(-t)| < 2\pi/3$,

$$A_i(t) \sim \frac{1}{\sqrt{\pi}(-t)^{1/4}} \sin\left[\frac{2}{3}(-t)^{3/2} + \frac{\pi}{4}\right],$$

$$B_i(t) \sim \frac{1}{\sqrt{\pi}(-t)^{1/4}} \cos\left[\frac{2}{3}(-t)^{3/2} + \frac{\pi}{4}\right].$$

Appendix to Chapter 4

A. GREEN'S IDENTITIES AND THEOREM

Consider the divergence theorem applied to vector $\bar{A}$:

$$\int_V \nabla \cdot \bar{A}\, dV = \int_S \bar{A} \cdot d\bar{s}. \tag{4A-1}$$

Letting $\bar{A} = u \nabla v$, where u and v are scalar fields, we get

$$\int_V \nabla \cdot (u \nabla v)\, dV = \int_S u \nabla v \cdot d\bar{s} = \int_S u \frac{\partial v}{\partial n} dS, \tag{4A-2}$$

where $\partial/\partial n$ is the outward normal derivative. Using the vector identity on the left side of (4A-2),

$$\nabla \cdot (u \nabla v) = \nabla u \cdot \nabla v + u \nabla^2 v, \tag{4A-3}$$

we get *Green's first identity*,

$$\int_V (\nabla u \cdot \nabla v + u \nabla^2 v)\, dV = \int_S u \frac{\partial v}{\partial n} dS. \tag{4A-4}$$

Now we interchange u and v,

$$\int_V (\nabla u \cdot \nabla v + v \nabla^2 u)\, dV = \int_S v \frac{\partial u}{\partial n} dS. \tag{4A-5}$$

Subtracting (4A-5) from (4A-4), we get *Green's second identity*, which is also called *Green's theorem.*

$$\int_V (u\nabla^2 v - v\nabla^2 u)\, dV = \int_S \left(u\frac{\partial v}{\partial n} - v\frac{\partial u}{\partial n}\right) dS. \tag{4A-6}$$

The two-dimensional equivalents of Green's first and second identities are

$$\int_a (\nabla_t u \cdot \nabla_t v + v\nabla_t^2 u)\, da = \int_l v\frac{\partial u}{\partial n} dl, \tag{4A-7}$$

$$\int_a (u\nabla_t^2 v - v\nabla_t^2 u)\, da = \int_l \left(u\frac{\partial v}{\partial n} - v\frac{\partial u}{\partial n}\right) dl, \tag{4A-8}$$

where a is the area surrounded by a closed boundary curve l. The one-dimensional equivalents are

$$\int_{x_1}^{x_2} \left(\frac{\partial u}{\partial x}\frac{\partial v}{\partial x} + v\frac{\partial^2 u}{\partial x^2}\right) dx = v\frac{\partial u}{\partial x}\Big|_{x_1}^{x_2}, \tag{4A-9}$$

$$\int_{x_1}^{x_2} \left(u\frac{\partial^2 v}{\partial x^2} - v\frac{\partial^2 u}{\partial x^2}\right) dx = \left(u\frac{\partial v}{\partial x} - v\frac{\partial u}{\partial x}\right)\Big|_{x_1}^{x_2}. \tag{4A-10}$$

B. BESSEL FUNCTIONS $Z_\nu(x)$

Bessel functions $Z_\nu(z)$ are the solutions of Bessel's differential equation:

$$\left[z^2\frac{d^2}{dz^2} + z\frac{d}{dz} + (z^2 - \nu^2)\right] Z_\nu(z) = 0.$$

J_ν, N_ν, $H_\nu^{(1)}$, and $H_\nu^{(2)}$ are called the Bessel, Neumann, and Hankel functions of the first kind, and the Hankel function of the second kind, respectively. They are related by the following:

$$H_\nu^{(1)}(z) = J_\nu(z) + jN_\nu(z),$$
$$H_\nu^{(2)}(z) = J_\nu(z) - jN_\nu(z). \tag{4B-1}$$

If ν is a noninteger, J_ν and $J_{-\nu}$ are independent. When $\nu = n =$ integer, Z_n is proportional to Z_{-n}.

$$Z_{-n}(z) = (-1)^n Z_n(z). \tag{4B-2}$$

For real $\nu \geq 0$, we have for $|z| << 1$:

$$J_\nu(z) \approx \frac{1}{\Gamma(\nu+1)}\left(\frac{z}{2}\right)^\nu,$$

$$N_\nu(z) \approx -\frac{\Gamma(\nu)}{\pi}\left(\frac{2}{z}\right)^\nu,$$

$$J_0(z) \approx 1 - \left(\frac{z}{2}\right)^2, \tag{4B-3}$$

$$N_0(z) \approx -\frac{2}{\pi} \ln \frac{2}{\gamma z},$$

where $\Gamma(\nu)$ is the gamma function, $\Gamma(n) = (n-1)!$, and $\gamma = 1.781072418$, Euler's constant. For $|z| >> 1, |z| >> \nu$,

$$J_\nu(z) \approx \left(\frac{2}{\pi z}\right)^{1/2} \cos\left(z - \frac{\nu\pi}{2} - \frac{\pi}{4}\right),$$

$$N_\nu(z) \approx \left(\frac{2}{\pi z}\right)^{1/2} \sin\left(z - \frac{\nu\pi}{2} - \frac{\pi}{4}\right). \tag{4B-4}$$

Appendix to Chapter 5

A. DELTA FUNCTION

The delta function is defined by $\delta(\bar{r}, \bar{r}') = 0$ whenever $\bar{r} \neq \bar{r}'$ and $\int_V \delta(\bar{r} - \bar{r}')\, dV = 1$ when V includes $\bar{r}'$.

Rectangular Coordinate

$$\delta(\bar{r} - \bar{r}') = \delta(x - x')\delta(y - y')\delta(z - z'),$$

where $\delta(x - x') = 0$ whenever $x \neq x'$ and $\int \delta(x - x') = 1$, and thus $\delta(x - x')$ has a dimension of $(\text{length})^{-1}$.

Cylindrical Coordinate

$$\delta(\bar{r} - \bar{r}') = \frac{\delta(\rho - \rho')\delta(\phi - \phi')\delta(z - z')}{\rho},$$

where

$$\delta(\rho - \rho') = 0 \qquad \text{whenever} \qquad \rho \neq \rho',$$
$$\delta(\phi - \phi') = 0 \qquad \text{whenever} \qquad \phi \neq \phi',$$
$$\delta(z - z') = 0 \qquad \text{whenever} \qquad z \neq z'$$

and

$$\int \delta(\rho - \rho')\, d\rho = 1,$$

$$\int \delta(\phi - \phi')\, d\phi = 1,$$

$$\int \delta(z - z')\, dz = 1.$$

Note that $dV = \rho\, d\rho\, d\phi\, dz$.

Spherical Coordinate

$$\delta(\bar{r} - \bar{r}') = \frac{\delta(r - r')\delta(\theta - \theta')\delta(\phi - \phi')}{r^2 \sin\theta},$$

where

$$\delta(r - r') = 0 \qquad \text{whenever} \qquad r \neq r',$$
$$\delta(\theta - \theta') = 0 \qquad \text{whenever} \qquad \theta \neq \theta',$$
$$\delta(\phi - \phi') = 0 \qquad \text{whenever} \qquad \phi \neq \phi',$$

and

$$\int \delta(r - r')\, dr = 1,$$

$$\int \delta(\theta - \theta')\, d\theta = 1,$$

$$\int \delta(\phi - \phi')\, d\phi = 1,$$

$$dV = r^2 \sin\theta\, dr\, d\theta\, d\phi.$$

The delta function $\delta(r - r')$ has the following important characteristics:

$$\int_{V_0} f(\bar{r})\delta(\bar{r} - \bar{r}')\, dV = f(\bar{r}'),$$

for an arbitrary function $f(\bar{r})$, whenever V_0 includes $\bar{r}'$. The delta function may be thought of as a limiting case of a rectangular pulse $f(x)$ of height h and width W with a unit area as the width approaches zero.

$$\delta(x) = \lim_{W \to 0} f(x), \qquad \text{keeping} \quad hW = 1.$$

The exact shape of the function $f(x)$ is unimportant. More rigorously, the delta function must be interpreted in terms of the theory of distribution.

We note the following characteristics:

$$\int_{-\epsilon}^{\epsilon} \delta(x)\, dx = 1$$

$$\int_{0}^{\epsilon} \delta(x)\, dx = \tfrac{1}{2},$$

$$\delta(-x) = \delta(x),$$

$$\int_{-\epsilon}^{\epsilon} f(x)\delta'(x)\, dx = f(x)\delta(x)\Big|_{-\epsilon}^{\epsilon} - \int_{-\epsilon}^{\epsilon} f'(x)\delta(x)\, dx = -f'(0).$$

Similarly,

$$\int f(x)\delta^{(n)}(x)\, dx = (-1)^n f^{(n)}(0),$$

$$\delta[g(x)] = \sum_{n=1}^{N} \frac{\delta(x - x_n)}{|g'(x_n)|},$$

where x_n are the zeros of $g(x)$ $[g(x_n) = 0]$.

Appendix to Chapter 6

A. STRATTON–CHU FORMULA

To prove (6-110), (6-111), and (6-112), we start with the vector Green's theorem:

$$\int_V (\bar{Q}\cdot\nabla\times\nabla\times\bar{P} - \bar{P}\cdot\nabla\times\nabla\times\bar{Q})\,dV = \int_s (\bar{P}\times\nabla\times\bar{Q} - \bar{Q}\times\nabla\times\bar{P})\cdot d\bar{S}. \quad \text{(6A-1)}$$

We let

$$\begin{aligned} \bar{P} &= \hat{a}G(\bar{r},\bar{r}'), \\ \bar{Q} &= \bar{E}(\bar{r}), \end{aligned} \quad \text{(6A-2)}$$

where $\hat{a}$ is a constant unit vector and G is the scalar Green's function. We also have Maxwell's equations

$$\begin{aligned} \nabla\times\bar{E} &= -j\omega\mu\bar{H} - \bar{J}_m, \\ \nabla\times\bar{H} &= j\omega\epsilon\bar{E} + \bar{J}, \\ \nabla\cdot\bar{E} &= \frac{\rho}{\epsilon}, \qquad \nabla\cdot\bar{H} = 0, \end{aligned} \quad \text{(6A-3)}$$

where we included the magnetic current density $\bar{J}_m$.

Now we apply Green's theorem to the volume V_1 surrounded by S_t consisting of S_1, S, and S_∞ (Fig. 6A-1). The surface S_1 is the surface of a small sphere centered at $\bar{r}'$. (Later $\bar{r}'$ will be identified as the observation point and $\bar{r}$ and $\bar{r}'$ will be interchanged.) We wish to transform (6A-1) into the form

$$\hat{a}\cdot\int_{V_1}\cdots dV = \hat{a}\cdot\int_{S_t}\cdots dS,$$

and obtain

$$\int_{V_1}\cdots dV = \int_{S_t}\cdots dS.$$

For this purpose, we examine the left side of (6A-1). We note that

$$\begin{aligned}
\nabla\times\nabla\times\bar{P} &= \nabla\times\nabla\times(\hat{a}G),\\
&= \nabla(\nabla\cdot\hat{a}G) - \nabla^2(\hat{a}G),\\
&= \nabla(\hat{a}\cdot\nabla G) + \hat{a}k^2G,\\
\bar{E}\cdot\nabla(\hat{a}\cdot\nabla G) &= \nabla\cdot(\bar{E}(\hat{a}\cdot\nabla G)) - \hat{a}\cdot\nabla G(\nabla\cdot\bar{E}),\\
\int_{V_1}\nabla\cdot(\bar{E}\hat{a}\cdot\nabla G)\,dV &= \int_{S_t}\hat{a}\cdot(\nabla G)\bar{E}\cdot\hat{n}\,dS.
\end{aligned}$$

Therefore, we get

$$\int_{V_1}\bar{Q}\cdot\nabla\times\nabla\times\bar{P}\,dV = \hat{a}\cdot\int_{S_t}(\nabla G)(\bar{E}\cdot\hat{n})\,dS + \hat{a}\cdot\int_{V_1}\left(k^2G\bar{E} - \frac{\rho}{\epsilon}\nabla G\right)dV.$$

Similarly, we get

$$\bar{P}\cdot\nabla\times\nabla\times\bar{Q} = \hat{a}G\cdot\nabla\times\nabla\times\bar{E} = \hat{a}\cdot G(k^2\bar{E} - j\omega\mu\bar{J} - \nabla\times\bar{J}_m).$$

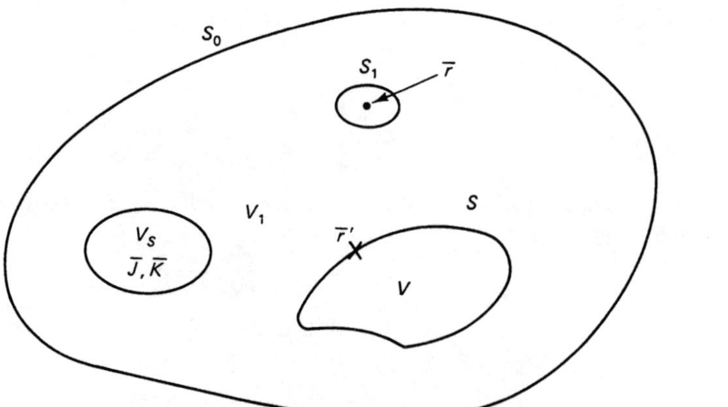

Figure 6A-1 Proof of Stratton–Chu formula.

We also note that

$$G\nabla\times\bar{J}_m = \nabla\times(G\bar{J}_m) + \bar{J}_m\times\nabla G$$

$$\int_{V_1}\nabla\times(G\bar{J}_m)\,dV = \int_{S_t}\hat{n}\times G\bar{J}_m\,dS.$$

Therefore, the left-hand side of (6A-1) becomes

$$\begin{aligned}\text{L.H.} &= \hat{a}\cdot\int_{V_1}\left(-\frac{\rho}{\epsilon}\nabla G + j\omega\mu\bar{J} + \bar{J}_m\times\nabla G\right)dV \\ &\quad + \hat{a}\cdot\int_{S_t}[\nabla G(\bar{E}\cdot\hat{n}) + \hat{n}\times(G\bar{J}_m)]\,dS.\end{aligned} \tag{6A-4}$$

Now we examine the right-hand side of (6A-1). We note that

$$\begin{aligned}\bar{P}\times\nabla\times\bar{Q}\cdot\hat{n} &= \hat{a}G\times\nabla\times\bar{E}\cdot\hat{n} \\ &= \hat{a}G\times(-j\omega\mu\bar{H} - \bar{J}_m)\cdot\hat{n} \\ &= \hat{a}\cdot[\hat{n}\times(j\omega\mu G\bar{H} + G\bar{J}_m)], \\ \bar{Q}\times\nabla\times\bar{P}\cdot\hat{n} &= \bar{E}\times\nabla\times(\hat{a}G)\cdot\hat{n} \\ &= \hat{a}\cdot(\hat{n}\times\bar{E})\times\nabla G.\end{aligned}$$

Therefore, the right side of (6A-1) becomes

$$\text{R.H.} = \hat{a}\cdot\int_{S_t}[-(\hat{n}\times\bar{E})\times\nabla G + j\omega\mu G\hat{n}\times\bar{H} + \hat{n}\times G\bar{J}_m]\,dS. \tag{6A-5}$$

We now equate (6A-4) to (6A-5). We also interchange $\bar{r}$ and $\bar{r}'$ so that the source point is $\bar{r}'$ and the observation point is $\bar{r}$. We also use $\hat{n}' = -\hat{n}$. We then get

$$\int_{V_1}\bar{E}_v\,dV' + \int_{S_t}\bar{E}_s\,dS' = 0, \tag{6A-6}$$

where $\bar{E}_s$ is given in (6-113) and $\bar{E}_v$ is given by

$$\bar{E}_v = -\left(j\omega\mu G\bar{J} + \bar{J}_m\times\nabla' G - \frac{\rho}{\epsilon}\nabla' G\right), \tag{6A-7}$$

where $\bar{J} = \bar{J}(\bar{r}')$ and $\bar{J}_m = \bar{J}_m(\bar{r}')$.

Similarly, for magnetic fields, we use the symmetry of Maxwell's equations (6A-3) with the following interchanges of the field quantities:

$$\begin{aligned} &\bar{E}\to\bar{H} && \rho\to\rho_m, \\ &\bar{H}\to-\bar{E} && \rho_m\to-\rho, \\ &\bar{J}\to\bar{J}_m && \epsilon\to\mu, \\ &\bar{J}_m\to-\bar{J} && \mu\to\epsilon. \end{aligned} \tag{6A-8}$$

We then get

$$\int_{V_1} \bar{H}_v \, dV' + \int_{S_t} \bar{H}_s \, dS' = 0, \tag{6A-9}$$

where $\bar{H}_s$ is given in (6-113) and $\bar{H}_v$ is given by

$$\bar{H}_v = -\left(j\omega\epsilon G\bar{J}_m - \bar{J} \times \nabla' G - \frac{\rho_m}{\mu} \nabla' G \right). \tag{6A-10}$$

Note that the surface integral and the volume integral are in the same form if we define the following:

$$\begin{aligned} \bar{J}_s &= \text{surface current density} = \hat{n}' \times \bar{H}, \\ \bar{J}_{ms} &= \text{surface magnetic current density} = -\hat{n}' \times \bar{E}, \\ \rho_s &= \text{surface charge density} = \epsilon(\hat{n}' \cdot \bar{E}), \\ \rho_{ms} &= \text{surface magnetic charge density} = \mu(\hat{n}' \cdot \bar{H}). \end{aligned} \tag{6A-11}$$

Now we apply (6A-6) to the problem shown in Fig. 12-10. First, we consider the case when the observation point $\bar{r}$ is outside S (Fig. 6-15). Consider the integral over S_1.

$$\int_{S_1} \bar{E}_s \, dS' = -\int_{S_1} [j\omega\mu G\hat{n}' \times \bar{H} - (\hat{n}' \times \bar{E}) \times \nabla' G - (\hat{n}' \cdot \bar{E}) \nabla' G] \, dS'. \tag{6A-12}$$

S_1 is the surface of a small sphere of radius ϵ centered at $\bar{r}$. As $\epsilon \to 0$, the first term contains only G, which varies as ϵ^{-1} and the surface is $4\pi\epsilon^2$, and therefore, this term vanishes. The second and third terms, however, contain the gradient of G and thus $\nabla' G = -(4\pi\epsilon^2)^{-1} \hat{n}'$, and

$$\begin{aligned} \lim_{\epsilon \to 0} \int_{S_1} \bar{E}_s \, \mathrm{h} dS' &= \lim_{\epsilon \to 0} [(\hat{n} \times \bar{E}) \times \hat{n}' + \hat{n}' \cdot \bar{E}\hat{n}'] \left(-\frac{1}{4\pi\epsilon^2} \right) 4\pi\epsilon^2 \\ &= -\bar{E}(\bar{r}). \end{aligned} \tag{6A-13}$$

The surface S_t consists of S_1, S_∞, and S. The integral over S_∞ vanishes as the fields diminish satisfying the radiation condition. Thus we get

$$\bar{E}(\bar{r}) = \int_{V_1} \bar{E}_v \, dv' + \int_S \bar{E}_s \, dS'. \tag{6A-14}$$

Now the volume integral over the source can be identified as the incident wave $\bar{E}_i$ because if we remove the object, the integral over S disappears and $\bar{E}(\bar{r})$ should be equal to the incident wave $\bar{E}_i(\bar{r})$. Therefore, we get (6-111),

$$\bar{E}_i(\bar{r}) + \int_S \bar{E}_s \, dS' = \bar{E}(\bar{r}). \tag{6A-15}$$

Similarly, we get

$$\bar{H}_i(\bar{r}) + \int_S \bar{H}_s \, dS' = \bar{H}(\bar{r}).$$

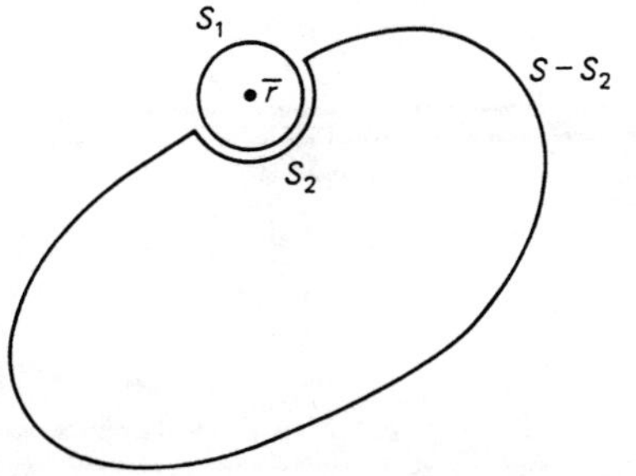

Figure 6A-2 Observation point on the surface S.

If the observation point $\bar{r}$ is on the surface S, we consider S, S_2 and $S - S_2$ (Fig. 6A-2). It is easy to see that

$$\int_{S_1} \bar{E}_s \, dS' = -\bar{E}(\bar{r}),$$

$$\int_{S_2} \bar{E}_s \, dS' = +\frac{1}{2}\bar{E}(\bar{r}). \tag{6A-16}$$

Therefore, we get

$$\bar{E}_i(\bar{r}) + \fint_S \bar{E}_s \, dS' = \frac{1}{2}\bar{E}(\bar{r}). \tag{6A-17}$$

Similarly,

$$\bar{H}_i(\bar{r}) + \fint_S \bar{H}_s \, dS' = \frac{1}{2}\bar{H}(\bar{r}),$$

where $\fint_S$ means the Cauchy principal value of the integral (over $S - S_2$ as $\epsilon \to 0$).

If the observation point $\bar{r}$ is inside S, we can immediately obtain (6-112) by noting that the volume integral in (6A-6) is $\bar{E}_i(\bar{r})$ and the surface integral is only on the surface S.

Appendix to Chapter 7

A. PERIODIC GREEN'S FUNCTION

Consider a periodic structure with period l_1 in the x direction and l_2 in the y direction, and consider one cell defined by $0 < x < l_1$, $0 < y < l_2$, and $-\infty \leq z \leq +\infty$. Within this cell, Green's function G satisfies the following:

$$(\nabla^2 + k^2)G = -\delta(x - x')\delta(y - y')\delta(z - z'). \tag{7A-1}$$

Expanding G in Floquet series, we write

$$G = \sum_m \sum_n g_{mn}(z, z')e^{-j\alpha x - j\beta y}, \tag{7A-2}$$

where $\alpha = \alpha_0 + 2m\pi/l_1$ and $\beta = \beta_0 + 2n\pi/l_2$. Substituting this into (7A-1) and noting that

$$\delta(x - x') = \sum_m \frac{e^{-j\alpha(x - x')}}{l_1}, \tag{7A-3}$$

we get

$$G = \sum_{m=-\infty}^{\infty} \sum_{n=-\infty}^{\infty} \frac{e^{-j\alpha(x - x') - j\beta(y - y') - j\gamma|z - z'|}}{2j\gamma l_1 l_2}, \tag{7A-4}$$

where $\alpha = \alpha_0 + 2m\pi/l_1$, $\beta = \beta_0 + 2n\pi/l_2$, and

$$\gamma = \begin{cases} (k^2 - \alpha^2 - \beta^2)^{1/2} & \text{if } k^2 > \alpha^2 + \beta^2 \\ -j(\alpha^2 + \beta^2 - k^2)^{1/2} & \text{if } k^2 < \alpha^2 + \beta^2 \end{cases}$$

B. VARIATIONAL FORM

Note that in general we write

$$\int_a^b dz \int_a^b dz' f^*(z) G(z, z', \beta) f(z') = 0, \tag{7B-1}$$

where G is Hermitian:

$$G(z, z', \beta) = G^*(z', z, \beta). \tag{7B-2}$$

If the true f is not known, we use an approximate function $f_a = f + \delta f$, where δf is the variation. The corresponding propagation constant is $\beta_a = \beta + \delta\beta + \cdots$. If we show that $\delta\beta = 0$, the variation of β is of the second order and the error in β_a is much smaller than the error in f_a. Thus we get the value of β close to the true value from (7B-1), even if we use a crude approximation for f.

Let us take the variation for (7B-1). We get

$$\int_a^b dz \int_a^b dz' [\delta f^*(z) G(z, z', \beta) f(z') + f^*(z) G(z, z', \beta) \delta f(z') + f^*(z) \frac{\partial G}{\partial \beta} \delta\beta f(z')] = 0,$$

and thus interchanging z and z' in the second term and using (7B-2), we get

$$\int_a^b dz \int_a^b dz' f^*(z) \frac{\partial G}{\partial \beta} f(z) + \int_a^b dz \int_a^b dz' [\delta f^*(z) G(z, z', \beta) f(z') + \delta f(z) G^*(z, z', \beta) f^*(z')] = 0. \tag{7B-3}$$

But at the correct value of $f(z)$,

$$\int_a^b dz' G(z, z', \beta) f(z') = 0,$$

and thus the second and third terms of (7B-3) vanish, proving that $\delta\beta = 0$.

C. EDGE CONDITIONS

In the neighborhood of a conducting wedge, electric and magnetic field components behave in a particular manner dependent on the wedge angle. Let us consider a conducting wedge whose angle is $(2\pi - \phi_0)$ (Fig. 7C-1). The field in the neigh-

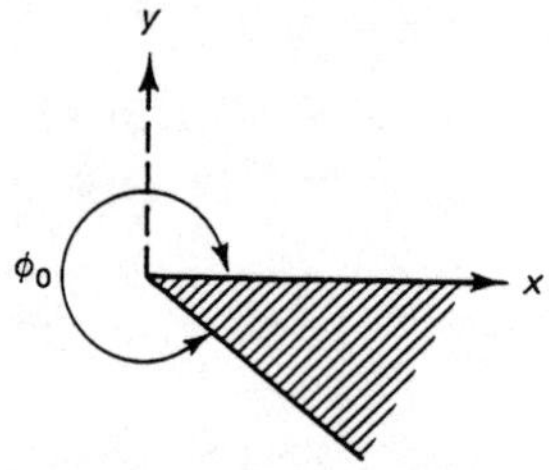

Figure 7C-1 Edge conditions.

borhood of the tip can be represented by two different modes: TM with E_z, H_ϕ, and H_ρ and TE with H_z, E_ϕ, and E_ρ. The TM modes may be given by

$$(\nabla^2 + k^2)E_z = 0 \tag{7C-1}$$

$$H_\phi = \frac{1}{j\omega\mu}\frac{\partial}{\partial\rho}E_z \quad \text{and} \quad H_\rho = -\frac{1}{j\omega\mu}\frac{1}{\rho}\frac{\partial}{\partial\phi}E_z \tag{7C-2}$$

A most general solution to (1) satisfying $E_z = 0$ at $\phi = 0$ and ϕ_0 is

$$E_z = \sum_{n=1}^{\infty} a_n Z_{\nu_n}(k\rho) \sin \nu_n \phi, \qquad \nu_n = \frac{n\pi}{\phi_0} \tag{7C-3}$$

and $Z_{\nu_n}(\rho)$ is an appropriate Bessel function. Noting that $\nu_n \neq 0$ and noninteger, $Z_\nu(\rho)$ can be either $J_\nu(k\rho)$ or $J_{-\nu}(k\rho)$. For small kp,

$$J_\nu(k\rho) \sim \rho^\nu, \qquad J_{-\nu}(k\rho) \sim \rho^{-\nu}$$

and thus taking the first term of (7C-3) for small $k\rho$, we get

$$E_z \sim \rho^\nu \sin \nu\phi \quad \text{or} \quad \rho^{-\nu} \sin \nu\phi,$$

$$H_\phi \sim \rho^{\nu-1} \sin \nu\phi \quad \text{or} \quad \rho^{-\nu-1} \sin \nu\phi, \tag{7C-4}$$

and

$$H_\rho \sim \rho^{\nu-1} \cos \nu\phi \quad \text{or} \quad \rho^{-\nu-1} \cos \nu\phi$$

Now the choice between the two sets above must be resolved. This is done by noting that the total energy in the vicinity of the edge must be finite.

The total power entering the region within radius ρ_0 is given by

$$P = \int_0^{\phi_0} \tfrac{1}{2}\bar{E} \times \bar{H}^* \cdot (-\hat{r})\rho_0\, d\phi$$
$$= \tfrac{1}{2}\int_0^{\phi_0} E_z H_\phi^* \rho_0\, d\phi. \tag{7C-5}$$

Using the first set of (7C-4), we get

$$P \sim (\text{const.})\rho^{2\nu}. \tag{7C-6}$$

In order that P be finite as $\rho \to 0$, we require that $\nu > 0$. On the other hand, if we use the second set of (7C-4), we get

$$P \sim (\text{const.})\rho_0^{-2\nu}. \tag{7C-7}$$

For P to be finite, ν must be negative, but this contradicts (7C-3), and therefore this must be rejected. Thus the behavior of the field in the vicinity of the edge is

$$\begin{aligned} E_z &\sim \rho^{\nu} \sin \nu\phi \\ H_\phi &\sim \rho^{\nu-1} \sin \nu\phi \\ H_\rho &\sim \rho^{\nu-1} \cos \nu\phi \quad \text{with} \quad \nu = \frac{\pi}{\phi_0}, \end{aligned} \tag{7C-8}$$

and the surface current density should behave as

$$J_z \sim \rho^{\nu-1}.$$

On the other hand, for TE modes, the same considerations lead to

$$\begin{aligned} H_z &\sim C_0 + C_1 \rho^{\nu} \cos \nu\phi, \\ E_\rho &\sim \rho^{\nu-1} \sin \nu\phi, \\ E_\phi &\sim \rho^{\nu-1} \cos \nu\phi \quad \text{with} \quad \nu = \frac{\pi}{\phi_0}, \end{aligned} \tag{7C-9}$$

and the surface current behaves as

$$J_\rho \sim C_0 + C_1 \rho^{\nu}.$$

Equations (7C-8) and (7C-9) represent the behavior of the field components in the neighborhood of the edge.

As an example, take a knife edge ($\phi_0 = 2\pi$). Then

$$E_z \sim \rho^{1/2} \qquad \text{and} \qquad H_z \sim C_0 + C_1 \rho^{1/2} \cos \nu\phi$$

but all the other field components become infinite at the edge. For example, E_ρ behaves as

$$E_\rho \sim \frac{1}{\rho^{1/2}} \sin \frac{\phi}{2}$$

and the current density on the surface becomes

$$J_z \sim H_\rho \sim \frac{1}{\rho^{1/2}}.$$

Even though the field and the current density themselves become infinite at the edge, the energy stays finite.

For a rectangular edge, $\phi_0 = 3\pi/2$, and thus

$$E_\rho \sim \frac{1}{\rho^{2/3}} \qquad \text{and} \qquad J_z \sim \frac{1}{\rho^{2/3}}.$$

Appendix to Chapter 8

A. MATRIX ALGEBRA

Consider a matrix given by

$$A = \begin{bmatrix} a_{11} & a_{12} & \cdots & a_{1n} \\ a_{21} & a_{22} & \cdots & a_{2n} \\ \vdots & & & \\ a_{m1} & a_{m2} & \cdots & a_{mn} \end{bmatrix}.$$

1. a_{ij} is called the element of the matrix A and to indicate the element a_{ij}, we write

$$A = (a_{ij}).$$

2. A has m rows and n columns and the *order* of the matrix is $m \times n$.
3. When $m = n$, A is called the *square matrix*.
4. Equality of matrices: $A = B$ means $a_{ij} = b_{ij}$. Addition of matrices: $A + B = C$ means $a_{ij} + b_{ij} = c_{ij}$. Matrix addition satisfies the commutative and associative laws:

$$A + B = B + A,$$

$$(A + B) + C = A + (B + C).$$

5. Multiplication by a scalar α:

$$\alpha A = B \qquad \text{means} \qquad \alpha a_{ij} = b_{ij}.$$

6. Multiplication by matrix:

$$AB = C \qquad \text{means} \qquad C_{ij} = \sum_{k=1}^{n} a_{ik} b_{kj}.$$

In order to have the product AB, the matrices must be conformable, which means that the number of columns of A must be equal to the number of rows of B. Thus, if the order of A is $m \times n$, the order of B must be $n \times p$ and the order of C becomes $m \times p$. The multiplication of matrices does not obey the commutative law, $AB \neq BA$. The associative law is valid, $(AB)C = A(BC)$.

7. A singular matrix is a square matrix with a zero determinant. A nonsingular matrix is a square matrix with a nonzero determinant.

8. The rank of the matrix is the highest order of a nonvanishing determinant in a matrix.

9. The reciprocal or inverse of A is given by

$$A^{-1} = \frac{(A_{ji})}{|A|},$$

where (A_{ji}) is the cofactor. The product of A and A^{-1} is the unit matrix

$$AA^{-1} = A^{-1}A = U.$$

The inverse of the product is given by

$$(AB)^{-1} = B^{-1}A^{-1}.$$

10. The transpose of A is given by

$$\tilde{A} = (a_{ji}).$$

11. A symmetric matrix is a matrix with $a_{ij} = a_{ji}$ and thus $\tilde{A} = A$.

12. An antisymmetric (or skew-symmetric) matrix is a matrix with $a_{ij} = -a_{ji}$, and thus $\tilde{A} = -A$ and all the diagonal elements are zero.

13. The transpose of the product is given by

$$(\widetilde{AB}) = \tilde{B}\tilde{A}.$$

14. Diagonal matrix: $a_{ij} = \delta_{ij} a_{ii}$, where

$$\delta_{ij} = \begin{cases} 1 & \text{if } i = j \\ 0 & \text{if } i \neq j \end{cases} \qquad \text{(Kronecker's delta).}$$

15. Unit matrix U: $a_{ij} = \delta_{ij}$. Zero matrix: $a_{ij} = 0$.

16. The adjoint matrix is the conjugate of the transpose:

$$A^{+} = \tilde{A}^{*}.$$

17. Orthogonal matrix A satisfies the following: $\tilde{A} = A^{-1}$ or $\tilde{A}A = U$.

18. Unitary matrix A satisfies: $A^{+} = A^{-1}$ or $A^{+}A = U$.

19. Hermitian matrix A satisfies: $A^{+} = A$. Skew-Hermitian matrix A satisfies: $A^{+} = -A$.

20. It is always possible to express A as a sum of symmetric and antisymmetric matrices:

$$A = \frac{A + \tilde{A}}{2} + \frac{A - \tilde{A}}{2},$$

or as a sum of Hermitian and skew-Hermitian matrices:

$$A = \frac{A + A^{+}}{2} + \frac{A - A^{+}}{2}.$$

Appendix to Chapter 10

A. FORWARD SCATTERING THEOREM (OPTICAL THEOREM)

Consider a linearly polarized wave $\bar{E}_i$ incident on an object and the incident Poynting vector $\bar{S}_i$:

$$\bar{E}_i = \hat{e}_i e^{-jk\bar{r}\cdot\hat{i}},$$
$$\bar{S}_i = \frac{1}{2}\bar{E}_i \times \bar{H}_i^* = \frac{|E_i|^2}{2\eta_0}\hat{i}, \qquad \eta_0 = \left(\frac{\mu_0}{\epsilon_0}\right)^{1/2}. \tag{10A-1}$$

The total field $\bar{E}$ and $\bar{H}$ are given by

$$\bar{E} = \bar{E}_i + \bar{E}_s,$$
$$\bar{H} = \bar{H}_i + \bar{H}_s, \tag{10A-2}$$

where $\bar{E}_s$ and $\bar{H}_s$ are the scattered wave.

First, we consider the total power P_a absorbed by the object. This is given by (Fig. 10A-1)

$$P_a = S_i \sigma_a = -\int_s \mathrm{Re}[\tfrac{1}{2}\bar{E} \times \bar{H}^*] \cdot d\bar{s}. \tag{10A-3}$$

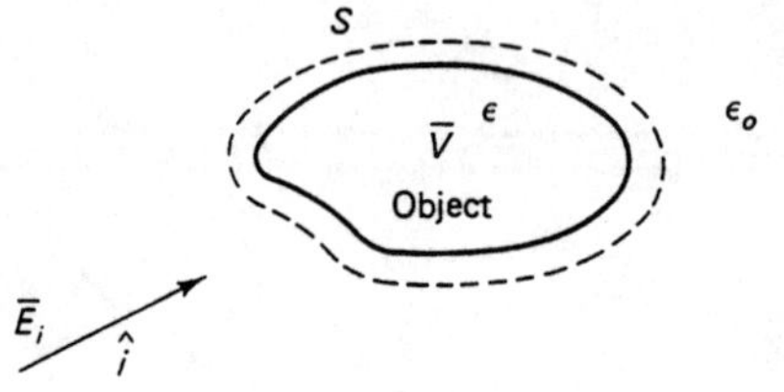

Figure 10A-1 Forward scattering theorem.

Now, we note that

$$\bar{E} \times \bar{H}^* = \bar{E}_i \times \bar{H}_i^* + \bar{E}_s \times \bar{H}_s^* + \bar{E}_i \times \bar{H}_s^* + \bar{E}_s \times \bar{H}_i^*. \tag{10A-4}$$

The scattered power P_s is given by

$$P_s = S_i \sigma_s = \int_s \mathrm{Re}[\tfrac{1}{2}\bar{E}_s \times \bar{H}_s^*] \cdot d\bar{s}. \tag{10A-5}$$

We substitute (10A-4) into (10A-3) and note (10A-5). We also use the following:

$$\begin{aligned}
&\int_s \bar{E}_i \times \bar{H}_i^* \cdot ds = 0,\\
&\mathrm{Re}(\bar{E}_i \times \bar{H}_s^*) = \mathrm{Re}(\bar{E}_i^* \times \bar{H}_s),\\
&\int_s \bar{E}_i \times \bar{H}_s^* \cdot d\bar{s} = \int_s \bar{E}_i \times \bar{H}^* \cdot d\bar{s},\\
&\int_s \bar{E}_s \times \bar{H}_i^* \cdot d\bar{s} = \int_s \bar{E} \times \bar{H}_i^* \cdot d\bar{s}.
\end{aligned} \tag{10A-6}$$

We then get

$$S_i(\sigma_a + \sigma_s) = -\int_s \mathrm{Re}\tfrac{1}{2}[\bar{E}_i^* \times \bar{H} + \bar{E} \times \bar{H}_i^*] \cdot d\bar{s}. \tag{10A-7}$$

Now noting

$$\begin{aligned}
\int_s \bar{E}_i^* \times \bar{H} \cdot d\bar{s} &= \int_v \nabla \cdot (\bar{E}_i^* \times \bar{H})\, dV,\\
\nabla \cdot (\bar{E}_i^* \times \bar{H}) &= \bar{H} \cdot \nabla \times \bar{E}_i^* - \bar{E}_i^* \cdot \nabla \times \bar{H}\\
&= \bar{H} \cdot (j\omega\mu_0 \bar{H}_i^*) - \bar{E}_i^* \cdot (j\omega\epsilon \bar{E}),\\
\nabla \cdot (\bar{E} \times \bar{H}_i^*) &= \bar{H}_i^* \cdot \nabla \times \bar{E} - \bar{E} \cdot \nabla \times \bar{H}_i^*\\
&= \bar{H}_i^* \cdot (-j\omega\mu_0 \bar{H}) - \bar{E} \cdot (-j\omega\epsilon_0 \bar{E}_i^*).
\end{aligned} \tag{10A-8}$$

We get

$$\begin{aligned}
S_i(\sigma_a + \sigma_s) &= -\mathrm{Re} \int_v \tfrac{1}{2}[-j\omega(\epsilon - \epsilon_0)]\bar{E} \cdot \bar{E}_i^*\, dV\\
&= -\mathrm{Im} \int_v \frac{\omega(\epsilon - \epsilon_0)}{2} E \cdot \bar{E}_i^*\, dV.
\end{aligned} \tag{10A-9}$$

However, the forward scattering amplitude $\bar{f}(\hat{i}, \hat{i})$ is given by

$$\bar{f}(\hat{i}, \hat{i}) = \frac{k^2}{4\pi} \int_V \frac{\epsilon - \epsilon_0}{\epsilon_0} \bar{E} e^{jk\bar{r}' \cdot \hat{i}} \, dV. \tag{10A-10}$$

Noting $\bar{E}_i$ in (10A-1), we get

$$\bar{f}(\hat{i}, \hat{i}) \cdot \hat{e}_i = \frac{k^2}{4\pi} \int_V \frac{\epsilon - \epsilon_0}{\epsilon_0} \bar{E} \cdot \bar{E}_i^* \, dV. \tag{10A-11}$$

Combining (10A-9) and (10A-11), we finally get the forward scattering theorem (optical theorem):

$$\sigma_a + \sigma_s = -\frac{4\pi}{k^2} \operatorname{Im} \bar{f}(\hat{i}, \hat{i}) \cdot \hat{e}_i. \tag{10A-12}$$

Appendix to Chapter 11

A. BRANCH POINTS AND RIEMANN SURFACES

Let us examine the solution (11-34) obtained in Chapter 11. In particular, it is important to investigate the contour of the integration in more detail. To do this, we should first note that the integrand contains $\lambda = \sqrt{k^2 - h^2}$, and this raises the question of whether we should take $\lambda = +\sqrt{k^2 - h^2}$ or $-\sqrt{k^2 - h^2}$. To study this question in more detail, we consider a more general complex function of a complex variable h:

$$W = f(h). \tag{11A-1}$$

If $f(h)$ is an analytic function within a region in the h plane, $f(h)$ is a single-valued function, and its value is uniquely determined by h (i.e., there is a one-to-one correspondence between W and h).

If $f(h)$ is a multivalued function of h, however, there are many different values of W for a given h. For example, letting $W = y$ (real) and $h = x$ (real), consider the case where

$$y^2 = x. \tag{11A-2}$$

For a given value of x, there are two values of y: $y_1 = +\sqrt{x}$ and $y_2 = -\sqrt{x}$ (Fig. 11A-1). We may call y_1 the first branch and y_2 the second branch. The point where these two branches meet is called the *branch point*.

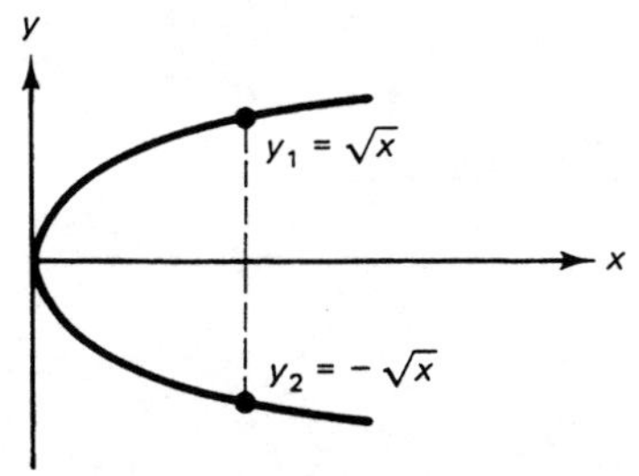

Figure 11A-1 Two branches.

By using these two branches y_1 and y_2, we can keep the different branches apart so that we know exactly which branch we are dealing with. When W is a complex function of a complex variable h, each branch is represented by a different complex plane. We call these different planes of a multivalued function the *Riemann surfaces*. On each Riemann surface, the function is single-valued and its value is uniquely determined.

Let us take an example:

$$W = \sqrt{h}. \tag{11A-3}$$

Let us represent h, a complex variable, in the following polar form:

$$h = re^{j\theta}.$$

Then W is given by

$$W_1 = r^{1/2} e^{j(\theta/2)}.$$

But this is not the only value of W. In the h plane, we can go around the origin and thus add 2π to θ without changing the value of h:

$$h = re^{j\theta} = re^{j(\theta + 2\pi)}.$$

But this second form of h gives a different value of W:

$$W_2 = r^{1/2} e^{j(\theta/2 + \pi)} = -W_1.$$

If another 2π is added to θ, W will go back to W_1.

$$h = re^{j(\theta + 4\pi)},$$

$$W = re^{j(\theta/2 + 2\pi)} = W_1.$$

Therefore, for a given h, there are two W's: W_1 and W_2. We note that to transfer from W_1 to W_2, we must go around the origin once in the h plane. The origin in this example is the branch point.

To describe this situation more clearly and to tell which of these two values W_1 and W_2, we are dealing with, we introduce the idea of the *branch cut*. In the h plane, we imagine a cut extending from the branch point $h = 0$ to infinity. We say that as long as we do not cross the branch cut, we are on the first branch (the first Riemann surface), and the value of the function is W_1. To get W_2, we must cross the branch cut. The branch cut can be drawn in any direction from the origin, and it need not be

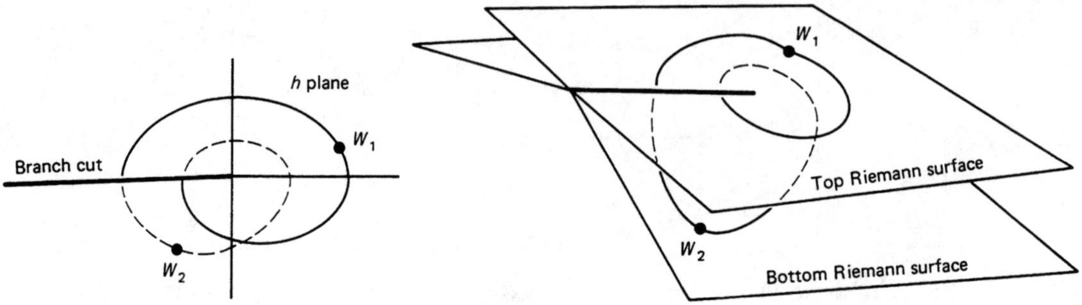

Figure 11A-2 Riemann surfaces.

a straight line. For convenience, we normally choose a cut along the negative real axis (Fig. 11A-2). According to this choice of the branch cut, we define

$$\sqrt{h} = \begin{cases} r^{1/2}e^{j(\theta/2)} = W_1 & \text{for } -\pi < \theta < \pi \\ -r^{1/2}e^{j(\theta/2)} = W_2 & \text{for } \pi < \theta < 3\pi \\ W_1 & \text{for } 3\pi < \theta < 5\pi. \end{cases}$$

This situation is pictured in the two Riemann surfaces in Fig. 11A-2. The top surface represents W_1 and the bottom surface W_2. These two surfaces are put together at the branch cut, indicating that when we cross the branch cut, W_2 must be used. As the branch cut is crossed the second time, we go back to the first Riemann surface and obtain W_1.

Another example of a multivalued function is

$$W = \ln h.$$

We note that for a given h,

$$h = re^{j\theta} = re^{j(\theta + 2n\pi)}.$$

We get an infinite number of W's:

$$W_n = \ln r + j(\theta + 2n\pi), \qquad n = 0, \pm 1, \pm 2, \ldots, \pm\infty.$$

Therefore, there are an infinite number of Riemann surfaces. As we go around the branch point $h = 0$, we move to the next Riemann surface. This is pictured in Fig. 11A-3.

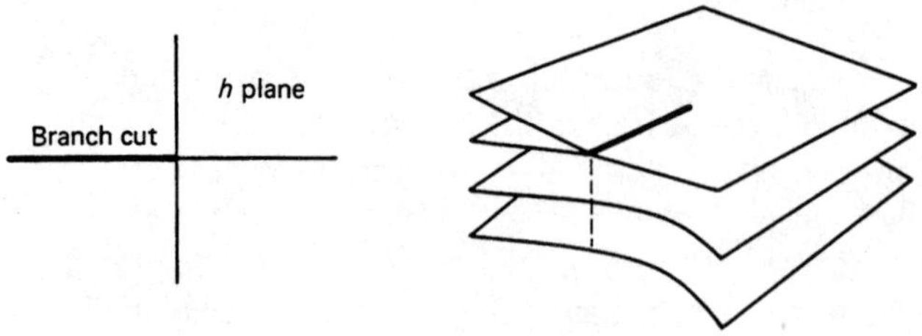

Figure 11A-3 $W = \ln h$.

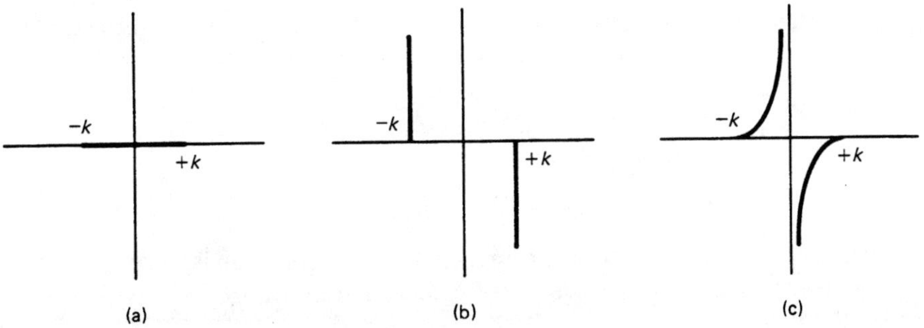

Figure 11A-4 Three branch cuts in the h plane.

Now consider $\lambda = \sqrt{k^2 - h^2}$. We note that there are two branch points $h = +k$ and $-k$. We may have the branch cuts as shown in Fig. 11A-4. The cut in Fig. 11A-4a is not acceptable on physical grounds. The cuts in both part (b) and part (c) are acceptable. Cut (b) is along the line $\operatorname{Re} h = \pm\operatorname{Re} k$ and cut (c) is along the curve $\operatorname{Im} \lambda = 0$.

B. CHOICE OF THE CONTOUR OF INTEGRATION AND THE BRANCH CUT

Equation (11-34) has the general form

$$I = \int_c C_n(h) H_n^{(2)}(\lambda\rho) e^{-jhz}\, dh, \qquad \lambda = \sqrt{k^2 - h^2}. \tag{11B-1}$$

Because of $\lambda = \sqrt{k^2 - h^2}$, the integrand has two branch points at $h = \pm k$.

In this section we show how the contour c in (11B-1) should be drawn and will discuss the reason for this choice. First, we note that due to the branch points, we can have two Riemann surfaces, and therefore we must choose the contour in such a manner that the inverse transform (11B-1) exists. This choice must be made on the basis of physical argument.

First, we note that the integrand in (11B-1) represents a wave with propagation constant h in the z direction. At a large distance from the origin, we approximate the Hankel function by

$$H_n^{(2)}(z) = \sqrt{\frac{2}{\pi z}}\, e^{-jz + j(2n+1)\pi/4}, \tag{11B-2}$$

and we write

$$I \approx \int_c C_n(h) \sqrt{\frac{2}{\pi\lambda\rho}}\, e^{j(2n+1)\pi/4 - j\lambda\rho - jhz}\, dh. \tag{11B-3}$$

If this equation is to represent a physical situation, the wave represented by the integrand must be an outgoing wave, and no wave should come back from infinity. This is the so-called *radiation condition*.

To satisfy this condition, the branch must be chosen so that λ is in the fourth quadrant of the complex plane,

$$\mathrm{Re}(\lambda) \geq 0 \qquad \text{and} \qquad \mathrm{Im}(\lambda) \leq 0, \tag{11B-4}$$

because the positive real part represents the outgoing phase progression, and the negative imaginary part represents the attenuation toward infinity.

$$e^{-j\lambda\rho} = e^{-j\lambda_r\rho + \lambda_i\rho}, \tag{11B-5}$$

where $\lambda = \lambda_r + j\lambda_i$ and $\lambda_r \geq 0$ and $\lambda_i \leq 0$. Now the choice of the path of integration and branch cut is made such that λ is always in the fourth quadrant.

Let us consider a point Q in the h-plane (Fig. 11B-1). We note that

$$\begin{aligned} h - k &= r_1 e^{j\theta_1}, \\ h + k &= r_2 e^{j\theta_2}. \end{aligned} \tag{11B-6}$$

Now consider λ:

$$\begin{aligned} \lambda_1 &= \sqrt{k^2 - h^2} = \sqrt{(k-h)(k+h)} \\ &= \sqrt{-r_1 r_2 e^{j(\theta_1 + \theta_2)}}. \end{aligned} \tag{11B-7}$$

Because of the square root, we can have two values for λ:

$$\begin{aligned} \lambda_1 &= \sqrt{r_1 r_2}\, e^{j(\theta_1 + \theta_2)/2 - j(\pi/2)}, \\ \lambda_2 &= \sqrt{r_1 r_2}\, e^{j(\theta_1 + \theta_2)/2 + j(\pi/2)}. \end{aligned} \tag{11B-8}$$

Let us first determine which of these two λ_1 or λ_2 satisfies the radiation condition. To do this, let us consider a point Q_1 on the contour between $+k$ and $-k$. At Q_1, $\theta_1 = \pi$ and $\theta_2 = 0$ and thus

$$\lambda_1 = \sqrt{r_1 r_2} \qquad \text{and} \qquad \lambda_2 = \sqrt{r_1 r_2}\, e^{j\pi}.$$

λ_2 is not in the fourth quadrant, therefore λ_1 must be chosen. Next consider the path from Q_1 to Q_2 going around the branch point at $+k$ as shown. Then at Q_2, $\theta_1 = 0$ and $\theta_2 = 0$, and therefore

$$\lambda_1 = \sqrt{r_1 r_2}\, e^{-j(\pi/2)},$$

which is in the fourth quadrant and satisfies the radiation condition.

On the other hand, if the path going from Q_1 to Q_2' is chosen, then at Q_2', $\theta_1 = 2\pi$ and $\theta_2 = 0$. Thus at Q_2',

$$\lambda_1 = \sqrt{r_1 r_2}\, e^{j(\pi/2)},$$

which is not in the fourth quadrant and does not satisfy the radiation condition.

In a similar manner, at Q_3, $\theta_1 = \pi$ and $\theta_2 = -\pi$ and therefore,

$$\lambda_1 = \sqrt{r_1 r_2}\, e^{-j(\pi/2)},$$

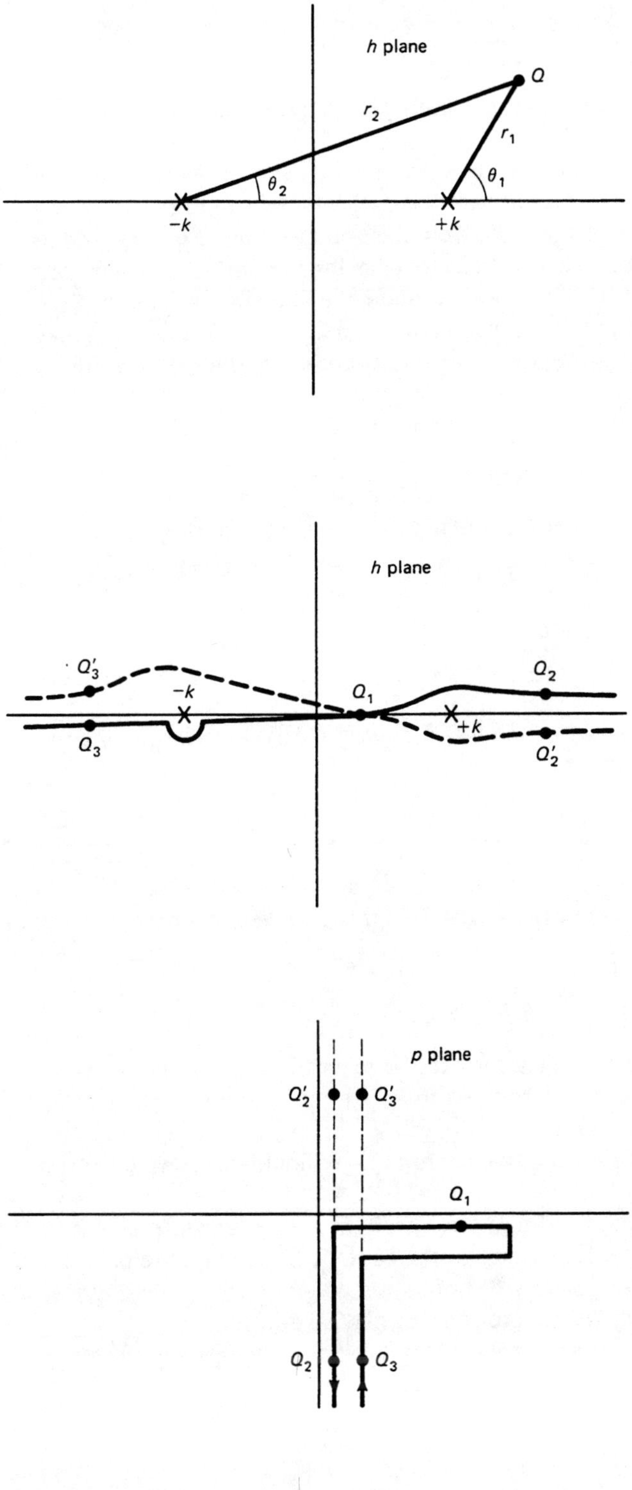

Figure 11B-1 Choice of contour.

while at Q_3',

$$\lambda_1 = \sqrt{r_1 r_2}\, e^{j(\pi/2)}.$$

Thus to satisfy the radiation condition, the path Q_3–Q_1–Q_2 must be chosen (Fig. 11B-1).

To ensure that the path of integration is as specified above, we draw branch cuts from the branch points and require that the path of integration not cross the cuts. We may have the cuts drawn vertically from the branch points (Fig. 11A-4b) or any other choice of the branch cuts is satisfactory, as long as the original contour does not cross them. However, when we evaluate the integral, we must often deform the path of integration in some convenient manner. Then since the path may move off the real axis of the h plane, one choice of branch cuts is more convenient than the other.

One convenient choice of the cuts is to draw the cuts along the curve

$$\text{Im}(\lambda) = 0. \tag{11B-9}$$

To clarify this situation, let us assume a small negative imaginary part of k:

$$k = k_r + jk_i. \tag{11B-10}$$

Since λ and h must satisfy the relation

$$\lambda^2 + h^2 = k^2, \tag{11B-11}$$

substituting (11B-10) and $\lambda = \lambda_r + j\lambda_i$, and $h = h_r + jh_i$ in (11B-11) and equating the real and imaginary parts of both sides, we obtain

$$\begin{aligned} \lambda_r^2 - \lambda_i^2 + h_r^2 - h_i^2 &= k_r^2 - k_i^2, \\ \lambda_r \lambda_i + h_r h_i &= k_r k_i. \end{aligned} \tag{11B-12}$$

Now let us consider the condition $\text{Im}(\lambda) = \lambda_i = 0$. In this case we must have

$$\begin{aligned} \lambda_r^2 + h_r^2 - h_i^2 &= k_r^2 - k_i^2, \\ h_r h_i &= k_r k_i. \end{aligned} \tag{11B-13}$$

Equation (11B-13) is an equation for a hyperbola going through the point $h = k$ (Fig. 11-5). We note that for the curve from $+k$ to A, $|k_r| > |h_r|$ and $|h_i| > |k_i|$, and this satisfies (11B-13) but for the curve from $+k$ to B, $|h_r| > |k_r|$ and $|h_i| < |k_i|$ and (11B-13) cannot be satisfied. Thus the curve for $\text{Im}(\lambda) = 0$ should be drawn from $+k$ to A.

Similarly, the curve for the condition $\text{Re}(\lambda) = 0$ should be drawn from $+k$ to B. We need to keep in mind that this choice of the contour is based on the physical consideration that the wave must satisfy the radiation condition. If other physical situations are considered, the choice of the contour may be different.

For example, if h and $\pm k$ are replaced by ω and $\pm\omega_0$, this represents a

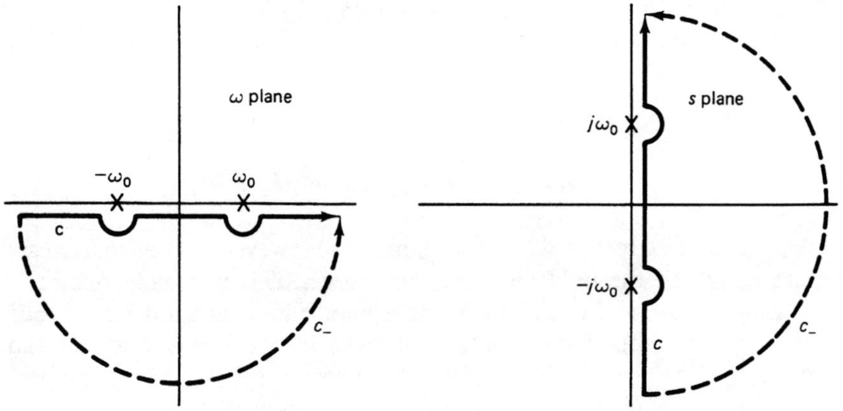

Figure 11B-2 Contour for the transient problem.

transient wave in a waveguide or a magnetoplasma. In this case, the wave may be given by

$$u(x,t) = \int_C A(\omega)e^{j\omega t - jk(\omega)x}\, d\omega,$$

where

$$k(\omega) = \left[\left(\frac{\omega}{c}\right)^2 - \left(\frac{\pi}{a}\right)^2\right]^{1/2}. \tag{11B-14}$$

The physical consideration in this case is that no wave should propagate faster than the light velocity, and thus $u(x,t) = 0$ for $t < x/c$. To ensure this condition, the contour C should be drawn to include all the singularities on one side (Fig. 11B-2).

Note that for $|\omega|$ large, we have

$$|e^{j\omega t - j(\omega/c)\sqrt{1-(\omega_p^2/\omega^2)}x}| \Rightarrow e^{-\omega_i(t-(x/c))},$$

with $\omega = \omega_r + j\omega_i$, and this goes to zero along $C_-(\omega_i < 0)$ for $t < x/c$ and thus, since there are no singularities between C and C_-, the integral along C is equal to the integral along C_-, which is zero. This is, of course, the same as the Laplace transform ($j\omega = S$).

For spatial frequency h, there is no preferred direction, so the function should exist for $x < 0$ as well as for $x > 0$. This is the basis for the choice in Figs. 11-4 and 11-5.

C. SADDLE-POINT TECHNIQUE AND METHOD OF STATIONARY PHASE

The complex integrals appearing in previous sections, such as

$$I_1 = \int_{-\infty}^{\infty} C_n(h) H_n^{(2)}(\sqrt{k^2 - h^2}\,\rho) e^{-jhz}\, dh, \tag{11C-1}$$

are in general extremely difficult to evaluate. However, this evaluation can be done quite easily for the field far from the source. This is usually referred to as the *radiation pattern* and is one of the most important characteristics of radiating systems. The technique used to obtain the radiation pattern is called the saddle-point technique. This technique is useful not only for particular cylindrical problems but for many other problems involving the evaluation of a contour integral. In this section we describe this technique.

Let us consider the following complex integral:

$$I = \int_c F(\alpha) \exp[Zf(\alpha)]\, d\alpha, \tag{11C-2}$$

where Z is a large positive real number and $f(\alpha)$ and $F(\alpha)$ are complex functions of the complex variable α. $F(\alpha)$ is a slowly varying function of α. Writing $f = f_r + jf_i$, we get

$$e^{Zf(\alpha)} = e^{Zf_r}(\cos Zf_i + j \sin Zf_i). \tag{11C-3}$$

Therefore, for a large Z, as f_i varies along the contour, the integrand oscillates very rapidly, and the evaluation of the integral would be extremely difficult.

However, there is a way to avoid this difficulty. If we deform the original contour in such a manner that the imaginary part f_i is constant along the new path, there is no rapid oscillation of the integrand. In this case, $\exp[Zf(\alpha)]$ varies slowly as $\exp(Zf_r)$ starts from zero, attains some value, and declines to zero again at the end of the contour.

The point in the complex plane where the real part becomes maximum should be the saddle point, as will be shown below. To study the behavior of $f(\alpha)$ near this point, let us first consider a point $\alpha = \alpha_s$ where

$$\frac{df(\alpha)}{d\alpha} = 0. \tag{11C-4}$$

We expand $f(\alpha)$ about this point,

$$f(\alpha) = f(\alpha_s) + (\alpha - \alpha_s)f'(\alpha_s) + \frac{(\alpha - \alpha_s)^2}{2!} f''(\alpha_s) + \cdots, \tag{11C-5}$$

and in the neighborhood of $\alpha = \alpha_s$, we can neglect the higher-order terms. Thus we get

$$f(\alpha) = f(\alpha_s) + \frac{(\alpha - \alpha_s)^2}{2!} f''(\alpha_s). \tag{11C-6}$$

We note that as the angle of $\alpha - \alpha_s$ increases, the corresponding increase of the angle of $f(\alpha) - f(\alpha_s)$ is twice that for $\alpha - \alpha_s$, and thus the angle $\pi/2$ in the α plane around α_s corresponds to π in the f plane.

The valley region, where $f_r(\alpha) < f_r(\alpha_s)$ in the f plane, is represented by two regions in the α plane about the saddle point. The mountain region $f_r(\alpha) > f_r(\alpha_s)$ is also represented by two regions in the α plane, and these four regions meet at the saddle point, each occupying the angle of $\pi/2$.

It is then clear that the point $\alpha = \alpha_s$ is the saddle point as shown in Fig. 11C-1. This is also clear considering that $f(\alpha)$ is an analytic function. An analytic function cannot have maxima or minima within the range of analyticity. This situation is mathematically identical to two-dimensional electrostatic problems. A two-dimensional electrostatic potential U and a flux function V can be represented by an analytic function $W = U + jV$ and U and V satisfy the Laplace equations, and it is clear that there is no maximum potential U in the region of analyticity. Only the singularity such as the electric charge can create maximum or minimum potentials. A point where $\partial f/\partial \alpha = 0$ is therefore a saddle point.

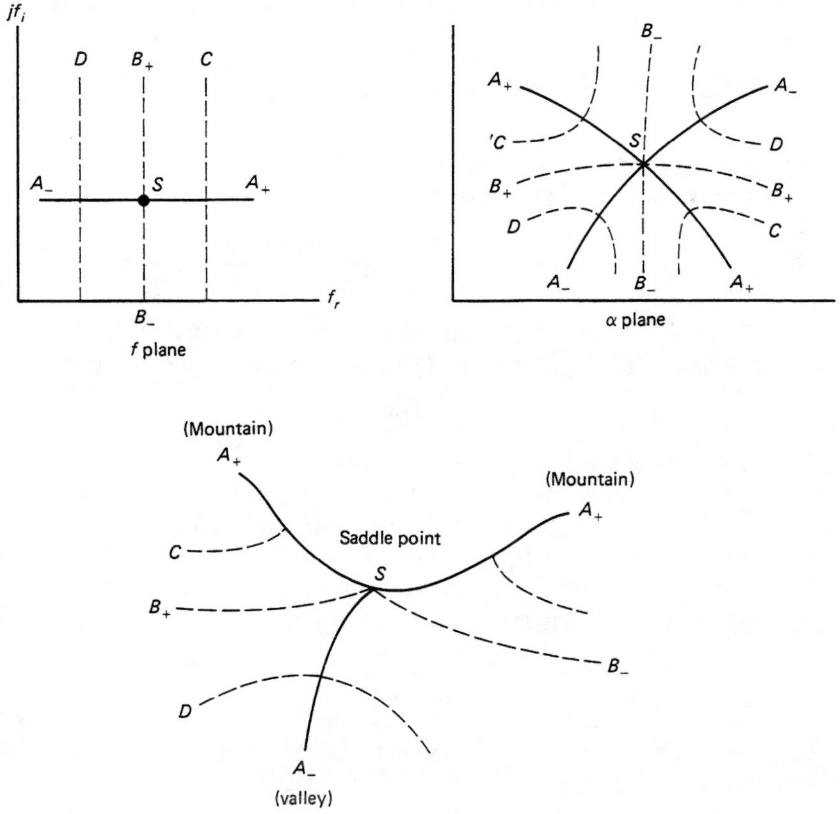

Figure 11C-1 Saddle point S.

We now wish to deform the original contour into the path along which the imaginary part f_i is constant. As can be seen from Fig. 11C-1, there are two perpendicular lines (SA_+ and SA_-) in the α plane along which f_i is constant. The path (SA_+) goes up the mountains on both sides of the saddle point, and the other (SA_-) goes down to the valleys on both sides. Obviously, the deformed path should be chosen to originate in one valley, cross the saddle point, and terminate in the other valley. This is the path A_-–S–A_-. As we note in Fig. 11C-1, this path of constant f_i is in fact the path along which the real part f_r decreases fastest from the saddle point. Thus the path is called the *steepest descent path*, and the saddle point technique is often called the *method of steepest descent*.

Let us evaluate the integral along this steepest descent contour (SDC):

$$I = \int_{\text{SDC}} F(\alpha)e^{Zf(\alpha)}\,d\alpha. \tag{11C-7}$$

We write

$$f(\alpha) = f(\alpha_s) + \frac{(\alpha-\alpha_s)^2}{2!}f''(\alpha_s),$$

where

$$\frac{df}{d\alpha} = 0 \qquad \text{at } \alpha = \alpha_s \text{ saddle point.}$$

First, we assume that $F(\alpha)$ is a slowly varying function, and that its value near α_s is almost constant. Then, approximately, we get

$$I = F(\alpha_s)e^{Zf(\alpha_s)}\int_{\text{SDC}} \exp\left[Z\frac{(\alpha-\alpha_s)^2}{2!}f''(\alpha_s)\right]d\alpha. \tag{11C-8}$$

To study the exponent in the integral, let us note that along the contour of the constant phase, the imaginary part should be constant and equal to its value at α_s:

$$f_i(\alpha) = f_i(\alpha_s). \tag{11C-9}$$

Therefore, along the path, we have

$$\text{Re}\,f(\alpha) = \text{Re}\,f(\alpha_s) + \frac{(\alpha-\alpha_s)^2}{2!}f''(\alpha_s).$$

Now we note that $\text{Re}\,f(\alpha)$ must decrease from the value at the saddle point $\text{Re}\,f(\alpha_s)$ and therefore the path must be chosen such that

$$\frac{(\alpha-\alpha_s)^2}{2!}f''(\alpha_s)$$

must be not only real, but also negative. This choice can be made by properly orienting the path. We let

$$\alpha - \alpha_s = se^{j\gamma},$$

where s is the distance measured from the saddle point and γ is the angle the path makes with the real axis. The exponent can be expressed by

$$\frac{Z(\alpha-\alpha_s)^2}{2} f''(\alpha_s) = Z\frac{s^2}{2} e^{j2\gamma} f''(\alpha_s)$$
$$= -\frac{Z[-e^{j2\gamma} f''(\alpha_s)]}{2} s^2.$$

We now choose γ such that

$$[-e^{j2\gamma} f''(\alpha_s)] = P$$

is real and positive. With this choice, P is equal to $|f''(\alpha_s)|$. Then the integral becomes

$$I = F(\alpha_s) e^{Zf(\alpha_s)} \int_{\text{SDC}} \exp\left(-\frac{ZPs^2}{2} + j\gamma\right) ds. \tag{11C-10}$$

This integral is in the form of the integral of a Gaussian curve, and for large Z, the integrand drops down very rapidly on both sides of $s = 0$. Therefore, the most contribution comes from the neighborhood of $s = 0$. Thus we approximate the integral by integrating over a small distance about $s = 0$.

$$I = F(\alpha_s) e^{Zf(\alpha_s) + j\gamma} \int_{-\epsilon}^{\epsilon} e^{-(ZP/2)s^2} ds.$$

After the change of variable from s to x,

$$\frac{ZP}{2} s^2 = x^2,$$

we get

$$I = F(\alpha_s) e^{Zf(\alpha_s) + j\gamma} \sqrt{\frac{2}{ZP}} \int_{\sqrt{ZP/2}\epsilon}^{\sqrt{ZP/2}\epsilon} e^{-x^2} dx.$$

For large Z, the limit of the integral can be extended to $-\infty$ and $+\infty$. Using

$$\int_{-\infty}^{\infty} e^{-x^2} dx = \sqrt{\pi},$$

we get finally

$$I = \int_c F(\alpha) e^{Zf(\alpha)} d\alpha$$
$$\cong F(\alpha_s) e^{Zf(\alpha_s) + j\gamma} \left[\frac{2\pi}{|f''(\alpha_s)|Z}\right]^{1/2}, \tag{11C-11}$$

where γ is chosen to make $[e^{j2\gamma} f''(\alpha_s)]$ real and negative, and $\partial f(\alpha_s)/\partial\alpha = 0$. Equation (11C-11) is the final result of the approximate evaluation of the integral based on the method of steepest descent (the saddle-point technique) valid for large Z.

There are two choices for γ. For example, if f'' has the angle of $\frac{1}{2}\pi$, then $2\gamma + \frac{1}{2}\pi = \pm\pi$, and we get $\gamma = \pi/4$ or $-3\pi/4$. The choice can be made by noting the path of integration (see Section 11-5).

We made several assumptions and approximations in deriving the result of the saddle-point integration. For example, we assumed that $F(\alpha)$ is a slowly varying function of α near $\alpha = \alpha_s$, and we approximated $F(\alpha)$ by $F(\alpha_s)$. We did not define what is meant by "slowly varying." We expanded $f(\alpha)$ in a Taylor's series and retained the first two terms. But what must be done if $f''(\alpha_s) = 0$? How do we justify taking the first two terms? Also, in one process we integrated from $-\epsilon$ to $+\epsilon$ under the assumption that the main contribution comes from this region. A more complete treatment must be done through the use of Watson's lemma.

Also, we did not account for the presence of singularities between the original contour and the deformed saddle point contour. If the poles are present between these two contours, they may give rise to "surface waves" and "leaky waves." Furthermore, when the pole is close to the saddle point, we need a modified saddle-point technique. Wave propagation over the earth excited by a dipole, which is often referred to as a Sommerfeld problem, requires this technique. Also, the existence of a branch point may give rise to lateral waves. These are discussed in Chapter 15.

The method of stationary phase is equivalent to the method of steepest descent. As seen in Fig. 11C-1, the steepest descent path is along the constant $f_i(SA_+, SA_-)$. However, we may go through the saddle point along constant $f_r(SB_+, SB_-)$. Then writing

$$I = F(\alpha_s)e^{Zf(\alpha_s)} \int e^{Z[(\alpha-\alpha_s)^2/2]f''(\alpha_s)}\, d\alpha, \tag{11C-12}$$

we choose the path such that

$$f_r(\alpha) = f_r(\alpha_s) = \text{constant}, \tag{11C-13}$$

and then we choose $\alpha - \alpha_s = se^{j\gamma}$ such that

$$\frac{Z(\alpha-\alpha_s)^2}{2} f''(\alpha_s) = j\left[Z\frac{s^2}{2}e^{j2\gamma - j(\pi/2)}f''(\alpha_s)\right]$$
$$= j\frac{Z}{2}Qs^2,$$

is purely imaginary. Then

$$I = F(\alpha_s)e^{Zf(\alpha_s)+j\gamma} \int e^{j(Z/2)Qs^2}\, ds,$$

which is evaluated to give

$$I \sim F(\alpha_s)e^{Zf(\alpha_s)+j\gamma+j(\pi/4)}\left[\frac{2\pi}{|f''(\alpha_s)|Z}\right]^{1/2}, \tag{11C-14}$$

where γ is chosen so that

$$[e^{j2\gamma - j(\pi/2)}f''(\alpha_s)]$$

is real.

Obviously, $\gamma = \gamma' + \pi/4$, and the method of steepest descent and the method of stationary phase give the same result. The method of stationary phase is used, for example, to study reflections from curved surfaces where the main contribution comes from the portion of the reflector where the phase variation is stationary.

The method of stationary phase is usually used when the integral is expressed in the following form:

$$I = \int_c F(\alpha) e^{jZg(\alpha)}\, d\alpha. \tag{11C-15}$$

The stationary phase point $\alpha = \alpha_s$ is then given by

$$\frac{\partial}{\partial \alpha} g(\alpha) = 0. \tag{11C-16}$$

The integral is then evaluated to give

$$I \cong F(\alpha_s) e^{jZg(\alpha_s) + j\gamma' \pm j(\pi/4)} \left[\frac{2\pi}{|g''(\alpha_s)|Z} \right]^{1/2}, \tag{11C-17}$$

where γ' is chosen such that

$$[e^{j2\gamma'} g''(\alpha_s)]$$

is real and positive (negative) for the upper (lower) sign in the exponent. Thus if $g'' > 0$, $\gamma' = 0$ and $+j\frac{1}{4}\pi$ must be used, and if $g'' < 0$, $\gamma' = 0$ and $-j\frac{1}{4}\pi$ must be used.

D. COMPLEX INTEGRALS AND RESIDUES

This appendix gives a short summary of important definitions and theorems for complex functions.

1. Analytic functions $w = f(z), z = x + jy$. If $f(z)$ possess a derivation at $z = z_0$ and in its neighborhood, then $f(z)$ is called the *analytic function*.

2. If $w = f(z) = u(x, y) + jv(x, y)$ is analytic, then u and v satisfy the *Cauchy–Riemann equation*:

$$\frac{\partial u}{\partial x} = \frac{\partial v}{\partial y}, \qquad \frac{\partial u}{\partial y} = -\frac{\partial v}{\partial x}. \tag{11D-1}$$

Combining these two, we get

$$\nabla^2 u = 0 \quad \text{and} \quad \nabla^2 v = 0. \tag{11D-2}$$

The real part and the imaginary part of an analytic function satisfy the two-dimensional Laplace equation.

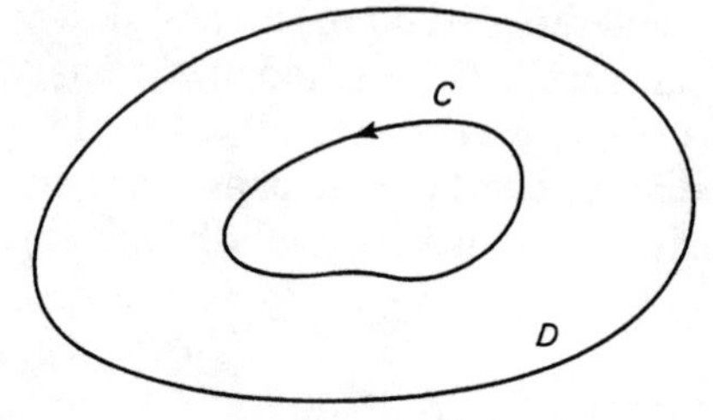

Figure 11D-1 Cauchy's integral theorem.

3. Cauchy's integral theorem. If $f(z)$ is analytic in a domain D, then for every closed path C in D (Fig. 11D-1)

$$\oint f(z)\,d\mathbf{z} = 0. \tag{11D-3}$$

4. Cauchy's integral formula. If $f(z)$ is analytic in D, then

$$f(z_0) = \frac{1}{2\pi j}\int_C \frac{f(z)}{z - z_0}\,d\mathbf{z},$$
$$f^{(n)}(z_0) = \frac{n!}{2\pi j}\int_C \frac{f(z)}{(z - z_0)^{n+1}}\,dz. \tag{11D-4}$$

5. If $f(z)$ is analytic in D, $f(z)$ can be expanded in a Taylor's series:

$$f(z) = f(a) + f'(a)(z - a) + f''(a)\frac{(z - a)^2}{2!} + \cdots. \tag{11D-5}$$

The radius of convergence of the series is the distance from a to the nearest singularity of $f(z)$.

6. Residue. If $f(z)$ has an isolated singularity at $z = a$, $f(z)$ can be expanded in a Laurent series,

$$f(z) = \sum_{n=0}^{\infty} c_n(z - a)^n + \frac{C_{-1}}{z - a} + \frac{C_{-2}}{(z - a)^2} + \cdots. \tag{11D-6}$$

This is valid in the domain $0 < |z - a| < R$, where R is the distance from a to the nearest singularity of $f(z)$. The coefficient C_{-1} is called the *residue* of $f(z)$ at $z = a$.

7. If $f(z)$ has a simple pole at a, the residue is given by

$$c_{-1} = \lim_{z \to a}(z - a)f(z)$$
$$= \frac{N(a)}{D'(a)}$$

where $f(z) = N(z)/D(z)$.

8. If $f(t)$ has the mth-order poles

$$f(z) = \sum_{n=0}^{\infty} c_n(z - a)^n + \frac{C_{-1}}{(z - a)} + \cdots + \frac{C_{-m}}{(z - a)^m},$$

then the residue is given by

$$c_{-1} = \frac{1}{(m-1)!} \lim_{z \to a} \frac{d^{m-1}}{dz^{m-1}} [(z-a)^m f(z)].$$

9. Residue theorem. If $f(z)$ is analytic inside a closed path C except for finite number of singular points $a_1, a_2, \ldots, a_m$, then

$$\int_C f(z)\, dz = 2\pi j \sum_{i=1}^{m} \text{residue at } a_i$$

The integral is taken counterclockwise.

Appendix to Chapter 12

A. IMPROPER INTEGRALS

If the integrand is unbounded at several points in the interval of integration, the integral is called "improper." Consider the following improper integral:

$$I = \int_a^b \frac{dx}{x^\alpha} \qquad \text{and } a < 0 \quad \text{and} \quad b > 0.$$

If $\alpha < 1$, the following limit exists and is finite:

$$\lim_{\epsilon_1 \to 0} \int_a^{-\epsilon_1} \frac{dx}{x^\alpha} + \lim_{\epsilon_2 \to 0} \int_{\epsilon_2}^b \frac{dx}{x^\alpha} = \frac{b^{1-\alpha} - a^{1-\alpha}}{1-\alpha}.$$

This limit is taken as the value of the integral I. The integral is said to be convergent and *weakly singular*. If $\alpha = 1$, then the following limit exists:

$$\lim_{\epsilon \to 0} \left(\int_a^{-\epsilon} \frac{dx}{x} + \int_\epsilon^b \frac{dx}{x} \right) = \lim_{\epsilon \to 0} \left(\ln \frac{\epsilon}{-a} + \ln \frac{b}{\epsilon} \right) = \ln \frac{b}{|a|}.$$

Note that in the above, we used the same ϵ for both sides of $x = 0$. This is essential to obtain the finite limit. The integral is said to be *singular in a Cauchy sense*, and the value above is called the *Cauchy principal value*. If $\alpha > 1$, the integral is said to be *strongly singular*.

The integral $I(\bar{r})$,

$$I(\bar{r}) = \int_V f(\bar{r}, \bar{r}')\, dV', \tag{12A-1}$$

is called the *improper integral* if $f(\bar{r}, \bar{r}')$ becomes unbounded at a point $\bar{r}' = \bar{r}$ within the volume V. The integral $I(\bar{r})$ is convergent if

$$\lim_{\nu \to 0} \int_{V-\nu} f(\bar{r}, \bar{r}')\, dV' \tag{12A-2}$$

exists, where ν is a small volume surrounding $\bar{r}' = \bar{r}$.

If we let $|\bar{r} - \bar{r}'| = R$,

$$\int_V \frac{dV'}{R^\beta} \quad \text{is convergent if } 0 < \beta < 3 \tag{12A-3}$$

and

$$\int_S \frac{dS'}{R^\alpha} \quad \text{is convergent if } 0 < \alpha < 2. \tag{12A-4}$$

Next consider the normal derivative

$$I_+ = \frac{\partial}{\partial n} \int_S f(\bar{r}')G_0(\bar{r}, \bar{r}')\, dS' \tag{12A-5}$$

as $\bar{r}$ approaches the surface S from the side $z > 0$. We let $S = (S - \sigma) + \sigma$, where σ is a small area shown in Fig. 12A-1. Then $G_0 \to R^{-1} = (z^2 + x^2 + y^2)^{-1/2}$ and

$$\frac{\partial}{\partial n}\left(\frac{1}{R}\right) = \frac{\partial}{\partial z}\left(\frac{1}{R}\right) = -\frac{1}{R^2}\frac{z}{R},$$

$$dS' = R^2\, d\Omega\, (R/z),$$

where we used $dS' = 2\pi r\, dr = 2\pi R\, dR$, $R = z/\cos\theta$, and $d\Omega = 2\pi \sin\theta\, d\theta$. Therefore, we get

$$\frac{\partial}{\partial n}\int_\sigma fG_0\, dS' = -\int_\sigma f(\bar{r})\frac{d\Omega}{4\pi} = -\frac{f(\bar{r})}{2},$$

$$I_+ = -\frac{f(\bar{r})}{2} + \int_{S-\sigma} f(\bar{r}')\frac{\partial}{\partial n}G_0(\bar{r}, \bar{r}')\, dS'. \tag{12A-6}$$

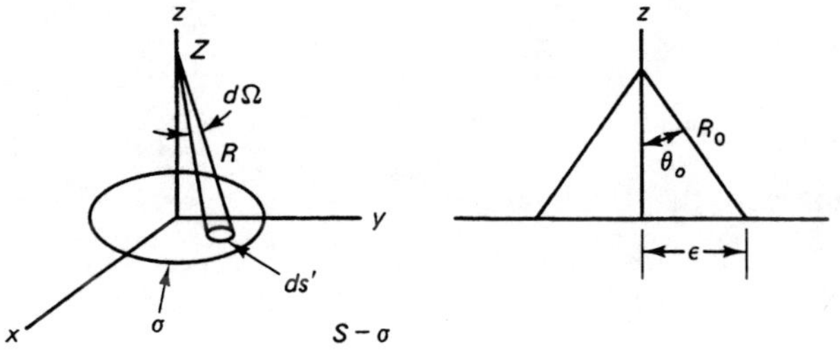

Figure 12A-1 Improper integral.

The integral over $S - \sigma$ is called the Cauchy principal value and is expressed as

$$⨍ f(\bar{r}')\frac{\partial}{\partial n}G_0(\bar{r},\bar{r}')\,dS'.$$

The integral is convergent because $(\partial/\partial n)G_0$ is zero near $\bar{r} = \bar{r}'$ and $z = 0$.

Note also that if $\bar{r}$ approaches the surface S from the side $z < 0$, we get

$$I_- = +\frac{f(\bar{r})}{2} + \int_{S-\sigma} f(\bar{r}')\frac{\partial}{\partial n}G_0(\bar{r},\bar{r}')\,dS'. \tag{12A-7}$$

Therefore, we get

$$I_+ - I_- = -f(\bar{r}). \tag{12A-8}$$

Consider the integral

$$I = \frac{\partial}{\partial n}\int_\sigma \frac{\partial}{\partial n'}G_0(\bar{r},\bar{r}')f(\bar{r}')\,dS', \tag{12A-9}$$

where σ is a small area with radius ϵ. Then using the results above and noting that $\partial/\partial n' = \partial/\partial z'$ rather than $\partial/\partial z$,

$$I = -\frac{\partial}{\partial n}\Omega f(\bar{r}), \tag{12A-10}$$

where Ω is the solid angle given by Fig. 12A-1.

$$\Omega = 2\pi(1 - \cos\theta_0) = 2\pi\left(1 - \frac{z}{R_0}\right).$$

Therefore, we get

$$I = -\frac{\partial}{\partial z}\Omega f = 2\pi\frac{\epsilon^2}{R_0^3}f, \tag{12A-11}$$

where we used

$$\frac{1}{2\pi}\frac{\partial\Omega}{\partial z} = -\frac{1}{R_0} + \frac{z}{R_0^2}\frac{\partial R_0}{\partial z}.$$

As $z \to 0$, I approaches $(2\pi/\epsilon)f(\bar{r})$, and thus I depends on the size of the area σ.

B. INTEGRAL EQUATIONS

Integral equations are classified as follows:

1. Fredholm integral equation of the first kind:

$$\int_a^b K(x,x')f(x')\,dx' = g(x), \tag{12B-1}$$

where $K(x,x')$ is a known function called the *kernel function*. $f(x')$ is the unknown and $g(x)$ is the given function. The solution depends greatly on the kernel function. Laplace transform is an example:

$$\int_0^{\infty} e^{-st} f(t)\, dt = \phi(s), \tag{12B-2}$$

where $\exp(-st)$ is the kernel. The solution is given by

$$f(t) = \frac{1}{2\pi j}\int_C \phi(s) e^{st}\, ds. \tag{12B-3}$$

For a more general kernel function, analytical solution is difficult to obtain.

2. Inhomogeneous Fredholm integral equation of the second kind:

$$\int_a^b k(x,x') f(x')\, dx' = f(x) + g(x). \tag{12B-4}$$

3. Homogeneous Fredholm integral equation of the second kind:

$$\int_a^b K(x,x') f(x')\, dx' = \lambda f(x). \tag{12B-5}$$

This constitutes an eigenvalue equation where λ is the eigenvalue.

4. In the above, if the upper limit b is replaced by the variable x, this is called the *Volterra integral equation*. For example, the Volterra integral equation of the first kind is given by

$$\int_a^x K(x,x') f(x')\, dx' = g(x).$$

Appendix to Chapter 14

A. STATIONARY-PHASE EVALUATION OF A MULTIPLE INTEGRAL I

$$I = \int_{-\infty}^{\infty} dx_1 \int_{-\infty}^{\infty} dx_2 \cdots \int_{-\infty}^{\infty} dx_N A(x_1 \cdots x_N) e^{if(x_1 x_2 \cdots x_N)}. \tag{14A-1}$$

First we find a stationary phase point $(x_{10}, x_{20}, x_{30}, \ldots, x_{N0})$ by satisfying N equations,

$$\frac{\partial f}{\partial x_1} = \frac{\partial f}{\partial x_2} = \frac{\partial f}{\partial x_3} = \cdots = \frac{\partial f}{\partial x_N} = 0. \tag{14A-2}$$

Then we expand f about this stationary-phase point.

$$f(x_1, x_2, \ldots, x_N) = f(x_{10}, x_{20}, \ldots, x_{N0}) + \frac{1}{2!}\left[(x_1 - x_{10})\frac{\partial}{\partial x_1} + (x_2 - x_{20})\frac{\partial}{\partial x_2} + \cdots + (x_N - x_{N0})\frac{\partial}{\partial x_N}\right]^2 f \Bigg|_{x_{10}, x_{20}, \ldots} + \text{higher-order terms}.$$

These higher-order terms contribute little to the integral unless the second derivatives of f are small. Furthermore, we assume that the amplitude $A(x_1 \cdots x_N)$ is a slowly varying function of $x_1 \cdots x_N$, and thus we approximate

$$A(x_1 \cdots x_N) \approx A(x_{10}, x_{20}, \ldots, x_{N0}).$$

Then we write, letting

$$x_1 - x_{10} = x_1', \quad x_2 - x_{20} = x_2', \quad \text{etc.},$$

$$I = A(x_{10}, x_{20}, \ldots, x_{N0}) e^{jf(x_{10}, x_{20}, \ldots, x_N)} \int_{-\infty}^{\infty} dx_1' \int_{-\infty}^{\infty} dx_2' \cdots \int_{-\infty}^{\infty} dx_N' e^{j(1/2)[T]}, \tag{14A-3}$$

where

$$[T] = \left(x_1' \frac{\partial}{\partial x_1'} + x_2' \frac{\partial}{\partial x_2'} + \cdots + x_N' \frac{\partial}{\partial x_N'} \right)^2 f.$$

We write $[T]$ in the following matrix form:

$$[T] = \tilde{x} F x$$

$$x = \begin{bmatrix} x_1' \\ \vdots \\ x_N' \end{bmatrix}, \qquad F = \begin{bmatrix} f_{11} & f_{12} & \cdots & f_{1N} \\ f_{21} & & & \\ \vdots & & & \\ f_{N1} & \cdots & \cdots & f_{NN} \end{bmatrix}, \tag{14A-4}$$

where

$$f_{ij} = \frac{\partial^2}{\partial x_i' \, \partial x_j'} f \bigg|_{x'_1 = 0, x'_2 = 0, \text{ etc.}}$$

Next we note that by the orthogonal transformation of X to Y,

$$X = PY \qquad \text{and} \qquad Y = \begin{bmatrix} y_1 \\ y_2 \\ \vdots \\ y_N \end{bmatrix}$$

We can convert $[T]$ into the following diagonal form:

$$\begin{aligned} [T] &= \tilde{X} F X \\ &= \tilde{Y} \tilde{P} F P Y \\ &= \tilde{Y} \alpha Y, \end{aligned} \tag{14A-5}$$

where

$$\alpha = \begin{bmatrix} \alpha_1^2 & 0 & 0 & 0 & 0 \\ 0 & \alpha_2^2 & 0 & 0 & 0 \\ 0 & 0 & \alpha_3^2 & 0 & 0 \\ 0 & 0 & 0 & \ddots & \alpha_N^2 \end{bmatrix}.$$

Thus we obtain

$$[T] = \alpha_1^2 y_1^2 + \alpha_2^2 y_2^2 + \cdots + \alpha_N^2 y_N^2. \tag{14A-6}$$

Also, note that the Jacobian of $x_1' \cdots x_N'$ with respect to $y_1 y_2 \cdots y_N$ is 1.

$$dx_1'\,dx_2' \cdots dx_N' = \frac{\partial(x_1', \ldots, x_N')}{\partial(y_1, \ldots, y_N)}\,dy_1 \cdots dy_N,$$

$$\text{Jacobian} = \frac{\partial(x_1' \cdots x_N')}{\partial(y_1 \cdots y_N)}$$

$$= \begin{vmatrix} \dfrac{\partial x_1'}{\partial y_1} & \cdots & \dfrac{\partial x_N'}{\partial y_1} \\ \vdots & & \\ \dfrac{\partial x_1'}{\partial y_N} & \cdots & \dfrac{\partial x_N'}{\partial y_N} \end{vmatrix} = |P| = 1.$$

Therefore, we obtain

$$\int_{-\infty}^{\infty} dx_1' \int_{-\infty}^{\infty} dx_2' \cdots \int_{-\infty}^{\infty} dx_N' e^{j(1/2)[T]} = \int_{-\infty}^{\infty} dy_1 \int_{-\infty}^{\infty} dy_2 \cdots \int_{-\infty}^{\infty} dy_N e^{j[\alpha_1^2 y_1^2 + \alpha_2^2 y_2^2 + \cdots + \alpha_N^2 y_N^2]}$$

$$= \frac{(2\pi)^{N/2} e^{jN(\pi/4)}}{\sqrt{\alpha_1^2 \alpha_2^2 \cdots \alpha_N^2}}. \tag{14A-7}$$

But since

$$\alpha_1^2 \alpha_2^2 \cdots \alpha_N^2 = |\alpha|$$
$$= |\tilde{P}FP|$$
$$= |\tilde{P}||F||P|$$
$$= |F|,$$

we obtain

$$I = A(x_{10}, x_{20}, \ldots, x_{N0}) e^{jf(x_{10}, x_{20}, \ldots, x_{N0})} \frac{(2\pi)^{N/2} e^{jN(\pi/4)}}{\sqrt{\Delta}}, \tag{14A-8}$$

where Δ is the determinant of $F = |F|$, and is called *Hesse's determinant*. For $N = 2$, we have

$$I = \int_{-\infty}^{\infty} dx_1 \int_{-\infty}^{\infty} dx_2 A(x_1 x_2) e^{jf(x_1 x_2)}$$
$$= A(x_{10} x_{20}) e^{jf(x_{10} x_{20})} \frac{(2\pi) e^{j(\pi/2)}}{\sqrt{f_{11} f_{12} - f_{12}^2}}, \tag{14A-9}$$

where x_{10}, x_{20} are given by

$$\frac{\partial f}{\partial x_1} = 0 \quad \text{and} \quad \frac{\partial f}{\partial x_2} = 0.$$

Appendix to Chapter 15

A. SOMMERFELD'S SOLUTION

Sommerfeld's 1909 paper contains an error in sign for $\sqrt{p_1}$ (which is essentially $-s_p$ instead of $+s_p$, and this $-s_p$ corresponds to the pole on the bottom Riemann surface in the α plane). If we deliberately change $\sqrt{p_1}$ to $-\sqrt{p_1}$ in (15-63), we get

$$\pi_{er} = 2\frac{e^{-jkR}}{4\pi R}[1 + j\sqrt{p_1\pi}\, e^{-p_1}\,\mathrm{erfc}(-j\sqrt{p_1})]. \qquad (15A\text{-}1)$$

Using

$$\mathrm{erfc}(-z) = 2 - \mathrm{erfc}(z),$$

we write

$$\pi_{er} = 2\frac{e^{-jkR}}{4\pi R}[1 - j\sqrt{p_1\pi}\, e^{-p_1}\,\mathrm{erfc}(+j\sqrt{p_1}) + 2j\sqrt{p_1\pi}\, e^{-p_1}],$$

which differs from the correct π in (15-63) as follows:

$$\pi_{er} = \pi + 2\frac{e^{-jkR}}{4\pi R}(2j\sqrt{p_1\pi}\, e^{-p_1}). \qquad (15A\text{-}2)$$

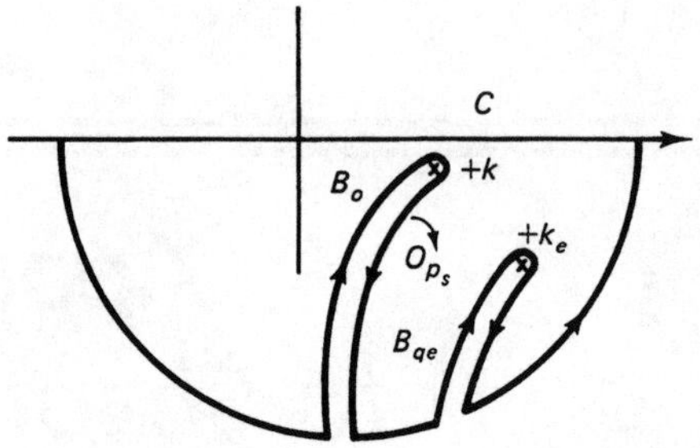

Figure 15A-1 Branch cut integration and residue.

It is easy to show that the foregoing difference between π_{er} and π is precisely equal to $-2\pi j$ (the residue at the Sommerfeld pole, s_p). Thus we write

$$\pi = \pi_{er} - (-2\pi j \cdot \text{residue at } s_p). \tag{15A-3}$$

Now, in his 1909 paper, Sommerfeld evaluated the integral in the following form:

$$\pi_{er} = B_q + B_{qe} + (-2\pi j \cdot \text{residue at } s_p), \tag{15A-4}$$

where B_q and B_{qe} are the branch cut integration along $\text{Im}\, q = 0$ and $\text{Im}\, q_e = 0$, respectively (Fig. 15A-1).

Now the term

$$(-2\pi j)(\text{residue at } s_p),$$

is precisely the Zenneck wave excited by the dipole source. But because of the sign error, the true field is π_{er} minus the Zenneck wave. Thus it *appeared* that the Zenneck wave contained in Sommerfeld's 1909 paper disappears if the correct sign is used. Therefore, it *appears* that the Zenneck wave does not exist.

As we pointed out earlier, this division (15A-4) is arbitrary, and the fact that one term exhibits the Zenneck wave characteristic does not mean that the wave exists. In fact, the question of the existence of a wave based on one term of the total wave is meaningless, because what is important is the total wave, not a portion of the wave. However, this sign error created controversies over the existence of the Zenneck wave for many decades (see Banos, 1966, p. 154). This error in sign was not detected until 1935 by Norton, even though Sommerfeld has the correct sign in his 1926 paper. The entire question of the controversies was finally resolved around 1950 through work of van der Waeden, Ott, and Banos and Wesley (see Banos, 1966).

B. RIEMANN SURFACES FOR THE SOMMERFELD PROBLEM

In the book the evaluation of the integral was performed in the α plane. It is instructive to see how the integration and the various Riemann surfaces may appear in the λ plane.

In the λ plane, the branch points appear at

$$\lambda = \pm k \qquad \text{and} \qquad \lambda = \pm k_e. \tag{15B-1}$$

Let us now draw the branch cuts from these branch points along

$$\operatorname{Im} q = \operatorname{Im} \sqrt{k^2 - \lambda^2} = 0,$$

and

$$\operatorname{Im} q_e = \operatorname{Im} \sqrt{k_e^2 - \lambda^2} = 0. \tag{15B-2}$$

As shown in Chapter 15, these branch cuts are hyperbolas passing through the branch points. It is obvious that the contour of the integration must be on the Riemann surface where

$$\operatorname{Im} q < 0 \qquad \text{and} \qquad \operatorname{Im} q_e < 0. \tag{15B-3}$$

In this case the wave attenuates as $z \to +\infty$ and $z \to -\infty$ and thus satisfies the *radiation condition.*

$$|e^{-jqz}| = e^{(\operatorname{Im} q)z} \to 0 \qquad \text{as } z \to +\infty,$$

$$|e^{jq_e z}| = e^{-(\operatorname{Im} q_e)z} \to 0 \qquad \text{as } z \to -\infty.$$

This surface is therefore called the *proper Riemann surface.*

In addition, however, there are three other Riemann surfaces on which the radiation condition is not satisfied, and therefore these are called the *improper Riemann surfaces*. We list these four Riemann surfaces:

I.	Proper Rieman surface:	$\operatorname{Im} q < 0, \operatorname{Im} q_e < 0$
II.	Improper Riemann surface:	$\operatorname{Im} q > 0, \operatorname{Im} q_e < 0$
III.	Improper Riemann surface:	$\operatorname{Im} q < 0, \operatorname{Im} q_e > 0$
IV.	Improper Riemann surface:	$\operatorname{Im} q > 0, \operatorname{Im} q_e > 0$

The first two belong to $\operatorname{Im} q_e < 0$ and the last two belong to $\operatorname{Im} q_e > 0$. Let us consider the first two surfaces. We first examine the complex q plane (Fig. 15B-1). We label each quadrant as B_2, B_1, T_2, and T_1. T_1 and T_2 belong to the top surface (I) and B_1 and B_2 belong to the bottom surface (II) in the λ plane. The corresponding α plane can be obtained by noting that

$$\lambda = k \sin \alpha \qquad \text{and} \qquad q = k \cos \alpha.$$

As we go around in the q plane from A to B to C to D, we can trace the corresponding locus in surfaces I and II of the λ plane and the α plane (Fig. 15B-1). Note that in the α plane, T_1, T_2, B_1, and B_2 are placed side by side, and the branch point $\lambda = +k$ in the λ plane is the regular point $\alpha = \pi/2$ in the α plane.

Now let us picture the four Riemann surfaces in the λ plane (Fig. 15B-2). Close examination of the Sommerfeld poles reveals that under this branch cut, the poles are on Riemann surfaces I and IV. In the α plane there are only two surfaces,

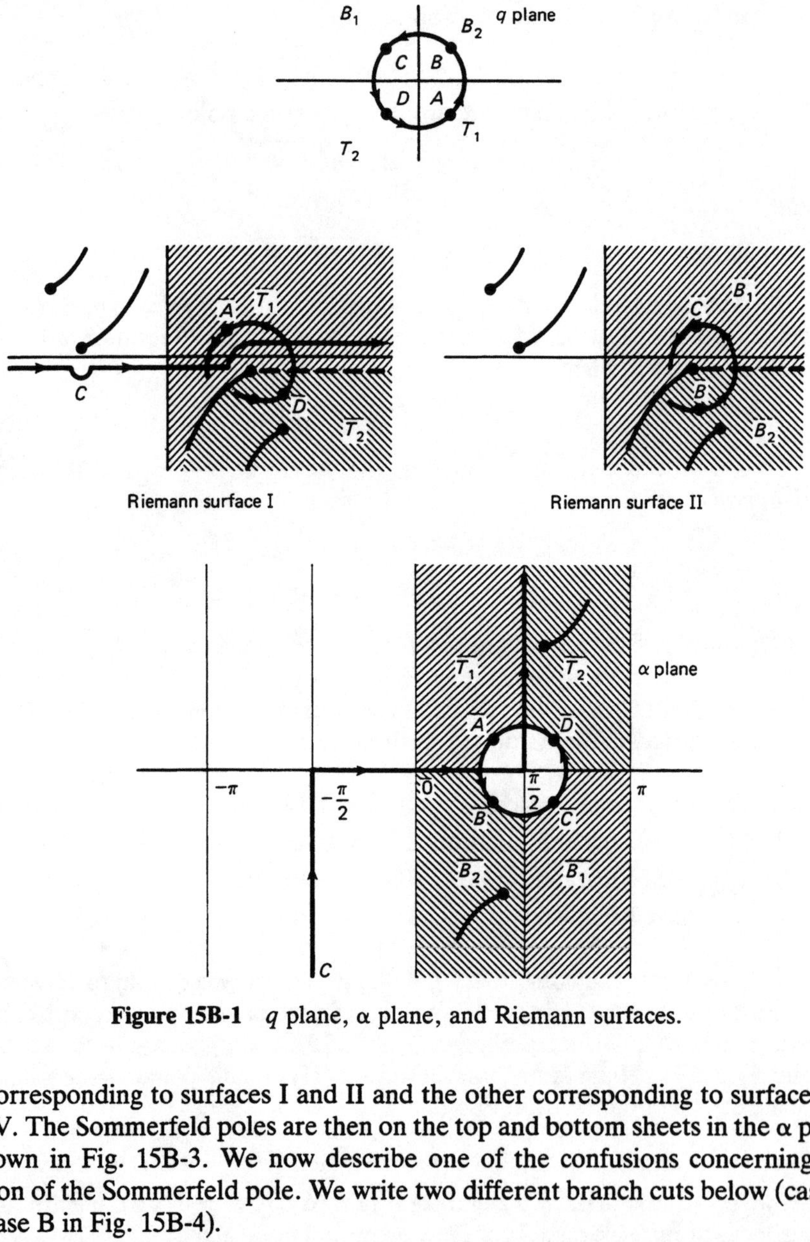

Figure 15B-1 q plane, α plane, and Riemann surfaces.

one corresponding to surfaces I and II and the other corresponding to surfaces III and IV. The Sommerfeld poles are then on the top and bottom sheets in the α plane as shown in Fig. 15B-3. We now describe one of the confusions concerning the location of the Sommerfeld pole. We write two different branch cuts below (case A and case B in Fig. 15B-4).

For case A the branch cuts are along $\operatorname{Im} q = 0, \operatorname{Im} q_e = 0$. For case B, the branch cuts are along $\operatorname{Re} \lambda = \operatorname{Re} k, \operatorname{Re} \lambda = \operatorname{Re} k_e$. Then for A, the pole is on I and IV, but for B, the pole is on II and III. Thus for A, the original integral along C is

$$\int_C = \int_{C_\infty} + \int_{B_{rA1}} + \int_{B_{rA2}} + 2\pi j \cdot \text{residue at } P_A,$$

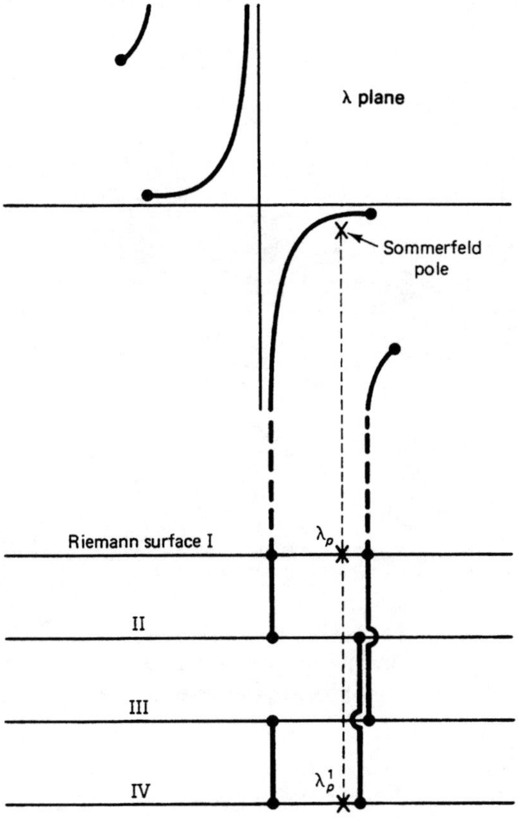

Figure 15B-2 Four Riemann surfaces and the Sommerfeld pole.

where B_{rA1} and B_{rA2} are the branch cut integrations. On the other hand, for case B,

$$\int_C = \int_{C_\infty} + \int_{B_{rB1}} + \int_{B_{rB2}} .$$

and there is no contribution from the pole P_B, because P_B is on II. Thus it *appears* that case A gives the extra term due to the pole, and case A and case B give two different answers. Note also that the residue has all the characteristics of the Zenneck wave and thus it *appears* that case A yields the Zenneck wave, whereas case B does not.

All these apparent contradictions are of course nonexistent. The branch cuts are drawn for convenience only, and both case A and case B should give the identical result. In fact, we can show that (the integral along C_∞ is zero) there is a difference between the branch cut integration for A and for B and this difference is exactly equal to the residue term, that is,

$$\int_{B_{rA1}} + \int_{B_{rA2}} + 2\pi j \cdot \text{residue} = \int_{B_{rB1}} + \int_{B_{rB2}} .$$

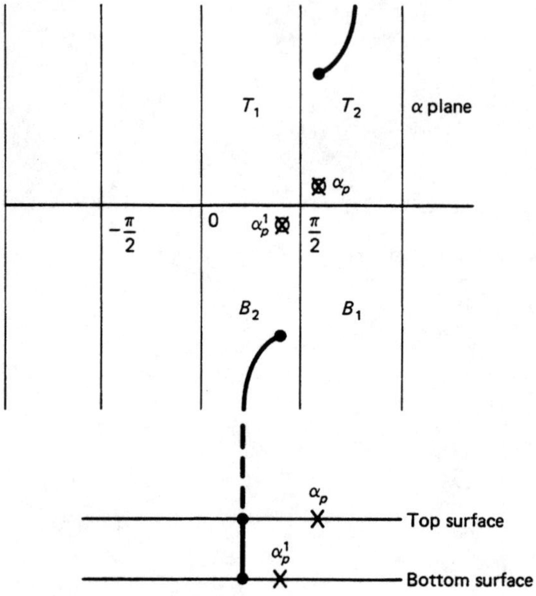

Figure 15B-3 Sommerfeld poles at α_p and α_p^1.

In the text, the integration was performed in the α plane, where Riemann surfaces I and II are placed side by side and thus the confusion between locations P_A and P_B never arises in the α plane.

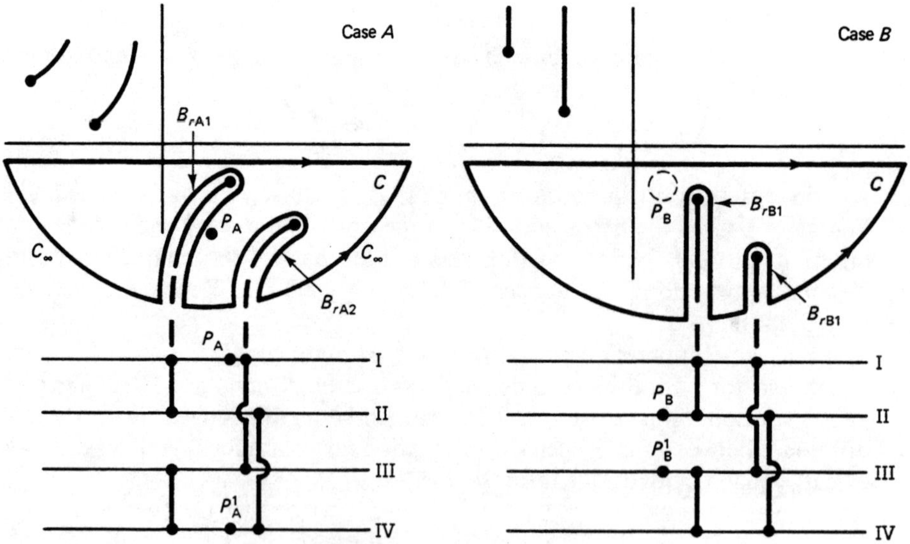

Figure 15B-4 Two different branch cuts.

C. MODIFIED SADDLE-POINT TECHNIQUE

Let us consider an integral

$$I_0 = \int_c F(\alpha) \exp[zf(\alpha)]\, d\alpha, \tag{15C-1}$$

where z is a large positive number and $F(\alpha)$ has a pole at α_p, which may be located at any part of the α plane.

We first find the saddle point $\alpha = \alpha_s$ from

$$\left.\frac{\partial f(\alpha)}{\partial \alpha}\right|_{\alpha_s} = 0, \tag{15C-2}$$

and write (15C-1) as follows:

$$I_0 = e^{zf(\alpha_s)} \int_c F(\alpha) e^{z[f(\alpha) - f(\alpha_s)]}\, d\alpha. \tag{15C-3}$$

Let us transform (15C-3) into the following form using the transformation from α to s:

$$f(\alpha) - f(\alpha_s) = -\frac{s^2}{2}, \tag{15C-4}$$

$$I_0 = e^{zf(\alpha_s)} \int_c F(s) e^{-z(s^2/2)} \left(\frac{d\alpha}{ds}\right) ds. \tag{15C-5}$$

The transformation (15C-4) is possible if the saddle point is an isolated one. We note that in (15C-5) the saddle point is at $s = 0$, and the steepest descent contour (SDC) is along the real axis of s. This is pictured in Fig. 15-8 using the following example:

$$\begin{aligned} f(\alpha) &= -j \cos(\alpha - \theta), \\ s &= 2e^{-j(\pi/4)} \sin\frac{\alpha - \theta}{2}, \\ \frac{d\alpha}{ds} &= \left(1 - j\frac{s^2}{4}\right)^{-1/2} e^{j(\pi/4)}. \end{aligned} \tag{15C-6}$$

Because of $d\alpha/ds$, two branch points (B_{r1} and B_{r2}) appear at $s = \pm 2e^{-j(\pi/4)}$ in the s plane, but these branch points are far from the saddle point and have little effect on the integral.

Now we evaluate (15C-5) for the following two cases:

1. *$F(\alpha)$ has no pole.* Let us first obtain the solution for a simpler case, where there is no pole in the integrand. In this case we can deform the contour from C to SDC:

$$\int_c ds = \int_{-\infty}^{\infty} ds. \tag{15C-7}$$

Now we expand $F(s)\, d\alpha/ds$ in a Taylor's power series:

$$F(\alpha)\frac{d\alpha}{ds} = \sum_{n=0}^{\infty} A_{2n} s^{2n} + \sum_{n=1}^{\infty} A_{2n-1} s^{2n-1}, \tag{15C-8}$$

where we separated the even and odd powers for convenience.

Substituting (15C-8) into (15C-5), we note that the odd terms of (15C-8) become zero. Thus

$$I_0 = e^{zf(\alpha_s)} \int_{-\infty}^{\infty} \sum_{n=0}^{\infty} A_{2n} s^{2n} e^{-z(s^2/2)}\, ds. \tag{15C-9}$$

The integration can be performed by using the gamma function:

$$\Gamma(x) = \int_0^{\infty} t^{x-1} e^{-t}\, dt. \tag{15C-10}$$

We also let

$$t = z\frac{s^2}{2}, \qquad x = n + \tfrac{1}{2}$$

and note that

$$\begin{aligned} \Gamma(\tfrac{1}{2}) &= \sqrt{\pi}, \\ \Gamma(n+\tfrac{1}{2}) &= \frac{1\cdot 3\cdot 5\cdots(2n-1)}{2^n}\sqrt{\pi}, \\ \int_{-\infty}^{\infty} s^{2n} e^{-z(s^2/2)}\, ds &= \frac{\Gamma(n+\tfrac{1}{2})2^{n+1/2}}{z^{n+1/2}} \\ &= \frac{1\cdot 3\cdot 5\cdots(2n-1)}{z^{n+1/2}}\sqrt{2\pi}. \end{aligned} \tag{15C-11}$$

We then obtain

$$I_0 = e^{zf(\alpha_s)} \sum_{n=0}^{\infty} A_{2n} \frac{1\cdot 3\cdot 5\cdots(2n-1)}{z^{n+1/2}}\sqrt{2\pi}. \tag{15C-12}$$

The first term,

$$e^{zf(\alpha_s)} A_0 \sqrt{\frac{2\pi}{z}},$$

is the result obtained from the usual saddle-point technique shown in Appendix 11C, and (15C-12) is the asymptotic series in inverse power of z. The series on the right side of (15C-12), which was obtained by term-by-term integration (15C-10), is in general divergent. However, it is an asymptotic series having the following characteristics.

If we take the partial sum

$$f_N = \sum_{n=0}^{N} A_n,$$

then the error $(I_0 - f_N)$ is smaller in absolute value than the first term neglected (A_{N+1}):

$$|I_0 - f_N| < |A_{N+1}|.$$

Usually, $|A_n|$ becomes smaller at first and then eventually gets larger and the series diverges.

Alternatively, I_0 in (15C-6) can be evaluated numerically along the real axis of s using the Gauss quadrature technique.

2. *$F(\alpha)$ has a pole.* Now we come to our main problem. When $F(\alpha)$ has a pole at α_p, or $s = s_p$ in the s plane, Taylor's expansion (15C-9) is valid only within a circle whose radius (called the *radius of convergence*) is the distance from the origin to $|s_p|$:

$$|s| < |s_p|.$$

If the pole is close to the saddle point, $|s_p|$ is very small. In order to extend the region of Taylor's expansion, we first write $F(s)\, d\alpha/ds$ as a sum of the term with a pole and the term that is regular at $s = s_p$:

$$F(s)\frac{d\alpha}{ds} = \frac{R_1(s_p)}{s - s_p} + G_1(s). \tag{15C-13}$$

$R_1(s_p)$ is the residue at $s = s_p$. Note also that the residue of the integrand $F(\alpha)\exp[zf(\alpha)]$ at the pole $\alpha = \alpha_p$ is given by

$$R(\alpha_p) = R_1(s_p)\exp[zf(\alpha_p)]. \tag{15C-14}$$

To show this, let $F(\alpha) = N(\alpha)/D(\alpha)$ and note that

$$R_1(s_p) = \left[\frac{N(s)}{\partial D/\partial s}\frac{d\alpha}{ds}\right]_{s=s_p} = \left.\frac{N(\alpha)}{\partial D/\partial \alpha}\right|_{\alpha=\alpha_p}.$$

The evaluation of the integral must now be done differently depending on whether the pole is below or above the original contour.

It is clear that the integral along the original contour C can now be deformed along the saddle-point contour (SDC). If the pole is located below the SDC, then

$$\int_C = \int_{\text{SDC}}. \tag{15C-15}$$

But if the pole is located above the SDC, we have

$$\int_C = \int_{\text{SDC}} - 2\pi j R(s_p), \tag{15C-16}$$

where $R(s_p)$ is the residue of $F(\alpha)\exp[-zf(\alpha)]$ at s_p shown in (15C-14). We combine (15C-15) and (15C-16) and write

$$\int_C = \int_{\text{SDC}} - 2\pi j R(s_p) U(\operatorname{Im} s_p), \tag{15C-17}$$

where $U(x)$ is a unit step function. Next we evaluate the integral along the SDC:

$$\int_{\text{SDC}} = e^{zf(\alpha_s)} \int_{-\infty}^{\infty} F(s) \frac{d\alpha}{ds} e^{-z(s^2/2)}\, ds \tag{15C-18}$$

$$= I_p + I_1,$$

where

$$I_p = e^{zf(\alpha_s)} \int_{-\infty}^{\infty} \frac{R_1(s_p)}{s - s_p} e^{-z(s^2/2)}\, ds, \tag{15C-19}$$

$$I_1 = e^{zf(\alpha_s)} \int_{-\infty}^{\infty} G_1(s) e^{-z(s^2/2)}\, ds. \tag{15C-20}$$

The evaluation of (15C-19) can be made as follows. Write

$$I_p = e^{zf(\alpha_s)} R_1(s_p) y(z), \tag{15C-21}$$

where

$$y(z) = \int_{-\infty}^{\infty} \frac{e^{-z(s^2/2)}}{s - s_p}\, ds. \tag{15C-22}$$

To evaluate $y(z)$ it is necessary to reduce the integral (15C-21) to a known integral. In this case because of the form of (15C-22), it is possible to convert (15C-22) into an error integral. To do this, we first multiply $y(z)$ by $\exp[zs_p^2/2]$ and take the derivative with respect to z.

$$\frac{\partial}{\partial z}[y(z)e^{z(s_p^2/2)}] = \int_{-\infty}^{\infty} \frac{s^2 - s_p^2}{2(s - s_p)} e^{-(z/2)(s^2 - s_p^2)}\, ds$$

$$= -\frac{s_p}{2} e^{z(s_p^2/2)} \int_{-\infty}^{\infty} e^{-z(s^2/2)}\, ds$$

$$= -\sqrt{\pi/2}\, s_p \frac{e^{z(s_p^2/2)}}{\sqrt{z}}.$$

We now integrate (15C-22) over z from $z = 0$ to z:

$$y(z)e^{z(s_p^2/2)} = y(0) - \sqrt{\pi/2}\, s_p \int_0^z \frac{e^{z(s_p^2/2)}}{\sqrt{z}}\, dz. \tag{15C-23}$$

Using the following definition of the error function,

$$\text{erf}(z) = \frac{2}{\sqrt{\pi}} \int_0^z e^{-t^2}\, dt,$$

(15C-23) becomes

$$y(z) = e^{-z(s_p^2/2)} y(0) + j\pi e^{-z(s_p^2/2)} \,\text{erf}(j\sqrt{z/2}\, s_p). \tag{15C-24}$$

However, $y(0)$ is given by

$$y(0) = \int_{-\infty}^{\infty} \frac{ds}{s - s_p} = \begin{cases} j\pi & \text{for } \operatorname{Im} s_p > 0 \\ 0 & \text{for } \operatorname{Im} s_p = 0. \\ -j\pi & \text{for } \operatorname{Im} s_p < 0 \end{cases} \tag{15C-25}$$

Thus we write

$$y(z) = e^{-z(s_p^2/2)}\left\{2\pi j U(\operatorname{Im} s_p) - j\pi\left[1 - \operatorname{erf}\left(j\sqrt{\frac{z}{2}}s_p\right)\right]\right\}. \tag{15C-26}$$

Using the complementary error function, erfc(z),

$$\operatorname{erfc}(z) = 1 - \operatorname{erf}(z),$$

we finally obtain I_p, using $f(\alpha_s) - (s_p^2/2) = f(\alpha_p)$,

$$I_p = R(s_p)\left[2\pi j U(\operatorname{Im} s_p) - j\pi \operatorname{erfc}\left(j\sqrt{\frac{z}{2}}s_p\right)\right]. \tag{15C-27}$$

The original integral I_0 in (15C-5) is, therefore, given by

$$I_0 = I_1 + I_p - 2\pi j R(s_p) U(\operatorname{Im} s_p). \tag{15C-28}$$

Note that the discontinuity in I_0 in (15C-27) will be exactly canceled by the last term of (15C-28), and thus we finally obtain the following.

(a) *The pole is located below the original contour in region A* (Fig. 15C-1).

$$\begin{aligned} I_0 &= \int_C F(\alpha) e^{zf(\alpha)}\, d\alpha \\ &= I_1 - j\pi R(s_p) \operatorname{erfc}\left(j\sqrt{\frac{z}{2}}s_p\right), \end{aligned} \tag{15C-29}$$

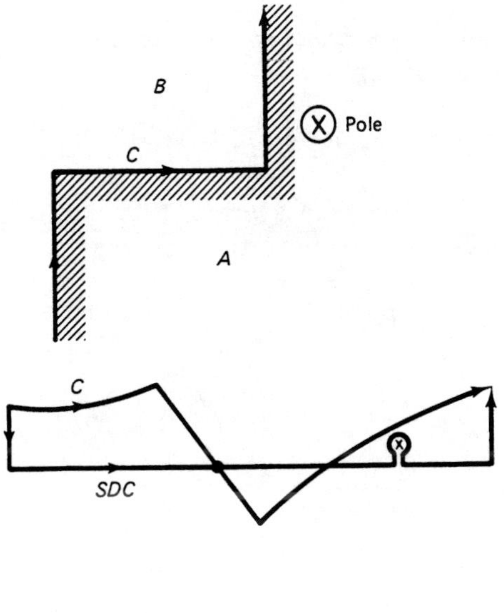

Figure 15C-1 Pole in region A.

where

$$I_1 = e^{zf(\alpha_s)} \int_{-\infty}^{\infty} \left[F(s)\frac{d\alpha}{ds} - \frac{R_1(s_p)}{s - s_p} \right] \exp\left(-z\frac{s^2}{2}\right) ds,$$

$$R(s_p) = R_1(s_p) \exp[zf(\alpha_p)]$$

$$= \text{residue of } F(\alpha) \exp[zf(\alpha)] \text{ at the pole } \alpha = \alpha_p,$$

$$f(\alpha) - f(\alpha_s) = -\frac{s^2}{2} \qquad \text{and } \alpha_s \text{ is the saddle point for } f(\alpha).$$

Note that the integrand for I_1 has no pole, and therfore this can be evaluated either by the expansion as in (15C-10) or numerically. The series expansion is given by

$$I_1 = e^{zf(\alpha_s)} \sum_{n=0}^{\infty} B_{2n} \left(\frac{2}{z}\right)^{n+1/2} \Gamma(n^{+1/2}), \tag{15C-30}$$

and B_{2n} is the coefficient of the expansion

$$\left[F(s)\frac{d\alpha}{ds} - \frac{R_1(s_p)}{s - s_p} \right] = \sum_{n=0}^{\infty} B_{2n} s^{2n} + \sum_{n=1}^{\infty} B_{2n-1} s^{2n-1},$$

$$B_0 = A_0 + \frac{R_1(s_p)}{s_p},$$

$$B_n = A_n + \frac{R_1(s_p)}{s_p^{n+1}}, \tag{15C-31}$$

$$A_0 = \left[F(s)\frac{d\alpha}{ds} \right]_{s=0},$$

$$A_n = \frac{1}{n!}\frac{d^n}{ds^n}\left[F(s)\frac{d\alpha}{ds} \right]_{s=0}.$$

(b) *The pole is located above the original contour (in region B)* (Fig. 15C-1). We follow a similar procedure and obtain

$$I_0 = I_1 + j\pi R(s_p) \operatorname{erfc}\left(-j\sqrt{\frac{z}{2}}s_p\right). \tag{15C-32}$$

The integral I_0 is given by (15C-29) if the pole is located below the original contour and by (15C-32) if the pole is located above the original contour.

Appendix to Chapter 16

A. HILBERT TRANSFORM

The Hilbert transform is defined by the following (Bracewell):

$$F_h(t) = \frac{1}{\pi}\int_{-\infty}^{\infty} \frac{f(t')}{t' - t}\, dt', \tag{16A-1}$$

where the integral is Cauchy's principal value. It is the convolution integral of $f(t')$ and $(-1)/(\pi t)$. The inverse Hilbert transform is given by

$$f(t) = -\frac{1}{\pi}\int \frac{F_h(t')}{t' - t}\, dt'. \tag{16A-2}$$

To show this, we take the Fourier transform of (16A-1). Noting that this is a convolution integral, we get

$$\begin{aligned} \overline{F}_h(s) &= \int F_h(t) e^{-ist}\, dt \\ &= K(s)F(s), \end{aligned} \tag{16A-3}$$

where

$$K(s) = -\!\!\!\!\!\!\int \frac{1}{\pi t} e^{-ist}\, dt,$$

$$F(s) = \int f(t) e^{-ist}\, dt.$$

The integral for $K(s)$ is Cauchy's principal value, and therefore

$$K(s) = \lim_{\epsilon \to 0} \left(-\frac{1}{\pi}\right)\left(\int_{-\infty}^{-\epsilon} + \int_{\epsilon}^{\infty}\right) \frac{e^{-ist}}{t}\, dt$$

$$= \lim \frac{1}{\pi} \int_{\epsilon}^{\infty} \frac{2i \sin st}{t}\, dt.$$

Noting that

$$\int_0^{\infty} \frac{\sin x}{x}\, dx = \frac{\pi}{2},$$

we get

$$K(s) = -\!\!\!\!\!\!\int \frac{1}{\pi t} e^{-ist}\, dt = i \operatorname{sgn}(s), \tag{16A-4}$$

where $\operatorname{sgn}(s) = 1$ if $s > 0$ and -1 if $s < 0$.

Taking the inverse transform, we get

$$-\frac{1}{\pi t} = \frac{1}{2\pi} \int i \operatorname{sgn}(s) e^{ist}\, ds. \tag{16A-5}$$

Taking the inverse Fourier transform of (16A-3), we write

$$F_h(t) = \frac{1}{2\pi} \int i \operatorname{sgn}(s) F(s) e^{ist}\, ds. \tag{16A-6}$$

On the other hand, from (16A-3) and (16A-4), we get

$$\begin{aligned} f(t) &= \frac{1}{2\pi} \int F(s) e^{ist}\, ds \\ &= \frac{1}{2\pi} \int [-i \operatorname{sgn}(s) \overline{F}_h(s)] e^{ist}\, ds. \end{aligned} \tag{16A-7}$$

This has a form identical to (16A-6), and therefore

$$f(t) = -\frac{1}{\pi} \int \frac{F_h(t')}{t' - t}\, dt'. \tag{16A-8}$$

B. RYTOV APPROXIMATION

Consider a wave field $U(\bar{r})$ satisfying the wave equation:

$$(\nabla^2 + k^2 n^2)U = 0. \tag{16B-1}$$

We let

$$U(\bar{r}) = \exp[\Psi(\bar{r})] \tag{16B-2}$$

and get Riccati's equation for $\Psi(\bar{r})$:

$$\nabla^2 \Psi + \nabla\Psi \cdot \nabla\Psi + k^2 n^2 = 0. \tag{16B-3}$$

Also we consider U_0 when $n = 1$.

$$\begin{aligned} U_0 &= \exp(\Psi_0), \\ \nabla^2 \Psi_0 + \nabla\Psi_0 \cdot \nabla\Psi_0 + k^2 &= 0. \end{aligned} \tag{16B-4}$$

We now let

$$\Psi = \Psi_0 + \Psi', \tag{16B-5}$$

and subtracting (16B-4) from (16B-3), we get

$$\nabla^2 \Psi' + 2\nabla\Psi_0 \cdot \nabla\Psi' = -[\nabla\Psi' \cdot \nabla\Psi' + k^2(n^2 - 1)]. \tag{16B-6}$$

Now using the identity

$$\nabla^2[U_0 \Psi'] = (\nabla^2 U_0)\Psi' + 2U_0 \nabla\Psi_0 \cdot \nabla\Psi' + U_0 \nabla^2 \Psi',$$

the left side of (16B-6) becomes

$$\frac{1}{U_0}[\nabla^2(U_0 \Psi') + k^2 U_0 \Psi'],$$

and we obtain

$$(\nabla^2 + k^2)(U_0 \Psi') = [\nabla\Psi' \cdot \nabla\Psi' + k^2(n^2 - 1)]U_0. \tag{16B-7}$$

This can be converted to the integral equation using Green's function G:

$$\Psi' = \frac{1}{U_0(\bar{r})}\int G(\bar{r} - \bar{r}')[k^2(n^2 - 1) + \nabla\Psi' \cdot \nabla\Psi']U_0(\bar{r}')\, dV'. \tag{16B-8}$$

The first iteration for Ψ' is obtained by letting $\nabla\Psi' = 0$ in the integral. This is called the *first Rytov solution* and is given by

$$\begin{aligned} U(\bar{r}) &= U_0(\bar{r}) \exp[\Psi_1(\bar{r})] \\ \Psi_1(\bar{r}) &= \frac{1}{U_0(\bar{r})}\int G(\bar{r} - \bar{r}')k^2(n^2 - 1)U_0(\bar{r}')\, dV'. \end{aligned} \tag{16B-9}$$

C. ABEL'S INTEGRAL EQUATION

Consider the following Volterra integral equation of the first kind, known as *Abel's integral equation*

$$g(t) = \int_0^t \frac{f(\tau)}{(t-\tau)^\alpha} d\tau, \qquad 0<\alpha<1, \tag{16C-1}$$

where $g'(t)$ is continuous. $g(t)$ is a given function of t and $f(\tau)$ is the unknown function to be determined. We will show that the solution to (16C-1) is given by

$$f(\tau) = \frac{\sin \pi\alpha}{\pi} \frac{g(0)}{\tau^{1-\alpha}} + \frac{\sin \pi\alpha}{\pi} \int_0^\tau \frac{g'(t)}{(\tau - t)^{1-\alpha}} dt. \tag{16C-2}$$

Furthermore, we will show that

$$\int_0^\tau f(\tau)\, d\tau = \frac{\sin \pi\alpha}{\pi} \int_0^\tau \frac{g(t)}{(\tau - t)^{1-\alpha}} dt. \tag{16C-3}$$

To prove this, we first note that $g(t)$ is a convolution integral of $K(\tau)$ and $f(\tau)$.

$$g(t) = \int_0^t K(t-\tau) f(\tau)\, d\tau, \tag{16C-4}$$

where $K(\tau) = \tau^{-\alpha}$.

We can solve this by using the Laplace transform:

$$\begin{aligned} \bar{g}(s) &= \mathcal{L}g(t), \\ \bar{K}(s) &= \mathcal{L}K(t), \\ \bar{f}(s) &= \mathcal{L}(f(t)). \end{aligned} \tag{16C-5}$$

We then get

$$\bar{g}(s) = \bar{K}(s)\bar{f}(s). \tag{16C-6}$$

Therefore, we get

$$f(t) = \mathcal{L}^{-1}\left[\frac{\bar{g}(s)}{\bar{K}(s)}\right].$$

The Laplace transform $\bar{K}(s)$ is given by

$$\bar{K}(s) = \mathcal{L}(\tau^{-\alpha}) = \frac{\Gamma(1-\alpha)}{s^{1-\alpha}} \qquad \text{for } \alpha<1. \tag{16C-7}$$

Therefore,

$$f(t) = \mathcal{L}^{-1}\left[\frac{s^{1-\alpha}}{\Gamma(1-\alpha)}\bar{g}(s)\right] = \frac{1}{\Gamma(1-\alpha)}\mathcal{L}^{-1}\left[\frac{1}{s^\alpha} s\bar{g}(s)\right]. \tag{16C-8}$$

We use the following to evaluate (16C-8):

$$\mathcal{L}^{-1}(s^{-\alpha}) = \frac{t^{\alpha-1}}{\Gamma(\alpha)}, \qquad \alpha > 0,$$
$$\mathcal{L}^{-1}(s\bar{g}(s)) = g'(t) + \delta(t)g(0), \tag{16C-9}$$
$$\frac{1}{\Gamma(1-\alpha)\Gamma(\alpha)} = \frac{\sin \pi\alpha}{\pi}.$$

We can then write (16C-8) in the following convolution form:

$$f(\tau) = \frac{\sin \pi\alpha}{\pi} \frac{g(0)}{\tau^{1-\alpha}} + \frac{\sin \pi\alpha}{\pi} \int_0^\tau \frac{g'(t)}{(\tau - t)^{1-\alpha}} dt. \tag{16C-10}$$

To show (16C-3), we integrate (16C-8) and get

$$\int_0^\tau f(\tau)\, d\tau = \frac{1}{\Gamma(1-\alpha)} \mathcal{L}^{-1}\left[\frac{1}{s^\alpha}\bar{g}(s)\right], \tag{16C-11}$$

and therefore this is given by the convolution integral (16C-3).

References

ABRAMOWITZ, M., and I. A. STEGUN, eds., *Handbook of Mathematical Functions*. Washington, D.C.: U.S. Government Printing Office, 1964.

ARVAS, E., and R. F. HARRINGTON, "Computation of the Magnetic Polarizability of Conducting Disks and the Electric Polarizability of Apertures," *IEEE Trans. Antennas Propag.*, 31 (1983), 719–724.

BAHL, I. J. and P. BHARTIA, *Microstrip Antennas,* Dedham, Mass, Artech House, 1980.

BALANIS, C. A., *Antenna Theory*. New York: Harper & Row, 1982.

———, *Advanced Engineering Electromagnetics*. New York: Wiley, 1989.

BANOS, A., JR., *Dipole Radiation in the Presence of a Conducting Half-Space*. Oxford: Pergamon Press, 1966.

BASSIRI, S., C. H. PAPAS, and N. ENGHETA, "Electromagnetic Wave Propagation through a Dielectric-Chiral Interface and through a Chiral Slab," *J. Opt. Soc. Am.*, A5 (1988), 1450–1459.

BAUM, C. E., "Emerging Technology for Transient and Broad-Band Analysis and Synthesis of Antennas and Scatterers," *Proc. IEEE,* 64 (1976), 1598–1616.

BOERNER, W.-M., ed., "Inverse Methods in Electromagnetic Imaging," *NATO-ASI Series C: Mathematical and Physical Sciences*. Dordrecht, The Netherlands: D. Reidel, 1985.

BOJARSKI, N. N., "A Survey of the Near-Field Far-Field Inverse Scattering Inverse Source Integral Equation," *IEEE Trans. Antennas Propag.*, 30 (1982a), 975–979.

———, "A Survey of the Physical Optics Inverse Scattering Identity," *IEEE Trans. Antennas Propag.*, 30 (1982b), 980–989.

BORN, M., and E. WOLF, *Principles of Optics*. Elmsford, N.Y.: Pergamon Press, 1970.

BOWMAN, J. J., T. B. A. SENIOR, and P. L. E. USLENGHI, *Electromagnetic and Acoustic Scattering by Simple Shapes*. Amsterdam: North-Holland, 1969.

BREBBIA, C.A., J. C. F. TELLES, and L. C. WROBEL, *Boundary Element Techniques*. New York: Springer-Verlag, 1984.

BREKHOVSKIKH, L. M., *Waves in Layered Media*. New York: Academic Press, 1960.

BREMMER, H., *Terrestrial Radio Waves*. Amsterdam: Elsevier, 1949.

BROOKNER, E., ed., *Radar Technology*. Dedham, Mass.: Artech House, 1977.

BUTLER, C. M., and D. R. WILTON, "General Analysis of Narrow Strips and Slots," *IEEE Trans. Antennas Propag.*, S28 (1980), 42–48.

———, Y. RAHAMT-SAMII, and R. MITTRA, "Electromagnetic Penetration through Apertures in Conducting Surfaces, *IEEE Trans. Antennas Propag.*, 26 (1978), 82–93.

CHENG, D. K., *Field and Wave Electromagnetics*. Reading, Mass.: Addison-Wesley, 1983.

COHEN, J., and N. BLEISTEIN, "A Velocity Inversion Procedure for Acoustic Waves," *Geophysics*, 44 (1979), 1077–1087.

COLLIN, R. E., *Field Theory of Guided Waves*. New York: McGraw-Hill, 1966.

———, and F. J. ZUCKER, eds., *Antenna Theory*. New York: McGraw-Hill, 1969.

DESCHAMPS, G. A., J. BOERSMA, and S.-W. LEE, "Three-Dimensional Half-Plane Diffraction: Exact Solution and Testing of Uniform Theories," *IEEE Trans. Antennas Propag.*, 32 (1984), 264–271.

DEVANEY, A. J., "A Filtered Back Propagation Algorithm for Diffraction Tomography," *Ultrason. Imaging*, 4 (1982), 336–350.

———, and R. P. PORTER, "Holography and the Inverse Source Problem: Part II," *J. Opt. Soc. Am.*, A2 (1985), 2006–2011.

ELLIOTT, R. S., *Electromagnetics*. New York: McGraw-Hill, 1966.

———, *Antenna Theory and Design*. Englewood Cliffs, N.J.: Prentice-Hall, 1981.

FELSEN, L. B., "Transient Electromagnetic Fields," in *Topics in Applied Physics*, Vol. 10. New York: Springer-Verlag, 1976.

———, and N. MARCUVITZ, *Radiation and Scattering of Waves*. Englewood Cliffs, N.J.: Prentice-Hall, 1973.

FINLAYSON, B. A., *The Method of Weighted Residuals and Variational Principles*. New York: Academic Press, 1972.

GHOSHAL, U. S., and L. N. SMITH, "Skin Effects in Narrow Copper Microstrip at 77K," *IEEE Trans. Microwave Theory Tech.*, 36 (1988), 1788–1795.

GOLDSTEIN, H., *Classical Mechanics*. Reading, Mass.: Addison-Wesely Publishing Company, 1981.

GRADSHTEYN, I. S., and I. M. RYZHIK, *Tables of Integrals, Series, and Products*. New York: Academic Press, 1965.

HANSEN, R. C., *Microwave Scanning Antennas*, Vols. 1, 2, and 3. New York: Academic Press, 1966.

———, ed., *Geometric Theory of Diffraction*. New York: IEEE Press, 1981.

HARDY, A., and W. STREIFER, "Coupled Mode Solutions of Multiwaveguide Systems," *IEEE J. Quantum Electron.*, 22 (1986), 528–534.

HARRINGTON, R. F., *Time-Harmonic Electromagnetic Fields*. New York: McGraw-Hill, 1961.

———, *Field Computation by Moment Methods*. New York: Macmillan, 1968.

HERMAN, G. T., "Image Reconstruction from Projections," in *Topics in Applied Physics*, Vol. 32. Berlin: Springer-Verlag, 1979.

HUYNEN, J. R., "Phenomenological Theory of Radar Targets," in *Electromagnetic Scattering*, ed. P. L. E. Uslenghi. New York: Academic Press, 1978.

IKUNO, H., and K. YASUURA, "Numerical Calculation of the Scattered Field from a Periodic Deformed Cylinder Using the Smoothing Process on the Mode-Matching Method," *Radio Sci.*, 13 (1978), 937–946.

ISHIMARU, A., *Wave Propagation and Scattering in Random Media*, Vols. 1 and 2. New York: Academic Press, 1978.

ITOH, T., "Special Domain Immitance Approach for Dispersion Characteristics of Generalized Printed Transmission Lines," *IEEE Trans. Microwave Theory Tech.*, 28 (1980), 733–736.

———, ed., *Planar Transmission Line Structures*. New York: IEEE Press, 1987.

———, *Numerical Techniques for Microwave and Millimeter Wave Passive Structures*. New York: Wiley, 1989.

JACKSON, J. D., *Classical Electrodynamics*. New York: Wiley, 1975.

JAHNKE, E., F. EMDE, and F. LOSCH, *Tables of Higher Functions*, 6th ed. New York: McGraw-Hill, 1960.

JAMES, G. L., *Geometrical Theory of Diffraction for Electromagnetic Waves*. Stevenage, Herts, England: Peter Peregrinus, 1976.

JOHNSON, C. C., and A. W. GUY, "Nonionizing Electromagnetic Wave Effects in Biological Materials and Systems," *Proc. IEEE*, 60 (1972), 692–718.

JONES, D. S., *The Theory of Electromagnetism*. New York: Macmillian, 1964.

———, *Methods in Electromagnetic Wave Propagation*. Oxford: Clarendon Press, 1979.

JORDAN, E. C., and K. G. BALMAIN, *Electromagnetic Waves and Radiating Systems*. Englewood Cliffs, N.J.: Prentice-Hall, 1968.

JULL, E. V., *Aperture Antennas and Diffraction Theory*. Stevenage, Herts, England: Peter Peregrinus, 1981.

KAK, A. C., "Computerized Tomography with X-Ray, Emission, and Ultrasound Sources," *Proc. IEEE*, 67 (1979), 1245–1272.

KANTOROVICH, L., and V. I. KRYLOV, *Approximate Methods of Higher Analysis*. New York: Interscience, 1958.

KELLER, J. B., "Geometric Theory of Diffraction," *J. Opt. Soc. Am.*, 52 (1962), 116–130.

KERKER, M., *The Scattering of Light and Other Electromagnetic Radiation*. New York: Academic Press, 1969.

KING, D. D., "Passive Detection," in *Radar Handbook*, ed. M. I. Skolnik. New York: McGraw-Hill, 1970, Chap. 39.

KLEINMAN, R. E., "Low Frequency Electromagnetic Scattering," in *Electromagnetic Scattering*, ed. P. L. E. Uslenghi. New York: Academic Press, 1978, Chap. 1.

KNOTT, E. F., and T. B. A. SENIOR, "Comparison of Three High-Frequency Diffraction Techniques," *Proc. IEEE*, 62 (1974), 1468–1474.

KONG, J. A., "Theorems of Bianisotropic Media," *Proc IEEE,* 60 1972), 1036–1046.

———, "Optics of Bianisotropic Media," *J. Opt. Soc. Am.,* 64 (1974), 1304–1308.

———, *Research Topics in Electromagnetic Wave Theory.* New York: Wiley-Interscience, 1981.

———, *Electromagnetic Wave Theory.* New York: Wiley, 1986.

KOSTINSKI, A. B., and W.-M. BOERNER, "On Foundations of Radar Polarimetry," *IEEE Trans. Antennas Propag.,* 34 (1986), 1395–1404.

KOUYOUMJIAN, R. G., and P. H. PATHAK, "A Uniform Geometric Theory of Diffraction for an Edge in a Perfectly Conducting Surface," *Proc. IEEE,* 62 (1974), 1448–1461.

KRAUS, J., *Radio Astronomy.* New York: McGraw-Hill, 1966.

LAKHTAKIA, A., V. V. VARADAN, and V. K. VARADAN, "Field Equations, Huygens' Principle, Integral Equations, and Theorems for Radiation and Scattering of Electromagnetic Waves in Isotropic Chiral Media," *J. Opt. Soc. Am.,* A5 (1988), 175–184.

LANDAU, L. M., and E. M. LIFSHITZ, *Electrodynamics of Continuous Media.* Reading, Mass.: Addison-Wesley, 1960.

LEE. S.-W., "Comparison of Uniform Asymptotic Theory and Ufimtsev's Theory of Electromagnetic Edge Diffraction," *IEEE Trans. Antennas Propag.,* 25 (1977), 162–170.

———, "Uniform Asymptotic Theory of Electromagnetic Edge Diffraction," in *Electromagnetic Scattering,* ed. P. L. E. Uslenghi. New York: Academic Press, 1978, pp. 67–119.

LEE, H. Y., and T. ITOH, "Phenomenological Loss Equivalence Method for Planar Quasi-TEM Transmission Lines with a Thin Normal Conductor or Superconductor," *IEEE Trans. Microwave Theory Tech.,* 37 (1989), 1904–1909.

LEE, S. Y., and N. MARCUVITZ, "Quasiparticle Description of Pulse Propagation in a Lossy Dispersive Medium," *IEEE Trans. Antennas Propag.,* 32 (1984), 395–398.

LEWIN, L., D. C. CHANG, and E. F. KUESTER, *Electromagnetic Waves and Curved Structures.* Stevenage, Hertz, England: Peter Peregrinus, 1977.

LEWIS, R. M., "Physical Optics Inverse Diffraction," *IEEE Trans. Antennas Propag.,* 17 (1969), 308–314.

LIVESAY, D. E., and K. M. CHEN, "Electromagnetic Fields Induced Inside Arbitrarily Shaped Biological Bodies," *IEEE Trans. Microwave Theory Tech.,* 22 (1974), 1273–1280.

LO, Y. T., and S.-W. LEE, eds., *Antenna Handbook.* New York: Van Nostrand Reinhold, 1988.

LO, Y. T., D. SOLOMON, and W. F. RICHARDS, "Theory and experiment on microstrip antennas," *IEEE Trans. Antennas Propag.,* 27, (1979), 137–145.

MA, M. T., *Theory and Application of Antenna Arrays.* New York: Wiley, 1974.

MAANDERS, E. J., and R. MITTRA, eds., *Modern Topics in Electromagnetics and Antennas.* Stevenage, Herts, England: Peter Peregrinus, 1977, Chap. 1.

MAGNUS, W., and F. OBERHETTINGER, *Special Functions of Mathematical Physics.* New York: Chelsea, 1949.

MAILLOUX, R. J., "Phased Array Theory and Technology," *Proc. IEEE,* 70 (1982), 246–291.

MARCUSE, D., *Light Transmission Optics.* New York: Van Nostrand Reinhold, 1982.

MARCUVITZ, N., *Waveguide Handbook.* MIT Radiation Laboratory Series, Vol. 10. New York: McGraw-Hill, 1951.

MENDELSSOHN, K., *The Quest for Absolute Zero*. New York: McGraw-Hill, 1966.

MITTRA, R., ed., *Computer Techniques for Electromagnetics*. Elmsford, N.Y.: Pergamon Press, 1973.

———, "Numerical and Asymptotic Techniques in Electromagnetics," in *Topics in Applied Physics,* Vol. 3. New York: Springer-Verlag, 1975.

MONTGOMERY, C. G., R. H. DICKE, and E. M. PURCELL, *Principles of Microwave Circuits,* MIT Radiation Laboratory Series, Vol. 8. New York: McGràw-Hill, 1948.

MOORE, J., and R. PIZER, *Moment Methods in Electromagnetics*. Chichester, England: Research Studies Press, 1984.

MORGAN, M. A., ed., *Finite Element and Finite Difference Methods in Electromagnetic Scattering*. New York: Elsevier, 1990.

MORSE, P. M. and H. FESHBACH, *Methods of Theoretical Physics*. McGraw-Hill Book Company, New York, 1953.

MORSE, P. M., and K. U. INGARD, *Theoretical Acoustics*. New York: McGraw-Hill, 1968.

NOBLE, B., *Methods Based on the Wiener-Hopf Technique*. Elmsford, N.Y.: Pergamon Press, 1958.

OGUCHI, T., "Electromagnetic Wave Propagation and Scattering in Rain and Other Hydrometers," *Proc IEEE,* 71 (1983), 1029–1078.

PORTER, R. P., and A. J. DEVANEY, "Holography and the Inverse Source Problem," *J. Opt. Soc. Am.,* 72 (1982), 327–330.

RAHMAT-SAMII, Y., and R. MITTRA, "Electromagnetic Coupling through Small Apertures in a Conducting Screen," *IEEE Trans. Antennas Propag.,* 25 (1977), 180–187.

RAMO, S., J. R. WHINNERY, and T. VAN DUZER, *Fields and Waves in Communication Electronics*. New York: Wiley, 1965.

RAY, P. S., "Broadband Complex Refractive Indices of Ice and Water," *Appl. Opt.,* 11 (1972), 1836–1844.

RUCK, G. T., D. E. BARRICK, W. D. STUART, and C. K. KRICHBAUM, *Radar Cross Section Handbook,* Vols. 1 and 2. New York: Plenum Press, 1970.

RUMSEY, V. H., "The Reaction Concept in Electromagnetic Theory," *Phys. Rev.,* Ser. 2, 94 (1954), 1483–1491.

RUNSEY, V. H., *Frequency Independent Antennas,* New York, Academic Press, 1966.

SCHELKUNOFF, S. A., *Applied Mathematics for Engineers and Scientists*. New York: Van Nostrand Reinhold, 1965.

SCOTT, C., *The Spectral Domain Method in Electromagnetics*. Norwood, Mass.: Artech House, 1989.

SENIOR, T. B. A., "Scattering by Resistive Strips," *Radio Sci.,* 14 (1979), 911–924.

———, and M. NOAR, "Low Frequency Scattering by a Resistive Plate," *IEEE Trans. Antennas Propag.,* 32 (1984), 272–275.

SHEN, L. C., and J. A. KONG, *Applied Electromagnetism*. Boston: PWS Publishers, 1987.

SKOLNIK, M. I., *Introduction to Radar Systems*. New York: McGraw-Hill, 1980.

———, ed., *Radar Handbook*. New York: McGraw-Hill, 1970.

SOMMERFELD, A., *Partial Differential Equations in Physics*. New York: Academic Press, 1949.

———,*Optics*. New York: Academic Press, 1954.

STARK, L., "Microwave Theory of Phase-Array Antennas—A Review," *Proc. IEEE,* 62 (1974), 1661–1701.

STEVENSON, A. F., "Solutions of Electromagnetic Scattering Problems as Power Series in the Ratio (Dimensions of Scatterer Wavelength)," *J. Appl. Phys.*, 24 (1953), 1134–1142.

STRATTON, J. A., *Electromagnetic Theory.* New York: McGraw-Hill, 1941.

STUTZMAN, W. L., and G. A. THIELE, *Antenna Theory and Design.* New York: Wiley, 1981.

TAI, C. T., *Dyadic Green's Functions in Electromagnetic Theory.* New York: Intext Educational Publishers, 1971.

TAMIR, T., ed., "Integrated Optics," in *Topics in Applied Physics,* Vol. 7. New York: Springer-Verlag, 1975.

———, and M. BLOK, ed., "Propagation and Scattering of Beam Fields," Special Issue, *J. Opt. Soc. Am.,* A3 (1986), 462–588.

TESCHE, F. M., "On the Analysis of Scattering and Antenna Problems Using the Singularity Expansion Technique," *IEEE Trans. Antennas Propag.*, 21 (1973), 52–63.

TSANG, L., and S. L. CHUANG, "Improved Coupled-Mode Theory for Reciprocal Anisotropic Waveguides," *J. Lightwave Technol.,* 6 (1988), 304–311.

———, A. ISHIMARU, R. P. PORTER, and D. ROUSEFF, "Holography and the Inverse Source Problem: Part III," *J. Opt. Soc. Am.,* A4 (1987), 1783–1787.

———, J. A. KONG, and R. T. SHIN, *Theory of Microwave Remote Sensing.* New York: Wiley, 1985.

UFIMTSEV, P. Y., "Comments on Comparison of Three High Frequency Diffraction Techniques," *Proc. IEEE,* 63 (1975), 1734–1737.

ULABY, F. T., R. K. MOORE, and A. K. FUNG, *Microwave Remote Sensing,* Vols. 1, 2, and 3. London: Addison-Wesley, 1981.

UNGER, H. G., *Planar Optical Waveguides and Fibers.* Oxford: Clarendon Press, 1977.

USLENGHI, P. L. E., ed., *Electromagnetic Scattering.* New York: Academic Press, 1978.

VAN BLADEL, J., *Electromagnetic Fields.* New York: McGraw-Hill, 1964.

———, "Low-Frequency Scattering by Hard and Soft Bodies," *J. Acoust. Soc. Am.,* 44 (1968), 1069–1073.

VAN DE HULST, *Light Scattering by Small Particles.* New York: Wiley, 1957.

VAN DUZER, T., and C. W. TURNER, *Principles of Superconductive Devices and Circuits.* New York: Elsevier, 1981.

WAIT, J. R., *Electromagnetic Radiation from Cylindrical Structures.* New York: Pergamon Press, 1959.

———, *Electromagnetic Waves in Stratified Media.* Oxford: Pergamon Press, 1962.

———, *Wave Propagation Theory.* Elmsford, N.Y.: Pergamon Press, 1981.

———, *Geo-Electromagnetism.* New York: Academic Press, 1982.

———, *Introduction to Antennas and Propagation.* Stevenage, Herts, England: Peter Peregrinus, 1986.

———, "Complex Resistivity of the Earth," in *Progress in Electromagnetics Research,* ed. J. A. Kong. New York: Elsevier, 1989, pp. 1–174.

WATERMAN, P. C., "New Formulations of Acoustic Scattering," *J. Acoust. Soc. Am.,* 45 (1969), 1417.

YAGHJIAN, A. D., "Electric Dyadic Green's Functions in the Source Region," *Proc. IEEE,* 68 (1980), 248–263.

YAMASHITA, E., and R. MITTRA, "Variational Method for the Analysis of Microstrip Line," *IEEE Trans. Microwave Theory Tech.,* 16 (1968), 251–256.

YASUURA, K., and Y. OKUNO, "Numerical Analysis of Diffraction from Grating by the Mode Matching Method with a Smoothing Procedure," *J. Opt. Soc. Am.,* 72 (1982), 847–852.

YEE, K. S., "Numerical Solution of Initial Boundary Value Problems Involving Maxwell's Equations in Isotropic Media," *IEEE Trans. Antennas Propag.,* 14 (1966), 302–307.

YEH, K. C., and C. H. LIU, *Theory of Ionospheric Waves.* New York: Academic Press, 1972.

ZIENKIEWICZ, O. C., *The Finite Element Method.* New York: McGraw-Hill, 1977.

Index

A

B

C

D

E

F

G

H

I

K

L

M

N

O

P

R

S

T

U

V

W

Y

Z